Environmental Science and Engineering

The ultimate goal of this series is to contribute to the protection of our environment, which calls for both profound research and the ongoing development of solutions and measurements by experts in the field. Accordingly, the series promotes not only a deeper understanding of environmental processes and the evaluation of management strategies, but also design and technology aimed at improving environmental quality. Books focusing on the former are published in the subseries Environmental Science, those focusing on the latter in the subseries Environmental Engineering.

Maulin Shah · Deepanwita Deka
Editors

Emerging Sustainable Technologies for Biofuel Production

Springer

Editors
Maulin Shah
Division of Applied and Environmental Microbiology
Industrial Waste Water Research Lab
Ankleshwar, Gujarat, India

Deepanwita Deka
Department of Botany
S. B. Deorah College
Affiliated to Gauhati University
Guwahati, Assam, India

ISSN 1863-5520 ISSN 1863-5539 (electronic)
Environmental Science and Engineering
ISBN 978-3-031-52166-9 ISBN 978-3-031-52167-6 (eBook)
https://doi.org/10.1007/978-3-031-52167-6

This Springer imprint is published by the registered company Springer Nature Switzerland AG
The registered company address is: Gewerbestrasse 11, 6330 Cham, Switzerland

Contents

Chapter 1
Introduction to Biofuel Production: A Step Towards Sustainable Energy

Mohammad Yusuf and Hussameldin Ibrahim

Abstract Biofuels are alternative fuels made from organic matter, such as crops (e.g., corn, sugarcane, soybeans) and waste products (e.g., cooking oil, animal fat). They are produced through processes such as fermentation, distillation, and refining, to be used in vehicles, power plants, and other applications as a substitute for fossil fuels. The production of biofuels aims to address two major global issues: energy security and climate change. Biofuels offer a domestically produced alternative to petroleum-based fuels, reducing dependence on foreign oil and improving energy security. Additionally, they have lower greenhouse gas emissions compared to fossil fuels, making them a potential solution to mitigate the effects of climate change. However, several challenges in biofuel production are also highlighted in the end, such as its impact on food prices, and land usage, etc. Bio-fuels are also in limelight in the aviation industry as well due to the social pressure to reduce carbon footprints. Overall, biofuel production represents an important and ongoing area of research and development as the world searches for sustainable energy solutions.

Keywords Biofuels · Energy · Greenhouse gas · Climate change · Sustainable energy

1.1 Introduction

In the current situation, fossil fuels serve as the primary source of energy, contributing approximately 80% overall, with the transport sector accounting for 58% of this share (Puricelli et al. 2021; Qureshi et al. 2023a). The sources of these fossil fuels and oil reserves are, however, quickly running out and are a major factor in the release of hazardous gases (Iqbal et al. 2021; Qureshi et al. 2023b). These emissions have detrimental consequences, including glacier retreat, biodiversity loss, climate

M. Yusuf (✉) · H. Ibrahim
Clean Energy Technologies Research Institute (CETRI), Process Systems Engineering, Faculty of Engineering and Applied Science, University of Regina, 3737 Wascana Parkway, Regina, SK S4S 0A2, Canada
e-mail: yusufshaikh.amu@gmail.com

M. Shah and D. Deka (eds.), *Emerging Sustainable Technologies for Biofuel Production*, Environmental Science and Engineering, https://doi.org/10.1007/978-3-031-52167-6_1

change, and increasing sea levels (Yusuf and Ibrahim 2023). Moreover, the high fossil fuel demand impacts global economic activities, leading to increased crude oil prices (Yusuf et al. 2021b, 2022). Unpredictable fuel consumption is mostly caused by the modern world's rapid pace, which is fueled by industrialization and motorization (Qureshi et al. 2021, 2022a).

To address these issues, extensive research is being conducted on alternative energy sources, particularly biofuels (Jahirul et al. 2012). Researchers are actively exploring the production of biofuels from sustainable biomass, as they offer an efficient alternative to non-renewable fuels (Show et al. 2018; Qureshi et al. 2022c). Biofuels have various benefits over petroleum fuels, including the ease with which they can be produced from biomass, sustainability and biodegradability, combustion that follows the carbon dioxide cycle, and environmental friendliness (Saurabh and Majumdar 2022; Qureshi et al. 2022b). The percentage of biofuels in the automotive market is anticipated to increase quickly over the next ten years because of these environmental advantages (Demirbas 2011). This growth will have a positive impact on the agricultural sector, leading to increased production and the generation of associated by-products.

Energy is essential for survival, and it has become crucial to maximize the use of limited resources as the demand for energy continues to rise rapidly. India stands as the world's third-largest consumer of primary energy and oil, following the United States and China (Fig. 1.1) (Maheshwari et al. 2021).

Biomass is recognized as the world's fourth-largest available energy resource. It functions as a sustainable and affordable energy storage medium that may be used when necessary. The yearly availability of biomass, which consists of agricultural

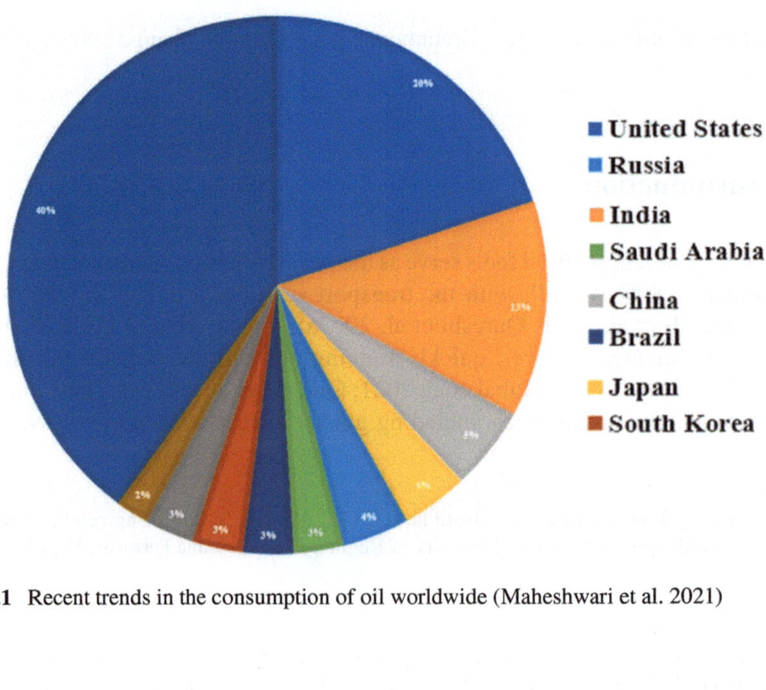

Fig. 1.1 Recent trends in the consumption of oil worldwide (Maheshwari et al. 2021)

and forestry biomass and has an estimated energy potential of 18,000 MW, is thought to be over 500 million metric tonnes in India alone (Gaurav et al. 2017). Although the mechanism of energy transfer is not yet fully understood, researchers are looking into ways to efficiently access the energy stored in biomass. Numerous types of biomass, such as waste from the industrial, forestry, agricultural, marine, and urban sectors, decompose uncontrollably, causing environmental problems and the unanticipated release of harmful gases. Through the use of green fuels, the optimal use of these renewable resources will directly reduce greenhouse gas emissions, reduce environmental pollution, and boost rural economies. The idea of "biorefineries" or "biomass refineries" has recently gained popularity. These facilities allow for the environmentally friendly production of essential energy (like direct electricity generation in combined cycles), liquid biofuels, and chemicals (like platform hydrocarbons derived from biomass using various processing techniques). Marine biomass, particularly seaweeds, offers a novel source of biofuel feedstock with high biomass productivity in a shorter amount of time, whereas the production of biofuels is mostly dependent on terrestrial plants, which have restrictions in terms of land availability for growth. Utilising these bioresources will enable the creation of renewable biomass-based energy (Hassan and Kalam 2013a).

Photosynthesis fuel is a potential area of biofuels, where plants and algae effectively absorb atmospheric CO_2 and store energy as biomass that may be transformed into a variety of energy sources. Without the creation of biomass, the photosynthetic microorganism cyanobacteria may convert CO_2 straight into ethylene, a fuel chemical. A research team at NREL has created genetically modified cyanobacteria for the hyper-production of ethylene, demonstrating the field's promise in the manufacture of biofuels.

1.1.1 Global Perspective

Globally, it has been observed that there are significant projections for the use of biofuels in future energy systems. The availability of bioresources for biofuel production will become a key determinant if the biofuel market realizes its full potential in the automotive industry. The production of biomethanol and bioethanol from biomass as alternatives to oil-based fuels can help the world move from an economy based on hydrocarbons to one based on carbohydrates. The implementation of biofuels has resulted in reduced oil imports, increased GDP, improved energy ratios, and enhanced per capita affluence. Each scenario offers numerous benefits, including the reduction of carbon dioxide emissions, job creation, economic growth, and improved energy efficiency in the production process (Mahapatra et al. 2021).

1.1.2 Biofuel Economy

Even with the high price of oil today, biofuels are still affordable compared to traditional fuels. Following a similar trend to the growth of the hydrocarbon economy throughout the previous century, the biofuel industry has been expanding quickly since the turn of the twenty-first century. Biofuels offer significant environmental benefits, and their effective integration into the transportation sector is expected to drive rapid growth in the near future. Both industrialized and developing countries consider biofuels as a relevant technology for various reasons (Hassan and Kalam 2013b).

Achieving national targets relies on public support, particularly in the transport sector, through the utilization of biofuels. To encourage the supply of feedstocks for biofuel production, support mechanisms such as biofuel obligations and tax reductions have been implemented.

In recent years, energy production has become a crucial driver of agricultural development, playing a vital role in economic growth. As a significant portion of the population resides in rural areas and relies on agricultural production, it is necessary to adopt development programs that integrate communities. These community-based initiatives enhance the socio-economic aspects of the country. Currently, it costs a lot to make synthetic ethanol from ethylene and methanol from natural gas. However, it is economically advantageous to produce bioethanol and biomethanol simultaneously from sugar juice, particularly in regions where hydroelectricity is more readily available and less expensive. Comparatively to the production of hydrogen from natural gas, coal, and biomass, the conversion of biomass into methanol is more expensive and less effective. The depletion of traditional energy sources such as coal, natural gas, and petroleum is frequently noted in research papers and evaluations, and the usage of ecologically benign biomass is recommended as the ideal substitute to meet energy demands.

1.2 Classification of Biofuels

Numerous classification techniques can be found in both literature and legislation due to the constantly evolving nature of biofuels. The division of biofuels into three generations is one that is frequently utilized. This classification takes into account the types of feedstock, their historical emergence in the market, and their level of maturity (Puricelli et al. 2021). The three generations are explained as follows.

First generation: These are traditional biofuels made from crops used as food and feed, such as maize and palm oil. First-generation biofuel production methods are well-established. However, issues develop because of rivalry in the food supply chain. Utilizing food and feed crops for energy generation may result in higher food costs and probable shortages, which has given rise to the "food vs. fuel" controversy.

Additionally, it's important to consider that climate change could negatively impact food crop yields in the future.

Second generation: These biofuels are derived from non-food and non-feed energy crops like straw and agricultural wastes. The production technologies for second-generation biofuels are relatively mature, but their potential is limited by complex processes and the limited availability of raw materials. Producing these biofuels is more complex and expensive compared to the first generation.

Third generation: This category comprises algae-based biofuels, with a concentration on microalgae or microbial algae. The most promising algae for producing biofuel are microalgae. Occasionally, the third generation is linked to other cutting-edge biofuels that haven't yet been commercialized, like those made from higher alcohols and cyanobacteria.

In some instances, a fourth generation of biofuels is mentioned in the literature. These biofuels are produced by genetically engineered microorganisms, photosynthetic microorganisms, or thermochemical processes using waste CO_2. This assessment, however, solely concentrates on the first, second, and third generations due to the lack of agreement regarding the classification of fourth-generation biofuels and their continued research. The elements affecting the evolution of feedstocks are depicted in Fig. 1.2.

It's crucial to understand that electro-fuels, often known as "e-fuels," are fuels produced from renewable power and CO_2 absorption and are not considered biofuels due to their non-biological nature. It should be noted that biogenic CO_2 can also be used to make e-fuels. For instance, electrolytic H_2 and O_2 can be used to increase biofuel production, and methane can be created from CO_2 absorbed during biogas upgrading. Due to the unclear classification and their hybrid nature, these types of fuels are not included in this chapter.

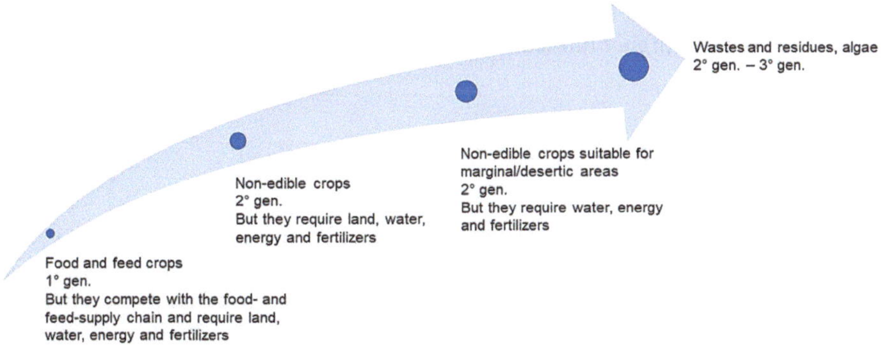

Fig. 1.2 The advancement of biofuel feedstocks (Puricelli et al. 2021)

1.2.1 Biodiesel

The fuel biodiesel is produced from lipid feedstocks like vegetable or animal oils. A procedure known as transesterification is used to produce fatty acid methyl ester (FAME), also known as biodiesel, from the oil to give the properties of the oil a closer resemblance to those of diesel. Compression-ignition engines (CIEs) may run on biodiesel without the need for any changes. However, there can be concerns regarding stability and decomposition, so biodiesel is typically mixed with traditional diesel in blends of 7–10% volume/volume (B7-B10) blends. It is necessary to modify the engines' operational settings, such as injection time, compression ratio, and injection pressure, in order to raise the mix percentages over this range. Captive fleets have been known to use blending grades of B20, B30, and even B100 throughout Europe (Ahmed et al. 2023).

1.2.2 HVO

Hydrotreated Vegetable Oil (HVO), a paraffinic diesel fuel made from biomass, is also referred to as "green diesel" in the industry. HVO is categorized as one of the "renewable diesels," along with Fischer–Tropsch diesel (FT-diesel). Next-generation diesel fuels made from biomass are referred to as "renewable diesel" in a wide sense. Thermochemical processes like pyrolysis and gasification and hydro processing (HVO) are the two primary classes of technologies utilized to manufacture renewable fuel. These fuels don't need to be blended before being used in conventional cars. HVO stands out because it lacks oxygen, aromatics, and sulfur, and it possesses superior properties compared to Fatty Acid Methyl Ester (FAME) and conventional diesel. It exhibits characteristics such as a higher cetane number, higher heating value, and improved oxidative stability. However, unlike FAME, HVO does not have the advantage of being highly lubricated (Hor et al. 2023).

1.2.3 Bioethanol

Liquid alcohol called bioethanol is created using a two-step procedure. Using yeasts, carbohydrates are fermented in the first phase to produce hydrous ethanol, which can include up to 95.6% ethanol and water. To produce anhydrous ethanol, the remaining water is eliminated in the second phase using methods such extractive distillation, membrane separation, or solid adsorbents. Based on the feedstock used, bioethanol is divided into three generations and frequently blended with petrol for use in spark-ignition engines. 10% volume/volume (E10) is the maximum amount of ethanol that can be blended with fuel in conventional automobiles in Europe. However, flexible fuel vehicles (FFVs), which make up a tiny portion (0.1%) of the EU's fleet of passenger cars, can run on ethanol blends of up to 85% volume/volume (E85) (Tse et al. 2021).

1.2.4 Synthetic Liquid Biofuels

Various thermochemical pathways can convert biomass into synthetic fuels suitable for transportation. Currently, there is no widely accepted term to categorize this group of biofuels. In this review, the authors propose the classification of "synthetic liquid biofuels" as the most appropriate. One commonly used thermochemical process is biomass gasification, which produces syngas. This syngas is then converted through a thermochemical pathway referred to as "biomass to liquids" (BtL). Synthetic diesel, sometimes referred to as syndiesel or Fischer–Tropsch diesel, is an illustration of a synthetic fuel created by this process. Syndiesel is manufactured from syngas using the Fischer–Tropsch process. Co-products of this process may include biogasoline, biokerosene, or both, depending on the catalysts and operating circumstances. Comparing synthetic biofuels to their fossil fuel equivalents, they display comparable or even better qualities. They can be utilized directly in existing engines without modification and contain little to no sulphur or aromatic chemicals. Gasification, pyrolysis, liquefaction, supercritical fluid extraction, and aqueous phase reforming are some of the thermochemical processes that can be used to produce synthetic biofuels. Additionally, the synthesis of biomethanol can be achieved through the steam reforming of biomethane (Ram and Salkuti 2023).

1.2.5 Biomethane

Vehicles powered by compressed natural gas, which made up 0.4% of the passenger car fleet in 2017, can be fueled using biomethane produced from biomass. Anaerobic digestion and methanation of biosyngas are the two processes that can produce biomethane. The first method is a well-established technology that creates biogas from organic matter. The biogas can then be further processed to produce biomethane, also known as upgraded biogas (UBG), by this process. In contrast, bio-synthetic natural gas (bio-SNG) is the term for biomethane that is created from biosyngas. However, the technique for producing SNG from the gasification of biomass is not yet commercially viable (Li et al. 2017).

1.2.6 Sustainable Aviation Fuel

Sustainable Aviation Fuels (SAF) provide a solution to reduce the emission of harmful gases and particles from aircraft exhaust, aiming to mitigate climate change by achieving overall net greenhouse gas emissions. By blending licensed substitute aviation fuel with conventional fuel, when compared to utilizing only conventional fuel, it is possible to reduce particle emissions directly behind the airplane by 50–70%. Currently, the ASTM D7566 standard permits the production of alternative aviation fuels using seven different processes, including agricultural waste, algae, biomass, and vegetable or animal FATS (Kurzawska 2021). Transitioning to renewable or low-carbon fuels is the primary path for the aviation industry to pursue. However, the widespread adoption of SAF poses significant challenges. It requires substantial investments in constructing new production facilities, achieving cost reductions across the value chain (including feedstock production, collection, and delivery), and obtaining ASTM certification (Chiaramonti 2019).

Although the immediate economic efficiency of SAF for reducing CO_2 emissions may be limited due to their cost, there are several opinions on the courtesy of this approach. Politics and society are immensely pressurizing on the aviation sector to cut CO_2 emissions. Regulatory measures, such as the blending mandate in the EU, impose potential costs on aviation users unless complementary support measures are implemented. The industry suggests allocating government revenues to SAF production to alleviate costs for users. The expenses associated with transitioning the aviation sector to SAF depend on the price difference between SAF and fossil fuels and the specified blending mandate quota. It is projected that the demand for SAF in the EU will reach approximately 47 million tons by 2050. However, regardless of who bears the costs, the switch to SAF will be costly, especially in the beginning when economies of scale and learning curve effects are still being fully realized (Grimme 2023).

1.2.7 Other Biofuels

Bio-DME: Dimethyl ether is an ether that can be used in compression-ignition engines (CIEs) as both a fuel additive and as a fuel source on its own. Either biomethanol is dehydrated or it is mixed with biosyngas to create bio-DME. DME can only be utilized in modified vehicles with a pressurized fuel chamber, similar to LPG vehicles, because it normally exists in a gaseous state (Ram and Salkuti 2023).

- *Bio-ETBE and bio-MTBE*: Bioethers are a type of fuel that is frequently added to other fuels to improve combustion efficiency. Isobutylene and bioethanol/biomethanol react to produce bio-MTBE and bio-ETBE, respectively.
- *Biohydrogen*: Fuel cells (FCs) are devices that use hydrogen (H_2) to produce energy from biomass. Biomass is gasified in the presence of water to produce H_2-rich biosyngas, which is then used to produce biohydrogen. The water–gas shift process is used to boost the H_2 content even more. The JRC mentions black liquor, farmed wood, and wood waste as suitable gasification materials. Biomethane reforming, bioethanol reforming, dark fermentation, and photo-fermentation are further processes for producing biohydrogen.

- *Biopropane/Bio-LPG*: Liquefied petroleum gas (LPG), which is also known as bio-LPG, has the potential to displace fossil LPG. It can be made either as a byproduct of HVO or through the one-stage catalytic synthesis of syngas generated from biomass gasification. Bio-LPG is already created as a byproduct in other biofuel synthesis techniques, but in relatively modest quantities.
- *Higher alcohols*: Compared to shorter-chain alcohols like ethanol and methanol, long-chain alcohols like biobutanol and biopentanol have better characteristics. They can be used in compression-ignition engines (CIEs) to blend elements with diesel. Butanol and pentanol can be produced utilising engineered microbes and cyanobacteria, biomass fermentation, biosyngas conversion, or biosynthesis from glucose. Recent evaluations on this subject show that ongoing scientific study is focused on lowering the production cost.
- *Mixed alcohols*: Chemical catalysis of biosyngas conversion can result in the production of a combination of methanol, ethanol, and butanol.

1.3 Challenges in the Biofuel Production

1.3.1 Economical Pretreatment

The following requirements should be met by a pretreatment method that is successful and economical: (a) the creation of reactive cellulosic fibres to aid in enzymatic breakdown, (b) the preservation of hemicelluloses and cellulose integrity, (c) the avoidance of the development of potential inhibitors for hydrolytic enzymes and fermenting microorganisms, (d) the reduction of energy requirements, (e) the cost of size reduction for feedstocks, (f) the material cost for building pretreatment reactors, (g), the generation of fewer residues, and Pretreatment techniques for lignocellulosic materials have been proposed, and they can be divided into four groups: "Physical pretreatment," "Physico-chemical pretreatment," "Chemical pretreatment," and "Biological pretreatment" (Khan et al. 2021).

1.3.2 Cost of Enzymes

In order to make biofuel production cost-effective, it is crucial to focus on developing efficient methods for hydrolytic enzyme production. The current high cost of commercial enzymes poses a significant challenge, making large-scale fuel production economically unviable for commercial purposes (Singh et al. 2022).

1.3.3 Fermentation Efficacy

Recent reports suggest that advancements in genetic engineering and novel screening technologies will lead to the development of bacteria and yeast strains capable of fermenting both glucose and xylose. Over the medium to long term, technological advancements will enhance the efficiency of fermentation processes, resulting in higher ethanol yields in less time. Additionally, these developments will also improve the resistance of the organisms, thereby reducing the need for extensive detoxification of the hydrolysate (Leong et al. 2021).

1.4 Sustainable Feedstock

1.4.1 Terrestrial Resource

The largest energy reserves are found in biomass, which includes a variety of terrestrial and marine resources. Due to its unique advantages, biomass has drawn a lot of interest from researchers. It is used as a major and plentiful renewable energy source on a global scale. Despite making up 14% of the world's primary energy, coal, petroleum, and natural gas are not sustainable sources of energy despite being the largest ones currently available. About 220 billion oven dry tonnes (odt) of terrestrial biomass are produced annually, according to estimates. Lignocellulose makes up the majority of plant biomass, which totals about 200×109 tonnes annually. The primary biomass has an energy output potential of $8–20 \times 109$ tonnes, which is four times the yearly global energy consumption (Brunerová et al. 2017).

It is frequently possible to obtain biomass resources locally, and developments in research and conversion techniques have made it possible to convert biomass into secondary energy carriers. The use of biomass, however, necessitates careful consideration of variables like spatial and temporal availability, transportation costs, and large-scale organic material storage. Sustainable biomass forms the basis for the creation of more environmentally friendly technologies for the generation of sustainable energy. Renewable energy sources such as solar and wind power are the first contemporary energy sources and support sustainable development.

It is frequently possible to obtain biomass resources locally, and developments in research and conversion techniques have made it possible to convert biomass into secondary energy carriers. The use of biomass, however, necessitates careful consideration of variables like spatial and temporal availability, transportation costs, and large-scale organic material storage. Sustainable biomass forms the basis for the creation of more environmentally friendly technologies for the generation of sustainable energy. Renewable energy sources such as solar and wind power are the first contemporary energy sources and support sustainable development.

1.4.2 Marine Resources

India, a tropical nation in South Asia (08.04–37.06 N and 68.07–97.25 E), is regarded as one of the world's most diversified nations. India stands out among the twelve nations—Mexico, Colombia, Ecuador, Peru, Brazil, Zaire, Madagascar, China, India, Malaysia, Indonesia, and Australia—that are renowned for having a high level of biodiversity. Together, these nations own more than 70% of the world's biodiversity. India specifically exhibits well-defined biodiversity across various levels, including species, habitat, and genetic diversity. Its extensive coastline, spanning approximately 7500 km, harbors a highly unique marine biota. The Indian coast is characterized by lateritic cliffs, rocky promontories, offshore stalks, long beaches, estuaries, lagoons, spits, and bars.

Compared to Australia and South Africa, India leads in the number of recorded specific and intraspecific seaweed taxa among countries located along the Indian Ocean. The Indian coast is known to host around 844 species of marine algae belonging to 217 genera, mainly distributed in inter-tidal and deep-water areas. However, recent reports indicate an increased number of genera and species of marine algae, with figures reaching 271 genera and 1153 species. The four groups of algae that make up this collection are Chlorophyceae, Phaeophyceae, Rhodophyceae, and Cyanophyceae. Particularly, the Andaman and Nicobar Islands and Tamil Nadu offer ideal conditions for prodigious seaweed development. Areas around Gujarat, Mumbai, Ratnagiri, Goa, Karwar, Kollam, Varkala, Vizhinjam in Kerala, and Chilka in Orissa are also abundant in seaweed beds. However, there hasn't been much research done on the chemical, biochemical, and genetic resources of Indian marine algae (Zaky 2021).

1.4.3 Sustainable Production

Nuclear and fossil fuel reserves are finite and concentrated in particular areas, and their depletion is happening quickly. In contrast, the distribution of renewable resources is uniform worldwide. More energy is being produced from renewable sources than is currently being used worldwide. However, due to issues with economics, equity, and the environment, it is uncertain whether current energy systems will endure (Hasan et al. 2023).

Various sustainability strategies and programs have been put into place for biofuels to protect the land, communities, and wildlife. NGOs, governmental bodies, and commercial organizations are all involved in these activities. To encourage the commercialization of biofuels, it is essential to mix governmental and private investments. For instance, Brazil and other nations have emphasized the generation of biofuel from crops and developed cooperative agreements. In the United States, government funding supports accelerated research and development in the biofuels market.

Among global energy producers, China ranks third, following Brazil and the US. The Chinese government is primarily emphasizing the industrial-scale production of ethanol. India has implemented a "National Biofuel Policy" with the aim of meeting 20% of diesel demand through biodiesel. Additionally, India has proposed replacing 10–20% of gasoline with bioethanol (Mahmood et al. 2022).

1.5 Conclusion

Biofuels are renewable fuel sources derived from organic matter, including crops like corn, sugarcane, and soybeans, as well as waste products such as cooking oil and animal fat. Through processes like fermentation, distillation, and refining, these fuels are created as an alternative to fossil fuels, finding applications in vehicles, power plants, and other areas. The primary objectives behind biofuel production are tackling global concerns related to energy security and climate change. Biofuels serve as a domestically sourced substitute for petroleum-based fuels, reducing reliance on foreign oil and enhancing energy security. Moreover, they exhibit lower emissions of greenhouse gases compared to fossil fuels, potentially offering a solution to combat climate change. Nonetheless, biofuel production also presents certain challenges, including its impact on food prices and land utilization. Despite these concerns, the development and exploration of biofuel production remain vital areas of ongoing research and development, as the world strives to discover sustainable energy solutions. Until electric transportation becomes widespread, biofuels offer potential as a substitute for fossil fuels, provided they are derived from advanced feedstocks. However, avoiding feedstocks linked to land use change is essential to prevent emissions comparable to fossil fuels. Shifting to new production technologies requires significant investments while improving methodologies and models in life cycle assessments (LCAs) is crucial for understanding global biofuel impacts. Further research on land use change and carbon debt is needed. Establishing a standardized approach akin to "Product environmental footprint category rules" would streamline LCAs by reducing variables.

References

Ahmed M, Alamgir K, Vo DN, Yusuf M, Haq A, Abdullah A, Aslam M, Patle DS, Ahmad Z, Ahmad E, Athar M (2023) Recent trends in sustainable biodiesel production using heterogeneous nanocatalysts: function of supports, promoters, synthesis techniques, reaction mechanism, and kinetics and thermodynamic studies. Energy Convers Manag 280:116821. https://doi.org/10.1016/j.enconman.2023.116821

Brunerová A, Malaťák J, Müller M, Valášek P, Roubík H (2017) Tropical waste biomass potential for solid biofuels production. Agron Res 15:359–368

Chiaramonti D (2019) Sustainable aviation fuels: the challenge of decarbonization. Energy Procedia 158:1202–1207. https://doi.org/10.1016/j.egypro.2019.01.308

Demirbas MF (2011) Biofuels from algae for sustainable development. Appl Energy 88:3473–3480. https://doi.org/10.1016/j.apenergy.2011.01.059

Gaurav N, Sivasankari S, Kiran GS, Ninawe A, Selvin J (2017) Utilization of bioresources for sustainable biofuels: a review. Renew Sustain Energy Rev 73:205–214. https://doi.org/10.1016/j.rser.2017.01.070

Grimme W (2023) The introduction of sustainable aviation fuels—a discussion of challenges, options and alternatives. Aerospace 10. https://doi.org/10.3390/aerospace10030218

Hassan MH, Kalam MA (2013a) An overview of biofuel as a renewable energy source: development and challenges. Procedia Eng 56:39–53. https://doi.org/10.1016/j.proeng.2013.03.087

Hassan MH, Kalam MA (2013) An overview of biofuel as a renewable energy source: development and challenges. Procedia Eng 56:39–53. https://doi.org/10.1016/j.proeng.2013.03.087

Hasan M, Abedin MZ, Bin Amin M, Nekmahmud M, Oláh J (2023) Sustainable biofuel economy: a mapping through bibliometric research. J Environ Manage 336:117644. https://doi.org/10.1016/j.jenvman.2023.117644

Hor CJ, Tan YH, Mubarak NM, Tan IS, Ibrahim ML, Yek PNY, Karri RR, Khalid M (2023) Techno-economic assessment of hydrotreated vegetable oil as a renewable fuel from waste sludge palm oil. Environ Res 220:115169. https://doi.org/10.1016/j.envres.2022.115169

Iqbal F, Abdullah B, Oladipo H, Yusuf M, Alenazey F, Nguyen TD, Ayoub M (2021) Recent developments in photocatalytic irradiation from CO_2 to methanol. In: Nguyen V-H, Vo D-VN, Nanda SBT-NP (eds) Nanostructured photocatal. from fundamental to practical applications. Elsevier, pp 519–540. https://doi.org/10.1016/B978-0-12-823007-7.00015-8

Jahirul MI, Rasul MG, Chowdhury AA, Ashwath N (2012) Biofuels production through biomass pyrolysis—a technological review. Energies 5:4952–5001. https://doi.org/10.3390/en5124952

Khan MAH, Bonifacio S, Clowes J, Foulds A, Holland R, Matthews JC, Percival CJ, Shallcross DE (2021) Investigation of biofuel as a potential renewable energy source. Atmosphere (Basel) 12. https://doi.org/10.3390/atmos12101289

Kurzawska P (2021) Overview of sustainable aviation fuels including emission of particulate matter and harmful gaseous exhaust gas compounds. Transp Res Procedia 59:38–45. https://doi.org/10.1016/j.trpro.2021.11.095

Leong HY, Chang C-K, Khoo KS, Chew KW, Chia SR, Lim JW, Chang J-S, Show PL (2021) Waste biorefinery towards a sustainable circular bioeconomy: a solution to global issues. Biotechnol Biofuels 14:87. https://doi.org/10.1186/s13068-021-01939-5

Li H, Mehmood D, Thorin E, Yu Z (2017) Biomethane production via anaerobic digestion and biomass gasification. Energy Procedia 105:1172–1177. https://doi.org/10.1016/j.egypro.2017.03.490

Mahapatra S, Kumar D, Singh B, Sachan PK (2021) Biofuels and their sources of production: a review on cleaner sustainable alternative against conventional fuel, in the framework of the food and energy nexus. Energy Nexus 4:100036. https://doi.org/10.1016/j.nexus.2021.100036

Maheshwari P, Haider MB, Kumar R (2021) Effect of water addition on CO_2 solubility of glycol-based deep eutectic solvents. SSRN Electron J. https://doi.org/10.2139/ssrn.3818321

Mahmood T, Hussain N, Shahbaz A, Mulla SI, Iqbal HMN, Bilal M (2022) Sustainable production of biofuels from the algae-derived biomass. Bioprocess Biosyst Eng. https://doi.org/10.1007/s00449-022-02796-8

Puricelli S, Cardellini G, Casadei S, Faedo D, van den Oever AEM, Grosso M (2021) A review on biofuels for light-duty vehicles in Europe. Renew Sustain Energy Rev 137:110398. https://doi.org/10.1016/j.rser.2020.110398

Qureshi F, Yusuf M, Abdullah B (2021) A brief review on hydrogen production to utilization techniques. In: 2021 third international sustainability and resilience conference: climate change, pp 66–71. https://doi.org/10.1109/IEEECONF53624.2021.9668127

Qureshi F, Yusuf M, Pasha AA, Khan HW, Imteyaz B, Irshad K (2022a) Sustainable and energy efficient hydrogen production via glycerol reforming techniques: a review. Int J Hydrogen Energy. https://doi.org/10.1016/j.ijhydene.2022.04.010

Qureshi F, Yusuf M, Kamyab H, Zaidi S, Junaid Khalil M, Arham Khan M, Azad Alam M, Masood F, Bazli L, Chelliapan S, Abdullah B (2022b) Current trends in hydrogen production, storage and applications in India: a review. Sustain Energy Technol Assess 53:102677. https://doi.org/10.1016/j.seta.2022.102677

Qureshi F, Yusuf M, Kamyab H, Vo DVN, Chelliapan S, Joo SW, Vasseghian Y (2022c) Latest eco-friendly avenues on hydrogen production towards a circular bioeconomy: currents challenges, innovative insights, and future perspectives. Renew Sustain Energy Rev 168:112916. https://doi.org/10.1016/j.rser.2022.112916

Qureshi F, Yusuf M, Arham Khan M, Ibrahim H, Ekeoma BC, Kamyab H, Rahman MM, Nadda AK, Chelliapan S (2023a) A state-of-the-art review on the latest trends in hydrogen production, storage, and transportation techniques. Fuel 340:127574. https://doi.org/10.1016/j.fuel.2023.127574

Qureshi F, Yusuf M, Ibrahim H, Kamyab H, Chelliapan S, Pham CQ, Vo D-VN (2023b) Contemporary avenues of the hydrogen industry: opportunities and challenges in the eco-friendly approach. Environ Res 229:115963. https://doi.org/10.1016/j.envres.2023.115963

Ram V, Salkuti SR (2023) An overview of major synthetic fuels. Energies 16. https://doi.org/10.3390/en16062834

Saurabh K, Majumdar R (2022) Fuels for sustainable transport in India BT—clean fuels for mobility. In: Di Blasio G, Agarwal AK, Belgiorno G, Shukla PC (eds) Springer Singapore, Singapore, pp 27–55. https://doi.org/10.1007/978-981-16-8747-1_3

Show KY, Yan YG, Lee DJ (2018) Biohydrogen production from algae: perspectives, challenges, and prospects. In: Biomass, biofuels, biochemicals: biofuels from Algae, 2nd edn. Elsevier, pp 325–343. https://doi.org/10.1016/B978-0-444-64192-2.00013-5

Singh R, Langyan S, Rohtagi B, Darjee S, Khandelwal A, Shrivastava M, Kothari R, Mohan H, Raina S, Kaur J, Singh A (2022) Production of biofuels options by contribution of effective and suitable enzymes: technological developments and challenges. Mater Sci Energy Technol 5:294–310. https://doi.org/10.1016/j.mset.2022.05.001

Tse TJ, Wiens DJ, Reaney MJT (2021) Production of bioethanol—a review of factors affecting ethanol yield. Fermentation 7. https://doi.org/10.3390/fermentation7040268

Yusuf M, Ibrahim H (2023) A comprehensive review on recent trends in carbon capture, utilization, and storage techniques. J Environ Chem Eng 11:116072. https://doi.org/10.1016/j.jece.2023.111393

Yusuf M, Alnarabiji MS, Abdullah B (2021) Clean hydrogen production technologies BT. In: Song GW, Liu JL, Bashir S (eds) Advances in sustainable energy: policy, materials and devices. Springer International Publishing, Cham, pp 159–170. https://doi.org/10.1007/978-3-030-74406-9_5

Yusuf M, Bazli L, Abdullah B (2022) Challenges and remediation for global warming to achieve sustainable development. In: Dubey AK, Narang SK, Srivastav AL, Kumar A, VBT-AI for RES García-Díaz (eds) Woodhead publishing series energy. Woodhead Publishing, pp 243–257. https://doi.org/10.1016/B978-0-323-90396-7.00017-1

Zaky AS (2021) Introducing a marine biorefinery system for the integrated production of biofuels, high-value-chemicals, and co-products: a path forward to a sustainable future. Processes 9. https://doi.org/10.3390/pr9101841

Chapter 2
Generations of Biofuel

L. Dharani, R. Umapriya, N. Arunkumar, M. Gokila, and R. Sakthi Shankar

Abstract In recent years, biofuel has gained popularity as a feasible substitute for conventional fuels like gasoline and diesel and as a renewable energy source. Depending on the complexity of their production procedures and the energy content of the fuel they create, the various types of biofuel are divided into generations. Fuels from the first generation include ethanol and biodiesel, which are produced by transforming organic material like sugarcane, corn, or soybeans into fuel. The second generation creates more sophisticated fuels like renewable diesel and drop-in aviation fuels that use cellulose and lignin as feedstocks to make a fuel that is comparable to those used in conventional transportation. The third generation is concentrated on utilising waste products to produce fuels, such as landfill gas and syngas, which are produced by transforming organic matter into a gas that may be used as fuel. Finally, the fourth generation is concentrated on the creation of advanced biofuels, such as hydrogen-based and algae biofuels. The current book chapter gives an overall idea about the various generations of biofuel and their source. The prospect for a larger-scale deployment of biofuel is becoming increasingly more enticing as the technology develops and the production costs become more affordable.

Keywords Biofuel · Fossil fuels · Technology · Organic waste · Diesel

2.1 Introduction

Fossil fuels, petroleum, coal, and methane are examples of conventional energy sources that are non-renewable (Singh et al. 2020). The explosive rise in global population has driven up consumer demand for energy. Consequently, the extensive usage of fossil fuels has resulted in their decline and depletion (Olabi et al. 2022). The twentieth century saw the recognition of climate change to be one of the greatest pressing problems that mankind is currently facing. Nearly eighty percent

L. Dharani (✉) · R. Umapriya · N. Arunkumar · M. Gokila · R. Sakthi Shankar
Department of Chemical Engineering, KPR Institute of Engineering and Technology, Coimbatore, India
e-mail: dharani.l@kpriet.ac.in

M. Shah and D. Deka (eds.), *Emerging Sustainable Technologies for Biofuel Production*, Environmental Science and Engineering, https://doi.org/10.1007/978-3-031-52167-6_2

of the energy is consumed through the burning of fossil fuels, which is a significant amount (Alalwan et al. 2019). Greenhouse gases (GHGs), which contribute to global warming and have a negative impact on the biosphere, escape into the atmosphere when fossil fuels are burned. Carbon dioxide (CO_2) emitted back into the atmosphere primarily a consequence of consuming fossil fuels increased by 1.6% in 2017 and is expected to rise considerably more in the coming years (Aron et al. 2020). The total greenhouse gas emissions (GHG) from roadways are predicted to rise from 54 gigatons of CO_2 equivalent to 87 gigatons of CO_2 equivalent by 2050, in accordance with the United Nations Environment Programme (UNDP) Emission Gap Report 2014. This could have a negative impact on the climate system, particularly global warming (Yusup et al. 2019). Development of sustainable energy and chemical sources has become increasingly important due to the depletion of fossil fuel supplies, the stiffening of environmental regulations, and political unrest (Ferreira et al. 2020). In order to replace fossil fuels, experts are working hard to find new and practical alternatives. The use of biofuels is now seen as a desirable replacement for fossil fuels (Olabi et al. 2022). A variety of chemical and biological processes are used to create biofuels, which are fuels made from solid biomass. Bioethanol, biodiesel, syngas, and biogas are typical forms of biofuel. In the United States, Brazil, the European Union, China, and India, biofuels have primarily been utilised to replace conventional transportation fuels (such petrol and diesel). It is possible to mix various ratios of bioethanol and biodiesel with traditional transportation fuels. As an illustration, "E5" (Gasparatos et al. 2013) is used to indicate a blend of 5% bioethanol and 95% petrol. Pyrolysis, anaerobic digestion, and dark fermentation are a few of the techniques used to transform biomass feedstock into biofuels. Pyrolysis converts biomass into bio-oil, which is then used to manufacture biofuels. However, there are a number of drawbacks to this procedure, including higher prices. In order to reduce the requirement for bio-oil modification, co-pyrolysis technology has been created. By employing the digestate as a feedstock for nutrients, biofuels, charcoal, algae farming, and polyhydroxyalkanoates (PHA), anaerobic digestion greatly aids in the development of a circular economy. Anaerobic bacteria obtained from substrates with high glucose content are used in dark fermentation, which creates biofuels without the use of light. First, second, third, and fourth generations of biofuels are divided into categories (Olabi et al. 2022). Sugar cane, vegetable oils, starch, soybean, palm, and maize oils are just a few of the edible feedstocks used to produce first-generation biofuel. Although deforestation has increased and food costs have grown as a result of the production of biofuels from edible feedstocks (Dias et al. 2008; Sakai et al. 2009). In order to address these problems, second-generation biofuels were made from the lipids of non-edible feedstocks, such as animal fats, wastes, and cooking oil (Jung et al. 2021). Second-generation feedstocks are less expensive than first-generation feedstocks, but they have contaminants that impede the conversion procedures needed to produce biofuel (Kim et al. 2022). Furthermore, second-generation lignocellulosic biomass can produce biofuels effectively; however, the pretreatment processes used to turn lignocellulosic biomass into biofuels produce a number of chemicals that prevent the growth of microorganisms and enzymes (Bhatia et al. 2020). In light of this, microalgae were selected as the feedstocks for third-generation biofuels (Mofijur et al. 2019). Lipids and biomass leftovers seen in microalgae and macroalgae species are transformed into biofuels in both the third and fourth generations (Olabi et al. 2022). Due to their reduced emissions of pollutants, less impacts on global warming, and ability to be renewed and sustainable, biofuels are becoming a more appealing alternative to fossil fuels (Mathimani et al. 2018; Saravanan et al. 2018). The annual release of 29 gigatonnes of CO_2 from fossil fuels is to blame. As a result, the total CO_2 emissions from fossil fuels amount to 35.3 billion tonnes (Abishek et al. 2014). The emission of oxygen from algae, on the other hand, which is utilised to produce biofuel, is substantial (10–45%), whereas the release of sulphur is noticeably low. Aside from that, algae-based biofuels are dependable, affordable, sustainable, and non-polluting (Gendy and El-Temtamy 2013). Up to 25% of the world's energy needs might be met by algae-based biofuels (Christenson and Sims 2011). These benefits have led to an emphasis in research on the production of biofuels from microalgae (Khan et al. 2018).

2.2 Generations of Biofuel

Bio feedstocks that contain fermentable sugar, fat, or carbohydrate are converted into biofuels through a fermentation process. Various kinds of energy, including heat, electricity, biogas, and liquid fuels, are produced by converting the biomass of the feedstocks. First, second, third, and fourth generation are the divisions used to classify the biological feedstocks used to produce biofuels (Aron et al. 2020) (Fig. 2.1).

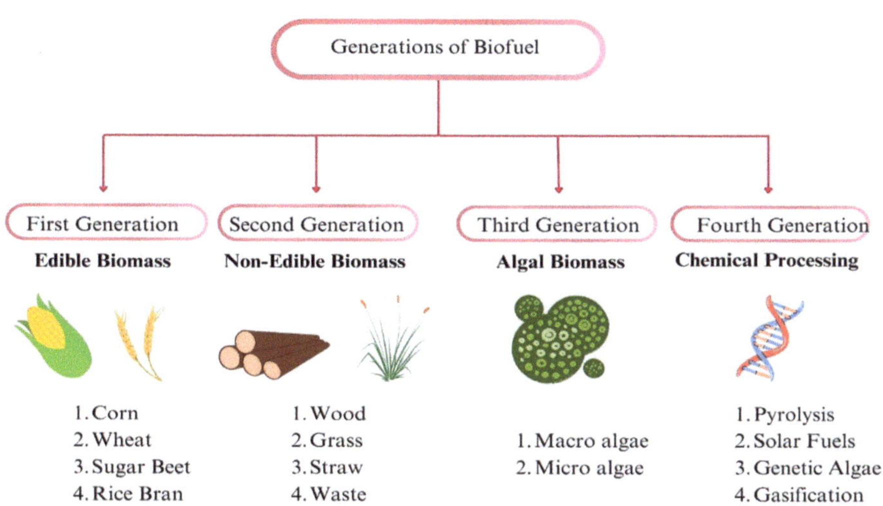

Fig. 2.1 Generations of biofuel

2.2.1 First Generation Biofuels

"First-generation biofuels" identifies those fuels that are derived solely from food crops, such as biofuels produced from biomass that comprises starch, sugar, and vegetable oils and fats. Although the structure of each generation of biofuel is the same, each generation is categorised according to the source from which it is made (Naqvi and Yan 2015). The process of producing first generation biofuels for commercial use is currently well developed. The first generation of biofuels use agricultural feedstocks to manufacture, as seen with ethanol produced from sugarcane and biodiesel made from vegetable oils, for example (Berni et al. 2014). Wheat, maize, and sugarcane are the most frequently utilised raw materials in the manufacturing of first-generation biofuel (Naqvi and Yan 2015). Significant volumes of saccharose can be harvested from sugar crops like sugarcane and sugar beetroot and fermented to produce bioethanol (Singh et al. 2010). Bioethanol and biodiesel are the two main first generation biofuel products. They both primarily serve as sources of bioenergy and have extremely similar chemical characteristics. Through blending with fossil fuels, biodiesel and bioethanol have been utilised as transportation fuel for millennia (Liew et al. 2014). From an economic and environmental point of view, sugarcane is a desirable feedstock for the manufacturing of ethanol, but due to weather and soil conditions, it is only grown in a few places (Naqvi and Yan 2015). Grain crops used as a feedstock make up a significant portion of the first-generation biofuel that has been commercialised in numerous countries. Bioethanol, a liquid biofuel that can be produced using a variety of conversion processes, can be produced using a range of different forms of biomass as feedstocks. Biomass that includes considerable levels of sugar or materials that can be converted into sugar can be fermented to make bioethanol. There are two different categories of feedstocks that can be used to produce first-generation bioethanol:

- Sugar beetroot, sweet sorghum, and sugar cane are examples of feedstocks that include sucrose.
- Starchy ingredients including grains of wheat, corn, barley, and rice (Hirani et al. 2018).

The first generation of biofuel is mostly produced from agricultural food crops due to competition over the use of premium farmland to grow food crops and the development for biofuel feedstock. Consequently, the worldwide population has experienced significant social and economic impacts as well as a "food versus fuel" conflict (Liew et al. 2014).

2.2.1.1 Corn

Particularly in the United States, corn is regarded as the main source of first-generation bioethanol; as of 2012, nearly 40% of the corn production there was used to make ethanol (Naqvi and Yan 2015). According to the efficiency of the production process, ethanol production in August 2012 ranged from 400 to 450 L per tonne of maize, with a market value of about US$338/t (Lee and Lavoie 2013). Through fermentation and distillation, ethanol is made from maize. By 2022, the US Government expects to create 36 billion gallons of biofuel, of which 15 billion gallons will be made from grains, including maize (Naqvi and Yan 2015). The pros and cons of employing maize as a feedstock to produce biofuels are listed in Table 2.1.

First generation bioethanol from maize requires pre-treating the starch to hydrolyze it into sugar prior to fermentation. The milling process reduces the raw material into smaller, more crystalline particles. The combination is then heated to high temperatures (140–180 °C) in order to liquefy the starch after these particles are combined using water and enzymes. During the saccharification procedure, an enzyme called gluco-amylase is introduced into the mixture to convert the molecules of starch into fermentable sugars. When yeast is added to the mixture, the carbohydrates are converted to ethanol and carbon dioxide (Berni et al. 2014).

Hydrolysis of polysaccharide: $(C_6H_{10}O_5)_n + nH_2O \rightarrow nC_6H_{12}O_6$

Ethanol fermentation: $C_6\ H_{12}O_6 \rightarrow 2C_2H_5OH + 2CO_2 + \text{Energy (ATP)}$ (Prasad et al. 2007)

In the absence of oxygen, this procedure is carried out. Water that is created as a byproduct of fermentation needs to be eliminated. This can be achieved by distilling the bioethanol solution to its purity is 95–96%, then dehydrating the bioethanol using a molecular filter (Kuhad et al. 2011). Due to the high cooking temperatures, which therefore imply high operating costs, starch-based biofuels prove less beneficial commercially than sugarcane-based ethanol. The first generation of ethanol is a fuel that can be blended with other fuels, such as petrol, or used alone (Berni et al. 2014).

Table 2.1 Advantages and disadvantages of biofuel obtained from corn

S. no	Advantages	Disadvantages
1	Existing facilities for growing, gathering, and processing	There will be an increase in the use of fertilisers and insecticides
2	Simple method to turn starch into ethanol	Increased demand for fertiliser and pesticides will result in higher expenses as well as soil and water contamination
3	There are no expenses associated with indirect land usage	Corn plays a crucial role in the food chain, and using it for biofuel will raise food prices
4	The facility's leftovers might also be utilised to make ethanol. There is a lot of maize resource available in the US	Low energy yield is only about 1.2

2.2.1.2 Sugarcane

A typical feedstock for the creation of biofuel is sugarcane. Being one of the major users is Brazil (Lee and Lavoie 2013). About 41 billion litres of bioethanol were reportedly produced globally in 2004. The USA (39%), Brazil (38%), and Asia (17%) are the three biggest manufacturers in the world. The USA produces 18 billion litres of bioethanol annually, making it the greatest producer in the world. With a capacity of 16.4 billion litres in 2004, Brazil ranked second among countries that generate bioethanol, providing about 18% of the country's need for vehicle fuel (Hirani et al. 2018). The method for turning sugarcane into ethanol is rather straightforward. The sucrose is removed from the sugarcane by crushing it in water, and the sucrose is then refined to create either raw sugar or ethanol. Increases in the price of sugar are problematic for the bioethanol industry, although being very beneficial for the producers (Lee and Lavoie 2013). Modern sugar mills combine the distillation portion with milling process that reduce size of cane into a smaller one and raw sugar refining. Sugarcane is washed, chopped, and shredded during the milling process using rotating knives. The juice from the crushed sugarcane is then filtered and pasteurised to remove any remaining micro-bacterial contaminants. The sugarcane is initially crushed before being broken up into smaller pieces. In addition, bioethanol distillery operations use the waste byproduct bagasse as a feedstock. The distillery facility qualifies as energy self-sufficient if it can generate all the heat and steam required for the distillery portion. Cane juice can be separated from the undesired non-alcoholic liquid called vinasse, which can then be burned or used as fertiliser. The juice syrup is composed of molasses and transparent crystals after evaporating and crystallising. The pasteurisation and fermentation steps necessary to produce ethanol are carried out after the molasses and crystals are separated using centrifuges. It typically takes 4–12 h to ferment one batch (Naqvi and Yan 2015). In addition to sugarcane and maize, sorghum, rice, sugar beetroot, wheat and cassava are utilised as major inputs for ethanol manufacture (Ravindranath et al. 2011). Additionally, the value of the by-products, such as the discarded grain left over from distillation that is utilised as livestock feed, is a net asset for the operation's overall economic balance (Lee and Lavoie 2013).

2.2.1.3 Vegetable Oil

First-generation biodiesel can replace traditional diesel made from fossil fuels because it is based on vegetable oil. Vegetable oil is made from vegetables, as its name suggests, and if it is consumed straight from the bottle, it is also referred to as virgin vegetable oil. Transesterification can be used to create first-generation biodiesel from vegetable oils such as soybean, palm, sunflower, and more (Kulkarni et al. 2006). Rapeseed oil, sunflower oil, and palm oil are the most widely used industrial feedstocks, particularly to produce biodiesel. Depending on the type and amount of fatty acids they contain, feedstocks can be categorised as saturated, mono-unsaturated, or poly-unsaturated. Process chemistry and economics have a role in deciding which oils to employ. Different vegetable oils (or fats) differ from one another for the process itself based on the quantity of free fatty acids linked to the triglycerides and the presence of other contaminants, such as compounds that could decrease the quality of the glycerine produced. Free fatty acids present in oil tend to interact unfavourably when combined with alkali catalyst employed for supporting the esterification step to produce soap, rendering the catalyst inactive. When available, less expensive feedstocks can be employed since feedstocks with comparable compositions can be used interchangeably in processes designed for those compositions (Girard and Fallot 2006). The primary ingredient in biodiesel, a liquid with a chemical structure like that of fossil fuel but containing no sulphur, is fatty acid methyl esters (FAMEs). Biodiesel is mostly made up of methyl or ethyl esters of fatty acids derived from plant and animal fats (Ren et al. 2010). Because first-generation biodiesel has about 9% lower energy density than petroleum-based diesel, it needs to be changed in diesel engines to prevent harm to rings, fuel lines, and rubber components (Lee 2011). The biodiesel can be used in any diesel engine without modifications in either its original state or in combination with conventional diesel. In any diesel engine without changes, the biodiesel can be utilised either in pure form or combined with fossil diesel (Demirbas and Demirbas 2011). The primary crops used to create biodiesel are soybeans in Brazil and the United States and rapeseed oil in Europe. Although palm oil continues to rule in other developing nations like Indonesia and Malaysia, jatropha is being cultivated more and more as a biofuel crop in several countries, including India and China. Additionally, a variety of species that produce tree-based oils, including Pongamia pinnata, are utilised to create the feedstock for biodiesel. Jatropha produces 1250 L/ha of biodiesel produced and palm oil produces 4594 L/ha/year of biodiesel (Ravindranath et al. 2011). In 2000, there were only a little under 0.8 billion litres of biodiesel produced globally. By 2005, that number had risen to 4 billion litres. Over 16 billion litres were produced overall in 2010. In 2011 and 2012, 22.4 billion and 22.5 billion litres of biodiesel, respectively, were produced, with output rising at a same steady rate. The USA is second in terms of the production of biodiesel in the world, producing more than 80% of it (Hirani et al. 2018). In Europe, 6 billion litres of biodiesel made from imported rapeseed and palm oil (about 2% of the demand for diesel fuel) were produced in 2007 (Naqvi and Yan 2015). In Malaysia and Indonesia, palm oil is widely available and is used for cooking 85–90% of the time. Feedstock costs account for about 80% of the total cost of making biodiesel from palm oil (Chew and Bhatia 2008). Making biodiesel is one of numerous processes being researched with the goal of producing fuels having properties that are extremely identical to those of diesel and kerosene. The most common biofuels generated throughout history are biodiesel and bioethanol. They are often created through fermentation and transesterification, respectively (Liew et al. 2014). Traditional biodiesel can be produced from animal fats, used cooking oil, and raw vegetable oils such those from soybean, rapeseed, palm, or sunflower. Vegetable oils that have been trans esterified with alcohols are the most popular type of biodiesel. Transesterification aims to alter the molecular makeup of vegetable oil in order to make it more physicochemically similar to mineral oil. During the transesterification process, vegetable oil is converted into methyl or ethyl esters of fatty acids, which yield the fatty acids that make up biodiesel. Both a short chain alcohol and an enzyme catalyst are required for the reaction to proceed. The two alcohols that are most frequently used in this kind of operation are methanol, that is toxic and generated from fossil fuels, and ethanol. When the fatty acid reacts with the short-chain alcohol, it produces water and the ester (biodiesel). In addition, glycerin, a byproduct mostly used in the cosmetics industry, is produced (Berni et al. 2014).

Process for Making Biodiesel

By extracting glycerin from vegetable fats and oils during the transesterification process, biodiesel is created. The following are some of the crucial production steps.

Feedstock Preparation

Virgin vegetable oils recycled yellow grease (vegetable oil), and tallow are the most common types of feedstock for biodiesel (Feofilova et al. 2010). To get rid of impurities like burnt food or dirt that was added during cooking or storage, used vegetable oil must be treated. Virgin vegetable oils are degummed to eliminate phospholipids, but they are not processed to a food-grade standard. Since water can hydrolyze triglycerides through the catalytic transesterification process, resulting in salts of the fatty acid and water is also removed from the feedstock (Knothe 2001).

Free Fatty Acid Treatment

Titration with a base solution is used to calculate the amount of free fatty acids, such as carboxylic acids, present in the vegetable oil feedstock before esterification. To produce glycerides and biodiesel, the acids must either be neutralised or esterified (Gao et al. 2008).

Esterification of Acids

The esterification process turns the surplus free fatty acids in the oil into biodiesel to increase the amount of the fuel produced. Once all pollutants and water content have been sufficiently filtered, the procedure works successfully. In order to generate biodiesel, the pretreatment oil is combined with a catalyst, such as sulfuric acid that has been dissolved in methanol, and heated. Following the esterification process, the mixture is eventually dewatered and fed onto the transesterification stage (Naqvi and Yan 2015).

Transesterification

Oils with less free fatty acids are delivered directly to the transesterification stage as feedstock (Knothe 2005). As a catalyst, the oil is combined with potassium hydroxide (KOH) that has been dissolved in methanol. Methyl esters (biodiesel), which are produced during the transesterification process, are also produced along with glycerol (Meher et al. 2006). When the reactants are mixed with a catalyst present, the reaction proceeds quickly. Because it is more affordable and reacts more quickly than other commercially available alcohols, methanol is a common ingredient in most biodiesel manufacturing facilities. It is significant to keep in mind that depending on the alcohol used, the grade of biodiesel (alkyl ester) varies (Knothe 2005). Between the reactants, which are the alcohol and the oil in the transesterification process, bases as well as acids are able to function as catalysts. If a base catalyst is utilised instead of an acid catalyst, the reaction does, however, proceed more quickly.

Methanol Recovery

Methanol is eliminated and recovered once the biodiesel and glycerol are removed as a result of the two layers becoming separated. This stops the reverse reaction. Before being employed again in the reactor for the transesterification process, the methanol is cleaned.

Refining Biodiesel

Glycerol, the leftover methanol, and the catalyst must be removed from the biodiesel during purification before it can be used as fuel. Before being stored, the biodiesel must first be dried. An extra distillation process is used at the production site to produce biodiesel fuel of excellent quality in terms of having no colour, no odour, and no sulphur (Naqvi and Yan 2015).

Triglycerides combine with a total of three molecules of methanol to create three molecules of methyl ester (biodiesel) along with a glycerol molecule, making them the ideal feedstock for biodiesel. Triglycerides are 100% triglyceride. Animal feed made from cake is one of the co- and by-products of biofuel production from oil crops, much like with starchy crops. The supply of animal feed is an issue in many developing nations, such as those in Africa, hence this by-product is crucial for those nations. Cotton seed cake used as animal feed in Burkina Faso is frequently or even equally priced with cotton seed oil. Many countries, including Brazil and India, are researching the enormous prospective of biodiesel made from biomass oil. In December 2004, Brazil's national biodiesel strategy got under way. By 2011, India planned to blend 20% biodiesel into its fuel. In-depth plans are being prepared to utilise waste areas in oilseed cultivation, including Jatropha curcas and other fascinating native species that are oil-rich, as they do not compete with food crops. The productivity data contains some errors and an inadequate information (Girard and Fallot 2006). The cost of oil from various vegetable sources varies on the international market. For instance, the market price for palm oil in August 2012 was US$931/t, whereas the price for soybean oil was US$1230/t. Another common feedstock used to produce biodiesel is canola oil, which has a market value of US$1180/t. Based on these calculations, it can be anticipated that each tonne of oil will produce 1000–1200 L of biodiesel, with a market price estimated from diesel at US$3.2077/gallon (US$0.85/L). The most significant element influencing the generation of biodiesel is the cost of the feedstock. Therefore, interest is growing in the utilisation of other less expensive sources such as old oils (US$331/ton) and oil from plants that are not edible, like jatropha (estimated between US$350 and US$500/ton) (Lee and Lavoie 2013). Using soy beans, castor oil, canola oil, barley, used cooking oil, and other crucial feedstock as the primary input, around 400,000 barrels of first-generation biodiesel were generated daily in 2011. Europe, South America, and North America are the three largest producers of biodiesel, accounting for around 45, 25, and 16% of global production, respectively. With respective shares of around 16, 11.7, 11.4,

13, and 8% of the global biodiesel output in 2011 (Naqvi and Yan 2015), the United States, Argentina, Brazil, Germany, and France all made major contributions. The environmental advantages of using biodiesel can be lost if the number of steps rises. It would be necessary to address these worries and challenges from an economic and ecological standpoint. When it comes to the energy needed to make fertilisers, distribute fertiliser, harvest seeds that contain oil, transport those seeds to a centralised processing facility, and other activities, the system may be energy heavy. By generating biodiesel locally rather than through a centralised production system, the situation that uses a lot of energy can be avoided (Naqvi and Yan 2015). Despite an increase, the demand for conventional diesel still outweighs the amount of biodiesel produced globally (Meher et al. 2006).

2.2.1.4 Soybeans

Soybeans are a widely grown crop, especially in North America, South America, and Asia. The United States is the world's largest soybean grower, producing around 32% of all soybeans. The second-largest soybean grower in the world is Brazil, which produces around 28% of all soybeans (Goldemberg and Guardabassi 2010). It is important to keep in mind that although though soybeans cost more than other food crops, they are nonetheless used extensively in the production of first-generation biofuels. Soybeans are often utilised to produce biodiesel rather than bioethanol (Naqvi and Yan 2015). The United States is less reliant upon diesel than Europe, where it makes up 70% of all transportation fuel usage, but it nonetheless produced 1.8 billion litres of biodiesel in 2007 and 4.2 billion litres in 2012, primarily using soybeans (Feofilova et al. 2010). There are many concerns in growing plants or vegetables for biofuel rather than for human consumption (Meher et al. 2006). Given that the major objective of agriculture is to supply food for the world's ever-increasing population, a collaborative search for solutions for satisfying the ethanol requirements for the transport industry is particularly vital at this time (Prasad et al. 2014). Productivity is projected to rise, nevertheless, as a result of improved farming methods and plant selection. There is no set estimate for the total amount of oil biomass that might be used for energy in each nation. In fact, a number of variables, including the quantity of land that is actually used for agriculture, the climate, the amount of seed oil produced, the need for fuel and food, rival uses, and even exchange rates, have a substantial impact on output, demand, exports, and prices (Girard and Fallot 2006).

2.2.2 *Second Generation Biofuels*

The second generation of biofuels does not use consumable sources, in contrast to the first generation, which only used easily accessible sugars, starches, or oils as their basic materials. They focus on a range of feedstocks and their constituent parts to examine a wider range of raw materials. The term "second generation biofuel" is defined with reference to conversion methods and feedstocks (Berni et al. 2014). Lignocellulosic materials, or wastes from first-generation operations, are used in the second generation along with the "nonedible" sections of the feedstocks. They include switchgrass, wood scraps, agave, sugar cane, and rice bagasse, as well as forestry byproducts and waste materials. The lignin structure must first be weakened and/or removed from such feedstocks using harsh thermochemical pretreatments, and then the cellulose polymer must be digested with a variety of cellulolytic enzymes to liberate glucose (Aparicio et al. 2020). The three main categories of biomass used to produce second-generation biofuels are typically: homogeneous, like white wood chips; quasi-homogeneous, like as agricultural and forestry residues and non-homogeneous such as municipal solid wastes (Lee and Lavoie 2013). In the end, these biofuels won't compete with food crops or won't compete with them much at all; instead, crop leftovers and other plant species that aren't crops will only be planted to produce bioenergy (Hirani et al. 2018). Because they don't directly compete with food crops, lignocellulosic biofuel feedstocks have the potential to be more plentiful and less expensive than agriculture feedstocks. Second generation biofuels can be used to develop and improve the production of biofuel from cellulose, and there are a number of potential advantages: Farmers' incomes could be enhanced via improved utilisation of byproducts, potential and yield in terms of toe/ha (tonnes of oil equivalent biofuel per hectare of land required) can be increased, disputes in the use of land over the cultivation of crops may be minimised as residues can be used or plantations can use set-aside land, and net GHG emissions are able to decrease. In addition to the possibilities to employ residues and municipal solid waste (MSW), there are opportunities for utilising set-aside land and less fertile soil as energy plants, particularly quickly growing tree species (Girard and Fallot 2006). Second generation biofuels are produced using biomass in an improved environmentally benign way that actually has no impact on CO_2 concentrations, if anything, they are even carbon negative (Naik et al. 2010). The generation of bioethanol from lignocellulosic biomass generated by different non-edible sources, which makes use of the full biomass, is one example. The source materials can be grasses (like switchgrass) and wood, as well as agricultural waste products like straw and stover, forestry waste products, or biomass crops like these. All these raw materials can be biochemically converted into biofuels by the application of enzymes or microbes, including genetically modified microorganisms made specifically for this purpose. The "fuel versus food" conundrum appears to be avoided by second generation biofuels since they employ distinct bio-conversion routes. The use of agricultural lands that may be used for growing food crops, however, can compete with them. Both thermochemical decomposition, which encompasses procedures like gasification, bio-carbonization, liquefaction, and thermal breakdown (pyrolysis), and biological digestion, which essentially refers to microbial digestion and fermentation, are employed to convert biomass. Although biological processing tends to be extremely specific as well as generates only a handful of discrete products with significant yield via catalysts that are biological in nature, while thermal conversion frequently yields multiple and complex products in a wide range of reaction times, and while inorganic catalysts are frequently used to

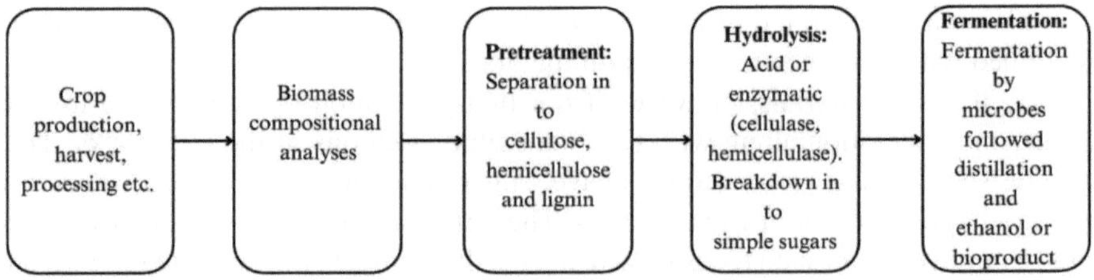

Fig. 2.2 Process of biofuel synthesis from lignocellulosic biomass

improve the product quality or spectrum. At the moment, molasses makes up 75% of the raw materials used to make ethanol in India. However, the supply of molasses is insufficient and uneven, which causes molasses prices to fluctuate greatly. It is unlikely that alcohol made solely from molasses will be available for blending with gasoline because India's prospects for expanding the country's sugarcane production beyond its current 4.0 million ha are bleak due to ongoing droughts, a lack of adequate rainwater supply, erratic summertime electricity availability, and alarming groundwater depletion. Also important and expensive are the related pollution control measures. As a result, ethanol production from lignocellulosic biomass provides an additional and alternative feedstock to sugarcane molasses, moulded grains, sugar beet, cassava, etc. It has been suggested to cultivate land intensively for novel energy crops in order to create lignocellulosic ethanol, and this is seen as a desirable choice by farmers (Rao et al. 2010) (Fig. 2.2).

2.2.2.1 Wood

Natural resources that are abundant and sustainable are essential for the functioning of technological civilizations as well as for the development of a worldwide economy that is sustainable through the production of ethanol to feed the global transportation sector. Lignocellulose has been crucial to the development of civilizations as a component of wood and paper products (Hirani et al. 2018). The majority of renewable energy used today is derived from woody biomass, which includes waste products from the forestry and wood-processing industries as well as actual wood. For many developing nations, it accounts for more than 60% of all primary energy use, and for the poorest African nations (Mali, Niger, etc.), it can even approach 80% or higher. When compared to other types of biomass, wood has a low level of contaminants and a relatively high density, both of which are interesting qualities. It therefore makes for the ideal feedstock for the manufacturing of biofuels. It is frequently utilised to produce various goods, such as paper, lumber, particle board, and fiber board, and as a result, it is one of the most expensive feedstocks (Girard and Fallot 2006). This is partially because of these qualities. Sawdust, wood chips, and wood pellets are examples of wood and wood byproducts created from hybrid poplar. These materials can be hydrolyzed to make sugar, which can then be fermented to create bioethanol. The percentages of cellulose, hemicelluloses, and lignin in wood range from 30–45%, 25–30%, while the remaining percentages are made up of protein and pectin. The effective utilisation of these three components, cellulose, hemicellulose, and lignin, would significantly contribute to the production of economically viable levels of bioethanol (Hirani et al. 2018). Waste wood from the wood industry, such as demolition wood, logging leftovers (tops, branches), process residues (off-cuts, sawdust), and logging residues from logging operations, have plenty of potential. These materials may be more affordable than logs. Depending on how effectively the industry they come from is run, these resources may or may not be available (Girard and Fallot 2006). The five critical steps in the biomass conversion processes that created bioethanol were the selection of the proper biomass, pretreatment, fermentation of hexoses and pentoses by various enzymes, and further processing (Hirani et al. 2018). To boost the production of bioethanol from lignocellulosic materials derived from plant waste, several initiatives are being conducted worldwide (Hirani et al. 2018). The best agricultural feedstocks for the generation of lignocellulosic ethanol are bagasse, sweet sorghum stover, corn stover, rice straw and sorghum stover, among others (Zinoviev et al. 2010; Rao et al. 2010).

2.2.2.2 Sorghum

Sweet sorghum, also known as Sorghum bicolor (L.) Moench., is an emerging energy crop with a rapid growth rate and the ability to collect vast volumes of biomass. It is adapted widely and is used as an alternative source of raw materials for the manufacturing of ethanol. With the help of its biomass, fuel-grade ethanol, glycols, speciality and bulk organic compounds, industrial alcohol, etc. can be produced. Different varieties of sweet sorghum had varying amounts of lignin (5.3–10.5%), hemicellulose (16.5–25.0%), and cellulose (30–35%). Additionally, farmers in India can keep using grains for food or animal feed and stalks for the production of lignocellulosic ethanol, thus cultivating sweet sorghum in dryland conditions does not present a risk to the security of food in the nation. Similar to sugarcane, sweet sorghum is a crop grown specifically for energy and has the capacity to accumulate sugars (10–20%) in its stalks. While its juice, which contains easily accessible "free soluble sugars," can be directly fermented to bioethanol, the leftover bagasse offers a further option for the generation of ethanol that is suitable for use in fuel. Bagasse, which is produced when the sugars in sweet sorghum are extracted, is readily available in considerable quantities (5–7 dry tons/ha) (Rao et al. 2010).

2.2.2.3 Switchgrass

The perennial herbaceous C4 grass species known as switchgrass is found in the prairies of North America. It excels as a bioenergy crop for preserving the soil of agricultural land because it has relatively high net energy benefits, requires relatively

Table 2.2 Cellulose, hemicellulose and lignin content in biomass for production of second generation biofuels

Cellulose, hemicellulose and lignin content in various sources of biomass			
Feedback	Cellulose (%)	Hemicellulose (%)	Lignin (%)
Wheat straw	38.2	24.7	23.4
Switchgrass	31.0	24.4	17.6
Corn stover	36.4	22.6	16.6
Sweet sorghum baggase	33.7	21.6	20.5
Poplar	49.9	20.4	18.1
Rice straw	34.2	24.5	23.4

little input contrasted with other perennial crops over high yields of biomass at environmentally friendly agronomic properties, and is known for its efficient soil carbon sequestration. It is one of the best biomass feedstocks for efficiently manufacturing bioethanol. The efficient production of bioethanol is expected to require 80–150 g/kg of dry switchgrass biomass (Table 2.2).

2.2.2.4 Miscanthus

This sterile hybrid grass is perennial and spreads by rhizomes. Compared to maize grown on the same land, it has the potential to create 2.5 times more ethanol when utilised as feedstock. Due to its perennial nature, it can thrive on poor soils, lacks competition with industrial crops, reduces soil erosion, and binds more carbon from the atmosphere onto the soil than can annual cash crops. Miscanthus species can be used to produce biofuel in the future because they have a high yield potential and noticeable lignin and carbohydrate content.

2.2.2.5 Crop Residues (Straw and Corn Stover)

Crop residues like maize stover, wheat straw, rice straw, and sugarcane bagasse are the most common crop by-products used as biomass around the world. Significant net energy advantages may result from both the reduction of greenhouse gas emissions and the use of these crop wastes to make bioethanol. Crop waste makes an easy-to-find, affordable feedstock for the manufacturing of bioethanol. When compared to bioethanol derived from grains, bioethanol produced from agricultural leftovers will provide a greater net energy benefit (Hirani et al. 2018).

2.2.2.6 Agricultural Wastes

Daily human population growth leads to the development of enormous amounts of diverse wastes. Difficulty with disposal has also emerged primarily as a consequence of the increased trash generation. It is more economically feasible to use waste to produce energy than to spend energy on getting rid of waste. An estimated 5% of biomass energy comes from agricultural waste. Aside from leaves, orchard trimmings, beets, corn, fruits, and sugarcane, agricultural biomass also includes rice husk, rice straw, and stalks. The food-related categories of crops are mentioned, as well as beets, maize, fruits, and sugarcane. Utilising agro waste allows for the production of numerous types of biofuels. Biohydrogen is one of the future's safest, cleanest, and nontoxic biofuels. When it is utilised as fuel, the only by-product is water, which contains no carbon and is not a pollution. At the moment, physicochemical processes that split fossil fuels virtually exclusively yield H_2. Thus, H_2 is a clean fuel that is created at high-temperature, high-pressure conditions from polluting and scarce sources, which releases greenhouse gases. As a result, it becomes crucial to use alternative sources to produce H_2 in a sustainable, environmentally friendly, and renewable manner. Biological processes, which are more energy-efficient and environmentally benign than physicochemical ones, could be one method for producing H_2 (Kour et al. 2019).

2.2.2.7 Second Generation Biofuel Technology for Lignocellulosic Biomass Conversion

There are three different categories of biomass resources: primary, secondary, and tertiary. Woody crops and crop residues are examples of primary supplies that are collected straight from the earth. Black liquor from the pulping process and sawdust in mills are examples of secondary sources. Tertiary sources include residues left over from previous consumers, such as animal fats and greases. All types of biomasses are capable of being utilised to produce biofuel, but most of it need to be pre-treated to change into a state that will be more conducive to conversion. After pretreatment, biomasses can be converted through physical, thermochemical, and biological conversion processes into a range of products with added value. These conversion procedures are covered in more detail in the following sections.

2.3 Pretreatment

The functionality of the conversion processes may be significantly impacted by the physical characteristics of the raw biomass. By ensuring that the biomass has a consistent size, shape, and bulk density, pretreatment is an essential step in getting the biomass ready for conversion. The conversion processes need for heat can be minimised using pretreatment procedures. Pretreatment methods employed in the business often fall into one of four categories: physical, biological, chemical, or combinatorial pretreatment (physiochemical and biochemical conversion). The physical treatment uses mechanical techniques including milling, chopping, grinding, and shredding to improve the specific surface area. After that, the biomass is dried to lower its moisture level in preparation for processing further down the line. To decompose the lignin, hemicellulose, and polyphenols, fungi (such as Phanerochaete chrysosporium, Phlebia radiata, etc.) are used in biological pretreatment. However, since these pretreatment procedures typically need for lengthy residence times, the industrial process is slowed down. To enhance the internal surface area of lignocellulosic biomass and alter its lignin structure, chemical pretreatment uses acids, alkalis, and organic solvents. Increased interior surface area would aid to increase product production and conversion efficiency. It should be noted that methods of combinatorial pretreatment seem to be more successful than other ways. The physiochemical and biochemical methods that work well include Ammonia Fiber/Freeze Explosion (AFEX) and Bioorganosolv.

Physical Conversion Processes

The physical processing of biomass through conversion processes can result in solid biofuel along with additional products. The three physical conversion methods that are applied most frequently are briquetting, pelletizing, and fibre extraction. Briquetting is a method for converting loose biomass wastes into fuel by compacting it into high density solid blocks, including sawdust and oil palm waste. Biomass is extruded and compressed into pellet form during the process of pelletizing. Cleansing the biomass to remove impurities and chipping it into uniform size are required before the pelletizing process. To soften the structure, hot water heated to about 100 °C must also be used to condition the biomass. Finally, the biomass is prepared for feeding into the pellet mill to begin the extrusion process. When compressed under great pressure, the biomass particles will combine to form a solid mass. They are sorted according to size after the extrusion process, where they are chilled to solidify. As an example, mesocarp fibre and empty fruit shells derived from palm trees are extracted from biomass wastes using the method known as fibre extraction. Cyclone separators can be used to extract fibre from the mesocarp, whilst crushing and shredding are used to obtain fibre from fruit brunches that have been removed of their fibre. Finally, a loose solid block of fibre is uniformly packed. The fibre is easily usable as a burning fuel, such as in a boiler for the production of energy, and an alternative natural fibre for a variety of uses (such as mattresses, fibre maps, etc.).

Thermochemical Processes

Thermochemical conversion of biomass, which also includes liquefaction, pyrolysis, gasification, and direct burning, yields energy in various forms. Following this, the intermediate is transformed to a variety of synthetic biofuels. In the section that follows, all four of the processes' fundamental principles are described in detail.

(a) Liquefaction—Heavy oil, which has a high viscosity and contains oils that cannot be dissolved in water, is the main product. Liquefaction requires temperatures between 350 and 370 °C and pressures between 5 and 20 MPa. The main fuel for liquefaction is biomass that contains a lot of lignin, such as sawdust. During the liquefaction process, the biomass is depolymerized, or disassembled into tiny, unstable, and reactive molecules. Re-polymerization of these molecules can create a liquid product containing a range of molecular weights. For the conversion into the heavy fuel oil, syngas, catalysts, and solvents (like phenol) are needed. To improve reaction kinetics and product yield, catalysts must be used during the liquefaction process.

(b) Pyrolysis—It involves thermal decomposition process done without oxygen that produces fuel gaseous products such charcoal and other fuels. Pyrolysis can be divided into two categories: slow pyrolysis and quick pyrolysis. While the synthesis of liquid fuels (bio-oil and biocrude) through fast pyrolysis occurs at temperatures between 400 and 700 °C and times of residence ranging from 1 to 5 s, slow pyrolysis is mostly employed to produce biochar at temperatures of 300–700 °C and residence times ranging from 30 to 200 s. Using additional processing techniques like hydrotreating and hydrocracking, naphtha and diesel range stock fuel can be made from bio-oils. Sulfur-containing compounds that are undesirable are removed, and the oxygen is converted into H_2O and CO_2. To produce clean naphtha and diesel range products, the hydrocracking technique breaks down large hydrocarbon molecules. The oil generated by thermal pyrolysis needs to be catalytically improved in order to produce liquid biofuel in this process. According to laboratory study, the steps of pyrolysis and hydrotreating can be combined and processed together by adding ZSM-5 (zeolite) catalyst. As a result, less energy is required for hydrotreating because the procedures in this case can be finished in a single phase.

(c) Gasification - Synthetic gas, sometimes known as syngas, is the common name for the gaseous product. To create fuels and intermediate chemicals, this syngas is predominantly used. These contaminants must be eliminated before the FT (Fischer Trophs) synthesis process of downstream catalytic upgrading. The long chain hydrocarbon product (such as petrol) is created by the FT process by converting the H_2 and CO. Additionally, the syngas can be used for alcohol synthesis, or catalytic upgrading, to create bioalcohol. It only takes a few seconds or minutes to produce bioalcohol. Also take note that the thermochemical technique to producing bioethanol is more efficient than the traditional biochemical route. The biological process (hydrolysis-fermentation) yields around 80% of the final bioethanol, whereas the gasification-alcohol synthesis pathway permits nearly 100% conversion within a short reaction time. This is a result of the hydrolysis limitation, which prevents all the lignin, cellulose, and hemicellulose from the biomass being fully degraded.

(d) Direct combustion—In this process, biomass undergoes combustion to generate heat in a biomass boiler. The heat generated is converted into power and steam by use of a boiler and steam turbine. There are two key elements for thermochemical processes that typically have a big impact on how well the process works. To guarantee an even temperature distribution inside the gasifier, it is first necessary to regulate the working temperature. Temperature variation may have an impact on energy consumption, biomass reactivity, and conversion efficiency (carbon conversion). Due to its ability to alter and optimise performance in response to changes in feedstock and market demand, the function of the catalyst in thermochemical processes, in addition to temperature control, is another crucial factor.

Biological Processes

The biological process of fermenting sugar results in the production of bioethanol once the cellulose and hemicellulose present in biomass have been converted into sugar. The primary energy source for producing bioethanol is lignocellulosic biomass, which is abundant in carbohydrates. There are essentially no greenhouse gas emissions produced during the preparation of lignocellulosic biomass, which is commonly available. This biological process requires pretreatment of the biomass before proceeding to the subsequent stage of enzymatic hydrolysis. Pretreatment is crucial for enhancing sugar yield because it may account for up to 60% of overall production expenses. Simultaneous Saccharification and Fermentation (SSF) is a method for producing large amounts of bioethanol from lignocellulosic biomass. This method combines fermentation with enzymatic hydrolysis to saccharify the biomass. High glucose concentrations must always be present to allow for the competitive restriction of sugar transport, which prevents the fermentation of pentoses like xylose. Syngas can be converted using either the thermochemical method or the biological approach. Through gasification at a high temperature, an interim gas or liquid gets produced through thermochemical conversion. Then, using a biocatalyst like fermentation bacteria, syngas can be converted into synthetic biofuel. In a water–gas shift process, CO from syngas is converted into H_2 with the help of a hydrogenogenic reaction. In addition, biochemical production of bioethanol is another option for using syngas. Saccharomyces cerevisiae and Zymomonas mobilis, in particular, are used in the production of bioethanol by microbial fermentation. Along with the production of syngas, anaerobic digestion also allows to produce H_2 through a dark fermentation route. Microalgal biomass serves as the source of energy for dark fermentation, which is carried out by obligate and facultative anaerobes among other microorganisms. The suppression of H_2-consuming organisms must be kept under control because it would lower the synthesis of H_2 if H_2 were consumed after it was created. Fermentation microorganisms exhibit certain advantages over traditional metal-based catalyst. The system must be thoroughly cleaned before the process begins in order to enhance bacterial activity. When all the operational factors that influence the production of bioethanol are considered, several indicators are acknowledged for enhancing cost-competitiveness. The price of raw materials and pre-treatment, syngas cleaning technology, high performance catalyst, enzyme, and off-gas recycling, heat integration for energy savings, and high value by-products are a few of the things that are considered (Liew et al. 2014).

2.3.1 Third Generation Biofuels

The term "third-generation biofuels" (TGB) refers to any sort of fuel that can be synthesised utilising algae as a source of biomass (Ganguly et al. 2021). The third generation of biofuel, which is already available and delivers more benefits than the first and second generations, is superior (Neto et al. 2019). TGB was created using a prospective feedstock that can be turned directly into energy (Ganguly et al. 2021). The third generation of biofuels uses microbial biomass, specifically lipids for the manufacture of biodiesel and carbohydrates for the production of bioethanol, which includes oleaginous microalgae, yeast, fungus, bacteria, and others. The term "oleaginous microbes" refers to any organism that can store lipids in amounts that do not exceed 20% of the dry weight of its cells (Srinivas et al. 2023). Algae converts solar energy into chemical energy with a high efficiency, giving it a significantly higher chance of manufacturing ethanol than the first and second generation biofuels (Liew et al. 2014) (Fig. 2.3).

2.3.1.1 Microalgae

Algae, one of the earliest life forms, can be seen throughout all known ecosystems that demonstrate remarkable diversity in a range of environmental conditions (Singh et al. 2011). Only a few of the many diverse kinds of microorganisms that

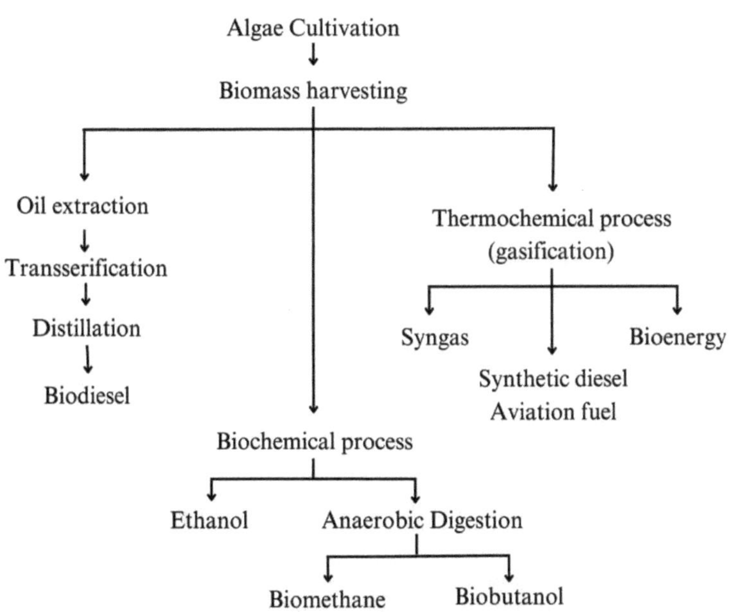

Fig. 2.3 Flow diagram of third generation biofuel production

go under the umbrella word "algae" include microalgae, macroalgae (seaweed), and cyanobacteria (formerly known as blue-green algae). Technology and the marine environment are just two of the many factors that might affect the production of biofuel from algae. The manufacture of biofuels from algae requires microorganisms with a lipid content. Algae come in more than 72,000 different species. Freshwater, sewage, or brackish water can all support the growth of algae. Algae offer a number of significant benefits when it comes to the production of biofuels, including the capacity to grow swiftly during harvesting cycles with the need for lower nutrients. Algal growth can occur naturally in a variety of environments. It is possible to change the rate of growth by increasing the levels of a number of growth factors, including as temperature, acidity, and enriched nutrients. Algae with a high energy production are envisioned as the primary source for third-generation biofuels. Algal biomass can be used to produce a wide range of products, such as ethanol, butanol, biodiesel, jet fuel, syngas, bio-oil, a number of chemical feedstocks, and organic fertilisers. Algal biomass combustion has the potential to produce power (Ganguly et al. 2021). Algae can be categorised into nine major groups, including cyanobacteria (Cyanophyceae), green algae (Chlorophyceae), diatoms (Bacillariophyceae), yellow-green algae (Xanthophyceae), golden algae (Chrysophyceae), red algae (Rhodophyceae), brown algae (Phaeophyceae), dinoflagellates (Dinophyceae), and "picoplankton" (Prasinophyceae and Eustigmatophyceae). Microalgae such as Spirulina, Chlorella, Dunaliella, and Haematococcus are being produced for commercial reasons to produce photosynthetically produced biomass ranging from a few tonnes to several hundred tonnes per year. They can grow 100 times more quickly than terrestrial plants and can double their biomass in less than a day. Obtaining biomass from microalgal plants has been a considerable challenge due to the tiny size and poor concentration under cultivation conditions. Microalgae are originally grown in the general production process with an optimal biomass concentration that ranges from 0.02 to 0.5% (Neto et al. 2019). Algae not only consume a lot of CO_2 for growth, but they also take in CO_2 from power plant emissions, convert biomass through photosynthesis, then release extra oxygen to the atmosphere. The risk of contamination is decreased by algae's contribution to the reduction of N and P levels in wastewater. The conversion of algal biomass into various forms of biofuels can be accomplished by a variety of techniques, including liquefaction, pyrolysis, gasification, extraction and transesterification, fermentation, and aneorbic digestion. Microalgae can generate large quantities of lipids, proteins, and carbohydrates that can be used to create biofuels in a short period of time. In contrast to higher plants, microalgae can withstand and use much higher levels of CO_2 (up to 150,000 ppmv (parts per million by volume)) in the environment. Microalgae can also release gases and soluble carbonates. Due to their highly effective CO_2 fixation, algal cells are in fact tiny biochemical factories with a higher apparent photosynthetic efficiency than terrestrial plants. Algae's potential to fix CO_2 was proposed as a method to remove CO_2 from power plant flue gases and, as a result, may serve to reduce GHG emissions. Three different ways to produce algae are: photoautotrophically, heterotrophically, and mixotrophically. All of these methods follow organic growth cycles. Up to 80% of the dry weight of several microalgal strains is made up of lipids. The most effective method for promoting microalgal lipid accumulation is nitrogen starvation, that gradually transforms the composition of lipids of free fatty acids into triacylglycerol (TAG), an additional valuable product for the production of biodiesel. The complete algal biomass or algal oil extracts can be converted into a variety of fuels, including biogas, liquid and gaseous transportation fuel, kerosene, ethanol, aviation fuel, and biohydrogen using appropriate processing technology. Algal biomass can be burned directly to produce heat and power through direct combustion, which involves chemical reactions to trans esterify the oil that has been recovered from the biomass of algae. As a result of their greater acceptance, numerous businesses are working to commercialise the first three conversion technologies. It is feasible to direct cellular activity towards the synthesis of specific products and even boost microalgae's processing capacity through the manipulation of metabolic pathways. The development of numerous transgenic algal strains with engineered photosynthesis, recombinant protein expression, and enhanced metabolism encourages the possibility of designer microalgae. A particular way for regulating microalgae uses specific environmental factors, like nutrients, to produce envisioned fluxes in metabolic processes (Singh et al. 2011).

2.3.1.2 Cyanobacteria

The production of biofuel is made possible by cyanobacteria's advantages of quick growth, high photosynthetic efficiency, genetic tractability, and genome accessibility. These factors also contribute to the release of a number of important biochemical compounds. The many types of biofuels produced at the time include biodiesel, bioethanol, biogas, biohydrogen, and bioelectricity. Acid content has a significant impact on the Arthrospira platensis bioethanol output. With 0.5 N HNO_3 and H_2SO_4, the greatest amount of bioethanol was produced. For the manufacture of biofuels, genetic engineering is very advantageous. Isobutanol is one of the possible options for a petrol alternative. It has been suggested that cyanobacterium uses carbon dioxide and solar energy. Engineered Synechococcus elongatus PCC 7942 may produce isopropanol in anaerobic environments. Additionally, Synechococcus elongates growth conditions under light were optimised, and aerobic conditions that boosted isopropanol synthesis were observed. The cyanobacterium Synechococcus sp. strain PCC 7002's mutation boosted H_2 production. Dihydrogen is produced by the monocellular, diazotrophic cyanobacterium Cyanothece sp. The production of hydrogen (H_2) is maximised by a combination of environmental and dietary factors. The cyanobacterium "Arthrospira maxima" had a moderately effective H_2 yield, as shown by the hydrogenase-mediated H_2 uptake reaction. One of the filamentous cyanobacteria, Spirulina platensis NIES-46, produced ethanol and hydrogen gas when grown under anaerobic circumstances. Anabaena cylindrica is a species of filamentous cyanobacteria, which are planktonic and produce biohydrogen. The synthesis of biohydrogen is influenced by a number of variables, including pH, ammonium, and ferric ions as well as light conditions that are either limited or enhanced. Electric energy is produced by photo-electrochemical cells from light energy. Cyanobacteria carry out the work by storing light energy captured during photosynthesis in the fuel cells (Kour et al. 2019).

2.3.2 *Fourth Generation Biofuels*

In fourth generation biofuel (FGB), genetically modified (GM) algae are employed to boost biofuel output. Even though GM algae biofuel has become a renowned alternative to fossil fuels, there are still several environmental and health risks that need to be taken into consideration. A successful commercialised production of FGB depends on an appraisal of these worries and the subsequent development of effective mitigation techniques to address them. To increase the production of biofuel, FGB is derived from GM algal biomasses. Increasing penetration of light, reducing photoinhibition, and increasing the efficiency of photosynthetic activity are common strategies used in the genetic modification of microalgae (Abdullah et al. 2019).

2.3.2.1 Genetically Modified Microalgae

Microalgae constitute an extensive category of eukaryotes and cyanobacteria that have a high ability for adaptation to several difficult environmental variables. Nitrogen and phosphorus, the two main nutrients, make up 10–20% of the biomass of microalgae. For their structure, storing food, pigments used for photosynthesis, cell membrane chemistry, and reproduction, algae use a variety of techniques. Not all the current microalgae are equally relevant in terms of their potential use in the production of biofuel. Thus, microalgae belonging to few classes have received the majority of attention in this field of study. The term "fourth generation biofuel" refers to the creation of biofuel by genetic modification of algae. Cyanobacteria, microalgae, and macroalgae are the sources of biomass for fourth generation biofuel. Cyanobacteria are prokaryotes, or organelles without membranes, which are members of the bacterial kingdom, often known as algae. The diatom family Bacillariophyceae, also referred to as Bacillariophyceae, is primarily unicellular and has silicate cell walls that contain chemicals that store carbohydrates and triglycerides (TAGs). Golden-brown Chrysophycea are a type of unicellular algae that can live in freshwater or the ocean. Oils and carbohydrates make up the bulk of Chrysophyceae storage structures, just like diatoms do. Yellow-green, unicellular eukaryotic algae called eustigmatophytes have a promising future as sources of biofuel. The majority of the cells of eustigmatophytes are tiny and contain significant levels of vital fatty acids. Proteins and starch are the primary storage components in the majority of Chlorophyceae algae. Both contained and uncontained systems can be used to grow GM microalgae. The difficulties brought on by each of these approaches, however, differ greatly from one another. Contamination and environmental exposure are kept to a minimum in the confined cultivation system, which has a more closely regulated environment. The confined cultivation system has a high initial cost, despite providing superior protection. Raceway ponds typically make up uncontained systems. The potential for the release of GM algae is substantially larger in uncontained culture systems than in contained ones because the latter are more vulnerable to leaks, animal interference, and aerosol dispersion. Additionally, operating costs are cheaper for open cultivation systems than for closed ones. Due to additional factors, such as user safety and diffusion into the environment, the majority of research has remained in lab settings. Therefore, it is essential to carefully consider the risks associated with using GM microalgae and to control the consequences on the ecosystem. The advantages of GM algae for the environment, including as CO_2 sequestration and assimilation, wastewater heavy metal bioremediation, and a decrease in greenhouse gas emissions, have been well researched. One way to enhance the penetration of light in packed cultures of microalgae is by using the truncated chlorophyll antenna of the chloroplast. It is feasible to improve the photosynthesis efficiency of GM microalgae by increasing the microalgae's absorbing spectrum range during photosynthesis. Increasing the quantity of light that reaches the microalgae culture by reducing the size of the chlorophyll antenna is one of the primary techniques used to genetically engineer microalgae. The manipulation of the pigments and reducing light absorption are other techniques. Metabolic engineering can also significantly raise the lipid or carbohydrate content of microalgae. Maximising the lipid and carbohydrate content of a microalgae biomass is the most attractive factor that can improve its effectiveness (Abdullah et al. 2019). Along with having a high level of adaptation to varied environments, they can, depending on the source of carbon that is available, get their necessities for photoautotrophic, heterotrophic, or mixotrophic development and activity. The microorganisms (cyanobacteria and microalgae) have a very high economic value for biotechnology as a result, and because of their exceptional qualities, they are widely used to produce biotechnological goods for a world dependent on oil (Moravvej et al. 2019) (Table 2.3).

2.4 Challenges

The potential of advanced biofuels is reliant on environmentally and socially appropriate technologies. The demand for biofuels is increasing quickly due to their attractive environmental benefits, which include: (i) renewable; (ii) lowering carbon dioxide indirectly while improving carbon fixation; (iii) facilitating regional economic growth; (iv) minimising air pollution (v) providing reliability in energy as well as decreasing the demand for imported oil; and (v) generating jobs for agriculturalists as well as technologically advanced employment. However, it is important to handle some of the major obstacles brought about by the economy, limited resources, conflicts over issues of food security, safety and security, demand for water, land usage, and different environmental worries that can have an impact on adjacent communities. Most of the time, a life-cycle assessment (LCA) can solve such problems. LCA can help in determining the potential long-term impacts from biofuel production upon the surroundings. Applying the LCA technique to advanced biofuels

Table 2.3 Advantages and Disadvantages of Generations of Biofuels

Biofuels	Advantages	Disadvantages
First generation biofuel	• Biodegradable • Energy security	• Competition of land use • Highest carbon footprint
Second generation biofuel	• No food com[petition • High-vale addded products produced • Energy security	• Complex processses are required • Conversion technologies are under development
Third generation biofuel	• High oil yield • No food and land competition • No toxic content • Energy security	• High processing cost • Production technology is under development • Difficult to harvest and process
Fourth generation biofuel	• Carbon negative biofuel • Energy security	• Lack of study on its practical performance in terms of technical and economic aspects • Still in research and development stage

requires first understanding farmer wants, feedstock type and choices, including changing the use of land (Ambaye et al. 2021). In addition, there are numerous other challenges that arise throughout the process of conversion, such as pH changes, ion and metal impediments, or the hydrolysis of waste products including acetic acid and levulinic acid that affect the volumetric efficiency and quantity of ethanol (Puri et al. 2012).

2.5 Future Prospects

There is a critical need for an established technology that can increase production and efficiency in an affordable method in order to get around these constraints. It's crucial to do cost studies at each level to evaluate the overall approach's viability in respect to efficiency as well as return on investment (Puri et al. 2012). Furthermore, the evolution of the many generations of biofuel must be considered when attempting to predict how the fusion of the various production systems will vary across the next 10 years:

- First-generation biofuels are being produced and utilised commercially to make biodiesel and bioethanol, but they need to be improved in light of GHG emissions

- Second-generation biofuel systems must be improved and optimised for scale-up in laboratories, and they may even need to be tested at a pilot scale
- Due in part to the large area needed to produce biodiesel from algae with high photosynthetic efficiency, this fuel has not found usage in industry. This third-generation biofuel must therefore be industrialised, which calls for study on bioprocess optimisation
- It is crucial to scale up those systems under various conditions because photo fermentation procedures are laborious and have only been evaluated in the laboratory
- The primary mechanism that enables CO_2 neutrality is methanogenesis, which uses gases as its primary raw materials. Considering this, the fifth-generation process is a novel method that has only been tried in a laboratory setting, but it has a good chance of being applied on a wide scale during the upcoming ten years (Porqueras et al. 2014)

By maximising the use of the output products, it is possible to create biofuels and other coproducts at the same time and so reduce production costs. A portion of the algal biomass feedstock can be used to make biofuel as well as a number of other valuable byproducts, including bioplastics and biofertilizer. Other byproducts include food supplements with nutritional value, animal feed, a variety of pharmaceutical ingredients used to make medicines, cosmetics, toiletries, and other products. As a result, the production cost can be reduced by offsetting the coproduct credit by cultivating highly productive algae species with the goal of creating both biofuel and other useful byproducts (Chowdhury et al. 2019).

2.6 Conclusion

Biofuels have gotten a lot of attention because of how safe and eco-friendly they are. Because it incorporates so many various scientific disciplines, including economics, biology, agronomy, ecology, microbiology, engineering, chemistry, etc., the research and development of biofuels is incredibly fascinating. Plants that aren't generally utilised to create biofuels or haven't yet been commercialised could be desirable feedstocks for these fuels because they are currently cost-effective (Malode et al. 2021). Utilising a sustainable power source in the transportation and various power and energy sectors is a priority for the nations that care about the environment. Biofuels have a significant potential to supplement current conventional energy sources (Chowdhury et al. 2019). Due to its enormous potential for resolving bioenergy issues and reducing emissions, algae biotechnology has recently gained growing attention. A new perspective for the quick advancement of photosynthetic organisms, particularly aquatic ones, has emerged thanks to the development of metabolic engineering of algae in the fields of bioenergy and biofuels, conservation of the environment, and reduction of emissions (Jing et al. 2011). If high yielding algae species can be found, modern production and harvesting techniques are used, and cutting-edge drying and oil extraction procedures are implemented, biofuel presents a real alternative to fossil fuel. Given the current state of the art, biofuel cannot completely replace fossil fuel, at least not in the near future (Alam et al. 2015).

Therefore, it can be inferred that algae biofuel could perfectly replace fossil fuel in the future if the cost of producing it is reduced using the following methods:

- Choosing the right algae species for the final products to make the most of biomass feedstock.
- Choosing suitable sites for the manufacturing and processing facilities.
- Creating the best production and harvesting methods possible.
- Inventing new methods for oil extraction and drying.
- Creating GMO algae that are highly productive (Chowdhury et al. 2019).

References

Abdullah B, Muhammad SAFS, Shokravi Z, Ismail S, Kassim KA, Mahmood AN, Md Maniruzzaman AA (2019) Fourth generation biofuel: a review on risks and mitigation strategies. Renew Sustain Energy Rev 107:37–50

Abishek MP, Patel J, Rajan AP (2014) Algae oil: a sustainable renewable fuel of future. Biotechnol Res Int 2014:1–8

Alalwan HA, Alminshid AH, Aljaafari HAS (2019) Promising evolution of biofuel generations. Subject review. Renewable Energy Focus 28:127–139

Alam F, Mobin S, Chowdhury H (2015) Third generation biofuel from Algae. Procedia Engineering 105:763–768

Ambaye TG, Vaccari M, Bonilla-Petriciolet A, Prasad S, van Hullebusch ED, Rtimi S (2021) Emerging technologies for biofuel production: a critical review on recent progress, challenges and perspectives. J Environ Manage 290:1–16

Aparicio E, Rodríguez-Jasso RM, Lara A, Loredo Trevino A, Aguilar CN, Kostas ET, Ruiz HA (2020) Biofuels production of third generation biorefinery from macroalgal biomass in the Mexican context: an overview. In: Sustainable seaweed technologies, Chapter 15, pp 393–446

Aron NSM, Khoo KS, Chew KW, Show PL, Chen W-H, Nguyen THP (2020) Sustainability of the four generations of biofuels—a review. Int J Energy Res 44:1–17

Berni MD, Dorileo IL, Prado JM, Forster-Carneiro T, Meireles MAA (2014) Advances in biofuel production. In: Biofuels production, Chapter 2, pp 11–58

Bhatia SK, Jagtap SS, Bedekar AA, Bhatia RK, Patel AK, Pant D, Rajesh Banu J, Rao CV, Kim YG, Yang YH (2020) Recent developments in pretreatment technologies on lignocellulosic biomass: effect of key parameters, technological improvements, and challenges. Biores Technol 300:1–13

Chew L, Bhatia S (2008) Catalytic processes towards the production of biofuels in a palm oil and oil palm biomass-based biorefinery. Biores Technol 99:7911–7922

Chowdhury H, Loganathan B, Mustary I, Alam F, Mobin SMA (2019) Algae for biofuels: the third generation of feedstock. In: Second and third generation of feedstocks, Chapter 12, pp 323–344

Christenson L, Sims R (2011) Production and harvesting of microalgae for wastewater treatment, biofuels, and bioproducts. Biotechnol Adv 29:686–702

Demirbas A, Demirbas MF (2011) Importance of algae oil as a source of biodiesel. Energy Convers Manage 52:163–170

Dias JM, Alvim-Ferraz MCM, Almeida MF (2008) Comparison of the performance of different homogeneous alkali catalysts during transesterification of waste and virgin oils and evaluation of biodiesel quality. Fuel 87:3572–3578

Feofilova EP, Sergeeva YE, Ivashechkin AA (2010) Biodiesel-fuel: content, production, producers, contemporary biotechnology (review). Appl Biochem Microbiol 46:369–378

Ferreira MFP, Oliveira BFH, Pinheiro WBS, Correa NF, Franca LF, Ribeiro NFP (2020) Generation of biofuels by slow pyrolysis of palm empty fruit bunches: optimization of process variables and characterization of physical-chemical products. Biomass Bioenerg 140:1–10

Ganguly P, Sarkhel R, Das P (2021) The second- and third-generation biofuel technologies: comparative perspectives. In: Sustainable fuel technologies handbook, Chapter 2, pp 29–50

Gao LJ, Xu B, Xiao GM, Lv JH (2008) Transesterification of palm oil with methanol to biodiesel over a KF/hydrotalcite solid catalyst. Energy Fuels 22:3531–3535

Gasparatos A, Stromberg P, Takeuchi K (2013) Sustainability impacts of first-generation biofuels. Anim Front 3:12–26

Gendy TS, El-Temtamy SA (2013) Commercialization potential aspects of microalgae for biofuel production: an overview. Egypt J Pet 22:43–51

Girard P, Fallot A (2006) Review of existing and emerging technologies for the production of biofuels in developing countries. Energy Sustain Dev 10:92–108

Goldemberg J, Guardabassi P (2010) The potential for first generation ethanol production from sugarcane. Biofuels, Bioprod Biorefin 4:17–24

Hirani AH, Javed N, Asif M, Basu SK, Kumar A (2018) A review on first- and second generation biofuel productions. Biofuels: Greenhouse Gas Mitigat Global 141–154

Jung S, Shetti NP, Reddy KR, Nadagouda MN, Park YK, Aminabhavi TM, Kwon EE (2021) Synthesis of different biofuels from livestock waste materials and their potential as sustainable feedstocks—a review. Energy Convers Manage 236:1–13

Khan M, Shin JH, Kim JD, Khan MI, Shin JH, Kim JD (2018) The promising future of microalgae: Current status, challenges, and optimization of a sustainable and renewable industry for biofuels, feed, and other products. Microb Cell Fact 17:1–21

Kim JY, Jung JM, Jung S, Park YK, Tsang YF, Lin KYA, Choi YE, Kwon EE (2022) Biodiesel from microalgae: recent progress and key challenges. Prog Energy Combust Sci 93:1–25

Knothe G (2001) Historical perspectives on vegetable oil-based diesel fuels. Inform 12:1103–1107

Knothe G (2005) Dependence of biodiesel fuel properties on the structure of fatty acid alkyl esters. Fuel Process Technol 86:1059–1070

Kour D, Rana KL, Yadav N, Yadav AN, Rastegari AA, Singh C, Negi P, Singh K, Saxena AK (2019) Technologies for biofuel production: current development, challenges, and future prospects. Pros Renew Bioprocess Fut Energy Syst Biofuel Biorefinery Technol 10:1–50

Kuhad RC, Gupta R, Khasa YP, Singh A, Zhang YHP (2011) Bioethanol production from pentosa sugars: current status and future prospects. Renew Sustain Energy Rev 15:4950–4962

Kulkarni M, Gopinath R, Meher LC, Dalai AK (2006) Solid acid catalyzed biodiesel production by simultaneous esterification and transesterification. Green Chem 8:1056–1062

Lee DH (2011) Algal biodiesel economy and competition among biofuels. Biores Technol 102:43–49

Lee RA, Lavoie (2013) From first- to third-generation biofuels: challenges of producing a commodity from a biomass of increasing complexity. Anim Front 3:6–11

Liew WH, Hassim MH, Ng DKS (2014) Review of evolution, technology and sustainability assessments of biofuel production. J Clean Prod 71:11–29

Lu J, Sheahan C, Fu P (2011) Metabolic engineering of algae for fourth generation biofuels production. Energy Environ Sci 4:2451–2466

Malode SJ, Keerthi Prabhu K, Mascarenhas RJ, Shetti NP, Aminabhavi TM (2021) Recent advances and viability in biofuel production. Energy Convers Manage 10:1–17

Mathimani T, Baldinelli A, Rajendran K, Prabakar D, Matheswaran M, van Leeuwen RP, Pugazhendhi A (2018) Review on cultivation and thermochemical conversion of microalgae to fuels and chemicals: process evaluation and knowledge gaps. J Clean Prod 208:1053–1064

Meher LC, Vidyasagar D, Naik SN (2006) Technical aspects of biodiesel production by transesterification—a review. Renew Sustain Energy Rev 10:248–268
Mofijur M, Rasul M, Hassan N, Nabi M (2019) Recent development in the production of third generation biodiesel from microalgae. Energy Procedia 156:53–58
Moravvej Z, Makarem MA, Rahimpour MR (2019) The fourth generation of biofuel. In: Second third generation of feedstocks, Chapter 20, pp 557–597
Naik SN, Goud VV, Rout PK, Dalai AK (2010) Production of first and second generation biofuels: a comprehensive review. Renew Sustain Energy Rev 14:578–597
Naqvi M, Yan J (2015) First-generation biofuels. Handb Clean Energy Syst 1–18
Neto JM, Komesu A, da Silva Martins LH, Goncalves VOO, de Oliveira JAR, Rai M (2019) Third generation biofuels: an overview. Sustain Bioenergy, Chapter 10, 283–298
Olabi AG, Alami AH, Alasad S, Aljaghoub H, Sayed ET, Shehata N, Rezk H, Abdelkareem MA (2022) Emerging technologies for enhancing microalgae biofuel production: recent progress, barriers, and limitations. Fermentation 8:1–21
Porqueras EM, Rittmann S, Herwig C (2014) Biofuels and CO_2 neutrality: an opportunity. Biofuels 3:413–426
Prasad S, Singh A, Joshi HC (2007) Ethanol as an alternative fuel from agricultural, industrial and urban residues. Resour Conserv Recycl 50:1–39
Prasad S, Amit K, Muralikrishna KS (2014) Biofuels production: a sustainable solution to combat climate change. Indian J Agric Sci 84:1443–1452
Puri M, Abraham RE, Barrow CJ (2012) Biofuel production: prospects, challenges and feedstock in Australia. Renew Sustain Energy Rev 16:6022–6031
Rao S, Seetharama N, Ratnavathi CV, Umakanth AV, Dalal M (2010) Second generation biofuel production from sorghum biomass. Presented at the 40th annual sorghum group meeting, Tamil Nadu Ag. University, pp 1–5
Ravindranath NH, Sita Lakshmi C, Manuvie R, Balachandra P (2011) Biofuel production and implications for land use, food production and environment in India. Energy Policy 39:5737–5745
Ren YL, Harvey A, Zakaria R (2010) Biorefining based on biodiesel production: chemical and physical characterisation of reactively extracted rapeseed. J Biobased Mater Bioenergy 4:79–86
Sakai T, Kawashima A, Koshikawa T (2009) Economic assessment of batch biodiesel production processes using homogeneous and heterogeneous alkali catalysts. Bioresour Technol 100:3268–3276
Saravanan AP, Mathimani T, Deviram G, Rajendran K, Pugazhendhi A (2018) Biofuel policy in India: a review of policy barriers in sustainable marketing of biofuel. J Clean Prod 193:734–747
Singh A, Olsena SI, Nigam PS (2011) A viable technology to generate third-generation biofuel. J Chem Technol Biotechnol 86:1349–1353
Singh A, Pant D, Korres NE, Nizami A-S, Prasad S, Murphy JD (2010) Key issues in life cycle assessment of ethanol production from lignocellulosic biomass: challenges and perspectives. Biores Technol 101:5003–5012
Singh D, Dilip Sharma SL, Soni SS, Sharma PK, Jhalani A (2020) A review on feedstocks, production processes, and yield for different generations of biodiesel. Fuel 262:1–15
Srinivas GLK, Singh D, Kumar S (2023) Transition of biofuels from the first to the fourth generation. In: Biofuels, Chapter 1, pp 3–17
Yusup S, Bokhari A, Trinh H, Shahbaz M, Patrick DO, Cheah KW, Azizan MT, Ramli A, Ameen M, Osman N, Shuhaili AFA, Singh HKG (2019) Emerging technologies for biofuels production. Biomass Biofuels Biochem 45–76
Zinoviev S, Muller-Langer F, Das P, Bertero N, Fornasiero P, Kaltschmitt M, Centi G, Miertus S (2010) Next-generation biofuels: survey of emerging technologies and sustainability issues. Chemsuschem 3:1106–1133

Chapter 3
Sustainable Technologies for Biofuel Production

A. L. Tecorralco-Bobadilla and P. X. Sotelo-Navarro

Abstract However, the increase in population makes it necessary to search for different biofuel production alternatives that will allow us to move towards less dependence on fossil fuels and reduce the environmental threats they cause. This chapter addresses the various technologies that exist to produce biofuels and discusses their sustainability. It is known that producing biofuels by different methods can mitigate the adverse effects on the environment. However, this must be demonstrated through sustainability analysis and addressed in the broad sense of the word "sustainable", which involves aspects of production costs and their relationship to economic viability, environmental impacts associated with biofuel production, as well as working conditions and remuneration of workers and the impact on society due to their use.

A. L. Tecorralco-Bobadilla · P. X. Sotelo-Navarro (✉)
Área de Investigación Tecnologías Sustentables, Universidad Autónoma Metropolitana, Av. San Pablo 180, Col. Reynosa Tamaulipas, 02200 Alcaldía Azcapotzalco, CP, México
e-mail: pearlo@azc.uam.mx; perla.sotelo@cinvestav.mx

P. X. Sotelo-Navarro
IxM-Investigadora por México CONACyT—CINVESTAV, DCTS, Av. Instituto Politécnico Nacional 2508, San Pedro Zacatenco, 07360 Gustavo A. Madero, México

M. Shah and D. Deka (eds.), *Emerging Sustainable Technologies for Biofuel Production*, Environmental Science and Engineering, https://doi.org/10.1007/978-3-031-52167-6_3

Graphical Abstract

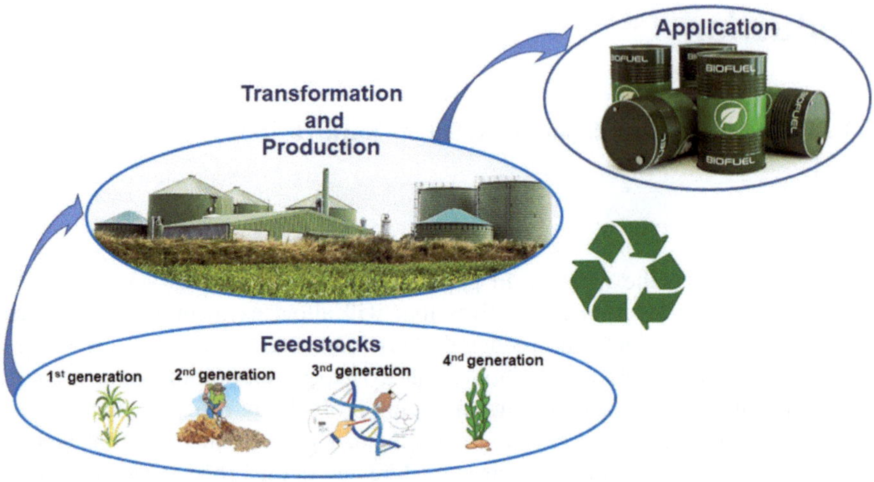

Keywords Biofuels · Sustainability · Environmental impact

3.1 Highlights

- Biofuel production significantly decreases environmental impacts compared to fossil fuels.
- The use of nanotechnology as an emerging technology can improve yields in the production of biofuels.
- Biomass can also be used to produce different chemical products.
- A significant challenge is to identify the best place to locate a biorefinery.

3.2 Introduction

With petroleum-based raw materials being the backbone of many large industrial products, their supply will likely run out soon, so we need alternative resources to secure raw materials. Expanding the base of these is essential to guarantee the global competitiveness today of any industry and country worth it is salt.

The growth of the world population and the demand for raw materials for industries is increasing, especially for the population's transport. Remember that research must remain constant concerning the exploration of fossil fuels and the production of oil and natural gas; biofuels have and will have more excellent study, research and economic relevance worldwide.

Biofuels are fuels from the use of biomass as raw material. If these biofuels are used for the propulsion of means of transport, they receive the more precise name of biofuels, although, on many occasions, they are referred to indistinctly in one way or another. It is accurate as transport fuels that its use is most widespread and where it is foreseeable that it will expand more in the future, given the significant dependence that the means of locomotion have today on oil.

Due to their chemical composition and the physical characteristics that this gives them, biofuels may be suitable in certain circumstances to replace gasoline in gasoline engines or diesel in diesel engines.

The substitution can occur ultimately or, what is more common, in a certain percentage of the mixture, as additives. Therefore, it is said that biofuels come from chemical reactions whose raw material is biomass, but… what is biomass? It is a comprehensive concept. This name refers to any organic matter derived from animal and vegetable residues or products, such as plants, wood, forestry and agricultural processes, or even biodegradable waste products from industrial or urban activity. Why use this biomass as a raw material to produce fuels? Because this is the only source of organic carbon available on Earth that is being produced continuously at an appreciable rate. Vegetable and animal matter is constantly being made on the planet; plants perform photosynthesis to feed themselves; they grow, reproduce, and new plants are born; animals feed on plants, produce, and reproduce. And so on. During photosynthesis, plants fix atmospheric CO_2, one of the gases responsible for the greenhouse effect (GHG, Greenhouse Gases), and take advantage of the sun's light energy. It is a fast and constant cycle of "creation" of matter, unlike fossil fuels (coal, oil, and natural gas), whose production-consumption balance is terribly unbalanced, which is why in a few decades the sources will run out. available reserves (approximately four times more oil is consumed than is discovered annually).

For this and other reasons, attention has long been focused on biomass as a possible substitute for fossil fuels in producing energy sources and certain chemical products. And because biomass is constantly being created and fixing atmospheric CO_2, it is often said that biofuels are "clean and renewable energies."

The above does not mean that biofuels are idyllic (despite the connotations that the prefix "bio" usually recalls) or that they do not substantially impact the environment, so one must be cautious. Now, it is true that, in principle, biofuels are neutral as regards their final use because the CO_2 released in their service has been previously fixed, as has been said, in photosynthesis. The absorbed offsets the emitted carbon dioxide, giving a balance close to zero. This balance is a great advantage compared to the enormous amount of carbon dioxide emitted by fossil fuels. Although they are not "perfect," they certainly beat fossil fuels not only in CO_2 emissions but in many other environmental aspects. For example, they do not emit sulfur dioxide (the leading cause of acid rain), and the concentration of emitted suspended particles, heavy metals, carbon monoxide, polycyclic aromatic hydrocarbons and volatile organic compounds is reduced.

Public acceptance through a good quality and environmental compatibility of new biofuel products and technologies should be the aim of this paper to highlight the need and urgency of these research activities for this change of raw materials, and the sustainability of biofuel production technologies must be proven.

3.3 The Global (Worldwide) Market for Biofuels

In 2020, there was a drop in the global use of fuel for transportation of 8.5% compared to 2019 due to international restrictions on the mobility of people and trade caused by the COVID-19 pandemic. The use of biofuels also decreased by 8.7% in the same period. The OECD-FAO projects an influence of support policies and the demand for fossil fuels on biofuel markets. In developing countries, blending targets are expected to increase to encourage global biofuel consumption in the medium term. Decreasing the demand for fossil fuels, as well as reducing political incentives in developing countries, will limit the expansion of biofuels. In nominal terms, it is projected that from 2022 to 2030, the international price of biofuels will increase; however, in real terms, it is expected to remain almost constant. Regulatory policies play an essential role in the prices of biofuels since they usually cover part of the production costs linked to the use of fossil fuels (OECD/FAO 2022).

Figure 3.1 shows a growth projection in global biofuel use for 2022–2030. According to the International Energy Agency (IEA) it is expected that the United States (US) and the European Union (EU) will decrease the total use of fuels, which will reflect a limited consumption of biofuels. A decrease in the use of palm oil-based biodiesel in the EU is expected, as the Renewable Energy Directive (RED) II classifies it as having a high risk of indirect land use change (ILUC). While for the US, sustained demand for biofuels is expected due to that country's Renewable Fuel Standard (RFS). On the other hand, domestic ethanol consumption will be limited due to the 10% ethanol blend (OECD/FAO 2022).

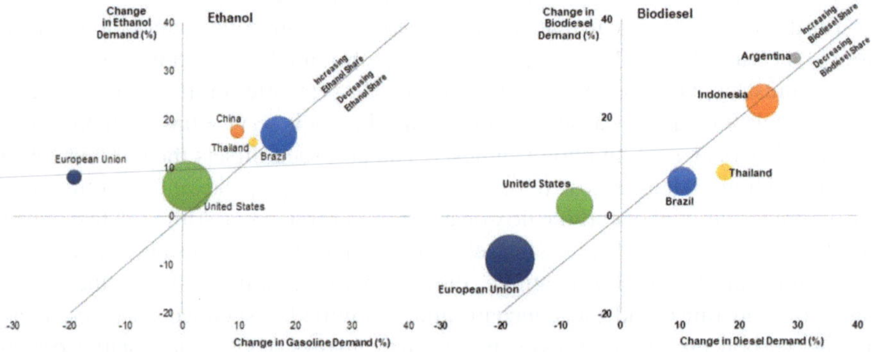

Fig. 3.1 Biofuel demand developments in major regions (modified OECD 2023)

Policy developments and fuel consumption trends play an important role in emerging economies. Total fuel consumption in Brazil is expected to grow in the forecast period, with balanced growth for ethanol and biodiesel. In China, corn stocks have decreased since 2017, so an E10 mandate (10% ethanol blend) is not expected to apply as proposed. Therefore, by 2030 it will maintain a mixing rate of 2% (OECD/FAO 2022).

For some emerging economies, regulations on biofuel blends are expected to change during the forecast period. Diesel and biodiesel usage in Indonesia is expected to increase. Along with the growth in fossil fuel demand, Indonesia encourages biodiesel use by implementing the B30 program (30% biodiesel blend) in 2020. This program allows assuming that during the projection, the blend rate will remain at 30%. The current mix rate in Argentina is 10%, and it is projected to be maintained. Sugarcane-based ethanol is expected to significantly contribute to India reaching a blend rate of 8%. The limited supply of food, mainly molasses, means that the projection is below the E20 planned to be reached by 2030 (OECD/FAO 2022).

Sugarcane and corn to produce ethanol and vegetable oils for biodiesel will remain traditional raw materials to produce biofuels worldwide. In Canada, the EU, Singapore and the US, biodiesel from used cooking oil will continue to play an important role. The reduction in the consumption of fossil fuels and GHG emissions is promoted in most countries with biofuel policies. The preceding will allow a national supply, leaving the participation of foreign trade low, anticipating a more significant decrease in the next decade. The world trade of ethanol will have a moderate reduction; however, that of biodiesel will decrease by 25%, reflected by the policies of the EU on biodiesel from palm oil. A decline in exports from Indonesia is expected to be remembered by an increase in domestic demand (OECD/FAO 2022).

The projection analyzed foresees that biofuels are produced from agricultural commodities, but a substantial increase in advanced or second-generation biofuels is not expected. Changes in regulatory frameworks or unforeseen technological advances may have deviations, implying projection uncertainty (OECD/FAO 2022).

3.4 Production and Consumption of Biofuels

Table 3.1 shows that biofuels (bioethanol and biodiesel) are produced from biomass. Currently, ethanol is made from corn, sugar cane, wheat, and molasses (60, 25, 3 and 2%, respectively), and the remaining 10% from other cereals (sugar beet or cassava). Vegetable oils are the main base of biodiesel; palm, soybean, and rapeseed with 30, 25 and 25%, respectively, and the remaining 25% from used cooking oil. The cellulosic raw materials on which advanced technologies are based do not represent a large part in the production of biofuels. Farmer support, GHG reduction and energy independence as primary objectives of national policies strongly influence the global biofuels sector (OECD/FAO 2022).

For 2030, global ethanol production of 132 and 50 million m^3 for biodiesel is expected due to increased Indonesian mandates. There is a significant variation in the

Table 3.1 Biofuel production ranking and a significant feedstock

Country	Production ranking (base period)		Major feedstock	
	Ethanol	Biodiesel	Ethanol	Biodiesel
United States	1 (48.2%)	2 (18.1%)	Corn	Soybean oil used as cooking oils
European Union	5 (4.8%)	1 (32.3%)	Sugar beet/wheat/maize	Rapeseed oil/Palm oil/ used cooking oils
Brazil	2 (26.7%)	4 (12.2%)	Sugarcane/corn	Soybean oil
China	3 (8.3%)	9 (2.3%)	Corn/cassava	Used cooking oils
India	5 (2.3%)	15 (0.5%)	Molasses	Used cooking oils
Canada	6 (1.6%)	13 (0.7%)	Corn/wheat	Canola oil/used cooking oil/soybean oil
Indonesia	20 (0.1%)	3 (15%)	Molasses	Palm oil
Argentina	8 (1.0%)	5 (5%)	Molasses/sugarcane/ maize	Soybean oil
Thailand	7 (1.4%)	7 (3.8%)	Molasses/cassava/sugar cane	Palm oil
Colombia	13 (0.44%)	11 (1.3%)	Sugarcane	Palm oil
Paraguayan	14 (0.42%)	19 (0.03%)	Corn/sugarcane	Jatropha

Notes

1. Numbers refer to the country ranking in global production; percentages refer to the production share of countries in the base period
2. In the OECD-FAO Agricultural Outlook (2021–2030), biodiesel includes renewable diesel (Hydrotreated Vegetable Oil or HVO), although these are different products

Source OECD/FAO (2021), "OECD-FAO Agricultural Outlook", OECD Agriculture statistics (database), http://dx.doi.org/https://doi.org/10.1787/agr-outl-data-en

raw materials used between countries (2022–2030). Despite the growing sensitivity towards sustainability, it is expected that the production of biofuels will continue to be from traditional feedstocks in many countries (OECD/FAO 2022).

3.5 Bioprocesses to Produce Biofuels with a Perspective on Biorefineries

With the idea of providing new knowledge and guidelines to meet future energy needs through the bioprocessing of bioresources, various researchers have developed advances and identified opportunities for biorefineries (Dragone et al. 2020; Fallahi et al. 2021; Usmani et al. 2021, 2022). Biorefineries have been conceived to valorize biomass, converting it into various value-added products and bioenergy to achieve a sustainable bioeconomy (Usmani et al. 2022).

The key to economic and ecosystem sustainability may be to use diverse biomass, including by-products (lignocellulosic biomass) and wastes (agricultural, food, municipal solids) and other renewable biological resources, and process them by biorefinery (Gupta et al. 2022).

Biorefining biomass from sustainable and eco-innovative bioprocess systems is aligned with a series of sustainable development goals (SDGs). Emphasizing that biomass is of biogenic origin, releasing CO_2 through any biochemical route does not contribute to an increase in atmospheric CO_2 (Kostas et al. 2021).

It is possible to mitigate climate change, help the circular economy and economically strengthen several industrial sectors by coupling bioprocesses in a well-optimized biorefinery that can obtain more than one bioproduct (Kostas et al. 2021).

The supply of environmentally friendly chemicals, energy, and fuels to sustain economic growth can be achieved with continuous improvement of energy crops, synthetic biology, aquaculture, biorefining and conversion technologies. However, the biorefinery industry still faces several challenges. The complex nature of lignocellulosic refining as a raw material has yet to be rationally profitable, coupled with the expensive processes of pretreatment and enzymatic hydrolysis of cellulose. This area has clear economic and technical barriers even though new cellulosic biomass bioprocesses have been developed.

3.6 Lignocellulosic Biorefinery

Lignocellulosic residues are made up of lignin, hemicellulose, and cellulose present in the plant cell wall, for which the biomass, mainly agricultural and forestry (UNAM 2021), are the primary raw material in biorefineries; each of these compounds has characteristics and different functions (Adewuyi 2022).

Lignin: Between 5 and 30% is found in the lignocellulosic residue; it is necessary to be removed to take advantage of the hemicellulose and cellulose because, due to its structure, it does not provide sugar in the production of fuels, it is hydrophobic, which can be related with which material has resistance to degradation.

Hemicellulose: 15–35% corresponds to this polymer; through the action of the hemicellulose enzyme, it is possible for acid or basic hydrolysis to simple sugar to occur. This sugar will be a raw material for biofuel production.

Cellulose: From 20 to 30% of the lignocellulosic material; in addition, it can be present up to 50% in the wall of the plant cell; the cellulase enzyme is in charge of degrading cellulose into glucose to be a substrate in the production of biofuel.

Therefore, each material can vary in the amount of lignin, hemicellulose and cellulose that it has in its structure (Table 3.2) and with it, the decision of what to use as a substrate in the production of biofuels, in addition to the most suitable bioprocess for its use sustainable.

Table 3.2 Biomass materials and their chemical composition (Irmak 2019)

Biomass materials	Cellulose (%)	Hemicellulose (%)	Lignin (%)
Switchgrass	33.8	28.4	16.6
Miscanthus (grass)	47.7	24.6	12.3
Poplar (hardwood)	52.1	27.4	15.9
Oak (hardwood)	40.4	35.9	24.1
Pine (softwood)	46.0	25.5	20.0
Spruce (softwood)	45.5	22.9	27.9
Corn stove (agricultural waste)	38.5	24.5	18.5
Rice husks (agricultural waste)	32.1	20.6	17.7
Corn bran (a by-product of milling)	20.5	65.3	1.6

It is crucial to carry out a pretreatment of this waste for its maximum use and thus avoid the formation of toxic compounds that may interfere with the production of biofuels.

There are different methods for the pretreatment of lignocellulosic waste, which can be Nauman Aftab et al. (2019).

- **Physical**: grinding, microwaves, mechanical extrusion, pyrolysis, pulsed electric field (PEF).
- **Chemicals**: acid, alkaline, organic solvents, ionic liquids, and ozonolysis.
- **Thermochemical**: Wet Oxidation, Ammonia Fiber Expansion (AFEX), Carbon Dioxide Explosion, Liquid Hot Water (LHW), Steam Explosion, and Sulfite Pretreatment to remove recalcitrance from lignocellulose, better known as SPORL.
- **Biological**: occurs due to the activity of microorganisms such as bacteria and fungi or enzymes, improving yield.

The processing of lignocellulosic material or biomass consists mainly of three steps: removal of lignin for the use of hemicellulose and cellulose, hydrolysis of carbohydrate polymers to generate free sugars and, finally, fermentation of free sugar to biofuel (Adewuyi 2022).

Biorefineries are the facilities in which biomass processing is carried out to obtain biofuels, which seek to be sustainable, so the raw material plays a significant role.

Biofuels are classified into first, second, third and fourth generation (Fig. 3.2), and the most significant variation is the raw material (Cavelius et al. 2023).

First generation: it is produced from edible raw material, or its origin is cropped, such as wheat, sugar cane or corn, which have a high starch and sucrose content; its fermentation is mainly to produce bioethanol, in the case of biodiesel, it is possible to obtain it from soybean oil. Biomass does not require intensive treatment; however, large tracts of land and high-water consumption are used (Cavelius et al. 2023).

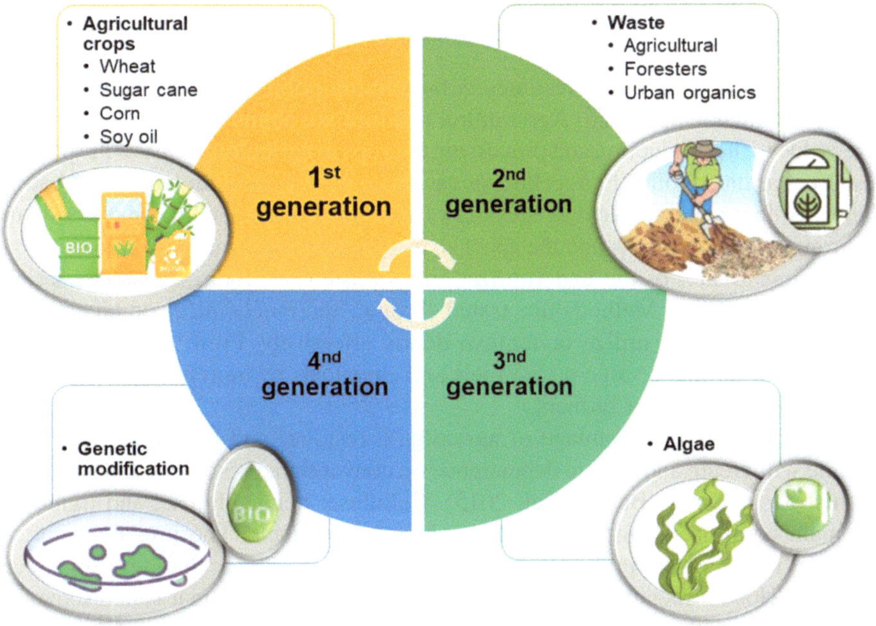

Fig. 3.2 Classification of biofuels

Second generation: they are mainly produced from the forest, and agricultural residues; it is also possible to use biomass such as trees or shrubs that can be cultivated for this purpose and have the characteristic of being materials with high lignocellulosic content, the leading biofuel obtained is bioethanol and to a lesser extent biobutanol. It has the advantage that it valorizes waste, so it is possible to integrate it into a circular economy system, in addition to the fact that there is no competition between allocating a product to food or biofuel; however, the available biomass is not high and requires intensive treatment (Kowalski et al. 2022).

Third generation: they are advanced biofuels that are produced from algae. This biomass reduces competition for land use for food crops, as with first-generation biofuels. Algae have the advantage of rapidly reproducing; however, nutrients such as nitrogen and phosphorus are required, which could cause alterations in the ecosystem due to their prolonged use and accumulation (Rafa et al. 2021).

Fourth generation: genetic engineering interacts in the production of these biofuels since genetic modifications of the microorganisms present are carried out to optimize the growth rate or interact with different carbon sources to produce a more significant amount of biofuel or improve its quality, which is its best advantage; in addition, there are times when it is possible to use CO_2 and synthesis gas directly. However, it also has the disadvantage that this type of production requires greater regulation or control of the process, mainly associated with biosafety (Cavelius et al. 2023).

3.7 Chemicals and Fuels from Agro-Waste

The were using agricultural residues in biorefining points to a reasonable solution with a sustainable approach. Agricultural residues are abundant at low or even free prices due to the production and processing of crops in agriculture. One of the primary resources as unexploited lignocellulosic raw material is agricultural residues. Among them are plant materials, leaves, and straws after harvest. Depending on the crop, the environmental conditions and the species, its characteristics may vary.

Burning in situ, or the underutilization of waste done today, generates serious pollution problems. Methods are tested through thermochemical or biochemical conversion routes to utilize or remove debris effectively. From these residues, it is possible to produce value-added fuels and chemicals by biorefining technologies and generate income for farmers.

Effectively taking advantage of agricultural residues and generating additional income will create and reduce dependence on conventional resources and low waste of this raw material (Beller et al. 2015). Cellulose, hemicellulose, and lignin are the main components for obtaining biochemical products. The important part of the plant is the sugars present as structural or storage polysaccharides. Sugars are always in close association with lignin. In the biorefinery, the sugar is first separated from the plant without damaging it to eliminate other components later. Glucose and hexoses are simple sugars predominantly used for biofuels such as ethanol. Lactic acid derived from glucose is a fundamental molecule in chemistry (Bouaid et al. 2010). Lactate esters are used as green solvents in industries, for example. It is possible to obtain acrylic acid and 1,2-propanediol as high-value products produced by further transforming lactic acid. An important derivative of glucose is succinic acid produced by chemical inducers (Ramesh et al. 2019).

In this sense, biorefining will play an increasingly important role in the supply of fuels, energy, and chemical products (1,4-butanediol, 2,3-butanediol, 1,2-propanediol, 3-hydroxypropionic acid, 1,3-propanediol, n -propanol, acetic acid, acetone, acrylic acid, adipic acid, n-butane, butyric, cadaverine, ethanol, fumaric acid, lactic acid, malic acid, propionic acid, glucaric acid, itaconic acid, isobutanol, isoprene, isopropanol, putrescine, styrene and succinic acid to mention a few) for sustainable economic growth with minimal impact on the environment by continuing developments and breakthroughs in new energy crops, synthetic biology for cell engineering, aquaculture, and conversion technologies (Hossain 2017).

3.8 Biofuels from Lignocellulosic Biomass Rich in Proteins and Lipids

Among the primary fuels produced and consumed is biodiesel, which seeks to replace diesel, made mainly from vegetable oil, cooking oils and animal fats through esterification (Alleman 2016). Bioethanol is produced through an alcoholic fermentation process. Biomass, such as rice straw, wheat, oats, sugarcane bagasse, and some types of trees, such as cherry trees, are favorable for producing this biofuel because it has a higher cellulose content than lignin and hemicellulose (Malik et al. 2022).

On the other hand, biogas is formed by anaerobic digestion, which indicates that it mainly occurs in the absence of oxygen and in four stages: hydrolysis, acidogenesis, acetogenesis, and methanogenesis, and different genera of microorganisms interact in each one. The composition of this biofuel is mainly methane (40–70%), carbon dioxide (15–60%) and, to a lesser extent, other gases such as hydrogen, nitrogen, and hydrogen sulfide.

Residues of animal origin, such as bovine and pig manure, those of vegetable origin, such as residues of corn, wheat and potato leaves, agro-industrial, forestry and even domestic are some of the substrates suitable for obtaining biogas; however, to obtain a quality The specific percentage of the composition also depends on other factors such as temperature, pH, nutrients, among others, which promote microbial activity (MINENERGIA et al. 2011).

There needs to be more research on protein-rich biowaste (PRB) used in biomethanation and the microbial communities that interact in the processes. However, until now, the production of biofuels by lipids such as used oils and carbohydrates such as lignocellulosic biomass has been very well studied.

Proteins for bioenergy production are an important raw material; these are polymers of amino acids connected by peptide bonds. Approximately 100 million tons of PRB are produced annually. Around 300 million tons of meat are produced annually; of these, slaughterhouse or kitchen waste represents 30%, and the pig, sheep, livestock, fishing, and poultry industries are a source of PBR. LPBCs (e.g., tanneries, shellfish, slaughterhouse, and kitchen wastes) are considered hazardous and can cause airborne diseases, odors and GHG emissions if left untreated. Through anaerobic digestion (AD), transesterification or fermentation, PRBs can produce biofuels. The high organic content for biomethanization of PBR of plant or animal origin can contribute to improved industrialization for biogas production (Ling et al. 2022).

Microbial acclimation, oxygen circulation, pretreatment, micro-aeration, nanoparticle treatment, bioaugmentation, and enhanced nutrient supply can enhance biogas production. A balance of the carbon/nitrogen (C/N) ratio by co-digesting PRBs can increase biomethanization (Ling et al. 2022).

Different studies have shown that lignocellulosic biomass has enough protein present, which can replace the fossil fuel market by promoting the profitable production of biofuels. Lignocellulosic biomass from protein-rich plant residues can be pretreated to obtain these proteins and converted into biofuels with the pretreated biomass.

3.9 Nanotechnology to Produce Biofuels

Economic and technological limitations exist in the technologies to process biomass (Masran et al. 2016). Therefore, interest in emerging nanotechnologies that improve current technologies and provide new alternatives to meet the world's energy needs has been boosted. Nanotechnology refers to the design of materials and phenomena at the nanoscale (1 and 100 nm). Bulk materials have very different properties from nanomaterials or nanosystems. Quantum mechanics primarily controls the nanoworld, with very different rules than classical physics. These differences cause the behavior of nanomaterials to be very different from bulk materials and therefore have properties (Nanotechnology 2014).

The critical bottleneck for lignin-derived biofuels is the need for more technology to convert biomass to liquid fuels efficiently. Developing efficient technological systems to solve the problem is an urgent need. Advances in nanotechnology that make it possible to understand and control molecular chemistry promise the potential development of efficient technologies to produce biofuels from biomass. Nanomaterials such as ZnO, TiO_2, SnO_2, Fe_3O_4, carbon, graphene and fullerene present unique properties that have allowed their application in the production of biofuels. Enzymes such as cellulases and hemicelluloses immobilized on nanomaterials have produced bioethanol (Antunes et al. 2017). The cost of the process can be reduced with this strategy, allowing to recycle the immobilized enzymes. The release of nanoparticles into the environment can pose a threat to the environment and human health (Gupta et al. 2022). Detecting and mitigating environmental impacts due to the production of nanomaterials must be considered with a life cycle approach (Keivani Nahr et al. 2015).

The application of nanotechnology to produce biofuels from lignocellulosic biomass still has few studies because the same nanotechnology is still under development. Table 3.3 summarizes some works that use nanotechnology to produce biofuels from lignocellulosic biomass. The spent tea, a solid residue, has been converted into biofuel with nanocatalytic catalysts in gasification, with the subsequent transesterification of the liquid fraction obtaining biodiesel. In contrast, the fermentation of the solid fraction with Aspergillus niger has been used to produce bioethanol (Mahmood and Hussain 2010).

Simultaneous solid state saccharification and fermentation (SSF) reactions of microcrystalline cellulose had higher ethane yields when cellulase is physisorbed on silica nanoparticles than when the enzyme is free. This yield shows that the immobilization of the enzyme on silica increases the results in the conversion by SSF of lignocellulosic material to ethanol (Lupoi and Smith 2011).

Aspergillus fumigatus cellulase immobilized on MnO_2 nanoparticles combined with yeast in SSF improve hydrolysis and efficiency to produce bioethanol in various organic wastes. The yield in the free enzyme was 18 g/L, while with the immobilized enzyme, it was 22 g/L. The immobilization also allowed to reuse of the cellulase and favored the operation in a greater range of pH and temperature (Cherian et al. 2015).

Table 3.3 Biofuels produced using nanotechnology

Feedstock	Biocatalyst	Products	References
Methanol and oleic acid	A solid acid catalyst, sulfonated-multiwalled carbon nanotubes (s-MWCNTs)	Biodiesel	Shu et al. (2009)
Spent tea *(Camellia sinensis)*	Cobalt nanoparticles	Biodiesel (40.8%) Bioethanol (57.5%)	Mahmood and Hussain (2010)
Microcrystalline cellulose	Cellulase physisorbed on silica nanoparticles	Bioethanol	Lupoi and Smith (2011)
Anaerobic sludge blanket (UASB)	Carbon Nanotubes	Biohydrogen (2.45 mol H_2/mol glucose)	Liu et al. (2012)
Jackfruit waste and sugarcane leaves	Cellulase immobilized on MnO_2 nanoparticles	Bioethanol (21.96 g/L)	Cherian et al. (2015)
Waste-activated sludge	Nano zero-valent iron (nZVI) and Fe_2O_3	Biogas (increased methane production by 120%)	Wang et al. (2016)
Sesbania aculate	Cellulase bound magnetic nanoparticles	Bioethanol (5.31 g/L)	Baskar et al. (2016)
Jatropha oil	Nanocatalysts CaO and Al_2O_3	Biodiesel	Hashmi et al. (2016)
Soybean oil	Magnetic nanocatalysts of $Ni_{0.5}Zn_{0.5}Fe_2O_4$ doped with Cu	Biodiesel	Dantas et al. (2017)
Anaerobic sludge	Immobilized *Enterobacter aerogenes* on functionalized multi-walled carbon nanotube (MWCNT-COOH)	Biohydrogen (2.2 moL H_2/mol glucose)	Boshagh et al. (2019)
Waste cooking oils	NP catalysts such as La_2O_3-SBA-15, MeO-SBA-15, ZnO-SBA-15	Biodiesel	Cao et al. (2019)
Jatropha curcus	Magnetic nanoparticles ($CaSO_4$ /Fe_2O_3-SiO_2)	Biodiesel	Teo et al. (2019)

Using magnetic nanoparticles attached to cellulase in the hydrolysis of Sesbania proved efficient in producing bioethanol. Maximum production of 5.3 g/L of bioethanol was achieved under optimal conditions of 1.5% p/v of catalyst, 4% p/v of biomass and 30 °C temperature (Baskar et al. 2016).

The use of microalgae to produce third-generation biofuels has also been reported to have high potential. Biodiesel production employing algae is promising since the algal biomass forms up to 50% of oil. Also, carbohydrates from algae are used in the production of bioethanol. Even the use of microalgae has several challenges to face, such as the water requirements that allow the algae to grow, the high cost of cultivation and harvesting, and the low concentration of CO_2 in the atmosphere before turning the algae into biofuels (Kim et al. 2013; Rashid et al. 2014). Some of these problems can be solved with nanotechnology. Catalyst supports or nanomaterial catalysts have been used to accumulate lipids, extract them, and achieve transesterification also in cultivation and harvesting (Lee et al. 2015). Cellulose and lipase enzymes immobilized on magnetic nanoparticles have been used to produce biodiesel from Chlorella salina, a marine microalga. Under optimized conditions of 2 g of each enzyme and 60% water, they yielded 94%. The use of the immobilized enzymes in 10 successive cycles was achieved efficiently (Duraiarasan et al. 2016).

There are still concerns to be resolved about applying nanoparticles to produce biofuels using microalgae, including efficiency, profitability, and environmental recycling.

Safety concerns towards health and the environment still need to be fully understood to promote nanotechnology despite appearing with an exciting option to produce biofuels in an efficient, economically viable and environmentally respectful manner.

In biofuel research, nanotechnology and nanomaterials are promising tools that provide quality products with cost-effective techniques. Unique properties such as crystallinity, adsorption capacity and stability, high surface area/volume ratio, unique attributes, size, and catalytic activity provide multiple advantages to nanoparticles (NPs) over other forms of biofuel synthesis. Nanocatalysts such as carbon nanotubes and metal oxide nanoparticles have additional properties that support high potential recovery in biofuel production. Pyrolysis, anaerobic digestion, gasification, and hydrogenation combined with nanotechnology have shown utility in the synthesis of biofuels.

3.10 Biofuels from Microorganisms

Microorganisms such as bacteria, yeasts and fungi are the source of lipids and enzymes in the production of biofuels, providing substrates such as fatty acids and alcohols. Still, they can also be a catalyst in the process.

It is possible to use microorganisms that come from different sources, such as wastewater, manure, and brewer fermentation residues, among others and that are used in the production of biodiesel due to their high lipid content; some may be Cho and Park (2018):

- **Oily microalgae**: *Chlorella pirenoidosa, Neochloris oleoabundans, Nanochloropsis sp., Scenedesmus bijuga, Scenedesmus bijuga,*
- **Oily yeasts**: *Cryptococcus curvatus, Rhodotorula glutinis, Lipomyces starkeyi, Lipomyces starkeyi, Yarrowia lipolytica,* among others.
- **Oily bacteria**: *Gordonia sp., Bacillus subtilis, Rhodococcus opacus.*

- **Oil fungi**: *Mucor circinelloides, Mortierella alpina, Cunninghamella echinulate, Mortierella elongata*

There are various enzymatic sources for the production of biodiesel. Each microorganism generates a different percentage yield of biofuel; for example, in the case of biodiesel: *Rhizomucor miehei* (87.8%), *Aspergillus oryzae* (94%), *Aspergillus oryzae* (95%), *Rhizopus oryzae* (84.6%), *Pseudomonas cepacian* (78%) *and Aspergillus niger* (80%), the origin of these microorganisms can be from different types of oils (Wang et al. 2021).

Different microorganisms have been found that are favorable in the production of bioethanol, some yeasts such as *Saccharomyces cerevisiae, Rhizopus, Mucor* and *Rhizomucor* fungi, bacteria such as *E. coli* and *Zymomonas mobilis* (Malik et al. 2022) and some algae such as *Alaria esculenta, Fucus vesiculosus, Laminaria digitata, Pelvetia canaliculate, Enteromorpha intestinalis,* among others (Greetham et al. 2018).

The central bacterial populations interacting in biogas production are fermentative, hydrogen-producing acetogenic, homoacetogenic, hydrogenotrophic methanogenic, and acetoclastic methanogenic. It is possible to mention by stage of the process the main genres that participate; in hydrolysis stand out *Bacteroides, Lactobacillus, Megasphaera, Bifidobacterium Propionibacterium, Sphingomonas, Sporobacterium,* Clostridium, Paenibacillus and Ruminococcus are found in the acidogenesis stage, *Cytophaga-Flavobacterium-Bacteroides, Acetobacter, Roseococcus* have been identified in the acetogenic stage *Acidicaldus, Craurococcus and in the methanogenic stage, mainly the archaea Methanobacterium, Methanospirillum hungatii,* and *Methanosarcina* are found (MINENERGIA et al. 2011).

There is great interest in using algae to produce biofuels such as bioethanol or biodiesel, which is why it is of great interest for developing marine biorefineries. Some species of algae include *Chlorella Vulgaris* (C. *Vulgaris), Tetraselmis* sp., *Kappaphycus alvarezii* and *Eucheuma denticulatum* (Zhou et al. 2022). In addition to the intention of promoting mutations in the species to improve their performance by increasing the production of lipids, as is the case with the *Chlamydomonas reinhardtii strain, Nannochloropsis salina, Nannochloropsis oceanica, Thalassiosira pseudonana, Nannochloropsis gaditana, Volvox carteri, Phaeodactylum tricornutum, Dunaliella salina* and *Picochlorum celeri* (Ruffing et al. 2022).

Cyanobacteria are another type of microorganism in which interest has arisen in their modification with the same objective of improving or producing more biofuel, some examples of which are *Synecococus 7942, Synecocystis 6803* for bioethanol production, for isobutanol the *Sinecocystis species 6803* and *Synecococcus 7942,* among others (Mund et al. 2022).

Currently, biotechnology and genetic engineering play an essential role in the production of biofuels and the microorganisms that interact, either for their identification or even in genetic modification for better use and therefore improve the quality of the final product, which can be renewable, in addition to contributing to the reduction of waste, as well as greenhouse gas emissions.

3.11 Strategies to Improve Enzymes Through Solid State Fermentation

The process of fermentation on a solid surface is known as solid-state fermentation (SSF); in the absence of "free" water, the moisture is absorbed into a solid matrix. Due to its lower complexity and more remarkable resemblance to the natural environment of the microorganisms, SSF presents advantages over other fermentation techniques. A disadvantage of this fermentation is that it is difficult to estimate the biomass concentration. Increased product concentration, lower manufacturing costs, direct application of the product, and reduced energy requirements make SSF a potential technique to produce various enzymes, such as lipase, cellulase, and tannase. Cellulase production can be improved in solid-state fermentation using multiple grades of lignocellulosic substrates, the microorganisms involved, the culture, and process parameters such as moisture content and temperature, water activity, nutrient diffusion, substrate particle size, pH, temperature, surfactants, and bioreactor designs. A significant problem related to the transfer of oxygen to the microorganisms, which, in turn, depends on the configuration, the size and the agitation/aeration system used in the reactor, occurs in the submerged fermentation that occupies a different place regarding various types of fermentations. An important parameter to characterize the oxygen transfer is K (oxygen transfer coefficient), whose value estimates how much oxygen is transferred by the equipment regardless of the volume of the reactor. Antioxidant production in SSF noted that pomace tends to increase antioxidant activity, converging on β-glucosidase activity with an increase. As shown in different studies, P. floridensis is an important organism used to produce lingo cellulolytic enzymes. The consecutive advance in the in vitro digestibility of wheat straw has been carried out to a greater extent (Rudakiya 2019).

There is a well-established fact that solid-state fermentation is one of the promising technologies for producing many industrially essential enzymes. Tannase production began in microbiology's early years, but more research is still needed to unravel many unknown facts about its efficient extraction and commercial production. The output of antioxidants with the help of SSF has led to the opening of many avenues, including the prevention of some deadly oxidation-related diseases, such as cancer. SSF is preferred over other fermentation technologies, such as submerged fermentation (SmF), for several reasons. One is the requirement of a lower amount of energy for the supply of oxygen so that the system can cope with the high oxygen demand.

3.12 Production of Second-Generation Biofuels: The State of the Art

The most important challenge in second-generation bioethanol production is obtaining the substrate since it must be a massive production. In addition to low cost, these do not come from sources destined for food (Malode et al. 2022). The optimization of this process is constantly sought.

Goh et al. (2022) mixed cooking oil residue with Jatropha esterified oil to carry out *the* transesterification process and obtain biodiesel that complied with ASTM parameters, such as acid value, calorific value, and viscosity, kinematics and oxidation stability. They emphasize that it is possible to improve the performance of this fuel by mixing it with commercial diesel and antioxidants.

Sugarcane bagasse is an abundant residue and constant study; when mixed with the imidazole catalyst, it has generated 100% enzymatic conversion to produce bioethanol, which is reflected in an increase of 37% of this biofuel, which glimpses the viability of biorefineries of this residue on an industrial scale (Goh et al. 2022).

Second-generation bioethanol from lignocellulosic biomass residues is a sustainable solution to declining oil reserves, environmental deterioration due to GHG emissions and national security. One of the best strategies can be the development and application of more efficient organisms in association with the intelligent integration of various processes in an integrated biorefinery approach, where many products can be obtained in addition to bioethanol alone can also include applications of consolidated bioprocessing.

3.13 Life Cycle Analysis Applied to the Production of Biofuels

Bioenergy as a renewable energy option is very relevant worldwide. Bioenergy can be carbon neutral and significantly reduce GHG emissions compared to fossil fuels. Several reviews of life cycle assessments (LCA) of biofuels have been published, where heat and power contribute 90% of total biomass energy consumption. Few studies focus on biomass heat and power generation, only addressing a few impact categories. No LCA review of biomass heat and electricity included a detailed quantitative assessment. There needs to be more quantitative description, insufficient consideration of impact categories, and a lack of care for heat generation in bioenergy LCA reviews.

In general, the optimal functional unit has yet to reach a consensus, the ideal distribution of the environmental impact between the co-products, the definition of the system's limit, or how to model the biomass carbon cycle. It has been concluded that a higher level of transparency and harmonization of the preparation of LCAs for electricity and biomass heat is needed to improve the comparability of such assessments.

To know the sustainability of a process, in this case, the production of biofuels, it is important to see the sustainability of the process, verifying compliance with technical, economic, and environmental criteria. Environmental sustainability is generally analyzed under the life cycle approach (Table 3.4).

Most of the LCA studies for different raw materials of biological origin used in the production of biofuels have shown a significant decrease in the associated environmental impacts, indicating that it is the way forward to sustain the world's energy demand sustainably.

3.14 Challenges and Prospects

The production of biofuels is an issue that arises from the need to replace the consumption of fossil fuels with fuels that have a low environmental impact throughout their life cycle and that are renewable.

As mentioned, the raw material for biofuels comes from organic sources; therefore, one of the main challenges facing production is obtaining it since it will be necessary not to consume products that can generate a high negative impact on the environment or other raw materials are consumed excessively, such as first generation biofuels or in the case of second-generation biofuels, resolve the issue of logistics regarding the selection, collection and transportation of organic waste for the location from a biorefinery.

As technology and biotechnology advance, the challenges involved in the production of these biofuels can be solved, where a more significant application of nanotechnology is glimpsed to stimulate and improve production processes, as well as synergy with raw materials according to with the generation of biofuel to optimize production and have a sustainable process.

It is important to mention that the production of biofuels is not an isolated process; what is meant is that it can also be part of a primary process and be integrated as an element that promotes zero waste generation or be incorporated into the application of the circular economy.

For now, the issues to be resolved are:

- What is the best raw material to produce biofuels?
- What biofuel to produce?
- What quality and cost must it have to compete with its fossil equivalent?
- Where to locate a biorefinery?

Among other questions, however, one should consider factors such as geographic location, raw material availability, costs associated with production and distribution, target market, and those that arise during decision-making or implementation of public policies in producing biofuels.

Table 3.4 LCA studies on the production of biofuels

Feedstock	Biofuel	Impact categories	Results	References
Corn stover (residue) and lignin-rich stream (waste)	Biofuels	Global warming Depletion of fossil resources eutrophication	Similar global warming between the two cases Corn stubble has 28% more potential for acidification and depletion of fossil fuels Similar eutrophication in both cases	Zoppi et al. (2023)
Sugarcane bagasse	Bioethanol	Cumulative energy demand, and water footprint, global warming potential (GWP), Abiotic Depletion Potential (ADP), Marine Aquatic Ecotoxicity (MAETP), Eutrophication Potential (EP), Ozone Layer Depletion (ODP), Photochemical Oxidation Potential (POP), Land Competition (LC), Human Toxicity Potential (HTP), Fresh Water Ecotoxicity Potential (FAETP), Terrestrial Ecotoxicity Potential (TETP)	Biofuel uses in vehicles (70%), and biofuel production (25%) stages present the highest contributions for the GWP impact category, quantified as ~26.7 kg CO_2-eq/L. When a cradle-to-gate perspective was modeled, this amount was significantly reduced (~8.4 kg CO_2-eq/L). For another impact categories evaluated, the highest contributions corresponded to the biofuel production (55–95% for ADP, MAETP, EP, ODP, and POP), and raw material extraction stages (70–100%: LC, HTP, FAETP and TETP)	Santoyo-Castelazo et al. (2023)

(continued)

Table 3.4 (continued)

Feedstock	Biofuel	Impact categories	Results	References
Second-generation feedstocks: Castor (Ricinus communis), Croton Megalocarpus, and Jatropha	Biodiesel (B10)	Global warming Human toxicity fossil depletion Ecotoxicity in freshwater Freshwater eutrophication Formation of photochemical oxidants particle formation	Compared to fossil diesel, a decrease of between 4 and 9% was observed in all impact categories on biodiesel production	Tibesigwa et al. (2023)
	Biodiesel (B100)		A decrease of 36 and 94% was observed in all impact categories on biodiesel production compared to fossil diesel; however, it can only be applied as a mixture for B10 and other combinations	
Wastewater Treatment Sludge (WWTS)	Biodiesel	Global warming potential and depletion of fossil resources mostly	The impact categories most affected by the systems are climate change (with a contribution from 30 to 36% in the seven scenarios) and the depletion of fossil resources (with a contribution from 23 to 31% in the seven scenarios)	Notarnicola et al. (2023)

References

Adewuyi A (2022) Underutilized lignocellulosic waste as sources of feedstock for biofuel production in developing countries. Front Energy Res 10:142. https://doi.org/10.3389/FENRG.2022.741570/BIBTEX

Alleman RLM (2016) Biodiesel handling and use guide. In: Biodiesel handling and use guide, 5th edn, Issue November

Antunes FAF, Gaikwad S, Ingle AP, Pandit R, dos Santos JC, Rai M, da Silva SS (2017) Bioenergy and biofuels: nanotechnological solutions for sustainable production, 3–18. https://doi.org/10.1007/978-3-319-45459-7_1

Baskar G, Naveen Kumar R, Heronimus Melvin X, Aiswarya R, Soumya S (2016) Sesbania aculeate biomass hydrolysis using magnetic nanobiocomposite of cellulase for bioethanol production. Renew Energy 98:23–28. https://doi.org/10.1016/j.renene.2016.04.035

Beller HR, Lee TS, Katz L (2015) Natural products as biofuels and bio-based chemicals: Fatty acids and isoprenoids. Nat Prod Rep 32(10):1508–1526. https://doi.org/10.1039/c5np00068h

Boshagh F, Rostami K, Moazami N (2019) Biohydrogen production by immobilized enterobacter aerogenes on functionalized multi-walled carbon nanotube. Int J Hydrogen Energy 44(28):14395–14405. https://doi.org/10.1016/j.ijhydene.2018.11.199

Bouaid A, Martínez M, Aracil J (2010) Biorefinery approach for coconut oil valorisation: a statistical study. Biores Technol 101(11):4006–4012. https://doi.org/10.1016/j.biortech.2009.12.074

Cho HU, Park JM (2018) Biodiesel production by various oleaginous microorganisms from organic wastes. https://doi.org/10.1016/j.biortech.2018.02.010

Cao X, Li L, Shitao Y, Liu S, Hailong Y, Qiong W, Ragauskas AJ (2019) Catalytic conversion of waste cooking oils for the production of liquid hydrocarbon biofuels using in-situ coating metal oxide on SBA-15 as heterogeneous catalyst. J Anal Appl Pyrolysis 138(November 2018):137–144. https://doi.org/10.1016/j.jaap.2018.12.017

Cavelius P, Engelhart-Straub S, Mehlmer N, Lercher J, Awad D, Brück T (2023) The potential of biofuels from first to fourth generation. PLOS Biol 21(3):e3002063. https://doi.org/10.1371/JOURNAL.PBIO.3002063

Cherian E, Dharmendirakumar M, Baskar G (2015) Immobilization of cellulase onto MnO_2 nanoparticles for bioethanol production by enhanced hydrolysis of agricultural waste. Cuihua Xuebao/Chinese J Catal 36(8):1223–1229. https://doi.org/10.1016/S1872-2067(15)60906-8

Dantas J, Leal E, Mapossa AB, Cornejo DR, Costa ACFM (2017) Magnetic nanocatalysts of $Ni_{0.5}Zn_{0.5}Fe_2O_4$ doped with Cu and performance evaluation in transesterification reaction for biodiesel production. Fuel 191:463–471. https://doi.org/10.1016/j.fuel.2016.11.107

Dragone G, Kerssemakers AAJ, Driessen JLSP, Yamakawa CK, Brumano LP, Mussatto SI (2020) Innovation and strategic orientations for the development of advanced biorefineries. *Bioresource Technology, 302*(December 2019), 122847. https://doi.org/10.1016/j.biortech.2020.122847

Duraiarasan S, Razack SA, Manickam A, Munusamy A, Syed MB, Ali MY, Ahmed GM, Mohiuddin MS (2016) Direct conversion of lipids from marine microalga C. salina to biodiesel with immobilised enzymes using magnetic nanoparticle. J Environ Chem Eng 4(1):1393–1398. https://doi.org/10.1016/j.jece.2015.12.030

Fallahi A, Farzad S, Mohtasebi SS, Mandegari M, Görgens JF, Gupta VK, Lam SS, Tabatabaei M, Aghbashlo M (2021) Sustainability assessment of sugarcane residues valorization to biobutadiene by exergy and exergoeconomic evaluation. Renew Sustain Energy Rev 147(April). https://doi.org/10.1016/j.rser.2021.111214

Goh B, Tung Chong C, Chyuan Ong H, Milano J, Halim Shamsuddin A, Jiat Lee X, Ng J-H (2022) Strategies for fuel property enhancement for second-generation multi-feedstock biodiesel. https://doi.org/10.1016/j.fuel.2022.123178

Greetham D, Zaky A, Makanjuola O, Du C (2018) A brief review on bioethanol production using marine biomass, marine microorganism and seawater. Curr Opin Green Sustain Chem 14:53–59. https://doi.org/10.1016/J.COGSC.2018.06.008

Gupta VK, Pandey A, Koffas M, Mussatto SI, Khare S (2022) Biobased biorefineries: Sustainable bioprocesses and bioproducts from biomass/bioresources special issue. Renew Sustain Energy Rev 167(May). https://doi.org/10.1016/j.rser.2022.112683

Hashmi S, Gohar S, Mahmood T, Nawaz U, Farooqi H (2016) Biodiesel production by using CaO-Al_2O_3 nano catalyst. Int J Eng Res Sci 2(3):2395–6992. https://pdfs.semanticscholar.org/0695/119726ef54d7f35415c12edeb0be82673db2.pdf

Hossain GS, Liu L, Du GC (2017) Industrial bioprocesses and the biorefinery concept. In: Current developments in biotechnology and bioengineering: bioprocesses, bioreactors and controls, pp 3–27. https://doi.org/10.1016/B978-0-444-63663-8.00001-X

Irmak S (2019) Challenges of biomass utilization for biofuels. In: Biomass for bioenergy—recent trends and future challenges. IntechOpen, pp. 1-12. https://doi.org/10.5772/intechopen.77410

Keivani Nahr F, Mokarram RR, Hejazi MA, Ghanbarzadeh B, Sowti Khiyabani M, Zoroufchi Benis K (2015) Optimization of the nanocellulose based cryoprotective medium to enhance the viability of freeze dried Lactobacillus plantarum using response surface methodology. Lwt 64(1):326–332. https://doi.org/10.1016/j.lwt.2015.06.004

Kim J, Yoo G, Lee H, Lim J, Kim K, Kim CW, Park MS, Yang JW (2013) Methods of downstream processing for the production of biodiesel from microalgae. Biotechnol Adv 31(6):862–876. https://doi.org/10.1016/j.biotechadv.2013.04.006

Kostas ET, Adams JMM, Ruiz HA, Durán-Jiménez G, Lye GJ (2021) Macroalgal biorefinery concepts for the circular bioeconomy: a review on biotechnological developments and future perspectives. Renew Sustain Energy Rev 151(August). https://doi.org/10.1016/j.rser.2021.111553

Kowalski Z, Kulczycka J, Verhé R, Desender L, De Clercq G, Makara A, Generowicz N, Harazin P (2022) Second-generation biofuel production from the organic fraction of municipal solid waste. Front Energy Res 10(August):1–15. https://doi.org/10.3389/fenrg.2022.919415

Lee YC, Lee K, Oh YK (2015) Recent nanoparticle engineering advances in microalgal cultivation and harvesting processes of biodiesel production: a review. Biores Technol 184:63–72. https://doi.org/10.1016/j.biortech.2014.10.145

Ling Z, Thakur N, El-Dalatony MM, Salama ES, Li X (2022) Protein biomethanation: insight into the microbial nexus. Trends Microbiol 30(1):69–78. https://doi.org/10.1016/j.tim.2021.06.004

Liu Z, Lv F, Zheng H, Zhang C, Wei F, Xing XH (2012) Enhanced hydrogen production in a UASB reactor by retaining microbial consortium onto carbon nanotubes (CNTs). Int J Hydrogen Energy 37(14):10619–10626. https://doi.org/10.1016/j.ijhydene.2012.04.057

Lupoi JS, Smith EA (2011) Evaluation of nanoparticle-immobilized cellulase for improved ethanol yield in simultaneous saccharification and fermentation reactions. Biotechnol Bioeng 108(12):2835–2843. https://doi.org/10.1002/bit.23246

Mahmood T, Hussain ST (2010) Nanobiotechnology for the production of biofuels from spent tea. Afr J Biotech 9(6):858–868. https://doi.org/10.5897/ajb09.1555

Malik K, Sharma P, Yang Y, Zhang P, Zhang L, Xing X, Yue J, Song Z, Nan L, Yujun S, El-Dalatony MM, Salama ES, Li X (2022) Lignocellulosic biomass for bioethanol: insight into the advanced pretreatment and fermentation approaches. Ind Crops Prod 188:115569. https://doi.org/10.1016/J.INDCROP.2022.115569

Malode SJ, Gaddi SAM, Kamble PJ, Nalwad AA, Muddapur UM, Shetti NP (2022) Recent evolutionary trends in the production of biofuels. Mater Sci Energy Technol 5:262–277. https://doi.org/10.1016/J.MSET.2022.04.001

Masran R, Zanirun Z, Bahrin EK, Ibrahim MF, Lai Yee P, Abd-Aziz S (2016) Harnessing the potential of ligninolytic enzymes for lignocellulosic biomass pretreatment. Appl Microbiol Biotechnol 100(12):5231–5246. https://doi.org/10.1007/s00253-016-7545-1

MINENERGIA M de E, PNUD P de las NU para el D, FAO O de las NU para la A y la A, GEF GEF (2011) Manual de biogás

Mund NK, Liu Y, Chen S (2022) Advances in metabolic engineering of cyanobacteria for production of biofuels. Fuel 322. https://doi.org/10.1016/J.FUEL.2022.124117

Nanotechnology (2014) No Title. http://science.howstuffworks.com/nanotechnology1.htm

Nauman Aftab M, Iqbal I, Riaz F, Karadag A, Tabatabaei M (2019) Different pretreatment methods of lignocellulosic biomass for use in biofuel production. In: Biomass for bioenergy—recent trends and future challenges. IntechOpen. https://doi.org/10.5772/intechopen.77410

Notarnicola B, Tassielli G, Renzulli PA, Di Capua R, Astuto F, Mascolo G, Murgolo S, De Ceglie C, Curri ML, Comparelli R, Dell'Edera M (2023) Life cycle assessment of UV-C based treatment systems for the removal of compounds of emerging concern from urban wastewater. Sci Total Environ 857(August 2022):159309. https://doi.org/10.1016/j.scitotenv.2022.159309

OECD/FAO (2022) OECD-FAO agricultural outlook 2022–2031

Rafa N, Ahmed SF, Badruddin IA, Mofijur M, Kamangar S (2021) Strategies to produce cost-effective third-generation biofuel from microalgae. Front Energy Res 9(September):1–11. https://doi.org/10.3389/fenrg.2021.749968

Ramesh D, Muniraj IK, Thangavelu K, Karthikeyan S (2019) Chemicals and fuels production from agro residues: a biorefinery approach. Springer International Publishing. https://doi.org/10.1007/978-3-319-94797-6_3

Rashid N, Ur Rehman MS, Sadiq M, Mahmood T, Han JI (2014) Current status, issues and developments in microalgae derived biodiesel production. Renew Sustain Energy Rev 40:760–778. https://doi.org/10.1016/j.rser.2014.07.104

Rudakiya DM (2019) Strategies to improve solid-state fermentation technology. In: New and future developments in microbial biotechnology and bioengineering: from cellulose to cellulase: strategies to improve biofuel production, August, pp 155–180. https://doi.org/10.1016/B978-0-444-64223-3.00010-2

Ruffing AM, Davis RW, Lane TW (2022) Advances in engineering algae for biofuel production. https://doi.org/10.1016/j.copbio.2022.102830

Santoyo-Castelazo E, Santoyo E, Zurita-García L, Camacho Luengas DA, Solano-Olivares K (2023) Life cycle assessment of bioethanol production from sugarcane bagasse using a gasification conversion process: bibliometric analysis, systematic literature review and a case study. Appl Therm Eng 219(PA):119414. https://doi.org/10.1016/j.applthermaleng.2022.119414

Shu Q, Zhang Q, Xu G, Wang J (2009) Preparation of biodiesel using s-MWCNT catalysts and the coupling of reaction and separation. Food Bioprod Process 87(3):164–170. https://doi.org/10.1016/j.fbp.2009.01.004

Teo SH, Islam A, Chan ES, Thomas Choong SY, Alharthi NH, Taufiq-Yap YH, Awual MR (2019) Efficient biodiesel production from Jatropha curcus using $CaSO_4/Fe_2O_3$-SiO_2 core-shell magnetic nanoparticles. J Clean Prod 208(2019):816–826. https://doi.org/10.1016/j.jclepro.2018.10.107

Tibesigwa T, Iezzi B, Lim TH, Kirabira JB, Olupot PW (2023) Life cycle assessment of biodiesel production from selected second-generation feedstocks. Cleaner Eng Technol 13:100614. https://doi.org/10.1016/J.CLET.2023.100614

Usmani Z, Sharma M, Awasthi AK, Lukk T, Tuohy MG, Gong L, Nguyen-Tri P, Goddard AD, Bill RM, Nayak SC, Gupta VK (2021) Lignocellulosic biorefineries: the current state of challenges and strategies for efficient commercialization. Renew Sustain Energy Rev 148(May):111258. https://doi.org/10.1016/j.rser.2021.111258

Usmani Z, Sharma M, Diwan D, Tripathi M, Whale E, Jayakody LN, Moreau B, Thakur VK, Tuohy M, Gupta VK (2022) Valorization of sugar beet pulp to value-added products: a review. Bioresource Technol 346(October 2021):126580. https://doi.org/10.1016/j.biortech.2021.126580

Wang T, Zhang D, Dai L, Chen Y, Dai X (2016) Effects of metal nanoparticles on methane production from waste-activated sludge and microorganism community shift in anaerobic granular sludge. Sci Rep 6(February):1–10. https://doi.org/10.1038/srep25857

Wang H, Peng X, Zhang H, Yang S, Li H (2021) Microorganisms-promoted biodiesel production from biomass: a review. Energy Convers Manage X 12. https://doi.org/10.1016/J.ECMX.2021.100137

Zhou Y, Remón J, Gracia J, Jiang Z, Pinilla JL, Hu C, Suelves I (2022) Toward developing more sustainable marine biorefineries: a novel 'Sea-Thermal' process for biofuels production from microalgae. Energy Convers Manage 270. https://doi.org/10.1016/J.ENCONMAN.2022.116201

Zoppi G, Tito E, Bianco I, Pipitone G, Pirone R, Bensaid S (2023) Life cycle assessment of the biofuel production from lignocellulosic biomass in a hydrothermal liquefaction—aqueous phase reforming integrated biorefinery. Renew Energy 206:375–385. https://doi.org/10.1016/J.RENENE.2023.02.011

Chapter 4
Biofuel Production: An Ecofriendly Way to Produce Sustainable Energy

Deepshikha Datta, Rajat Kanti Ghosal, Soheli Biswas, Esha Mandal, and Bimal Das

Abstract Sustainable biofuel production is an essential component of a broader strategy to address climate change, minimize greenhouse gas emissions, reduce water and energy consumption, and promote the use of non-food feedstocks and decrease dependence on fossil fuels. The current work explores the potential of sustainable technologies for biofuel production. It discusses the environmental impacts of traditional biofuel production methods and highlights the need for sustainable alternatives. The review deals with various sustainable technologies, including second-generation biofuels, algae-based biofuels, and waste-to-energy systems. The benefits and challenges of each technology are examined, and potential future developments are discussed. The range of sustainable technologies for biofuel production, including the use of genetically modified organisms to increase yield and efficiency, the development of new feedstocks such as microalgae and cellulosic biomass, and the implementation of advanced conversion processes such as hydrothermal liquefaction and gasification. Various sustainable technologies for biofuel production include advanced feedstock cultivation and harvesting methods, biorefinery processes that convert biomass into various biofuels and other valuable products, and efficient conversion processes that utilize waste streams and reduce emissions are discussed. It serves as a useful resource for researchers, practitioners, and policymakers interested in advancing in the field of bioenergy.

Keywords Sustainable biofuel · Feedstocks · Sustainable technologies · Bioenergy · Emissions · Conversion

D. Datta
Department of Chemistry, Brainware University, Barasat, Kolkata 700125, West Bengal, India

R. K. Ghosal · S. Biswas
Department of Biotechnology, Brainware University, Barasat, Kolkata 700125, West Bengal, India

E. Mandal
Department of Biotechnology, National Institute of Technology, Durgapur 713213, India

B. Das (✉)
Department of Chemical Engineering, National Institute of Technology, Durgapur 713213, India
e-mail: bimal.das@che.nitdgp.ac.in

M. Shah and D. Deka (eds.), *Emerging Sustainable Technologies for Biofuel Production*, Environmental Science and Engineering, https://doi.org/10.1007/978-3-031-52167-6_4

4.1 Introduction

A number of biochemical, physical, and thermochemical methods are used to create the energy-dense molecules that make up sustainable biofuel production techniques from biomass derived from plants and microbes. Greenhouse gas (GHG) emissions from transportation are growing in a rapid division than from any other sources (Sims 2014). The sector is largely dependent over non-renewable fuels, that is considered for around 97% of total transport fuels in 2018 (Kusch-Brandt 2019). For usage in the transportation, industrial, residential, commercial, and power generating sectors, they represent a workable substitute for traditional fossil fuels. In contradict to fossil fuel energy origins, bio-fuels provide energy indemnity, lower GHG discharges, boost the economics, and promote rustic communities through generating employment opportunities and local harvesting merchantry opportunities (Sims 2014). To decrease non-renewable fuels, along with mitigating environmental changes, biofuels are widely observed as a hopeful substitute medium in transport facilities. Bio-fuels have been in application from the very beginning of the automotive transportation system (Michael et al. 2011). Biofuels are divided into multiple generations based on the feedstock. To encourage and improve sustainable biofuel production, a number of elements, including feedstock, effective production methods, technological preparedness, and market-driven regulations, must be taken into account. By the 1940s, biofuels were considered a viable transportation fuel, and various bio-ethanol compositions were widely applied as fuel energy in the USA regions or, Europe etc. (Michael et al. 2011). By the time of the oil emergency in the 1970s, most of the continents turned towards commercial renewable biofuel energy generation; but Brazil was the only country who has begun to generate ethanol on a bigger scale under the National Ethanol Program "Proálcool" (Sims 2014; Soccol et al. 2005). In the period of 1990s, when crude oil costs were at a top level and concerning energy safety worldwide, the United States and most of the countries in Europe has displayed policies to boost up bio-fuel generation (Sims 2014; Elbehri et al. 2013). Bio-fuel production has attracted attention from all over the world over the last few decennaries with the prosperity in climate change mitigation policies and tactics to decrease greenhouse gas generation from the transportation systems. Since then, over 50 countries started bio-fuel production programs and put biofuel blending objectives into their own fuel production tanks (Nikolina 2016). Most famous programs are the Renewable Fuel Standard (RFS) in the USA and the Renewable Energy Directive (RED) in Europe (EPA 2010; Kupiec 2021; Jeswani et al. 2020).

Biofuels have both beneficial and drawback sites in terms of impact on climate, economy, and hold down capability in the market (Azapagic and Stichnothe 2011). One side there are, reducing greenhouse gas generations, energy safety and rustic prosperity, which are vital factors that control biofuel generation worldwide. And the other side, perturbations over the increased biofuel generation, like high-placed arc over food-supply costs, the presumption of increased greenhouse gas generation by straight and inverse LUC (land use change) from bio-fuel generation, along with the risk of land corrosion, crops, water supplies, and ecology (Bringezu 2009). To make biofuels economically viable and competitive with fossil fuels, market interventions, financial help, grants, and subsidies must be supported. Application of 1st generation feedstocks, like corn starch, partly became a problematical outpouring, mainly due to comparison with food generation and perturbations about the use of lands for fuel production, which were used for cultivation. The increasing need for cultivational products threatens to rise deforestation and the usage of those lands which has great biodiversity, to fulfil biofuel need, along with the amalgamated use of fresh water, manures and pesticides with bad environmental effects. Few of these problems might be solved with the help of 2nd generation raw materials; howsoever, sustainable economy of few 2nd generation bio-fuels continues to be questionable in nowadays economic references, mainly due to lesser oil cost (Roy and Dutta 2013; Stephenson et al. 2010; Daystar et al. 2015). 3rd generation (e.g., algal) bio-fuels might also decrease the problem of food supplement issues and land usage, though micro-algae can grow over cultivable soil and in sewage, salt or salty water, and they grow up very quickly. Howsoever, the generation of biofuel from micro-algae is strength-severe and currently impractical from an economical viewpoint (Passell et al. 2013). In order to embolden the improvement of sustainable bio-fuels, regulatory strategies like the Renewable Energy Directive and the Renewable Fuel Standard provide for different sustainable measures for bio-fuels. Greenhouse gas generation during the life cycle is one of the prime measurements. The Renewable Energy Directive requires biofuels for minimum GHG generation (50%) compare to their non-renewable substitutes for plants operating by October 2015 (Kupiec 2021). The Renewable Fuel Standard needs generators of developed bio-fuels to decrease greenhouse gas generation (at a 50% minimal level), whilst standard bio-fuels must acquire at least 20% decreasement in greenhouse gas generation (EPA 2010). Environmental alternation impacts concerned with greenhouse gas generation and another perspective of biofuel sustainability should be assessed over a life-cycle cornerstone using Life Cycle Assessment (LCA) to get off burden changing from one portion of the life cycle to another supply chain (Jeswani et al. 2020).

This chapter investigates different traditional biofuel production techniques, their environmental impacts, different sustainable techniques and modern techniques of biofuel production and also aims to review energy production from different wastes (bagasse, waste oil, waste water).

4.2 Traditional Bio Fuel Production Techniques

Sustainable and renewable energy resources include bio fuels. However, they do emit carbon dioxide into the environment, they do so very sparingly since they can cut greenhouse emissions by approximately 30%. People are now turning to bio fuel to satisfy their everyday demands in an effort to create a more environmentally friendly future and do away with the need for fossil fuels and other conventional energy sources. As a result, many nations use bio fuel for transportation. Moreover, it is employed to produce energy. Several impoverished and emerging nations' rural areas will utilize bio fuel for cooking as well. Overall, there are several applications for this specific fuel.

Biomass, which is used as a source of clean energy in and of itself, is used to create bio fuel. Through a process, we can turn biomass into bio fuel. If we have the necessary tools and supplies, we can even complete these tasks in our homes. The various methods of changing biomass into microalgal fuels are shown in Table 4.1.

All the traditional techniques follow the following common pathways for the feed stock to produce (Fig. 4.1).

Filtering from the biomass:

The filtering procedure is used to remove unwanted particles from the biomass. Here, we take the used vegetable oil and heat it to a specific temperature. The waste materials will naturally separate from the main mixture once the liquid has been heated. The final step is to filter it using ordinary filter paper.

Table 4.1 Various methods for changing the biomass into microalgal fuels

Biomass	Conversion route	Process	Product
Microalgal biomass	Biochemical	Fermentation	Bioethanol
		Anaerobic digestion	Biogas
		Photo biological water splitting	Biohydrogen
	Thermochemical	Combustion	Heat and Electricity
		Liquefaction	Bio-oil
		Gasification	Syngas, Bio-Oil
		Pyrolysis	Syngas
	Chemical	Trance Esterification	Biodiesel

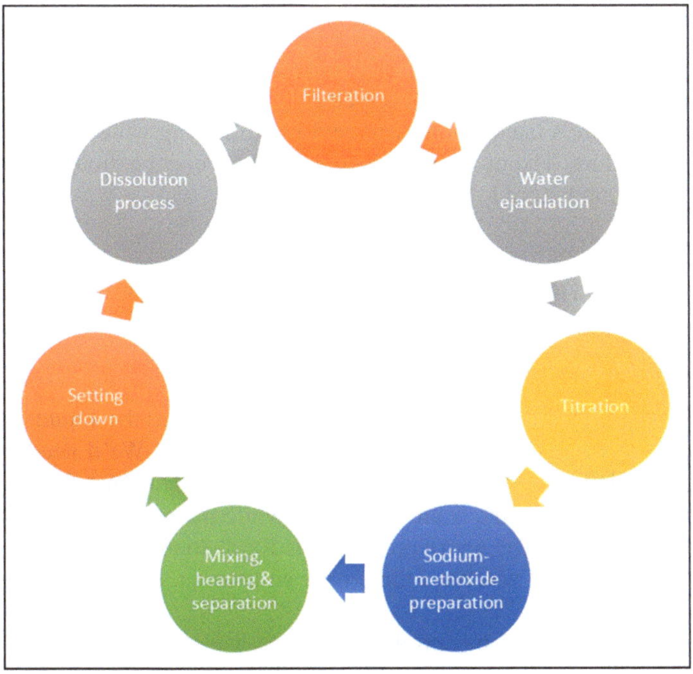

Fig. 4.1 Traditional technique for biofuel production

Water ejaculation process:

The leftover gangue must next be cleaned of water. The procedure will be delayed if the water is allowed to persist in the mixture. We can speed up the reaction significantly by getting rid of all the water. Heating the mixture steadily for a while at 212 °F is the simplest technique to eliminate water from the mixture.

Titration of the mixture:

To establish how much chemical catalyst will be required, the mixture is titrated. All chemical reaction needs a catalyst, which is an essential element. It essentially decides how quickly and how significant thing we'll get. As a result, this phase is crucial in the production of biofuel.

Sodium meth-oxide preparation:

In this phase, we combine sodium hydroxide and methanol (18–20% of the used vegetable oil). As a result, sodium meth-oxide is produced, serving as a catalyst for the subsequent process. It streamlines the entire reaction process and aids in carrying out synthesis reactions on the reagents. A crucial component in this manufacturing process is sodium methoxide. It is regarded as a common ingredient that is utilized to speed up the response and provide superior outcomes.

Mixing, heating and separation process:

The residue is then heated to between 120 and 130 °F. Following that, we thoroughly mix it. The mixture will be distributed equally through this procedure. This will facilitate the mixture's later cooling and settling, following which we may start the separation procedure. The mixing and heating process may be thought of as the last stage of preparation before to separation process.

Setting down process:

The liquid is allowed to cool and settle when the mixing process is finished, and then the bio fuel is getting extracted.

Dissolution process:

The biofuel may be removed from the mixture's top once the liquid has cooled. Similar to oil in water, it will be observed floating on top. We'll need to take out the glycerine below it in order to extract the biofuel. This may be achieved by simply emptying the gasoline from the bottom while maintaining stability. Finally, the biofuel is prepared. The entire procedure is for a small-scale operation, as was previously stated. Yet, if you have the proper equipment, supplies, and setup, you can scale it up as needed.

It should be noted that the production process also uses chemical catalysts, such as lye. Ultrasonics, however, are currently being investigated by scientists and researchers as additional catalysts. Recent research suggests combining chemicals and ultrasonic can increase fuel production while shortening processing time. Also, this improves how well biomass is used. Nowadays, a range of industries, including the renewable energy sector, are employing ultrasonic systems and technologies, like Coltraco. Although the technology has recently gained greater popularity in other domains, it shouldn't be long before it is actively employed in the production of biofuels and other renewable energy sources (Zafar 2022). There are some major biofuel production companies present, which is briefly described in Table 4.2.

4.3 Environmental Impact of Traditional Technique

Many recognized adverse environmental impacts of petroleum extraction, transportation, refining, and burning include the destruction of delicate biological ecosystems and significant greenhouse gas (GHG) emissions. Although replacing petroleum-based fuels with biofuels might lessen the country's reliance on foreign oil and perhaps lessen overall environmental impact, biofuels do have certain environmental costs. Each phase of a biofuel's life cycle involves the use of non-renewable resources and produces emissions that have an impact on the land, the air, and the water. So, it would be necessary to assess the environmental advantages and drawbacks of biofuels and petroleum-based fuels over the course of their respective life cycles in order for policymakers to determine which trade-offs are tolerable (in Fig. 4.2 biofuel

Table 4.2 Details of major biofuel production facilities using organic wastes from various companies. Reproduced with permission from Stephen and Periyasamy (2018)

Company	Country	Biofuel types	Installed capacity (million litres per year)
High plains bio energy LLC (Carsanba et al. 2018)	United States	Biodiesel from animal fats (Lucia and Grisolia 2018)	114
Urban X renewable group (Carsanba et al. 2018)		Renewable diesel from waste frying oil	
DuPont Danisco (Carsanba et al. 2018)		Celluloid ethanol from waste biomass (Lucia and Grisolia 2018)	0.95
Abengoa (Carsanba et al. 2018)			95
Aemetis (Lucia and Grisolia 2018)			208
Weyland (Carsanba et al. 2020)	Norway		0.2 million
Iogen corporation (Reddy et al. 2020)	Canada		
GranBio/GraalBio (Reddy et al. 2020)	Brazil		82
Gushan environmental energy (Carsanba et al. 2018)	China	Biodiesel from recycled cooking oil (Carsanba et al. 2018)	
Fujian Zhongde energy co. limited (Carsanba et al. 2018)		Biodiesel from waste vegetable oil (Carsanba et al. 2018)	
Nextgeneration biomass to liquid diesel	Singapore	Biodiesel from residual animal fat (Adeniyi et al. 2018)	
Biogasol (Carsanba et al. 2018)	Denmark	Celluloid ethanol from wheat straw (Carsanba et al. 2018)	5 million
Abengoa (Carsanba et al. 2018)	Spain	Celluloid ethanol from waste biomass (Carsanba et al. 2018)	19 million
Ethtec (Carsanba et al. 2018)	Australia	Celluloid ethanol from woody residues (Carsanba et al. 2018)	

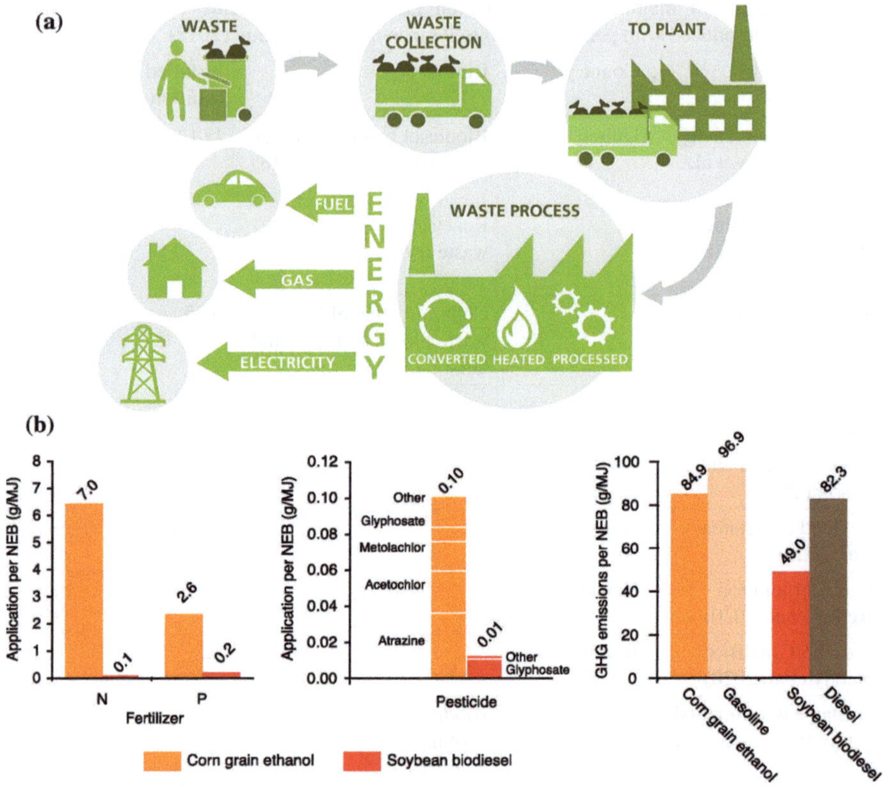

Fig. 4.2 **a** Waste-to-energy (WtE) or energy-from-waste (EfW) (Peters et al. 2023) and **b** Bio fuel life cycle environmental effects (Zhou et al. 2013)

cycle's environmental effects have been shown). However, decades of research on this subject have shown that some methods of producing biofuels from specific feedstocks offer distinct advantages over others and thus have greater potential for providing environmental benefits over petroleum-derived fuels. There is no simple or single way to compare biofuels and petroleum-derived fuels over their full life cycles and over their entire suite of environmental effects. However, certain phases of the life cycle of bio fuels have more of an impact on the environment than others, and as a result, they should get special consideration in aiming tactics for improving environmental results (National Research Council 2012).

4.3.1 Climate Change Possibilities

The use of traditional fossil fuels such as coal, oil, and natural gas has been identified as a major contributor to climate change. When these fuels are burned, they release large amounts of greenhouse gases, primarily carbon dioxide (CO_2), into the atmosphere. These gases trap heat from the sun, causing the Earth's temperature to rise and leading to a wide range of negative impacts on the environment and human health (IPCC A 2013). The burning of traditional fossil fuels such as coal, oil, and gas is the primary source of greenhouse gas emissions that contribute to climate change (Fig. 4.3). When these fuels are burned, they release carbon dioxide and other greenhouse gases into the atmosphere, which trap heat and contribute to global warming (Epstein et al. 2011). The burning of coal is the largest source of carbon emissions globally. The extraction and transportation of oil and gas also contribute significantly to greenhouse gas emissions, as do leaks from pipelines and other infrastructure (Epstein et al. 2011). In addition, the process of extracting and processing these fuels can lead to other environmental problems, such as air and water pollution, habitat destruction, and ecosystem degradation. The continued use of traditional fuels at current rates is likely to exacerbate climate change and its impacts. To mitigate these risks, there is a growing need for a transition to renewable energy sources, such as solar, wind, and hydropower. These sources do not emit greenhouse gases and have a much lower impact on the environment (Owusu and Asumadu-Sarkodie 2016). Governments, businesses, and individuals are increasingly recognizing the importance of reducing dependence on traditional fuels and investing in renewable energy. Many countries have set ambitious targets for transitioning to clean energy, and there is growing public support for policies and actions that promote sustainability and climate action (Gielen et al. 2019).

The impact of traditional fuels on climate change is significant and has been widely studied by scientists. The Intergovernmental Panel on Climate Change (IPCC) has concluded that human activities, including the burning of fossil fuels, are the primary cause of global warming since the mid-twentieth century (Carslaw et al. 2018). If we continue to rely on traditional fuels at current rates, the impacts of climate change are

Fig. 4.3 Various effects of climate change due to use of traditional fuels

expected to become increasingly severe, including rising sea levels, more frequent and severe weather events, and increased risk of food and water shortages (Gitz et al. 2016). To avoid the worst impacts of climate change, it is essential to transition to cleaner and renewable energy sources, such as solar, wind, and geothermal power. This transition will require significant investments in technology, infrastructure, and policy changes, but it is necessary to reduce our dependence on traditional fuels and mitigate the effects of climate change (Gielen et al. 2019).

Biofuels have been proposed as a potential solution to mitigate the negative impacts of climate change by reducing greenhouse gas emissions. Biofuels are produced from organic matter such as crops, agricultural waste, and forest residues, which are then converted into liquid fuels such as ethanol and biodiesel (Jeswani et al. 2020). The impact of biofuels on climate change depends on several factors, such as the type of feedstock used, the production process, and the transport and distribution of the biofuels. In general, biofuels have the potential to reduce greenhouse gas emissions compared to fossil fuels because they are renewable and have a lower carbon footprint (Hanaki and Portugal-Pereira 2018). However, the production of biofuels by traditional techniques can also have negative environmental and social impacts if not done sustainably. For example, if forests are cleared to grow feedstock crops, the carbon stored in the trees is released into the atmosphere, resulting in a net increase in greenhouse gas emissions. Similarly, the use of fertilizers and pesticides in feedstock production can lead to water pollution and biodiversity loss (Malode et al. 2021).

4.3.2 *Energy Use*

Energy consumption throughout the life cycle of biofuels has been analysed using a variety of measures, including net energy ratio, primary, secondary, or cumulative energy demand, and fossil energy use (Arvidsson et al. 2012). Even though energy security and reducing reliance on fossil fuels are significant goals of these policies, in addition to mitigating climate change, many national policies on bio fuels are focused on the use of fossil fuels. As shown in Fig. 4.4, the majority of research show that first- and second-generation biofuels use less than 0.5 MJ MJ^{-1} of fossil energy. The energy density of first-generation bio fuels ranges from 0.04–0.86 MJ MJ^{-1} to 0.57–0.87 MJ MJ^{-1}, despite the fact that there are several distinct types of bio fuels. Negative values are produced by energy credits for byproducts like electricity and warmth. Various agricultural practises, feed stock yield, conversion technology, and allocation strategies are a few of the causes of these variances. The assumptions made on the type of energy (fossil or biomass) employed in the conversion process have an impact on the outcomes as well.

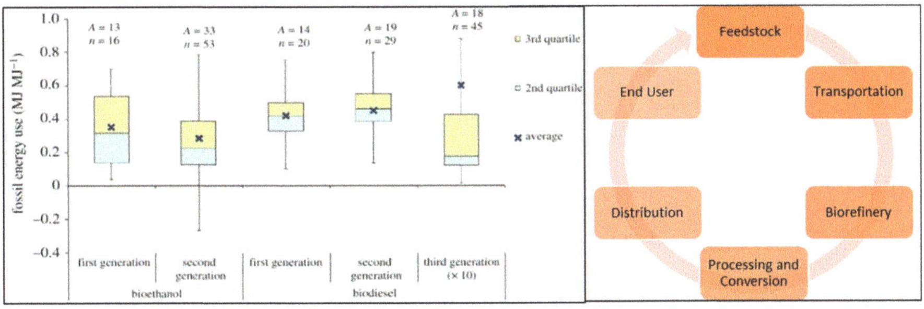

Fig. 4.4 Fossil energy utilization along the lifespan of biofuels (Reijnders and Huijbregts 2008)

4.3.3 Water Use

Particularly for first-generation bio fuels, the manufacturing of feed stocks might involve significant water usage (Escobedo and Calderón 2021; Correa et al. 2017). This is problematic because the irrigation water needed for some feed stocks may compete with water utilized for other uses, such food production. By 2050, the total amount of water consumed worldwide could significantly increase due to increased agricultural biomass production for biofuels (Hammond and Li 2016). In regions already experiencing water scarcity, this additional water demand could significantly worsen the overall environmental effects of bio fuels.

Water usage is often not included in LCA evaluations of bio fuels, despite the fact that several studies have expressly focused on this aspect of bio fuel production. The majority of them provide data on the amount of water consumed, including the amounts of green (soil moisture) and blue (surface) water. Even if the amount of water used for a particular product is the same, this is insufficient to evaluate the local environmental consequences of water consumption since they heavily depend on the region's water supply and the distinct characteristics of the hydrological cycle (Arodudu et al. 2020). The bulk of agricultural goods need a lot of water overall when green water is taken into account. Since green water may in reality have little influence on the regional hydrological cycle, including it might overstate the genuine impact of water demand for bio fuels (Teswani and Azapagic 2011; Gerbens-Leenes et al. 2009).

The use of bio fuels in blue water varies greatly depending on the underlying crop and nation, according to more recent study (Berger et al. 2015) that looked at the water footprint of first-generation bio fuels utilised in Europe (Fig. 4.5). The usage of less water in the production of bio ethanol from wheat and sugar beetroot is due to the fact that many countries grow crops with little or no irrigation. In contrast, Portugal utilises 86 m^3 GJ to make bio ethanol from maize (Berger et al. 2015).

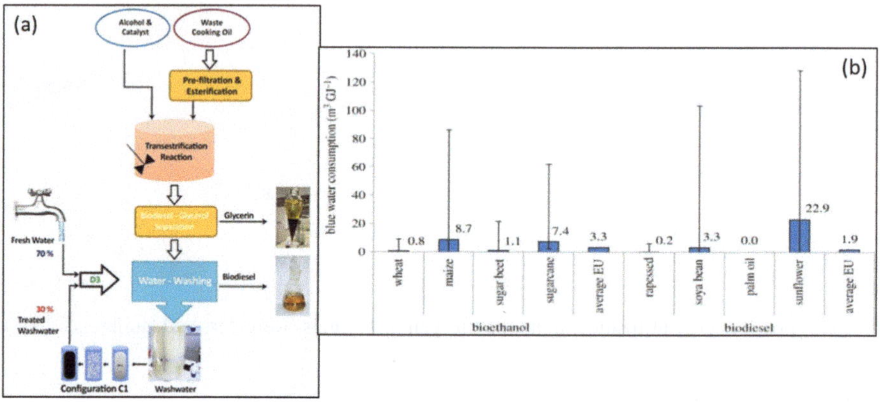

Fig. 4.5 Shows how much blue water is used in Europe to produce bio fuels (Jeswani et al. 2020) and the utilization of water for production of biofuel (Jaber et al. 2015)

4.3.4 Biodiversity

Biofuels have the potential to contribute to the loss of biodiversity through habitat loss and degradation, excessive nutrient loads and other kinds of pollution, over-exploitation and unsustainable land use, as well as the breeding of invasive alien species for use as feed stocks (Escobedo and Calderón 2021). Biodiversity is impacted by the usage of feed stock, the scale of production, management strategies, and LUC (Correa et al. 2017).

Local biodiversity may be directly threatened by intensive farming practises and the use of agrochemicals in the production of feed stock for first-generation bio fuels (Liu et al. 2014; Elshout et al. 2019). By directly destroying animal habitats like tropical rain forests, LUC brought on by increasing bio fuel production increases the danger of biodiversity loss (Elbehri et al. 2013).

It is believed that second-generation bio fuels are less harmful to biodiversity than first-generation bio fuels, and they may even be advantageous (Eisentraut 2010). This is true for plant-based lignocellulosic feed stocks since they have a lengthy life cycle, require little in the way of fertilizers and pesticides, and require little in the way of human involvement while they are growing. For instance, SRC willow planted on a broad scale can help various bird species, butterflies, and floral plants (Rowe et al. 2009). Moreover, if feed stocks are grown on degraded land, the variety of species may increase. Similar to this, permanent grasslands that are exploited to produce biomass may increase the diversity of bird species, including migratory ones. Large energy crop mono cultures, however, have the potential to harm local biodiversity, especially through the loss of habitat and the spread of invasive species. Certain invasive characteristics may be seen in eucalyptus, switchgrass, and some Miscanthus species (Woods 2008).

It is anticipated that the use of agricultural and forestry waste as a biofuel feed-stock would be less detrimental to biodiversity than the use of special energy crops

(Rowe et al. 2009). Reducing the amount of decomposing wood, a niche habitat, and disturbing wildlife because of greater access to the forest are two effects of using forestry wastes on biodiversity. A further issue would be the excessive removal of agricultural waste from fields, since this may promote weed growth and increase pesticide usage, which would negatively impact local biodiversity. The large-scale production of algae poses a major danger to coastal biodiversity through the invasion of algal species into coastal shallow habitats as mud flats, salt marshes, mangroves, sea grass bed, and coral reefs (Woods 2008).

While biodiversity loss is one of the current major environmental issues, LCA assessments of bio-energy systems seldom include it as an effect category (Flude 2019). One of the requirements in sustainability certification programmers is the preservation of biodiversity or the avoidance of biodiversity loss due to bio fuels. Yet, it is challenging to define and quantify systems that support biodiversity, and it is also difficult to monitor biodiversity loss (Jeswani et al. 2020).

4.3.5 *Other Environmental Impacts*

In the LCA study on biofuels, the extra environmental effects have been calculated using a variety of impact assessment approaches. It is difficult to assess them and provide a relevant range of outcomes for different bio fuels as a result. The research's focus differs as well, with some studies emphasising the cradle-to-grave system border and others the cradle-to-gate system barrier. The conclusions of the latter studies also rely on assumptions about the type of vehicle that burns biofuels. However, a lot of studies show that the advantages of lower GHG emissions from bio fuels compared to fossil fuels are given up for other impacts, such eutrophication and acidification (Jeswani et al. 2020). In Table 4.3 a brief discussion has been given about different biofuels, their feedstocks, acidification and eutrophication level.

4.4 Sustainable Techniques

Adopting more sustainable biofuel production systems, which currently include waste from sustainable sources, local perennial crops and microalgae production systems produced on low-biodiversity or degraded land, can reduce the magnitude of a range of socio-economic and environmental impacts that first-generation biofuels, mainly in terms of reducing competition with food production and biodiversity. Further different sustainable techniques to produce biofuel, challenges of those technologies, and future developments are discussed below.

Table 4.3 Biofuel types and their feedstocks according to the first, second and third generation

Biofuel type	Feedstock	Acidification	Eutrophication	Source
First generation				
Bio ethanol	Corn	1.4–3	4.4–20	Yang et al. (2012)
	Wheat	3	5	Belboom et al. (2015)
	Sugar beet	1.4–1.8	6–15	Bessou et al. (2013)
	Sugarcane	2	2.8	Cavalett et al. (2013)
Bio diesel	Rape seed	1.3–1.7	3.1–5	Iriarte et al. (2012), Gonzalez-Garcia et al. (2013)
	Soya bean	1.3–1.7	4–5	Panichelli et al. (2009)
	Palm oil	1.3	14	Arpornpong et al. (2015)
Second generation				
Bio ethanol	Short rotation coppice	0.45	1.2	González-García et al. (2012)
	Switchgrass	1.1	3.2	Bai et al. (2010)
	Straw	1.6–3	2–3.6	Wang et al. (2013)
Bio diesel	Used cooking oil	0.2	0.63	Ripa et al. (2014)
	Jatropha	1	1	Kalaivani et al. (2014)
Third generation				
Bio diesel	Algae	2.6–3	2.1–3.2	Kalaivani et al. (2014)

4.4.1 Different Sustainable Techniques

Generally, four different types of sustainable techniques can be used to produce biofuel, which is elaborately described in Table 4.4.

4.4.2 Challenges of Technologies

4.4.2.1 Direct Use and Blending

Animal fat or vegetable oil gives sufficient power. But it has some problems due to its unacceptable properties, so it cannot be used in DI engines without any modification. To avoid such problems, alternative fuel sources are directly blended with conventional fossil fuels (Mendhe et al. 2015; Atabani and da Silva César 2014; Rodionova et al. 2017).

Table 4.4 Sustainable techniques to produce biofuels

Techniques	Description	References
Direct use and blending	• Animal fat or vegetable oil can be used as fuel in direct injection engines; it has a good heating value and could give sufficient power • To improve the fuel quality, reduce fossil fuel consumption, etc., alternative fuel sources are directly blended, so it is also preferable as a most convenient way to use an alternative fuel such as biofuels • The bio-oil and diesel blends will be in different ratios like 10:1,10:2, 10:3, etc.	Rajalingam et al. (2016), Mendhe et al. (2015), Atabani and da Silva César (2014), Rodionova et al. (2017)
Transesterification process	• The transesterification process is a reaction between triglycerides in vegetables and alcohol which yields biodiesel (a monobasic ester) and glycerol • In this process, some catalysts are also used to increase the speed of the reaction and the quality of the outcome product. Like- a. *Base catalyst transesterification process*: 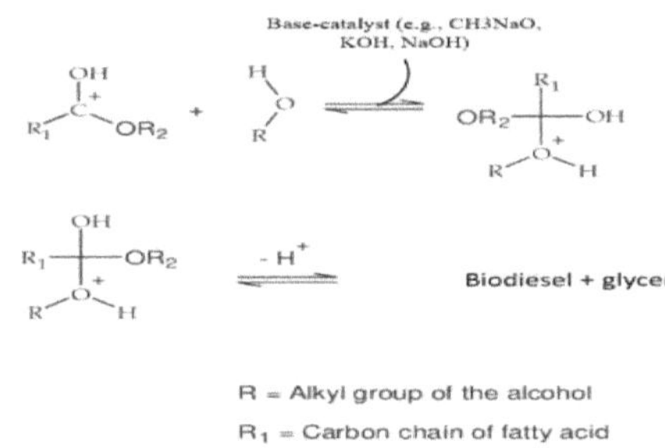 To separate out biodiesel and glycerol, the solution should place in a separating stock for 12–24 h. The upper layer is biodiesel and the lower will be the glycerol b. *Acid catalyst esterification process*: 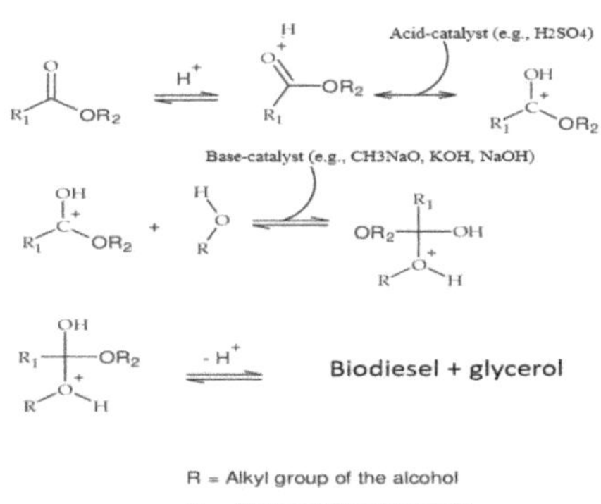 This process also can be used to convert triglycerides into biodiesel but it will take more time so it is not preferable	Bullen et al. (2006), Ibrahim et al. (2019), Venkanna and Venkataramana Reddy (2009), Rajalingam et al. (2016), Dhar et al. (2012), Ragit et al. (2011), Subramaniam et al. (2013), Canakci and Van Gerpen (2001), Yuji et al. (2002), Fonseca et al. (2013)

(continued)

Table 4.4 (continued)

Techniques	Description	References
Thermal cracking (pyrolysis)	Thermal cracking is a process of converting the complex structure of hydrocarbons into its simplest structure with or without a catalyst. Due to this process, the density and viscosity of the oil will reduce a. *Pyrolysis process to produce biodiesel* 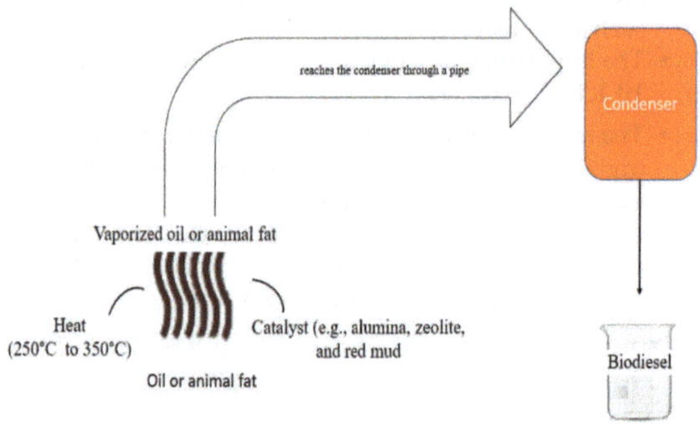 	Ong et al. (2014), Canakci and Van Gerpen (1999), Silitonga et al. (2014), Parawira (2010), Parvizsedghy and Sadrameli (2015), Jahirul et al. (2012), Vivek and Gupta (2004)
Microemulsion	• The microemulsion is defined as thermodynamically stable, isotropic liquid mixtures of oil, water, and surfactant (compounds that lower the surface tension of a liquid, the interfacial tension between two liquids) • Generally, the alcohol used to increase the volatile property of oil, it reduces the smoke. The microemulsion process is also used to get a good spry property when injected into the engine by nozzle	Parawira (2010), Vivekpantidar et al. (2014), Yusuf et al. (2011), Koh and Gazi (2011), Schenk et al. (2008)

4.4.2.2 Transesterification Process

In the transesterification process, the amount and types of catalysts are decided by the amount of free fatty acid present in the feedstock oil. The higher amount of free fatty acid is unfavourable for biodiesel production which leads to the formation of soap and biodiesel yield efficiency.

Base catalyst transesterification process:

Sodium methoxide is the most efficient catalyst but it is not economic. The reaction temperature (50–60 °C) should also be maintained in order to get the highest biodiesel yield. The reaction temperature should be below of methanol boiling point. Otherwise, methanol is wasted because of vaporization. The process, stirring at 1300 rpm is carried throughout the process for dynamic mixing to enhance the reaction rate. Nearly one hour will take place to complete the reaction (Bullen et al. 2006; Ibrahim et al. 2019; Venkanna and Venkataramana Reddy 2009; Rajalingam et al. 2016; Dhar et al. 2012; Ragit et al. 2011; Subramaniam et al. 2013; Canakci and Van Gerpen 2001; Yuji et al. 2002).

Acid catalyst esterification process:

Some feedstock oil has more than 1% free fatty acid; this free fatty oil cannot be converted into biodiesel. So, the yield of biodiesel from the transesterification process will be very low. In this process the alcohol will react with oil using acid catalyst and produces the biodiesel and water, the water must be removed immediately because the water will lead to the formation of soap while in base catalyst transesterification process. Here phosphoric acid or sulphuric acid is used as a catalyst. The product obtained from acid catalyst is use to produce biodiesel through base case catalyst transesterification process (Ong et al. 2014; Canakci and Van Gerpen 1999; Silitonga et al. 2014).

4.4.2.3 Thermal Cracking (Pyrolysis)

In vegetable oil as an alternative fuel these two properties are affecting the atomization of engine. Collected fuel through this process has direct use in diesel engine without any type of alternation. The thermal cracking biodiesel plant has a reactor with safety valve, drain pipe, temperature indicator, etc., the oil or animal fat need to be converted into the biodiesel is placed inside the reactor, then heat is applied to the reactor (Parvizsedghy and Sadrameli 2015; Jahirul et al. 2012; Vivek and Gupta 2004).

4.4.2.4 Microemulsion

This process will solve the problem of viscosity and some other atomization properties of oil (Parawira 2010; Vivekpantidar et al. 2014; Yusuf et al. 2011). Alkyl nitrate will be the cetane number improver. If micro-emulsified diesel is used in diesel engines, some problems will arise such as incomplete combustion, carbon deposit, and nozzle failure (Yusuf et al. 2011).

4.4.3 Future Development

- Rapid advancements in biofuel science have reduced the cost per gallon from $400,000 to $6. Soon that number will be even lower, making it competitive with today's fossil fuels. That's a key step toward replacing gasoline, diesel, and jet fuel in our cars, trucks, and planes (Schenk et al. 2008).
- Producing affordable and reliable jet fuel poses the most significant challenge for biofuels researchers, but scientists at the Energy Biosciences Institute have discovered a method that could make it easier and cheaper to produce jet fuel (a type of aviation fuel designed for use in aircraft powered by gas-turbine engines) from biomass (Stanton and Gustafson 2019).

- A brand-new ARPA-E (Advanced Research Projects Agency-Energy) project will use robots to study the characteristics and DNA of sorghum, a plant with a high sugar content ideal for producing biofuel. The information will be gathered in a database that scientists developing biofuels can use to help maximize new strains of the plant (Blaustein 2012).
- Over the past century, global energy demand has primarily depended on fossil fuels. This contributed to the depletion of non-renewable resources. In addition, the degradation of the environment has been correlated to the development and increasing civilization. For example, land pollution is directly related to increasing land commercialization and urbanization. The advantages presented by biofuels make them a viable alternative to conventional fossil fuels. Biofuels present a substantial role in combating climate change and securing the planet's future energy needs. However, the subject of change is often the major obstacle to the fuel industry (Escobar et al. 2009). Companies that have long been comfortable with readily available fossil fuels are still dissuaded from using non-renewable resources. There is an urgent need for countries to produce ecological and economic energy sources. This can be achieved through government, social, and business policy-making collaborations. To sum up, the long-term goal is to develop an alternative and sustainable energy source for future consumption. To achieve this, everyone must ensure that different stakeholders (social, environmental, business, and government) are advantageous and sustainable to one another. Thus, biofuels are on their way to becoming the future's answer to meeting the world's energy security (Klymchuk et al. 2020).

4.5 Waste to Energy

Bioenergy is a source of energy from the organic material that plants make up, known as biomass. Biomass contains carbon that is absorbed by plants during photosynthesis. When this biomass is used to produce energy, the carbon is released during combustion and simply returned to the atmosphere, making modern bioenergy a promising fuel with near-zero emissions. Here, with few waste products such as bagasse, waste oil, and wastewater; waste to biofuel energy production procedures have been discussed briefly.

4.5.1 From Bagasse

One of the most undertaken husbandry biomass energy sources is sugarcane. Sugar cane produces mainly two types of biomasses—sugar cane and bagasse. Sugarcane trash is the field residue left after harvesting the sugarcane stalk, while bagasse is the fibrous residue left after grinding the sugarcane, with a moisture content of 45–50%

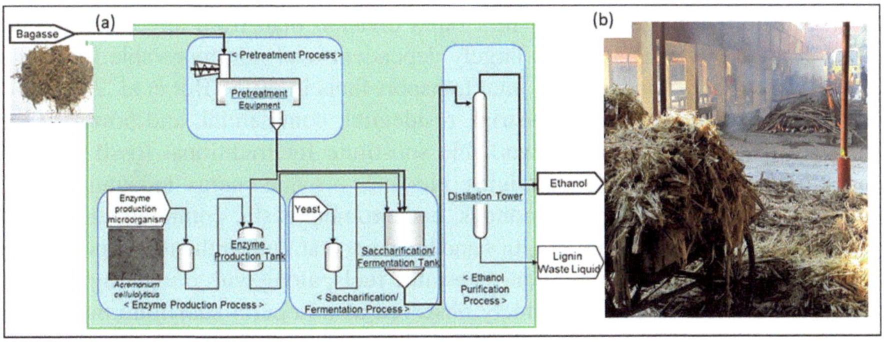

Fig. 4.6 **a** Energy/biofuel production from bagasse and **b** Collection of bagasse from sugarcane (Dias et al. 2009; Bot et al. 2021)

and consisting of a combination of solid fibers with mild and plane parenchymatous fiber (core) tissue with high hygroscopicity.

The pulp predominantly holds cellulose, hemi-cellulose, pentosanes, lignin, sugars, wax, and minerals. The amount gained differs from 20 to 40% per sugarcane and is mainly related to the fraction of fiber in the sugarcane and the purity of the stocked sugarcane, which in turn rely upon the harvesting culture (Abdalla et al. 2022).

The composition of bagasse depends on the variety and maturity of the sugar cane, as well as on harvesting methods and the efficiency of sugar processing. The pulp is usually burned in furnaces to produce steam for energy production. Pulp is also becoming an attractive raw material for bioethanol production (Fig. 4.6) (Chandel et al. 2012).

It is also used as a raw material for paper production and as a raw material for cattle. Pulp's value as a fuel depends greatly over its calorific value, which in turn relies upon its combination, mainly with regard to water content and sugarcane's calorific value, that is specially dependent upon sucrose content.

Moisture content is the main factor that determines calorific value, i.e., the lower the moisture content, the higher the calorific value. A good crushing process helps to produce content of 45% which has low humidity, while a moisture content of 52% will indicate poor milling efficiency. Most mills produce bagasse with a moisture content of 48%, and most boilers are designed to burn bagasse at about 50% moisture (Pan et al. 2022).

4.5.2 *From Waste Oil*

A number of biochemical, physical, and thermochemical methods are used to create the energy-dense molecules that make up sustainable biofuel production techniques from biomass derived from plants and microbes. Greenhouse gas (GHG) emissions from transportation are growing in a rapid division than from any other sources (Dias et al. 2009). The sector is largely dependent over non-renewable fuels, that is considered for around 97% of total transport fuels in 2018 (Bot et al. 2021). For usage in the transportation, industrial, residential, commercial, and power generating sectors, they represent a workable substitute for traditional fossil fuels. In contradict to fossil fuel energy origins, bio-fuels provide energy indemnity, lower GHG discharges, boost the economics, and promote rustic communities through generating employment opportunities and local harvesting merchantry opportunities (Dias et al. 2009). To decrease non-renewable fuels, along with mitigating environmental changes, bio-fuels are widely observed as a hopeful substitute medium in transport facilities. Bio-fuels have been in application from the very beginning of the automotive transportation system (Marcucci et al. 2021). Biofuels are divided into multiple generations based on the feedstock. To encourage and improve sustainable biofuel production, a number of elements, including feedstock, effective production methods, technological preparedness, and market-driven regulations, must be taken into account. By the 1940s, biofuels were considered a viable transportation fuel, and various bio-ethanol compositions were widely applied as fuel energy in the USA regions or, Europe etc. (Marcucci et al. 2021). By the time of the oil emergency in the 1970s, most of the continents turned towards commercial renewable biofuel energy generation; but Brazil was the only country who has begun to generate ethanol on a bigger scale under the National Ethanol Program "Proálcool" (Dias et al. 2009; Amorim et al. 2018). In the period of 1990s, when crude oil costs were at a top level and concerning energy safety worldwide, the United States and most of the countries in Europe has displayed policies to boost up bio-fuel generation (Dias et al. 2009; Srinivas and Satyarthi 2012). Bio-fuel production has attracted attention from all over the world over the last few decennaries with the prosperity in climate change mitigation policies and tactics to decrease greenhouse gas generation from the transportation systems. Since then, over 50 countries started bio-fuel production programs and put biofuel blending objectives into their own fuel production tanks (Baroutian et al. 2013). Most famous programs are the Renewable Fuel Standard (RFS) in the USA and the Renewable Energy Directive (RED) in Europe (Tian-Feng et al. 2013; Bhatia et al. 2020; Liu et al. 2016).

Biofuels have both beneficial and drawback sites in terms of impact on climate, economy, and hold down capability in the market (Leandro et al. 2019). One side there are, reducing greenhouse gas generations, energy safety and rustic prosperity, which are vital factors that control biofuel generation worldwide. And the other side, perturbations over the increased biofuel generation, like high-placed arc over food-supply costs, the presumption of increased greenhouse gas generation by straight and inverse LUC (land use change) from bio-fuel generation, along with the risk of land corrosion, crops, water supplies, and ecology (Jayakody et al. 2018). To make biofuels economically viable and competitive with fossil fuels, market interventions, financial help, grants, and subsidies must be supported. Application of 1st generation feedstocks, like corn starch, partly became a problematical outpouring, mainly due to comparison with food generation and perturbations about the use of lands for fuel production, which were used for cultivation. The increasing need for cultivational products threatens to rise deforestation and the usage of those lands which has great biodiversity, to fulfil biofuel need, along with the amalgamated use of fresh water, manures and pesticides with bad environmental effects. Few of these problems might be solved with the help of 2nd generation raw materials; howsoever, sustainable economy of few 2nd generation bio-fuels continues to be questionable in nowadays economic references, mainly due to lesser oil cost (Kehrein et al. 2020; Mehariya et al. 2021; Jha and Schmidt 2021). 3rd generation (e.g., algal) bio-fuels might also decrease the problem of food supplement issues and land usage, though micro-algae can grow over cultivable soil and in sewage, salt or salty water, and they grow up very quickly. Howsoever, the generation of biofuel from micro-algae is strength-severe and currently impractical from an economical viewpoint (Dragone et al. 2010). In order to embolden the improvement of sustainable bio-fuels, regulatory strategies like the Renewable Energy Directive and the Renewable Fuel Standard provide for different sustainable measures for bio-fuels. Greenhouse gas generation during the life cycle is one of the prime measurements. The Renewable Energy Directive requires biofuels for minimum GHG generation (50%) compare to their non-renewable substitutes for plants operating by October 2015 (Bhatia et al. 2020). The Renewable Fuel Standard needs generators of developed bio-fuels to decrease greenhouse gas generation (at a 50% minimal level), whilst standard bio-fuels must acquire at least 20% decreasement in greenhouse gas generation (Tian-Feng et al. 2013). Environmental alternation impacts concerned with greenhouse gas generation and another perspective of biofuel sustainability should be assessed over a life-cycle cornerstone using Life Cycle Assessment (LCA) to get off burden changing from one portion of the life cycle to another supply chain (Liu et al. 2016).

This chapter investigates different traditional biofuel production techniques, their environmental impacts, different sustainable techniques and modern techniques of biofuel production and also aims to review energy production from different wastes (bagasse, waste oil, waste water) (Fig. 4.7).

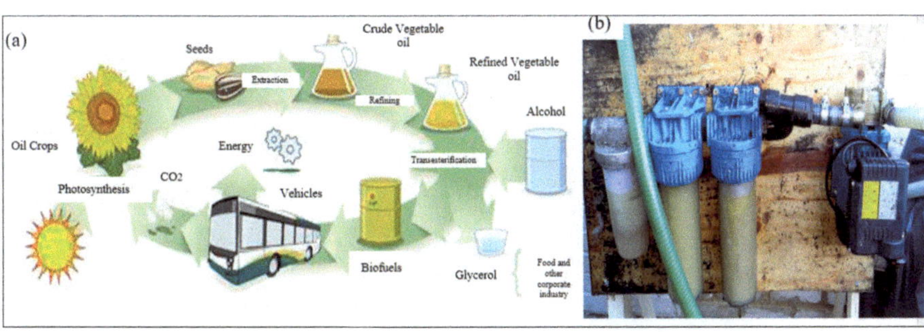

Fig. 4.7 **a** Energy/biofuel production from waste oil and **b** Waste vegetable oil filtration setup (Baroutian et al. 2013; Tian-Feng et al. 2013)

4.5.3 From Wastewater

The world's population consistency in emerging nations is particularly high as a result of rapid urbanisation and industrialisation. Because of numerous human activities (anthropogenic actions), this unsustainable population growth has resulted in the production of vast amounts of wastewater. Large volumes of sewage are produced by households, municipal corporations (MCs), urban local bodies (ULBs), and industry and dumped into nearby water bodies and streams as well as other sources besides any alteration, leading to water contamination (Fig. 4.8). In addition to a number of water-borne diseases, such bad sewage treatment causes energy waste for sewage treatment and a lack of fresh water. However, each of these major issues is capable of providing a remedy. Almost all the wastewater produced by industrial and WWTP facilities is enriched in various bio-polymers, including starch, lactose, glucose-lignocellulose, proteins, lipids, and fats. Dark fermentation (DF) and anaerobic digestion (AD) transform these biopolymers into sustainable biofuels including ethanol, butanol, biodiesel, biogas, hydrogen, methane, and biomethane, among others (Bhatia et al. 2020).

It is well-known that bioethanol is being researched as a renewable liquid fuel. Gasohol and Diesohol may be created using bioethanol either alone or in conjunction with other liquid fuels. Bioethanol is often produced by fermenting simple carbohydrates like glucose and fructose in an environment with no oxygen (anaerobic conditions). This fermentation is carried out by some yeasts, such as *Saccharomyces* sp., and other bacteria, such as *Zymomonas* sp. (Amorim et al. 2018; Liu et al. 2016). Numerous companies, including those in the sugar, food, meat, pulp, and paper sectors, have produced wastewater that is high in carbohydrates (glucose, fructose, lactose, etc.) that can be fermented to produce bioethanol. However, contemporary difficulties pertain to the utilisation of waste streams in which organic carbon is absent as simple sugars, by chemical or biological pre-treatment, or by novel microbes that ferment a wide variety of organic components (Bhatia et al. 2020). Procedures

Fig. 4.8 **a** Energy/biofuel production from wastewater and **b** Algae turning wastewater into biofuel (Mehariya et al. 2021; Jha and Schmidt 2021)

for pulp pre-treatment are now the subject of extensive research. A further energy-intensive distillation is now needed to process the lower yield of ethanol generated by fermentation (usually 10% (v/v)). The distillation process used in conventional ethanol facilities will use more than 30% of the thermal energy required to produce the fuel (Leandro et al. 2019; Jayakody et al. 2018; Kehrein et al. 2020).

4.6 Modern Techniques of Biofuel Production

4.6.1 Genetically Modified Organism

Biofuels being prospective replacements to conventional fossil fuels, more or less, are dependent upon various biological sources. Classified into varied generations, the first-generation biofuels significantly acknowledge natural, financial and political concerns, requiring arable agricultural lands, leading to diminished terrains for animal and human feedstock production. Subsequently, the 2nd generation sourced from different non-food items, such as bio-cellular mass, straw, manure etc., did not provoke industrial profitability, thereby concerning the development of 3rd generation biofuels. Overcoming the weaknesses of 1st and 2nd generation biofuels, the 3rd generation biofuels apparently were dependent upon microalgae, which proved to be delivered 20 times more biofuel than traditional crops, assimilate carbon dioxide and relatively simple processing (Adeniyi et al. 2018; Dragone et al. 2010). The eventual introduction of 4th generation biofuels sourced from genetically modified organisms (algae biomass) to achieve enhanced biofuel production with improved light penetration, increased photosynthetic efficiency, reduction in photoinhibition and expanding the absorption spectrum of microalgae account to mainstream strategies for increasing biofuel efficiency (Tandon and Jin 2017; Buitrón et al. 2017; Wolf et al. 2018).

Extensive research has been carried out in this regard, with more specification towards genetic modifications. This exploitation of genetically modified organisms targeting biofuel production started in 1996 (Abdullah et al. 2019). Genetically modified plants like soyabeans, rapeseeds and maize have been accounted with occupying global area of 73.3×10^{13}, 7×10^{13} and 46.8×10^{13} m^2 respectively (Ullah et al. 2018). However, further developments to ensue upscaling, reduction of production costs, were targeted towards commercialization. The latest biofuel generation, the fourth-generation biofuels, nevertheless encompassed the utilization of multiple types of sugars to higher lipid synthesis or increased carbon fixation and enhanced photosynthesis (Cavelius et al. 2023a). Different strategies like engineering native producers' metabolic pathways (improving growth rates, utilising various carbon sources, directing the metabolic flux towards biofuel production, and increasing production titers) and reconstructing naturally occurring producers' metabolic pathways in more genetically accessible model organisms like *E. coli*, *B. subtilis*, *P. putida* and *S. cerevisiae* (Cavelius et al. 2023a). In addition, to streamline product recovery, overexpression of specific membrane transporters has also been practiced to avoid accumulation and toxicity. Membrane transporters in *E. coli*, have been successfully incorporated to excrete n-alkanes (Malik et al. 2021; Foo and Leong 2013). The antennae systems of algae have been influenced with techniques like CRISPR-Cas9 to increase photosynthetic efficiency (Javed et al. 2019). Fourth generation biofuel procedures focusing upon optimized cyanobacteria for the production of ethanol, butanol, isobutanol etc. have been successfully realized, with production of bioethanol titers up to 5.5 g/L (Mund et al. 2022; Zahra et al. 2020; Liu et al. 2019). Furthermore, techniques like random mutagenesis by UV light, fast neutron irradiation and chemical agents have been utilized for production of biodiesel by *Y. lipolytica*, *R. toruloides*, *L. starkeyi* and *C. oleaginosus* (Awad et al. 2020; Cavelius et al. 2023b). The recent advancement via electrobiofuels to establish new-to-nature hybrid systems, which are directed to use carbon sources and renewable electricity to produce commodity chemicals and biofuels have been affiliated with higher proton efficiency of modern photovoltaic systems with sustainable production of biofuel and increased process effectiveness (Eva-Mari 2016).

4.6.2 Microalgae and Cellulosic Biomass

As described in the prior paragraphs, genetic engineering has played an important role in development of biofuels. Concerning the fourth generation of biofuels, microalgae has proved to be a prominent renewable source of biofuel. They also gained popularity than lignocellulosic biomass because of their biochemical makeup, which includes high levels of protein, lipids, and carbohydrates. Cultivation of algal biomass for biofuels has been advantageous in terms of faster biomass feedstock, CO_2 sequestration etc. Incorporating genetic engineering techniques like modifications of genes involved in fatty acid biosynthesis and triacylglycerol (TAG) pathways, Kennedy pathway, have also been successful in elevating lipid content (Sun et al. 2019; Park et al. 2019). Moderation of the expression of ACCase in *S. quadricuada*, by Gomma et al. has been seen to result in 1.6 fold increase in fatty acid accumulation. A 24, 36 and 39.6% increase in fatty acid production was observed by enhancing the TAG biosynthesis pathway in *H. pluvialis*, *N. oceanica*, and *Schizochytrium* respectively (Park et al. 2019). 11.3 and 600% increments in lipid accumulation was demonstrated by Yan et al. and Blazeck et al. when they increased the acetyl CoA supply in *Schizochytrium* sp. and *Yarrowia lipolytica* (Shokravi et al. 2021; Yao et al. 2016). Most reports have also concurred that microalgae can grow faster than other biomass feedstocks, even though each species has a unique growth rate and time to double (Blazeck et al. 2014). In addition, techniques like ZFNs and TALENs to strains like *P. tricornutum* and *N. salina* (Ho et al. 2014; Daboussi et al. 2014; Serif et al. 2017), CRISPR to *C. reinhardtii*, *N. oceanica*, *P. ellipsoidea* and *P. tricornutum* (Beacham et al. 2015; Shin et al. 2016; Jiang and Weeks 2017; Wang et al. 2016), have also been related with microalgae strains to increase biofuel efficiency. Further discussion regarding additional genetic modification techniques for enhancing the resistance of microalgal culture to bacteria, fungi, zooplankton, and viruses have been described by Li and Tsai (2009), Lofgren et al. (2008) and Gressel et al. (2010) respectively (Kasai et al. 2015; Nymark et al. 2016).

4.7 Advanced Conversion Process

The usage of biofuels over existing fuels offer several advantages, including safe storage, easier transport, biodegradable, promising sustainable energy source, etc. Several processes have been implicated to obtain biofuels from raw materials, like chemical, biochemical, thermochemical, etc. Biochemical methods like fermentation, anaerobic digestion, esterification, transesterification, have been some of the common methods for biofuel production. In fact thermochemical processes like gasification, pyrolysis, combustion, and hydrothermal liquefaction, have also been associated with biofuels. Out of these, processes like gasification and hydrothermal liquefaction (HTL) are major advanced conversion processes, practiced in today's world.

4.7.1 Hydrothermal Liquefaction

Hydrothermal liquefaction (HTL) is one of the most promising thermochemical liquefaction methods, it produces bio-crude from organic matter in hydrothermal settings, with temperatures ranging from 250 to 450 °C and pressures between 100 and 300 bar. It is capable of efficiently converting lignocellulosic to organic waste without restriction on lipid content (Löfgren et al. 2008). During the liquefaction process, the biomass undergoes a number of depolymerization events, including as hydrolysis, dehydration, and decarboxylation, which result in the production of insoluble products like bio-crude oil or bio-carbon. Other by-products are also created, primarily acids or phenols, which are soluble organic compounds or gases (CO_2, CO, H_2, or CH_4) (Gressel et al. 2010; Grande et al. 2021). Despite the fact that HTL can be related to reaction temperature, the amount of hydrogen remains constant while the amount of oxygen decreases as the process temperature increases because a higher temperature promotes deoxygenation.

Wet biomass from algae (proteins and lipids) and dry lignocellulosic biomass (made of cellulose, hemicellulose, and lignin) are the biomass types that are most frequently used in hydrothermal liquefaction. Evidence from many sources has demonstrated the significance of lipids, proteins, and carbohydrates in the conversion of HTL to biofuel, with lipids being the most effective (Grande et al. 2021).

The biocrude produced by HTL, although being an effective process with enormous potential, typically contains heteroatoms such oxygen, nitrogen, and sulphur, making it incompatible for transport and optimisation (Castello et al. 2018). While oxygen lowers the calorific value and energy density of biocrude, nitrogen and sulphur actually destroy equipment and contaminate the atmosphere (Xu et al. 2018). Additionally, the biocrude is unstable under harsh circumstances due to the substantial quantity of oxygenated chemicals present, rendering it prone to condensation or polymerization (Miguel Mercader et al. 2011). Although decarboxylation, and dehydration processes are used in hydrotreating to remove oxygen, further study is needed to investigate better options.

4.7.2 *Gasification*

Biomass gasification can be considered as one of the one of the significant and flexible method to process bio-feedstock to produce biofuels (usually in the form of syngas). Expanding research and implementation opportunities for biofuel gasification are being driven by the process' superior flexibility, both in terms of the feedstock type and the energy generation or fuel production options. Consisting of majorly five stages, namely, drying, pyrolysis, oxidation, reduction and cracking, this resultant syngas directly is used as a fuel to generate heat, electricity etc. based on configuration, gasifiers are classified into three main types: fluidized bed, fixed bed and entrained flow (Shahabuddin et al. 2020). Fixed-bed gasifiers are perfect for using biofuel on a small scale. Biofuel and Fuel Derived from Waste are treated using fluidized bed gasifiers (FDW). These fluidized bed gasifiers are further separated into bubbling and circulating types. While bubbling type beds are used to treat FDW, circulating type beds are more frequently used for biogas. Last but not least, entrained flow gasifiers are used to process coal and, in some cases, biofuel (Sikarwar et al. 2016).

4.8 Conclusion

Sustainable biofuel production offers a promising solution for addressing climate change and reducing greenhouse gas emissions. This review highlights the environmental impacts of traditional biofuel production and emphasizes the need for sustainable alternatives. The various sustainable technologies discussed in this work offer potential solutions to many of the challenges that are faced during biofuel production. Sustainable alternatives to traditional biofuel production methods, including second-generation biofuels, algae-based biofuels, and waste-to-energy systems, have been discussed, along with advanced feedstock cultivation and harvesting methods and efficient conversion processes that utilize waste streams and reduce emissions. However, there are also several challenges to be addressed, including the development of new feedstocks, the implementation of advanced conversion processes, and the need for more efficient harvesting and conversion methods. Overall, this review serves as a valuable resource for researchers, practitioners, and policymakers interested in advancing the field of bioenergy and promoting sustainable biofuel production.

References

Abdalla ME, Abdalla SA, Taqvi SAA, Naqvi SR, Chen W-H (2022) Investigation of biomass integrated air gasification regenerative gas turbine power plants. Energies 15(3):741

Abdullah B, Muhammad SAFAS, Shokravi Z, Ismail S, Kassim KA, Mahmood AN, Aziz MMA (2019) Fourth generation biofuel: a review on risks and mitigation strategies. Renew Sustain Energy Rev 107:37–50

Adeniyi OM, Azimov U, Burluka A (2018) Algae biofuel: current status and future applications. Renew Sustain Energy Rev 90:316–335

Arodudu O, Holmatov B, Voinov A (2020) Ecological impacts and limits of biomass use: a critical review. Clean Technol Environ Policy 22(8):1591–1611

Arpornpong N, Sabatini DA, Khaodhiar S, Charoensaeng A (2015) Life cycle assessment of palm oil microemulsion-based biofuel. Int J Life Cycle Assess 20:913–926

Arvidsson R, Fransson K, Froling M, Svanstrom M, Molander S (2012) Energy use indicators in energy and life cycle assessments of biofuels: review and recommendations. J Clean Prod 31:54–61

Atabani AE, da Silva César A (2014) Renew Sustain Energy Rev 37:644–655

Awad D, Younes S, Glemser M, Wagner FM, Schenk G, Mehlmer N, Brueck T (2020) Towards high-throughput optimization of microbial lipid production: from strain development to process monitoring. Sustain Energy Fuels 4(12):5958–5969

Azapagic A, Stichnothe H (2011) Life cycle sustainability assessment of biofuels. In: Handbook of biofuels production, pp 37–60

Bai Y, Luo L, van der Voet E (2010) Life cycle assessment of switchgrass-derived ethanol as transport fuel. Int J Life Cycle Assess 15:468–477

Baroutian S, Aroua MK, Raman AAA, Shafie A, Ismail RA, Hamdan H (2013) Blended aviation biofuel from esterified Jatropha curcas and waste vegetable oils. J Taiwan Inst Chem Eng 44(6):911–916

Beacham TA, Macia VM, Rooks P, White DA, Ali ST (2015) Altered lipid accumulation in Nannochloropsis salina CCAP849/3 following EMS and UV induced mutagenesis. Biotechnol Rep 7:87–94

Belboom S, Bodson B, Léonard A (2015) Does the production of Belgian bioethanol fit with European requirements on GHG emissions? Case of wheat. Biomass Bioenergy 74:58–65

Berger M, Pfister S, Bach V, Finkbeiner M (2015) Saving the planet's climate or water resources? The trade-off between carbon and water footprints of European biofuels. Sustainability 7:6665–6683

Bessou C, Lehuger S, Gabrielle B, Mary B (2013) Using a crop model to account for the effects of local factors on the LCA of sugar beet ethanol in Picardy region, France. Int J Life Cycle Assess 18:24–36

Bhatia RK, Sakhuja D, Mundhe S, Walia A (2020) Renewable energy products through bioremediation of wastewater. Sustainability 12(18):7501. https://doi.org/10.3390/su12187501

Blaustein R (2012) Can biology transform our energy future? ARPA-E infuses innovative research ventures with fresh funds. Bioscience 62(2):115–119

Blazeck J, Hill A, Liu L, Knight R, Miller J, Pan A, Otoupal P, Alper HS (2014) Harnessing Yarrowia lipolytica lipogenesis to create a platform for lipid and biofuel production. Nat Commun 5(1):3131

Bot BV, Sosso OT, Tamba JG, Lekane E, Bikai J, Ndame MK (2021) Preparation and characterization of biomass briquettes made from banana peels, sugarcane bagasse, coconut shells and rattan waste. Biomass Convers Biorefin 1–10

Bringezu S (ed) (2009) Towards sustainable production and use of resources: assessing biofuels. UNEP/Earthprint

Buitrón G, Carrillo-Reyes J, Morales M, Faraloni C, Torzillo G (2017) Biohydrogen production from microalgae. In: Microalgae-based biofuels and bioproducts, pp 209–234

Bullen RA, Arnot TC, Lakeman JB, Walsh FC (2006) Biofuel cells and their development. Biosens Bioelectron 21(11):2015–2045

Canakci M, Van Gerpen J (1999) Am Soc Agric Eng 42(5):1203–1210

Canakci M, Van Gerpen J (2001) Am Soc Agric Eng 44(6):1429–1436

Carsanba E, Papanikolaou S, Erten H (2018) Production of oils and fats by oleaginous microorganisms with an emphasis given to the potential of the nonconventional yeast Yarrowia lipolytica. Crit Rev Biotechnol 38(8):1230–1243

Carsanba E, Papanikolaou S, Fickers P, Erten H (2020) Lipids by Yarrowia lipolytica strains cultivated on glucose in batch cultures. Microorganisms 8(7):1054

Carslaw K, Lee L, Regayre L, Johnson J (2018) Climate models are uncertain, but we can do something about it. Eos 99

Castello D, Pedersen TH, Rosendahl LA (2018) Continuous hydrothermal liquefaction of biomass: a critical review. Energies 11(11):3165

Cavalett O, Chagas MF, Seabra JEA, Bonomi A (2013) Comparative LCA of ethanol versus gasoline in Brazil using different LCIA methods. Int J Life Cycle Assess 18:647–658

Cavelius P, Engelhart-Straub S, Mehlmer N, Lercher J, Awad D, Brück T (2023a) The potential of biofuels from first to fourth generation. PLoS Biol 21(3):e3002063

Cavelius P, Engelhart-Straub S, Heieck K, Pilz M, Melcher F, Brück T (2023) Agricultural biocatalysis: from waste stream to food and feed additives. In: Agricultural biocatalysis, pp 133–182

Chandel AK, da Silva SS, Carvalho W, Singh OV (2012) Sugarcane bagasse and leaves: foreseeable biomass of biofuel and bio-products. J Chem Technol Biotechnol 87(1):11–20

Correa DF, Beyer HL, Possingham HP, Thomas-Hall SR, Schenk PM (2017) Biodiversity impacts of bioenergy production: microalgae vs. first generation biofuels. Renew Sustain Energy Rev 74:1131–1146

Daboussi F, Leduc S, Maréchal A, Dubois G, Guyot V, Perez-Michaut C, Amato A, Falciatore A, Juillerat A, Beurdeley M, Voytas DF (2014) Genome engineering empowers the diatom Phaeodactylum tricornutum for biotechnology. Nat Commun 5(1):3831

Daystar J, Treasure T, Gonzalez R, Reeb C, Venditti R, Kelley S (2015) The NREL biochemical and thermochemical ethanol conversion processes: financial and environmental analysis comparison. BioResources 10:5096–5116. https://doi.org/10.15376/biores.10.3.5096-5116

De Amorim ELC, Dos Amorim SNC, Macê do WV, Batista EA (2018) Biohydrogen and bioethanol production from agro-industrial wastewater. In: Ahamed MI, Hasan SW (eds) Organic pollutants in wastewater II. Materials Research Foundations, Millersville, PA, USA, pp 122–137

De Miguel Mercader F, Koehorst PJJ, Heeres HJ, Kersten SR, Hogendoorn JA (2011) Competition between hydrotreating and polymerization reactions during pyrolysis oil hydrodeoxygenation. AIChE J 57(11):3160–3170

Dhar A, Kevin R, Agarwal AK (2012) Fuel Process Technol 97:118–129

Dias MOS, Ensinas AV, Nebra SA, Filho RM, Rossell CEV, Maciel MRW (2009) Production of bioethanol and other bio-based materials from sugarcane bagasse: integration to conventional bioethanol production process. Chem Eng Res Des 87(9):1206–1216

Dragone G, Fernandes BD, Vicente AA, Teixeira JA (2010) Third generation biofuels from microalgae

Eisentraut A (2010) Sustainable production of second-generation biofuels: potential and perspectives in major economies and developing countries

Elbehri A, Segerstedt A, Liu P (2013) Biofuels and the sustainability challenge: a global assessment of sustainability issues, trends and policies for biofuels and related feedstocks. Food and agriculture organization of the United Nations (FAO)

Elshout PMF, van Zelm R, van der Velde M, Steinmann Z, Huijbregts MAJ (2019) Global relative species loss due to first-generation biofuel production for the transport sector. GCB Bioenergy 11:763–772

EPA (Environmental Protection Agency) (2010) Renewable Fuel Standard Program (RFS2) regulatory impact analysis. Assessment and Standards Division, Office of Transportation and Air Quality. EPA-420-R-10-006; February

Epstein PR, Buonocore JJ, Eckerle K, Hendryx M, Stout BM III, Heinberg R, Clapp RW, May B, Reinhart NL, Ahern MM, Doshi SK, Glustrom L (2011) Full cost accounting for the life cycle of coal in "Ecological Economics Reviews." Ann N Y Acad Sci 1219:73–98

Escobar JC, Lora ES, Venturini OJ, Yáñez EE, Castillo EF, Almazan O (2009) Biofuels: environment, technology and food security. Renew Sustain Energy Rev 13(6–7):1275–1287

Escobedo MJ, Calderón AC (2021) Biomasa microalgal con alto potencial para la producción de biocombustibles. Scientia Agropecuaria 12(2):265–282

Eva-Mari A (2016) From first generation biofuels to advanced solar biofuels. Ambio 45(1):24–31

Flude A (2019) Appendix C: Copyright permissions. In: Bhutan's development nexus: economic analysis of the complex interplay between carbon neutrality, gross national happiness and economic growth, p 203

Fonseca MG, Batista LN, Silva VF, Pissurno ECG, Soares TC, Jesus MR, Cruz GF (2013) Advanced techniques in soybean biodiesel. In: El-Shemy H (ed) Soybean-bio-active compounds. InTech, Rijeka, Croatia, pp 23–47. https://doi.org/10.5772/52990

Foo JL, Leong SSJ (2013) Directed evolution of an E. coli inner membrane transporter for improved efflux of biofuel molecules. Biotechnol Biofuels 6(1):1–12

Gerbens-Leenes W, Hoekstra AY, van der Meer TH (2009) The water footprint of bioenergy. Proc Natl Acad Sci USA 106:10 219–10 223

Gielen D, Boshell F, Saygin D, Bazilian MD, Wagner N, Gorini R (2019) The role of renewable energy in the global energy transformation. Energy Strat Rev 24:38–50

Gitz V, Meybeck A, Lipper L, Young CD, Braatz S (2016) Climate change and food security: risks and responses. Food and Agriculture Organization of the United Nations (FAO) Report, vol 110, pp 2–4

González-García S, Iribarren D, Susmozas A, Dufour J, Murphy RJ (2012) Life cycle assessment of two alternative bioenergy systems involving *Salix* spp. biomass: bioethanol production and power generation. Appl Energy 95:111–122

Gonzalez-Garcia S, Garcia-Rey D, Hospido A (2013) Environmental life cycle assessment for rapeseed-derived biodiesel. Int J Life Cycle Assess 18:61–76

Grande L, Pedroarena I, Korili SA, Gil A (2021) Hydrothermal liquefaction of biomass as one of the most promising alternatives for the synthesis of advanced liquid biofuels: a review. Materials 14(18):5286

Gressel J, Chen O, Danon M, TransAlgae Ltd (2010) Method and system for protection and cross protection of algae and cyanobacteria from virus an bacteriophage infections. U.S. Patent Application 12/584,572

Hammond GP, Li B (2016) Environmental and resource burdens associated with world biofuel production out to 2050: footprint components from carbon emissions and land use to waste arisings and water consumption. Gcb Bioenergy 8(5):894–908

Hanaki K, Portugal-Pereira J (2018) The effect of biofuel production on greenhouse gas emission reductions. In: Biofuels and sustainability: holistic perspectives for policy-making, pp 53–71

Ho SH, Ye X, Hasunuma T, Chang JS, Kondo A (2014) Perspectives on engineering strategies for improving biofuel production from microalgae—a critical review. Biotechnol Adv 32(8):1448–1459

Ibrahim TK, Mohammed MK, Al Door WHA, Al-Sammarraie AT, Basrawi F (2019) Study of the performance of the gas turbine power plants from the simple to complex cycle: a technical review. J Adv Res Fluid Mech Therm Sci 57(2):228–250

IPCC A (2013) Climate change 2013: the physical science basis. Contribution of working group I to the fifth assessment report of the intergovernmental panel on climate change, p 1535

Iriarte A, Rieradevall J, Gabarrell X (2012) Transition towards a more environmentally sustainable biodiesel in South America: the case of Chile. Appl Energy 91:263–273
Jaber R, Shirazi MMA, Toufaily J, Hamieh AT, Noureddin A, Ghanavati H, Ghaffari A, Zenouzi A, Karout A, Ismail AF, Tabatabaei M (2015) Biodiesel wash-water reuse using microfiltration: toward zero-discharge strategy for cleaner and economized biodiesel production. Biofuel Res J 2(1):148–151
Jahirul MI, Rasul MG, Chowdhury AA, Ashwath N (2012) Energies 5:4952–5001
Javed MR, Noman M, Shahid M, Ahmed T, Khurshid M, Rashid MH, Ismail M, Sadaf M, Khan F (2019) Current situation of biofuel production and its enhancement by CRISPR/Cas9-mediated genome engineering of microbial cells. Microbiol Res 219:1–11
Jayakody LN, Liu J-J, Yun EJ, Turner TL, Oh EJ, Jin Y-S (2018) Direct conversion of cellulose into ethanol and ethyl-β-d-glucoside via engineered Saccharomyces cerevisiae. Biotechnol Bioeng 115:2859–2868
Jeswani HK, Chilvers A, Azapagic A (2020) Environmental sustainability of biofuels: a review. Proc R Soc A 476(2243):20200351. https://doi.org/10.1098/rspa.2020.0351
Jha P, Schmidt S (2021) State of biofuel development in sub-Saharan Africa: how far sustainable? Renew Sustain Energy Rev 150:111432
Jiang WZ, Weeks DP (2017) A gene-within-a-gene Cas9/sgRNA hybrid construct enables gene editing and gene replacement strategies in Chlamydomonas reinhardtii. Algal Res 26:474–480
Kalaivani K, Ravikumar G, Balasubramanian N (2014) Environmental impact studies of biodiesel production from *Jatropha curcas* in India by life cycle assessment. Environ Prog Sustain Energy 33:1340–1349
Kasai Y, Oshima K, Ikeda F, Abe J, Yoshimitsu Y, Harayama S (2015) Construction of a self-cloning system in the unicellular green alga Pseudochoricystis ellipsoidea. Biotechnol Biofuels 8:1–12
Kehrein P, Van Loosdrecht M, Osseweijer P, Garfí M, Dewulf J, Posada J (2020) A critical review of resource recovery from municipal wastewater treatment plants–market supply potentials, technologies and bottlenecks. Environ Sci Water Res Technol 6(4):877–910
Klymchuk O, Khodakivska O, Kovalov B, Brusina A, Benetyte R, Momotenko I (2020) World trends in bioethanol and biodiesel production in the context of sustainable energy development. Int J Glob Environ Issues 19(1–3):90–108
Koh MY, Gazi TIM (2011) Renew Sustain Energy Rev 15(5):2240–2251
Kupiec B (2021) Legal status of a renewables self-consumer under Polish law in light of Directive (EU) 2018/2001 of the European Parliament and of the Council of 11 December 2018 on the promotion of the use of energy from renewable sources
Kusch-Brandt S (2019) Urban renewable energy on the upswing: a spotlight on renewable energy in cities in REN21's. "Renewables 2019 global status report"
Leandro MJ, Marques S, Ribeiro B, Santos H, Fonseca C (2019) Integrated process for bioenergy production and water recycling in the dairy industry: selection of Kluyveromyces strains for direct conversion of concentrated lactose-rich streams into bioethanol. Microorganisms 7:545
Li SS, Tsai HJ (2009) Transgenic microalgae as a non-antibiotic bactericide producer to defend against bacterial pathogen infection in the fish digestive tract. Fish Shellfish Immunol 26(2):316–325
Liu J, Dantoft SH, Würtz A, Jensen PR, Solem C (2016) A novel cell factory for efficient production of ethanol from dairy waste. Biotechnol Biofuels 9:33
Liu X, Miao R, Lindberg P, Lindblad P (2019) Modular engineering for efficient photosynthetic biosynthesis of 1-butanol from CO2 in cyanobacteria. Energy Environ Sci 12(9):2765–2777
Liu Y, Xu Y, Zhang F, Yun J, Shen Z (2014) The impact of biofuel plantation on biodiversity: a review. Chin Sci Bull 59:4639–4651
Löfgren SE, Miletti LC, Steindel M, Bachere E, Barracco MA (2008) Trypanocidal and leishmanicidal activities of different antimicrobial peptides (AMPs) isolated from aquatic animals. Exp Parasitol 118(2):197–202
Lucia U, Grisolia G (2018) Cyanobacteria and microalgae: thermoeconomic considerations in biofuel production. Energies11:156

Malik S, Shahid A, Liu CG, Khan AZ, Nawaz MZ, Zhu H, Mehmood MA (2021) Developing fourth-generation biofuels secreting microbial cell factories for enhanced productivity and efficient product recovery; a review. Fuel 298:120858

Malode SJ, Prabhu KK, Mascarenhas RJ, Shetti NP, Aminabhavi TM (2021) Recent advances and viability in biofuel production. Energy Convers Manag X 10:100070

Marcucci SM, Araki CA, da Silva LS, Zanin GM, Arroyo PA (2021) Influence of the chain length of the fatty acids present in different oils and the pore diameter of the support on the catalytic activity of immobilized lipase for ethyl ester production. Braz J Chem Eng 38:511–522

Mehariya S, Goswami RK, Verma P, Lavecchia R, Zuorro A (2021) Integrated approach for wastewater treatment and biofuel production in microalgae biorefineries. Energies 14(8):2282

Mendhe PM, Baig MM, Madane CD (2015) Vol 2, Special Issue-1

Michael K, Steffi N, Peter D (2011) The past, present, and future of biofuels—biobutanol as promising alternative. In: dos Santos MA (ed) Biofuel production—recent developments and prospects. InTech, Rijeka, Croatia, pp 451–486

Mund NK, Liu Y, Chen S (2022) Advances in metabolic engineering of cyanobacteria for production of biofuels. Fuel 322:124117

National Research Council (2012) Renewable fuel standard: potential economic and environmental effects of US biofuel policy. National Academies Press

Nikolina SAJN (2016) International Renewable Energy Agency (IRENA)

Nymark M, Sharma AK, Sparstad T, Bones AM, Winge P (2016) A CRISPR/Cas9 system adapted for gene editing in marine algae. Sci Rep 6(1):24951

Ong HC, Masjuki HH, Mahlia TMI, Silitonga AS, Chong WT, Yusaf T (2014) Energy 69:427–445

Owusu PA, Asumadu-Sarkodie S (2016) A review of renewable energy sources, sustainability issues and climate change mitigation. Cogent Eng 3(1):1167990

Pan S, Zabed HM, Wei Y, Qi X (2022) Technoeconomic and environmental perspectives of biofuel production from sugarcane bagasse: current status, challenges and future outlook. Ind Crops Prod 188:115684

Panichelli L, Dauriat A, Gnansounou E (2009) Life cycle assessment of soybean-based biodiesel in Argentina for export. Int J Life Cycle Assess 14:144–159

Parawira W (2010) Biodiesel production from Jatropha curcas: a review. Sci Res Essays 5(14):1796–1808

Park S, Nguyen THT, Jin E (2019) Improving lipid production by strain development in microalgae: strategies, challenges and perspectives. Biores Technol 292:121953

Parvizsedghy R, Sadrameli SM (2015) Int J Chem Mol Nucl Mater Metall Eng 9(7)

Passell H et al (2013) Algae biodiesel life cycle assessment using current commercial data. J Environ Manage 129:103–111. https://doi.org/10.1016/j.jenvman.2013.06.055

Peters R, Berlekamp J, Tockner K, Zarfl C (2023) RePP Africa—a georeferenced and curated database on existing and proposed wind, solar, and hydropower plants. Sci Data 10(1):16

Ragit SS, Mohapatra SK, Kundu K, Gill P (2011) Biomass Bioenergy 35:1138–1144

Rajalingam A, Jani SP, Senthil Kumar A, Adam Khan M (2016) Production methods of biodiesel. J Chem Pharm Res 8(3):170–173

Reddy CV, Reddy KR, Harish VVN, Shim J, Shankar MV, Shetti NP, Aminabhavi TM (2020) Metal-organic frameworks (MOFs)-based efficient heterogeneous photocatalysts: synthesis, properties and its applications in photocatalytic hydrogen generation, CO_2 reduction and photodegradation of organic dyes. Int J Hydrog Energy 45(13):7656–7679

Reijnders L, Huijbregts MAJ (2008) Biogenic greenhouse gas emissions linked to the life cycles of biodiesel derived from European rapeseed and Brazilian soybeans. J Clean Prod 16:1943–1948

Ripa M, Buonauio C, Melllino S, Fiorentino G, Ulgati S (2014) Recycling waste cooking oil into biodiesel: a life cycle assessment. Int J Perform Eng 10:347–356

Rodionova MV, Poudyal RS, Tiwari I, Voloshin RA, Zharmukhamedov SK, Nam HG, Zayadan BK, Bruce BD, Hou HJ, Allakhverdiev SI (2017) Biofuel production: challenges and opportunities. Int J Hydrog Energy 42(12):8450–8461

Rowe RL, Street NR, Taylor G (2009) Identifying potential environmental impacts of large-scale deployment of dedicated bioenergy crops in the UK. Renew Sustain Energy Rev 13:271–290

Roy P, Dutta A (2013) Life cycle assessment of ethanol derived from sawdust. Bioresour Technol 150:407–411. https://doi.org/10.1016/j.biortech.2013.08.057

Schenk PM, Thomas-Hall SR, Stephens E, Marx UC, Mussgnug JH, Posten C, Kruse O, Hankamer B (2008) Second generation biofuels: high-efficiency microalgae for biodiesel production. Bioenergy Res 1:20–43

Serif M, Lepetit B, Weißert K, Kroth PG, Bartulos CR (2017) A fast and reliable strategy to generate TALEN-mediated gene knockouts in the diatom Phaeodactylum tricornutum. Algal Res 23:186–195

Shahabuddin M, Alam MT, Krishna BB, Bhaskar T, Perkins G (2020) A review on the production of renewable aviation fuels from the gasification of biomass and residual wastes. Biores Technol 312:123596

Shin SE, Lim JM, Koh HG, Kim EK, Kang NK, Jeon S, Kwon S, Shin WS, Lee B, Hwangbo K, Kim J (2016) CRISPR/Cas9-induced knockout and knock-in mutations in Chlamydomonas reinhardtii. Sci Rep 6(1):1–15

Shokravi H, Shokravi Z, Heidarrezaei M, Ong HC, Koloor SSR, Petrů M, Lau WJ, Ismail AF (2021) Fourth generation biofuel from genetically modified algal biomass: challenges and future directions. Chemosphere 285:131535

Sikarwar VS, Zhao M, Clough P, Yao J, Zhong X, Memon MZ, Shah N, Anthony EJ, Fennell PS (2016) An overview of advances in biomass gasification. Energy Environ Sci 9(10):2939–2977

Silitonga AS, Ong HC, Mahlia TMI, Masjuki HH, Chong WT (2014) Energy Procedia 61:480–483

Sims RE (2014) Renewable energy and climate change mitigation: an overview of the IPCC Special Report. In: Weather matters for energy, pp 91–110

Soccol CR, Vandenberghe LPS, Costa B, Woiciechowski AL, de Carvalho JC, Medeiros ABP, Francisco AM, Bonomi LJ (2005) Brazilian biofuel program: an overview. J Sci Ind Res 64:897–904

Srinivas D, Satyarthi JK (2012) Challenges and opportunities in biofuels production

Stanton BJ, Gustafson RR (2019) Advanced hardwood biofuels northwest: commercialization challenges for the renewable aviation fuel industry. Appl Sci 9(21):4644

Stephen JL, Periyasamy B (2018) Innovative developments in biofuels production from organic waste materials: a review. Fuel 214:623–633. https://doi.org/10.1016/j.fuel.2017.11.042

Stephenson AL, Dupree P, Scott SA, Dennis JS (2010) The environmental and economic sustainability of potential bioethanol from willow in the UK. Bioresour Technol 101:9612–9623. https://doi.org/10.1016/j.biortech.2010.07.104

Subramaniam D, Murugesan A, Avinash A (2013) Int J Energy Environ (IJEE) 4(5):859–870

Sun XM, Ren LJ, Zhao QY, Ji XJ, Huang H (2019) Enhancement of lipid accumulation in microalgae by metabolic engineering. Biochim Biophys Acta (BBA) Mol Cell Biol Lipids 1864(4):552–566

Tandon P, Jin Q (2017) Microalgae culture enhancement through key microbial approaches. Renew Sustain Energy Rev 80:1089–1099

Teswani HK, Azapagic A (2011) Water footprint: methodologies and a case study for assessing the impacts of water use. J Clean Prod 19:1288–1299

Tian-Feng CAI, Hui-Peng LI, Hua Z, Ke-Jian LIAO (2013) Purification of crude glycerol from waste cooking oil based biodiesel production by orthogonal test method. China Petrol Process Petrochem Technol 15(1):48

Ullah K, Sharma VK, Ahmad M, Lv P, Krahl J, Wang Z (2018) The insight views of advanced technologies and its application in bio-origin fuel synthesis from lignocellulose biomasses waste, a review. Renew Sustain Energy Rev 82:3992–4008

Venkanna BK, Venkataramana Reddy C (2009) Bioresour Technol 100:5122–5125

Vivek, Gupta AK (2004) Biodiesel production from Karanja oil. J Sci Ind Res 63:39–47

Vivekpantidar, Chandra A, Singh M, Kale RK (2014) J Sci Ind Res 63:461–464

Wang L, Littlewood J, Murphy RJ (2013) Environmental sustainability of bioethanol production from wheat straw in the UK. Renew Sustain Energy Rev 28:715–725

Wang Q, Lu Y, Xin Y, Wei L, Huang S, Xu J (2016) Genome editing of model oleaginous microalgae Nannochloropsis spp. by CRISPR/Cas9. Plant J 88(6):1071–1081
Wolf BM, Niedzwiedzki DM, Magdaong NCM, Roth R, Goodenough U, Blankenship RE (2018) Characterization of a newly isolated freshwater Eustigmatophyte alga capable of utilizing far-red light as its sole light source. Photosynth Res 135:177–189
Woods J (2008) Sustainable biofuels: prospects and challenges. The Royal Society, London
Xu D, Lin G, Guo S, Wang S, Guo Y, Jing Z (2018) Catalytic hydrothermal liquefaction of algae and upgrading of biocrude: a critical review. Renew Sustain Energy Rev 97:103–118
Yang Y, Bae J, Kim J, Suh S (2012) Replacing gasoline with corn ethanol results in significant environmental problem-shifting. Environ Sci Technol 46:3671–3678
Yao L, Cengic I, Anfelt J, Hudson EP (2016) Multiple gene repression in cyanobacteria using CRISPRi. ACS Synth Biol 5(3):207–212
Yuji S, Yomi W, Akio S, Yoshio T (2002) J Mol Catal B Enzym 17:133–42
Yusuf NNAN, Kamarudin SK, Yaakub Z (2011) Energy Convers Manag 52(7):2741–2751
Zafar S (2022) Breaking down the process of biofuel production
Zahra Z, Choo DH, Lee H, Parveen A (2020) Cyanobacteria: review of current potentials and applications. Environments 7(2):13
Zhou W, Min M, Hu B, Ma X, Liu Y, Wang Q, Shi J, Chen P, Ruan R (2013) Filamentous fungi assisted bio-flocculation: a novel alternative technique for harvesting heterotrophic and autotrophic microalgal cells. Sep Purif Technol 107:158–165

Chapter 5
Evolution of Sustainable Technologies for Production of Biofuel

Pooja Sharma, Himanshi Chaudhary, Sanjukta Vidyant, and Shubha Dwivedi

Abstract In the current scenario, the overconsumption of non renewable fossil fuels and anthropogenic stress by human interference may fulfill the global energy demand but affect the sustainability of environment and threatened the environment with the dispossession and diminishing of fossil fuels. This leads the researchers to formulate new concepts and technology to fulfill energy demand by the production of Biofuels like bio-oil, biodiesel, bioethanol, and syngas from the plant based material like peel waste, microalgae and macro algae and microbes based material. This chapter focuses to using electric transport, solar power, carbon capture and storage technologies for sustainable development of environment and for the production of E20 fuel blend of ethanol and fossil based fuel. This will mark beneficial for agriculture, transportation, growth of country, sustainability of environment, energy demand, controlled pollution and flags green mobility.

Keywords Sustainable technology · Biofuel · Green mobility · E20 fuel · New technologies

5.1 Introduction

In 2022, united nation survey and published data of increasing human population that emerge as the novel trouble for sustainability of earth (FAO and UNEP 2021; Li et al. 2019; Moravvej et al. 2019). The anthropogenic activity of human serves as major threat for accomplishing the goals of sustainable environment by causing pollution, degradation of soil by digging and dumping waste into pits (Barreto 2018), uses of fossil fuel for transport vehicles [98%] (Gasparatos et al. 2013), loss of biodiversity (Mohr and Raman 2013), acid rain (Carriquiry et al. 2011), incineration

P. Sharma · S. Vidyant · S. Dwivedi (✉)
Department of Biotechnology, IIMT University, Meerut, Uttar Pradesh 250001, India
e-mail: shubhadwivedi5@gmail.com

H. Chaudhary
Department of Microbiology, IIMT University, Meerut, Uttar Pradesh 250001, India

M. Shah and D. Deka (eds.), *Emerging Sustainable Technologies for Biofuel Production*, Environmental Science and Engineering, https://doi.org/10.1007/978-3-031-52167-6_5

of fossil fuels that leads for the emanation of green gases (Carbon monoxide, nitrogen dioxide, carbon dioxide, nitrogen monoxide, sulphur dioxide (Carriquiry et al. 2011) hydrocarbons) causes global warming (Hirani et al. 2018), increase carbon foot prints (Abas et al. 2015; Mohammadi et al. 2011; Moravvej et al. 2019), emancipation of mercury lead elements in water and atmosphere by burning coal (Carriquiry et al. 2011).

Over years, the global energy demand is accomplish by using non renewable resources like coals (Dangi et al. 2018), petroleum, natural gas and crude oil produced from organic matter [plant, animal] by undergoing extreme sedimentation process, into fuels under extreme pressure in addition to temperature and extracted in the form of gaseous, solid and liquid state (Pittman et al. 2011a). Non renewable resources are used years ago as they are reasonably priced, produced naturally, easily available, helps in economic growth for worldwide development and considered as the ultimate as well as primary source of energy production although its fails to accomplish the goals of sustainable development as they cannot replenish therefore using renewable sources considered as the outstanding solution for energy crisis (Dangi et al. 2018; Gasparatos et al. 2013; Pittman et al. 2011a). The export of natural gas, coal and petroleum is controlled by center east countries like Saudi Arab, Iran, Iraq and Qatar that have the reserve that raise the market value and prices all over countries. and production of natural gases, coal and petroleum are located in the center east countries like (Carriquiry et al. 2011) in According to world energy outlook evaluation Renewable energy sources can be replenished, Cost efficient, sustainable, potentially stable, less polluted, highly efficient, plenteous in environment. Renewable resources like Windmill, solar mill, geothermal, tidal, biofuel (Bioethanol, Biodiesel) emerge as outstanding alternative resource to produce power and channel electricity worldwide (United Nations Department of Economic and Social Affairs, Population Division 2022).

Sustainable development is an advance strategy with the intention to meet the requirements of the current without jeopardizing future generation's ability to meet their own needs. In other words, it aims to hit equilibrium amid community, cost-effective, and ecological factors in order to build a more sustainable future (Havlík et al. 2011). The United Nations' SDGs, a sequence of 17 goals endorsed by all UN member states in 2015, are frequently associated with sustainable development. Among these goals, goal 7th and 13th are the abolition of poverty and hunger, the promotion of clean energy and sustainable cities, the conservation of biodiversity and ecosystems, and the provision of safe drinking water and sanitation (Gasparatos et al. 2013; Havlík et al. 2011). Sustainable development necessitates a multifaceted strategy that considers the interdependence of economic, social, and environmental systems. This means that policies and activities must meet numerous goals rather than just one (Havlík et al. 2011; Naqvi and Yan 2015). A sustainable development strategy, for example, might promote renewable energy as resources of reducing greenhouse gas emissions and combating climate change while also creating jobs and fostering economic growth. To achieve sustainable development, all parties, including governments, businesses, civil society organizations, and individuals, must be involved. This necessitates collaboration and communication in order to define common goals and develop equitable, inclusive, and successful solutions (Gasparatos et al. 2013; Xiao and Gao 2011). Overall, sustainable development is a critical notion for assuring a better future for everybody while also safeguarding the earth and its resources for future generations (FAO and UNEP 2021; Gasparatos et al. 2013; Havlík et al. 2011; Xiao and Gao 2011).

5.1.1 Biofuels: Generation and Types

Biofuel is a energy containing compounds that are formed from sustainable organic resources like plants and mammal waste like wood, agricultural crops waste (Xiao and Gao 2011). It is environmentally friendly alternative that be capable of be used to replace remnant fuels like coal, petrol, diesel and natural gas. Biofuels available in a variety of forms, including ethanol, biodiesel, and biogas, biochar, biohydrogen, bioalcohol (Chye et al. 2018). Ethanol is primarily derived from corn or sugarcane and can be mixed with petrol for use in automobiles. Biodiesel is derivative starting vegetable oils or animal fats and is suitable for utilize in diesel engines (Xiao and Gao 2011). Biogas is produced from organic materials such as sewage or manure and be capable of be utilized to produce high temperature or power. Bio-oil, also known as pyrolysis oil, is formed from heating biomass in the nonexistence of oxygen (Hubbert 1949). Biofuels have the impending to lessen greenhouse gas emissions and dependence on fossil fuels, but they also have some drawbacks, such as land-use competition, water usage, and potential unconstructive impact on biodiversity (Pittman et al. 2011b). Additionally, the production of some biofuels can generate significant amounts of greenhouse gases, depending on the creation, development and the nature of biomass used (Lee and Lavoie 2013). The renewable energy that is resulting from organic matter, such as plants or plant waste called as biofuel can be used as an alternative to fossil fuels. There are different types of biofuels involved to replace fossil fuel and evolve as a best alternative source for energy consumption (Pittman et al. 2011a; Lü et al. 2011; Liang et al. 2012).

1. Production of bioethanol: Ethanol is the bulk and usually produced biofuel and is typically made from corn, sugarcane, or other high-sugar crops. The process involves converting the sugars in the plant material into alcohol through fermentation.
2. Biodiesel production: Biodiesel is typically made from vegetable oils or animal fats, which are changed into petroleum that can be used in diesel engines. The process involves production of fatty acid methyl esters [FAME] by reacting the oils or fats.
3. Biogas production: biogas production is done through the anaerobic digestion of organic materials, such as agricultural waste, food waste and sewage. The method involves infringementing down the organic matter using bacteria to produce a mixture of methane and carbon dioxide, which can be used at the same time as a fuel (Carriquiry et al. 2011).

4. Cellulosic ethanol production: Cellulosic ethanol is made starting non-food sources, such as wood wastes, farming waste, or grasses. The process involves breaking down the complex sugars in the plant material using enzymes to produce ethanol.
5. Algal biofuel production: Algal biofuels are produced from microalgae, which are grown in ponds or tanks and harvested to extract the oils or carbohydrates that can be converted into biofuels (Singh et al. 2014; Teng and Chen 2019).

This chapter focuses on using different types of biofuels greater than fossil fuels to fulfill the increasing demand of energy followed through the production of biofuel to accomplish the demand of sustainable goals.

5.2 Evolution of Technologies for Biofuel Production

The three most important types of biofuels generation technologies are first-generation, second-generation, and third-generation. First-generation biofuels are through food crops such as sugarcane, corn and soybeans, and have been criticized for contributing to food scarcity and lashing up food prices (Liew et al. 2014). The fuels of second-generation are prepared from non-food crops and waste materials such as agricultural residues, forestry residues, and municipal unyielding waste. The fuels of third-generation are prepared from algae as well as other aquatic plants (Mao et al. 2015; Shafiee and Topal 2009). The Fig. 5.1 shows the various generations, feedstock and products produced as the biofuels.

5.2.1 Technology for First Generation Biofuel Production

First generation biofuels are created by using feedstock of agricultural wastes and crops like corn, sugarcane, sugar beet, soya bean, rapeseed, and peanut that contain starch, sugar and oil compounds. The two mainly frequent biofuel are Biodiesel and Bioethanol that have similar properties but produced using different feedstock and technology. Nowadays, the biodiesel and bioethanol are generally blended with fossil fuels and used for transportation purpose that considerably increases biofuel production rate world widely which requires agricultural lands for the production of feedstock like sugarcane, corn, etc. (Nanda et al. 2018). The overproduction of feedstock in limited and specific areas causes extreme violation and degradation of lands, overproduction and production of green house gas and loss of habitat in specific areas. The fabrication technology of First generation biofuel is transesterification and fermentation process for biodiesel as well as bioethanol creation (Hirani et al. 2018).

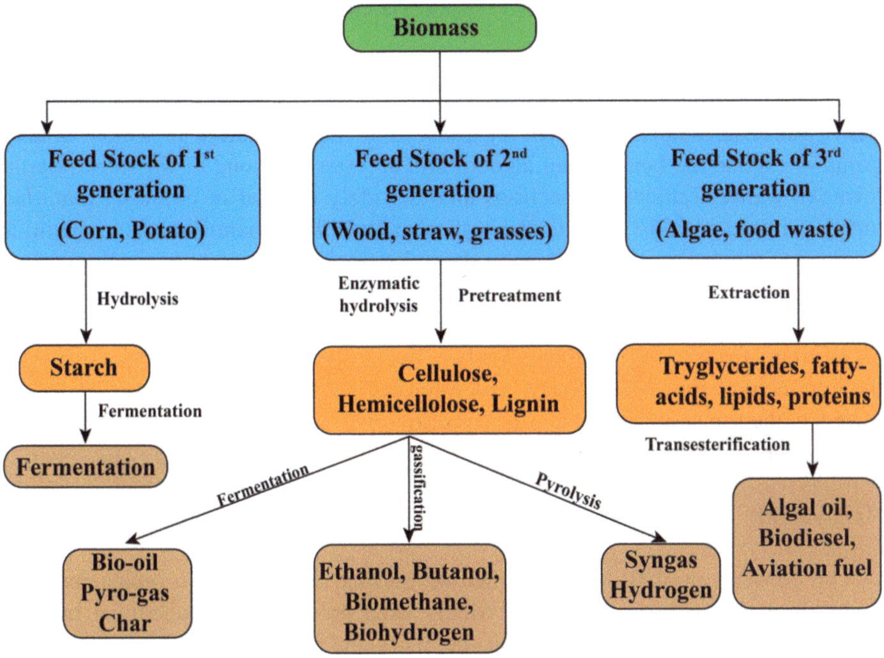

Fig. 5.1 The figure depicts the various generation and feedstock that undergo pretreatment processes like hydrolysis, enzymatic hydrolysis and extraction process that further undergo fermentation, gasification and transesterification to produce bio-oil, pyro-gas, char, ethanol, biomethane, bihydrogen, syngas and biodisesl

5.2.1.1 Transesterification

Triglycerides retort chemically through alcohol [e.g. biomethanol] to produce biodiesel [esters of fatty acids] and glycerol during the transesterification development. This process ought to be performed in the occurrence of a catalyst or else enzyme, or below supercritical conditions. It is usually catalyzed by an acid or base catalyst and takes place at high temperatures (Chye et al. 2018). The technique is extensively employed in the production of biodiesel, in which vegetable oils or animal fats are transformed into fatty acid methyl esters [FAMEs] that can be used as a diesel fuel replacement. An alcohol molecule reacts with an ester molecule in the reaction, resulting in the production of a new ester and an alcohol molecule. For example, three molecules of methanol react by way of one molecule of triglyceride to structure three molecules of FAME and one molecule of glycerol during the transesterification of triglycerides [chief constituent of vegetable oils and animal fats (Liew et al. 2014; Solomon 2010).

Types of Transesterification

Alkyl Transesterification

Alkyl transesterification is a type of transesterification reaction in which an alkyl group is exchanged between an alcohol and an ester. It is one of the most popular forms of transesterification reactions and is widely utilized in biodiesel manufacturing (Chye et al. 2018). An alcohol (often methanol or ethanol) reacts in the midst of an ester (such as a triglyceride) in the existence of a catalyst (typically a strong base such as sodium or potassium hydroxide). The ester is disassembled into its constituent elements (often a glycerol molecule and three fatty acid chains), and an alkyl group from the alcohol (typically a methyl or ethyl group] replaces one of the fatty acid chains to generate a new ester molecule) (Lee and Lavoie 2013). For example, the transesterification of triglycerides with methanol that would proceed to the reaction as follows:

$$\text{Triglyceride} + 3\ \text{Methanol} \rightarrow 3\ \text{Fatty Acid Methyl Esters [FAME]} + \text{Glycerol}$$

The consequential FAMEs can then be purified and used as a replacement for fuel. Alkyl transesterification be a widely used process in the chemical industry and has many applications in organic chemistry and biochemistry. This method is a relatively undemanding and competent technique for the production of esters and is an important step in the production of biodiesel (Teng and Chen 2019).

Aromatic Transesterification

Aromatic transesterification is a form of transesterification response in which an aromatic group is exchanged between an alcohol and an ester. Aromatic transesterification is a slower reaction that often requires more intense reaction conditions than alkyl transesterification, which includes the exchange of an alkyl group (Pittman et al. 2011b). An alcohol (such as methanol or ethanol) reacts by an ester (such as an aromatic ester) in the presence of a strong acid (sulfuric acid) or a strong base (sodium or potassium hydroxide) catalyst. The aromatic ester is disassembled into its constituent pieces, and the aromatic group from the alcohol replaces one of the aromatic ester's groups to generate a new ester molecule (Lee and Lavoie 2013). An aromatic group is exchanged between an alcohol and an ester in this type of transesterification. This is a slower reaction than alkyl transesterification and requires more intense reaction conditions. For example, in the transesterification of methyl benzoate with methanol, the reaction would proceed as follows:

$$\text{Methyl Benzoate} + \text{Methanol} \rightarrow \text{Methyl 3 - Methoxybenzoate} + \text{Methanol}$$

The outcome is a novel ester molecule with an aromatic group attached, methyl 3-methoxybenzoate. Aromatic transesterification is a typical method for synthesizing pharmaceuticals and other chemical molecules. It can be an effective method for changing the chemical properties of aromatic esters, leading to the synthesis of novel molecules with distinct chemical and physical properties. It is, however, a slower and more complicated reaction to perform than alkyl transesterification (Schenk et al. 2008).

Heterocyclic Transesterification

A heterocyclic group is exchanged between an alcohol and an ester. Heterocyclic compounds are organic compounds with at least one non-carbon atom (such as nitrogen, oxygen, or sulphur) within their ring structure. An alcohol (such as methanol or ethanol) is reacting through an ester (such as a heterocyclic ester) in the presence of a catalyst (such as an acid or a base) in this reaction. The heterocyclic ester is disassembled, and the heterocyclic group from the alcohol replaces one of the groups on the heterocyclic ester to generate a new ester molecule (Singh et al. 2014). In the transesterification of ethyl isocyanate with methanol, for example, the reaction would proceed as follows:

$$\text{Ethyl Isocyanate} + \text{Methanol} \rightarrow \text{Methyl Carbamate} + \text{Ethanol}$$

The resulting methyl carbamate molecule is a novel ester compound with a heterocyclic group. Heterocyclic transesterification is widely employed in pharmaceutical chemical synthesis, particularly in the creation of carbamate esters and other heterocyclic esters. It is an effective method for changing the chemical properties of heterocyclic compounds, and it can lead to the synthesis of novel molecules with distinct chemical and physical features (Pittman et al. 2011b).

Enzymatic Transesterification

Enzymatic transesterification is a sort of transesterification reaction in which enzymes, rather than chemical catalysts, act as catalysts. Enzymes are biological catalysts that can rapidly and specifically enhance chemical processes. To generate a new ester molecule, an alcohol (such as methanol or ethanol) reacts with an ester (such as a triglyceride) in the existence of an enzyme catalyst (lipase) (FAO and UNEP 2021).

The enzyme catalyses the ester's breakdown into its constituent parts before facilitating the transfer of the alcohol group to the ester to generate a new ester molecule. For example, the enzymatic transesterification of triglycerides with methanol, the reaction would proceed as follows:

$$\text{Triglyceride} + 3\ \text{Methanol} \rightarrow 3\ \text{Fatty Acid Methyl Esters [FAME]} + \text{Glycerol}$$

This reaction is catalyzed by an enzyme called lipase, which speeds up the break of triglycerides into fatty acids and glycerol and then enhances the transfer of methyl groups from methanol to fatty acids to generate FAMEs. Enzymatic transesterification is a possible alternative to classic chemical transesterification reactions, especially for biodiesel generation. It has a amount of advantages, including lower energy consumption, less waste formation, and more control over reaction conditions. However, enzymes are more expensive than standard chemical catalysts, and reaction conditions are more sensitive to factors such as pH and temperature (Schenk et al. 2008; Singh et al. 2014).

Acid-Catalyzed Transesterification

Acid-catalyzed transesterification is a form of transesterification reaction where an acid catalyst, like sulfuric acid or hydrochloric acid used to speed up the effect between an alcohol and an ester. By protonating the carbonyl oxygen atom, the acid catalyst activates the ester molecule, building it more susceptible to nucleophilic attack by the alcohol. An alcohol [such as methanol or ethanol] is reacted with an ester [such as a fatty acid ester] in the presence of an acid catalyst in this reaction (Singh et al. 2014; Teng and Chen 2019). The carbonyl oxygen atom in the ester is protonated by the acid catalyst, which then conducts a nucleophilic substitution reaction with the alcohol. The end result is a new ester molecule containing an alkyl group from the alcohol and a carbonyl group from the catalyst (Pittman et al. 2011b; Teng and Chen 2019). The acid catalyst protonates the carbonyl oxygen atom in the fatty acid ester, making it more vulnerable to methanol's nucleophilic assault. The end result is a mix of FAMEs and glycerol. In the synthesis of biodiesel from vegetable oils or animal fats, acid-catalyzed transesterification is typically used (Nada and Alrikabi 2014). However, it has a number of disadvantages, including the risk of acid corrosion, the invention of enormous amounts of waste, and the formation of unwanted byproducts (FAO and UNEP 2021; Xiao and Gao 2011). Furthermore, the utilize of acid catalysts can cause the ester products to degrade, lowering the total yield of the process. For example, in the acid-catalyzed transesterification of triglycerides with methanol, the reaction would proceed as follows:

$$\text{Triglyceride} + 3\ \text{Methanol} \rightarrow 3\ \text{Fatty Acid Methyl Esters [FAME]} + \text{Glycerol}$$

Base-Catalyzed Transesterification

Base-catalyzed transesterification is a reaction in which a base catalyst like sodium or potassium hydroxide is used to aid the reaction amid an alcohol and an ester. By deprotonating the carbonyl oxygen atom, the base catalyst activates the ester molecule, making it more vulnerable to nucleophilic assault by the alcohol. An alcohol [such as methanol or ethanol] is reacted with an ester [such as a triglyceride] in the presence of a base catalyst in this reaction. The ester is deprotonated by the base catalyst, which then undergoes a nucleophilic substitution reaction with the alcohol (FAO and UNEP 2021; Schenk et al. 2008). The resultant ester molecule has an alkyl group from the alcohol and a carbonyl group from the original ester. A strong base catalyst, such as sodium hydroxide or potassium hydroxide, catalyses this process (Gajraj et al. 2018). The base catalyst deprotonates the carbonyl oxygen atom in the fatty acid ester, leaving it more vulnerable to methanol's nucleophilic attack

(Shahzad 2015). The end result is a mix of FAMEs and glycerol. In the synthesis of biodiesel from vegetable oils or animal fats, base-catalyzed transesterification is typically utilized. It outperforms acid-catalyzed transesterification in various ways, including greater yields, less corrosion, and fewer undesirable byproducts (Lee and Lavoie 2013). The employment of base catalysts, on the other hand, may result in saponification of the ester products and the generation of soaps, lowering the largely capitulate of the reaction. Furthermore, the reaction conditions may be more sensitive to variables such as temperature and pressure (Pittman et al. 2011b; Teng and Chen 2019). For example, in the base-catalyzed transesterification of triglycerides with methanol, the reaction would proceed as follows:

$$\text{Triglyceride} + 3\text{Methanol} \rightarrow 3\ \text{Fatty Acid Methyl Esters [FAME]} + \text{Glycerol}$$

Non-catalytic Transesterification

A method in which esterification or transesterification reactions are carried out without the aid of a catalyst is known as non-catalytic transesterification. The Trans-esterification reactions typically include the replacement of an alcohol group in one molecule with an ester group in an additional to create a new ester and alcohol. This method is frequently employed to create biodiesel (renewable fuel derived from vegetable oils or animal fats). Non-catalytic transesterification involves conducting the process devoid of a catalyst at high temperatures and pressures. The transesterification reaction takes place without the aid of a catalyst when the reactants are heated to high temperatures and pressures. After that, the products are separated and refined using a variety of techniques, including distillation. The absence of catalysts makes non-catalytic transesterification more advantageous than catalytic transesterification in terms of cost and environmental impact. The need for greater reaction temperatures and longer reaction durations, which can result in higher energy consumption and lower product yields, are some of its drawbacks (Singh et al. 2014; Teng and Chen 2019).

Microwave Aided Transesterification

Microwaves are employed in the microwave aided transesterification process to hasten the reaction that turns animal or vegetable lipids into biodiesel. To speed up the transesterification reaction, the oil or else fat is heated along with an alcohol and a catalyst while being exposed to microwave radiation. A chemical process called transesterification involves replacing the alcohol group of an ester molecule with another alcohol to create a new ester. In the process of making biodiesel, the alcohol group of a vegetable oil or animal fat is exchanged with an alcohol like methanol or ethanol to create methyl or ethyl esters, respectively (FAO and UNEP 2021; Gasparatos et al. 2013). Compared to conventional transesterification methods, microwave aided transesterification has a number of benefits, such as quicker reaction times, higher yields, and less energy use. The use of microwaves can also improve reaction efficiency by lowering the production of unwanted byproducts. The necessity for specialized equipment and the potential for localized hot patches that could lead to thermal degradation of the biodiesel product are two issues that come with microwave assisted transesterification. To assure the generation of high-quality biodiesel using this process, rigorous optimization of the reaction conditions is necessary (Schenk et al. 2008; Singh et al. 2014).

Ultrasound-Aided Transesterification

A chemical reaction method called ultrasound-assisted transesterification is used to turn triglycerides, like vegetable oil, into biodiesel. In order to create glycerol and biodiesel, the transesterification process requires the exchange of ester groups between a triglyceride and an alcohol in the presence of a catalyst. By speeding up the reaction and boosting the yield of biodiesel, ultrasound waves are utilised to improve the transesterification process. Cavitations bubbles are produced by the ultrasonic waves, and when they collide, they produce intense pressure and warmth locally (Lee and Lavoie 2013; Pittman et al. 2011b). These circumstances speed up the process and encourage the production of biodiesel by making it easier for the triglycerides and alcohol to break down. In comparison to traditional procedures, using ultrasonic for transesterification has a number of benefits, including faster reaction times, less energy use, and higher biodiesel yields. Low temperatures and pressures make ultrasound-assisted transesterification less likely to have unfavourable side effects, while also requiring less expensive machinery. Overall, ultrasound-assisted transesterification offers a more effective and environmentally friendly option to conventional methods for the generation of biodiesel (Schenk et al. 2008; Xiao and Gao 2011).

5.2.1.2 Fermentation

First-generation biofuels are those that are created through the fermentation of food crops like corn, sugarcane, and soybeans. Microorganisms like yeast or bacteria are employed in the fermentation process to turn the sugars in crops into ethanol or other biofuels. Pretreatment, hydrolysis, and fermentation are frequently included in the fermentation process. The complex carbohydrates in the crops are transformed into simple sugars during pretreatment so they may be quickly fermented (Schenk et al. 2008; Singh et al. 2014). Typically, physical or chemical techniques like grinding, heating, or adding acids or enzymes are used to accomplish this. After the sugars are released, a procedure known as fermentation transforms them into ethanol or other biofuels. Microorganisms consume the sugars during fermentation to create ethanol and carbon dioxide as byproducts. After being cleaned up, the ethanol is combined with petrol to create a biofuel that can be used in regular combustion engines. First-generation biofuels have the benefit of being made from widely accessible food crops, but they also have a number of disadvantages (Pittman et al. 2011b; Teng and Chen 2019). The production of food crops for biofuels can compete with food production, leading to food shortages and higher prices. Additionally, the fermentation process requires large amounts of water and energy, contributing to environmental problems such as water pollution and greenhouse gas emissions.

Steps of Fermentation

Feedstock Preparation

In the fermentation process used to produce biofuel, the preparation of the feedstock is a crucial stage. By dissolving complicated carbohydrates or other compounds into simpler sugars that can be quickly ingested by microorganisms like yeast or bacteria, feedstock pretreatment aims to make the raw materials more accessible for fermentation (Rodionova et al. 2017; Schenk et al. 2008). The various methods involved in stock preparation are:

(a) **Enzymatic Hydrolysis**: Enzymatic hydrolysis is the process of breaking down complex carbohydrates into simpler sugars by adding enzymes to the feedstock. This technique is frequently applied to feedstock like wood chips, switch grass, or maize Stover.
(b) **Acid Hydrolysis**: To convert complex carbohydrates into simpler sugars, acid hydrolysis entails treating the feedstock with an acid, such as sulfuric acid. This technique is frequently applied to feedstock like corn or sugarcane.
(c) **Heat Treatment**: Heating the feedstock causes complex carbohydrates to disintegrate into simpler sugars. This technique is frequently applied to feedstock like corn or wheat.

Pretreatment

In the fermentation process used to produce biofuels, pretreatment is a crucial step, especially for feedstocks that contain complex polysaccharides, lignocellulosic materials, or other substances that are challenging to ferment (Singh et al. 2014; Teng and Chen 2019). In order for yeast or bacteria to more easily ferment these complicated molecules, pretreatment is used to convert them into simpler sugars. Here are a few typical pretreatment techniques used in the creation of biofuels:

(a) **Physical Pretreatment**: Crushing, grinding, or milling the feedstock can increase surface area and decrease particle size, making it more palatable to microorganisms during fermentation. Physical pretreatment may also include mechanical or thermal processing, such as steam explosion or high-pressure homogenization, to alter the feedstock's molecular structure.
(b) **Chemical Pretreatment**: To break down lignin or complex carbohydrates, chemical pretreatment techniques add chemicals like acids, bases, or solvents to the feedstock. To improve the breakdown of complex molecules, chemical

pretreatment might also involve adding enzymes or microbial cultures to the feedstock.

(c) **Biological pretreatment**: In biological pretreatment, the feedstock's complex polysaccharides or lignocellulosic components are broken down by microorganisms like fungi or bacteria. Agricultural waste or forestry wastes are examples of feedstocks for which this technique may be useful.

(d) **Combination Pretreatment**: To obtain a more thorough breakdown of complex chemicals, combination pretreatment entails employing various pretreatment techniques, such as a combination of physical

Fermentation

The process of fermentation is the conversion of sugars or other organic compounds into alcohol, acids, or gases by microorganisms like yeast or bacteria. A crucial step in the manufacture of ethanol and other kinds of biofuels, fermentation is important in the context of biofuel production (Teng and Chen 2019). The general steps in the fermentation process for making biofuels are as follows:

(a) **Inoculation**: The initial stage entails adding the microbes to the feedstock, such as corn, sugarcane, or other plants, which have been cleaned and processed in order to facilitate fermentation.

(b) **Fermentation**: Under carefully controlled temperature and pH conditions, the prepared feedstock is combined with water and the microorganisms. The microorganisms use the feedstock's sugars and other organic components as food during fermentation, turning them into ethanol or other biofuels as well as carbon dioxide and other byproducts.

(c) **Monitoring and Control**: The fermentation process is carefully watched over and managed to create the best possible environment for the activity and growth of the microorganisms. To maximize the effectiveness and production of fermentation, variables including temperature, pH, and nutrient levels are carefully regulated.

(d) **Harvesting**: After fermentation is finished, a process known as distillation is used to separate and purify the combination of biofuel and other byproducts. To make ethanol or other biofuels more concentrated, the biofuel is separated from the water and other byproducts and concentrated (Lee and Lavoie 2013).

Separation and Distillation

In order to distinguish the biofuel from other fermentation byproducts, separation and distillation are crucial phases in the manufacturing of biofuels, notably ethanol (Schenk et al. 2008; Xiao and Gao 2011). The general procedures for separation and distillation for the production of biofuel are as follows:

Separation: To separate the combination of biofuel, water, and other fermentation byproducts, particles and contaminants are removed using centrifugation or other techniques.

Distillation: The biofuel, which is mostly ethanol, is subsequently separated from water and other byproducts by distilling the remaining mixture. The combination is heated until the ethanol vaporizes since ethanol has a lower boiling point than water. The vapour is then condensed and collected.

Rectification: The ethanol produced during distillation is further purified through a procedure called rectification in order to obtain high-quality ethanol for use as a fuel. To get rid of any impurities and raise the concentration of the ethanol, this includes repeatedly distilling and condensing it.

Dehydration: The purified ethanol can then be dried out to remove any last traces of water and boost its concentration even more. Azeotropic distillation or other techniques that employ additional chemicals to extract the residual water are often used to achieve this.

5.2.1.3 Current Scenario of First Generation Biofuel

Over the precedent few decades, there has been an augment in demand for first-generation biofuels, which are prepared from victuals crops like corn, sugarcane, and vegetable oils. This demand has been sparked by a number of factors including worries concerning climate change, energy security, and the desire to lessen reliance on fossil fuels (Singh et al. 2014; Teng and Chen 2019). The most widely produced types of biofuels in 2019 were ethanol and biodiesel, which accounted for around 65% of global biofuel output, according to the International Energy Agency's [IEA] Bioenergy Report 2020. The research further mentions that laws and incentives meant to encourage the adoption of renewable energy sources are projected to cause the demand for biofuels to continue rising in the upcoming years. First-generation biofuel production and use, however, have come under fire for a number of reasons. One key issue is the possible conflict between food crops and biofuels for resources and land, which might drive up food costs and worsen food poverty in some areas. Additionally, the manufacturing of first-generation biofuels has been associated to deforestation, changes in land use, and other negative environmental effects, which can partially balance the environmental advantages of utilizing biofuels in place of fossil fuels (FAO and UNEP 2021; Xiao and Gao 2011). The growth of second and third-generation biofuels are made from non-food sources such algae, forestry waste, and agricultural wastes, has changed in response to some of these worries. These biofuels could potentially provide more ecologically friendly and sustainable energy sources (Schenk et al. 2008; Singh et al. 2014; Teng and Chen 2019).

5.2.2 Technology for Second Generation Biofuel Production

Advanced biofuels known as second-generation biofuels are made from non-food sources like community concrete waste, farming waste, and wooded area residues. Second-generation biofuels, which are thought to be more sustainable than first-generation biofuels derived from food crops like corn and sugarcane, do not compete with the food supply chain (Leong et al. 2018). Second-generation biofuels are produced by dissolving the cellulose and complex sugars in non-food sources into simpler sugars, which are subsequently transformed into biofuels including ethanol, butanol, and biodiesel. This procedure, called cellulosic biofuel manufacturing, uses cutting-edge technology including fermentation and enzyme hydrolysis (FAO and UNEP 2021; Xiao and Gao 2011). Second-generation biofuels are now able to be produced on a large-scale for commercial use thanks to recent advancements in research and development (Nanda et al. 2018). The word "second-generation biofuels" refers to biofuels made from non-food sources as feedstocks, such as bioethanol in addition to biodiesel. In end, biofuels won't compete with food crops or won't compete as much, and crop scraps and additional non-crop plant genus will only exist planted for the creation of bioenergy (Havlík et al. 2011; Schenk et al. 2008). The process of creating bioethanol from lignocelluloses involves several steps. Pretreatment of the processed biomass often involves an amalgamation of heat and chemical treatment. Pretreatment modified natural cellulose fibers to encourage cellulose hydrolysis. The most favored commercial microbe for the manufacture of bioethanol, Saccharomyces, ferments the sugar produced by the enzymatic hydrolysis. Simultaneous saccharification and fermentation, or SSF, is a technique whereby enzymatic hydrolysis and fermentation are accepted out simultaneously. This decreased end product buildup and improved the enzymatic hydrolysis process' efficiency. SSF is high, though, depending on how suitable the biomass is and whether there are any available comestible enzymes or yeast strains (Ewunie et al. 2020; Nanda et al. 2018; Teng and Chen 2019).

5.2.2.1 Approaches of Second Generation Biofuel

Thermo Chemical Pathway

The "thermo" technique refers to particular processes that include heating biomass with little or no oxidizing agent. All of those processes result in the separation of biomass into three fractions: biochar, a solid, and syngas, a gas typically made up of "CO", "H", short-chain alkanes, and CO_2 (Chye et al. 2018). Pyrolytic oil and bio oil are the two current names for the liquid fraction produced by these processes. Torrefaction occurs when biomass is treated by the side of low down temperatures (250–350 °C) with no oxygen, and the main conversion product is charcoal. Thermo chemical routes are a collection of procedures that use heat and chemical reactions to transform biomass feedstock into biofuels (FAO and UNEP 2021). Each of these pathways involves a different thermo chemical process, such as pyrolysis, gasification, or liquefaction, and each has advantages and drawbacks of its own.

Pyrolysis

In order to create bio-oil, char, and gas, pyrolysis, a thermo chemical process, requires heating biomass feedstocks without the presence of oxygen. The biomass is heated in between 400 and 600 °C in a reactor vessel during the process, which causes it to disintegrate into its constituent elements. The biomass is quickly heated to a high temperature during pyrolysis, which causes it to break down into a liquid, a solid, and a gas. The resultant liquid, also referred to as bio-oil or pyrolysis oil, is a richly energetic dark brown or black liquid. The solid byproduct, known as char, is a carbon-rich substance that can be utilized as a fuel or as a soil supplement. Different reactor types, including as fixed-bed reactors, fluidized-bed reactors, and entrained-flow reactors, can be used for pyrolysis (Xiao and Gao 2011). Depending on the type of feedstock utilized, the desired result, and the process circumstances, each reactor type has pros and cons. The bio-oil created by pyrolysis has a number of potential uses, such as fuel for the production of heat and power, feedstock for the refinement of crude oil into transportation fuels, and a supply of chemicals for the manufacture of polymers, resins, and other goods. However, the bio-oil has a number of difficulties that restrict its commercial viability, such as its high acidity, instability, and propensity to degrade over time (Schenk et al. 2008).

Gasification

Biomass feedstocks are heated in the presence of a gasifying agent, such as air, oxygen, or steam, to produce syngas [synthesis gas], a gas combination. Gasification is a thermo chemical process. In a gasifier, where the biomass is often subjected to high pressures and temperatures (typically between 700 and 1,200 °C), the process typically takes place (Teng and Chen 2019). Depending on the type of feedstock utilized and the parameters of the process, the biomass is transformed into a combination of carbon monoxide, hydrogen, carbon dioxide, and other gases during gasification. The syngas can be converted further into transportation fuels like ethanol and synthetic diesel or utilized as a fuel for the production of heat and power. Different types of gasifies, including as fixed-bed, fluidized-bed, and entrained-flow gasifies, can be used for gasification. Depending on the feedstock utilized, the desired product, and the processing circumstances, each type of gasifier has pros and cons. In comparison to other methods, gasification is a prospective thermo chemical approach for the amalgamation of biofuels from biomass feedstocks (Pittman et al. 2011b). To make the process more commercially viable, additional research and development are required to improve it and lower its costs. In comparison to alternative thermo chemical routes, gasification has a number of benefits, including as higher thermal efficiency, greater feedstock flexibility, and fewer emission of greenhouse gases and other pollutants. In order to further cut greenhouse gas emissions, gasification can be combined with other techniques like carbon capture and storage (Liew et al. 2014).

Liquefaction

The thermo chemical process of liquefaction includes heating biomass feedstocks in the presence of a solvent or a catalyst to produce bio-oil or biocrude, a liquid fuel. Hydrothermal liquefaction, solvolysis, and rapid pyrolysis are just a few of the methods that can be used to carry out the procedure. When biomass is liquefied, the complex organic molecules in the biomass are broken down into simpler ones by being heated to high temperatures (usually between 250 and 500 °C) and pressure in the presence of a solvent or catalyst. The resulting bio-oil has a chemical structure akin to crude oil from petroleum, is dark brown or black in colour, and contains a significant amount of energy. The bio-oil that results from liquefaction has a variety of possible uses, such as fuel for the production of heat and electricity, a feedstock for the refinement of biofuels, and a source of chemicals for the manufacture of polymers, resins, and other goods. However, the bio-oil has a number of difficulties that restrict its commercial viability, such as its high acidity, instability, and propensity to degrade over time. Liquidification has the advantage of processing a variety of feedstocks, including wet biomass such as algae and sewage sludge, which is challenging to treat using alternative thermo chemical processes. Additionally, liquefaction can provide a lot of bio-oil, often between 30 and 70% of the dry weight of the feedstock (Solomon 2010).

Biochemical Pathway

A technique utilized in the creation of second-generation biofuels is the "bio" route approach. Second-generation biofuels are derived from non-food sources such agricultural waste, forestry byproducts, and energy crops as opposed to first-generation biofuels, which are made from food crops like corn or sugarcane. Using numerous biological processes, the "bio" route approach dissects complex organic matter into simpler parts that can be turned into biofuels. The four processes of this method are typically pretreatment, hydrolysis, fermentation, and distillation. The pretreatment stage involves cleaning the biomass to get rid of any contaminants and make it more receptive to enzymes. Enzymes are then utilized in the hydrolysis process to convert the biomass' complex polysaccharides and other compounds into simple sugars. In order to make ethanol or other biofuels, the resultant sugar solution is subsequently fermented by microorganisms like bacteria or yeast (Bhuiya et al. 2016). The biofuels are then cleaned and distilled to remove any last traces of contaminants. Comparing the "bio" route approach to first-generation biofuels, there are a number of benefits. As it uses non-food sources that do not compete with the production of food, it is more sustainable. Additionally, it emits fewer greenhouse gases and has less of an adverse effect on the environment. However, the "bio" pathway approach also has some challenges, such as the high cost of pretreatment and the low yields of biofuels from some types of biomass. Nevertheless, research is ongoing to optimize this approach and improve the efficiency and economic viability of second-generation biofuel production (Neto et al. 2019; Teng and Chen 2019).

Current Scenario of Second Generation Biofuel

As nations throughout the world explore for ways to lessen their dependency on fossil fuels and reduce their carbon footprint, the demand for second-generation biofuels is rising. First-generation biofuels have come under fire for their effects on food prices and land use, while second-generation biofuels are considered as a more environmentally responsible and long-term replacement (Neto et al. 2019). The International Energy Agency [IEA] published a research predicting further growth in the market for biofuels, with second-generation biofuels contributing significantly to this growth. According to the analysis, the demand for biofuels may rise by about 25% by 2025, with second-generation biofuels accounting for about two-thirds of this growth (Li et al. 2019). The demand for second-generation biofuels is anticipated to rise as a result of some nations setting goals to use more biofuels in their energy mix (Abdullah et al. 2019; Liew et al. 2014). The European Union, for instance, has set a goal of using 14% renewable energy in transportation by 2030, which will necessitate a major rise in the usage of second-generation biofuels. In order to lower its carbon emissions, the aviation industry is likewise looking into using second-generation biofuels. The International Air Transport Association [IATA] has set a goal of achieving carbon-neutral growth in the aviation industry by 2020, which is anticipated to drive demand for biofuels in the future. Several airlines have already conducted test flights using biofuels made from waste and other non-food sources.

5.2.3 Technology for Third Generation Biofuel Production

Biofuels of the third generation are those produced using non-food sources like algae, jatropha, and other feedstocks. Because they are generated utilizing cutting-edge technologies and have the potential to be more efficient and sustainable than first- and second-generation biofuels, these biofuels are also referred to as advanced biofuels. Third-generation biofuels, such as those derived from algae, are an excellent example. Algae are a viable source of renewable energy since they can be produced in huge quantities and grow quickly (Abdullah et al. 2019; Solomon 2010). Algae can also be cultivated on non-arable terrain and do not compete for resources with food crops. Another potential feedstock for third-generation biofuels is jatropha, a tropical shrub that produces seeds rich in oil. Jatropha can be produced on marginal topography and doesn't need a lot of water or fertilizer, unlike corn and sugarcane, which are frequently used to make first- and second-generation biofuels (Achten et al. 2008; Alaswad et al. 2015; Nanda et al. 2018). In general, third-generation biofuels have the potential to be more cost-, energy-, and sustainably-efficient than earlier biofuel generations. Simple plants known as algae range in size from microscopic [microalgae] to enormous seaweeds [macroalgae], from little green spots to superior cyanobacteria and glittering diatoms. Algae can be categorized into nine major groups, including yellow-green algae [*Xanthophyceae*], cyanobacteria [*Cyanophyceae*], diatoms [*Bacillariophyceae*], green algae [*Chlorophyceae*], red algae [*Rhodophyceae*], brown algae [*Phaeophyceae*], golden algae *[Chrysophyceae*], dinoflagellates [*Dinophyceae*]. However, additional study and development are required to commercialize these technologies and make them competitive with conventional fossil fuels (Alaswad et al. 2015; Bhuiya et al. 2016; Solomon 2010).

For the production of third-generation biofuels including bioethanol, biodiesel, and biogas, algae can be utilized as a feedstock. For third-generation biofuels, algae can be used in the following ways:

(a) **Production of bioethanol**: Algae may be used to create bioethanol, a renewable and sustainable fuel that can be mixed with petrol. The fermentation of sugars obtained from algal biomass can result in the production of algae-based bioethanol.
(b) **Production of biodiesel**: Algae can also be utilized to create biodiesel, a renewable and clean-burning substitute for traditional petroleum-based diesel. The process of extracting lipids (oils) from algal biomass and transesterifying them into biodiesel yields algae-based biodiesel.
(c) **Biogas production**: Algae can also be utilized to create biogas, a renewable fuel that can be used for transportation, heating, and electricity production. The anaerobic digestion method, in which microorganisms break down organic materials and release biogas as a byproduct, can use algae biomass as a substrate.
(d) **Carbon capture and utilization**: Algae can also be used for carbon capture and utilization [CCU], which is a process by which algae take atmospheric carbon dioxide [CO_2] and turn it into biomass. Biofuels like bioethanol and biodiesel can then be made from the biomass (Schenk et al. 2008; Xiao and Gao 2011).

The biomass (i.e., the organic material) derived from microalgae is referred to as micro algal biomass. Microalgae are tiny photosynthetic organisms that can use the process of photosynthesis to change light energy into organic matter. One of the most promising sources of renewable biomass for a variety of uses, these organisms are found in a variety of aquatic habitats such as seas, lakes, rivers, and ponds. A variety of goods, including biofuels, dietary supplements, animal feed, and pharmaceuticals, can be made from microalgae biomass. Due to their quicker growth rates, higher photosynthetic efficiencies, and less need for land, microalgae biomass production is substantially more efficient than traditional crop-based biomass production. Although microalgae biomass has great potential, there are still a number of technical and financial issues that must be resolved before it can be widely used as a source of biomass. These difficulties include boosting micro algal cultivation productivity, creating efficient methods for harvesting and processing, and enhancing the financial viability of producing micro algal biomass. However, research in this area is moving forward quickly, and interest in using micro algal biomass for sustainable development is rising (Solomon 2010).

5.2.3.1 Methods of Production of Third Genertaion Biofuel

Biochemical Conversion of Micro Algal Biomass

The process of employing micro algal biomass as a feedstock for the creation of different biochemicals, such as biofuels, chemicals, and pharmaceuticals, is known as biochemical conversion of microalgae biomass. The process of turning micro algal biomass into simpler molecules that can be used to create a variety of products with added value requires the employment of microbes, enzymes, or other catalysts to break down the complex organic compounds present in the biomass. Microalgal biomass can be converted biochemically via a variety of methods, such as enzymatic hydrolysis, anaerobic digestion, and fermentation. Utilizing microorganisms, fermentation entails breaking down complex organic chemicals found in biomass into simpler ones like ethanol, butanol, and other organic acids. Utilizing microorganisms to break down biomass in the absence of oxygen to produce biogas and other byproducts is known as anaerobic digestion. Enzymatic hydrolysis uses enzymes to convert the biomass' complex organic molecules into simple sugars, which can then be used as a feedstock for the creation of biofuels and other products with added value (Liew et al. 2014).

In comparison to typical crop-based biomass, the biochemical conversion of micro algal biomass provides a number of benefits, such as faster growth rates, higher photosynthetic efficiencies, and a need for less area. Furthermore, a number of technologies, such as open ponds, photo bioreactors, and closed systems, can be used to culture microalgae, giving producer's choice over both the quantity and quality of the biomass they can generate. The development of cost-effective and efficient conversion technologies, the optimization of micro algal cultivation and harvesting, and the creation of markets for the value-added products are just a few of the technical and financial challenges associated with the biochemical conversion of micro algal biomass. Nevertheless, the biochemical conversion of micro algal biomass offers a lot of potential for the generation of sustainable and renewable biofuels as well as other value-added goods with further research and development (Abdullah et al. 2019; Solomon 2010).

Anaerobic Digestion of Microalgal Biomass

In the absence of oxygen, organic matter is broken down via anaerobic digestion, creating biogas that can be used as a sustainable energy source. Microalgae have a high biomass productivity and nutritional content, making them a suitable feedstock for anaerobic digestion (Xiao and Gao 2011).

There are multiple steps involved in the anaerobic digestion of micro algal biomass. The biomass is first gathered and cleaned up to get rid of any undesirable impurities. The organic material is then combined with water and sent to an anaerobic digester, where microorganisms degrade it in the absence of oxygen. Methane and carbon dioxide are combined to produce biogas, which can either be burned as fuel or utilized to generate power. The content of the biomass, the temperature and pH of the digester, and the type of microorganisms utilized are just a few of the variables that might impact how effectively micro algal biomass is an aerobically digested. Typically, a temperature range of 25–40 °C, a pH range of 6.5–8.5, and the utilization of mesophilic or thermophilic bacteria are the best conditions for anaerobic digestion of micro algal biomass. In comparison to other forms of biomass, anaerobic digestion of micro algal biomass has a number of benefits, including high productivity, minimal land use requirements, and the potential to yield high-value byproducts like fertilizer and animal feed. The process is not without its difficulties, though, including the requirement for efficient harvesting and processing techniques and the possibility for nutritional imbalance in the digester (Schenk et al. 2008; Xiao and Gao 2011).

Photo Biological Hydrogen Production

Using photosynthetic bacteria, photo biological hydrogen production creates hydrogen gas from water and sunshine. This procedure has the potential to be a sustainable and renewable replacement for current hydrogen generation techniques, which rely on non-renewable fossil fuels. Green algae and cyanobacteria are the most prevalent microorganisms employed in the photobiological hydrogen generation process. Through the process of photosynthesis, these bacteria are able to use sunlight to transform carbon dioxide and water into oxygen and organic compounds. Some of these microbes are also capable of producing hydrogen gas as a byproduct of their metabolic processes under specific conditions during digestion. Growing the photosynthetic microorganisms in an appropriate growth medium under regulated environmental factors including temperature, light intensity, and nutrient availability is the first step in the photobiological hydrogen production process. Through photosynthesis, the microbes produce oxygen and organic chemicals when they are exposed to light. Researchers can also stimulate the microbes to create hydrogen gas as a byproduct of their metabolic processes by changing the ambient circumstances. The process of producing hydrogen by photobiological means has a number of benefits. It does not rely on non-renewable fossil fuels and is a sustainable and renewable process. Second, it creates no greenhouse gases or other pollutants; therefore it has a negligibly low impact on the environment. However, there are a number of difficulties with the synthesis of photobiological hydrogen. These include the necessity for complex equipment and cultivation techniques, the challenge of regulating the metabolism and growth of the microorganisms, and the comparatively low effectiveness of the procedure in comparison to other hydrogen generation techniques (Gasparatos et al. 2013; Schenk et al. 2008; Xiao and Gao 2011).

Overall, photo biological hydrogen generation holds great potential as an environmentally friendly and renewable means of producing hydrogen, and further study in this area may result in important advances in the science.

Thermochemical Conversion of Microalgal Biomass

Thermo chemical techniques for micro algal conversion involve heating the biomass to high temperatures and pressures without oxygen in order to transform microalgae into biofuels and other high-value products. Several distinct thermo chemical processes, including pyrolysis, gasification, and liquefaction, may be used in this process. In comparison to alternative conversion techniques, the utilization of thermo chemical procedures for micro algal conversion has a number of benefits. First, a variety of high-value goods, such as biofuels, chemicals, and fertilizers, can be made using it. Second, it allows for flexibility in the choice of feedstock because these products can be made using a number of micro algal species. Thirdly, it can be utilized to transform microalgae into goods that are more transportable and storage-friendly because to their higher energy density compared to the original biomass (Bhuiya et al. 2016; Solomon 2010).

The thermo chemical conversion of microalgae is not without its difficulties, though. These include the process' high energy needs, the demand for advanced infrastructure and equipment, and the possibility of adverse environmental effects including greenhouse gas emissions.

Micro Algal Biomass Conversion Using Pyrolysis

Pyrolysis is a thermo chemical method for converting micro algal biomass into bio-oil, gas, and char by heating the biomass in the absence of oxygen. Different reactor designs and operating conditions can be used to perform pyrolysis, yielding various product yields and quality.

When micro algal biomass is heated to temperatures between 300 and 800 °C during pyrolysis, it breaks down into several compounds. A complex mixture of oxygenated hydrocarbons, the resulting bio-oil can be utilized as a fuel or chemical feedstock. The gas fraction, which can be utilized as fuel or processed further to create chemicals, is mostly made up of hydrogen, carbon monoxide, carbon dioxide, and other gases. Micro algal pyrolysis has a number of advantages over conventional conversion processes. First off, it is a reasonably easy procedure that can be completed using inexpensive tools. It is also more environmentally friendly than other methods that utilize a lot of water because it doesn't need water or other solvents. Thirdly, it can be utilized to create a variety of goods that can be used as chemical feedstock or as fuels (Schenk et al. 2008; Xiao and Gao 2011).

However, pyrolysis of microalgae is not without its difficulties. The high ash content of microalgae presents one difficulty since it can cause slag and reactor fouling by causing the growth of ash. The presence of nitrogen in microalgae, which might result in the generation of NO_x and N_2O during the burning of the bio-oil, presents another difficulty. Pyrolysis is a promising process for turning microalgae into bio-oil, gas and char, and it has the potential to aid in the creation of sustainable chemicals and fuels. To improve the procedure and solve the difficulties involved in the pyrolysis of microalgae, more study is required (Solomon 2010).

Microalgal Biomass Conversion Using Gasification

Thermo chemical conversion of micro algal biomass into synthetic gas [syngas] in the presence of a constrained amount of oxygen or steam is known as gasification of microalgae. To achieve varying yields and product quality, this process can be run utilizing various reactor designs and operational settings. The micro algal biomass is heated during gasification to temperatures between 700 and 1200 °C, and a small quantity of oxygen or steam is added to the reactor to partially oxidize the biomass. Hydrogen, carbon monoxide, carbon dioxide, and other gases combine to form syngas, which can be used as a fuel or chemical feedstock. The char is a solid byproduct that can be activated or used as a soil amendment (Solomon 2010).

Microalgae may be gasified, which provides a number of benefits over conventional conversion processes. The first application is to create high-quality syngas that can be utilized as a fuel or chemical feedstock. Second, it is a comparatively clean process with low pollutant emissions. Thirdly, it can be utilized to create a variety of goods that can be used as chemical feedstocks or as fuels. Microalgae gasification does, however, come with significant difficulties. The high ash content of microalgae presents one difficulty since it can cause slag and reactor fouling by causing the growth of ash. The potential for tar development in the syngas, which might lower the effectiveness of downstream operations, is another difficulty (Bhuiya et al. 2016).

Microalgal Biomass Conversion Using Liquefaction

Liquefaction of microalgae is a thermo chemical process that transforms the biomass of the algae into a liquid bio-oil under conditions of high pressure and temperature. To achieve varying yields and product quality, this process can be run utilizing various reactor designs and operational settings. The micro algal biomass is heated to temperatures between 250 and 400 °C under high pressure [up to 30 MPa] to liquefy it. A solvent or catalyst may also be used. The complex organic molecules in the biomass are broken down into smaller pieces by the solvent or catalyst, which combine with the liquid phase to generate a bio-oil. The bio-oil, a complex mixture of oxygenated hydrocarbons, can serve as a feedstock for fuel-related chemicals. Microalgae liquefaction has a number of advantages over conventional conversion processes. First, it can be used to create a liquid bio-oil with properties resembling those of petroleum crude oil, which can then be refined to create a variety of goods, such as petrol, diesel, and chemicals. Second, it is a comparatively clean process with low pollutant emissions. Thirdly, it allows for flexibility in product quality and yield since a range of solvents and catalysts can be used to carry out the process. The liquefaction of microalgae is not without its difficulties, though. One difficulty is the process' high expense, which is caused by the need for high pressure and temperature as well as the usage of pricey solvents and catalysts (Xiao and Gao 2011).

Transesterification of Microalgal Biomass

Transesterification is a technique used to convert micro algal biomass into biodiesel by converting the lipids found in microalgae. A chemical process known as transesterification entails the exchange of ester groups between an alcohol and a fatty acid. In the case of microalgae, the procedure entails the extraction of lipids from the microalgae cells and the subsequent reaction of the recovered lipids with an alcohol, such as methanol or ethanol, in the presence of a catalyst, usually an acid or a base. Either a homogeneous or heterogeneous catalyst can be used to complete the transesterification process. Acids like sulfuric acid or hydrochloric acid as well as bases like sodium hydroxide or potassium hydroxide are frequently used as homogeneous catalysts. Typically, heterogeneous catalysts are solid substances like zeolites and metal oxides. The mixture is split into two layers after the transesterification procedure, with the top layer being the biodiesel and the bottom layer being the byproduct glycerol (Bhuiya et al. 2016; Solomon 2010). The biodiesel made from microalgae has qualities that are comparable to those of conventional biodiesel made from vegetable oils and can be used to replace fossil diesel. Microalgae can be produced on non-arable soil utilizing seawater or wastewater as a supply of nutrients, minimizing the conflict between food and fuel production. This is one benefit of employing microalgae for the production of biodiesel. Additionally, compared to other biomass sources, microalgae have a high lipid content, which makes them a desirable feedstock for the generation of biodiesel. The use of transesterification to convert micro algal biomass is not without its difficulties, though. The high cost of growing microalgae on a large scale, the low yield of biodiesel from microalgae, and the requirement to optimize the process to lessen the environmental impact of the production process are a few of these. Nevertheless, there is still a need for more study and development in this area, and improvements in technology and process optimization are anticipated to increase the viability of transesterification of micro algal biomass as a long-term source of biofuels (Gasparatos et al. 2013; Nada and Alrikabi 2014; Solomon 2010).

Microalgal Biomass Conversion Using Direct Combustion

Another technique for converting micro algal biomass is direct combustion, which includes burning micro algal biomass to generate heat and energy. Dry micro algal biomass, which is normally gathered and dried to a moisture level of under 10%, is burned during the process.

A significant portion of the organic material in micro algal biomass, such as lipids, proteins, and carbohydrates, can be directly burned to produce energy. The heat produced by burning the microalgal biomass is utilized to create steam, which can then be used to power a steam turbine to produce energy (Güney 2019; Singh and Gu 2010). The resulting electricity can subsequently be used to run buildings, businesses, or factories. There are various benefits of burning micro algal biomass directly. First off, microalgae are a potentially desirable source of biomass for energy generation since they can be produced on non-arable soil. Second, garbage can be disposed of while also producing renewable energy by using wastewater or other waste streams as a source of nutrients for growing microalgae (Gasparatos et al. 2013). Last but not least, the combustion of micro algal biomass emits a lot less greenhouse gas than that of fossil fuels, making it a greener choice for energy production. Microalgal biomass combustion does come with certain difficulties, though. For instance, the collection and drying of microalgae biomass can be costly and energy-intensive, and the burning process can emit pollutants into the atmosphere such particulate matter, sulphur dioxide, and nitrogen oxides. Overall, the direct combustion of micro algal biomass is a viable route for the production of renewable energy, but more research

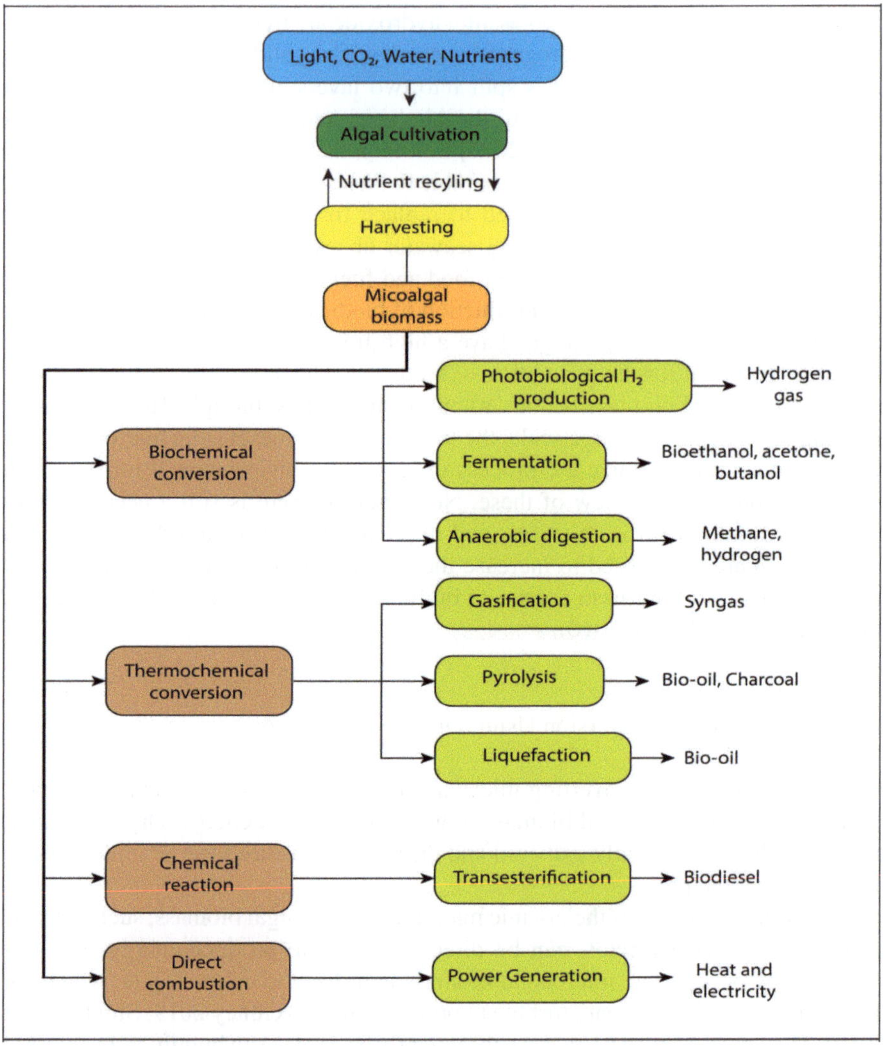

Fig. 5.2 This indicates the culturing and harvesting of algal biomass that undergo conversion by the process of thermo chemical, biochemical, chemical reactions and direct combustion through various technologies to produce biodiesel, bio-oil, charcoal, acetone, bioethanol and generation of electricity

and development are required to optimize the method and overcome its difficulties (Bhuiya et al. 2016; Mao et al. 2015). The Fig. 5.2 indicates the production of Biofuel using Micro algal biomass.

5.2.4 Technology for Fourth Generation Biofuel Production

Advanced biofuels known as fourth generation biofuels come from non-food sources such algae, wood, grasses, and agricultural waste. These biofuels can be used directly in current engines without requiring any modifications, which is why they are also referred to as “drop-in” fuels.

The fact that fourth-generation biofuels are more environmentally friendly and have a smaller carbon footprint than traditional biofuels is one of their main benefits. Algae-based biofuels, for instance, can be cultivated in wastewater or on land that is unfit for other crops, reducing competition for arable land and lowering the environmental effect of farming (Gajraj et al. 2018). The potential for fourth-generation biofuels to be more energy-efficient than traditional biofuels is another advantage. As an illustration, some biofuels made from algae can be created with a net energy gain, which means that the energy used to make the fuel is lower than the energy it supplies (Solomon 2010).

All things considered, fourth-generation biofuels have a lot of potential to lessen our dependency on fossil fuels and slow down climate change. However, before they can be generated on a wide scale and be competitive with conventional fuels, there are still substantial technological and financial obstacles that must be removed. Examples of fourth-generation biofuels include the following:

(a) Biofuels made from algae can be grown in ponds or tanks and include biodiesel, bioethanol, and biogas. Algae-based biofuels have the potential to be less carbon-intensive and more energy-efficient than conventional biofuels.
(b) Cellulosic biofuels: Non-food sources including switchgrass, corn stover and wood chips are used to make cellulosic biofuels. Because they utilise agricultural waste and other non-food sources, they have the potential to be more environmentally friendly and less carbon-intensive than conventional biofuels.
(c) Biofuels derived from organic waste, such as food scraps, paper trash, and sewage sludge, are known as waste-based biofuels. Because they make use of waste materials that would otherwise be thrown away, they can be a more economical and environmentally friendly method of producing biofuels.
(d) Biofuels produced by genetically modified organisms [GMOs]: Researchers are striving to create microorganisms with improved biofuel production capabilities. These biofuels, which use microorganisms created to produce biofuels more quickly, have the potential to be more cost- and environmentally-friendly than conventional biofuels.

5.2.4.1 Algae-Based Fourth Generation Biofuel

Fourth-generation biofuels made from algae are known as algae-based biofuels. Algae are photosynthetic microorganisms that can flourish in freshwater, saltwater, and wastewater, among other settings. Algae are grown in ponds or tanks, exposed to sunlight, and fed nutrients like nitrogen and phosphorus to create algae-based biofuels

(Leong et al. 2018). The algae take in carbon dioxide from the atmosphere as they expand and transform it into organic chemicals. When the algae have fully developed, they are harvested and processed to remove the lipids, which can then be used to make biodiesel. Using algae as a feedstock for biofuels has a number of benefits (Naik et al. 2010). In order to reduce competition for arable land and lessen the environmental impact of farming, algae can be produced on terrain that is unsuitable for other crops. Additionally, wastewater can be used to cultivate algae, lowering manufacturing costs and reducing the environmental impact of waste disposal. However, making biofuels from algae comes with its own set of difficulties. One is that the energy-intensive nature of producing algae and harvesting their lipids may reduce some of the environmental advantages of using biofuels (Bhuiya et al. 2016; Solomon 2010).

5.2.4.2 Cellulosic Based Fourth Generation Biofuel

Fourth-generation biofuels of the cellulosic variety are made from non-food sources such forestry waste, agricultural waste, and special energy crops. These resources contain complex carbohydrates that can be broken down and transformed into biofuels, such cellulose and hemicellulose. The biomass feedstock must first be gathered and transported to a processing facility where it is processed to make it more palatable to enzymes in order to manufacture cellulosic biofuels. The cellulose fibres may need to be broken down and the lignin removed using a mix of mechanical, thermal, and chemical processes. Lignin is a complex polymer that can obstruct the conversion process. After pretreatment, enzymes are added to the feedstock to break down the cellulose and hemicellulose into simple sugars, which can then be fermented to produce biofuels like ethanol or butanol. In rare circumstances, a distillation or purification phase may come after the fermentation process to further refine the biofuel output. Cellulosic biofuels have the potential to be produced from non-food sources, which reduces the competition for arable land and has a little influence on food costs. Because they utilise waste materials and non-food crops as feedstocks, cellulosic biofuels have the potential to be more environmentally friendly and less carbon-intensive than traditional biofuels (Sims et al. 2010). However, making cellulosic biofuels is not without its difficulties. One reason is that the conversion procedure may be difficult and expensive, needing specialized enzymes and machinery. Additionally, it might be challenging to obtain constant production and pricing due to the variability in feedstock supply and consistency. Despite these difficulties, efforts are being made to create and explore cellulosic biofuels in order to make them more affordable and scalable. Cellulosic biofuels may be able to play a significant role in helping to meet the growing demand for environmentally friendly and low-carbon transportation fuels (Bhuiya et al. 2016; Solomon 2010).

5.2.4.3 Biofuels from Genetically Modified Organisms [GMOs]

Biofuels made from genetically altered microbes or plants are known as biofuels from genetically modified organisms (GMOs) (Mohammadi et al. 2011). These species have undergone modifications to improve their capacity to make biofuels, whether by boosting yield, raising stress tolerance, or refining metabolic pathways. The bacterium *Escherichia coli* is one example of a genetically modified microorganism used to make biofuels; it has been manipulated to make ethanol from plant sugars more effectively than wild-type strains. *Saccharomyces cerevisiae,* yeast that has been altered to create cutting-edge biofuels like isobutanol, is another example. Additionally, plants that have been genetically altered are being researched as feedstock for biofuels (Ewunie et al. 2020). For instance, scientists are aiming to create crops that are more resilient to pests, drought, and other challenges, which might enhance yields and lessen the impact of farming on the environment (Barreto 2018). Additionally, some researchers are investigating the use of non-food crops as feedstocks for biofuels, such as *switchgrass* and *miscanthus,* which could lessen rivalry with food crops and the effect on food prices. GMO use in the manufacture of biofuel is not without criticism, though (Demirbas and Balat 2006). Some detractors contend that genetically modified organisms endanger the environment and public health and that it is unclear what will happen to modified creatures after they are released into the wild. GMOs may be subject to intellectual property restrictions, which might restrict access to the technology and raise costs for farmers and consumers (Abas et al. 2015; Mohr and Raman 2013; Moravvej et al. 2019). Despite these reservations, research and development on genetically modified organisms for the production of biofuels continues. These efforts are aimed at increasing yields, lowering costs, and reducing the environmental effect of biofuels. GMOs are one of many technologies being investigated in this quickly developing industry, yet it is unclear whether or not they will play a big part in the production of biofuels in the future (Mao et al. 2015; Shafiee and Topal 2009).

5.2.5 Future Aspects of Sustainable Technologies for Biofuel Production

With continued research and development targeted at enhancing the efficiency, sustainability, and scalability of biofuels, the future of sustainable technologies for the production of biofuels appears hopeful. Future biofuel technologies will prioritize a number of critical areas, including;

(a) Improvements in feedstock development: Researchers are attempting to create feedstocks for biofuels in more effective and sustainable ways. In addition to using waste resources like crop residues, forestry debris, and municipal solid waste, this also entails the development of non-food crops that may be produced on marginal land.

(b) New conversion techniques: To increase the effectiveness and sustainability of biofuel production, new conversion techniques are being developed. For instance, scientists are investigating the application of thermo chemical processes like gasification and pyrolysis, which can transform biomass into biofuels and other high-value products.
(c) Microbial engineering: In order to manufacture biofuels more effectively and sustainably, scientists are striving to modify microorganisms like yeast and bacteria. This includes the creation of genetic engineering methods and synthetic biology tools that can improve metabolic pathways and raise yields.
(d) Integration of biorefineries: Biorefineries that can make a variety of products from biomass, such as biofuels, chemicals, and materials, are currently being developed. Utilizing all components of the biomass feedstock can boost the economic viability of biofuels and reduce waste.
(e) Carbon capture and utilization: To lessen the carbon impact of biofuels, researchers are investigating the use of carbon capture and utilization technologies. The use of BECCS technologies, which can capture and store carbon dioxide emissions from the generation of bioenergy, is one example (FAO and UNEP 2021; Shafiee and Topal 2009; Teng and Chen 2019).

5.2.6 Conclusion

In conclusion, sustainable biofuel production technologies are essential to ensuring that biofuels can act as a sustainable replacement for fossil fuels. Some of the environmentally friendly methods for producing biofuels include second-generation biofuels, algae biofuels, cellulosic ethanol, biogas, and waste-to-energy technologies. These innovations might lower greenhouse gas emissions, cut waste, and advance energy independence. It is crucial to remember that the sustainability of biofuel production is reliant on elements including the source of feedstock, land utilization, and water usage. Therefore, to make sure that the production of biofuel is actually sustainable, it is crucial to assess the environmental, social, and economic effects. In general, sustainable biofuel production can significantly contribute to climate change mitigation and the advancement of a more sustainable energy system.

References

Abas N, Kalair A, Khan N (2015) Review of fossil fuels and future energy technologies. Futures 69:31–49

Abdullah B, Syed Muhammad SAF, Shokravi Z, Ismail S, Kassim KA, Mahmood AN, Aziz MMA (2019) Fourth generation biofuel: a review on risks and mitigation strategies. Renew Sustain Energy Rev 107:37–50

Achten WMJ, Verchot L, Franken YJ, Mathijs E, Singh VP, Aerts R, Muys B (2008) Jatropha bio-diesel production and use. Biomass Bioenergy 32:1063–1084

Alaswad A, Dassisti M, Prescott T, Olabi AG (2015) Technologies & developments of third generation biofuel production. Renew Sustain Energy Rev 51:1446–1460
Barreto RA (2018) Fossil fuels, alternative energy and economic growth. Econ Model 75:196–220
Bhuiya MMK, Rasul MG, Khan MMK, Ashwath N, Azad AK (2016) Prospects of 2nd generation biodiesel as a sustainable fuel—Part: 1 selection of feedstocks, oil extraction techniques and conversion technologies. Renew Sustain Energy Rev 55:1109–1128
Carriquiry MA, Du X, Timilsina GR (2011) Second generation biofuels: economics and policies. Energy Policy 7(39):4222–4234
Chye JTT, Jun LY, Yon LS, Pan S, Danquah MK (2018) Biofuel production from algal biomass. In: Konur O (ed) Bioenergy and biofuels. CRC Press, Taylor & Francis Group, Boca Raton, FL, pp 87–117
Dangi AK, Sharma B, Hill RT, Shukla P (2018) Bioremediation through microbes: systems biology and metabolic engineering approach. Crit Rev Biotechnol 1–20
Demirbas MF, Balat M (2006) Recent advances on the production and utilization trends of bio-fuels: a global perspective. Energy Convers Manag 47:2371–2381
Ewunie GA, Morken J, Lekang OI, Yigezu ZD (2020) Factors affecting the potential of Jatropha curcas for sustainable biodiesel production: a critical review. Renew Sustain Energy Rev 110500
FAO and UNEP (2021) Global assessment of soil pollution—summary for policy makers. Rome, FAO
Gajraj RS, Singh GP, Kumar A (2018) Third-generation biofuel: algal biofuels as a sustainable energy source. In: Kumar et al (eds) Biofuels: greenhouse gas mitigation and global warming, A. Springer [India] Pvt. Ltd, pp 307–325
Gasparatos A, Stromberg P, Takeuchi K (2013) Sustainability impacts of first-generation biofuels. Anim Front 3(2):12–26
Güney T (2019) Renewable energy, non-renewable energy and sustainable development. Int J Sustain Dev World Ecol 1–9
Havlík P, Schneider UA, Schmid E, Böttcher H, Fritz S, Skalský R, Obersteiner M (2011) Global land-use implications of first and second generation biofuel targets. Energy Policy 39(10):5690–5702
Hirani AH, Javed N, Asif M, Basu SK, Kumar A (2018) A review on first-and second-generation biofuel productions. Biofuels: greenhouse gas mitigation and global warming, pp 141–154. https://doi.org/10.1007/978-81-322-3763-1_8
Hubbert MK (1949) Energy from fossil fuels. Science 109(2823):103–109
Lee RA, Lavoie J-M (2013) From first-to third-generation biofuels: challenges of producing a commodity from a biomass of increasing complexity. Anim Front 3(2):6–11
Leong W, Lim J, Lam M, Uemura Y, Ho Y (2018) Third generation biofuels: a nutritional perspective in enhancing microbial lipid production. Renew Sustain Energy Rev 91:950–961
Li P, Sakuragi K, Makino H (2019) Extraction techniques in sustainable biofuel production: a concise review. Fuel Process Technol 193:295–303
Liang S, Xu M, Zhang T (2012) Unintended consequences of bioethanol feedstock choice in Chine. Biores Technol 125:312–317
Liew WH, Hassim MH, Ng DKS (2014) Review of evolution, technology and sustainability assessments of biofuel production. J Clean Prod 71:11–29
Lü J, Sheahan C, Fu P (2011) Metabolic engineering of algae for fourth generation biofuels production. Energy Environ Sci 4:2451–2466
Mao X, Jiang R, Xiao W, Yu J (2015) Use of surfactants for the remediation of contaminated soils: a review. J Hazard Mater 285:419–435
Mohammadi M, Najafpour GD, Younesi H, Lahijani P, Uzir MH, Mohamed AR (2011) Bioconversion of synthesis gas to second generation biofuels: a review. Renew Sustain Energy Rev 15:4255–4273
Mohr A, Raman S (2013) Lessons from first generation biofuels and implications for the sustainability appraisal of second generation biofuels. Energy Policy 63:114–122

Moravvej Z, Makarem MA, Rahimpour MR (2019) The fourth generation of biofuel. Second and third generation of feedstocks, pp 557–597

Nada Kh MA Alrikabi (2014) Renewable energy types. J Clean Energy Technol 2:1

Naik SN, Goud VV, Rout PK, Dalai AK (2010) Production of first and second generation biofuels: a comprehensive review. Renew Sustain Energy Rev 14:578–597

Nanda S, Rana R, Sarangi PK, Dalai AK, Kozinski JA (2018) A broad introduction to first-, second-, and third-generation biofuels. In: Recent advancements in biofuels and bioenergy utilization, pp 1–25

Naqvi M, Yan J (2015) First-generation biofuels. In: Handbook of clean energy systems, pp 1–18

Neto JM, Komesu A, da Silva Martins LH, Gonçalves VOO, de Oliveira JAR, Rai M (2019) Third generation biofuels: an overview. Sustain Bioenergy 283–298

Pittman JK, Dean AP, Osundeko O (2011a) The potential of sustainable algal biofuel production using wastewater resources. Biores Technol 102(17):17–25

Pittman JK, Dean AP, Osundeko O (2011b) The potential of sustainable algal biofuel production using wastewater resources. Biores Technol 102:17–25

Rodionova MV, Poudyal RS, Tiwari I, Voloshin RA, Zharmukhamedov SK, Nam HG et al (2017) Biofuel production: challenges and opportunities. Int J Hydrog Energy 42(12):8450–8461

Schenk PM, Thomas-Hall SR, Stephens E, Marx UC, Mussgnug JH, Posten C et al (2008) Second generation biofuels: high efficiency microalgae for biodiesel production. BioEnergy Res 01:20–43

Shafiee S, Topal E (2009) When will fossil fuel reserves be diminished? Energy Policy 37(17):181–189

Sims REH, Mabee W, Saddler JN, Taylor M (2010) An overview of second generation of biofuel technologies. Biores Technol 101:1570–1580

Singh B, Guldhe A, Rawat I, Bux F (2014) Towards a sustainable approach for development of biodiesel from plant and microalgae. Renew Sustain Energy Rev 29:216–245

Singh J, Gu S (2010) Commercialization potential of microalgae for biofuels production. Renew Sustain Energy Rev 14:2596–2610

Solomon BD (2010) Biofuels and sustainability. Ann N Y Acad Sci 1185:119–134

Teng Y, Chen W (2019) Soil microbiomes—a promising strategy for contaminated soil remediation: a review. Pedosphere 29(3):283–297

Shahzad U (2015) The need for renewable energy sources. ITEE J (2012–2015)

United Nations Department of Economic and Social Affairs, Population Division (2022) World population prospects 2022: summary of results. UN DESA/POP/2022/TR/NO. 3

Xiao G, Gao L (2011) First generation biofuels. INTECH Open Access Publisher. ISBN 9533074787

Chapter 6
Innovative Approaches for Sustainable Biodiesel Production and Integration in Circular Economy

Divyesh Bhisikar, Shishir Mendhekar, Ashootosh Mandpe, and Sonam Paliya

Abstract The demand for clean and renewable energy has increased in recent years, and biodiesel has become an environmentally beneficial alternative to traditional fossil fuels. However, using traditional feedstocks such as vegetable and waste oils has several drawbacks, including high costs and the need for labour and land. To overcome these challenges, recent research has focused on using oleaginous microorganisms to produce biodiesel from lignocellulosic substrates. This approach has proven to be highly efficient and cost-effective, with these microorganisms' producing lipids with a composition resembling that of vegetable oils in stressful conditions. This book chapter offers a comprehensive overview of the various feedstocks used in biodiesel production, including non-edible oils, vegetable oils, oleaginous microalgae, fungus, yeast, and bacteria. It also breaks down the biodiesel production process into several steps, including pre-treatment, saccharification, oleaginous microbe-mediated fermentation, lipid extraction, transesterification, and biodiesel purification. Furthermore, the chapter discusses the integration of biofuels in the circular economy and highlights the relevance of biotechnological applications in ensuring the sustainability of biofuels and biorefineries. By embracing the use of oleaginous microorganisms and lignocellulosic substrates, biodiesel production can become a sustainable and cost-effective solution for reducing greenhouse gas emissions and promoting a cleaner and more sustainable future.

D. Bhisikar
Department of Data and Analytics, PGP Glass Pvt. Ltd., Vadodara 390012, India

S. Mendhekar
Department of Microbiology, Kurukshetra University, Kurukshetra 136119, India

A. Mandpe (✉)
Department of Civil Engineering, Institute of Technology Indore, Indore, Madhya Pradesh 453552, India
e-mail: as_mandpe@iiti.ac.in

S. Paliya
Department of Biosciences and Biomedical Engineering, Indian Institute of Technology Indore, Indore 453552, India

M. Shah and D. Deka (eds.), *Emerging Sustainable Technologies for Biofuel Production*, Environmental Science and Engineering, https://doi.org/10.1007/978-3-031-52167-6_6

Keywords Biodiesel · Biorefineries · Circular economy · Liquid biofuel · Renewable energy · Transesterification

6.1 Introduction

Energy is a crucial resource for the long-term development of humanity. The world is currently facing an energy crisis, which has become one of the most pressing global issues. Fuels play an important part in addressing our energy demands since they may be burnt to create huge amounts of energy. Many parts of daily life rely on fuel, notably the movement of products and people. According to recent statistics, fossil fuels, including petroleum, coal, and natural gas, are the primary sources of energy worldwide, accounting for 80% of total energy production. These fuels are commonly used in various sectors, including manufacturing, transportation, and power generation, among others. The use of diesel-powered machinery is common in manufacturing and transportation industries that heavily rely on fossil fuels such as gasoline, diesel, and petrol to power private vehicles, buses, trucks, and ships. This widespread dependence on non-renewable energy sources is a significant factor in daily life. However, the production of local crude oil is struggling to keep up with the increasing global population, leading to concerns about energy security and availability. The lack of alternative energy sources has also contributed to the ongoing energy crisis, making it a pressing issue that requires immediate attention. Fossil fuels are derived from ancient animals and microbes, and their creation takes millions of years. Because of this, they are classified as non-renewable sources of energy. When the price of oil spikes, it often leads to economic recessions and global conflicts. Despite the widespread dependence on non-renewable energy sources, renewable energy has now become the most affordable power source in most regions worldwide. Renewable energy technologies are becoming increasingly affordable, with solar-powered electricity costs falling by 85% between 2010 and 2020. Wind and solar energy costs have dropped significantly. This makes renewable energy more appealing, particularly in low- and middle-income nations, where new electricity will be most needed. As a result, low-carbon energy has a clear chance to play a substantial part in satisfying this demand and offering a sustainable alternative to non-renewable energy sources.

Biodiesel is an eco-friendly alternative to petroleum diesel, produced in the United States from renewable sources. Biodiesel as an automobile fuel not only improves air quality and the environment but also enhances safety. As fossil fuel resources have decreased and environmental concerns have increased, biodiesel has surged in favour (Chintagunta et al. 2021b). Biodiesel has grown in popularity in recent years as fossil fuel resources have depleted and environmental concerns have grown.

This chapter explores the history and contemporary advancements in biodiesel technology, including different types of biodiesel, its properties, processing techniques, and economic factors. Because of the detrimental environmental effect of petroleum-based diesel engines and the dwindling supply of fossil fuel resources, biodiesel has become a viable diesel fuel alternative. Any natural oil or fat may be chemically combined with an alcohol, such as ethanol or methanol, to produce biodiesel. Methanol has long been the most often used alcohol in commercial biodiesel production. Several studies have shown that biodiesel fuel produced from vegetable oil may be safely used in diesel engines (Jönsson and Martín 2016). Biodiesel, in reality, has a similar energy density as normal diesel fuel. When soybean oil and methanol are transesterified in the presence of acid catalysts, biodiesel can be created. Because biodiesel and petroleum-based diesel have identical combustion properties, biodiesel has emerged as one of the most feasible and sustainable renewable fuels for cars.

Biodiesel is a viable long-term fuel derived from waste oils, plant oils, microbial lipids, and animal fats (Jeevan Kumar et al. 2017b; Banerjee et al. 2019). The high cost of non-edible and vegetable oils, the high demand for land, and the low yield of oil have rendered these feedstocks useless for production of biodiesel (Martin et al. 2019). Generally, these feedstocks account for 75% of the total cost of biodiesel production (Subramaniam et al. 2010). Developing cost-effective feedstocks, in conjunction with the careful use of waste substrates, might be a viable alternative for cutting process costs. Furthermore, encouraging the circular economy process is becoming increasingly popular in order to reduce resource waste. Oleaginous microorganisms (OMs) may boost lipid accumulation using low-cost feedstocks (Garlapati et al. 2020).

Furthermore, using OMs may contribute to realizing the potential of the circular economy and developing cost-effective approaches. Crude glycerol, a biodiesel industry waste, can be employed as a carbon source for lipid accumulation (Chintagunta et al. 2021b). These lipids can then be recycled by being transformed into fatty acid methyl esters (FAMEs) and glycerol. Imagining this process might show its viability and, as a result, cost-effective biodiesel synthesis utilizing agro-residues, glycerol, and other waste substrates (Chandel et al. 2020).

The cultivation of oleaginous microalgae (OMs) is simple and does not require a large quantity of area or acreage. Furthermore, it leads to increased lipid accumulation with optimal lipid composition in shorter incubation durations. Furthermore, OMs may absorb ambient CO_2 and use it in lipid synthesis, aiding in carbon sequestration. The purpose of this chapter is to encourage the use of lignocellulosic biomass (LCB) and oleaginous microalgae (OMs) for biodiesel production by maximizing their potential (Kumar et al. 2017). Circular economy processes can also include the generation of additional biofuels, which include bioethanol and biodiesel. Oil crops like soybean and rapeseed, oil trees like palm oil and Chinese pistachio, and other sources like waste cooking oil and animal fat may all be utilized to make biodiesel. Fatty acid methyl ester (FAME) is a particularly important biodiesel product because of its renewable and environmentally advantageous properties. The raw materials of biodiesel are natural and renewable, as all of the oils used are produced from plants or animal fat, making them biodegradable and non-toxic. Traditional vegetable oil utilization solutions, such as direct mixing and micro emulsion, do not address issues like carbon deposits, lubricant contamination, and high-temperature pyrolysis cracking (Kalscheuer et al. 2006). The essential process factors in this type of operation are the catalyst quantity, reaction temperature, alcohol/vegetable-oil ratio, mixing intensity (RPM), catalyst type, and raw oils used. Ester exchange, on the other hand, is a more complicated method of converting triglycerides to monoesters. The ester exchange transesterification process reduces vegetable oil viscosity while keeping its heat value. Transesterification is a chemical reaction that occurs through a catalyst between triglycerides and short-chain alcohol to produce monoesters. Long-branched-chain triglycerides are converted to monoesters and glycerine in this process.

6.2 Feedstocks for Biodiesel Production

The cost of feedstocks has been highlighted as the most crucial element determining the biodiesel industry's overall economic viability (Zhang et al. 2003). The cost of feedstocks represents approximately 70–75% of the total cost of biodiesel production. Therefore, it is crucial to consider the cost of raw materials when producing sustainable biodiesel. In order to achieve an efficient biodiesel production process, it is necessary to use low-cost feedstocks, keep production costs low, and operate on a large scale. While vegetable oils from edible plants are the most commonly used feedstocks, there are ethical concerns about using food as fuel, as well as the high cost of food. As a result, research into low-cost non-food feedstocks for biodiesel synthesis is ongoing (Haas et al. 2006). In order for non-food feedstocks to be effectively utilized, several criteria need to be met, including a year-round supply of raw materials, the ability to produce value-added products for biochemical, a favourable low agricultural inputs fatty acid composition, such as fertilizers, high oil content, water, pesticides, and acclimatizing to local conditions such as latitude, and soil type (Patel et al. 2017). The most effective alternative to petroleum diesel is biodiesel, made from feedstocks that fulfill most of the abovementioned requirements. Various feedstocks or waste substrates for biodiesel synthesis have been researched in recent years, and they may be divided into four categories: animal fat, waste or recycled oil, vegetable oil (edible and non-edible), and oleaginous microorganisms (OMs) (Sampath Kumar et al. 2019).

6.2.1 Vegetable Oils (Edible/Non-edible)

According to Demirbas et al. (2016), vegetable oil has been identified as a viable biodiesel feedstock due to its ease of availability, transportability, sustainability, high combustion ability, low sulphur content, and eco-friendliness. However, because of its unsaturated hydrocarbon chain, vegetable oil is very viscous and reactive, which are key drawbacks of this feedstock (Demirbas 2009a). Vegetable oils, both edible and non-edible, are included in the phrase "vegetable oil." The primary and most often utilized feedstock for biodiesel production is edible vegetable oil, which accounts for more than 95% of worldwide biodiesel output. Edible oils are common biodiesel feedstocks due to their ease of availability, ease of processing, and high biodiesel quality (Demirbas 2009b).

Rapeseed oil an edible vegetable oil is often used as a feedstock in the production of biodiesel in Canada and Europe, palm oil in Southeast Asia, sunflower oil in Europe, and soybean oil in the United States. Other culinary oils, such as linseed, cottonseed, coconut, sesame, peanut, maize, almond, and many more, have also been explored for biodiesel generation (Demirbas et al. 2016). Commercial biodiesel production using edible vegetable oils is limited due to their high production and trade costs, making them an unlikely choice for long-term feedstock solutions. Despite this, rapeseed and sunflower oil are the most often utilized edible vegetable oils for biodiesel generation, accounting for 84 and 13% of total output, respectively. Soybeans and palm oil also contribute to global biodiesel output at 2 and 1%, respectively. However, concerns related to edible oils, such as rising prices and the food versus fuel crisis, limit their long-term viability as a substrate for biodiesel production (Atabani et al. 2012). To fulfill market demand, environmental factors must be taken into account while producing biodiesel from edible oil-based feedstocks. This includes the requirement to boost feedstock production, which can result in deforestation, agricultural land usage, and soil deterioration. To overcome this issue, non-edible vegetable oils have been found as a possible biodiesel manufacturing option. Non-edible oils have various advantages over edible oil feedstocks, including the capacity to grow on unused land, lower production costs, and no food versus fuel conflict. Extensive study has been conducted in recent years on the generation of biodiesel from diverse non-edible oil sources (Jena et al. 2010; Wang et al. 2011; Sharma and Singh 2010).

Many non-edible oil-producing plants have been studied as potential biodiesel feedstocks. These include *Azadirachta indica* (neem), Castor *Ricinus communis* (castor), Cotton seed *Gossypium hirsutum* (cotton seed), *Hevea brasiliensis* (rubber), *Jatropha curcas* (jatropha), Jojoba *Simmondsia chinensis* (jojoba), Karanja *Pongamia pinnata* (karanja), Kusum *Schleichera oleosa* (kusum), Mahua *Madhuca indica* (mahua), Nahor *Mesua ferrea* (Nahor), *Moringa oleifera* (moringa), Soap nut *Sapindus mukorossi* (soap nut), *Simarouba glauca* (Simarouba), and Tobacco *Nicotiana tabacum* (tobacco). Non-edible vegetable oil is a promising substitute to edible oils as biodiesel feedstock. However, there are some constraints, like the long growth period, the need for abundant land for production, and variations in lipid conditions depending on the season, climate, plant species, and geography (Kumar et al. 2020a, b; Singh et al. 2020) (Fig. 6.1).

Disposing large amounts of waste oil is a major environmental issue in most countries. The production of biodiesel from waste oil has the potential to ease this issue. The presence of impurities like polymers, free fatty acids (FFAs), and other chemicals, on the other hand, complicates the conversion process. Furthermore, animal fat, a common source of biodiesel synthesis, is in low supply in relation to the global demand for biodiesel (Talavari et al. 2021).

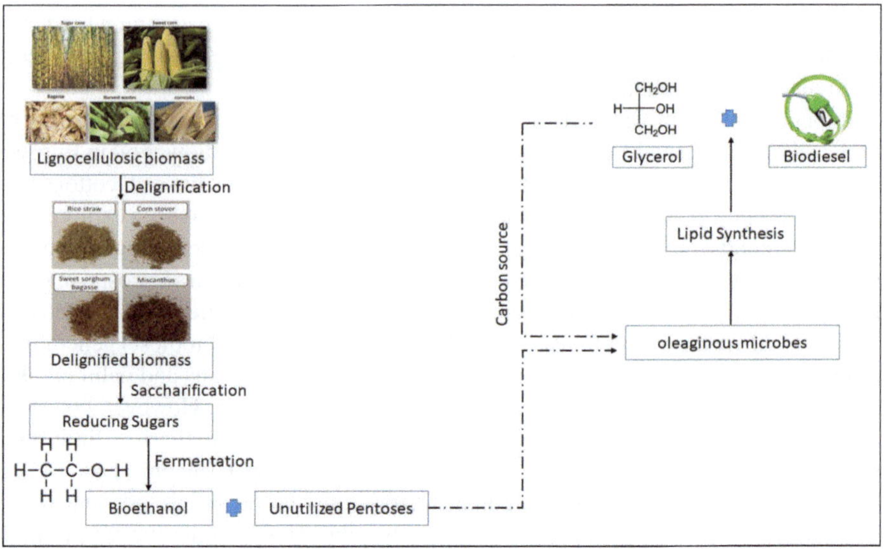

Fig. 6.1 Integrated biofuel generation from lignocellulosic biomass employing oleaginous microbes is shown schematically

6.2.2 Oleaginous Microalgae

Oleaginous microalgae are microorganisms that may accumulate over 20% of their dry mass as lipids under particular stress circumstances, such as excessive carbon and low nitrogen sources. In fact, with optimal stress requirements, they can accumulate up to 70% or more. This high lipid content makes them a promising feedstock for the production of biodiesel through transesterification. These bacteria can turn a variety of renewable resources into microbial oil, which may subsequently be utilized to make biodiesel (Meng et al. 2009; Pinzi et al. 2014). Single-cell oil, also known as microbial oil, is produced by a variety of microbes, including bacteria, yeast, and fungi. The filamentous fungus Mucor circinelloides was the first microbe used for the commercial production of single-cell oil in 1985, but its use in biodiesel production was not initially explored. The process for producing biodiesel from OMs involves culturing the microbe, harvesting the biomass, drying the biomass, extracting the lipid, and finally, transesterifying the lipid to produce biodiesel (Gujjala et al. 2017). Biodiesel produced from OM offers various advantages over other feedstocks, including the ability to overcome the food versus fuel issue because it does not need the utilization of edible plant or animal resources. Furthermore, lipid formation from OM requires shorter incubation durations than plant and animal resources and is not affected by seasonal, climatic, or geographical fluctuations. OM may also be used to transform low-cost agro-industrial waste and even municipal trash into microbial oil of comparable quality to a number of high-value lipids (Cho and Park 2018). Microalgae-based biodiesel is gaining popularity due to its potential to supply enough oil for world consumption, with biodiesel yields per hectare much exceeding those achieved from plant-based feedstocks. Furthermore, microalgae can assist in minimizing land use and food versus fuel issues and be farmed in ecosystems unsuited for energy crops. They can also help to mitigate the greenhouse impact by absorbing carbon dioxide from the atmosphere. Microalgae normally have a lipid concentration of 20–50% of their dry cellular weight, but under some conditions, this can reach up to 90% (Jin et al. 2015). Assuming a microalgal cell lipid content of 30% dry weight, the potential microalgal oil output varies between 4.5 and 7.5 t/ha/year (Table 6.1).

It is significantly greater than rapeseed, palm, soybean, and jatropha oil output, which are 1.4, 1.6, 0.4, 3.6, and 4.1 t/ha/year, respectively (Lam and Lee 2011). Microalgae are classified into three categories based on their feeding modes: autotrophic, heterotrophic, and mixotrophic, indicating that they have diverse metabolic processes. Heterotrophic microalgae use organic substances as carbon sources for growth, which results in better productivity and concentration. On the other hand, autotrophic microalgae receive carbon from carbon dioxide, which is

Table 6.1 Lipid production by oleaginous microalgae using lignocellulosic substrates (Lam and Lee 2011)

Oleaginous microalgae	Substrate	Lipid production	References
Schizochytrium sp.	Sugarcane bagasse hydrolysate	N.A. g/L 45.15%	Nguyena et al. (2018)
Schizochytrium mangrovei	Food waste hydrolysate	3.52 g/L 16.4%	Pleissner et al. (2013)
Schizochytrium limacinum	Sweet sorghum juice (squeezed by a mill)	2.15 4.95 g/L 55.3 ± 70.5%	Liang et al. (2010)
Nannochloropsis sp.	Palm oil mill effluent	3.2 g/L 11.0%	Cheirsilp et al. (2017)
Chlorella pyrenoidosa	Rice straw enzymatic hydrolysate	1.55 g/L 53.6%	Li et al. (2001)
Chlorella protothecoides	Cassava starch enzymatic hydrolysate, corn powder enzymatic hydrolysate	2.14 g/L 22 ± 55.2%	Lu et al. (2010)
Chlorella protothecoides	Sugarcane bagasse hydrolysate	5.8 g/L 34.0%	Mu et al. (2015)
Chlorella vulgaris	Wheat bran enzymatic hydrolysate	N.A. g/L 0.6 ± 0.9%	El-Sheekh et al. (2012)
Auxenochlorella protothecoides	Organosolv/steam explosion pretreated Birch biomass	5.7 g/L 66%	Patel et al. (2018)
Auxenochlorella protothecoides	Organosolv/steam explosion pretreated Spruce biomass	5.3 g/L 63%	Patel et al. (2018)

diminished by light energy and creates oxygen. The cost of oil recovery from biomass, biomass productivity, lipid content and, production scale are the key variables leading to the higher cost of microalgal oil production (Balat 2011). Higher biomass and lipid productivity are advantages of heterotrophic and mixotrophic development. Still, one of the fundamental restraints is the lack of an inexpensive organic carbon source, which is amongst the major contributing causes to the higher production costs of microalgae-based biodiesel. To overcome this problem, abundant lignocellulosic waste has been offered as a low-cost organic carbon source for microalgae culture.

Filamentous fungi and molds are considered oleaginous microorganisms, which are capable of storing up to 80% of their biomass as lipids. Various oleaginous fungal species have been investigated in recent years and may be cultivated utilizing sustainable carbon sources such as lignocellulosic materials. These species' lipids contain more unsaturated fatty acids than yeast. Oleaginous fungus have been found as a possible biodiesel feedstock competitor with lower output costs (biomass and oil) than microalgae due to their high lipid content and ability to grow in standard bioreactors. The application of fungal oil for biodiesel generation, on the other hand, is still in the laboratory. Around 30 yeast species have been identified as oleaginous, and study in this field is ongoing (Garay et al. 2016). Many of the identified oleaginous yeast species are capable of large-scale fermentation and have the potential for genetic improvement (Athenaki et al. 2018). Oleaginous yeasts are capable of storing up to 70% of their lipid under nutrient-limited conditions and have the advantage of faster growth compared to microalgae. The majority of their lipids consist of triglycerides. However, the high cost of the commercial substrate makes yeast oil production more expensive than vegetable oil production. Therefore, the selection of oleaginous yeast species capable of thriving on low-cost substrates is crucial in reducing production costs (Leesing and Karraphan 2011).

Several experiments have been conducted to investigate the potential of using lignocellulosic waste as an affordable carbon source for yeast oil production using various yeast species. Sreeharsha and Mohan (2020) also evaluated the potential of oleaginous yeast for lipid synthesis. However, despite these efforts, the commercial availability of yeast oil has not yet been achieved.

6.2.3 Fungi

Improving the cost-effectiveness of producing sustainable aviation fuels can be achieved by identifying microorganisms that are capable of producing significant quantities of single-cell oils (SCOs) that have similar compositions to traditional lipids (Athenaki et al. 2018). The filamentous fungus Mucor circinelloides has the distinction of being the first microorganism used in the industrial production of polyunsaturated fatty acids (PUFAs) (Čertík et al. 2012). Single-cell oils generated from oleaginous microorganisms like filamentous fungus have a high level of unsaturation, making them attractive as feedstock for biodiesel generation. The high manufacturing costs, however, have hampered the large-scale industrialization of microbial single-cell oil (SCO) production. Although advances in fermentation technology and genetically engineered oleaginous strains have increased microbial lipid synthesis's competitiveness, the cost of manufacturing microbial lipids remains prohibitively expensive. The cost of microbial lipids must be decreased in order for them to be economically feasible for usage in biodiesel, drop-in biofuels, and other goods (Abghari and Chen 2014). Combining synthesised high value compounds from agro-industrial waste and lignocellulosic biomass as feedstocks with microbial oil production is one potential approach for decreasing the cost of polyunsaturated fatty acid (PUFA) production (Ratledge 2004).

The filamentous fungus can yield PUFAs from low-cost substrates because they secrete a range of hydrolyzing enzymes that allow them to metabolize complicated substrates (Gmoser et al. 2019). The use of filamentous fungi as SCO producers efficiently utilizes low-cost substrates like lignocellulosic and agro-industrial waste, making the microbial lipid synthesis process more cost-effective (Martínez et al. 2015). Single-cell oils derived from filamentous fungus are appealing feedstocks for sustainable production of biodiesel because they yield SCOs with characteristics similar to traditional vegetable oils. Furthermore, biodiesel created from microbial oils may be combined in any proportion with petroleum fuel, making it a versatile and sustainable alternative to regular diesel fuel (Bhanja et al. 2014). Because of the numerous benefits of filamentous fungi-derived SCOs for third-generation biodiesel synthesis, screening attempts to find promising oleaginous strains have increased (Abu-Elreesh and Abd-El-Haleem 2014; Zeid et al. 2019). The capacity of a fungal strain to generate significant levels of total lipids, particularly triacylglycerols (TAGs), and the precise fatty acid composition ideal for biodiesel synthesis are the major criteria to consider while selection of a fungal strain for production of biodiesel (Bardhan et al. 2019). Oleaginous fungal strains that yield oils high in monounsaturated fatty acids (MUFA) and saturated fatty acids (SFA) are suitable for biodiesel production; whereas, polyunsaturated fatty acids (PUFAs) are preferred for pharmaceuticals and nutraceuticals (Patel et al. 2017). Lipids with a high concentration of PUFAs and more than two double bonds are more prone to auto-oxidation, making them more unstable and technically inappropriate for biodiesel generation. As a result, PUFA-rich lipids are chosen for nutraceuticals and medicines over biodiesel production (Table 6.2).

Lipids utilized as feedstock for biodiesel synthesis should be rich in MUFAs like palmitoleic acid and oleic acid, low in SFAs like palmitic acid and stearic acid, and low in PUFAs like linoleic acid and γ-linolenic acid. Fatty acid profiles have a significant impact on various properties of like melting point, heating value/calorific value, cetane number, oxidative stability, and kinematic viscosity, all of which affect the performance of the created biodiesel (Miglio et al. 2013).

To guarantee that the chosen strain is acceptable for biodiesel production, the screening process for choosing a suitable fungal strain for biodiesel synthesis must be carried out with utmost caution. Fungal strains that generate lipids with desired characteristics can metabolize low-cost substrates, develop quickly, and are easy to harvest are deemed more attractive due to their potential to considerably cut production costs. Oleaginous bacteria synthesize fatty acids in two ways: de novo

Table 6.2 Lipid production by oleaginous molds using lignocellulosic substrates (Chaturvedi and Verma 2013)

Oleaginous molds	Substrate	Lipid production	References
Phanerochaete hrysosporium ATCC 24725	Wheat bran, corn straw and glucose mixture	>40%	Liu et al. (2019)
Mucor circinelloides Q531	Mulberry branches	42.434.01 mg/gds 28.8 ± 2.85%	Qiao et al. (2018)
Mucor circinelloides	Hydrolyzed whey permeates	32%	Chan et al. (2020)
Mortierella isabellina	Corn stover enzymatic hydrolysate, sweet sorghum	0.016 ± 0.11 mg/gds 29.47 ± 38.36%	Economou et al. (2011)
Microsphaeropsis sp.	Wheat straw	80 mg/gds 10.2%	Abghari and Chen (2014)
Colletotrichum sp.	Rice straw + wheat bran	68.2 mg/gdsa N.A. %	Dey et al. (2011)
Aspergillus oryzae	Wheat straw + bran	62.9 mg/gdsa N.A. %	Tamano et al. (2013)
Alternaria sp.	Rice straw + wheat bran	60.3 mg/gdsa N.A. %	Dey et al. (2011)

(on hydrophilic substrates) and ex novo (on hydrophobic substrates) (Chaturvedi and Verma 2013). A deeper understanding of the cellular mechanisms that drive lipid metabolism in filamentous fungus is necessary to increase lipid yields and fatty acid profiles. Despite the continuous lipid accumulation route, there is considerable heterogeneity across oleaginous species and even strains of the similar species. The underlying reasons for these variations are unknown. Fortunately, genetic information is now publicly available in the scientific literature, and new research has revealed new insights into the regulation of lipid metabolism in oleaginous microorganisms.

6.2.4 *Yeast*

In laboratory experiments, yeast has been shown to accumulate lipids in the form of triglycerides when grown in a medium with low nitrogen and high carbon content. To study this process, researchers designed a custom medium that optimized glucose, inorganic nitrogen, and organic nitrogen concentrations to promote maximum lipid formation. They also investigated the effect of incubation time on lipid accumulation. The fermentation process was initiated by inoculating the specialized medium with 10% of a 24-h-old culture grown in GYE broth (Somerville et al. 2010). Through these experiments, researchers were able to better understand the conditions and timing that lead to optimal lipid production in yeast. The mechanisms involved in

TAG production by oleaginous yeast are highly regulated and systematic. To improve lipid synthesis output approaches such as gene overexpression, knockout genes, and multigene strategies are commonly used. TAGs may be accumulated as lipid droplets by oleaginous yeast, accounting for up to 70% of their biomass, but non-oleaginous yeasts such as Saccharomyces cerevisiae and Candida can only acquire 10% lipid content (Hossain and Badr 2007). Non-oleaginous yeast increases their mannan and glucan content in a nitrogen-limited medium with an extra carbon supply, whereas oleaginous yeast converts the surplus carbon source into lipids. When nitrogen is decreased from the culture media, oleaginous yeast activates adenosine monophosphate deaminase, catalyzing the conversion of AMP into inosine 5′-monophosphate and ammonium. The decrease in AMP content also causes isocitrate dehydrogenase inactivation, which disturbs the tricarboxylic acid cycle metabolic pathway. ATP: citrate lyases cleave citrate in the cytoplasm of oleaginous yeast, and citrate is transferred from the mitochondria to the cytosol via the malate/citrate translocase system. Acetyl-Co generated in this process is subsequently transformed into malonyl-Co A with the aid of acetyl-Co A carboxylase, resulting in the de-novo synthesis of lipids with 14 to 16-carbon long fatty acid chains (Table 6.3).

Furthermore, it has been shown that the fatty acid composition of oleaginous yeast can be altered by environmental or physical stress. For instance, under nitrogen-limited conditions, oleaginous yeast can accumulate a high lipid content in its cellular compartment (Sitepu et al. 2014). Different fatty acids. For instance, *R. toruloides* synthesizes large amounts of palmitoleic acid, which is an omega-7 monounsaturated fatty acid. *Y. lipolytica*, on the other hand, is capable of producing odd-chain fatty acids, such as heptadecanoic acid and nonadecanoic acid. *Cryptococcus curvatus*, in contrast, is a good source of lauric acid, which is a medium-chain fatty acid. The fatty acid composition of the lipids synthesized by oleaginous yeast is influenced by the type of carbon source utilized, as well as the culture conditions employed. As such, the Manipulation of growth conditions or modification of the metabolic pathways involved in lipid production can be utilized to create particular fatty acid profiles in oleaginous yeast, with applications in the pharmaceutical, food, and biofuel sectors.

6.2.5 Bacteria

The other approach is to genetically engineer oleaginous bacteria to efficiently utilize lignocellulosic biomass as a carbon source. However, the complexity of lignocellulosic biomass deconstruction and utilization by microorganisms presents significant challenges that require extensive research. One promising approach is to utilize consolidated bioprocessing (CBP), which integrates microbial fermentation and enzymatic hydrolysis into a single process. This approach aims to produce biofuels or biochemicals from lignocellulosic biomass in an environmentally sustainable and cost-effective manner. Despite the difficulties, using lignocellulosic biomass as a source of carbon for oleaginous bacteria is a viable strategy for the long-term synthesis of microbial lipids (Huber 2008).

Table 6.3 Lipid production by oleaginous yeast using lignocellulose substrates (Sitepu et al. 2014)

Oleaginous yeast	Substrate	Lipid production	References
Yarrowia lipolytica	Sugarcane bagasse and rice bran hydrolysate	5.2 ± 6.68 g/L 48 ± 58.5%	Tsigie et al. (2012)
Vishniacozyma psychrotolerans	Groundnut shell hydrolysate	46%	Deeba et al. (2017)
Trichosporon cutaneum	Corncob acid and enzymatic hydrolysates, diluted acid pretreated and biodetoxified corn stover	9.8 ± 12.3 g/L 32.1 ± 40%	Gao et al. (2014)
Rhodotorula paludigenum	Corncob hydrolysate	3.29 g/L 58%	Chaiyaso et al. (2019)
Rhodotorula glutinis	Corncob acid hydrolysate; Wheat straw acid hydrolysate	1.4 ± 5.5 g/L 11.86 ± 36.4%	Liu et al. (2015)
Rhodotorula glutinis	Cassava bagasse hydrolysate	10.42 g/L 51%	Liu et al. (2018)
Rhodosporidium toruloides	Jerusalem artichoke, cassava starch enzymatic hydrolysate	14 ± 39.6 g/L 43.3 ± 63.4%	Zhao et al. (2011)
Pichia kudriavzevii	Rice husk hydrolysate	28.57 ± 0.009%	Ananthi et al. (2019)
Naganishia albida	Biowaste hydrolyzed by microbes	13.5 g/L 20%	Sathiyamoorthi et al. (2019)
Meyerozyma guilliermondii	Sugarcane bagasse hydrolysate	37.99 ± 0.003%	Ananthi et al. (2019)
Lipomyces starkeyi	Corncob acid hydrolysate, sugarcane bagasse acid hydrolysate	8.1 g/L 26.9 ± 55%	Xavier et al. (2017)
Cutaneotrichosporon dermatis	Corn stover hydrolysate	20.36 g/L 56%	Yu et al. (2020)
Cutaneotrichosporon cutaneum	Corn stover hydrolysate	4–5 g/L	Wang et al. (2016)
Cryptococcus vishniaccii MTCC232	Paper mill sludge	7.8 g/L 53.4%	Deeba et al. (2016)
Cryptococcus curvatus	Corncob hydrolysate, sweet sorghum bagasse enzymatic hydrolysate	10.83 g/L 61 ± 73.26%	Chang et al. (2015)
Candida phangngensis PT1-17	Switchgrass hydrolysate	9.8 g/L	Quarterman et al. (2018)

Enzymatic catalysis is highly specific, and it can achieve high conversion rates with low amounts of methanol, and it does not produce undesirable by-products. However, enzymatic catalysis has several limitations that limit its commercial viability, including high enzyme prices, enzyme inactivation by methanol, long reaction times, and the need for the use of costly co-solvents. Novel catalytic approaches, such as microwave-assisted transesterification, ultrasonic-assisted transesterification, and supercritical methanol transesterification, have been developed to overcome

Table 6.4 Lipid production by oleaginous bacteria using lignocellulose substrates (Zeid et al. 2019)

Oleaginous microbes	Substrate	Lipid production	References
Rhodococcus opacus PD630	Orange waste	83%	Voss and Steinbuchel (2001)
Rhodococcus opacus	Oxygen-pretreated Kraft lignin	14.20%	Wei et al. (2015)
Rhodococcus opacus	Effluent from lignocellulosic pretreatment	26.9	Goswami et al. (2017)
R. opacus Xsp8	Kraft hardwood (hydrolysate)	45.80%	Kurosawa and Sinskey (2013)
Gordonia sp. DG	Orange waste	71%	Gouda et al. (2008)

the limitations of traditional processes. These methods have shown higher conversion rates, shorter reaction times, lower energy requirements, and reduced waste generation. Despite their benefits, these innovative processes are still in the research and development phase and require further investigation before they can be implemented on a commercial scale. Overall, the development of a cost-effective, environmentally friendly, and scalable process for biodiesel production remains a challenge for the industry. Although enzymatic manufacturing is a green and energy-efficient process, it is expensive and susceptible to environmental factors and deactivation, thus limiting its industrialization potential. By examining the recent progress in this field, this review hopes to contribute to the realization of efficient and cost-effective enzymatic biodiesel production on a commercial scale (Table 6.4).

6.3 Biodiesel Production Process

6.3.1 Pre-production

The most potential substitutes for fossil fuels are thought to be lignocellulosic materials (Shaheena et al. 2019) due to their production via biological photosynthesis from readily available atmospheric CO_2, water, and sunlight. This biomass represents a long-term source of organic carbon, comprising over 90% of global plant biomass and yielding more than 200 billion tonnes of biomass annually (Kuhad and Singh 2008). The resources and classification of LCBs are not well-defined and can vary based on the source. Generally, LCBs are classified as herbaceous grasses, forestry residues, agricultural residues and weeds, and industrial and urban wastes. These sources can include a wide range of materials such as corn Stover, wheat straw, sawdust, and municipal solid waste (Kumar et al. 2020a, b). Agricultural waste is shown to be the most abundant lignocellulosic resource (Zeid et al. 2019). Forest residues, such as branches, bark, and sawdust, are the primary sources of woodland residues. LCBs are also found in industrial and urban waste materials such as paper waste, wood waste, and municipal solid waste. These wastes are generated in enormous amounts and are frequently disposed of in landfills, producing environmental issues. LCBs can help reduce waste and promote sustainable development by turning these wastes into value-added goods like biofuels (Sampath Kumar et al. 2021).

To diminish recalcitrance in the conversion of lignocellulosic biomass to microbial oil, a pre-treatment step is necessary. Chemical, physical, and biological approaches are examples of pre-treatment processes. Prior to pretreatment, the biomass must be milled, which can add up to 5% to the entire cost. Pre-treatment, on the other hand, generates byproducts such as furan aldehydes, aliphatic carboxylic acids, and aromatic carboxylic acids, which can be hazardous to the development of microorganisms engaged in lipid synthesis bioprocessing (Himmel et al. 2007; Zhou et al. 2011).

6.3.2 Saccharification

Saccharification is the process of breaking down lignocellulosic biomass into fermentable sugars, which is typically preceded by pre-treatment to increase efficiency. Chemical and enzymatic processes are commonly used to hydrolyze the cellulose part of biomass. Chemical hydrolysis uses acid at high temperatures; however, this approach can produce toxic compounds and result in low sugar yields. Enzymatic hydrolysis occurs at low temperatures, pressures, and pH levels, resulting in a high yield of fermentable sugar with minimal by-product formation and no equipment damage. This process, however, necessitates a large quantity of complex enzymes, such as cellulase and hemicellulase, which catalyze the conversion of cellulose and hemicellulose. Endoglucanases, exoglucanases, and β-glycosylases are components of these enzymes. Efficient microbial strains and cost-effective procedures for enzyme production are necessary to make this process economically viable (Kucharska et al. 2020).

6.3.3 Fermentation

Oleaginous microbe-mediated fermentation is a fermentation technology that uses oleaginous microbes, which are microorganisms capable of accumulating high amounts of lipids in their cells, to transform organic resources such as sugars or agricultural waste into lipids. These lipids can then be extracted and utilized as a biofuel source or as a feedstock in the manufacture of chemicals, cosmetics, and other items. Examples of oleaginous microorganisms include yeast, fungi, and microalgae, which absorb organic matter and transform it into lipids through metabolic pathways during fermentation. The fermentation parameters, such as pH, temperature, and nutrient availability, can be adjusted to improve the process. One advantage of this method is its ability to use diverse feedstocks, including waste materials. However, the lipid extraction process can be challenging as it requires the separation of lipids from biomass and other biological components. Despite its potential as a sustainable and effective approach for generating lipids and other compounds from various organic sources, further research is needed to enhance the method and overcome its limitations (Chintagunta et al. 2021b; Jeevan Kumar et al. 2017a).

6.3.4 Lipid Extraction

A variety of microorganisms, such as *Mortierella isabellina*, *Chlorella* sp., *Rhodosporidium toruloides*, *Mycobacterium tuberculosis*, *Gordonia* sp., and others, are known to utilize the hydrolysate as a carbon source produced through the saccharification process (Kumar et al. 2020a, b). Oleaginous microorganisms (OMs) have a restricted capacity to directly use the polysaccharides contained in biomass because they lack cellulase and xylanase activity. To serve as a carbon source for the bacteria, polysaccharides must be hydrolyzed into fermentable sugars like glucose and xylose. Glucose and other carbohydrates can be converted to pyruvate by OMs, which is then used in the production of lipids. The de novo biosynthesis of OMs is comprised of three physiological phases, with the growth phase under balanced conditions being one of them. The oleaginous phase occurs in nitrogen-limited circumstances, after the carbon molecules in the growth medium have been depleted, and is a reserved lipid turnover phase. Glycolysis and the pentose phosphate pathway transform the carbon supply into cell mass, but a deficiency in nutrients causes oil build-up. Along with nutrient content, variables such as temperature, pH, culture agitation, and dissolved oxygen also contribute to lipid accumulation.

A decrease in carbon and nitrogen availability causes an increase in the expression of AMP deaminase (AMPD) in an organism. This enzyme is in charge of converting adenosine monophosphate (AMP) to inosine monophosphate (IMP) and ammonia. As a result, AMP levels and NAD+ (NADP+)-dependent isocitrate dehydrogenase (ICDH) activity drop. Meanwhile, isocitrate accumulates in the mitochondria and is isomerized to citrate, which is subsequently transferred into the cytoplasm in exchange for malate in the presence of aconitase. Only mitochondria, however, synthesize ATP citrate lyase, an enzyme that transforms citrate into acetyl-CoA and oxaloacetate. Acetyl-CoA serves as a precursor for fatty acid synthesis, whereas oxaloacetate is transformed into malate and delivered back into the mitochondria via malate dehydrogenase.

Separate Hydrolysis and Lipid generation (SHLP) separate the hydrolysis and lipid generation procedures into two different phases. However, because glucose inhibits cellulase, the generation of simple sugars is decreased. Simultaneous Saccharification and Lipid Production (SSLP), on the other hand, involves the simultaneous production of saccharification and lipids in the same reactor. Microbes digest the sugars generated during hydrolysis, which serves to lessen the inhibitory impact of glucose. Finally, in Consolidated Bioprocessing (CBP), cellulase, hydrolysis, and lipid synthesis all take place in the same phase.

Simultaneous Saccharification and Lipid Production (SSLP) is regarded as a more cost-effective technique than Separate Hydrolysis and Lipid Production (SHLP) since the process is accomplished in a single vessel, lowering the overall cost of equipment and labour. However, one significant disadvantage of SSLP is that it necessitates a temperature compromise since the temperature conditions for hydrolysis and lipid formation differ. This can reduce process efficiency and lower ultimate product yield (Akadiri et al. 2012). There are several ways to increase the lipid output of OMs cultivated on lignocellulosic hydrolysate.

6.3.5 *Transesterification*

To extract lipids from bacteria, it is necessary to rupture the cell wall. Bead milling, high-pressure homogenization, osmotic shock, subcritical water hydrolysis, ultrasound, pulsed electric field, microwaves, autolysis, enzymatic hydrolysis, and chemical hydrolysis are all used to disturb cells. These techniques are used to split apart the cells and liberate the lipids, allowing for further processing. The technique of choosing is determined by several criteria, including the type of bacteria, the characteristics of the lipids, and the size of the production (Khot et al. 2020). In addition, nanoscalpels, hydroxyl radicals, and photocatalytically active amino clay-conjugated TiO_2 are employed to disturb cells (Chintagunta et al. 2021a). To extract lipids from OMs, various methods such as solvent extraction, increased temperature and pressure solvent extraction, pressing and solvent-integrated extraction, supercritical extraction, electroporation-assisted extraction, ultrasound-assisted extraction, and ionic liquid co-solvent extraction are used. In contrast, the cost of lipid extraction can account for up to 90% of the cost of biodiesel synthesis. As a result, green solvent and green extraction technologies are being developed to make the process more cost-effective by reducing solvent usage while saving time and energy. These technologies aim to lessen the environmental impact of the extraction process while also cutting costs.

Although oleaginous biomass is high in lipids, it is also high in proteins and carbs. Soybeans, for example, include 20% fat, 20% carbs, and approximately 40% protein (Maddi et al. 2018). Non-lipid components of lipid-rich microalgae strains can be processed to provide value-added goods like fuel and methane. While the leftover biomass from lipid extraction can be added to soil, dried algal biomass can be pyrolyzed to create oils that can be used as transportation fuels. In order to produce fuel, gasoline, and methane, wet biomass may also be treated hydrothermally and anaerobically. There are no phenolic oligomers in the final bio-oil since neither the algal biomass nor the leftovers contain lignin. On the other hand, proteins in algal biomass/remnants will be converted into pyrroles, indoles, and poly-heteroaromatic chemicals that must be removed before burning to prevent catalyst poisoning or nitrogen oxide emissions.

Pyrolysis of algal biomass in the presence of zeolite, on the other hand, causes denitrification, deoxygenation, and the generation of aromatic hydrocarbons. Aromatics are valuable molecules that contribute to the higher octane rating of petrol. As a result of this process, bio-oil with less nitrogen and oxygen-containing molecules and more aromatic hydrocarbons is produced. This method can raise the value of bio-oil while lowering the process's environmental effect.

After extracting lipids from algal biomass, they need to be converted into simple alkyl esters, also known as biodiesel, to reduce their viscosity. This is done through a process called transesterification. Transesterification can be catalyzed by chemicals or enzymes or can occur without any catalyst (Kumar et al. 2019). Chemical-catalyzed reactions can be either heterogeneous or homogeneous. Heterogeneous reactions use solid acid or alkali catalysts like sulfated zirconia or calcium-based metal oxides (Aransiola et al. 2014). The use of liquid acid or alkali catalysts, such as hydrochloric acid, sulfuric acid, sodium hydroxide, or potassium hydroxide, is required for homogenous processes. However, the alkali-based transesterification process is not suitable for high FFA feedstocks because soap production may occur. Transesterification is a critical step in converting algal biomass into biofuels and high-value compounds (Ong et al. 2014).

6.3.6 Biodiesel Purification

The ability of oleaginous bacteria to manufacture specific compounds from a range of carbon sources is leading to an increase in their attractiveness (Kumar et al. 2020a, b). Manipulation of the OMs strains is required to produce high lipid productivity and yield for industrial-scale biofuel generation, industrial chemicals, and biorefinery. Synthetic biology, metabolic engineering (ME), genome editing, and genetic components can all be combined to accomplish this. Strategies relevant to ME of OMs include the production of various fatty acid-derived compounds such as FFAs, TAGs, FAMEs, and FALs, as well as acetyl CoA-derived products like terpenoids and poly-3-hydroxybutyrate (PHB). Additionally, the production of industrial compounds like erythritol, succinic acid, α-ketoglutarate, citric acid, and itaconic acid can also be achieved through ME. Furthermore, the utilization of low-cost substrates such as inulin, LCB, molasses, starch, and glycerol can be explored to reduce the cost of production (Fig. 6.2).

The numerous programs seek to develop feasible methods for producing biofuels or biorefinery chemicals, with the ultimate goal of maximizing feedstock utilization in the fermentation process. Many strains, however, are unable to thrive in low-cost carbon sources; hence OMs must be engineered to successfully use these sources. Xylose Cellobiose, from hemicelluloses, starch, inulin, molasses, and glycerol, a byproduct of the biodiesel industry, is being investigated by researchers. Furthermore, the creation of model systems containing specific industrial chemicals may result in the construction of biorefineries. Despite encouraging outcomes, oleaginous yeast and microalgae modification for ME is still in its early phases (Shi and Zhao 2017).

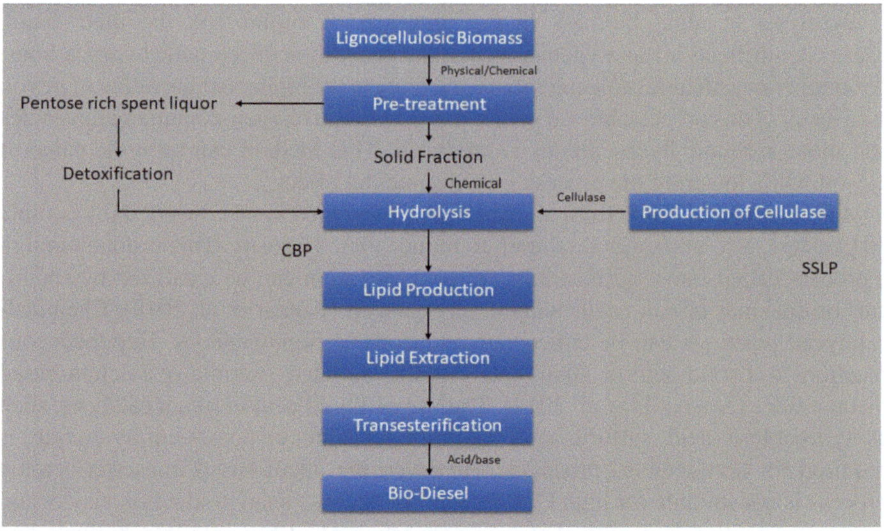

Fig. 6.2 Schematic diagram of lipid biosynthesis from lignocellulosic biomass using oleaginous microorganisms

Synthetic biology technology has been used to engineer strain cycles using the Design-Build-Test-Learn (DBTL) methodology. Other yeasts besides *Yarrowia lipolytica* and *R. toruloides* have benefited from the development of novel genetic engineering techniques thanks to recent advances in synthetic biology research (Park et al. 2018; Xiong et al. 2021).

In nonconventional yeasts such as *R. toruloides* and *Y. lipolytica*, recent research has concentrated on discovering critical genetic components such as constitutive, inducible, targeted, and repressible promoters. Modifying tandem copies of upstream activation sequences (UASs) has been proven in studies to improve hybrid promoters for higher expression in *Y. lipolytica* (Nora et al. 2019). Furthermore, genome editing approaches such as clustered regularly interspersed short palindromic repeats (CRISPR) and Cas protein have been investigated in these yeasts to improve lipid accumulation efficiency while minimizing off-target consequences (Abdel-Mawgoud and Stephanopoulos 2020; Yang et al. 2019). Researchers have been exploring ways to improve genetic engineering for increased lipid accumulation in oleaginous microorganisms (OMs). Koh et al. (2014) developed a Cre-loxp recombination system for marker-free transformation in *R. toruloides* and *Y. lipolytica*, which disrupted the Ku70 expressing gene and increased homologous frequency. Additionally, bacterial transcriptional factors have been combined with genetically encoded biosensors relevant to the malonyl-CoA and flavonoid pathways in *Y. lipolytica* to improve the stability and production of oleaginous engineered strains (Blazeck et al. 2011; Otoupal et al. 2019). The study on genetic engineering and biorefinery for oleaginous microorganisms (OMs) reveals the possibility of establishing commercially viable methods to increase lipid accumulation. Researchers have been able to optimize lipid accumulation in OMs like *Yarrowia lipolytica* and *Rhodosporidium toruloides* using techniques such as CRISPR-Cas mediated genome editing and the use of bacterial transcriptional factors and biosensors. These advances in genetic engineering, along with biorefinery, have the potential to provide more efficient and cost-effective methods of manufacturing biofuels and high-value chemicals.

6.4 Advantages of Using Oleaginous Microorganisms and Lignocellulosic Substrates

Using oleaginous microorganisms and lignocellulosic substrates to create biofuels and other useful products has various advantages. Here are a few examples:

1. Diverse feedstocks: Oleaginous microorganisms may consume a diverse variety of feedstocks, such as lignocellulosic substrates, agricultural waste, and even industrial effluents. This is advantageous because it permits the use of materials that would otherwise be wasted, hence decreasing waste and environmental effects. Lignocellulosic substrates are particularly appealing because they are plentiful and inexpensive, and they may be obtained from a range of sources, including agricultural wastes, forestry waste, and municipal solid waste.
2. High lipid productivity: Oleaginous microorganisms are extremely effective lipid producers, with some strains collecting more than 80% of their biomass as lipids. Because of their high lipid productivity, they can create huge volumes of lipids per unit of substrate eaten, making them a cost-effective and sustainable source of lipids to produce biofuels and other goods.
3. Low land use: Unlike typical biofuel feedstocks, which need a lot of arable lands, oleaginous microorganisms may be produced in much smaller places, such as bioreactors. This is advantageous because it allows to produce of biofuels and other goods without competing with food production, as well as reducing the burden on natural ecosystems caused by land-use change.
4. Reduce greenhouse gas emissions: Using oleaginous microorganisms with lignocellulosic substrates can help reduce greenhouse gas emissions in a variety of ways. For starters, it can reduce the quantity of rubbish that winds up in landfills, where it decomposes and generates methane, a powerful greenhouse gas. Second, it can provide a renewable energy source that is not reliant on fossil fuels, which are a major source of greenhouse gas emissions. Finally, using biofuels obtained from oleaginous microorganisms can assist in reducing the carbon footprint of transportation, which is an important contributor to the emission of greenhouse gases.

5. Improved economic value: By converting waste materials into valuable products such as biofuels, chemicals, and food additives, the employment of oleaginous bacteria and lignocellulosic substrates might provide new economic opportunities. This is advantageous because it has the potential to create new employment and industries while also lowering the economic expenses connected with garbage disposal and the usage of fossil fuels.

Overall, using oleaginous bacteria with lignocellulosic substrates has the potential to be a sustainable and effective means of producing biofuels and other useful products with various environmental and economic benefits. Nevertheless, further study is needed to improve the process and solve problems such as the expense of fermentation equipment and resource rivalry among various microbial species.

6.4.1 Cost Effectiveness

The economic efficacy of employing oleaginous microorganisms with lignocellulosic substrates to make biofuels and other products is determined by various factors, including feedstock cost, conversion process efficiency, and market demand for the final product. These are some of the elements that may have an impact on the cost-effectiveness of this approach:

1. Feedstock cost: Feedstock cost is an important aspect in assessing the economic effectiveness of utilizing oleaginous microorganisms with lignocellulosic substrates. Lignocellulosic substrates are usually less expensive than traditional biofuel feedstocks such as maize and soybeans; however, the cost might vary based on the feedstock's source and availability. The cost of processing and preparing the material for fermentation might also have an influence on the final cost.
2. Conversion efficiency: Another essential aspect in assessing the economic effectiveness of employing oleaginous microorganisms with lignocellulosic substrates is conversion efficiency. Factors such as the kind of microbe utilized, fermentation conditions, and downstream processing procedures can all have an influence on efficiency.
3. Market demand: The economic efficiency of utilizing oleaginous microorganisms and lignocellulosic substrates can be influenced by market demand. If there is a large demand for biofuels or other goods made from oleaginous microorganisms, the higher selling price may balance the expense. Yet, if market demand is low, the expense of the procedure may be prohibitively expensive.
4. Production scale: Production scale is another essential aspect in assessing the economic effectiveness of employing oleaginous bacteria with lignocellulosic substrates. Because of the higher cost of equipment and the reduced efficiency of smaller fermentation systems, small-scale manufacturing may be more expensive. But, when the manufacturing scale rises, the cost per unit of output might fall.

This cost benefit is substantial, and it has the potential to make biodiesel manufacturing more economically viable. Oleaginous bacteria may produce large amounts of lipids, which can then be removed and turned into biodiesel. Biodiesel costs can fluctuate greatly based on a variety of factors, such as government policy, global oil prices, and market demand. Nonetheless, biodiesel generated from oleaginous microorganisms and lignocellulosic substrates has the potential to be more cost competitive than biodiesel produced from standard vegetable oils, particularly if feedstock costs and conversion process efficiency can be reduced further. In conclusion, because of the cheaper cost of the feedstock, high lipid productivity, and developments in fermentation and extraction technologies, utilizing oleaginous microorganisms with lignocellulosic substrates in biodiesel synthesis can be cost-viable. The market price of biodiesel, on the other hand, continues to be an important factor for assessing the overall cost efficiency of the process.

6.4.2 Sustainability

Biodiesel is a sustainable fuel made from several feedstocks such as vegetable oil, animal fat, and used cooking oil. Nevertheless, these feedstocks have supply, cost, and environmental effect limits. The utilization of oleaginous microorganisms and lignocellulosic substrates in biodiesel synthesis offers the potential to overcome some of these constraints and increase the biodiesel industry's sustainability.

Oleaginous microorganisms, such as algae and fungus, may accumulate huge quantities of lipids, which can be turned into biodiesel. Lignocellulosic substrates are high in cellulose, and hemicellulose can be employed as a low-cost and plentiful feedstock for oleaginous microorganism growth. By reusing waste materials that would otherwise be wasted, the utilization of lignocellulosic substrates can further lessen the environmental effect of biodiesel production.

The method of producing biodiesel from oleaginous microorganisms and lignocellulosic substrates generally consists of three major steps:

1. Oleaginous microorganism growth on lignocellulosic substrates: Oleaginous microorganisms are cultured in bioreactors or open ponds on lignocellulosic substrates. The microbes use cellulose and hemicellulose in the substrate to generate lipids, which accumulate in their cells throughout this stage.
2. Microorganism harvesting: Once the microbes have collected enough lipids, they are harvested and separated from the substrate. Harvesting can be accomplished using a variety of processes, including centrifugation, filtering, and flocculation.
3. Lipids to biodiesel conversion: The lipids are transformed to biodiesel by transesterification, which involves mixing the lipids with an alcohol, such as methanol or ethanol, in the presence of a catalyst, such as sodium hydroxide or potassium hydroxide. The resulting fatty acid methyl esters (FAMEs) combination can be processed and used as biodiesel.

The employment of oleaginous microorganisms with lignocellulosic substrates in the synthesis of biodiesel has various benefits over conventional feedstocks, including:

1. Environmental sustainability: Using lignocellulosic substrates as a feedstock lowers rivalry between food and fuel production and makes use of waste materials, lowering the environmental effect of biodiesel production.
2. Lignocellulosic substrates are plentiful and inexpensive, making them an appealing feedstock for biodiesel synthesis.
3. High lipid productivity: Oleaginous microorganisms produce lipids that may be turned into biodiesel with high efficiency, resulting in a high biodiesel production per unit of feedstock used.
4. Lowering greenhouse gas emissions: Using biodiesel made from oleaginous microorganisms and lignocellulosic substrates can lower greenhouse gas emissions by minimizing waste and utilizing renewable feedstocks.

However, there are several drawbacks to using oleaginous bacteria with lignocellulosic substrates in biodiesel synthesis. These are some examples:

1. Production costs: The cost of fermentation equipment and the process of isolating microorganisms from the substrate might be significant, affecting the method's economic feasibility.
2. Resource competition: The development of oleaginous microorganisms on lignocellulosic substrates might be hampered by competition for nutrients and oxygen with other microbial species.
3. Technical constraints: The process of producing biodiesel from oleaginous microorganisms and lignocellulosic substrates is still in its early stages and requires more research and development to refine and enhance.

6.4.3 *Efficiency*

Biodiesel is a non-toxic, renewable fuel that may be made from a variety of sources, including vegetable oils, animal fats, and algae. Oleaginous microorganisms, such as yeasts and microalgae, can also be employed to create lipids for biodiesel production. Agricultural waste and forestry residues, for example, can be utilized as a low-cost and plentiful supply of carbon for the development of these microorganisms. Here are some specifics on how oleaginous bacteria and lignocellulosic substrates might boost biodiesel production efficiency:

1. High lipid productivity: Oleaginous microorganisms have high lipid productivity, which means they can create a considerable quantity of lipids per unit of substrate eaten. Certain microalgae species, for example, can collect more than 50% of their dry weight in lipids. Because of the high lipid productivity, a big volume of biodiesel may be generated from a little amount of substrate, lowering total costs and enhancing biodiesel production efficiency.

2. Reduce feedstock costs: Lignocellulosic substrates are a low-cost and plentiful source of carbon for oleaginous microorganism growth. Agricultural wastes such as corn stover and sugarcane bagasse, as well as forestry residues such as wood chips and sawdust, are examples of these substrates. Utilizing lignocellulosic substrates as a feedstock for biodiesel production helps lower feedstock costs, which are a significant component of overall biodiesel production costs.
3. Less rivalry for food resources: Conventional biodiesel sources, such as animal fats and vegetable oils, compete for land and resources with food production. Because these substrates do not compete with food production, using lignocellulosic substrates and oleaginous microorganisms as a source of biodiesel can help to lessen competition.
4. Lower environmental effect: Using lignocellulosic substrates and oleaginous microorganisms in biodiesel synthesis might lessen the process's environmental impact. Lignocellulosic substrates, for example, are a waste product of agriculture and forestry, and utilizing them as a feedstock for biodiesel synthesis can help to minimize waste and greenhouse gas emissions related to their disposal. Moreover, biodiesel made from oleaginous microorganisms has a lower carbon footprint than fossil fuels, lowering greenhouse gas emissions and helping to mitigate climate change.
5. Production Flexibility: The utilization of oleaginous microorganisms and lignocellulosic substrates in biodiesel synthesis might increase production flexibility. Oleaginous microorganisms may thrive in a variety of conditions, including freshwater, saltwater, and wastewater, enabling biodiesel synthesis in regions where standard biodiesel production is not feasible. Furthermore, independent of seasonal fluctuations, the utilization of lignocellulosic substrates can provide a continuous and stable source of feedstock for biodiesel synthesis.

The employment of oleaginous microorganisms with lignocellulosic substrates in biodiesel synthesis has the potential to boost efficiency, lower costs, and lessen the process's environmental effects. However, further study is needed to enhance the process and solve its hurdles, such as selecting the most appropriate microbes and developing cost-effective techniques for lipid extraction and biodiesel generation.

6.4.4 Composition Resembling That of Vegetable Oils

Oleaginous microorganisms and lignocellulosic substrates with lipid compositions like vegetable oils are viable sources of lipids for biodiesel generation. Depending on the species, strain, and growing circumstances, the lipid concentration and composition of oleaginous microorganisms can vary greatly. They do, however, typically include a blend of fatty acids, glycolipids, and other lipids comparable to those found in vegetable oils.

1. Yeasts: The capacity to collect lipids is recognized in several yeast species, including *Rhodosporidium toruloides*, *Yarrowia lipolytica*, and *Candida* sp.

These lipids are mostly made up of fatty acids, the most numerous of which are palmitic acid (C16:0), linoleic acid (C18:2), and oleic acid (C18:1). Depending on the strain and growing circumstances, the lipid content of these yeasts can range from 20 to 80% of their biomass.

2. Filamentous fungi: Certain filamentous fungi, like *Mortierella isabellina* and *Cunninghamella echinulata*, are known to generate lipids. The lipids generated by these fungi are predominantly constituted of saturated and unsaturated fatty acids, the most numerous of which are palmitic acid (C16:0), stearic acid (C18:0), oleic acid (C18:1), linoleic acid (C18:2).
3. Lignocellulosic substrates: Lignocellulosic substrates such as wood chips, sawdust, and maize stover can also be employed as a source of lipids in the manufacture of biodiesel. These substrates contain cellulose, hemicellulose, and lignin, which can be degraded into sugars and fermented by oleaginous microbes. The resultant lipids are predominantly made up of fatty acids, the most numerous of which are palmitic acid (C16:0) and oleic acid (C18:1).

Generally, oleaginous microorganisms and lignocellulosic substrates have lipid compositions like vegetable oils, making them a suitable source of lipids for biodiesel generation. Further research is needed, however, to optimize the process and enhance lipid yields and compositions.

6.5 Integration of Biofuels in the Circular Economy

The incorporation of biofuels into the circular economy entails producing biofuels from renewable resources and waste products, therefore lowering greenhouse gas emissions and establishing a more sustainable energy system. The circular economy is a system in which resources are utilized for as long as feasible, waste is reduced, and materials are recycled and reused in order to establish a closed loop of resource usage. These are some ideas for incorporating biofuels into the circular economy:

1. Use of waste materials: Biofuels may be made from waste resources such as agricultural leftovers, forestry debris, and municipal solid waste. These materials would otherwise be thrown or burned, increasing greenhouse gas emissions and complicating waste management. We can transform garbage into a lucrative resource and lessen the environmental effect of waste management by utilizing it as a feedstock for biofuels.
2. Bio-based material manufacturing: Biofuels may also be utilized as a feedstock for the manufacture of bio-based products like bioplastics and biochemicals. These materials can be used in place of fossil-based materials, lowering the carbon footprint of numerous products and fostering a more sustainable economy.
3. Usage of renewable energy: Renewable energy sources such as solar and wind power may be used to power biofuel production. This decreases the usage of fossil fuels in the manufacturing process, hence lowering greenhouse gas emissions.

4. Circular supply chain: Biofuel production may be incorporated into a circular supply chain in which waste materials are utilized as feedstock, and the produced biofuels and byproducts are reused in other operations. This creates a closed loop of resource use and reduces the need for virgin materials.
5. Carbon capture and utilization: Biofuels may also be utilized in carbon capture and utilization (CCU) operations, which absorb and utilize carbon dioxide as a feedstock in the manufacture of biofuels or other goods. This has the potential to cut greenhouse gas emissions while also creating a more sustainable energy system.

Delignification, saccharification, and fermentation are the three key stages in the production of bioethanol. Following delignification, the primarily enzymatically destroyed lignin can be used as antioxidants, glucose biosensors, reusable adsorbents, silver nanoparticles, electric double-layer capacitors, and other uses (Kumar et al. 2020a, b). Pentoses are generally wasted during the fermentation process.

6.5.1 Explanation of the Circular Economy and Its Components

The circular economy is a concept that seeks to establish a closed-loop system in which resources are used if feasible, waste is reduced, and materials are recycled and reused. This principle may be used to produce biodiesel in a variety of ways, including:

Biodiesel may be made from a variety of feedstocks, including used cooking oil and animal fats. Biodiesel production may help to waste reduction and the creation of a closed-loop system in which trash is exploited as a resource by employing waste materials as a feedstock.

Use of by-products: Glycerol is produced as a by-product of biodiesel manufacturing. Glycerol may be converted into a variety of products, including bio-based chemicals and materials. Glycerol, for example, may be used to make propylene glycol, a chemical that is used to make antifreeze, cosmetics, and other items.

Renewable energy sources such as solar and wind power can be used to power biodiesel manufacturing. Biodiesel manufacturing may minimize greenhouse gas emissions and contribute to a more sustainable energy system by using renewable energy to power the production process. Biodiesel may be recycled and reused as a fuel for a variety of purposes, including transportation, heating, and power generation. Spent biodiesel may be collected, cleaned, and re-used as a fuel. This closes the resource cycle and lowers the requirement for virgin resources.

6.5.2 *Importance of Integrating Biofuels in Circular Economy*

Biofuels are fuels made from organic elements that are renewable, such as crops, agricultural and forestry waste, and algae. They have gained popularity as a viable alternative to fossil fuels due to their ability to minimize greenhouse gas emissions and reliance on non-renewable resources. Including biofuels in a circular economy has various advantages, including:

Waste and emissions reduction: Waste is reduced, and resources are reused as much as feasible in a circular economy. Waste biomass, such as agricultural leftovers, forestry waste, and municipal solid waste, may be converted into biofuels, which can help decrease waste and greenhouse gas emissions. Helping local agriculture and rural development: By producing biofuels from locally obtained feedstocks, farmers and rural communities may gain new revenue prospects. This can aid economic growth by reducing reliance on imported fossil fuels.

Improving energy security: Biofuels can assist in reducing reliance on fossil fuels, improving energy security, and decreasing vulnerability to price volatility and supply interruptions.

Innovating and developing technology: The creation and manufacturing of biofuels necessitate major technological innovation, which can lead to the development of new and more efficient processes and products. This can assist in boosting the green economy and offer new job possibilities. Promoting sustainable transportation: Biofuels may be employed in the transportation industry as a sustainable alternative to fossil fuels. They can be combined with petrol and diesel or used in automobiles as pure biofuels, lowering emissions and improving air quality. The incorporation of biofuels into a circular economy is critical for a variety of reasons, including lowering greenhouse gas emissions, boosting sustainable agriculture, and enhancing energy security. In this response, we will go through the significance of incorporating biofuels into a circular economy in further depth.

For starters, the usage of biofuels can help to reduce greenhouse gas emissions from the transportation sector, which is a major source of global carbon dioxide emissions.

Biofuels are made from renewable biomass, such as agricultural waste, forestry leftovers, and energy crops, which absorb CO_2 from the atmosphere as they develop. When biofuels are burned, carbon dioxide is released, but this is compensated by the carbon dioxide absorbed during biomass growth. In this way, biofuels can provide a significant reduction in greenhouse gas emissions compared to fossil fuels.

Second, biofuel manufacturing can help to encourage sustainable agriculture. Several biofuels are made from crops like maize, sugarcane, and soybeans, which may be cultivated on land that would otherwise be unsuitable for food crops. This can assist in relieving strain on food crops while also promoting sustainable land use practices. Moreover, biofuel production may give economic advantages to rural areas by producing jobs and sustaining local economies.

6.6 Relevance of Biotechnological Applications in Ensuring the Sustainability of Biofuels and Biorefineries

Biotechnological applications are critical to the long-term viability of biofuels and biorefineries. Biotechnology can assist in enhancing biofuel production efficiency and minimizing environmental impact, as well as enabling the use of a broader range of feedstocks. In this answer, we will go through the importance of biotechnological applications in assuring the long-term viability of biofuels and biorefineries in greater depth.

Increasing Efficiency: By inventing more efficient fermentation methods, biotechnology can assist in enhancing the efficiency of biofuel production, resulting in better yields and lower production costs. The introduction of genetically engineered microbes, for example, can assist in boosting ethanol production productivity. Enzyme engineering, on the other hand, can increase the activity and stability of enzymes employed in biofuel generation, resulting in better conversion rates and reduced enzyme consumption.

Biotechnology can also assist in lessening the environmental effect of biofuel production by decreasing waste and lowering the usage of hazardous chemicals. Bioreactors, for example, may trap and utilize CO_2 emissions from biofuel production, lowering the process's carbon footprint. Moreover, the use of biocatalysts in biofuel production might eliminate the need for hazardous chemicals, reducing pollution.

Using a Broader Range of Feedstocks: Biotechnology allows for the use of a broader range of feedstocks for biofuel production, such as lignocellulosic biomass and algae. Lignocellulosic biomass, such as agricultural waste and forestry residues, is abundant and affordable, but due to its complicated structure, it is difficult to convert into biofuels. Biotechnology can aid in the breakdown of the complex structure of lignocellulosic biomass, making it more suitable for conversion into biofuels. Similarly, algae may be cultivated on non-arable land and in wastewater and transformed into biofuels by biotechnological methods.

In conclusion, biotechnological applications are critical for ensuring the sustainability of biofuels and biorefineries. Biotechnology can improve the efficiency of biofuel production, reduce the environmental impact of the process, enable the utilization of a wider range of feedstocks, and support the development of biorefineries. It is important to continue to invest in biotechnological research and development to further improve the sustainability and viability of biofuels and biorefineries. Biotechnological applications play a critical role in ensuring the sustainability of biofuels and biorefineries. Biofuels are renewable energy sources that are derived from biomass, such as plant material, waste, or algae. Biorefineries are facilities that use biomass to produce a variety of products, including biofuels, chemicals, and materials.

One of the key ways that biotechnological applications can contribute to the sustainability of biofuels and biorefineries is by improving the efficiency of biomass conversion. Biotechnology can be used to develop and optimize enzymes that can break down biomass into sugars, which can then be fermented to produce biofuels.

These enzymes can be produced using genetically engineered microorganisms or by using directed evolution to improve the properties of existing enzymes. This can increase the efficiency of biomass conversion, reducing the amount of biomass and other resources needed to produce biofuels. Biotechnological applications can also contribute to the sustainability of biofuels and biorefineries by improving the economics of biomass conversion. Biotechnology can be used to develop and optimize microorganisms that can produce high-value chemicals and materials from biomass. This can help to diversify the revenue streams of biorefineries, making them more economically viable and reducing their dependence on biofuels alone.

Finally, biotechnological applications can contribute to the sustainability of biofuels and biorefineries by supporting sustainable agriculture practices. Biotechnology can be used to develop crops that are more resilient to pests and diseases, require less water and fertilizer, and can grow on marginal lands. This can help to reduce the environmental impact of agriculture, as well as improve the economic viability of biofuels and biorefineries.

In conclusion, biotechnological applications are critical for ensuring the sustainability of biofuels and biorefineries. Biotechnology can help to improve the efficiency, environmental performance, and economics of biomass conversion, as well as support sustainable agriculture practices. As such, biotechnological applications will continue to be an important area of research and development for the biofuels and biorefineries industries.

6.7 Conclusion

Moreover, biofuels have the potential to play a significant role in the transition to a more sustainable and low-carbon energy system. Biofuels have various advantages over fossil fuels, including lower greenhouse gas emissions, better agricultural sustainability, and increased energy security. Nevertheless, the production and use of biofuels pose a number of obstacles, including land use disputes, food security issues, and significant environmental consequences.

To guarantee that biofuels are produced and utilized in a sustainable and responsible way, the full lifespan of the biofuel must be considered from production to end users. This involves ensuring that the biomass used to make the biofuel is sourced sustainably, limiting resource usage during production, and encouraging the recycling and reuse of biofuel by-products.

6.7.1 Summarize the Advantages of Utilizing Oleaginous Microorganisms and Lignocellulosic Substrates for Biodiesel Synthesis

Oleaginous bacteria and lignocellulosic substrates have various advantages in the synthesis of biodiesel, including:

High lipid content: Oleaginous microorganisms like algae and yeast may collect a lot of lipids (fats and oils) in their cells, which can be turned into biodiesel. Because they require less land, water, and other resources than traditional oilseed crops, they are a highly efficient source of biodiesel feedstock.

Lignocellulosic substrates, such as agricultural waste, forestry residues, and energy crops, can be utilized as feedstock for oleaginous bacteria, offering a low-cost and sustainable biomass source for biodiesel generation. This might also help to decrease waste and encourage circular economy principles.

Decreased greenhouse gas emissions: When compared to fossil fuel, biodiesel generated from oleaginous microorganisms and lignocellulosic substrates has the potential to greatly reduce greenhouse gas emissions. This is since the carbon in biodiesel is derived from atmospheric CO_2, which is stored by biomass during growth, resulting in a net reduction in CO_2 emissions.

Biodiesel made from oleaginous microorganisms and lignocellulosic substrates can help to improve energy security by lowering reliance on imported fossil fuels. This can aid in the promotion of local and regional economic growth as well as the improvement of energy availability in remote or rural locations.

Co-product potential: The synthesis of biodiesel from oleaginous microorganisms and lignocellulosic substrates might result in the production of lucrative co-products such as animal feed, fertilizer, and biogas. This has the potential to improve the overall economic feasibility and sustainability of biodiesel production.

Finally, using oleaginous microbes with lignocellulosic substrates for biodiesel synthesis has various advantages, including high lipid content, diverse substrate supplies, lower greenhouse gas emissions, enhanced energy security, and the possibility for co-products. These advantages make them an appealing alternative for producing low-carbon biodiesel, which can aid in the transition to a more sustainable and resilient energy system.

6.7.2 Implications for Promoting a Cleaner and More Sustainable Future

Biofuels, such as biodiesel made from oleaginous microorganisms and lignocellulosic substrates, have the potential to promote a cleaner and more sustainable future. Here are some of the ways that biofuels may help to create a more sustainable and low-carbon energy system.

Emissions of greenhouse gases have been reduced: When compared to fossil fuels, biofuels have the potential to considerably cut greenhouse gas emissions. As compared to fossil diesel, the usage of biodiesel generated from oleaginous microorganisms and lignocellulosic substrates can cut carbon dioxide emissions by up to 85%.

Renewable energy: Because biofuels are manufactured from renewable feedstocks like plant biomass and microorganisms, they are a more sustainable alternative to fossil fuels. Moreover, the utilization of lignocellulosic substrates for biofuel production can aid in waste reduction and the promotion of a circular economy.

Energy security: By lowering reliance on imported fossil fuels, biofuels can assist in boosting energy security. Biofuel production can potentially generate new economic prospects and jobs, particularly in rural regions.

The development of biotechnological applications for biofuel production, such as genetic engineering and metabolic engineering, can lead to the creation of new and novel technologies that can be used in other sectors.

International cooperation: Biofuel production and consumption may also foster international cooperation and collaboration, notably in the areas of energy security and climate change mitigation. It is crucial to emphasize, however, that the production and use of biofuels involve a number of obstacles and possible environmental implications, such as land use disputes, food security issues, and potential negative effects on biodiversity and ecosystems. As a result, it is critical to guarantee that biofuels are produced and utilized in a sustainable and responsible way, taking into account the biofuel's whole lifespan, from production to end users.

Overall, the usage of biofuels, particularly those derived from oleaginous microorganisms and lignocellulosic substrates, has the potential to significantly contribute to a cleaner and more sustainable future. It has the potential to help reduce greenhouse gas emissions, promote a circular economy, and promote sustainable agriculture and rural development. Nevertheless, in order to fully appreciate these benefits, biofuels must be produced and utilized in a sustainable and ethical way.

References

Abdel-Mawgoud AM, Stephanopoulos G (2020) Improving CRISPR/Cas9-mediated genome editing efficiency in Yarrowia lipolytica using direct tRNA-sgRNA fusions. Metab Eng 62:106–115. https://doi.org/10.1016/J.YMBEN.2020.07.008

Abghari A, Chen S (2014) Yarrowia lipolytica as an oleaginous cell factory platform for production of fatty acid-based biofuel and bioproducts. Front Energy Res 2:21. https://doi.org/10.3389/fenrg.2014.00021

Abu-Elreesh G, Abd-El-Haleem D (2014) Promising oleaginous filamentous fungi as biodiesel feed stocks: screening and identification. Eur J Exp Biol 4:576–582

Akadiri PO, Chinyio EA, Olomolaiye PO (2012) Design of a sustainable building: a conceptual framework for implementing sustainability in the building sector. Buildings 2:126–152. https://doi.org/10.3390/buildings2020126

Ananthi V, Prakash GS, Chang SW, Ravindran B, Nguyen DD, Vo DN et al (2019) Enhanced microbial biodiesel production from lignocellulosic hydrolysates using yeast isolates. Fuel 256:115932. https://doi.org/10.1016/j.fuel.2019.115932

Aransiola EF, Ojumu TV, Oyekola OO, Madzimbamuto TF, Ikhu-Omoregbe DIO (2014) A review of current technology for biodiesel production: state of the art. Biomass Bioenerg 61:276–297. https://doi.org/10.1016/J.BIOMBIOE.2013.11.014

Atabani AE, Silitonga AS, Badruddin IA, Mahlia TMI, Masjuki HH, Mekhilef S (2012) A comprehensive review on biodiesel as an alternative energy resource and its characteristics. Renew Sustain Energy Rev 16:2070–2093. https://doi.org/10.1016/J.RSER.2012.01.003

Athenaki M, Gardeli C, Diamantopoulou P, Tchakouteu SS, Sarris D, Philippoussis A, Papanikolaou S (2018) Lipids from yeasts and fungi: physiology, production and analytical considerations. J Appl Microbiol 124:336–367. https://doi.org/10.1111/JAM.13633

Balat M (2011) Potential alternatives to edible oils for biodiesel production—a review of current work. Energy Convers Manag 52:1479–1492. https://doi.org/10.1016/J.ENCONMAN.2010.10.011

Banerjee R, Kumar SPJ, Mehendale N, Sevda S, Garlapati VK (2019) Intervention of microfluidics in biofuel and bioenergy sectors: technological considerations and future prospects. Renew Sustain Energy Rev 101:548–558. https://doi.org/10.1016/J.RSER.2018.11.040

Bardhan P, Gohain M, Daimary N, Kishor S, Chattopadhyay P, Gupta K, Chaliha C, Kalita E, Deka D, Mandal M (2019) Microbial lipids from cellulolytic oleaginous fungus Penicillium citrinum PKB20 as a potential feedstock for biodiesel production. Ann Microbiol 69:1135–1146. https://doi.org/10.1007/S13213-019-01494-3/TABLES/4

Bhanja A, Minde G, Magdum S, Kalyanraman V (2014) Comparative studies of Oleaginous fungal strains (Mucor circinelloides and Trichoderma reesei) for effective wastewater treatment and bio-oil production. Biotechnol Res Int 2014:1–7. https://doi.org/10.1155/2014/479370

Blazeck J, Liu L, Redden H, Alper H (2011) Tuning gene expression in Yarrowia lipolytica by a hybrid promoter approach. Appl Environ Microbiol 77:7905–7914. https://doi.org/10.1128/AEM.05763-11

Čertík M, Adamechová Z, Laoteng K (2012) Microbial production of γ-linolenic acid: submerged versus solid-state fermentations. Food Sci Biotechnol 21:921–926. https://doi.org/10.1007/S10068-012-0121-2/METRICS

Chaiyaso T, Manowattana A, Techapun C, Watanabe M (2019) Efficient bioconversion of enzymatic corncob hydrolysate into biomass and lipids by oleaginous yeast Rhodosporidium paludigenum KM281510. Prep Biochem Biotechnol 49:545–556. https://doi.org/10.1080/10826068.2019.1591985

Chan LG, Dias FFG, Saarni A, Cohen JL, Block DE, Taha AY, ... Bell JMLNM (2020) Scaling up the bioconversion of cheese whey permeate into fungal oil by mucor circinelloides. J Am Oil Chem Soc 97(7):703–716. https://doi.org/10.1002/aocs.12372

Chandel AK, Garlapati VK, Jeevan Kumar SP, Hans M, Singh AK, Kumar S (2020) The role of renewable chemicals and biofuels in building a bioeconomy. Biofuels Bioprod Biorefining 14:830–844. https://doi.org/10.1002/BBB.2104

Chang Y-H, Chang K-S, Lee C-F, Hsu C-L, Huang C-W, Jang H-D (2015) Microbial lipid production by oleaginous yeast Cryptococcus sp. in the batch cultures using corncob hydrolysate as carbon source. Biomass Bioenergy 72:95–103. https://doi.org/10.1016/j.biombioe.2014.11.012

Chaturvedi V, Verma P (2013) An overview of key pretreatment processes employed for bioconversion of lignocellulosic biomass into biofuels and value added products. 3 Biotech 3:415–431. https://doi.org/10.1007/S13205-013-0167-8

Cheirsilp B, Thawechai T, Prasertsan P (2017) Immobilized oleaginous microalgae for production of lipid and phytoremediation of secondary effluent from palm oil mill in fluidized bed photobioreactor. Bioresour Technol 241:787–794. https://doi.org/10.1016/j.biortech.2017.06.016

Chintagunta AD, Kumar A, Jeevan Kumar SP, Verma ML (2021a) Contribution of metallic nanomaterials in algal biofuel production:331–353. https://doi.org/10.1007/978-3-030-53065-5_9

Chintagunta AD, Zuccaro G, Kumar M, Kumar SPJ, Garlapati VK, Postemsky PD, Kumar NSS, Chandel AK, Simal-Gandara J (2021b) Biodiesel production from lignocellulosic biomass using oleaginous microbes: prospects for integrated biofuel production. Front Microbiol 12:2080. https://doi.org/10.3389/FMICB.2021.658284/BIBTEX

Cho HU, Park JM (2018) Biodiesel production by various oleaginous microorganisms from organic wastes. Bioresour Technol 256:502–508. https://doi.org/10.1016/J.BIORTECH.2018.02.010

Deeba F, Pruthi V, Negi YS (2016) Converting paper mill sludge into neutral lipids by oleaginous yeast Cryptococcus vishniaccii for biodiesel production. Bioresour Technol 213:96–102. https://doi.org/10.1016/j.biortech.2016.02.105

Deeba F, Pruthi V, Negi YS (2017) Fostering triacylglycerol accumulation in novel oleaginous yeast Cryptococcus psychrotolerans IITRFD utilizing groundnut shell for improved biodiesel production. Bioresour Technol 242:113–120. https://doi.org/10.1016/j.biortech.2017.04.001

Demirbas A (2009a) Progress and recent trends in biodiesel fuels. Energy Convers Manag 50:14–34. https://doi.org/10.1016/J.ENCONMAN.2008.09.001

Demirbas A (2009b) Production of biodiesel fuels from linseed oil using methanol and ethanol in non-catalytic SCF conditions. Biomass Bioenergy 33:113–118. https://doi.org/10.1016/J.BIOMBIOE.2008.04.018

Demirbas A, Bafail A, Ahmad W, Sheikh M (2016) Biodiesel production from non-edible plant oils. Energy Explor Exploit 34:290–318. https://doi.org/10.1177/0144598716630166

Dey P, Banerjee J, Maiti MK (2011) Comparative lipid profiling of two endophytic fungal isolates—Colletotrichum sp. and Alternaria sp. having potential utilities as biodiesel feedstock. Bioresour Technol 102(10):5815–5823. https://doi.org/10.1016/j.biortech.2011.02.064

Economou CN, Aggelis G, Pavlou S, Vayenas DV (2011) Single cell oil production from rice hulls hydrolysate. Bioresour Technol 102:9737–9742. https://doi.org/10.1016/j.biortech.2011.08.025

El-Sheekh MM, Bedaiwy MY, Osman ME, Ismail MM (2012) Mixotrophic and heterotrophic growth of some microalgae using extract of fungal-treated wheat bran. Int J Rec Org Waste Agric 1:121–129. https://doi.org/10.1186/2251-7715-1-12

Gao Q, Cui Z, Zhang J, Bao J (2014) Lipid fermentation of corncob residues hydrolysate by oleaginous yeast Trichosporon cutaneum. Bioresour Technol 152:552–556. https://doi.org/10.1016/j.biortech.2013.11.044

Garay LA, Sitepu IR, Cajka T, Chandra I, Shi S, Lin T, German JB, Fiehn O, Boundy-Mills KL (2016) Eighteen new oleaginous yeast species. J Ind Microbiol Biotechnol 43:887–900. https://doi.org/10.1007/S10295-016-1765-3

Garlapati VK, Chandel AK, Kumar SPJ, Sharma S, Sevda S, Ingle AP, Pant D (2020) Circular economy aspects of lignin: towards a lignocellulose biorefinery. Renew Sustain Energy Rev 130:109977. https://doi.org/10.1016/J.RSER.2020.109977

Gmoser R, Sintca C, Taherzadeh MJ, Lennartsson PR (2019) Combining submerged and solid state fermentation to convert waste bread into protein and pigment using the edible filamentous fungus N. intermedia. Waste Manag 97:63–70. https://doi.org/10.1016/J.WASMAN.2019.07.039

Goswami L, Namboodiri MMT, Kumar RV, Pakshirajan K, Pugazhenthi G (2017) Biodiesel production potential of oleaginous Rhodococcus opacus grown on biomass gasification wastewater. Renew Energy 105:400–406. https://doi.org/10.1016/j.renene.2016.12.044

Gouda MK, Sanaa H, Omar Linda M, Aouad (2008) Single cell oil production by Gordonia sp. DG using agro-industrial wastes. World J Microbiol Biotechnol 24(9):1703–1711. https://doi.org/10.1007/s11274-008-9664-z

Gujjala LKS, Kumar SPJ, Talukdar B, Dash A, Kumar S, Sherpa KC, Banerjee R (2017) Biodiesel from oleaginous microbes: opportunities and challenges. 10:45–59. https://doi.org/10.1080/17597269.2017.1402587

Haas MJ, McAloon AJ, Yee WC, Foglia TA (2006) A process model to estimate biodiesel production costs. Bioresour Technol 97:671–678. https://doi.org/10.1016/J.BIORTECH.2005.03.039

Himmel ME, Ding SY, Johnson DK, Adney WS, Nimlos MR, Brady JW, Foust TD (2007) Biomass recalcitrance: engineering plants and enzymes for biofuels production. Science 315:804–807. https://doi.org/10.1126/SCIENCE.1137016

Hossain KA, Badr O (2007) Prospects of renewable energy utilisation for electricity generation in Bangladesh. Renew Sustain Energy Rev. https://doi.org/10.1016/j.rser.2005.12.010

Huber GW (2008) Breaking the chemical and engineering barriers to lignocellulosic bio-fuels: next generation hydrocarbon biorefineries. A research roadmap for making lignocellulosic bio-fuels a practical reality

Jeevan Kumar SP, Srinivas Gujjala LKS, Dash A, Talukdar B, Banerjee R (2017a) Biodiesel production from lignocellulosic biomass using oleaginous microbes. Lignocellul Prod Ind Appl 65–92. https://doi.org/10.1002/9781119323686.CH4

Jeevan Kumar SP, Vijay Kumar G, Dash A, Scholz P, Banerjee R (2017b) Sustainable green solvents and techniques for lipid extraction from microalgae: a review. Algal Res 21:138–147. https://doi.org/10.1016/J.ALGAL.2016.11.014

Jena PC, Raheman H, Prasanna Kumar GV, Machavaram R (2010) Biodiesel production from mixture of mahua and simarouba oils with high free fatty acids. Biomass Bioenerg 34:1108–1116. https://doi.org/10.1016/J.BIOMBIOE.2010.02.019

Jin M, Slininger PJ, Dien BS, Waghmode S, Moser BR, Orjuela A, da Costa Sousa L, Balan V (2015) Microbial lipid-based lignocellulosic biorefinery: feasibility and challenges. Trends Biotechnol 33:43–54. https://doi.org/10.1016/J.TIBTECH.2014.11.005

Jönsson LJ, Martín C (2016) Pretreatment of lignocellulose: formation of inhibitory by-products and strategies for minimizing their effects. Bioresour Technol 199:103–112. https://doi.org/10.1016/J.BIORTECH.2015.10.009

Kalscheuer R, Stölting T, Steinbüchel A (2006) Microdiesel: Escherichia coli engineered for fuel production. Microbiology 152:2529–2536. https://doi.org/10.1099/MIC.0.29028-0

Khot M, Raut G, Ghosh D, Alarcón-Vivero M, Contreras D, Ravikumar A (2020) Lipid recovery from oleaginous yeasts: perspectives and challenges for industrial applications. Fuel 259:116292. https://doi.org/10.1016/J.FUEL.2019.116292

Koh CMJ, Liu Y, Moehninsi Du M, Ji L (2014) Molecular characterization of KU70 and KU80 homologues and exploitation of a KU70-deficient mutant for improving gene deletion frequency in Rhodosporidium toruloides. BMC Microbiol 14. https://doi.org/10.1186/1471-2180-14-50

Kucharska K, Słupek E, Cieśliński H, Kamiński M (2020) Advantageous conditions of saccharification of lignocellulosic biomass for biofuels generation via fermentation processes. Chem Pap 74:1199–1209. https://doi.org/10.1007/s11696-019-00960-1

Kuhad RC, Singh A (2008) Lignocellulose biotechnology: current and future prospects. 13:151–172. https://doi.org/10.3109/07388559309040630

Kumar D, Singh B, Korstad J (2017) Utilization of lignocellulosic biomass by oleaginous yeast and bacteria for production of biodiesel and renewable diesel. Renew Sustain Energy Rev 73:654–671. https://doi.org/10.1016/J.RSER.2017.01.022

Kumar NS, Geetha RS, Vasu AT, Ramu DV (2019) Performance assessment of biodiesel transesterified from Labeo rohita visceral oil using IC engine. Curr Trends Biotechnol Pharm 13:348–356

Kumar B, Bhardwaj N, Agrawal K, Chaturvedi V, Verma P (2020) Current perspective on pretreatment technologies using lignocellulosic biomass: an emerging biorefinery concept. Fuel Process Technol 199:106244. https://doi.org/10.1016/j.fuproc.2019.106244

Kumar KK, Deeba F, Sauraj Negi YS, Gaur NA (2020) Harnessing pongamia shell hydrolysate for triacylglycerol agglomeration by novel oleaginous yeast Rhodotorula pacifica INDKK. Biotechnol Biofuels 13. https://doi.org/10.1186/S13068-020-01814-9

Kurosawa K, Sinskey AJ (2013) Engineering xylose metabolism in triacylglycerol producing Rhodococcusopacus for lignocellulosic fuel production. Biotechnol Biofuels 6: 134–147. https://doi.org/10.1186/1754-6834-6-134

Lam MK, Lee KT (2011) Renewable and sustainable bioenergies production from palm oil mill effluent (POME): win-win strategies toward better environmental protection. Biotechnol Adv 29:124–141. https://doi.org/10.1016/J.BIOTECHADV.2010.10.001

Leesing R, Karraphan P (2011) Kinetic growth of the isolated oleaginous yeast for microbial lipid production. Afr J Biotechnol 10:13867–13877. https://doi.org/10.5897/AJB10.2162

Li P, Miao X, Li R, Zhong J (2001) In situ biodiesel production from fast-growing and high oil content Chlorella pyrenoidosa in rice straw hydrolysate. J Biomed Biotechnol 1–8. https://doi.org/10.1155/2011/141207

Liu L, Chen J, Lim P-E, Wei D (2018) Enhanced single cell oil production by mixed culture of Chlorella pyrenoidosa and Rhodotorula glutinis using cassava bagasse hydrolysate as carbon source. Bioresour Technol 255:140–148. https://doi.org/10.1016/j.biortech.2018.01.114

Liu L, Song J, Li Y, Li P, Wang, H (2019) Robust and cost-saving static solid cultivation method for lipid production using the chlamydospores of Phanerochaete chrysosporium. Biotechnol Biofuels Bioprod 12(1). https://doi.org/10.1186/s13068-019-1464-1

Liu Y, Wang Y, Liu H, Zhang JA (2015) Enhanced lipid production with undetoxified corncob hydrolysate by Rhodotorula glutinis using a high cell density culture strategy. Bioresour Technol 180:32–39. https://doi.org/10.1016/j.biortech.2014.12.093

Lu Y, Zhai Y, Liu M, Wu Q (2010) Biodiesel production from algal oil using cassava (Manihot esculenta Crantz) as feedstock. J Appl Phycol 22(5):573–578. https://doi.org/10.1007/s10811-009-9496-8

Maddi B, Viamajala S, Varanasi S (2018) Pyrolytic fractionation: a promising thermochemical technique for processing Oleaginous (Algal) biomass. ACS Sustain Chem Eng 6:237–247. https://doi.org/10.1021/ACSSUSCHEMENG.7B02309/SUPPL_FILE/SC7B02309_SI_001.PDF

Martin NR, Kelley P, Klaski R, Bosco A, Moore B, Traviss N (2019) Characterization and comparison of oxidative potential of real-world biodiesel and petroleum diesel particulate matter emitted from a nonroad heavy duty diesel engine. Sci Total Environ 655:908–914. https://doi.org/10.1016/J.SCITOTENV.2018.11.292

Martínez EJ, Raghavan V, González-Andrés F, Gómez X (2015) New biofuel alternatives: integrating waste management and single cell oil production. Int J Mol Sci 16:9385–9405. https://doi.org/10.3390/IJMS16059385

Meng X, Yang J, Xu X, Zhang L, Nie Q, Xian M (2009) Biodiesel production from oleaginous microorganisms. Renew Energy 34:1–5. https://doi.org/10.1016/J.RENENE.2008.04.014

Miglio R, Palmery S, Salvalaggio M, Carnelli L, Capuano F, Borrelli R (2013) Microalgae triacylglycerols content by FT-IR spectroscopy. J Appl Phycol 25:1621–1631. https://doi.org/10.1007/S10811-013-0007-6/METRICS

Mu J, Li S, Chen D, Xu H, Han F, Feng B, Li Y (2015) Enhanced biomass and oil production from sugarcane bagasse hydrolysate (SBH) by heterotrophic oleaginous microalga Chlorella protothecoides. Bioresour Technol 185:99–105. https://doi.org/10.1016/j.biortech.2015.02.082

Nguyena HC, Sub C-H, Yub Y-K, Huong DTM (2018) Sugarcane bagasse as a novel carbon source for heterotrophic cultivation of oleaginous microalga Schizochytrium sp. Ind Crops Prod 121:99–105. https://doi.org/10.1016/j.indcrop.2018.05.005

Nora LC, Wehrs M, Kim J, Cheng JF, Tarver A, Simmons BA, Magnuson J, Harmon-Smith M, Silva-Rocha R, Gladden JM, Mukhopadhyay A, Skerker JM, Kirby J (2019) A toolset of constitutive promoters for metabolic engineering of Rhodosporidium toruloides. Microb Cell Fact 18. https://doi.org/10.1186/S12934-019-1167-0

Ong HR, Khan MR, Chowdhury MNK, Yousuf A, Cheng CK (2014) Synthesis and characterization of CuO/C catalyst for the esterification of free fatty acid in rubber seed oil. Fuel 120:195–201. https://doi.org/10.1016/J.FUEL.2013.12.015

Otoupal PB, Ito M, Arkin AP, Magnuson JK, Gladden JM, Skerker JM (2019) Multiplexed CRISPR-Cas9-based genome editing of Rhodosporidium toruloides. mSphere 4. https://doi.org/10.1128/MSPHERE.00099-19

Park YK, Nicaud JM, Ledesma-Amaro R (2018) The engineering potential of Rhodosporidium toruloides as a workhorse for biotechnological applications. Trends Biotechnol 36:304–317. https://doi.org/10.1016/J.TIBTECH.2017.10.013

Patel A, Arora N, Mehtani J, Pruthi V, Pruthi PA (2017) Assessment of fuel properties on the basis of fatty acid profiles of oleaginous yeast for potential biodiesel production. Renew Sustain Energy Rev 77:604–616. https://doi.org/10.1016/J.RSER.2017.04.016

Patel A, Matsakas L, Rova U, Christakopoulos P (2018) Heterotrophic cultivation of Auxenochlorella protothecoides using forest biomass as a feedstock for sustainable biodiesel production. Biotechnol Biofuels 11(1):1–16. https://doi.org/10.1186/s13068-018-1173-1

Pinzi S, Leiva D, López-García I, Redel-Macías MD, Dorado MP (2014) Latest trends in feedstocks for biodiesel production. Biofuels Bioprod Biorefining 8:126–143. https://doi.org/10.1002/BBB.1435

Pleissner D, Lam WC, Sun Z, Lin, CSK (2013) Food waste as nutrient source in heterotrophic microalgae cultivation. Bioresour Technol 137:139–146. https://doi.org/10.1016/j.biortech.2013.03.088

Qiao W, Tao J, Luo Y, Tang T, Miao J, Yang Q (2018) Microbial oil production from solid-state fermentation by a newly isolated oleaginous fungus Mucor circinelloides Q531 from mulberry branches. Royal Soc Open Sci 5(11):180551. https://doi.org/10.1098/rsos.180551

Quarterman JC, Slininger PJ, Hector RE, Dien BS (2018) Engineering Candida phangngensis—an oleaginous yeast from the Yarrowia clade—for enhanced detoxification of lignocellulose-derived inhibitors and lipid overproduction. FEMS Yeast Res 18:foy102. https://doi.org/10.1093/femsyr/foy102

Ratledge C (2004) Fatty acid biosynthesis in microorganisms being used for single cell oil production. Biochimie 86:807–815. https://doi.org/10.1016/J.BIOCHI.2004.09.017

Sampath Kumar NS, Sruthi Geetha R, Thirumala Vasu A, Vijaya Ramu D (2019) Performance assessment of biodiesel transesterified from labeo rohita visceral oil using IC engine. Curr Trends Biotechnol Pharm 13:348–356

Sampath Kumar NS, Sarbon NM, Rana SS, Chintagunta AD, Prathibha S, Ingilala SK, Jeevan Kumar SP, Sai Anvesh B, Dirisala VR (2021) Extraction of bioactive compounds from Psidium guajava leaves and its utilization in preparation of jellies. AMB Expr 11. https://doi.org/10.1186/s13568-021-01194-9

Sathiyamoorthi E, Kumar P, Kim BS (2019) Lipid production by Cryptococcus albidus using biowastes hydrolysed by indigenous microbes. Bioprocess Biosyst Eng 42(5):687–696. https://doi.org/10.1007/s00449-019-02073-1

Shaheena S, Chintagunta AD, Dirisala VR, Sampath Kumar NS (2019) Extraction of bioactive compounds from Psidium guajava and their application in dentistry. AMB Expr 9. https://doi.org/10.1186/S13568-019-0935-X

Sharma YC, Singh B (2010) An ideal feedstock, kusum (Schleichera triguga) for preparation of biodiesel: optimization of parameters. Fuel 89:1470–1474. https://doi.org/10.1016/J.FUEL.2009.10.013

Shi S, Zhao H (2017) Metabolic engineering of Oleaginous yeasts for production of fuels and chemicals. Front Microbiol 8. https://doi.org/10.3389/FMICB.2017.02185

Singh D, Sharma D, Soni SL, Sharma S, Kumar Sharma P, Jhalani A (2020) A review on feedstocks, production processes, and yield for different generations of biodiesel. Fuel 262:116553. https://doi.org/10.1016/j.fuel.2019.116553

Sitepu IR, Garay LA, Sestric R, Levin D, Block DE, German JB, Boundy-Mills KL (2014) Oleaginous yeasts for biodiesel: current and future trends in biology and production. Biotechnol Adv 32:1336–1360. https://doi.org/10.1016/J.BIOTECHADV.2014.08.003

Somerville C, Youngs H, Taylor C, Davis SC, Long SP (2010) Feedstocks for lignocellulosic biofuels. Science 329:790–792. https://doi.org/10.1126/SCIENCE.1189268

Sreeharsha RV, Mohan SV (2020) Obscure yet promising Oleaginous yeasts for fuel and chemical production. Trends Biotechnol 38:873–887. https://doi.org/10.1016/J.TIBTECH.2020.02.004

Subramaniam R, Dufreche S, Zappi M, Bajpai R (2010) Microbial lipids from renewable resources: production and characterization. J Ind Microbiol Biotechnol 37:1271–1287. https://doi.org/10.1007/S10295-010-0884-5

Talavari R, Hosseini S, Moradi G (2021) Low-cost biodiesel production using waste oil and catalyst. Waste Manag Res J Sustain Circ Econ 39:250–259. https://doi.org/10.1177/0734242X20935174

Tamano K, Bruno KS, Karagiosis SA, Culley DE, Deng S, Collett JR, et al (2013) Increased production of fatty acids and triglycerides in Aspergillus oryzae by enhancing expressions of fatty acid synthesis-related genes. Appl Microbiol Biotechnol 97:269–281. https://doi.org/10.1007/s00253-012-4193-y

Tsigie YA, Wang CY, Kasim NS, Diem QD, Huynh LH, Ho QP, Truong CT, Ju, YH (2012) Oil production from Yarrowia lipolytica Po1g using rice bran hydrolysate. J Biomed Biotechnol 2012:378384. https://doi.org/10.1155/2012/378384

Voss I, Steinbuchel A (2001) High cell density cultivation of Rhodococcusopacus for lipid production at a pilot-plant scale. Appl Microbiol Biot 55:547–555. https://doi.org/10.1007/s002530000576

Wang JQ, Zhang H, Bao J (2016) Inhibitor degradation and lipid accumulation potentials of oleaginous yeas Gao t Trichosporon cutaneum using lignocellulose feedstock. Bioresour Technol 218:892–901. https://doi.org/10.1016/j.biortech.2016.06.130

Wang R, Hanna MA, Zhou WW, Bhadury PS, Chen Q, Song BA, Yang S (2011) Production and selected fuel properties of biodiesel from promising non-edible oils: Euphorbia lathyris L., Sapium sebiferum L. and Jatropha curcas L. Bioresour Technol 102:1194–1199. https://doi.org/10.1016/J.BIORTECH.2010.09.066

Wei Z, Zeng G, Huang F, Kosa M, Huang D, Ragauskas AJ (2015) Bioconversion of oxygen-pretreated kraft lignin to microbial lipid with oleaginous Rhodococcus opacus DSM 1069. Green Chem 17:2784–2789. https://doi.org/10.1039/C5GC00422E

Xavier MCA, Coradini ALV, Deckmann AC, Franco TT (2017) Lipid production from hemicellulose hydrolysate and acetic acid by Lipomycesstarkeyi and the ability of yeast to metabolize inhibitors. Biochem Eng J 118:11–19. https://doi.org/10.1016/j.bej.2016.11.007

Xiong X, Xia Y, Qiao J (2021) Editorial: physiology, application, and bioengineering of Oleaginous microorganisms. Front Microbiol 12. https://doi.org/10.3389/FMICB.2021.650957

Yang Z, Edwards H, Xu P (2019) CRISPR-Cas12a/Cpf1-assisted precise, efficient and multiplexed genome-editing in Yarrowia lipolytica. Metab Eng Commun 10. https://doi.org/10.1016/J.MEC.2019.E00112

Yu Y, Xu Z, Chen S, Jin M (2020) Microbial lipid production from dilute acid and dilute alkali pretreated corn stover via Trichosporon dermatis. Bioresour Technol 295:122253. https://doi.org/10.1016/j.biortech.2019.122253

Zeid A, Zaher E, Allah K, Ahmed R (2019) Screening and optimization of some Egyptian soil-born fungi for lipids production as possible source for biofuel. Egypt J Exp Biol 1. https://doi.org/10.5455/egyjebb.20190709123823

Zhang Y, Dubé MA, McLean DD, Kates M (2003) Biodiesel production from waste cooking oil: 2. Economic assessment and sensitivity analysis. Bioresour Technol 90:229–240. https://doi.org/10.1016/S0960-8524(03)00150-0

Zhou CH, Xia X, Lin CX, Tong DS, Beltramini J (2011) Catalytic conversion of lignocellulosic biomass to fine chemicals and fuels. Chem Soc Rev 40:5588–5617. https://doi.org/10.1039/C1CS15124J

Chapter 7
Plant Based Biofuels: A Sustainable Energy Source

Deepshikha Datta, Dhriti Barman, Rajat Kanti Ghosal, Esha Mandal, and Bimal Das

Abstract Plant-based biofuels are renewable energy sources that are derived from organic matter, such as agricultural crops, forest residues, or algae and are considered as an alternative to traditional fossil fuels, as they are renewable and have a lower carbon footprint. The present work provides an overview of plant-based biofuels, including their production, potential, and challenges. The work further focuses on the different types of plant-based feedstocks used for biofuel production, such as sugarcane, corn, soybeans, and palm oil, and the different biofuel production processes, including fermentation and transesterification. The potential benefits of plant-based biofuels, such as reducing greenhouse gas emissions, enhancing energy security, and providing economic opportunities for farmers are elaborately discussed. The chapter concludes by discussing the current state of plant-based biofuel production, ongoing research, and future opportunities for sustainable and environmentally responsible biofuels. Overall, this book chapter provides a comprehensive overview of the plant-based biofuels landscape, which can inform and guide future policy and research efforts towards sustainable energy production. Ongoing research and development is focused on improving the efficiency and sustainability of plant-based biofuels, as well as developing new feedstocks and production processes. Finally, the work highlights the current state of research and development, and the potential for future advances in plant-based biofuel technology.

Keywords Plant-based biofuel · Feedstocks · Challenges · Policy · Economic opportunities

---

D. Datta
Department of Chemistry, Brainware University, Barasat, Kolkata, West Bengal 700125, India

D. Barman · R. K. Ghosal
Department of Biotechnology, Brainware University, Barasat, Kolkata, West Bengal 700125, India

E. Mandal
Department of Biotechnology, National Institute of Technology, Durgapur 713213, India

B. Das (✉)
Department of Chemical Engineering, National Institute of Technology, Durgapur 713213, India
e-mail: bimal.das@che.nitdgp.ac.in

M. Shah and D. Deka (eds.), *Emerging Sustainable Technologies for Biofuel Production*, Environmental Science and Engineering, https://doi.org/10.1007/978-3-031-52167-6_7

7.1 Introduction

Biofuels are liquid or gaseous fuels made from specific types of plant materials. It is one form of fuel that offers a lot of promise for our energy future since it is both renewable and ecologically benign. Some biofuels can offer renewable alternatives to fossil fuels, such as petrol. Biofuel is therefore sustainable. Typically, biofuels are made from plant components that cannot be consumed by people, such wood chips, grass, and maize stalks. Biomass Any organic (carbon-containing) material derived from living things, such as plants, is referred to as biomass in general. Cellulose, hemicellulose, and lignin are the three primary components that make up plant biomass. is another term for the plant materials that are used to produce biofuels, and it includes plants and plant wastes including grasses, maize stalks, and wood chips. Scientists may dissect and transform plant cells into sustainable fuels or chemicals by harvesting and processing biomass. In order to speed up this process, scientists are employing sophisticated chemistry to create biofuel from plants that are already living, rather than waiting a million years for nature to transform them into fossil fuels.

7.2 Carbon Footprint

The total quantity of carbon dioxide (CO_2) emissions produced by a person or another organization during all their activities (e.g., building, corporation, country, etc.) (Fig. 7.1). It comprises direct emissions from the production of energy used to power the consumption of products and services, as well as indirect emissions from the burning of fossil fuels in manufacturing, heating, and transportation. Moreover, the idea of a carbon footprint frequently considers other greenhouse gas emissions as well, including those from methane, nitrous oxide, or chlorofluorocarbons (CFCs).

The Canadian ecologist William Rees and the Swiss-born regional planner Mathis Wackernagel created the ecological footprint idea at the University of British Columbia in the early 1990s. It is related to and descended from the carbon footprint hypothesis. The ecological footprint is the total amount of land required to sustain a specific activity or population. It covers factors that have an impact on the environment, such water use and the space used for food production. On the other hand, a carbon footprint is frequently calculated in tonnes of CO_2 or CO_2 equivalent each year (Wiedmann and Minx 2008).

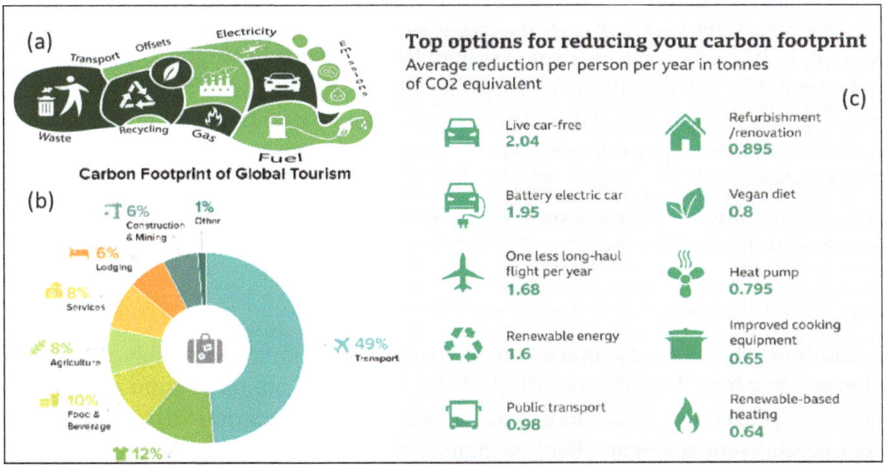

Fig. 7.1 Carbon footprinting (Wiedmann and Minx 2008); **a** renewable bio fuel source, **b** carbon footprint of global tourism, **c** top options for reducing your carbon footprint

7.3 Different Types of Plant Based Biofuels

Biofuel is a form of fuel made from biomass (that is from plants and animal material). Biofuels may be created from sources other than plants, such as industrial waste having a biological origin, and are quickly available, in contrast to fossil fuels, which require time to develop due to the slow geological process. Moreover, compared to fossil fuels, biofuel is a cleaner source of energy and emits less pollutants and greenhouse gases. Biodiesel and bio-alcohol are the two basic types of biofuel. Although biodiesel is produced from the oil already present in such plants, bio-alcohol is produced when microorganisms break down the starch in plants. The primary crops used to produce biofuel are listed below.

7.3.1 Agricultural Crops

a. *Corn*

The primary source of ethanol-based biofuel is thought to be corn. Sugar, which is the primary ingredient in ethanol, is abundant in corn. By a brewing procedure that is comparable to making beer, the sugar-rich maize is converted into ethanol. The kernels are first pulverized, and then they are combined with yeast and warm water. The use of kernels reduces the cost of fermenting compared to using a full maize plant. The mixture ferments as a result of the yeast, yielding ethanol. The resulting ethanol is then used by blending it with gasoline.

b. *Sugarcane*

In addition to maize, sugarcane is a significant source of ethanol, which is usually thought to be less expensive than gasoline. Brazil has made progress in recent years in lessening its reliance on fossil fuels and increasing the use of biofuel. Sugarcane planting has been promoted by the government among its farmers. Producing ethanol from sugarcane is approximately six times less expensive than doing it from corn. The harvesting of sugarcane, which entails burning the field, results in the release of significant volumes of greenhouse gases into the atmosphere despite the crop being grown with less pesticides.

c. *Soyabean*

In addition to being used to make tofu, tacos, and shampoo, soybeans may also be used to make gasoline. The majority of biodiesels in the US are made from soybean oil. Pure soybean biodiesel or a mixture of diesel and biodiesel can be used to power heavy commercial vehicles in an effective manner. In addition to being more ecologically friendly, soybean biodiesel generates more energy than maize ethanol. Up to 6 L of biodiesel may be made from a bushel of soybeans. Nevertheless, compared to sunflower seeds and canola, its oil content is lower.

d. *Palm oil*

The palm tree is used to produce the energy-efficient biofuel known as palm oil. Compared to petrol, this form of biofuel emits less carbon dioxide into the atmosphere, making it ecologically beneficial. It's interesting to note that diesel-powered automobiles can run on palm oil without needing to be converted. Along with being widespread near the shore, palm trees are also well-known for their fruits. One of the main contributors to the economy of Malaysia and Indonesia has been palm oil. Yet, it is thought to be environmentally harmful to remove land in order to cultivate the crop (Lopes and Steidle Neto 2011).

7.3.2 Forest Residues

A various kind of trees, including branches, leaves, bark, and trunk pieces, make up forest leftovers. They mostly consist of the remains of trees that have been removed. The supply of forest might be an issue since, in the case of a mega type reactor, the production of remaining forest products would not come from the same location or continually. Despite its limited availability, it should produce bio crude that has the same consistency as woody biomass. However, due to the varied components of the residues, the pre-treatment required for forest residues can be slightly different from that required for woody biomass. Therefore, certain parts need to be chipped, powdered, and ground.

However, bio-crude produced from forest waste should be similar to bio-crude produced from woody biomass. In general, bark would make up a sizable fraction of the total weight of forest trash, and the water content may be slightly higher than the

Table 7.1 Results for various feed stock sources (Letcher 2022)

Feedstock	C (%)	H (%)	N (%)	O (%)	S (%)	HHV (MJ kg^{-1})	References
Microalgae (*Nannochloropsis* sp.)	43.3	6	6.4	25.1	0.5	NR	Liu et al. (2011)
Macro algae	31.6	4.85	0.9	34.2	2.44	17.6	Bardhan et al. (2022)
Woody biomass (Aspen wood)	50.4	6.19	0.19	43.2	NR	NR	Lopes and Steidle Neto (2011)
Straw	37.1	5.6	0.5	44.7	0.3	17.5	Lopes and Steidle Neto (2011)
Manure	49.4	6.46	3.33	39.8	1.05	13.3	Letcher et al. (2022)
Sewage sludge	23–40	3.8–5.9	2.5–7.9	18.8–23.5	0.8–1	NR	(Wiedmann and Minx 2008)

woody biomass. Biocrude yields from forest waste were reported to be 22% in the literature, less than straw and woody biomass. Some illustrative analytical results for various feedstock sources are shown in Table 7.1 (Letcher 2022).

7.3.3 Algae

Aquatic organisms called algae use carbon dioxide (CO_2) and light to build biomass. Macroalgae and microalgae are the two types of algae. Macroalgae, which are measured in inches, are the massive, multicellular algae that are usually seen growing in ponds. These larger algae can develop in a variety of ways. The largest multicellular algae are referred to as seaweed; an example is the huge kelp plant, which may reach lengths of more than 100 feet. Microalgae, on the other hand, are unicellular, tiny algae that are measured in micrometres and frequently grow in suspension in water.

Microalgae have long been regarded as potential feasible sources for the production of biofuel due to their relatively high oil content and fast biomass synthesis. Microalgae grow far more fast than terrestrial crops. Algal mass culture may be done on non-arable fields using saline water that is not drinkable and waste water. As a result, more academics, entrepreneurs, and the general public are becoming interested in employing micro algae as a replacement for bio diesel, bio fuel and feed stock.

a. *Oil from Algae*

Fig. 7.2 Oil production from algae—**a** Photobioreactors; **b, c** Closed-loop system; **d** Open pond system

The process employed in an oil press is quite similar to one of the techniques to extract oil from algae. Around 75% of the total oil produced by the algae plant is obtained using this approach, which is the easiest and most popular way to extract oil from algae.

The hexane solvent approach is another popular technique. This procedure can produce up to 95% of the available oil from algae when used in conjunction with the oil press technique. It makes use of a two-step procedure. Using an oil press is the first option. The remaining algae is then combined with hexane, washed, and filtered to get rid of any remaining chemicals in the oil, continuing the process after halting there (Fig. 7.2).

The supercritical fluid approach, which is applied less often, may extract up to 100% of the algae's available oil. To alter carbon dioxide's composition into a liquid and a gas, pressure and heat are used. After that, it is combined with the algae, which entirely transforms into oil. While it can produce 100% of the available oil, this alternative isn't very common due to the large availability of algae and the additional effort and equipment needed (Holechek et al. 2022). In this process the remaining algae is then combined with hexane, washed, and filtered to get rid of any remaining chemicals in the oil, continuing the process after halting there.

b. *Algae Production for Biodiesel*

The techniques used to encourage algae growth in a certain way to produce the maximum oil are more varied than the extraction procedures. Growing algae for bio diesel has a very varied process and approach compared to nearly universal extraction methods. There are three main approaches of producing algae, and bio diesel producers have made great efforts to modify and optimise these approaches.

- *Open-Pond Agriculture*:

Open-pond cultivation is the most organic method of cultivating algae for the manufacture of biodiesel and one of the simplest procedures to comprehend. This strategy, as its name suggests, involves growing algae on open ponds in an effort to increase output, especially in regions of the world that are extremely warm and sunny. While being the most straightforward method of manufacture, it has significant disadvantages, including a relatively high risk of contamination. This method can effectively boost algae production, however controlling water temperature can be exceedingly challenging. Another uncontrollable component, the weather, is more of a factor in this strategy than it is in others (Holechek et al. 2022).

- *Vertical Growth*:

A vertical growth system, often known as a closed-loop production system, is another way to cultivate algae. This method was developed by biofuel businesses who wanted to create algae more quickly and effectively than they could with pond growth. Algae is vertically grown in clear plastic bags that are stacked and coated for weather protection. These sacks provide exposure to sunlight from various angles. The additional light is not insignificant because the clear plastic bag provides sufficient exposure to boost output rates. It goes without saying that there will be more oil to extract the more algae are produced. Moreover, the vertical growth approach separates algae from pollution, in contrast to the open pond method that exposes algae to it (Holechek et al. 2022).

- *Closed-Tank Bioreactor Plants*:

Closed-tank bioreactor facilities, which grow algae within to improve already high oil production levels, are the third type of extraction used by biodiesel producers. Large, rounded barrels are used in indoor plant construction that may almost perfectly support the growth of algae. In these barrels, algae may be managed to develop to its highest potential, even to the point of daily harvesting. It should come as no surprise that this approach produces a lot of algae and oil for biodiesel. In close proximity to energy plants, closed bioreactor facilities can be created to recycle excess carbon dioxide rather than contaminating the air.

The closed container and closed-pond procedures are still being refined by producers of biodiesel, and some have even created a version called as fermentation. With this method, algae that "eats" sugar is grown in confined containers to promote growth. Because it gives producers total environmental control, fermentation is appealing to them. Its independence from weather and other comparable environmental factors is another benefit. Nonetheless, this technique has scientists considering environmentally friendly ways to get enough sugar to promote algae production (Holechek et al. 2022).

7.4 Production Potential of Various Sources for Plant-Based Biofuels

The transition from fossil fuels to biofuels may have a number of benefits. In contrast to fossil fuels, which are made from limited resources, biofuels are produced from renewable feedstocks. Therefore, their production and consumption may theoretically go on forever.

Cornstarch and cane sugar have been the main ingredients used to produce ethanol for the past 20 years. In 2014, cornstarch was used to produce 14 billion gallons of ethanol in the United States (Fig. 7.3). Since, the remaining corn crop is used for human food and animal feed. Currently, only approximately 40% of the corn grown in the USA is used to produce ethanol, and it is unlikely that this percentage will increase in the future. Brazil generated 7.3 billion gallons of ethanol from cane sugar in 2014. Together, Brazil and the United States generate more than 90% of the ethanol consumed worldwide. Brazil and USA are considered the greatest producer of biofuel (mainly bioethanol) from sugar cane in the world (Singhania et al. 2008; Kumar et al. 2015).

About 4 billion gallons (15.1 billion L) of production were produced in 2005, and over 14 billion gallons (53 billion L) were produced in 2014 (Singhania et al. 2008). There are several concerns about the unfavorable effects and sustainability of Brazil's rapidly expanding sugarcane-based ethanol industry. The removal of lead compounds

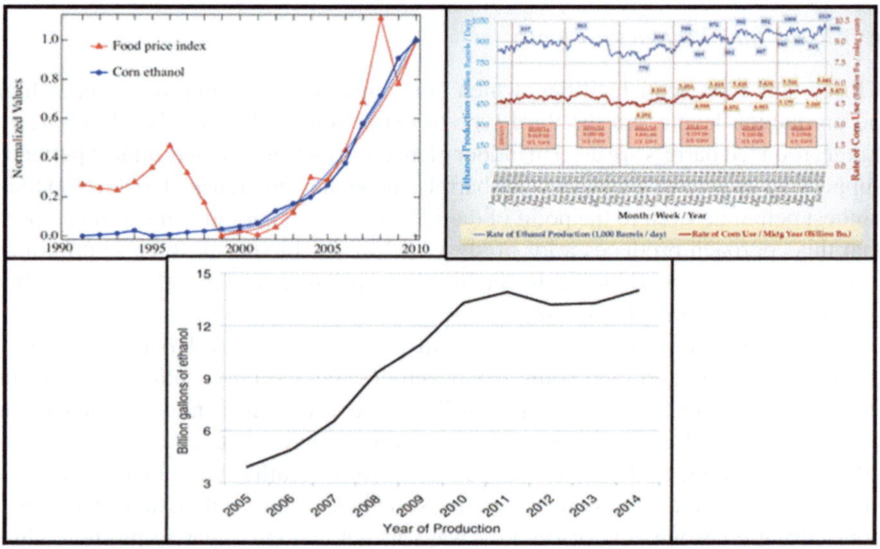

Fig. 7.3 Volumes of ethanol produced from corn starch in the USA over the past nine years—**a** Corn ethanol and food price index according to years 1990–2010 **b** Ethanol production according to months, weeks and years **c** Gallons of ethanol produced according to years 2005–2014 (Singhania et al. 2008)

from petrol and the decline in noxious emissions are benefits. As sugarcane ethanol is a sustainable fuel and very minimal amounts of fossil fuels are needed to produce it, there is also a reduction in CO_2 emissions. These beneficial effects can be seen most clearly in the improvement of the air quality in urban areas, as well as in rural areas where mechanized green cane harvesting is being adopted to replace sugarcane burning.

Negative consequences could include the degradation or damage to soils due to the use of chemicals and soil decarbonization, destruction or damage of high-biodiversity areas, deforestation, contamination or depletion of water resources, a reduction in food security due to competition between the production of food and fuel, and deterioration of working conditions in the fields.

Recent attention has been focused on biodiesel, a renewable fuel that is also pollution-free. Because the carbon in exhaust from vehicles was recently repaired from atmospheric carbon, it is carbon neutral. The oils and fats that can be used to produce biodiesel include virgin vegetable oils like soybean, canola, and camelina, old cooking oils, and animal fats like lard or tallow. The global biodiesel industry has expanded significantly over the past few years 2, but since 2008 there has been a decline due to the global economic downturn. Over the years, Europe and then the USA have produced the most biodiesel. 6 billion gallons of production were produced globally in 2012 (Fig. 7.4) (Singhania et al. 2008).

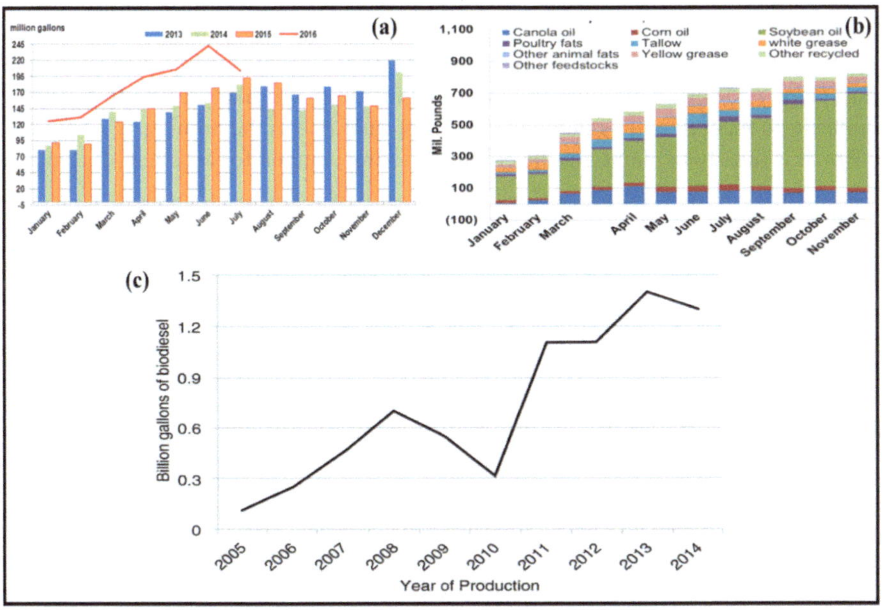

Fig. 7.4 Volumes of biodiesel produced in the USA for the past nine years using all sources of oil—**a** Million gallons of biodiesel produced from years 2013–2016 and months January–December **b** Feedstocks used according to months in biodiesel production **c** Gallons of biodiesel production from 2005 to 2014 (Kumar et al. 2015)

Production climbed from 112 million gallons in 2005 to nearly 1.3 billion gallons in 2014. During the economic slump from 2008 to 2010, there was a significant decline in production (Singhania et al. 2008). According to the assumption that if all 600 million cars in the globe ran on bioethanol, then a total of 1,059 billion L of bioethanol would be required annually. We can calculate how much biodiesel would be needed each year for the entire planet if the 600 million identical vehicles already on the road were biodiesel vehicles. We estimate that each car uses 17 km/L of biodiesel on average and that each person travels 15,000 km annually. 600 million of these vehicles would use 530 billion L of biodiesel annually. Currently, the world generates less biodiesel than is required by humans per year (based on data from 2008; EMO 2011). Table 7.2 shows that around 136 Mtonnes (160 billion L) of vegetable oil were produced globally in 2009, or roughly 1.176 L/kg. There wouldn't be enough biodiesel to run all the cars on the earth, even if all the vegetable oil produced were utilised for that purpose. Palm oil, soybean, rapeseed, and sunflower oil made up 86% of the world's total production of vegetative oils. Only palm oil made up around 35% of the total, and palm oil is only grown in tropical regions. This highlights the necessity for humans to create new oil crops and breed existing ones (Liu et al. 2011).

The USA and Brazil being the leading countries to produce millions of gallons of biofuel worldwide. The table depicts data from the year 2011–2013 (Table 7.3). Table 7.4 shows the region along with the production rate of biodiesel worldwide according to the feed stocks used to produce them.

According to data from 2016, the world produced 100.2 billion L of bioethanol. The global production as well as consumption of bioethanol are likely to rise to about 134.5 billion L by 2024, with annual production increasing steadily. The above pie-chart (Fig. 7.5) shows that two-thirds of this increase will likely come from Brazil, primarily to meet domestic demand. The continual rise in the number of FFVs sold in Brazil is mostly to blame for the country's rising demand for bioethanol. Hence, it

Table 7.2 The essential oil crops and their oil extraction in 2009 with plants (Liu et al. 2011)

Oil crops	Production (Million MTs)	Percentage
Soybeans	213,976,284	33.72
Oil Palm (Fruit)	172,840,720	27.24
Cotton seed	67,803,392	10.69
Coconuts	54,254,232	8.55
Rapeseed	48,907,026	7.71
Groundnuts (in shell)	37,228,389	5.87
Sunflower seed	30,595,462	4.82
Sesame seed	3,312,986	0.52
Linseed	2,900,587	0.46
Castor beans	1,392,592	0.22
Safflower seed	717,778	0.11
Mustard seed	592,026	0.09
Total	634,521,474	100

Table 7.3 World bio-fuel production (Kumar et al. 2015)

Continent	Millions of gallons		
	2011	2012	2013
North and central America	14,401	13,768	13,300
South America	5,772	5,800	5,920
Brazil	5,573	5,577	6,267
Europe	1,168	1,139	1,371
Asia	890	952	545
China	555	555	696
Canada	462	449	523
Rest of World	126	113	727

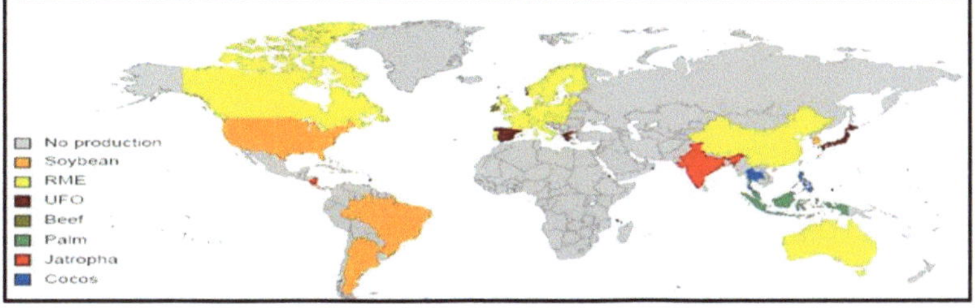

Table 7.4 World biodiesel production (Kumar et al. 2015)

Region	Million litres	Feed stocks
EU	9848	Rapeseed, soy oil, palm and tallow
USA	1682	Soybean , tallow, canola, palm
Brazil	1386	Soybean, tallow, other vegetable oils
Argentina	1250	Soybean
China	191	Waste vegetable oils
Asia	1687	Palm, soybean, waste vegetable oils
Rest of world	395	Tallow, rapeseed, palm, soybean, waste vegetable oils
World	16436	

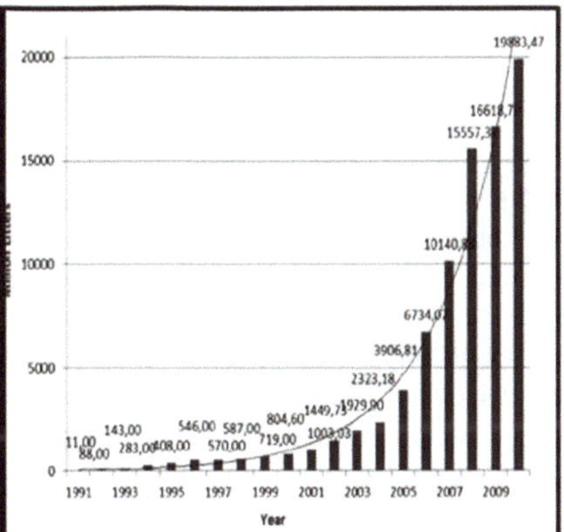

is expected that the USA and Brazil will continue to be the two largest producers of bioethanol, followed by the European Union and China. Around 40% of the world's bioethanol output comes from sugar cane and sugar beetroot, while approximately 60% comes from feedstocks containing starch. Grain (mainly wheat) and sugar beetroot are the most viable renewable raw sources for bioethanol manufacturing in Europe. Moreover, bioethanol was produced in France from wine surplus (Bušić et al. 2018).

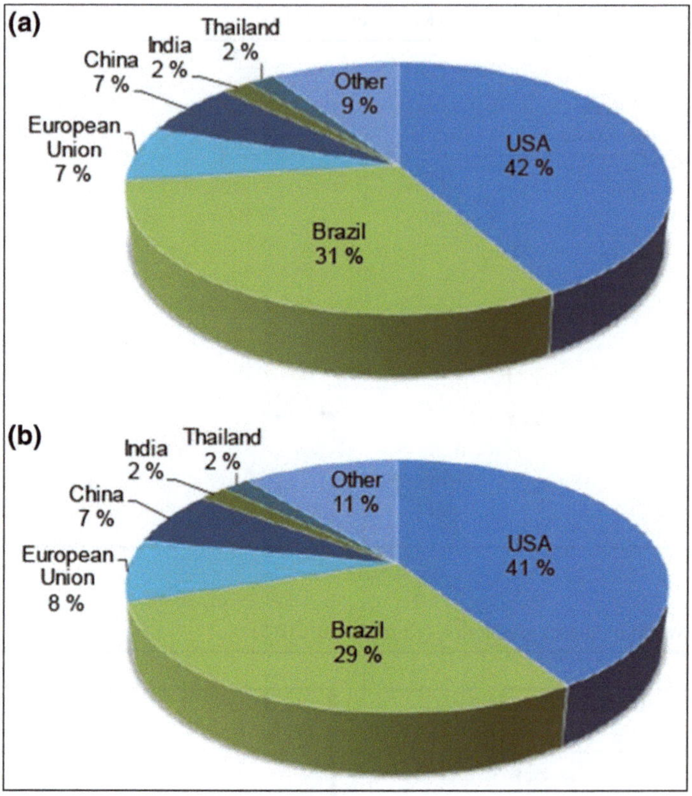

Fig. 7.5 Predictions for global bioethanol production and consumption **a** at present **b** by 2024 (Bušić et al. 2018)

Depending on the type of feedstock used, the cost of raw materials can account for between 40 and 75% of the entire cost of producing bioethanol. In Brazil, the cost of producing bioethanol from sugar cane is between \$0.20 and \$0.30 USD/L. Bioethanol made from sugar beetroot and maize had the lowest production costs in the United States and the European Union, at 0.30 and 0.53 USD/L, respectively. Depending on the price of the feedstock, the costs for wheat, sweet sorghum, or cassava in China range from 0.28 to 0.46 USD/L. In India, the cost of producing bioethanol from raw materials including sugar is approximately 0.44 USD/L, while the cost from raw materials containing lignocellulose ranges from 0.80 to 1.20 USD/L, depending on the type of feedstock used (Bušić et al. 2018).

7.5 Production techniques of Plant based biofuels

(a) *Fermentation*—It is a process in which microbes use chemical pathways to change carbohydrates into a new product (Taveira et al. 2021). The fermentation process is considered to be one of the most crucial methods in Biofuel Production. Different process configurations, such as SSF (Simultaneous Saccharification and Fermentation), SHF (Separate Hydrolysis and Fermentation), SSCF (Simultaneous Saccharification and Co-fermentation), and CBP (Consolidated Bio-Processing (CBP) can be used to ferment hydrolyzed biomass into bioethanol and value-added products/chemicals (Fig. 7.6) (Chandel et al. 2018).

Various types of fermentation processes take place during biofuel production. For batch fermentation, the Melle-Boinot technique is typically used to produce bioethanol. It involves the broth (pH adjustment, 14–22% by mass), sterilizing it, and then fermenting the yeast. The yeast is reused for the second fermentation in order to achieve higher cell densities, and the liquid portion of the fermented broth moves on to the ethanol separation stage, which is frequently distillation. The yeast cells are removed from the broth, which is subsequently distilled, during the fed-batch process, which calls for low beginning substrate concentrations. Separated yeast cells are often employed for a fresh fed-batch phase of bioethanol production after the bioreactor and broth preparation operations. Due to its ability to produce the highest volumetric productivity per unit of bioprocess, this is the most extensively applied manufacturing

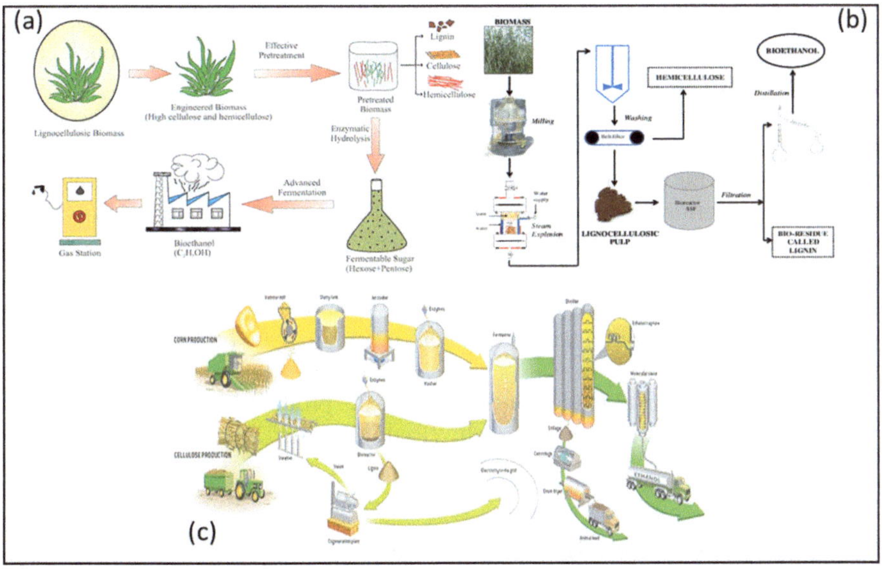

Fig. 7.6 Processes for making bioethanol from lignocellulosic biomass—**a** Production of bioethanol from biomass by pretreatment, enzymatic hydrolysis and advanced fermentation **b** Production of biodiesel from biomass by washing, distillation and filtration **c** Biodiesel production from corn and cellulose production (Lamichhane et al. 2021)

process in Brazil for the generation of bioethanol. The improvement of the feeding procedure is essential for raising the yield and productivity of the bioprocess in this mode of operation. In the repeated (or multiple) batch fermentation, the use of flocculating yeast strain plays the key role. The clarified broth is withdrawn from the bioreactor when the initial batch procedure is complete and the yeast cells settle (flocculate) on the bottom. The bioreactor is then filled with an equivalent volume of fresh broth for the subsequent batch, which leads to high cell concentrations and lessened ethanol inhibition. As a result of the buildup of yeast inhibitory chemicals in the broth, these batches can be repeated until the activity and viability of the yeast cells are lost, at which point fresh inoculum must be prepared for system reinoculation (Bušić et al. 2018).

(b) *Transesterification*—Triacyl glycerides from a variety of feedstocks, including non-edible vegetable oils, oil seeds, animal fats or tallow, used cooking oil, and microbial lipids or single cell oil (from algae, filamentous fungi, oleaginous yeast, and bacteria), are transformed into biodiesel, which is made from fatty acid methyl esters in the presence of alcohols like methanol or ethanol, during the process known as transesterification. This process is frequently used to reduce the consistency of feedstock or non-edible oil and increase its compatibility with fuel to a level similar to diesel and oil derived from fossil fuel (Bardhan et al. 2022).

Trans-esterification involves two methods: Catalytic and Non-catalytic. It is generally accepted that catalytic trans-esterification has two drawbacks, the first of which is that the processes are time-consuming and require separating the biodiesel from a mixture of oil, alcohol, catalyst, and saponified contaminants. However, since no catalyst is required throughout the supercritical trans-esterification process, which hinders the production of soap or saponification, purifying biodiesel is much easier (Baskar et al. 2019).

Preparing the catalyst and combining it with alcohol is the first step. The second step is the transesterification reaction, which occurs when the catalyst/alcohol mixture reacts with the fatty acid (triglycerides). The ideal temperature range for the transesterification reaction is between 40 and 60 °C, and as the temperature rises, the reaction rate rises. The third step involves separating glycerol from biodiesel, and the fourth step involves adding water to both the biodiesel and glycerol in order to purify both, particularly any glycerol that may have remained in the biodiesel. For the creation of neutralized glycerol, the acid must be applied to glycerol (Fig. 7.7) (Suali and Sarbatly 2012).

Compared to biodiesel made conventionally, the produced biodiesel is nontoxic, low in power, and environmental friendly (Reddy et al. 2022).

7.6 Benefits of Plant Based Biofuels

Several challenges posed by politics, society and industry have postured to represent challenges, thus influencing the competitiveness of plant-based biofuels to be an eventual source of fuel in the coming years. Global demand of food is projected

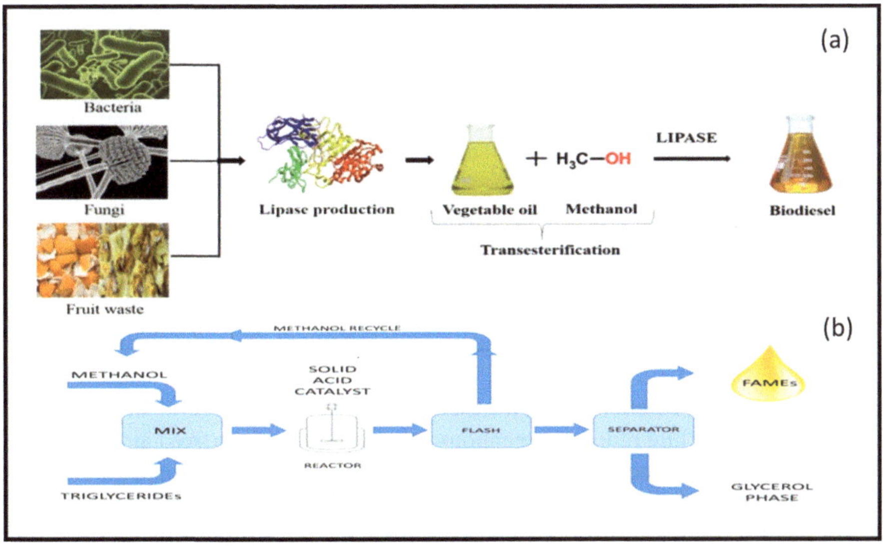

Fig. 7.7 Different steps involved in biodiesel production—**a** Biodiesel from bacteria, fungi and fruit waste **b** Separation of glycerol and FAMEs from the mixture of methanol and triglycerides for biodiesel production (Bhan and Singh 2020)

to rise twice by the mid of 21st century, due to population boosts, their increasing protein consumption and growing per capita income (Tilman et al. 2011). Declining water table and extreme weather conditions have also posed a threat to the agricultural sector. Although the automobile industry has started to switch to Electric vehicles (EVs), a massive amount of energy production for electricity and petrochemicals, still relies on coal and petroleum.

Plants as biofuels cater several benefits, including maintaining a neutral carbon balance by reducing the emissions of greenhouse gases, enhancing food and energy resilience, and imperative to become a platform for both economy and employment. The worldwide energy demand is currently fulfilled by burning mostly oil, natural gas, and coal and the trend of consuming these non-renewable fossil fuels is increasing day by day.

Also, the present release of greenhouse gases from the burning of fossil fuels, primarily carbon dioxide, has had a significant impact on the global ecosystem. Consequently, the biofuels industry has grown quickly as a result of the need to find low-carbon, renewable fuels as a crucial part of the strategy to reduce greenhouse gas emissions., including the plant-based biofuel industry (Hoefnagels et al. 2010). The biofuel production not only takes into account the environmental issues but also cares for the economy of the nation, as such, it contributes to the environment, energy security and economy as follows:

1. **Renewable**: Plant-based biofuels are derived from renewable sources, such as agricultural crops and waste materials. This means that they can be produced

indefinitely without depleting finite resources, unlike traditional fossil fuels (Kralova and Sjöblom 2010).

2. **Reduced carbon emissions**: Burning plant-based biofuels produces fewer greenhouse gas emissions than fossil fuels. Biofuels can also help reduce overall carbon emissions by absorbing CO_2 from the atmosphere during plant growth (Hill et al. 2006).
3. **Energy security**: By reducing dependence on foreign oil, plant-based biofuels can enhance energy security and reduce the impact of price volatility in the fossil fuel market (Hill et al. 2006).
4. **Job creation**: Biofuel production requires significant labor inputs, creating employment opportunities in rural areas where agriculture is a major industry (Doshi et al. 2016).
5. **Environmental benefits**: Plant-based biofuels offer a range of environmental benefits, including reduced greenhouse gas emissions, improved air quality, and reduced dependence on fossil fuels. These benefits can help mitigate the impacts of climate change and improve the overall health of our planet (Hill et al. 2006).
6. **Economic benefits**: Biofuels can provide economic benefits to farmers, rural communities, and countries that produce them. By creating a new market for crops, biofuels can provide additional revenue streams for farmers and create jobs in rural areas. Biofuel production can also reduce dependence on foreign oil, which can provide economic benefits at the national level (Hill et al. 2006).
7. **Energy security**: Biofuels can enhance energy security by diversifying our energy sources and reducing our reliance on foreign oil. This can help to stabilize energy prices and reduce the risk of supply disruptions (Shogren et al. 2019).
8. **Health benefits**: By reducing emissions from transportation, biofuels can improve air quality and reduce the risk of respiratory illnesses. Additionally, because biofuels are renewable and sustainable, they can reduce exposure to harmful pollutants associated with fossil fuel extraction and production (Doshi et al. 2016).

Advocating to plant-based biofuels ensures reduction in agri-waste, decrease in reliance on fossil fuels and regulating GHG emissions (Lange et al. 2021; Turan et al. 2022). There has been a significant improvement in the understanding of the energetic life-cycle performance of plant-based biofuels in context of GHGs, despite the fact that it is difficult to determine how well plant biofuels perform in terms of greenhouse gas emissions and the use of fossil fuels due to the numerous, sometimes not known parameters and influences, as well as methodological concerns (Hoefnagels et al. 2010). Owing to the benefits in the genre of energy security, manufacturing plant-based biofuels require less non-renewable energy, when compared with petroleum-based materials. They have also been proven to be beneficial for decreasing climate change and improving energy security by supplying renewable and sustainable energy sources (Shogren et al. 2019).

They also play an important part in sustaining and boosting the economy. Converting agricultural commodities to chemicals or other fuels usually take place in rural areas, so as to minimize transportation costs and also to enhance job opportunities. Growth in these industries further lead to development in these areas, thus encouraging rural investment (Shogren et al. 2019).

7.7 Future Prospects

Plant based biofuels has been one of the giant steps in context of sustainable development. Research has been aiming at the development of hybrid crops with higher fermentable starch content. Furthermore, having a favourable energy density, and being easily portable, these kinds of biofuels can either serve as a bridge fuel during the switchover from fossil fuels to more advanced renewable energy sources in the future. In fact, yield of plant-based biofuels can be enhanced by taking giant steps in the agro-industry, by development of highly fermentable hybrid strains, improvising enzyme efficiencies and enhancing separation and fermentation technology, and reducing the cost and duration (Ho et al. 2014). Metabolic engineering comes into play over here, which will enable plants/microorganisms to work like efficient cell factories, thereby increasing the production of biofuels (Hasheminejad et al. 2011; Majidian et al. 2018). The potential for increasing biofuel production may be increased by using cutting-edge techniques like high-throughput genome sequencing, very sensitive transcriptomics and proteomics, various metabolomic technologies, and fluxomic (Medema et al. 2014; Alonso-Gutierrez et al. 2015; Tang et al. 2013). Recently, the usage of bio-electrochemical systems has emerged to be a technology, efficient for biofuel production, which might be used in context of plant-based biofuels (Ahmad et al. 2022). Anaerobic digestion in integration of bio-electrochemical systems, may also contribute to increase in plant-based biofuel production, in the near future, thereby increasing overall performance (Ahmad et al. 2022). Synthetic promoters, genome editing, chromosomal rearrangement, and synthetic plastid genomics are also a few approaches that can be used to increase biofuel.

7.8 Conclusion

Plant-based biofuels have emerged as a promising alternative to traditional fossil fuels, offering numerous potential benefits such as reduced carbon footprint, enhanced energy security, and economic opportunities for farmers. This book chapter has provided an extensive overview of plant-based biofuels, including their production, feedstocks, and different biofuel production processes. Ongoing research and development efforts are aimed at improving the efficiency and sustainability of plant-based biofuels, and exploring new feedstocks and production methods. The current state of plant-based biofuel production and future opportunities for sustainable and environmentally responsible biofuels have been discussed, providing valuable insights that can inform and guide future policy and research efforts towards sustainable energy production.

References

Ahmad S, Iqbal K, Kothari R, Singh HM, Sari A, Tyagi VV (2022) A critical overview of upstream cultivation and downstream processing of algae-based biofuels: opportunity, technological barriers and future perspective. J Biotechnol 351:74–98

Alonso-Gutierrez J, Kim EM, Batth TS, Cho N, Hu Q, Chan LJG, Petzold CJ, Hillson NJ, Adams PD, Keasling JD, Martin HG (2015) Principal component analysis of proteomics (PCAP) as a tool to direct metabolic engineering. Metabolic Eng 28:123–133

Bardhan P, Deka A, Bhattacharya SS, Mandal M, Kataki R (2022) Economical aspect in biomass to biofuel production. In: Value-chain of biofuels. Elsevier, pp 395–427

Baskar G, Kalavathy G, Aiswarya R, Selvakumari IA (2019) Advances in bio-oil extraction from nonedible oil seeds and algal biomass. In: Advances in eco-fuels for a sustainable environment. Woodhead Publishing, pp 187–210

Bhan C, Singh J (2020) Role of microbial lipases in transesterification process for biodiesel production. Environ Sustain 3:257–266

Bušić A, Marđetko N, Kundas S, Morzak G, Belskaya H, IvančićŠantek M, Komes D, Novak S, Šantek B (2018) Bioethanol production from renewable raw materials and its separation and purification: a review. Food Technol Biotechnol 56(3):289–311

Chandel AK, Garlapati VK, Singh AK, Antunes FAF, da Silva SS (2018) The path forward for lignocellulose biorefineries: bottlenecks, solutions, and perspective on commercialization. Bioresour Technol 264:370–381

Doshi A, Pascoe S, Coglan L, Rainey TJ (2016) Economic and policy issues in the production of algae-based biofuels: a review. Renew Sustain Energy Rev 64:329–337

Hasheminejad M, Tabatabaei M, Mansourpanah Y, Javani A (2011) Upstream and downstream strategies to economize biodiesel production. Bioresour Technol 102(2):461–468

Hill J, Nelson E, Tilman D, Polasky S, Tiffany D (2006) Environmental, economic, and energetic costs and benefits of biodiesel and ethanol biofuels. Proc Natl Acad Sci 103(30):11206–11210

Ho SH, Ye X, Hasunuma T, Chang JS, Kondo A (2014) Perspectives on engineering strategies for improving biofuel production from microalgae—a critical review. Biotechnol Adv 32(8):1448–1459

Hoefnagels R, Smeets E, Faaij A (2010) Greenhouse gas footprints of different biofuel production systems. Renew Sustain Energy Rev 14(7):1661–1694

Holechek JL, Geli HM, Sawalhah MN, Valdez R (2022) A global assessment: can renewable energy replace fossil fuels by 2050? Sustainability 14(8):4792

Kralova I, Sjöblom J (2010) Biofuels–renewable energy sources: a review. J Dispers Sci Technol 31(3):409–425

Kumar S, Manukonda S, Satyanarayana PV, Ramanjaneyulu AV (2015) Agricultural crops as potential source of bio-fuel production. Int J Econ Plants 2(2):83–88

Lamichhane G, Acharya A, Poudel DK, Aryal B, Gyawali N, Niraula P, Phuyal SR, Budhathoki P, Bk G, Parajuli N (2021) Recent advances in bioethanol production from lignocellulosic biomass. Int J Green Energy 18(7):731–744

Lange L, Connor KO, Arason S, Bundgård-Jørgensen U, Canalis A, Carrez D, Gallagher J, Gøtke N, Huyghe C, Jarry B, Llorente P (2021) Developing a sustainable and circular bio-based economy in EU: by partnering across sectors, upscaling and using new knowledge faster, and for the

benefit of climate, environment & biodiversity, and people & business. Front Bioeng Biotechnol 1456
Letcher T (2022) Comprehensive renewable energy
Liu C, Sun C, Bernardes MAS (2011) The future crops for biofuels. Econ Eff Biofuel Prod 25:38
Lopes DC, Steidle Neto AJ (2011) Potential crops for biodiesel production in Brazil: a review. World J Agric Sci 7(2):206–217
Majidian P, Tabatabaei M, Zeinolabedini M, Naghshbandi MP, Chisti Y (2018) Metabolic engineering of microorganisms for biofuel production. Renew Sustain Energy Rev 82:3863–3885
Medema MH, Cimermancic P, Sali A, Takano E, Fischbach MA (2014) A systematic computational analysis of biosynthetic gene cluster evolution: lessons for engineering biosynthesis. PLoS Comput Biol 10(12):e1004016
Reddy CN, Vineetha Y, Priyanka A, Shalini A, Mishra B, Rajasri Y, Swapna V (2022) Techno-economic and environmental impact analysis of biofuels produced from microalgal biomass. Biofuels Bioenergy 687–712
Shogren R, Wood D, Orts W, Glenn G (2019) Plant-based materials and transitioning to a circular economy. Sustain Prod Consum 19:194–215
Singhania RR, Parameswaran B, Pandey A (2008) 1 Plant-based biofuels. In: Handbook of plant-based biofuels, p 1
Suali E, Sarbatly R (2012) Conversion of microalgae to biofuel. Renew Sustain Energy Rev 16(6):4316–4342
Tang J, Chen Q, Xu L, Zhang S, Feng L, Cheng L, Xu H, Liu Z, Peng R (2013) Graphene oxide–silver nanocomposite as a highly effective antibacterial agent with species-specific mechanisms. ACS Appl Mater Interfaces 5(9):3867–3874
Taveira IC, Nogueira KMV, Oliveira DLGD, Silva RDN (2021) Fermentation: humanity's oldest biotechnological tool. Front Young Minds 9:1–7
Tilman D, Balzer C, Hill J, Befort BL (2011) Global food demand and the sustainable intensification of agriculture. Proc Natl Acad Sci 108(50):20260–20264
Turan V, Aydın S, Sönmez O (2022) Production, cost analysis, and marketing of bioorganic liquid fertilizers and plant nutrition enhancers. In: Industrial microbiology based entrepreneurship: making money from microbes. Springer Nature, Singapore, pp 193–198
Wiedmann T, Minx J (2008) A definition of 'carbon footprint.' Ecol Econ Res Trends 1(2008):1–11

Chapter 8
Plant Based Biofuels: Sustainable Solution to Fuel Industry

Sanjukta Vidyant, Pooja Sharma, Himanshi Chaudhary, and Shubha Dwivedi

Abstract The need for the development of sustainable fuel alternatives have been boosted by the rapid rise in the depletion of fossils fuels, environment degradation, and industrialization. The global development patterns for biofuel technology have been developing in recent decades, Biofuels, which are created from renewable, organic (carbon containing) elements like plant matter and animal waste are possible replacements for fossils fuels. These substances together referred to as biomass, include agricultural products and trash, algae, special energy crops, and forestry byproducts. Sugar crops like sugarcane, starch crops like maize and sorghum, oilseed crops like soybean and canola and cellulose plants like corn stalks, grasses and leftover wood chips are typical agricultural plants that are utilized as biomass. In this we will discussed that how fossil fuels have long been replaced by biofuels and biofuel crops. Instead of the extremely slow geological processes involved in the creation of fossil fuels, biofuel is a low carbon fuel that is made from biomass.

Keywords Biofuels · Biomass · Fossil fuels · Environmental degradation · Renewable energy

8.1 Introduction

Due to an upsurge in gasoline prices, the quick depletion of fossil fuels, environmental damage caused by those fuels, and changes in the global climate, biofuel has attracted a lot of attention. Fossil fuel emissions are a significant factor in the rising levels of CO2 in the atmosphere that cause global warming (Allakhverdiev et al. 2009; Razzak et al. 2013; Voloshin et al. 2015). The effective and sustainable development of renewable energy sources is of global interest. The creation

S. Vidyant · P. Sharma · S. Dwivedi (✉)
Department of Biotechnology, IIMT University, Uttar Pradesh, Meerut-250001, India
e-mail: shubhadwivedi5@gmail.com

H. Chaudhary
Department of Microbiology, IIMT University, Uttar Pradesh, Meerut-250001, India

M. Shah and D. Deka (eds.), *Emerging Sustainable Technologies for Biofuel Production*, Environmental Science and Engineering, https://doi.org/10.1007/978-3-031-52167-6_8

of novel technologies and activities is necessary for a sustainable transition from nonrenewable to renewable resources (Dragone et al. 2010). One of the most prevalent biopolymers in the world, cellulose is a possible feedstock for the production of biofuels, which might replace the usage of fossil fuels and lower greenhouse gas emissions (Levin et al. 2006). After coal, natural gas, and oil, biomass is a primary resource, especially in developing nations. It is widely recognized, that the main biofuels are biodiesel, triglycerides, fatty acids, lipids, carbohydrates, ethanol, alcohols, cellulose, or the biomass of living things. Several species of algae, bacteria, or yeast can produce these fuels (Poudyal et al. 2015; Slade and Bauen 2007; histi 2007; Carlsson et al. 2007). The use of various feedstock's such as energy crops (such as miscanthus and sorghum), agricultural residues (such as straw, biogases, nuts, and rice husks), forestry residues (such as wood chips and bark), solid wastes like sewage and municipal waste, and algae has resulted in a significant increase in the consumption of biomass (Azapagic 2014; Demirbas 2009; Heiman 2016). The principal chemical building block for the synthesis of bioethanol and biomethanol is sugar (Dias et al. 2009). More biomass would be utilized, which would lessen reliance on nonrenewable energy sources while also promoting a sustainable and carbon–neutral energy supply. Biofuels have become an increasingly important source of renewable energy, and among them, plant-based biofuels have gained significant attention. Raw materials for the manufacture of biofuel can be obtained from a variety of sources, including waste fossil fuels, energy or crops grown in agriculture, etc. (Chum et al. 2012; Yuan et al. 2008). There is a significant growth in energy demand nowadays due to the development of cutting-edge technologies, which results in an excessive use of fossil fuels (Hafeez et al. 2020). Because of its positive social, economic, and environmental effects, renewable energy is being seen as a potential replacement (Potters et al. 2010). However, people continued to use fossil fuels rather than plant resources as primary sources of energy since they were easily accessible and economically viable(Mizik and Gyarmati 2021).The concept of using biofuels has existed for centuries. Biological materials were used to produce energy earlier in human history (Mizik and Gyarmati 2021).Biofuels can be used either in their pure form or as fuel additives. Additionally, biofuels are typically separated into bioethanol and biodiesel. In the transportation sector, liquid biofuels can be used as an alternative to conventional fuels, accounting for about 18% of total primary energy consumption (Wang et al. 2007). Currently, bioethanol accounts for around 80% of liquid biofuel production, with biodiesel producing the remaining 20% (Figueiredo et al. 2009).The biofuels' oxygen concentration sets them apart from other petroleum feedstock's. Compared to petroleum products, which have no oxygen, they range from 10 to 45% in oxygen content. When compared to petroleum sources, biofuels also have lower levels of sulphur and nitrogen. The adoption of the Sustainable Development Goals was a significant move, and several nations have since set comparable goals, notably the USA, which intends to replace 20% of its road transport fuel by biofuel by the year 2022 to overcome this problem (Law 2007; Commission and (EC), Directive 2009; Transforming our world 2030). There are many ways to make biofuels, including biological, chemical, and physical processes (Hajilary et al. 2019). Fuels are produced using substances that are biological in nature, namely plants, microbes,

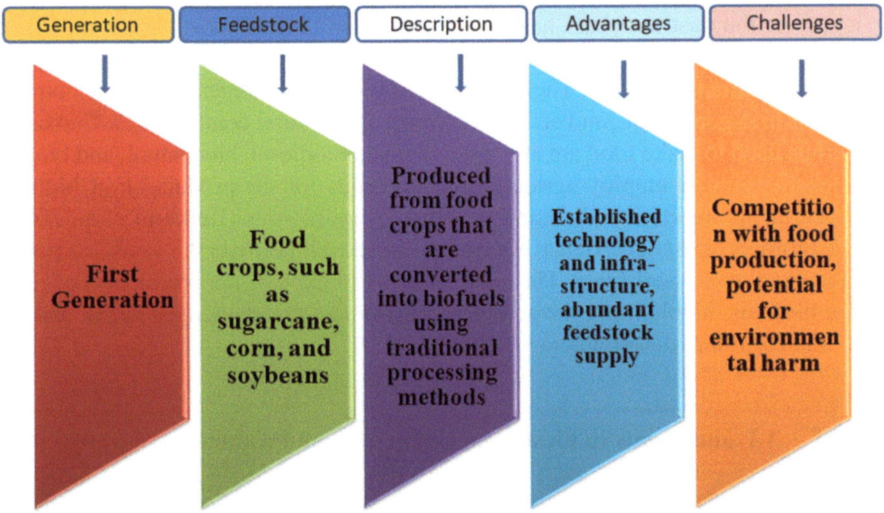

Fig. 8.1 Diagrammatic representation of First generation biofuel

livestock, and pollutants at this time. All biofuels share a fundamental, sustainable source. They differ from petroleum and natural gas, which depend on historical process of photosynthesis in that they convert solar energy into chemical energy in a "present-day" manner through photosynthesis. The implementation of innovative energy guidelines in various nations, particularly the United States, that demand for more sustainable, greener fuels for transportation, has been a major factor in the growth of the usage of biofuels over the past ten years. Due to growing demands for substituting petroleum-based fuels with renewable ones in order to reduce the production of greenhouse gases and combat climate change, biofuels have grown to become the most produced and used renewable fuel globally.

Based on the feedstock used to produce them, biofuels are broadly categorized. They fall into the following categories based on the feedstock: They are categorized as First, Second, Third and Fourth (Wu and Miao 2014; Dahman et al. 2019) (Fig. 8.1).

8.1.1 Production of First Generation Biofuel

First-generation biofuels, also known as conventional biofuels, since they are derived from using well-known technology and procedures like fermentation, distillation, and transesterification (Jeswani et al. 2020; Kulkarni et al. 2006). This generation of biofuel production uses traditional feedstock's such as corn, sugarcane, vegetable oils, and animal fats (Evans 2008). Since the process is primarily concerned with producing fuel and discarding the remaining non-fuel material as waste, membranes are not necessary in these processes (Oumer et al. 2018). The sharp increase in oil prices over the past ten years has also made liquid biofuels cost-competitive with petroleum-based transportation fuels, which has sparked a worldwide upsurge in research and manufacturing. The majorities of biofuels generated today are conventional, or first-generation, and are derived from agricultural commodities. Currently, materials used to make food are also used to make biodiesel, bioethanol, and biogas. It is challenging to employ agricultural food crops for the production of biofuels because of the trend towards rising demand for edible oils (Kulkarni et al. 2006). These biofuels have been extensively studied and developed over the years as a potential alternative to fossil fuels. This chapter aims to explore the production of plant based biofuels, highlighting the advancements made, the challenges faced, and the potential for future development.

8.1.1.1 Advancements in First Generation Biofuel Production

First-generation biofuels have witnessed significant advancements in production processes, making them more efficient and viable. The primary types of first-generation biofuels are bioethanol and biodiesel as shown in Fig. 8.2.

Considering biofuels release fewer greenhouse gases than conventional fuels, blending them with existing fuels is an efficient strategy to cut transportation-related carbon dioxide emissions. Plant-based biofuels are made from renewable biomass, such as crops and agricultural waste, and can be used to power vehicles, generate electricity, and heat buildings. In this essay, we will discuss the use and applications of plant-based biofuels and their benefits (Fig. 8.3).

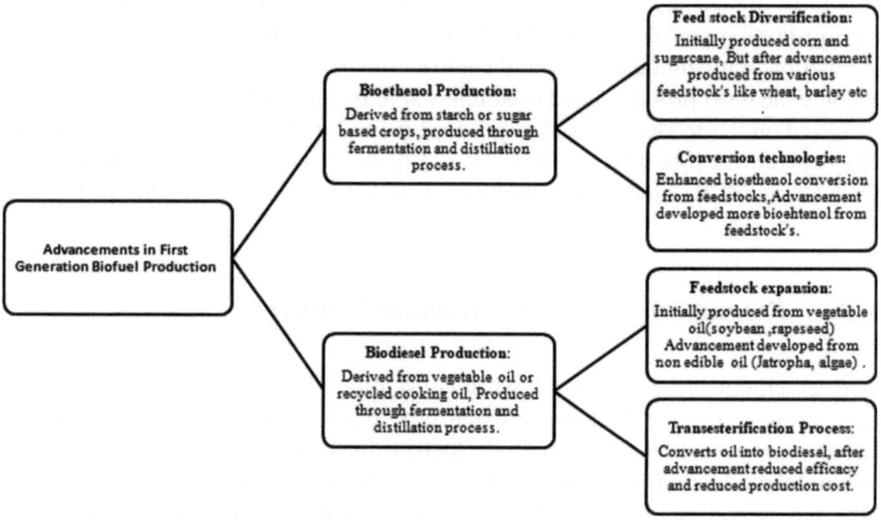

Fig. 8.2 Flowchart representation of Advancement in first generation biofuel production and its type

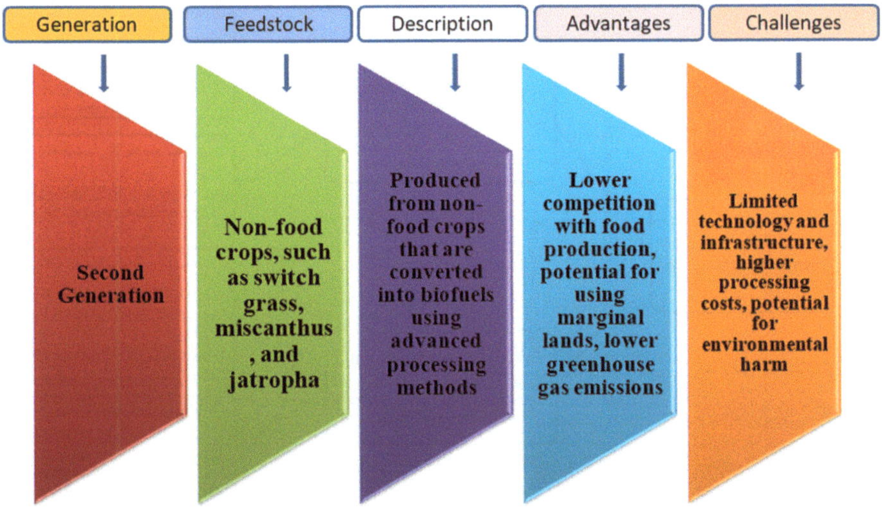

Fig. 8.3 Diagrammatic representation of Second generation biofuel

8.1.2 Production of Second Generation Biofuel

In the quest for sustainable renewable energy sources, second-generation biofuels have emerged as a promising alternative to conventional fossil fuels. These advanced biofuels are derived from non-food feedstock's such as lignocelluloses biomass and algae, offering several advantages over first-generation biofuels. This essay explores the production of second-generation biofuels, highlighting their advancements, environmental benefits, and the challenges faced in their implementation. Due to significant restrictions in the production of first-generation biofuels, second-generation biofuel technologies have been creating (Evans 2008).

8.1.2.1 Advancements in Second-Generation Biofuel Production

Second-generation biofuels utilize non-food feedstock's, enabling a more sustainable and environmentally friendly approach to biofuel production (Evans 2008). Other types of crops, such as switch grass, wheat straw, oats, etc. that bear small grain, but are not consumed as food, can also be extracted (Inderwildi and King 2009). Key advancements in second-generation biofuel production in shown in Fig. 8.4.

The fact that second-generation biofuels are made from non-food feedstock's like specialized energy crops (like Miscanthus, switch grass, short rotation coppice (SRC), and other lignocellulosic plants), agricultural and forestry waste, and other waste products (like municipal solid waste) is one of their key characteristics (Fig. 8.5).

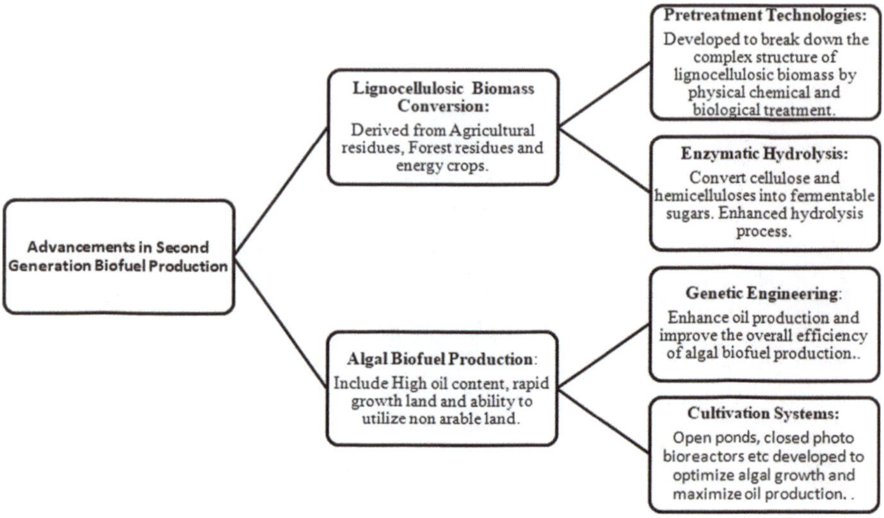

Fig. 8.4 Flow chart representation of Second generation biofuel production

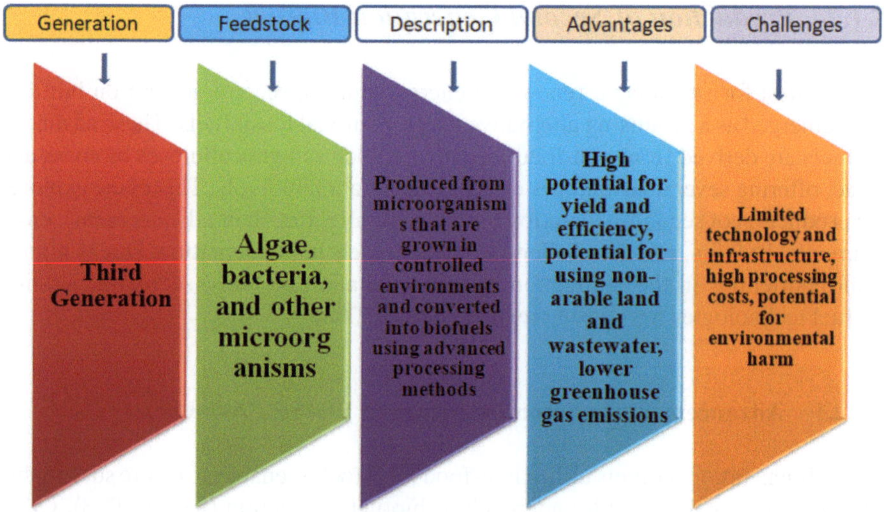

Fig. 8.5 Diagrammatic representation of Third generation biofuel

8.1.3 Production of Third Generation Biofuel Production

As the demand for sustainable and renewable energy sources continues to grow, third-generation biofuels have emerged as a promising solution. These advanced biofuels, derived from non-food feedstock's such as algae and waste biomass, offer significant advantages over first and second-generation counterparts. This essay explores the production of third-generation biofuels, highlighting their advancements, environmental benefits, and the challenges faced in their implementation. Algae can provide up to 30 times more energy per acre in third-generation biofuels than land crops like soybeans (Inderwildi and King 2009; Hartman 2008). Algae fuel still faces challenges, such as the requirement for equal mixing during production, which if accomplished by agitation, could slow the growth of the biomass (Bulkowska et al. 2016). A number of companies and governmental organizations are funding initiatives to lower start-up and ongoing costs and make the production of algal fuel profitable (Parry et al. 2007).

8.1.3.1 Advancements in Third-Generation Biofuel Production

Third-generation biofuels represent a significant leap forward in biofuel production, utilizing novel feedstock's and advanced technologies. Key advancements in third-generation biofuel production is shown in Fig. 8.6.

While third-generation biofuels offer significant environmental benefits, their large-scale implementation requires careful consideration of potential ecological impacts. Assessments of water usage, land use changes, and biodiversity preservation should be integral parts of the development.

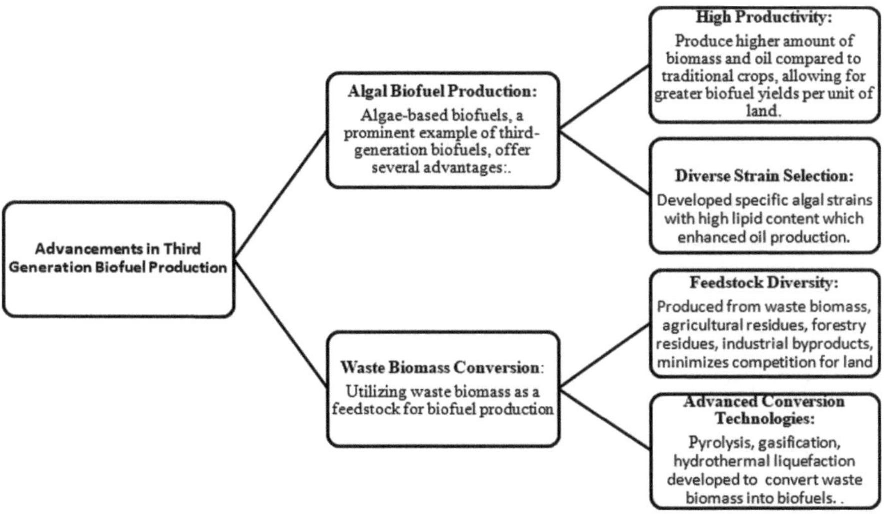

Fig. 8.6 Flow chart representation of Advancement and types of product derive from Third generation biofuel

8.1.4 Production of Fourth Generation Biofuel Production

The cause of the issue, also known as global climate change, is primarily attributed to human activities to satisfy the world's insatiable energy needs. These activities have led to the production and consumption of energy, which have brought about the impending dangers of global warming, including increased temperatures, precipitation, snow cover, rising sea levels, and extreme weather. As the world seeks innovative solutions to combat climate change and transition to sustainable energy sources, fourth-generation biofuels have emerged as a cutting-edge technology. Also known as advanced or synthetic biofuels, these next-generation fuels are produced from non-food biomass, algae, and even carbon dioxide captured from industrial emissions. This essay explores the production of fourth-generation biofuels, highlighting their advancements, environmental benefits, and the challenges faced in their implementation. the technique of synthetic biology to redesign microbes that photosynthesize with the capacity to readily manufacture some specific goods from CO2, which contributes to about 60% of the world's warming temperatures, is one such possibility. Sustainable developments in both the production and application of the future fuels and high-value products strongly depend on the collaboration of scientists and engineers in the related fields with significant advances (Parry et al. 2007; Resch et al. 2008).

8.1.4.1 Advancements in Fourth-Generation Biofuel Production

Fourth-generation biofuels represent a revolutionary approach to renewable energy production, leveraging advanced technologies and unconventional feedstock's. Key advancements in fourth-generation biofuel production are shown in Fig. 8.7.

Fourth-generation biofuels offer numerous environmental benefits, making them a promising solution for a sustainable future. It represents a groundbreaking advancement in renewable energy production.

8.2 Various Types of Biofuel

Biofuels are considered as an alternative to fossil fuels because they can help reduce greenhouse gas emissions and dependence on non-renewable resources. Biofuels can be categorized into several types based on their production methods are shown in Table 8.1.

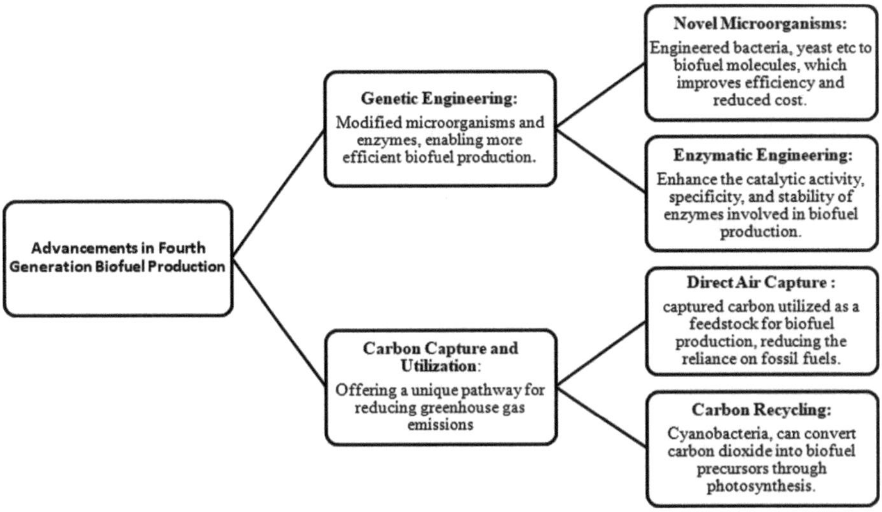

Fig. 8.7 Flow chart representation of advancement and types of product derived from fourth generation biofuel

Table 8.1 It depicts the types of Biofuels and their production method

Types of biofuel	Production method
Ethanol	Fermentation of sugar crops (e.g., corn, sugarcane) or cellulose biomass (e.r., switch grass,wood chips)
Bio-diesel	Transesterification of vegetable oils (e.g., soybean, canola) or animal fats (e.g.,tallow, grease)
Biogas	Anaerobic digestion of organic materials (e.g., sewage, agricultural waste, food scraps)
Bio-methanol	Gasification or fermentation of biomass feed-stocks (e.g., forest residues, agricultural waste)
Bio-ether	Etherification of alcohols (e.g., ethanol, methanol) with isobutylene or tertiary butyl alcohol
Green diesel	Hydro-treatment or hydrogenation of vegetable oils or animal fats, similar to petroleum diesel refining
Bio-hydrogen	Fermentation of biomass by certain bacteria or algae, followed by extraction of hydrogen gas
Algal biofuels	Cultivation and extraction of oils from algae, which are then converted into biodiesel or other fuel types
Solid biofuels	Combustion or gasification of biomass in the form of wood pellets, agricultural residues, or energy crops
Biobutanol	Fermentation of various feed stocks (e.g., corn, sugar-beets) using specific bacteria
Bio-jet fuel	Conversion of vegetable oils, animal fats or biomass into a suitable jet fuel substitute

8.3 Harnessing Nature's Power: Plants in Biofuel Production

Cellulose, hemicelluloses, and lignin are the three substances that make up the walls of plant cells. The sugar building blocks found in the cellulose and hemicelluloses can be used as an energy source. To provide the plant the structural support it needs to survive, the lignin, a fibrous molecule, binds all of the molecules together into a compact structure. The fibrous lignin must be broken down to have access to the sugars needed to produce ethanol. The first phase in turning plants into fuel is called "deconstruction," which involves rupturing the plant cell wall. Biofuels, derived from plant-based feedstocks, have emerged as a viable renewable energy source, providing a sustainable alternative to fossil fuels. Plants play a vital role in biofuel production as they possess the unique ability to convert solar energy into chemical energy through photosynthesis. This essay explores how plants are utilized in biofuel production, highlighting different feedstock's, conversion processes, and their environmental implications. Plant-based biofuels are derived from various sources, including sugar crops, such as sugarcane and sugar beet, and oil crops, such as rapeseed, soybeans, and palm oil. These crops are processed to extract oils or fermentable sugars, which are then converted into biofuels. The most common plant-based biofuels are ethanol and biodiesel.

Ethanol is a biofuel produced from the fermentation of sugars or starches, primarily from crops such as corn, wheat, and sugarcane. It is commonly used as a gasoline additive, improving fuel efficiency and reducing emissions. Ethanol is also used as a standalone fuel in flex-fuel vehicles, which can run on either gasoline or ethanol blends. These crops contain high levels of sugar, which can be converted into ethanol by yeast fermentation. Additionally, cellulose biomass, including switch grass and wood chips, can also be used as feedstock through a process called cellulosic ethanol production.

Biodiesel, on the other hand, is made from vegetable oils or animal fats, such as soybean oil, canola oil, and recycled cooking oil. Such oils are chemically reacted with an alcohol (usually methanol or ethanol) and a catalyst to produce biodiesel. Biodiesel can be blended with petroleum diesel and used in conventional diesel engines without any modification. It is a cleaner-burning fuel that reduces greenhouse gas emissions and particulate matter. Biodiesel is produced through a process called transesterification. This process breaks down the oils or fats into fatty acid methyl or ethyl esters, which can be used as a diesel fuel substitute.

8.3.1 Process of Biofuel Production from Plants

The process of biofuel production from plants typically involves several steps. Here's a stepwise overview of the process is shown in this flowchart (Fig. 8.8)..

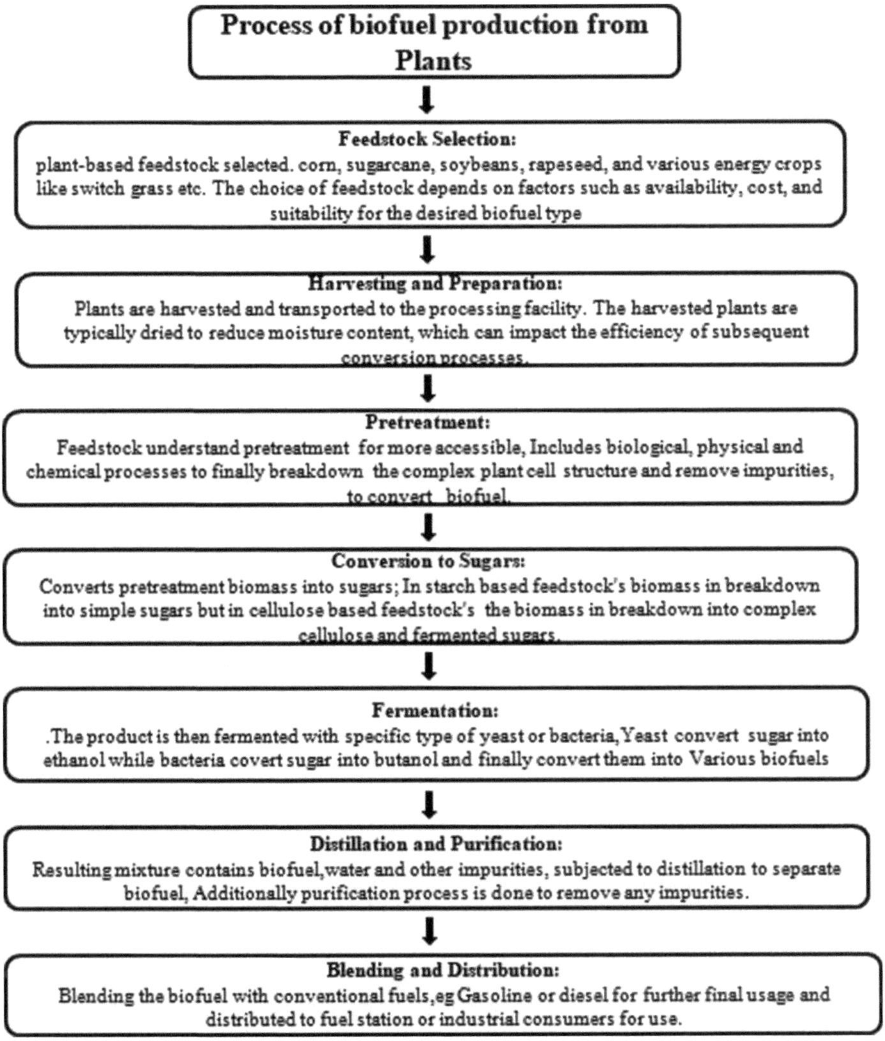

Fig. 8.8 Step wise process of biofuel production from plants

It's important to note that different biofuel production processes can vary based on the type of feedstock, the specific biofuel being produced, and the technological advancements in the field. Additionally, emerging technologies, such as advanced biofuels or synthetic biology, may introduce new steps or modify existing ones to improve biofuel production efficiency.

8.4 Plant-Based Feedstock's for Biofuel Production

Plants offer a wide range of feedstock's that can be used to produce biofuels, including.

8.4.1 *Starch and Sugar Crops*

Bioethanol for use in petrol engines can be produced by fermenting biological feedstock's that include significant amounts of sugar or components that can be turned into sugar, such as starch or cellulose (Balat and Balat 2009; Mojović et al. 2007; Baras et al. 2002). Biofuel production using starch and sugar crops is primarily focused on the production of ethanol. Crops such as corn, sugarcane, and wheat are rich in starch and sugar, making them suitable feedstocks for bioethanol production. These crops undergo fermentation processes, where enzymes break down the starch or sugar into simple sugars, which are then converted into ethanol through microbial fermentation. Traditional raw materials for ethanol production include molasses. When sugar is made from sugar beetroot (beetroot molasses) or sugarcane (cane molasses), it is obtained as a byproduct. Molasses was essentially the only raw material used to make ethanol for a very long period. But today, a variety of additional carbohydrate sources are available for the generation of biofuels (Mohanty et al. 2009; Mojovic et al. 2009). Starch crops like corn, wheat, and sorghum, as well as sugar crops like sugarcane and sugar beets, are commonly utilized for this purpose (Gnansounou et al. 2005). Here's an overview of the biofuel production process using starch and sugar crops:

- **Feedstock Preparation**: Starch and sugar crops are harvested and transported to the processing facility. The crops are cleaned and any impurities, such as dirt or debris, are removed.
- **Milling and Grinding**: The harvested crops are subjected to milling or grinding to break them down into smaller particles (Aprodu and Banu 2017). Dry milling and wet milling are the two milling methods used in the creation of starch ethanol. This process increases the surface area and improves the efficiency of subsequent enzymatic or chemical reactions.
- **Cooking**: The milled or ground crops are cooked with water to soften the starch granules and facilitate their conversion into sugars. Heating the mixture at high temperatures helps break down the starch molecules into simpler sugars.
- **Enzymatic Hydrolysis**: After cooking, the mixture undergoes enzymatic hydrolysis. Enzymes, such as amylases, are added to the mixture to catalyze the conversion of starch into fermentable sugars (Sakwa 2018). The amount of glucose available for ethanol fermentation is determined by how well starch is hydrolyzed, making it a crucial component in the conversion of starch to ethanol's overall efficiency (Cripwell et al. 2019). Amylases break down the long chains of starch molecules into shorter chains and eventually into glucose. The plant material's

composition is changed during pretreatment to speed up the enzymatic breakdown of hemicelluloses as well as cellulose and increase yields (Harmsen 2010; Bjerre et al. 2000; Allen et al. 2001).
- **Fermentation**: The obtained sugar solution is cooled and transferred to fermentation tanks. Yeast, such as Saccharomyces cerevisiae, is added to the tanks. The yeast consumes the sugars in the solution and produces ethanol as a byproduct through the process of anaerobic fermentation (Nikolić et al. 2009a, 2009b). The fermentation process typically takes several days. Optimizing the fermentation stage with regard to the key factors of temperature, pH, medium composition, mixing, aeration, and infection control is crucial for the effective generation of ethanol (Mojović et al. 2007).
- **Distillation**: After fermentation, the resulting mixture, known as the "beer," contains ethanol, water, residual sugars, and other impurities. The beer undergoes distillation to separate the ethanol from the mixture. Distillation involves heating the mixture to vaporize the ethanol, which is then condensed and collected as a concentrated ethanol solution.
- **Dehydration**: The ethanol solution obtained from distillation may still contain a significant amount of water. Dehydration processes, such as molecular sieves or azeotropic distillation, are employed to remove excess water and increase the ethanol concentration.
- **Denaturing (Optional)**: In some cases, ethanol produced for biofuel purposes is denatured to make it unsuitable for human consumption and exempt from certain taxes. Denaturing involves adding small amounts of chemicals, such as gasoline or methanol, to the ethanol.
- **Blending and Distribution**: The final step involves blending the ethanol with gasoline to create ethanol-gasoline blends, such as E10 (10% ethanol, 90% gasoline) or E85 (85% ethanol, 15% gasoline). These blends can be used as transportation fuels and distributed to fuel stations for use in flex-fuel vehicles.

Within the next 20 years, it is anticipated that bioethanol, which is made from renewable biomass like sugar, starch, or lignocellulosic materials, will be one of the most prevalent renewable biofuels used in the transportation sector. Although lignocelluloses processing, one of the most promising second-generation biofuel technologies, is given priority in the manufacture of ethanol globally, the use of lignocellulosic feedstock for fuel ethanol is still being improved. At the industrial level, feedstock's based on sugar and starch is still common and, so far, economically advantageous. It's important to note that the biofuel production process from starch and sugar crops can vary depending on factors such as feedstock characteristics, regional practices, and advancements in technology. Additionally, ongoing research and development efforts aim to improve the efficiency and sustainability of biofuel production from these crops.

8.4.1.1 Some Important Examples of Plants Containing Starch Used in Ethanol Production

Plant-based biofuels are derived from a variety of sources, including crops and agricultural waste. The main plant sources for biofuel production are:

1. **Sugarcane**—Sugarcane is a primary source of biofuel, particularly ethanol. It is grown in tropical and subtropical regions and can be processed to produce both sugar and ethanol. In contrast to corn, which supplies starch, sugarcane produces sugar that can be easily fermented into ethanol that calls for fermentation after heating (Mojović et al. 2007). Ethanol produced might be used for transportation and crop harvesting, using less fossil fuel overall unit production of ethanol. The sugarcane plant is a popular biofuel because of its exceptionally high efficiency and excellent prospects for power in the residue (No 2019). The production of biofuel from sugarcane and other types of feedstock's as shown in Fig. 8.9 below.
2. **Corn**—Corn is the most common crop used for biofuel production in the United States which is regarded as the primary source of first-generation bioethanol; as of 2012, about 40% of the corn production there was utilized to make ethanol. It is processed to produce ethanol, which is used as a gasoline additive (Li and Yang 2016). The US government has set a goal of producing 36 billion gallons

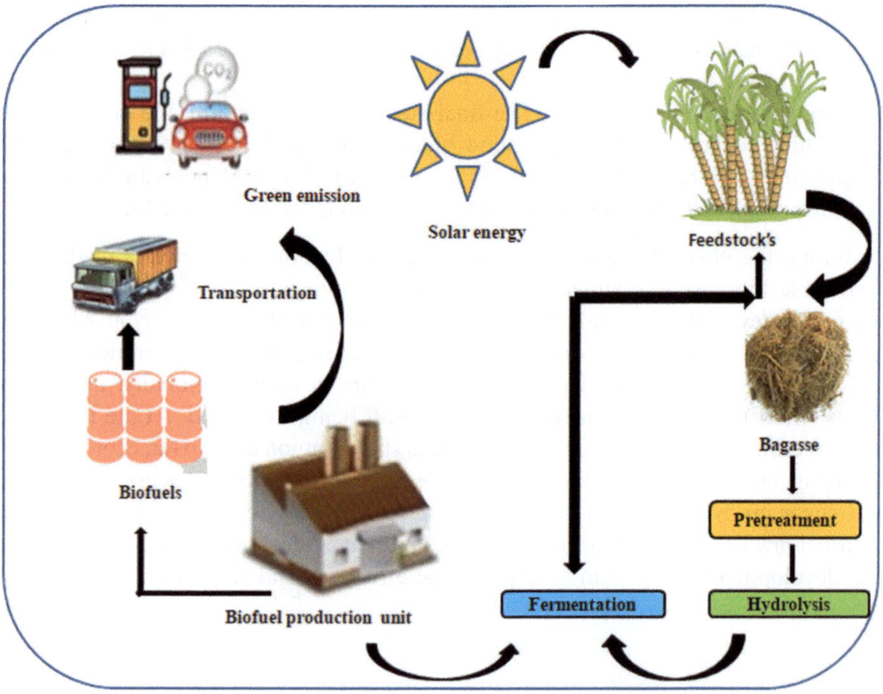

Fig. 8.9 Diagrammatic represents depicts biofuel production from feedstock's

of biofuel by 2022, with 15 billion gallons of that amount coming from grains like maize (Khoo et al. 2020).

8.4.2 Oilseeds

Oilseeds like soybean, rapeseed, and sunflower seeds contain high levels of vegetable oil. These oils can be extracted and processed into biodiesel through a chemical reaction called transesterification. The vegetable oil reacts with an alcohol, typically methanol, to produce biodiesel and glycerin as a byproduct. Biofuel production with oilseed crops, such as soybeans, canola, sunflower, and palm, primarily focuses on the production of biodiesel. These oil-rich crops contain triglycerides that can be converted into biodiesel through a process called transesterification. Here's an overview of the biofuel production process using oilseed crops.

- **Feedstock Preparation**: Oilseed crops are harvested and transported to the processing facility. The crops are cleaned to remove impurities like dirt, stones, and plant debris.
- **Seed Crushing**: The oilseed crops undergo seed crushing to extract the oil. The seeds are mechanically pressed or subjected to solvent extraction methods to separate the oil from the seed. Solvent extraction involves using organic solvents, such as hexane, to dissolve the oil and separate it from the solid residue.
- **Oil Refining** (Optional): Depending on the quality and purity requirements of the biodiesel, the extracted oil may undergo refining processes to remove impurities like free fatty acids, phospholipids, and color pigments. Refining steps may include degumming, neutralization, bleaching, and deodorization.
- **Transesterification**: The extracted and refined oil is converted into biodiesel through a process called transesterification. Transesterification involves reacting the oil with an alcohol, typically methanol or ethanol, in the presence of a catalyst (usually sodium hydroxide or potassium hydroxide). The reaction breaks down the triglycerides in the oil into fatty acid methyl or ethyl esters (biodiesel) and produces glycerin as a byproduct.
- **Separation and Washing**: After the transesterification reaction, the mixture is allowed to settle to separate the biodiesel from glycerin and any remaining impurities. The biodiesel is then washed with water or other agents to remove any traces of catalyst, soap, or impurities.
- **Dehydration (Optional)**: Biodiesel produced from oilseed crops may undergo a dehydration process to remove excess water, as water can negatively affect the quality and stability of the biodiesel. Dehydration methods may involve using adsorbents or molecular sieves.
- **Quality Testing**: The biodiesel is tested for quality parameters such as flash point, cloud point, viscosity, acidity, and purity to ensure it meets the required specifications and regulatory standards.
- **Blending and Distribution**: The final step involves blending the biodiesel with conventional diesel fuel to create biodiesel blends suitable for use in diesel

engines. Common blends include B5 (5% biodiesel, 95% diesel) and B20 (20% biodiesel, 80% diesel). The blended biodiesel is distributed to fuel stations for use in vehicles equipped with diesel engines.

8.4.2.1 Some Important Examples of Plants Containing Vegetable Oils Produce Biofuel

1. **Soybeans**—Soybeans are the common crop used for biofuel production, primarily to produce biodiesel. The plants' height varies greatly depending on the climate and genetic differences. United States, Brazil, Argentina, China, and India are the top soybean-producing nations. Due to the seeds' high protein composition, their oil content is only about 18%. Biodiesel made from soybeans is a cleaner-burning fuel that reduces greenhouse gas emissions and particulate matter (Romano and González Suárez 2009; Brasileira and de Pesquisa 2009).
2. **Rapeseed**—Rapeseed does well in soils with poor fertility but high sulphur content. It can be cultivated as a winter cover crop and has a high oil production (40–50%), allowing for double farming and rotations of crops. It is also used to produce biodiesel, particularly in Europe. It is a fast-growing crop that can be grown in colder climates. In the European Union, it is the most significant raw ingredient for the production of biodiesel. However, several Central and South American nations had technological restrictions for planting and harvesting, mostly because there was a lack of proper knowledge about fertilization, handling, and storage of seeds (the seeds are very minute and require specialized agricultural apparatus). Additionally, its utilization has been constrained by low costs compared to wheat, which serves as its primary rival for crop rotation, and by poor yield per unit area. Rapeseed flour is utilized as a protein supplement in cow feeds because it has a higher nutritional value than soybean flour. (Agricultura and Pecuária 2009).
3. **Palm Oil**—Palm oil is a widely used source of biofuel in Asia, particularly in Indonesia and Malaysia. A tropical plant with a 20–25 m height range and a 25-year lifespan is the oil palm (Will and the U.S. Produce 2022, 2014). The fruit yields two types of oil: palm kernel oil, which is extracted from the fruit's nut, and palm oil proper, which is derived from the pulp. The fruit yields two types of oil: palm kernel oil, which is extracted from the fruit's nut, and palm oil proper, which is derived from the pulp. Over the past several years, there has been a steady rise in the demand for palm oil on a global scale. This oil is used for cooking, as a component of margarine, as an ingredient in butter, and as an additive for baked goods. It is crucial to note that pure palm oil is semisolid at room temperature (20–22 °C) and is frequently combined with other vegetable oils, occasionally partially hydrogenated, in applications. However, there are concerns about the environmental impacts of palm oil production, including deforestation and habitat destruction (Agudelo Santamaría and Benjumea Hernández 2004).
4. **Jatropha**—Jatropha is a non-food crop that can be grown on marginal lands and used to produce biodiesel. Jatropha is a bush which thrives in dry climates.

The most well-known cultivar, *Jatropha curcas*, needs little water and other care, making it suitable for warm climates with low fertility. It is particularly attractive for its potential to provide a sustainable source of biofuel without competing with food crops. Oil yield varies depending on the extraction technique; it ranges from 28 to 32% for presses and up to 52% for solvent extraction. Jatropha oil is toxic, and the seeds are poisonous. Jatrophic acid and the globulin curcasin are responsible for the toxicity. Yield is influenced by weather, soil, rainfall, and care taken during planting and harvesting. After three to four years, jatropha plants start to produce, and they live for around 50 years. (Wise 2012).

5. **Sunflower**—Safflower is a good dry-land adapter. Despite having a poor grain yield per hectare, seeds have a high oil content of between 30 and 40%. It has economic possibilities for dry areas as a result. Safflower is currently used to make bread, oil, and bird feed. Safflower oil is of excellent quality and contains little cholesterol. In addition to being used to make food for humans, it is also used to make lacquers, soaps, paints, and other coating materials. There are two types: oleic acid, which is high in monounsaturated fatty acids, and linoleic acid, which is high in polyunsaturated fatty acids neither variety contains a lot of saturated fatty acids. It is significant to note that safflower oil is refined using traditional techniques without the addition of antioxidants and is extracted using hydraulic presses instead of solvents.

The creation of novel process methods to manufacture high-quality transportation fuels from biologically produced materials is necessary for the widespread use of biofuels in the future raw materials. It's important to note that different oilseed crops may have variations in their oil content, composition, and processing methods. Additionally, the biodiesel production process may undergo specific modifications depending on regional practices and advancements in technology. The cetane value of green diesel is greater, and it has good cold flow characteristics. Additionally, it has great storage stability and is totally compatible with the typical mixture of diesel fuels made from petroleum. Ongoing research also aims to explore alternative methods, such as enzymatic or microbial processes, to improve the efficiency and sustainability of biodiesel production from oilseed crops.

8.4.3 Lignocellulosic Biomass

Lignocellulosic biomass, derived from non-food sources such as agricultural residues, forestry waste, and dedicated energy crops like switch grass and miscanthus, can be converted into biofuels (Shelley and Alcoholic fuels. Boca Raton 2006). This biomass contains cellulose, hemicelluloses, and lignin (Lee et al. 2007; Huber and Corma 2007). Various technologies like pretreatment, enzymatic hydrolysis, and fermentation or thermo chemical conversion processes are employed to break down the complex structure of lignocellulosic biomass and produce bioethanol, biogas, or bio-oil (Shelley and Alcoholic fuels. Boca Raton 2006; Naik et al. xxxx). Biofuel production with lignocellulosic biomass involves converting the complex carbohydrate structure of non-edible plant materials, such as agricultural residues (e.g., corn Stover, wheat straw), forest residues, and dedicated energy crops (e.g., switch grass, miscanthus), into biofuels. The process typically includes the following steps:

- **Feedstock Preparation**: Lignocellulosic biomass is harvested and collected, and any large impurities like rocks or branches are removed. The biomass is then processed to reduce particle size through chipping, grinding, or shredding. This increases the surface area and facilitates subsequent processing.
- **Pretreatment**: Lignocellulosic biomass has a complex structure that requires pretreatment to make it more accessible for enzymatic or microbial conversion. Pretreatment methods include physical, chemical, or biological processes. These processes remove lignin, break down hemicelluloses, and increase the porosity of the biomass, making it easier to access the cellulose.
- **Enzymatic Hydrolysis**: After pretreatment, the biomass is subjected to enzymatic hydrolysis. Enzymes, typically celluloses and hemicelluloses, are added to break down the cellulose and hemicelluloses into simple sugars (glucose, xylose, etc.). This step requires optimal conditions of temperature, pH, and enzyme concentration to maximize sugar yields.
- **Fermentation**: The obtained sugar solution from enzymatic hydrolysis is then fermented by microorganisms, such as yeast or bacteria. Yeast fermentation is commonly used to convert sugars into ethanol, while bacteria like Clostridium may be employed for the production of other biofuels like butanol. The microorganisms consume the sugars and produce biofuels along with byproducts like carbon dioxide.
- **Separation and Purification**: After fermentation, the biofuel mixture contains the desired biofuel (e.g., ethanol), water, residual sugars, and other impurities. The mixture is subjected to separation techniques such as distillation, evaporation, or membrane filtration to separate the biofuel from the water and impurities. Additional purification steps may be performed to remove any remaining impurities and increase the biofuel concentration.
- **Post-Processing (Optional)**: Depending on the desired biofuel type and quality standards, additional post-processing steps may be conducted. These steps can include dehydration, rectification, or esterification to remove excess water, adjust fuel properties, or enhance stability.
- **Blending and Distribution**: The final biofuel product is blended with conventional fuels or used as a standalone fuel, depending on its properties and intended application. Biofuels derived from lignocellulosic biomass can be blended with gasoline or diesel to create biofuel blends suitable for use in existing vehicles and infrastructure.

8.4.3.1 Some Important Examples of Plants Containing Vegetable Oils Produce Biofuel

Lignocellulosic biomass refers to plant-based materials that contain cellulose, hemicellulose, and lignin, which can be used as feedstock's for producing lignocellulosic biofuels. Here are a few examples of lignocellulosic biomasses:

1. **Agricultural Residues**: Crop residues left after harvest can serve as lignocellulosic biomass. Examples include corn stover (stalks, leaves, and cobs), wheat straw, rice straw, sugarcane bagasse, and peanut shells.
2. **Forest Residues**: Various byproducts from forestry operations can be utilized as lignocellulosic biomass, such as logging residues (branches, bark, and treetops), sawdust, wood chips, and forest thinning.
3. **Energy Crops**: Certain dedicated energy crops are specifically grown for their lignocellulosic biomass. Examples include switch grass, miscanthus, giant reed, and willow. These crops can be cultivated on marginal lands and have high biomass productivity.
4. **Agricultural Wastes**: Waste materials from agricultural processes can be used as lignocellulosic biomasses. This includes straw from rice, wheat, and barley harvests, corn cobs, husks, and stalks, as well as sugarcane bagasse and other residues left after processing crops.
5. **Algal Biomass**: Certain species of algae can also be considered lignocellulosic biomasses. Algae have high growth rates and can be cultivated in large quantities. After extraction of lipids for biodiesel production, the remaining algal biomass can be used for lignocellulosic biofuel production.
6. **Municipal Solid Waste**: Organic components of municipal solid waste, such as yard waste, food scraps, and paper waste, can be utilized as lignocellulosic biomass for biofuel production.

These examples highlight the diverse range of biomass sources that can be used to produce lignocellulosic biofuels. The choice of biomass depends on factors like availability, cost, regional suitability, and sustainability considerations. It's important to note that lignocellulosic biofuel production is still undergoing research and development to improve efficiency and reduce costs. Various technological advancements, such as consolidated bioprocessing and advanced fermentation techniques, are being explored to enhance the commercial viability of lignocellulosic biofuel production.

8.4.4 Algae

The world's fastest-growing vegetation is algae. They don't need a lot of fresh water to grow (Singh and Gu 2010; Pittman et al. 1725; Christenson and Sims 2011). They are also more oil-rich and capable of greater photosynthetic efficiency than other raw materials or oil crops, and they have a quick doubling time (about 1224 h), a simple structure, and a short doubling time. Algae are microorganisms that can efficiently convert sunlight, carbon dioxide, and nutrients into biomass, often rich in lipids or oils. Algal biomass can be harvested and processed to extract lipids, which are then converted into biofuels such as biodiesel or renewable diesel. Algae's high growth rates, ability to grow in diverse environments, and potential utilization of non-arable land and wastewater make them a promising feedstock for biofuel production. Biofuel production with algae, specifically microalgae, offers great potential due to their high oil content and rapid growth rates. Microalgae produce more oil than corn, soybean, canola, coconut, and other similar crops (Khan et al. 2009). Algae can produce different types of biofuels, including biodiesel, bioethanol, and biogas. Here's an overview of the biofuel production process using algae:

- **Strain Selection and Cultivation**: The first step is to select suitable algae strains that have high oil content and can thrive in the specific environmental conditions (Jiang et al. 2011; Yoo et al. 2010). Algae can be cultivated in open ponds, closed photo bioreactors, or other cultivation systems. They require sunlight, nutrients (such as nitrogen and phosphorus), and carbon dioxide for photosynthesis and growth (Vergarafernandez et al. 2008). The final innovation is the creation of useful byproducts from waste, such as biogas, biobutanol, omega 3 oil for animal feed, and cosmetics (Nakas 1983; Belarbi et al. 2000; Besada et al. 2009; Spolaore et al. 2006).
- **Algae Growth and Harvesting**: Algae cultivation involves providing optimal conditions for their growth, including temperature, pH, and nutrient supply. Algae multiply through cell division and form dense populations. Once the algae reach the desired biomass concentration, they are harvested. Harvesting methods include centrifugation, filtration, flocculation, or flotation to separate the algae from the growth medium (Tredici 2010; Eppink et al. 2018).
- **Lipid Extraction**: Algae contain lipid-rich cells that can be converted into biofuels. The harvested algae biomass undergoes lipid extraction to separate the lipids from the rest of the cellular components. Extraction methods include mechanical pressing, solvent extraction, or supercritical fluid extraction. The goal is to obtain a lipid-rich extract known as algal oil (Popoola and Yangomodou 2006).
- **Transesterification (Biodiesel Production)**: If the aim is to produce biodiesel, the extracted algal oil is subjected to transesterification. Transesterification involves reacting the algal oil with an alcohol, typically methanol or ethanol, in the presence of a catalyst (such as sodium hydroxide or potassium hydroxide). This reaction converts the algal oil into fatty acid methyl or ethyl esters, which are the main components of biodiesel.
- **Fermentation (Bioethanol or Biogas Production)**: Alternatively, the extracted algae biomass can be subjected to fermentation processes to produce bioethanol or biogas. For bioethanol production, enzymes are used to break down the complex carbohydrates (such as starch or cellulose) present in the algae into simple sugars,

which are then fermented by yeast or bacteria to produce ethanol. Biogas production involves anaerobic digestion, where bacteria break down the algae biomass to produce methane and carbon dioxide.

- **Distillation and Purification**: After the transesterification or fermentation process, the biofuel mixture undergoes distillation to separate the desired biofuel from impurities and byproducts. Additional purification steps, such as dehydration or rectification, may be performed to remove any remaining impurities and increase the biofuel concentration.
- **Post-Processing (Optional)**: Depending on the biofuel type and quality standards, additional post-processing steps may be conducted. These steps can include filtration, centrifugation, or further purification to ensure the biofuel meets the required specifications.
- **Blending and Distribution**: The final biofuel product can be blended with conventional fuels, such as gasoline or diesel, to create biofuel blends suitable for use in existing infrastructure and vehicles. The blended biofuel is then distributed to fuel stations or industrial consumers for use.

8.4.4.1 Some Important Examples of Algae that Produce Biofuel

1. **Brown algae**: It is a macro algae. Due to presence of xanthophylls they are brown in color. They contain antifungal, antifungal, anti tumor activity (Burton 2009). In today's industries, brown algae are utilized to produce energy (as a substitute for fossil fuels) and biodiesel, but green algae can also be used (Gamal 2011).
2. **Green algae**: it is also macro algae which are also known as chlorophyta and due to the presence of chlorophyll it is green in color. They are in high demand for the production of biofuel because they have high concentrations of sugar and cellulose. It contain anti bacterial, anti fungal, anti inflammatory, immunosuppressive activity (Awad 2000)
3. **Red algae**: A macro algae family algae known as rhodophyta. Due to the presence of phycoerythrin and phycothcyanin pigment it is red in color. Its cell wall consists of cellulose and agar and it contains anti viral, anti microbial, anti inflammatory activity and cytotoxic activity (Choi et al. 2000).
4. **Diatoms**: these are microalgae, used as energy source. They are mostly grown for high-value components like pigments and proteins (Burton 2009).
5. **Blue-green algae**: These are similar to bacteria, family of microalgae, known as cyanobacteria contain antibacterial, antifungal, antiviral and cytotoxic activity (Burton 2009).

A wide variety of biofuels, including biodiesel, bioethanol, and biobutanol, can be produced from algae, including very high concentrations of lipids or carbohydrates, depending on the species and growing circumstances (Edwards 2008). It's worth noting that algae-based biofuel production is still in the development stage, and various challenges, including scalability, cost-effectiveness, and strain productivity, need to be addressed for widespread commercial implementation. Ongoing research aims to improve cultivation techniques, optimize lipid extraction methods, and enhance the overall efficiency and sustainability of algae-based biofuel production.

8.5 Challenges of Plant Harvested Biofuel Production

Despite the progress made in plants harvested biofuel production, it faces several challenges that need to be addressed for sustainable development as we look ahead, the future scope of plant-based biofuels is vast, but it also faces several challenges that need to be addressed for successful implementation and widespread adoption.

1. **Food Versus Fuel Debate**: The use of edible crops for biofuel production has raised concerns regarding food security and increased competition for agricultural land. Striking a balance between fuel production and food supply is essential.
2. **Land and Water Resources**: Expanding biofuel production requires significant land and water resources. This may lead to deforestation, habitat destruction, and increased water usage, potentially causing environmental degradation.
3. **Greenhouse Gas Emissions**: While biofuels are considered carbon–neutral, the production processes, including cultivation, processing, and transportation, may contribute to greenhouse gas emissions. The net carbon savings from first-generation biofuels can be limited, highlighting the need for further improvements.

Plant-based biofuels hold significant promise as a renewable and environmentally friendly energy source. However, realizing their full potential requires addressing challenges related to feedstock availability, technological advancements, sustainability concerns, economic viability, policy frameworks, and public perception. By overcoming these challenges, the future scope of plant-based biofuels can contribute to a more sustainable and resilient energy future.

8.6 Advantages and Applications of Plant Based Biofuel Production

Despite the challenges, plants based biofuels continue to play a vital role in the transition towards renewable energy sources. To ensure their sustainable development, the following areas hold promise for future advancements:

- They are renewable and sustainable, as they are derived from crops that can be grown every year.
- Plant-based biofuels have emerged as a promising alternative to fossil fuels, offering reduced greenhouse gas emissions, energy security, and the potential for rural development.

- Biofuels made from plants are more environmentally friendly than fossil fuels, as they emit fewer greenhouse gases and pollutants, contributing to lower levels of air pollution and improving overall air quality. Third, they can contribute to energy security, reducing dependence on foreign oil and stabilizing energy prices.
- They are used as transportation fuels in the form of ethanol and biodiesel.
- They are also used to generate electricity through combustion or co-firing with coal in power plants.
- In addition, they can be used in heating systems for residential and commercial buildings, providing an alternative to traditional heating oil.

8.7 Future Aspects of Plants Harvested Biofuel Production

1. **Nanotechnology**: The importance of using nanotechnology to improve biofuel production in numerous industries is growing (Jang et al. 2011; Kokand et al. 2021). Due to its high surface area, it plays a crucial function in the macro algal biorefinery by synthesizing nanoparticles for moving components within the algal biomass. Immobilizing cellulose with nanoparticles is a novel technique for improving the production of biofuel from macro algae, which could reduce the need for hydrolyzing enzymes [91].
2. **Expansion of Feedstock Availability**: The future scope of plant-based biofuels relies on the availability of sufficient and sustainable feedstock. While current biofuel production primarily utilizes food crops such as corn and sugarcane, future efforts must focus on developing advanced technologies to produce biofuels from non-food biomass, such as agricultural residues, algae, and dedicated energy crops. This expansion of feedstock availability will reduce competition with food production and help ensure long-term sustainability.
3. **Technological Advancements**: Continued research and development are crucial for advancing plant-based biofuel technologies. This includes improving the efficiency of biofuel conversion processes, developing new catalysts, and optimizing cultivation techniques for high-yield energy crops. Additionally, advancements in genetic engineering and synthetic biology may enable the production of biofuels with enhanced properties and reduced environmental impacts.
4. **Sustainability and Environmental Concerns**: While plant-based biofuels offer the potential to mitigate climate change, there are concerns about their overall sustainability and environmental impacts. The future scope of biofuels should focus on implementing robust sustainability criteria, such as land-use planning, water management, and avoiding deforestation. It is essential to ensure that the production of biofuels does not result in negative consequences, such as increased land degradation, water scarcity, or biodiversity loss.

5. **Economic Viability and Market Integration**: For plant-based biofuels to reach their full potential, they must become economically competitive with fossil fuels. Future efforts should concentrate on reducing production costs through technological advancements, scale-up of production facilities, and supportive policies such as tax incentives and subsidies. Additionally, integrating biofuels into existing fuel infrastructure and establishing a well-functioning market will be crucial for their widespread adoption.
6. **Policy and Regulatory Frameworks**: The future scope of plant-based biofuels depends heavily on supportive policy frameworks. Governments should establish long-term, stable policies that provide incentives for research, development, and commercialization of biofuels. Furthermore, regulatory frameworks should ensure that biofuels meet sustainability criteria and provide clear guidelines for feedstock sourcing, production processes, and emissions standards.
7. **Public Perception and Acceptance**: The successful implementation of plant-based biofuels also relies on public acceptance and perception. Education and awareness programs can help address misconceptions and highlight the environmental benefits of biofuels. Engaging stakeholders, including farmers, local communities, and environmental organizations, in the decision-making process can foster acceptance and facilitate the development of sustainable biofuel value chains.
8. **Advanced Conversion Technologies**: Research and development efforts should focus on improving conversion technologies to enhance the efficiency of biofuel production processes. This includes exploring novel fermentation techniques, enzyme engineering, and catalytic processes.
9. **Second and Third-Generation Biofuels**: Second-generation biofuels, produced from lignocellulosic biomass, and third-generation biofuels, derived from algae, offer the potential for higher yields, reduced environmental impact, and improved sustainability compared to first-generation biofuels. The use of plant-based biofuels offers several benefits.

Plant-based biofuels have gained significant attention as a renewable and sustainable source of energy. Ethanol and biodiesel are the most common plant-based biofuels and have various applications, including transportation, electricity generation, and heating. While there are challenges associated with the use of plant-based biofuels, they offer several benefits, including lower greenhouse gas emissions, improved air quality, and energy security. As technology advances and production processes become more efficient, plant-based biofuels are likely to become an increasingly important source of renewable energy.

It's worth noting that the exact classification of biofuels into these generations can be somewhat subjective, and there is ongoing research into new and innovative feedstocks and processing methods. In addition to these crops, agricultural waste such as corn stover, wheat straw, and sugarcane biogases can also be used to produce biofuels. Overall, there are many plant sources for biofuel production, each with its own advantages and disadvantages. The choice of plant source depends on a variety of factors, including climate, geography, and availability of resources.

8.8 Conclusion

In conclusion, plant-based biofuel production presents a promising pathway towards a more sustainable and environmentally friendly energy sector. The administration of gasoline as a diesel fuel substitute and in blends has numerous benefits. By harnessing the power of renewable biomass resources, we have the opportunity to reduce our reliance on fossil fuels, mitigate climate change, and promote rural development. However, the successful implementation of plant-based biofuels requires addressing various challenges. Vegetable oils can be derived from a variety of oil crops that are grown in settings that range widely, some of which are unsuitable for conventional agricultural agriculture. Oils for the manufacturing of biodiesel can be produced from microalgae growing in ponds and photo biological reactors. Additionally, used cooking oils and fat byproducts from the meat processing sector may be utilized in the creation of biodiesel. No matter the scale of production, the stages of the production process are the same, albeit equipment variations may be substantial. Following the treatment of the beginning ingredients, the transesterification reaction yields a mixture of fatty acids containing methyl esters (FAME) with glycerin as a byproduct (often with methanol and a basic catalyst). To meet the specifications defined by global biodiesel standards, the combination of methyl ester compounds needs to be extracted from the glycerin and purified.

Expanding the availability of diverse and sustainable feedstock's, such as agricultural residues and energy crops, is crucial for long-term viability. Technological advancements in conversion processes and genetic engineering can enhance efficiency and improve the properties of biofuels. Sustainability considerations must be carefully integrated into the production cycle, ensuring responsible land use, water management, and biodiversity preservation.

Additionally, creating a favorable economic environment through cost reduction strategies, market integration, and supportive policies will drive the widespread adoption of plant-based biofuels. Policymakers play a vital role in establishing clear regulations, sustainability criteria, and long-term incentives to foster research, development, and commercialization efforts. Public perception and acceptance can be nurtured through education, awareness, and stakeholder engagement, fostering a positive outlook on the benefits of biofuels.

By overcoming these challenges, plant-based biofuel production can contribute significantly to a more sustainable energy landscape, reducing greenhouse gas emissions, enhancing energy security, and promoting rural economies. It represents a key component of the broader transition towards a low-carbon future, offering a viable alternative to fossil fuels while mitigating environmental impacts. With continued innovation, collaboration, and commitment, plant-based biofuels can play a vital role in shaping a more sustainable and resilient energy sector for generations to come.

References

Agudelo Santamaría JR, Benjumea Hernández PN (2004) Biodiesel de aceite crudo de Palma colombiano: aspectos de su obtención y utilización

Allakhverdiev SI, Kreslavski VD, Thavasi V, Zharmukhamedov SK, Klimov VV, Nagata T, et al. (2009) Hydrogen photoproduction by use of photosynthetic organisms and biomimetic systems. Photochem Photobiol Sci 8:148e56

Allen SG, Schulman D, Lichwa J, Antal MJ, Laser M, Lynd LR. (2001) A comparison between hot liquid water and steam fractionation of corn fiber. Ind Eng Chem Res 40:293441. Available from: https://doi.org/10.1021/ie990831h.

Aprodu I, Banu I (2017) Milling, functional and thermo-mechanical properties of wheat, rye, triticale, barley and oat. J Cereal Sci 77:42–48. https://doi.org/10.1016/j.jcs.2017.07.009

Awad NE. Biologically active steroid from the green alga Ulva lactuca. Phytother Res 2000; 14:6413. Available from: https://doi.org/10.1002/1099-1573(200012)14:8 , 641: aid-ptr668 . 3.0. co;2-r

Azapagic A (2014) Sustainability considerations for integrated biorefineries. Trends Biotechnol 32(1):1–4

Balat M, Balat H (2009) Appl Energy 86:2273–2282

Baras J, Gaćeša S, Pejin D (2002) Chem Ind 56:89–105

Belarbi E-H, Molina E, Chisti Y (2000) Retracted: a process for high yield and scaleable recovery of high purity eicosapentaenoic acid esters from microalgae and fish oil. Process Biochem 35:95169. (00)00126–6 Available from: https://doi.org/10.1016/s0032-9592

Besada V, Andrade JM, Schultze F, Gonza´lez JJ. Heavy (2009) metals in edible seaweeds commercialised for human consumption. J Marine Systems. 75:30513. Available from: https://doi.org/10.1016/j.jmarsys.2008.10.010

Bjerre AB, Olesen AB, Fernqvist T, Plo¨ger A, Schmidt AS (2000) Pretreatment of wheat straw using combined wet oxidation and alkaline hydrolysis resulting in convertible cellulose and hemicellulose. Biotechnol Bioeng. 49(5):56877

Bulkowska K, Gusiatin ZM, Klimiuk E, Pawlowski A, Pokoj T (2016) Biomass for biofuels. London, United Kingdom: Taylor and Francis Group. ISBN: 978–1–138–02631–5

Burton T (2009) Sustainable energy Ireland. Sustainable energy Ireland: 188

Carlsson AS, van Beilen JB, Moller R, Clayton D. (2007) In: Bowles D, € editor. Micro- and macro-algae: utility for industrial applications, outputs from the EPOBIO project. Newbury (UK): University of York: CPL Press. p 1e82

Choi JS, Park HJ, Jung HA, Chung HY, Jung JH, Choi WC. (2000) A cyclohexanonyl bromophenol from the red alga Symphyocladia latiuscula. J Nat Prod. 63:17056. Available from: https://doi.org/10.1021/np0002278

Christenson L, Sims R (2011) Production and harvesting of microalgae for wastewater treatment, biofuels, and bioproducts. Biotechnol Adv. 29:686702. Available from: https://doi.org/10.1016/j.biotechadv.2011.05.015

Chum H, Faaij APC, Moreira J, Edenhofer O (2012) Renewable energy sources and climate change mitigation. In: Girardin LO, Roman M, Special report of the Intergovernmental Panel on Climate Change. Cambridge: Cambridge University Press, p 209–331. Available from: http://rael.berkeley.edu/old_drupal/sites/default/files/IPCC 2011—Special Report on Renewable Energy Sources and Climate Change Mitigation.pdf. Accessed in: Feb. 18, 2016

Cripwell RA et al. (2019) Construction of industrial *Saccharomyces cerevisiae* strains for the efficient consolidated bioprocessing of raw starch Biotechnol. Biofuels

Dahman Y, Syed K, Begum S, Roy P, Mohtasebi B (2019). Biofuels. biomass, Biopolymer-Based materials, and bioenergy, 277–325. https://doi.org/10.1016/b978-0-08-102426-3.00

Demirbas A (2009) Political, economic and environmental impacts of biofuels: a review. Appl Energy. 86:S108e17

Dhillon R, von Wuehlisch G (2013) Mitigation of global warming through renewable biomass. Biomass Bioenergy 48:7589

Dias MOS, Ensinas AV, Nebra SA, Filho RM, Rossell CEV, Maciel MRW (2009) Production of bioethanol and other bio-based materials from sugarcane bagasse: integration to conventional bioethanol production process. Chem Eng Res Des. 87:1206e16
Dragone G, Fernande B, Vicente AA, Teixeira JA (2010) Third generation biofuels from microalgae. In: Mendez-Vilas A, editor. Current research, technology and education topics in applied microbiology and microbial biotechnology. Formatex. p 1355e66
Edwards MR (2008) Green algae strategy: end oil imports and engineer sustainable food and fuel. ISBN-10: 1440421846; ISBN-13: 978–1440421846
Ehhalt D, Prather M, Dentener F, Derwent R, Dlugokencky EJ, Holland E et al. (2001) Atmospheric chemistry and greenhouse gases. Pacific Northwest National Lab. (PNNL), Richland, WA
Empresa Brasileira de Pesquisa Agropecuária (2009) http://www.embrapa.br/. Accessed 6 November
Eppink MHM, Barbosa MJ, Wijffels RH (2018) Biorefining of microalgae: production of high-value products, bulk chemicals and biofuels. Biorefinery. Available from: https://doi.org/10.1515/energyo.0114.00001.
European Commission (EC), Directive 2009/28/EC, The promotion of the use of energy from renewable sources, 2009. https://www.inforse.org/europe/EU_res-directive.htm (accessed on 11 July 2021)
Evans G. International biofuels strategy project. Liquid transport biofuels—technology status report, NNFCC 08–017: National Non-Food Crops Centre; 2008. Available from http://www.globalbioenergy.org/uploads/media/0711_NNFCC__Liquid_Transport_Biofuels_Technology_Status_Report.pdf
Figueiredo M.K.-K, Romeiro GA, ďAvila LA, Damasceno RN, Franco AP (2009) The isolation of pyrolysis oil from castor seeds via a low temperature conversion (LTC) process and its use in a pyrolysis oil–diesel blend, Fuel 88 2193–2198, https://doi.org/10.1016/j.fuel.2009.05.025
Gamal AAE (2011) Biological importance of marine algae. Handbook of marine macroalgae: 135. https://doi.org/10.1002/9781119977087.ch1
Gnansounou E, Bedniaguine D, Dauriat A (2005) In: Proce- edings of the 7th IAEE European energy conference. Bergen, Norway
Greene CH, Pershing AJ (2007) Climate drives sea change. Sci. -New York Then Washington- 315(5815):1084
Hafeez S, Al-Salem SM, Manos G, Constantinou A (2020) Fuel production using membrane reactors: a review. Environ Chem Lett 18:1477–1490. https://doi.org/10.1007/s10311-020-01024-7
Hajilary N, Shirazian S, Rezakazemi M (2019) Biofuel types and membrane separation, Environ. Chem. Lett. 17 1–18 https://doi.org/10.1007/s10311-018-0777-9
Harmsen HW (2010) Literature review of physical and chemical pretreatment processes for lignocellulosic biomass energy:149
Hartman EA (2008) promising oil alternative: algae energy, Washington Post, 2008. Available from: http://www.washingtonpost.com/wp-dyn/content/article/2008/01/03/AR2008010303907.html?noredirect 5 on
Heiman K (2016) Novel approaches to microalgal and cyanobacterial cultivation for bioenergy and biofuel production. Curr Opin Biotechnol. 38:183e9
Histi Y (2007) Biodiesel from microalgae. Biotechnol Adv 25:249e306
Huber GW, Corma A (2007) Synergies between bio- and oil refineries for the production of fuels from biomass. Angew Chem Int Ed 46:7184–7201
Inderwildi OR, King DA (2009) Quo vadis biofuels? Energy Environ Sci. 2:343. Available from: https://doi.org/10.1039/b822951c
Jang J-S, Cho Y, Jeong G-T, et al. (2011) Optimization of saccharification and ethanol production by simultaneous saccharification and fermentation (SSF) from seaweed, Saccharina japonica. Bioprocess Biosyst Eng. 35(1–2):11–18
Jeswani HK, Chilvers A, Azapagic A (2020) Environmental sustainablility of biofuels: a review. Royal Society Publishing 476(2243):2020035. https://doi.org/10.1098/rspa.2020.0351

Jiang L, Luo S, Fan X, Yang Z, Guo R (2011) Biomass and lipid production of marine microalgae using municipal wastewater and high concentration of CO_2. Appl Energy 88:333641. Available from: https://doi.org/10.1016/j.apenergy.2011.03.043.

Khan SA, Rashmi, Hussain MZ, Prasad S, Banerjee U (2009) Prospects of biodiesel production from microalgae in India. Renew Sustain Energy Rev. 13:236172. Available from: https://doi.org/10.1016/j.rser.2009.04.005.

Khoo KS, Chia WY, Tang DYY, et al. (2020) Nanomaterials utilization in biomass for biofuel and bioenergy production. Energies. 13(4):892. [Google Scholar]

Kokand AK, Chew KW, Manickam S, et al. (2021) Emerging algal nanotechnology for high-value compounds: a direction to future food production. Trends Food Sci Technol. 116:290–302

Kulkarni M, Gopinath R, Meher LC, Dalai AK (2006) Solid acid catalyzed biodiesel production by simultaneous esterification and transesterification. Green Chem 8:1056–1062

Ministerio da Agricultura Pecuária e Abastecimento do Brazil http://www.agricultura.gov.br/. Accessed 6 November 2009

Lee S, Speight JG, Loyalka SK (2007) Hand book of alternative fuel technologies. CRC Taylor and Francis Group, USA

Levin D, Islam R, Cicek N, Sparling R (2006) Hydrogen production by Clostridium thermos cellum 27405 from cellulosic biomass substrates. Int J Hydrogen Energy 31:1496503. Available from: https://doi.org/10.1016/j.ijhydene.2006.06.015

Li S, Yang X (2016). Biofuel production from food wastes. Luque R, Lin CSK, Wilson K, Clark J. In Handbook of biofuels production (pp. 617–653). Duxford, England: Woodhead Publishing. https://doi.org/10.1016/b978-0-08-100455-5.00020-5

McCarthy JJ (2001) Climate Change 2001: Impacts, Adaptation, and Vulnerability: Contribution of Working Group II to the Third Assessment Report of the Intergovernmental Panel on Climate Change. Cambridge University Press

Mizik T, Gyarmati G (2021) Economic and sustainability of biodiesel production—a systematic literature review. Clean Technol 3:19–36. https://doi.org/10.3390/cleantechnol3010002

Mohanty SM, Behera S, Swain MR, Ray RC (2009) Appl Energy 86:640–644

Mojovic L, Pejin D, Grujic O, Markov S, Pejin J, Rakin M, Savic D (2009) Progress in the production of bioethanol on starch-based feedstocks. Chem Ind Chem Eng Q 15(4):211–226. https://doi.org/10.2298/ciceq0904211m

Mojović L, Pejin D, Lazić M (2007) in: Bioethanol as fuel—state of the art and perspectives, Faculty of Tecnology, Leskovac, pp. 1–149 (in Serbian)

Naik SN, Goud VV, Rout PK, Jacobson K, Dalai AK. Characteisation of Canadian biomass for alternate renewable biofuel. Renewable Energy; in Press

Nakas MS (1983) System development for linked-fermentation production of solvents from algal biomass. Appl Environ Microb 46:101723

Nikolić S, Mojović L, Pejin D, Rakin M, Vučurović V (2009a) Food Technol. Biotechnol. 47:83–88

Nikolić S, Mojović L, Rakin M, Pejin D, Nedović V (2009b) J Chem Technol Biotechnol 84:497–503

No SY (2019) Production of liquid biofuels from biomass, Application of Liquid Biofuels to Internal Combustion Engines, Green Energy and Technology, Springer. Singapore. https://doi.org/10.1007/978-981-13-6737-3_1

Oumer AN, Hasan MM, Baheta AT, Mamat R, Abdullah AA (2018) Bio-based liquid fuels as a source of renewable energy: a review. Renew Sustain Energy Rev 88:82–98. https://doi.org/10.1016/j.rser.2018.02.022

Parry M, Canziani O, Palutikof J, van der Linden PJ, Hanson CE (2007) Climate Change 2007: Impacts. Cambridge University Press, Cambridge, Adaptation and Vulnerability

Pittman JK, Dean AP, Osundeko O (2011) The potential of sustainable algal biofuel production using wastewater resources. Bioresour Technol. 102:1725. Available from: https://doi.org/10.1016/j.biortech.2010.06.035.

Popoola TOS, Yangomodou OD (2006) Extraction, properties and utilization potentials of cassava seed oil. Biotechnology (Faisalabad). 5:3841. Available from: https://doi.org/10.3923/biotech.2006.38.41.

Potters G, Goethem DV, Schutte F (2010) Promising biofuel resources: lignocellulose and algae, Nat. Education 3(9):14 https://www.nature.com/scitable/topicpage/promising-biofuel-resources-lignocellulose-and-algae-14255919/

Poudyal RS, Tiwari I, Najafpour MM, Los DA, Carpentier R, Shen J-R, et al. (2015) Current insights to enhance hydrogen production by photosynthetic organisms. In: Stolten D, Emonts B, editors. Hydrogen science and engineering; p 461e87. Wiley-VCH Books

Razzak SA, Hossain MM, Lucky RA, Bassi AS, de Lasa H (2013) Integrated CO_2 capture, waste water treatment and biofuel production by microalgae culturing-a review. Renew Sustain Energy Rev. 27:622e53.

Resch G, Held A, Faber T, Panzer C, Toro F, Haas R (2008) Potentials and prospects for renewable energies at global scale. Energy Policy 36(11):40484056

Romano SD, González Suárez E (2009) Biocombustibles líquidos en Iberoamérica, 1st edn. Ediciones Cooperativas, Buenos Aires Secretaría de Agricultura, Ganadería, Pesca y Alimentos, República Argentina http://www.sagpya.mecon.gov.ar/new/0-0/agricultura/otros/estimaciones/infgeneral.php. Accessed 6 November 2009

Sakwa L et al (2018) Consolidated bioprocessing of raw starch with *Saccharomyces cerevisiae* strains expressing fungal alpha-amylase and glucoamylase combinations FEMS Yeast Res

Shelley M (2006) Alcoholic fuels. Boca Raton. FL 33487–2742, USA: CRC Taylor and Francis Group

Singh J, Gu S (2010) Commercialization potential of microalgae for biofuels production. Renew Sustain Energy Rev. 14:2596610. Available from: https://doi.org/10.1016/j.rser.2010.06.014

Slade R, Bauen A (2013) Micro-algae cultivation for biofuels: cost, energy balance, environmental impacts and future prospects. Biomass Bioenergy 53:29e38

Spolaore P, Joannis-Cassan C, Duran E, Isambert A (2006) Commercial applications of microalgae. J Biosci Bioeng. 101:8796. Available from: https://doi.org/10.1263/jbb.101.87

Tredici MR (2010) Photobiology of microalgae mass cultures: understanding the tools for the next green revolution. Biofuels. 1:14362. Available from: https://doi.org/10.4155/bfs.09.10.

United State Congress, Public Law 110–140, Energy Independence and Security Act of 2007, December 19, 2007. https://www.epa.gov/laws-regulations/summary-energy-independence-and-security-act

United Nations (UN), Transforming our world: the 2030 Agenda for Sustain. https://sustainabledevelopment.un.org/post2015/transformingourworld/publication

Vergarafernandez A, Vargas G, Alarcon N (2008) Velasco A. Evaluation of marine algae as a source of biogas in a two-stage anaerobic reactor system. Biomass Bioenergy. 32:33844. Available from: https://doi.org/10.1016/j.biombioe.2007.10.005

Voloshin RA, Kreslavski VD, Zharmukhamedov SK, Bedbenov VS, Ramakrishna S, Allakhverdiev SI (2015) Photoelectrochemical cells based on photosynthetic

Wang Y, Ou S, Liu P, Zhang Z (2007) Preparation of biodiesel from waste cooking oil via two-step catalyzed process. Energy Convers Manage 48(1):184–188. https://doi.org/10.1016/j.enconman.2006.04.016

World watch Institute (2014) How Will the U.S. Produce 36 Billion Gallons of Biofuel by 2022? http://www.worldwatch.Org/node/5600 (accessed 11 August 2014)

Wise, T (2012) the Cost to Developing Countries of U.S. Corn Ethanol Expansion. Global Development and Environment Institute working paper no. 12 – 02, http://www.ase.tufts.edu/gdae/Pubs/wp/12-02WiseGlobalBiofuels.pdf

Wu H, Miao X (2014) Biodiesel quality and biochemical changes of microalgae Chlorella pyrenoidosa and Scenedesmus obliquus in response to nitrate levels. Bioresour Technol. 170:4217. Available from: https://doi.org/10.1016/j.biortech.2014.08.017.

Yoo C, Jun S-Y, Lee J-Y, Ahn C-Y, Oh H-M (2010) Selection of microalgae for lipid production under high levels carbon dioxide. Bioresour Technol. 101. Available from: https://doi.org/10.1016/j.biortech.2009.03.030

Yuan JS, Tiller KH, Al-Ahmad H, Stewart NR, Stewart Jr. CN (2008) Plants to power: bioenergy to fuel the future. Trends in Plant Science, 13:8, p 421–429 Zych, D. The viability of corn cobs as a bioenergy feedstock. University of Minnesota, 2008. https://doi.org/10.3389/fpls.2013.0006

Chapter 9
Plant-Based Biofuels

Akshita Kanwar, Parmeshwar Vitthal Gavande, and Arun Goyal

Abstract The largest share of organic carbon is stored on Earth in the form of plant biomass which contains a range of carbohydrate polymers such as cellulose and hemicellulose along with lignin, to make up the lignocellulose. Crude oil-based fuels being both non-renewable and hazardous to the environment, plant-based biofuels promise hope for the ever-increasing energy demands of the future. Biofuel includes solid, liquid and gas fuels which are primarily obtained from photosynthetic organisms. Hence, the burning of biofuels is a carbon–neutral process. Plant-based biofuels can be majorly categorized as lignocellulosic bioethanol, biodiesel, biohydrogen and bio-oil. Lignocellulosic bioethanol is generated through systematic deconstruction of plant biomass which involves processes like efficient mechanical and chemical pre-treatment of selected biomass followed by its enzymatic hydrolysis during the saccharification. The biomass saccharified into simple C-6 or C-5 sugars (such as D-glucose or D-xylose, respectively) is then fermented to bioethanol followed by its collection through distillation or fractionation. Another major biofuel, Biodiesel is obtained from plant materials like seeds by extraction of plant oil followed by transesterification. This chapter reviews the different areas of plant-based biofuels, their economic viability, challenges and future prospects.

Keywords Biofuel · Biomass · Lignocellulose · Bioethanol · Biodiesel · Renewable energy

A. Kanwar · A. Goyal (✉)
Centre for the Environment, Indian Institute of Technology Guwahati, Guwahati 781039, Assam, India
e-mail: arungoyl@iitg.ac.in

P. V. Gavande · A. Goyal
Department of Biosciences and Bioengineering, Indian Institute of Technology Guwahati, Guwahati 781039, Assam, India

M. Shah and D. Deka (eds.), *Emerging Sustainable Technologies for Biofuel Production*, Environmental Science and Engineering, https://doi.org/10.1007/978-3-031-52167-6_9

9.1 Introduction

Climate Change could well place mankind in perils if no solution is sought. The global temperature is increasing year after year. The temperature at the surface of the earth and the proximal atmospheric layers is rising due to anthropogenic activities which have led to the global warming. Petroleum fuels used in vehicles cause the release of unburnt hydrocarbons, soot, volatile organic compounds, aldehydes, polycyclic aromatic hydrocarbons, CO_x, SO_x, NO_x, and Particulate Matter (PM) 2.5 particles (fine particles of size 2.5 μm) into the atmosphere (Sandaka and Kumar 2023). The persistence of these particles in the atmosphere leads to various secondary effects like the creation of ozone gas (O_3) near the surface of the earth, smog, and an increment in the concentration of Green House Gases (GHG's) (Sandaka and Kumar 2023). GHG's have commenced a ripple effect of climate change, global warming and the hole in the ozone layer in the upper atmosphere. According to Intergovernmental Panel on climate change (IPCC), GHG emissions can have detrimental implications on the world economy, food security and global health. The United Nations Framework Convention on Climate Change (UNFCCC) has targeted to achieve environmental decarbonisation and a reduction in the global warming at 1.5 °C at the earliest (Durga et al. 2022). The global mean surface temperature (GMST) for 2006–2015 period was 0.87 °C higher than the GMST for 1850–1900 period. The average temperature on the earth during the twentieth century was 14 °C. Since the pre-industrial era, there has been an increment of more than 1 °C in the average annual temperature till 2022. Many nations have agreed to attain net-zero carbon emissions between the years 2050 and 2070 (Durga et al. 2022). This has necessitated the development of green energy technologies that are sustainable, clean, and environment-friendly. Figure 9.1 represents the neutral carbon cycle (Sandaka and Kumar 2023). Carbon neutrality represents net-zero carbon emissions which is attained by an offset of total carbon (carbon dioxide gas or GHG's) produced directly or indirectly over a definite time period through carbon sequestration (Chen et al. 2022). Reduction in carbon dioxide (CO_2) emissions and removal of carbon from the atmosphere (negative emissions) are the two approaches towards carbon neutrality. Decarbonisation of the transport sector by employing biofuels for transportation can reduce (CO_2) emissions, while afforestation can curb the GHG emissions.

Fossil fuels account for 29 gigatons of carbon dioxide gas emissions yearly (Prasad et al. 2019). Fossil fuels are on the verge of exhaustion as they are non-renewable sources of energy. Moreover, the burning or combustion of fossil fuels causes global warming and major environmental issues. Crude oil (Petroleum) is the major fuel source which accounts for 32.9% of the global energy consumed; moreover, 40% of the world's electricity is supplied by coal. Renewable sources account for 18% of global energy consumption (Prasad et al. 2019). The production of ethanol and biodiesel is expected to expand to 134.5 and 39 billion litres, respectively, by 2024. Originally, ethanol was used by Nikolas Otto as a motor fuel in the internal combustion engine (ICE) in the year 1897. In the late twentieth century, countries like Brazil

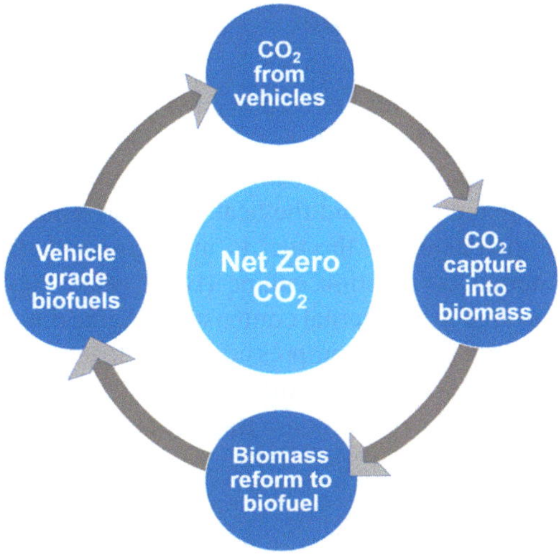

Fig. 9.1 Carbon–neutral cycle of biofuels attained by net zero carbon emissions

and the United States of America produced biofuel from biomass like maize and sugarcane.

9.2 Biofuel

The human population is on the rise that is now driving away the global supply of fuel energy. Biofuels are the best alternative to the fossil fuels. Biofuels can be produced from numerous varieties of biomass, for example, agricultural wastes, crops, municipal or domestic waste, algae etc. Biofuel made from biomass is less polluting with lower greenhouse gas emissions (Demirbas 2007). Biofuel refers to the solid, liquid, or gaseous fuel obtained from biomass. Liquid biofuels include biomethanol, bioethanol, biodiesel and Fischer–Tropsch diesel, while, biohydrogen and biomethane are gaseous fuels (Demirbas 2007). Biofuels have many advantages over fossil fuels. Biomass is easily and locally available. Biomass has no net carbon releases. Biofuels have an environment-friendly potential. Biofuel is a biodegradable (CO_2 emitted after burning fuel is utilised by plants causing no harm to environment) and sustainable source of energy. The combustion of biofuels represents a closed carbon-dioxide cycle. Biofuel has 10–45% of oxygen levels, while petroleum has none (Demirbas 2007). The sulphur and nitrogen levels of biofuel are significantly lower leading to reduced emissions of pollutants which would otherwise cause acid rain (Puppan 2002). Moreover, compared to other renewable energy resources like wind, water, solar, thermal etc., biomass can supply solid, liquid, and gaseous fuels, which can be stored, transported and availed distantly. Biomass has the capability to provide supply chain security (Demirbas 2007).

9.2.1 Bioethanol

Bioethanol is also termed as ethyl alcohol or grain alcohol with the chemical formula as $CH_3–CH_2–OH$. It is a liquid fuel with a higher-octane number which enables greater engine performance. The flammability limit of bioethanol is high; moreover, the flame speed and the heat of vaporization is higher than that of gasoline. These properties allow the fuel to undergo combustion without causing explosion in an internal combustion engine (Balat et al. 2008). Bioethanol is advantageous over gasoline in an internal combustion engine due to properties like shorter burning time, a higher ratio of compression and a leaner air-to-fuel ratio (Lean -burn), leading to theoretical efficiency (thermodynamic efficiency) as the excess oxygen in the engine emits less hydrocarbons (Balat et al. 2008).

9.2.2 Biodiesel

It is the best alternative for diesel used in diesel engines. Sources of biodiesel include plants (seeds) like *Zanthoxylum bungeanum, Guizotia abyssinica, Phoenix dactylifera, Datura stramonium, Terminalia belerica* etc. (Singh et al. 2014). Biodiesel is the methyl ester of fatty acids and often called as FAME (Fatty acid methyl esters). Biodiesel has numerous environmental and ecological benefits as against ordinary diesel used for power generation. Biodiesel has the ability to mitigate greenhouse gases, reduce air contamination, decompose the fuel effectively and leave no carbon unburnt (Georgogianni et al. 2008). Biodiesel is safe in terms of energy as its circulation and dispersion are local, reducing credibility on imported fuels. Moreover, biodiesel is economically viable. The fuel has the potential to provide sustainability and create nationwide industrial employment, besides generating enormous agricultural growth and profits (Riaz et al. 2022).

9.3 Plant Biomass Production

Biofuels comprise energy-enriched chemicals which are directly obtained from biological processes or indirectly obtained from biomass of living organisms like plants or algae by chemical conversion. The characteristic feature distinguishing plants from other renewable energy sources is the ability to photosynthesise (Rodionova et al. 2017). According to Kamen (1963), photosynthesis is defined as a serial process where electromagnetic energy from the sun light is transformed into chemical energy, which is used to carry out numerous biosynthetic processes (Kamen 2013). During the process of photosynthesis, light energy is absorbed by the green photosynthetic pigment of plants called chlorophyll (Björn et al. 2009). Chlorophyll utilises light energy to reduce carbon dioxide into sugars with the help of water.

There are two phases of photosynthesis: The light-dependent phase and the light-independent phase. The light-dependent phase is the first step or the primary process of photosynthesis in which a reducing power (NADPH) is generated in the presence of light (Rodionova et al. 2017). The light-independent phase is the second step in which the reduction of carbon dioxide into carbohydrates takes place in the dark with the help of reducing power. On striking the chlorophyll molecule, the visible light photon releases an electron of chlorophyll (on the porphyrin ring) in the outer orbital of the molecule, which is the excited state (Björn et al. 2009). The electron remains in its usual position in the ground state. The red-light photon excites an electron from the ground state to the first singlet state (S1). The blue light photon of higher energy than the red photon, excites the electron from the ground level molecular orbital to the second singlet state (S2). The energy of absorbed photons is manifested in the form of heat, fluorescence, and phosphorescence (Björn et al. 2009). Only a small part of the absorbed energy appears as chemical energy (ATP and NADPH), which is used in the dark reactions of photosynthesis for carbon dioxide assimilation into carbohydrates. The product synthesis phase of photosynthetic carbon dioxide assimilation is associated with producing various products like sugars, organic acids, fatty acids, and amino acids (Rodionova et al. 2017).

9.4 Classification of Biofuels Based on the Type of Bioresource

Biofuels are categorised into four generations: First, second, third and fourth generation. The first-generation biofuel utilises edible crops like starch and sugar containing sugarcane and corn. This raises the concern regarding 'Food versus Fuel' as the global supply of food is affected. Moreover, the first-generation biofuels require more land, water and fertilisers, which significantly increases the cost of production (Prasad et al. 2019). Second-generation biofuel uses inedible biomass like lignocellulose, including inexpensive wood, crop residues, sawdust, etc. Hence, second-generation biofuel employs better renewable alternative resources (Alalwan et al. 2019). The cellulosic biomass, generally the agricultural residues are abundantly available and are economically viable. The third-generation biofuel employs aquatic feedstock, like algal biomass. Algae have the ability to carry out photosynthesis. They can capture carbon dioxide to produce oxygen (Rodionova et al. 2017). In one of the studies towards the environmental implications of various technological processes for biodiesel production, a Life Cycle Inventory (LCI) was compiled (Martins et al. 2018). Results from the process inputs revealed that the cultivation of algal biomass demands a considerable amount of nutrients in the form of nitrogen. A significant amount of freshwater is required to cool the production system during summer (thermoregulation). Hence, the most water-exhaustive step of the system is algal biomass cultivation. Nevertheless, in the study, the water was reused at a rate of 90%, leading to a decrement in the water footprint. As algae consumed carbon dioxide for its growth, a considerable decrement in the carbon footprint of the process was also observed, which led to the conclusion that the process has the potential to mitigate global warming (Branco-Vieira et al. 2020). For biodiesel production from microalgae, the most energy-intensive steps include biomass drying, disruption of cells by ball milling technology, and biomass harvesting by pumping the biomass culture followed by centrifugation. The model suggested that a total of 181 tons of algal biomass can be produced, generating 15 tons of biodiesel annually. 12 kg of dried algal biomass is required to produce 1 kg of biodiesel (Branco-Vieira et al. 2020). However, the oil generated by algae is more volatile at high temperatures due to its high unsaturation and is more likely to degrade due to increase in viscosity (Prasad et al. 2019). The fourth-generation biofuels employ bioengineered micro-organisms like genetically modified crops or algae (Alalwan et al. 2019). These biofuels are in the stage of development. The microorganisms are genetically altered to consume a more significant amount of environmental carbon dioxide than what the fossil fuels emit on burning.

9.5 Plant Biofuels Based on Biomass Containing Starch and Sucrose

Several crops containing carbohydrates have been employed as feedstock for bioethanol production. Bioethanol can be produced from sugar containing crops like sugar cane, sweet sorghum, beet root and fruits or starch containing crops such as wheat, rice and corn (Almodares and Hadi 2009). Sweet sorghum grows rapidly, possesses high sugar accumulation ability and adapts well to surroundings. Moreover, it possesses higher biomass potential as the stalk along with grains and fibre containing cellulose can be used for biofuel production. Sweet sorghum has a high efficiency of photosynthesis owing to its C4 pathway (Anglani 1998). The main sugars present in the kernels of sorghum are glucose and fructose (monosaccharides), sucrose and maltose (disaccharides) and raffinose (trisaccharide). For ethanol production, the sweet sorghum juice is extracted from the stalk by milling followed by screening and sterilization by heating the juice at 100 °C (Almodares and Hadi 2009). Thereafter, the sweet sorghum juice is clarified by adjusting the pH and removing the impurities (Quintero et al. 2008). The juice is filtered using a rotary vacuum filter, subjecting to evaporation for concentration, wherein the syrup is converted into ethanol by an yeast *Saccharomyces cerevisiae* by fermentation (Almodares and Hadi 2009). In the process, reducing sugar is converted into ethanol, carbon dioxide and small quantities of aldehydes, ketones and glycerol. The alcohol (ethanol) obtained is further subjected to distillation and dehydration leading to concentration up to 95% (v/v). Agriculture and Agri-Food Canada (AAFC, Ottawa) has developed hybrid corn possessing a high content of stalk sugar which is referred to as 'sugarcorn' (Reid and Morrison 2015). Bioethanol was produced using the juice extracted from these rapidly-growing energy crops through fermentation by microbes. Sucrose, glucose and fructose accounted for about 80% of total sugar (Gomez-Flores et al. 2018). The sugarcorn juice contained 145 g/L of total carbohydrates. The sugarcorn juice was supplemented with 3.0 g/L of yeast extract for the growth of *Saccharomyces cerevisiae* that gave 45.6 g/L of ethanol after 72 h fermentation. The ethanol yield of 0.41 g per gram of carbohydrates was obtained.

9.6 Biodiesel Production from Plant Based Oil

Biodiesel is produced from plant seed oil by employing four methods namely, transesterification, pyrolysis, microemulsion and hydrocarbon blending (Riaz et al. 2022). The transesterification reaction is employed widely for biodiesel production which converts the triglycerides (fats) into methyl esters in the presence of catalyst. Transesterification is a reversible three step reaction where one mole of ester is released at every individual step. The triglyceride gives rise to diglyceride, subsequently the diglyceride forms a monoglyceride and finally the monoglyceride forms glycerol on reaction with alcohol with the release of an ester at every individual step as shown in Fig. 9.2 (Hincapié et al. 2011). Stoichiometrically, 1 mol of triglyceride reacts with 3 mol of short chain alcohol (methanol or ethanol) to form 3 mol of alkyl esters and 1 mol of glycerol. The catalysts employed for transesterification are categorised as homogeneous such as sodium hydroxide and potassium hydroxide catalysts and heterogenous catalysts such as sulfonated graphine and calcium oxide. There are two methods for carrying out the transesterification reactions: Conventional and in situ (Riaz et al. 2022). The conventional method uses refined oil (pre-extracted oil from vegetable seed) for the reaction with an alcohol and as such there is a stage of oil extraction prior to the transesterification. The plant seeds are heated at 85 °C in oven, followed by maceration with extracting solvent (petroleum ether). The solid coarse sand sized product is transferred to the Soxhlet extractor for extraction of oil. The solvent is removed from the oil by using rotary evaporator at 40 °C. The plant oil, methanol and NaOH are refluxed together and heated at 60 °C, besides stirring by a low frequency ultrasonicator. When the plant oil is completely converted to biodiesel, the reaction is stopped and the mixture is allowed to stand. Phase separation allows the formation of two layers: Upper layer of ester mixture and lower layer of glycerine (Hincapié et al. 2011). However, the in situ method of transesterification uses the plant seeds directly to participate in the reaction with the alcohol without extracting the oil from seeds to produce fatty acid methyl esters (Georgogianni et al. 2008). The plant seeds after maceration are transferred to the reaction flask in which methanol and NaOH are added. This step is followed by heating and stirring. After the complete plant oil conversion, the reaction is stopped. The anoxygenic thermal transformation of a substance into another by heating alone or employing a catalyst is called pyrolysis. In the process of pyrolysis, the long chain hydrocarbons of the plant oil are broken down to short chain hydrocarbons which can be condensed to produce biofuel. The products formed by the process of pyrolysis of plant oil at a temperature range of 500–600 °C include biochar (solid fuels), bio-oil (Liquid fuel)

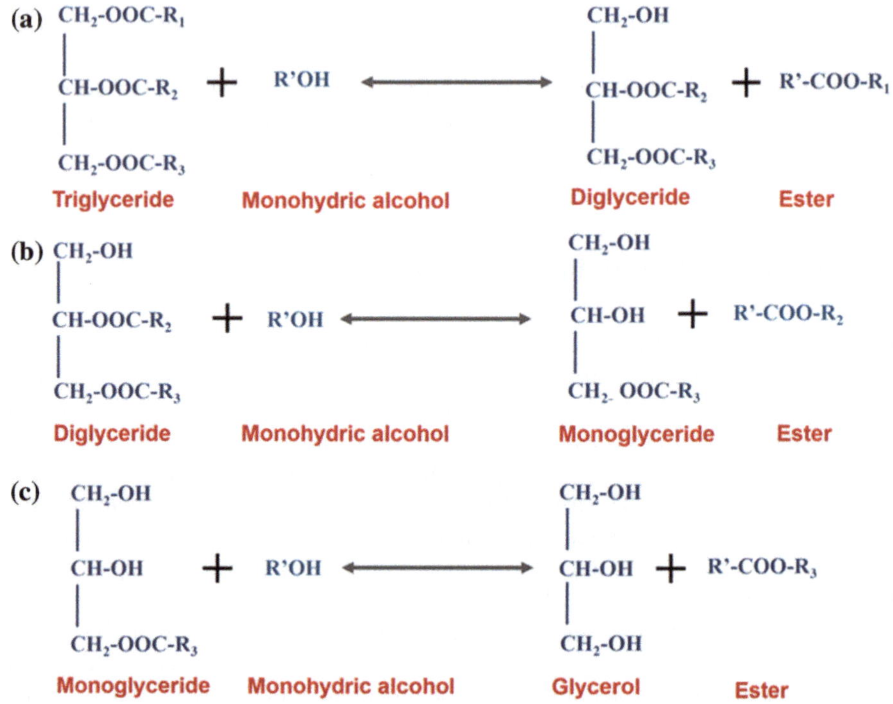

Fig. 9.2 Transesterification reaction for biodiesel production. **a** Triglyceride reacts with monohydric alcohol to form diglyceride and ester. **b** The diglyceride reacts with another mole of alcohol to form monoglyceride and ester. **c** Finally, the monoglyceride reacts with alcohol to form glycerol and mono alkyl ester

and biogas (gas fuel) (Riaz et al. 2022). Microemulsions involve the emulsification of the extracted plant oil (via hydraulic press with pressure up to 100 MPa) with emulsifying agents like alcohol by mixing. The microemulsions have less viscosity and are thermodynamically stable. Microemulsions have three components, oil-phase, aqueous-phase and surfactant which forms clear and stable fluids. The technique of hydrocarbon blending (dilution) involves the mixing of alternative or renewable fuels such as biodiesel with the contemporary fossil fuels like diesel (Riaz et al. 2022). Blending the extracted plant oil with diesel reduces viscosity and enhances volatility of the biofuel. The use of vegetable oils blended with diesel requires significant modifications in the engine, such as replacement of pipes prone to corrosion.

Linseed oil (*Linum usitatissimum)* has been used for biodiesel production as the plant has non-edible seeds with a considerable amount of oil in the range of 30–40% (Bahadorian et al. 2023). The seeds have substantial unsaturated and polyunsaturated fatty acid content (Taherkhani and Sadrameli 2018). The biodiesel production was optimised by employing in situ transesterification of linseed oil along with the conversion of linseed residue into biochar, bio-oil and biogas in a pyrolysis reactor (Bahadorian et al. 2023). Parameters of transesterification used were particle size of seeds (0.2–0.3 mm), ratio of solvent (n-Hexane)-solid (seed) (9.14 ml/g) and the ratio of co-solvent acetone-alcohol ethanol (0.52 ml/ml). Other parameters like temperature (55 °C) and time (180 min) of reaction, rate of agitation (800 rpm) and concentration of the KOH catalyst (based on amount of oil; 5%) were kept constant. The biodiesel production 960 g/L with 96.8% purity was obtained (Bahadorian et al. 2023). *Madhuca indica* is an indigenous tree inhabiting in the temperature range, 2–46 °C with an annual rainfall requirement in the range, 750 mm-1875 mm. The flowers of the tree are used for the production of alcohol, while the seeds are used for producing oil (Kapilan 2022). The use of *Madhuca indica* tree-oil for commercial biodiesel production was studied in order to establish the technical and economic feasibility of biodiesel. The molecular weight of Madhuca oil is found out to be 872.4 g/mole and its composition in terms of fatty acids was: Arachidic acid (1.25%), Eicosenoic acid (1.72%), Linoleic acid (24.52%), Myristic acid (1.69%), oleic acid (39.75%) Palmitic acid (13.47%) and stearic acid (17.15%) (Sudalai et al. 2023). Transesterification of the oil was done using a heterogenous catalyst which was a functionalised industrial waste dolomite (copper-modified-dolomite catalyst). The optimised reaction conditions for transesterification were as follows: Temperature 75 °C, reaction time 6 h, copper-modified-dolomite catalyst concentration 3% (w/w) and methanol/oil molar ratio 15:1; an average biodiesel yield of 96.9% was obtained. It was observed that on increasing the concentration of copper-modified-dolomite catalyst, the yield of fatty acid methyl esters (FAME) and the oil conversion efficiency increased remarkably. Nevertheless, neither the oil conversion nor the FAME yield was improved by increasing the concentration of the catalyst beyond 3% (v/v) (Sudalai et al. 2023). This is explained by the fact that adding more catalysts would increase the rate of reaction and not the yield of the product referred to as reaction equilibrium constraint. This reaction equilibrium constraint is due to increased viscosity of the reaction mixture leading to poor diffusion across the three-phase methanol-oil-catalyst systems (Faruque et al. 2020). The physicochemical properties of the biodiesel obtained from *M. indica* fall within the range of American Society of Testing and Materials (ASTM) and European (EN) standards (Sudalai et al. 2023). In a study, biodiesel was produced from *Jatropha curcas* oil in the presence of different catalysts prepared from the fruit peel, stem and rhizome of post-harvested banana plant (*Musa champa*) (Basumatary et al. 2023). Cost effectiveness of both the catalyst and biodiesel production was justified as the catalysts prepared from the post-harvest plant were environment-friendly, innocuous, renewable, economical, inexhaustible and certainly accessible. The peel of *Musa champa* was calcined at 550 °C for 2 h. The catalyst CMCP-550 exhibited best results with a biodiesel yield of 98%. The optimised conditions for the reaction include molar ratio of methanol to oil (MRMO) of 9:1, CMCP-550 catalyst dose of 5 wt.%, temperature at 65 °C and a reaction time of 10 min (Basumatary et al. 2023).

Employing a single feedstock (edible or non-edible vegetable oil) for biodiesel production has several demerits like soil fertility losses, mono-cultivation and increased cost of cultivation. The cultivation of single feedstock for oil production coincides with food crisis worldwide in terms of high cost of food items (Ong et al. 2019). Moreover, biodiesel obtained from single feedstock has been known to affect diesel engine and fuel injection systems, besides tank corrosion and storage and combustion issues. Nevertheless, mixing of different oil feedstocks overcomes the above-mentioned problems. The chemical composition and acid value of feedstock change by mixing oil in different proportions (Riaz et al. 2022). Several studies have been carried out on the mixing of oils from different feedstocks. In one such study, biodiesel was produced from the mixture of five oil feedstocks: Sunflower, Jatropha, Canola, Pongamia and Soybean. The quinary oil mixture was taken in an equivalent ratio (Brahma and Basumatary 2023). Calcined materials (thermal treatment at high temperature without melting the material) prepared from rhizome, trunk and peel of *Musa chinensis* post-harvest were used as heterogenous base catalysts for transesterification. The calcined peel catalyst was the most efficient, yielding 95% biodiesel at the optimum conditions of 9:1 MRMO (Molar ratio of methanol to oil), 5 wt.% catalyst dose, temperature 65 °C and 11 min of reaction time. The catalyst was recycled up to three times with a biodiesel production of 900 g/L (Brahma and Basumatary 2023). Table 9.1 shows the various feedstock employed for biodiesel production.

9.7 Plant Lignocellulosic Biomass-Based Biofuel

73×10^6 tons of global dry squandered crops have the potential to generate 49 gallons per year of bioethanol (Prasad et al. 2019). Figure 9.3 shows various processes to obtain biofuel from biomass.

9.7.1 Lignocellulosic Biomass Characteristics

Lignocellulose constitutes about 50% of the world's total biomass possessing a high energy value (Demirbas 2007). Cellulose, hemicellulose and lignin are the three main components of lignocellulosic biomass, besides small amounts of waxes, minerals, pectin and proteins. Cellulose, hemicellulose and lignin constitute about 40–60%, 20–40% and 10–25% respectively, of the total dry biomass (Prasad et al. 2019). Cellulose with the molecular formula $(C_6H_{10}\,O_5)_n$ is the most abundant polysaccharide on earth. Cellulose is constituted by insoluble crystalline microfibrils which are recalcitrant to enzymatic hydrolysis (Schwarz 2001). A Cellulose molecule consists of about 1,250–12,500 β-D-glucose residues, where individual D-glucose molecules are connected by β-1, 4-glycoside linkages (Chatterjee et al. 2015). Cellulose is a linear polymer with no coiling or rod-shaped structures which leads to crystallization of the molecule. About 30 discrete cellulose molecules assemble to form larger units called protofibrils which in turn form larger units termed as microfibrils (Lakhundi et al. 2015). The interchain and intrachain hydrogen bonding amongst the parallel layers of cellulose lead to the formation of firmly packed microfibrils. Hence, the

Table 9.1 Feedstock for biodiesel production

Feed-stock	Catalyst	Biodiesel Production (g/L)	Biodiesel properties	Ref
Trachyspermum ammi (Caraway, non-edible seed oil)	Heterogenous green nano catalyst of manganese oxide from *Colchicum aitchisoni*, Catalyst resusability of 5 cycles	940	Acid value 0.124 mgKOH/g Flash point 99 °C Kinematic Viscosity 4.2 cSt at 40 °C Density 0.814 kg/m^3 at 40 °C Sulphur content 0.0001 wt%	Ahmad et al. (2023))
Phoenix sylvestris (Date-palm, non-edible oil)	Anhydrous acid catalyst H_2SO_4 followed by second stage alkali transesterification	920	Acid value 1.6 mg KOH/g Flash point 180 °C Kinematic viscosity 4.1 cSt at 40 °C; Cetane number 57 Calorific value 38.6 MJ/kg	Bibin et al. (2023))
Madhuca indica (Mahua)	Copper-impregnated dolomite heterogenous catalyst	962	Dynamic Viscosity 2.89 cP at 40 °C Density 862 kg/m^3 at 25 °C Flash point 135 °C; Cloud point 4 °C; Pour point -8 °C	Sudalai et al. (2023))

(continued)

Table 9.1 (continued)

Feed-stock	Catalyst	Biodiesel Production (g/L)	Biodiesel properties	Ref
Linum usitatissimum (Linseed)	KOH (oil-based concentration, 5 wt%)	960	Acid value 0.37 mg KOH/g Dynamic Viscosity 4.55 cP at 25 °C Density 879 kg/m^3 at 25 °C Flash point 170 °C; Pour point-15 °C; Cloud point -6 °C	Bahadorian et al. (2023))
Catharanthus roseus (Periwinkle)	Barium-doped CaO nanocatalyst	920	Density 848 kg/m^3; Cloud point 6 °C; Pour point -1 °C; Flash point 178 °C; Fire point 191 °C Kinematic viscosity 2.16 cSt at 40 °C; Calorific value 29,652 kJ/kg Copper strip corrosion at 60 °C 1a-slightly tarnished; Fire point 191 °C	Karthikeyan et al. (2023))

cSt: Centistokes; cP: Centipoise

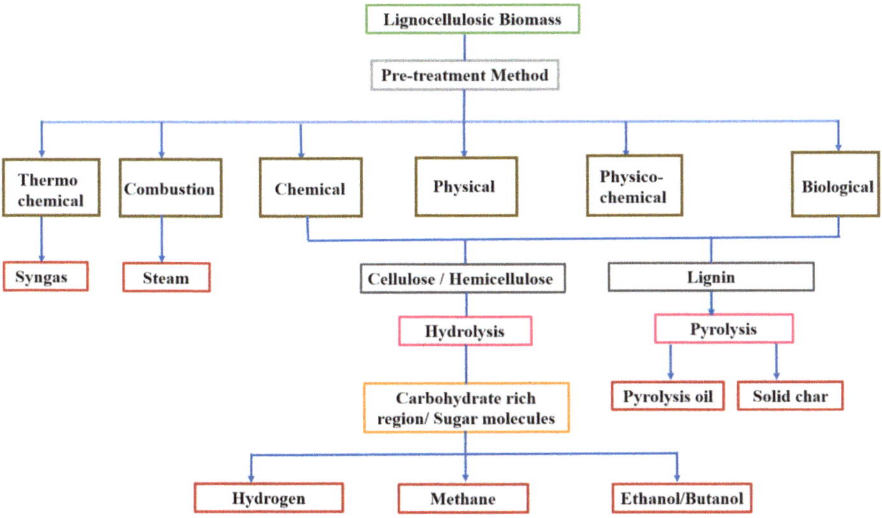

Fig. 9.3 Processing of lignocellulosic biomass for production of biofuels

structural forms of cellulose comprise ordered (Crystalline) and disordered (Amorphous) regions. In nature, the crystalline form of cellulose occurs as I_α and I_β. I_β form of crystalline cellulose molecules is more thermodynamically stable and is principally present in terrestrial plants rendering the biomass recalcitrant towards enzyme degradation (Prasad et al. 2019). I_α has a single chain triclinic structure whose unit cell parameters are not established firmly. I_α is metastable and can be converted to I_β entirely by hydrothermal pre-treatment in the presence of sodium hydroxide (Schwarz 2001). Hemicellulose is a heterogeneous biopolymer which is present in the primary and secondary plant cell walls (Álvarez et al. 2016). The different polysaccharides constituting the hemicelluloses are xylan, galactomannan, glucomannan, galactoglycomannans, xyloglucan, glucuronoarabinoxylans, mixed linkage containing glucans, arabinogalactan and arabinoglucuronoxylan (Sorieul et al. 2016). Lignin is a non-carbohydrate molecule available in the cell wall of plants in abundance (Prasad et al. 2019). It is composed of three basic units, namely *p*-coumaryl, coniferyl and sinapyl alcohols (Parthasarathi et al. 2011). Lignin forms cross-linkages with cellulose and hemi-cellulose polysaccharides resulting in a complex matrix (Prasad et al. 2019).

9.7.2 Processing of Lignocellulose

The features associated with the processing of lignocellulose to bioethanol include the depolymerisation of cellulose and hemicellulose molecules to soluble sugar molecules followed by fermentation of hexoses (C6 sugar) and pentoses (C5 sugar).

Moreover, the lignocellulosic feedstock should have a lower content of lignin which would lower the cost of bioethanol production as fewer and inexpensive chemicals will be required for lignin dissolution during pre-treatment (Demirbas 2007). Table 9.2 displays different plant biomass along with their respective pre-treatment methods for bioethanol production.

9.7.2.1 Biomass Dissolution

The first step in biofuel production from biomass includes fractionation of the plant cell wall. Pre-treatment involves the disruption of lignin and cellulose crystallinity, besides an increase in the lignocellulose porosity. The removal of lignin and hydrolysis of hemicellulose are classified as pre-treatment (Hamelinck et al. 2005). Several quintessential methods for lignocellulose pre-treatment are as follows:

i. Pre-treatment by physical methods: Mechanical pulverization (grinding, milling and chopping) of the biomass reduces the crystallinity of cellulose polymer. Particle size of up to 10–30 mm is obtained after chopping, and finally grinding reduces the particle size up to 0.2–2 mm. The method is useful in terms of accessible execution, biomass dewaterability, less energy utilization and negligible odour generation. The method, however, does not remove pathogens from the biomass (Sun and Cheng 2002).
ii. Thermal pre-treatment: This pre-treatment method removes the pathogen, besides reducing the viscosity and water content of the biomass. Thermal treatment affects the cell membrane of the biomass by disintegrating it leading to the breakdown of macromolecules to micromolecules, referred to organic matter solubilization (Amin et al. 2017). Various methods of heating are prevalent such as microwave, steam and electric methods. Microwave heating yields better biopolymer solubilisation rates as it polarizes the macromolecules. Thermal treatment of biomass at temperatures greater than 160 °C results in hemicellulose and lignin solubilization. However, phenolic compounds are released on thermal treatment which have repressive effect on the microbial consortia (Hendriks and Zeeman 2009).
iii. Steam explosion: In the process of biomass fractionation by steam explosion, a high pressure (5–50 atm) and temperature (160–260 °C) steam is passed into the sealed chamber containing the lignocellulosic feedstock, followed by the release of pressure after 5 min. This leads to the expansion of the steam within matrix of lignocellulose thereby detaching the fibres (Amin et al. 2017).
iv. Pyrolysis: Crops like *Miscantheus giganteus* and switch grass, besides plant materials like green waste, waste wood, saw dust, woody weeds, nut-shells, cotton trash and rice hulls are the biomass feedstock for pyrolysis. The biomass is heated at a temperature greater than 300 °C which leads to the direct decomposition of cellulose into gaseous products and char (in the form of residue). Oxygen enhances the process of pyrolysis. Lower temperature causes less degree of degradation (Millett et al. 1976).

Table 9.2 Different plant biomass and their pre-treatment methods for bioethanol production

Pre-treatment process	Mechanism of pre-treatment	Lignocellulosic Biomass	Compositional analysis after pre-treatment	Bio-ethanol production (g/L)	References
1. Physical Sonification	Ultrasound forms bubbles increasing cellulosic porosity; operating power 400 W, frequency 24 kHz	*Ipomoea carnea* (Bush morning glory)	Cellulose 71 wt%, Hemicellulose 8 wt%, Lignin 19 wt%	850	Muthuvelu et al. (2019)
2. Chemical Alkaline	NaOH Ruptures the H-bonds in cellulose; breaks the intercellular bond between cellulose and lignin; 5%, v/v peroxide at 50 °C for 1 h	*Brassica napus* (Rape seed straw)	Cellulose 59% of the total solid fraction, (Xylan) Hemicellulose 34% of the total solid fraction	5 g/100 g of raw material	Karagöz et al. (2012)
Acidic	Dilute sulphuric acid enhances the hydrolysis of amorphous cellulose; NH_4Cl concentration 0 – 0.4 mol/L, impregnation time 10–50 min. impregnation temperature 130 – 190 °C	*Eucalyptus* (Eucalyptus wood chips)	Cellulose 556 g/kg, Hemi-cellulose 18 g/kg, Lignin 337 g/kg	820	Yang et al. (2021)
3. Physico-chemical Steam explosion	Individual fibres get detached by heating with steam at high pressure, followed by sudden release of pressure; Acid catalysed (sulphuric acid, 0.25%, v/v) steam explosion in an autoclave (Temperature 121 °C, pressure 15 psi); solid loading 10% w/v	*Citrus* (Citrus peel)	Cellulose 35%, Hemi-cellulose 5%, Lignin 8%, Pectin 7%	180	John et al. (2017)
Ammonia Fibre Explosion Process	AFEX involves the breakage of ester and ether bonds between cellulose and hemicellulose; CO_2 added ammonia explosion, ammonia concentration 14.3%, CO_2 loading 2.2 MPa, Temperature 165 °C, residence time 70 min	(*Oryza sativa*) Rice straw	Cellulose 55% Hemi-cellulose 25%	970	Cha et al. (2014)

(continued)

Table 9.2 (continued)

Pre-treatment process	Mechanism of pre-treatment	Lignocellulosic Biomass	Compositional analysis after pre-treatment	Bio-ethanol production (g/L)	References
4. Biological Fungal	Less energy is consumed with a higher efficiency of hydrolysis; *Phanerochaete chrysosporium* secretes extraceelular peroxidases; Dilute acid is added to microbial pretreatment system	*Eichhornia crassipes* (Water hyacinth)	Cellulose 39%, Hemi-cellulose 18%, Lignin 5%	1.4	Zhang et al. (2018)
5. Integrated High-pressure microwave + 40% w/v NaCS	Output is substantiated, besides a reduction in time and chemicals; 40%, w/v sodium cumene sulfonate followed by pre-treatment under constant pressure of 156 psi, constant microwave generator power of 600 W for 60 min	*Triticum aestivum* (Wheat straw)	Cellulose 86% Dry Weight (DW), Hemi-cellulose 4% DW, Lignin 4% DW	54	Mikulski and Klosowski (2023)

v. Chemical pre-treatment: This method involves the degradation of the organic compounds by strong or weak acids, alkalis and oxidants. This method is applicable for leaves and yard wastes biomass with higher lignin content (Amin et al. 2017). The method is unsuitable for carbohydrate rich biomass. The reagents for the alkali-pre-treatment include hydroxides of calcium, ammonium, potassium and sodium. Cellulose swells in the presence of alkalis such as sodium hydroxide or lithium hydroxide. However, cellulose is amphiphilic in nature, it cannot be degraded completely by an alkali (Kim et al. 2016). The hydroxyl groups of the alkali are able to interact with the phenolic monomer groups in lignin thereby dissolving the polymer into smaller molecules. Concentrated acid solutions such as sulphuric acid and hydrochloric acid are able to dissolve cellulose completely. Acid solutions diluted to less than 5% leads to hemicellulose degradation at a moderate temperature (120–210 °C) and pressure (10 atm) (Amin et al. 2017).

vi. Ozonation: Ozone is a strong oxidising agent. It decomposes itself to form hydroxyl radicals. Ozone reacts with biomass both directly and indirectly. When the ozone reacts directly with biomass, it degrades the fermentable sugars rendering the reaction unrewarding (Carballa et al. 2007). Ozone reacts indirectly with the hydroxyl ions of biomass leading to the deterioration of the complex organic compounds such as fats, lipids and proteins (Ariunbaatar et al. 2014). Ozonation removes lignin from the biomass without producing any toxic products. (Ariunbaatar et al. 2014). Ozonolysis is carried out at normal pressures and room temperature. Ozonation of biomasses like sugarcane bagasse, green hay, peanut shells, wheat straw, pine shavings and cotton stalk have been done to degrade lignin and hemicellulose. Nevertheless, the use of ozone is uneconomical as large amount of ozone gas is needed for the pre-treatment process.

9.7.2.2 Enzymatic Saccharification of Pre-Treated Biomass for Bioethanol Production

In cellulose hydrolysis, the cellulose molecules are converted into glucose.

$$(C_6H_{10}O_5)_n + nH_2O \rightarrow nC_6H_{12}O_6$$

The reaction can be catalysed by employing concentrated acid, dilute acid, cellulases or hemicellulases (Hamelinck et al. 2005). Cellulases and hemicellulases depolymerise cellulose and hemicellulose respectively. Cellulases belong to the family of glycoside hydrolases, GH5, GH1 and GH8. The catalytic module of multimodular cellulases are associated with carbohydrate binding module(s) (CBMs), which paly important role in adhering to the polysaccharides, such as cellulose (Prasad et al. 2019). The catalytic module is often linked to other modular accessory domains. Cellulases are basically three types of enzymes, namely endoglucanase (EC 3.2.1.4), exoglucanase (EC 3.2.1.91) and β-glucosidase (EC 3.2.1.21).

Endoglucanases cleave the β-1,4-glycosidic bond internally by binding to the amorphous region of the cellulose molecule generating cello-oligosaccharides of various degrees of polymerization (DP) in the range, DP3—DP10. Exoglucanase binds the crystalline region of the cellulose molecule, hence it acts at the reducing as well as the non-reducing end of cellulose chain in a processive manner to release short chain cello-oligosaccharides or cellobiose. Finally, β-glucosidase hydrolyzes or cleaves cellobiose (soluble cello-oligosaccharides) to glucose (Schwarz 2001). Xylanases hydrolyse the xylan and are usually present in GH families 1, 2, 3, 4, 5, 10 and 11 (Prasad et al. 2019). Endo-β-1, 4-xylanase hydrolyses the β-1, 4 bond of the xylan backbone to give short chain xylo-oligomers.

These enzymes are able to deconstruct plant cell wall polysaccharides when present in complexed or non-complexed enzyme systems. The complexed systems are multienzyme complexes called cellulosomes which are present in fungi and anaerobic bacteria (Gray et al. 2006). The cellulosomes either remain free or attached to the cell surface. The CBMs present in cellulosome attach to the lignocellulosic biomass and its catalytic modules hydrolyse the polysaccharides and thus carry out its deconstruction into monomeric sugars like D-glucose and D-xylose (Schwarz 2001). On the other hand, individual enzymes are produced by aerobic microbes such as *Cellulomonas, Cellvibrio* and *Microbispora bispora* in a non-complex system and are secreted into the extracellular environment (Gray et al. 2006).

9.7.2.3 Fermentation of C5 and C6 Sugars

The monosaccharides, glucose and xylose present in the hydrolysates obtained after enzymatic saccharification of lignocellulose biomass are subjected to fermentation (Hamelinck et al. 2005). Fermentation is a biological process where micro-organisms such as yeast, bacteria or fungi convert the monomeric sugars obtained after hydrolysis to ethanol anaerobically. The glucose, a C6 sugar, is converted to ethanol and CO_2 (Eq. 9.1) by the enzyme Zymase produced by *Saccharomyces cerevisiae* (Prasad et al. 2019). Moreover, the five-carbon sugar (C5), xylose can also be converted to ethanol by xylose utilizing microorganisms such as *Scheffersomyces stipites* and *Candida shehatae*, as shown in Eq. 9.2 (Moysés et al. 2016). According to the fermentation reaction, a maximum of 0.51 kg of theoretical yield of ethanol can be obtained from glucose (Prasad et al. 2019).

$$\underset{\text{Glucose}}{C_6H_{12}O_6} \rightarrow \underset{\text{Bioethanol}}{2C_2H_5OH} + \underset{\text{Carbon - dioxide}}{2CO_2} \tag{9.1}$$

$$\underset{\text{(Xylose)}}{3C_5H_{10}O_5} \rightarrow \underset{\text{(Ethanol)}}{5C_2H_5OH} + \underset{\text{(Carbon dioxide)}}{5CO_2} \tag{9.2}$$

The yeast, *Saccharomyces cerevisiae* is employed in sugar-based and corn-based bioethanol industries. Nevertheless, various bacteria such as, *Corynebacterium glutamicum, Zymomonas mobilis, Clostridium thermocellum* etc. are also utilised for bioethanol production (Prasad et al. 2019). Table 9.3 gives description of various fermentation methods used for bioethanol production.

9.7.3 Case Study

In a study, watermelon waste was used to produce bioethanol and biogas. Pretreatment was done by mechanical method. The yeast *Wickerhamomyces sp. UFFS-CE-3.1.2* was employed for bioethanol production with the processed watermelon pulp (Scapini et al. 2023). The watermelon pulp had a total carbohydrate (sugars including sucrose, fructose and glucose) content of 76.9 g/L. Sequential batch fermentation was carried out with 30 cycles of cell recycling (Scapini et al. 2023). At the end of each individual process, the yeast cells were segregated from the fermentation broth and reinserted into a fresh fermentation medium. The maximum ethanol yield of 41 g/L was obtained in the sixth batch after 3 h (Scapini et al. 2023). Moreover, the residual organic part of the biomass was subjected to anaerobic digestion to produce biogas, a mixture of methane (CH_4) and carbon dioxide (CO_2). Total biogas (581 $mL_{N}.g_{VS}$) and methane (290 $mL_{CH4}.g_{VS}$) were produced from residual watermelon rinds. These values are expressed in standard millilitres (ml) per gram (g) of the volatile solids (fatty acids) added.

9.8 Production of Bio-Oil from Plant Lignocellulosic Biomass

Bio-oil is referred to as pyrolysis liquid, pyrolysis oil, bio-crude, wood liquid, wood oil or wood distillate. Bio-oil is a free-flowing dark brown liquid comprising a great amount of water, usually 15–35 wt%. It is an organic liquid mixture which contains organic compounds such as alcohols, acids, aldehydes, ketones, phenols, esters, ethers, furans, sugars, alkenes, nitrogen compounds and oxygenates (Velden et al. 2010). Bio-oil has a high oxygen content and low pH value (2–3.7). Therefore, it is corrosive, unstable and chemically reactive. Nevertheless, upgrading bio-oil via catalytic cracking technology and high-pressure hydro processing can render bio-oil applicable in engines. Many processes are being developed to transform the lignocellulosic biomass. Pyrolysis is one such simple and feasible technology encompassing a wide range of biomass (Demirbas 2007). Pyrolysis is defined as heating under low oxygen levels. Biomass pyrolysis is constituted of three phases. The first phase includes initial evaporation of the moisture content higher than the amount of moisture at equilibrium (free moisture); the second phase includes primary decomposition, which occurs at 200–400 °C. During this phase, biomass starts decomposing and forms solid char. The third phase includes secondary decomposition, such as oil

Table 9.3 Various enzymatic saccharification process configurations for fermentation

Process alternatives for fermentation	Technical aspects	Micro-organism	Cellulose consumed (% of feedstock)	Cellulose conversion efficiency (%)	Ref
Separate Hydrolysis and Fermentation (SHF)	• Enzymatic hydrolysis and hexose sugar (glucose) fermentation occur separately at individual optimum conditions. Low quantity of enzyme input; high ethanol yield • High investment cost due to set up of two reactors. Cellulase activity inhibited by glucose released in enzymatic hydrolysis of cellulose	Recombinant *Saccharomyces cerevisiae*	6	75	Prasad et al. (2019) (Hamelinck et al. 2005)
Simultaneous Saccharification and Fermentation (SSF)	• Cellulose hydrolysis and hexose sugar (glucose) fermentation occur in single reactor. Less cost investment due to the involvement of single reactor. Cellulase activity not inhibited by glucose due to simultaneous fermentation • Either enzymatic hydrolysis of cellulose or fermentation can be performed optimally	Recombinant *Saccharomyces cerevisiae*	6	80	Hamelinck et al. (2005)

(continued)

Table 9.3 (continued)

Process alternatives for fermentation	Technical aspects	Micro-organism	Cellulose consumed (% of feedstock)	Cellulose conversion efficiency (%)	Ref
Simultaneous Saccharification and Co-fermentation (SSCF)	• Simultaneous fermentation of C5 and C6 sugars in the same reactor. Low-cost investment • Microbes producing high ethanol grades in C5 and C6 sugar fermentation not yet extensively available	Recombinant *Zymomonas mobilis*	5	88	Prasad et al. (2019)
Consolidated BioProcessing (CBP)	• Enzymes, alongwith sugars and ethanol are produced by a single community of microbes in a single reactor. It is cost effective and energy efficient. Pre-treatment is not required • Conversion time is relatively high, ethanol yield and productivity is low, suitable microbes are not available	Co-cultures of *Clostridium thermocellum* and *Clostridium thermoshydro-sulfuricum*	4	90	Hamelinck et al. (2005)

cracking and polymerisation. Decomposition of hemicellulose, cellulose and lignin occurs at 250–350 °C, 325–400 °C and 300–550 °C, respectively (Velden et al. 2010).

The primary decomposition of cellulose can be understood by the Waterloo mechanism (Matsuoka et al. 2014).

i. Dehydration: Cellulose undergoes dehydration at a slow heating rate, temperature less than 250 °C (reaction constant k_1) to form char, carbon dioxide, carbon monoxide, water etc. (Matsuoka et al. 2014).
ii. Polymerisation: A low degree of cellulose polymerisation is also observed (reaction constant, k_2) to form active cellulose. Reducing ends of cellulose molecule play an important role in cellulose activation (Lédé 2012). The cellulose microfibrils consist of a crystalline region which contains disordered regions intrinsically. These regions, known as para crystalline regions, occur at fixed intervals along the microfibril direction longitudinally (Lédé 2012). A sharp reduction in the degree of polymerisation (DP) occurs at the interface of crystalline and amorphous regions of cellulose molecule during pyrolysis at low temperature. This initial reduction in DP of cellulose molecule results in the formation of reducing ends indicating that hydrolysis occurs in pyrolysis at a temperature higher than the boiling point of water (Gökçek et al. 2023). Under the pyrolysis conditions, the reducing ends are more labile than the repeating groups or the non-reducing ends of the cellulose. The reducing ends undergo discolouration, forming dark colour compounds and glycosylation at a temperature range of 140–260 °C. During thermal glycosylation, the aldehyde groups (reducing end) react with the alcohol in the pyrolysis reaction to produce oligo saccharides and water. Moreover, dehydration reaction of cellulose leads to the formation of conjugated double bond structures, imparting dark colour to the products (Lédé 2012). Reducing ends are intrinsically formed in smaller proportions between the cellulose crystalline regions, and hence the volatile products, besides water, do not eliminate efficiently during the process of pyrolysis (Gökçek et al. 2023). In this case, water leads to cellulose hydrolysis in the para crystalline region. These chain reactions occur in smaller areas and lead to the expansion of disrupted regions. This expansion results in the activation of crystallite surface cellulose molecules (Lédé 2012).
iii. Depolymerisation: Active cellulose depolymerises at a temperature range, 300–450 °C to form anhydro sugars such as levoglucosan (reaction constant k_3) (Matsuoka et al. 2014).
iv. Fragmentation: Active cellulose undergoes fragmentation to form carbonyls, acids and alcohol at an optimum temperature of 600 °C (reaction constant k_4) (Matsuoka et al. 2014).

Maximum bio-oil yields (60–70 wt%) are obtained by fast pyrolysis which includes some essential parameters such as, moderate temperature range of 450–650 °C, biomass particles size range of 2 mm, higher heating rates of 10^3–10^5 °C/s, shorter vapour residence time of less than 2 s and an expeditious pyrolytic vapour quenching to subdue secondary reactions of pyrolysis (Velden et al. 2010).

9.9 Bio-Hydrogen Production from Plant Lignocellulosic Biomass

Bio-hydrogen is a clean energy fuel with a sustainable potential having carbon-free emissions. Hydrogen has a high content of energy, 142 MJ/kg. The energy content of hydrogen is four times higher than that of oil in terms of lower heating values (LHV); LHV for hydrogen is 120 MJ/kg, while LHV for oil is 43.4 MJ/kg (Zhu et al. 2023). Investigations are being carried out to transform food waste into bio-hydrogen. Moreover, the combustion of hydrogen produces only heat and water indicating its eco-friendly nature. The combustion of hydrogen in an internal engine and fuel cell can be directly done. Hydrogen can be produced from renewable sources by employing the systems such as electrolysis, thermolysis, biochemical process, dark fermentation, photo fermentation and microbial electrolysis (Simmons et al. 2010). Dark fermentation is a bio-chemical process wherein biogas is generated from organic wastes in the absence of light and oxygen. The biogas contains hydrogen, carbon dioxide and by-products. Intermediates in the soluble form, like pyruvate and butyrate are converted to bio-hydrogen via microorganisms (*Clostridium* species). Photo-fermentation involves the conversion of organic substrates into hydrogen by employing anoxic photosynthetic organisms. The substrates are oxidised to produce protons, which in turn are reduced to hydrogen via photoenergy and reducing power availed by nitrogenase enzyme. This nitrogenase enzyme present inside the membrane of organic substrates convert nitrogen gas into ammonia and form hydrogen gas by proton reduction as shown in the reaction (Simmons et al. 2010).

$$N_2 + 8H^+ + 8e^- + 16\ ATP \rightarrow 2NH_3 + H_2 + 16\ ADP + 16\ Pi$$

In a study, bio-hydrogen was produced from apple pulp waste via dark fermentation. A batch-reactor with substrate to inoculum ratio (S/I) of 2 g Volatile solid (VS) per gram of VS was set up. 100 mg/L of magnetite was added to produce a maximum biohydrogen of 74 ml/gVS with bacterium *Sporolactobacillus* at 39% relative abundance (Zhu et al. 2023).

9.10 Challenges Involved for Plant-Based Biofuels and Their Future Prospects

9.10.1 Lignin Recalcitrance of Plant Biomass

Lignin is a phenolic molecule containing complex aromatic rings which forms inter-linkages with the plant polysaccharides present in lignocellulose. It is one of the major constituents of plant which provide mechanical strength. Lignin is difficult for biological hydrolysis and thus responsible for protection of plant polysaccharides from enzymatic hydrolysis. This characteristic of plant biomass is known as the lignin recalcitrance. Lignocellulosic plant biomass contains 7–20% of lignin (O'Neill and York, 2003). Lignin recalcitrance is the major reason for lignocellulosic polysaccharides being unavailable for efficient saccharification (27, Hood 2016). The ether bonds of lignin further crosslinked with the plant polysaccharide in the cell wall makes the plant lignocellulose heavily indigestible for carbohydrate active enzymes (CAZyme). Efforts are underway in generating plant varieties having altered forms of lignin ether bonds in the cell wall which are easy to break (Pollegioni et al. 2015). Downregulation of certain enzymes such as Caffeic-acid O-Methyl-Transferase (COMT) responsible for lignin biosynthesis of switchgrass resulted in the increased digestibility by CAZymes (Fu et al. 2011).

9.10.2 Enhancement of Plant Biomass Content

Sustainability of plant-based biofuels is possible only when there is a sufficient availability of plant biomass. Delaying the flowering period helps in notable additional growth of plant stem as observed in *Brachypodium distachyon* plant, wherein Phytochrome C was downregulated (27, Hood 2016). Moreover, certain *Qualitative Trait Loci* (QTLs) were reported in switchgrass plant which are responsible for plant height and biomass content (Serba et al. 2015).

9.10.3 Production of Indigenous Enzymes in Plant Systems

CAZymes are usually produced exogenously in the recombinant expression systems like strains of *E. coli* which are different from the natural source of the enzyme. However, the current approach of exogenous production of CAZymes is expensive as it requires control over all the protein expression and production conditions. So, research is underway to develop plant pockets, such as seeds in maize to accumulate the industrially relevant CAZymes required for lignocellulose conversion (Hood et al. 2011). Thus, the plant biomass itself will have the pocket containing enzymes which can degrade the lignocellulose.

9.10.4 Exploring Multifunctional Enzymes for Plant Biomass Deconstruction

The estimated price to make the bioethanol process cost-effective is $ 0.10 per gallon of bioethanol. However, billions of dollars have been spent in last 20 years to develop the efficient CAZymes like cellulases which has not turn out to be a breakthrough success. Lignocellulose contains an array of polysaccharides like cellulose, xylan, xyloglucan, mannan and pectin which has either C-6 or C-5 sugar (like D-glucose or D-xylose) as their building blocks. Thus, the research strategy has now slightly shifted from developing a singular CAZyme towards constructing or exploring multi-functional enzymes capable of deconstruction of more than one polysaccharide or perform more than one hydrolysis functions (Olson et al. 2015). These approaches either search natural multifunctional enzymes often present as a part of modules in complex cellulosome structures or they are developed through different mutation techniques like rational engineering-based site-directed mutagenesis. Nath *et. al.* in 2019 developed a bifunctional chimeric enzyme, *Ct*GH1-L1-*Ct*GH5-F194A from *Clostridium thermocellum* capable of performing endoglucanase as well as β-glucosidase activity on cellulosic substrate (Nath et al. 2019). Also, Gavande *et. al.* in 2022 reported a recombinant multifunctional cellulase enzyme, *Rf*GH5_4 from bacterium *Ruminococcus flavefaciens* which efficiently hydrolysed β-D-glucan, xyloglucan, galactomannan, glucomannan and xylan (Gavande et al. 2022). In a study, Sugarcane tops were pre-treated by alkali method using 10% (w/v) sodium hydroxide, pH 13 (Khaire et al. 2021). They used recombinant hydrolytic enzymes, *endo*-1,4-β-glucanase (*Ct*Xyn11A), cellobiohydrolase (CBH5A) and β-glucosidase (*Ht*Bgl1) from *Clostridium thermocellum* and obtained 0.19 g/g glucose of bioethanol yield. In another report, Sugarcane Bagasse (SCB) was pre-treated with 1% (w/v) NaOH followed by organosolv (85% (v/v) phosphoric acid at 50 °C for 1 h) treatment (Nath et al. 2021). The solid residue obtained after pre-treatment of SCB was enzymatically saccharified by bifunctional chimeric enzyme, *Ct*GH1-L1-*Ct*GH5-F194A (β-glucosidase, *Ct*GH1 and endoglucanase, *Ct*GH5-F194A) along with cellobiohydrolase, *Ct*CBH5A cloned from *Clostridium thermocellum.* Subsequently, this enzymatic hydrolysate was fermented by *Saccharomyces cerevisiae* at 30 °C and 150 rpm for 24 h. This Separate Hydrolysis and Fermentation (SHF) approach of fermentation gave bioethanol yield of 0.42 g/g of glucose. These approaches are promising to make the process of biofuel production cost-effective although they are still relatively at nascent stage.

9.11 Biofuel Based Economy

The biofuel economy is projected to rise in the twenty-first century (Moon et al. 2022). In comparison to the current oil prices, biofuels are expensive than conventional fuels. Public support system to encourage feedstock and biofuel supply is required which includes assistance for raw material cultivation and capital investments for biofuel processing, besides reductions in biofuel taxes. Furthermore, a short distance between the sources of raw material and the biofuel plant would aid in the reduction of carbon emissions as well as the total cost of transportation. The chief energy consuming sector includes transport. Biodiesel has the potential to replace fossil diesel, while bioethanol can be employed as a replacement for gasoline. Technology based on biofuel has the potential to generate more employment as compared to conventional (fossil fuels) fuel-based technologies (Demirbas 2008). In accordance with the International Energy agency (IEA), 5% and 8% of the accessible cropland in the European Union and USA, respectively is essential for gasoline displacement of 5%. Moreover, 15% and 13% of the accessible cropland in the European Union and USA, respectively is essential for diesel displacement of 5% (Demirbas 2008).

Plant lignocellulose waste can be economic, as it can be used for biofuel production which is otherwise burnt. Plant biomass is affordable and does not require additional croplands for plantations. Lignocellulosic biomass does not compete with food resources. The processing of plant biomass can be consolidated via innovative technologies based on biorefinery concept. Biorefinery is the integrated and sustainable refining of plant biomass into biofuels (biohydrogen, bioethanol, biodiesel) and various other commercial value-added products like lactic acid, succinic acid and levulinic acid. Besides, an integrated approach to biorefinery will be feasible wherein the lignocellulose processing can be integrated with various renewable technologies like solar, thermal or energy storage in the form of solar plant and grid respectively. Lignocellulosic biofuel will promote national economic growth by reducing import of fuels. The use of biofuel will restrict the expansion of coal and petroleum industry. Biofuel application will have a paradigm shift in various energy models and accompanying manufacturing and service sectors. Plant biomass has the potential to provide a continuous supply of energy.

References

Ahmad M, Zafar M, Bokhari A, Akhtar MS, Alshgari RA, Karami AM, Asif S (2023) Membrane reactor for production of biodiesel from nonedible seed oil of Trachyspermum ammi using heterogenous green nanocatalyst of manganese oxide. Chemosphere 322:138078

Alalwan HA, Alminshid AH, Aljaafari HAS (2019) Promising evolution of biofuel generations. Subject Review. Renewable Energy Focus 28:127–139

Almodares A, Hadi MR (2009) Production of bioethanol from sweet sorghum: A review. Afr J Agric Res 4(9):772–780

Álvarez C, Reyes-Sosa FM, Díez B (2016) Enzymatic hydrolysis of biomass from wood. Microb Biotechnol 9(2):149–156

Amin FR, Khalid H, Zhang H, Rahman SU, Zhang R, Liu G, Chen C (2017) Pretreatment methods of lignocellulosic biomass for anaerobic digestion. AMB Express 7:1–12

Anglani C (1998) Sorghum carbohydrates–A review. Plant Foods Hum Nutr 52:77–83

Ariunbaatar J, Panico A, Frunzo L, Esposito G, Lens PNL, Pirozzi F (2014) Enhanced anaerobic digestion of food waste by thermal and ozonation pretreatment methods. J Environ Manage 146:142–149

Bahadorian A, Sadrameli SM, Pahlavanzadeh H, Kashkouli MNI (2023) Optimization study of linseed biodiesel production via in-situ transesterification and slow pyrolysis of obtained linseed residue. Renewable Energy 203:10–19

Balat M, Balat H, Oz C (2008) Progress in bioethanol processing. Prog Energy Combust Sci 34(5):551–573

Basumatary B, Brahma S, Nath B, Basumatary SF, Das B, Basumatary S (2023) Post-harvest waste to value-added materials: Musa champa plant as renewable and highly effective base catalyst for Jatropha curcas oil-based biodiesel production. Bioresource Technology Reports, 101338

Bibin C, Sheeja R, Devarajan Y, Madhu S, Ponnappan VS, Natrayan L (2023) Environment impact assessment of agricultural diesel engines utilizing biodiesel derived from phoenix sylvestris oil. Environ Res 224:115432

Björn LO, Papageorgiou GC, Blankenship RE (2009) A viewpoint: Why chlorophyll a? Photosynth Res 99:85–98

Brahma S, Basumatary B (2023) Biodiesel production from quinary oil mixture using highly efficient Musa chinensis based heterogeneous catalyst. Fuel 336:127150

Branco-Vieira M, Costa D, Mata TM, Martins AA, Freitas MAV, Caetano NS (2020) A life cycle inventory of microalgae-based biofuels production in an industrial plant concept. Energy Rep 6:397–402

Carballa M, Manterola G, Larrea L, Ternes T, Omil F, Lema JM (2007) Influence of ozone pre-treatment on sludge anaerobic digestion: Removal of pharmaceutical and personal care products. Chemosphere, 67(7) : 1444–1452

Cha Y-L, Yang J, Ahn JW, Moon YH, Yoon YM, Yu GD, An GH, Choi IH (2014) The optimized CO_2-added ammonia explosion pretreatment for bioethanol production from rice straw. Bioprocess Biosyst Eng 37:1907–1915

Chatterjee S, Sharma S, Prasad RK, Datta S, Dubey D, Meghvansi, M.k., Vairale, M.G. and Veer, V. (2015) Cellulase enzyme based biodegradation of cellulosic materials: An overview. Cellulose 5(6):271–282

Chen L, Msigwa G, Yang M, Osman AI, Fawzy S, Rooney DW, Yap PS (2022) Strategies to achieve a carbon neutral society: A review. Environ Chem Lett 20(4):2277–2310

Demirbas A (2007) Progress and recent trends in biofuels. Prog Energy Combust Sci 33(1):1–18

Demirbas A (2008) Biofuels sources, biofuel policy, biofuel economy and global biofuel projections. Energy Convers Manag 49(8):2106–2116

Durga S, Evans M, Clarke L, Banerjee R (2022) Developing new pathways for energy and environmental decision-making in India: A review. Environ Res Lett 17(6):063004

Faruque MO, Razzak SA, Hossain MM (2020) Application of heterogeneous catalysts for biodiesel production from microalgal oil—A review. Catalysts 10(9):1025

Fu C, Mielenz JR, Xiao X, Ge Y, Hamilton CY, Rodriguez M Jr, Chen F, Foston M, Ragauskas A, Bouton J (2011) Genetic manipulation of lignin reduces recalcitrance and improves ethanol production from switchgrass. Proceedings of the National Academy of Sciences, U S A 108(9):3803–3808

Gavande PV, Nath P, Kumar K, Ahmed N, Fontes CMGA, Goyal A (2022) Highly efficient, processive and multifunctional recombinant endoglucanase RfGH5_4 from Ruminococcus flavefaciens FD-1 v3 for recycling lignocellulosic plant biomasses. Int J Biol Macromol 2022(209):801–813

Georgogianni KG, Kontominas MG, Pomonis PJ, Avlonitis D, Gergis V (2008) Conventional and in situ transesterification of sunflower seed oil for the production of biodiesel. Fuel Process Technol 89(5):503–509

Gökçek ÖB, Baş F, Muratçobanoğlu H, Demirel S (2023) Investigation of the effects of magnetite addition on biohydrogen production from apple pulp waste. Fuel 339:127475

Gomez-Flores R, Thiruvengadathan TN, Nicol R, Gilroyed B, Morrison M, Reid LM, Margaritis A (2018) Bioethanol and biobutanol production from sugarcorn juice. Biomass Bioenerg 108:455–463

Gray KA, Zha L, Emptage M (2006) Bioethanol. Curr Opin Chem Biol 10(2):141–146

Hamelinck CN, Hooijdonk GV, Faaij APC (2005) Ethanol from lignocellulosic biomass:Techno-economic performance in short-, middle-and long-term. Biomass Bioenerg 28(4):384–410

Hendriks A, Zeeman G (2009) Pretreatments to enhance the digestibility of lignocellulosic biomass. Biores Technol 100(1):10–18

Hincapié G, Mondragón F, López D (2011) Conventional and in situ transesterification of castor seed oil for biodiesel production. Fuel 90(4):1618–1623

Hood EE, Nelson P Powell R (2011) Plant biomass conversion. John Wiley and Sons. ISBN: 0813816947

John I, Yaragarla P, Muthaiah P, Ponnusamy K, Appusamy A (2017) Statistical optimization of acid catalyzed steam pretreatment of citrus peel waste for bioethanol production. Resour-Effic Technol 3(4):429–433

Kamen MD (2013). Primary processes in photosynthesis. Elsevier. 1. ISBN : 1483274454

Kapilan N (2022) Studies on Mahua Oil Biodiesel as the Sustainable Fuel in Compression Ignition Engine. Acta Technologica Agriculturae 25(2):85–91

Karagöz P, Rocha IV, Özkan M, Angelidaki I (2012) Alkaline peroxide pretreatment of rapeseed straw for enhancing bioethanol production by same vessel saccharification and co-fermentation. Biores Technol 104:349–357

Karthikeyan M, Sundararaman S, Deivasigamani P, Rajasimman M (2023) Synthesis and characterization of barium doped CaO heterogeneous nanocatalyst for the production of biodiesel from Catharanthus roseus seeds: Kinetics, optimization and performance evaluation. Environ Res 222:115336

Khaire KC, Moholkar VS, Goyal A (2021) Alkaline pretreatment and response surface methodology based recombinant enzymatic saccharification and fermentation of sugarcane tops. Biores Technol 341:125837

Kim JS, Lee YY, Kim TH (2016) A review on alkaline pretreatment technology for bioconversion of lignocellulosic biomass. Biores Technol 199:42–48

Lakhundi S, Siddiqui R, Khan NA (2015) Cellulose degradation: A therapeutic strategy in the improved treatment of Acanthamoeba infections. Parasit Vectors 8:1–16

Lédé J (2012) Cellulose pyrolysis kinetics: An historical review on the existence and role of intermediate active cellulose. J Anal Appl Pyrol 94:17–32

Martins AA, Marques F, Cameira M, Santos E, Badenes S, Costa L, Vieira VV, Caetano NS, Mata TM (2018) Water footprint of microalgae cultivation in photobioreactor. Energy Procedia 153:426–431

Matsuoka S, Kawamoto H, Saka S (2014) What is active cellulose in pyrolysis? An approach based on reactivity of cellulose reducing end. J Anal Appl Pyrol 106:138–146

Mikulski D, Klosowski G (2023) High-pressure microwave-assisted pretreatment of softwood, hardwood and non-wood biomass using different solvents in the production of cellulosic ethanol. Biotechnol Biofuels Bioprod 16(1):19

Millett MA, Baker AJ, Satter LD (1976) Physical and chemical pretreatments for enhancing cellulose saccharification. Dep Agric, Madison, Wisconsin.

Moon I, Yun WY, Sarkar B (2022) Effects of variable setup cost, reliability, and production costs under controlled carbon emissions in a reliable production system. Eur J Ind Eng 16(4):371–397

Moysés DN, Reis VCB, Almeida JRMD, Moraes LMPD, Torres FAG (2016) Xylose fermentation by Saccharomyces cerevisiae: Challenges and prospects. Int J Mol Sci 17(3):207

Muthuvelu KS, Rajarathinam R, Kanagaraj LP, Ranganathan RV, Dhanasekaran K, Manickam NK (2019) Evaluation and characterization of novel sources of sustainable lignocellulosic residues for bioethanol production using ultrasound-assisted alkaline pre-treatment. Waste Manage 87:368–374

Nath P, Dillon A, Kumar K, Sharma K, Jamaldheen SB, Moholkar VS, Goyal A (2019) Development of bi-functional chimeric enzyme (CtGH1-L1-CtGH5-F194A) from endoglucanase (CtGH5) mutant F194A and beta-1,4-glucosidase (CtGH1) from Clostridium thermocellum with enhanced activity and structural integrity. Biores Technol 282:494–501

Nath P, Maibam PD, Singh S, Rajulapati V Goyal A (2021). Sequential pretreatment of sugarcane bagasse by alkali and organosolv for improved delignification and cellulose saccharification by chimera and cellobiohydrolase for bioethanol production. 3 Biotechnology, 11, 1–16

Olson DG, Sparling R, Lynd LR (2015) Ethanol production by engineered thermophiles. Curr Opin Biotechnol 33:130–141

Ong HC, Milano J, Silitonga AS, Hassan MH, Wang CT, Mahlia TMI, Siswantoro J, Kusumo F, Sutrisno J (2019) Biodiesel production from Calophyllum inophyllum-Ceiba pentandra oil mixture: Optimization and characterization. J Clean Prod 219:183–198

Parthasarathi R, Romero RA, Redondo A, Gnanakaran S (2011) Theoretical study of the remarkably diverse linkages in lignin. J Phys Chem Lett 2(20):2660–2666

Pollegioni L, Tonin F, Rosini E (2015) Lignin-degrading enzymes. Fed Eur Biochem Soc J 282(7):1190–1213

Prasad RK, Chatterjee S, Mazumder PB, Gupta SK, Sharma S, Vairale MG, Datta S, Dwivedi SK, Gupta DK (2019) Bioethanol production from waste lignocelluloses: A review on microbial degradation potential. Chemosphere 231:588–606

Puppan D (2002) Environmental evaluation of biofuels. Period Polytech Soc Manag Sci 10(1):95–116

Quintero JA, Montoya MI, Sánchez OJ, Giraldo OH, Cardona CA (2008) Fuel ethanol production from sugarcane and corn : Comparative analysis for a Colombian case. Energy 33(3):385–399

Reid LM, Morrison, M.j., Zhu, X., Jindal, K.K. and Ma, B.L. (2015) High stalk sugar corn: A potential biofuel crop for Canada. Agron J 107(2):475–485

Riaz I, Shafiq I, Jamil F, Al-Muhtaseb AH, Akhter P, Shafique S, Park YK Hussain M (2022) A review on catalysts of biodiesel (methyl esters) production Catalysis Reviews, 1–53

Rodionova MV, Poudyal RS, Tiwari I, Voloshin RA, Zharmukhamedov SK, Nam HG, Zayadan BK, Bruce BD, Hou HJM, Allakhverdiev SI (2017) Biofuel production: Challenges and opportunities. Int J Hydrogen Energy 42(12):8450–8461

Sandaka BP, Kumar J (2023) Alternative vehicular fuels for environmental decarbonization: A critical review of challenges in using electricity, hydrogen, and biofuels as a sustainable vehicular fuel. Chem. Eng. J. Adv. 100442

Scapini T, Bonatto C, Dalastra C, Bazoti SF, Camargo AF, Junior SLA, Venturin B, Steinmetz RLR, Kunz A, Fongaro G (2023) Bioethanol and biomethane production from watermelon waste: A circular economy strategy. Biomass Bioenerg 170:106719

Schwarz W (2001) The cellulosome and cellulose degradation by anaerobic bacteria. Appl Microbiol Biotechnol 56:634–649

Serba DD, Daverdin G, Bouton JH, Devos KM, Brummer EC, Saha MC (2015) Quantitative Trait Loci (QTL) Underlying biomass yield and plant height in switchgrass. Bioenergy Research 8(1):307–324

Simmons BA, Loque D, Ralph J (2010) Advances in modifying lignin for enhanced biofuel production. Curr Opin Plant Biol 13(3):313–320

Singh B, Guldhe A, Rawat I, Bux F (2014) Towards a sustainable approach for development of biodiesel from plant and microalgae. Renew Sustain Energy Rev 29:216–245

Sorieul M, Dickson A, Hill SJ, Pearson H (2016) Plant fibre: Molecular structure and biomechanical properties, of a complex living material, influencing its deconstruction towards a biobased composite. Materials 9(8):618

Sudalai S, Vairaprakash P, Devanesan MG, Arumugam A (2023) Sustainable biodiesel production from Madhuca indica oil using a functionalized industrial waste as a catalyst: Ready to scale-up approach. Ind Crops Prod 193:116233

Sun Y, Cheng J (2002) Hydrolysis of lignocellulosic materials for ethanol production: A review. Biores Technol 83(1):1–11

Taherkhani M, Sadrameli SM (2018) An improvement and optimization study of biodiesel production from linseed via in-situ transesterification using a co-solvent. Renewable Energy 119:787–794

Van de Velden M, Baeyens J, Brems A, Janssens B, Dewil R (2010) Fundamentals, kinetics and endothermicity of the biomass pyrolysis reaction. Renewable Energy 35(1):232–242

Yang Q, Huo D, Han X, Gu C, Hou Q, Zhang F, Si C, Liu Z, Ni Y (2021) Improvement of fermentable sugar recovery and bioethanol production from eucalyptus wood chips with the combined pretreatment of NH4Cl impregnation and refining. Ind Crops Prod 167:113503

Zhang Q, Wei Y, Han H, Weng C (2018) Enhancing bioethanol production from water hyacinth by new combined pretreatment methods. Biores Technol 251:358–363

Zhu J, Song W, Chen X, Sun S (2023) Integrated process to produce biohydrogen from wheat straw by enzymatic saccharification and dark fermentation. Int J Hydrogen Energy 48(30):11153–11161

Chapter 10
Plant-Based Biofuels: A Sustainable Solution for Energy Production

Alok Kumar Panda, Vivek Manyapu, Ashootosh Mandpe, and Sameeksha Paliya

Abstract The excessive use of non-renewable energy sources has inflicted severe damage on the environment, prompting the need for renewable energy alternatives. While renewable energies like solar, wind, and hydropower hold potential, their resource requirements for manufacturing devices cannot be ignored. This chapter proposes biofuels as a promising solution to tackle the increasing demand for energy and depleting resources. Plant-based biofuels obtained from energy crops demonstrate higher efficiency for biofuel production than those derived from biowastes and algae. This chapter delves into the various processes involved in converting vegetable oils, crop byproducts, and wastes into biofuels. It also explores the recent techniques and new varieties of crop sources for biofuel production. The use of plant-based biofuels poses a lower environmental threat as they have a low net release of carbon dioxide and sulfur. Therefore, plant-based biofuels are the only form of renewable energy that can completely replace fossil fuels for our present and future energy constraints. Overall, this chapter highlights the significance of biofuels as a renewable energy alternative and their potential to mitigate the environmental impact caused by non-renewable energy sources. The incorporation of plant-based biofuels in our energy systems could significantly contribute to a sustainable future.

Keywords Biofuels · Crop byproducts · Energy crops · Plant-based sources · Renewable energy · Sustainability

---

A. K. Panda
Department of Agronomy, Sam Higginbottom University of Agriculture, Technology and Sciences, Prayagraj 211 007, India

V. Manyapu
Department of Tropical Agriculture and International Cooperation, National Pingtung University of Science and Technology, Pingtung 912, Taiwan

A. Mandpe (✉)
Department of Civil Engineering, Indian Institute of Technology Indore, Indore 453 552, India
e-mail: as_mandpe@iiti.ac.in

S. Paliya
Department of Chemical Engineering, IPS Academy, Institute of Engineering and Science, Indore 452 012, India

M. Shah and D. Deka (eds.), *Emerging Sustainable Technologies for Biofuel Production*, Environmental Science and Engineering, https://doi.org/10.1007/978-3-031-52167-6_10

10.1 Introduction

Energy is a necessary aspect of society and plays a significant role in raising the social and economic standards of life in society. With the passage of time, man has employed a variety of resources to create energy, ranging from wood to nuclear energy (Mirza et al. 2008). Wind, solar, hydro, tidal waves, biomass, and biodiesel are examples of renewable energy resources, while non-renewable energy resources comprise oil, gas, and coal (Ali et al. 2017). Recently, it has been observed that there has been a lot of interest in the topic of renewable energy. This is mostly due to the negative environmental consequences resulting from the entire reliance on non-renewable energy sources (fossil fuels). The main drawback of this energy source is the release of carbon dioxide (CO_2) and numerous harmful air contaminants into the environment from the burning of fossil fuels (coal, oil, natural gas) (Perera 2017).

The most serious environmental effects of non-renewable sources are acid rain and global warming. Acid rain occurs when gasses like Sulphur dioxide, nitrous oxide, and other pollutants are released into the atmosphere, which results in a chemical reaction with rainwater (Xu et al. 2015). According to the US electric power sector, mam-made sources account for approximately 67% of all US SO_2 emissions and 22% of total US NO emissions (Mohajan 2018). Because of this chemical mix, the rain turns somewhat acidic (i.e., 5.6), disrupting the nearby ecology (Burns et al. 2016). In terms of ecology, acid rain alters the acidity of lakes and streams, which may be highly destructive to fish and other aquatic species; it is also toxic to trees, weakening the forest ecosystem (Singh and Agrawal 2004).

Global warming is the term used to describe the rise in the Earth's internal surface air and ocean temperatures (Grover and Rajeshwari 2009). The average temperature of Earth increases due to certain greenhouse gasses like CO_2, methane, nitrous oxide, CFC, ozone, and water vapor, which trap the IR-radiation emitted by the earth, resulting in the greenhouse effect. Regrettably, modern human actions have exacerbated the natural greenhouse effect. According to IPCC 2021, human-induced greenhouse gases have caused around 1.1 °C increase in average temperature between 1850–1900, with global temperatures predicted to approach or surpass 1.5 °C during the next 20 years on average. In 2021, the USPA issued a study estimating overall greenhouse gas emissions from several sectors in the US, which found that around 28% of greenhouse gas emissions came from transportation, 25% from power production, 23% from industry, 13% from commercial and residential, and 10% from agricultural.

There is no additional energy from non-renewable sources indefinitely since it depletes once consumed (Ali et al. 2017). It is predicted that by 2100, global energy demand will be five times that of today (Holechek et al. 2022), whereas renewable sources have an unlimited supply and are recycled or replaced (Ahuja and Tatsutani, 2009), making them an ideal alternative to non-renewable energy sources.

Knowing that non-renewable resources would ultimately run out, the desire for an alternative has grown. Renewable sources may be preferable to non-renewable sources. The abundant supply of renewable energy is its most significant quality. It has no bounds and is a clean energy source that has a substantially lower environmental effect than traditional fossil energy technologies. This source involves solar-energy, wind-energy, hydropower, geothermal-energy, and biomass-energy.

Solar energy is an environmentally friendly non-consumable energy source. It can be produced using either photovoltaic (PV) or concentrated solar power (CSP) systems. It has incredibly been promised as a replacement for fossil fuels and coal (Liqun and Zhixin 2009). Wind energy is a clean energy resource that, given the right conditions, may contribute to conventional energy production and has a good impact on the environment (Umar and Abubakar 2014). Wind turbines are used to generate power with the help of blowing wind. When the wind turns a turbine's propeller-like blades around a rotor, a generator spins and produces electricity. Hydropower is electricity produced by harnessing the energy of flowing water. River and stream water can be stored in reservoirs behind dams and released as needed to drive turbines and generate energy. Dams and reservoirs may regulate the flow of water in rivers and streams from higher to lower altitudes, which can then be utilized to generate energy. Geothermal energy is defined as the heat energy held unusually in accessible depths of the earth's crust that may be economically profited (Akinbami 2001). This energy is pure and renewable, and it is utilized to generate power. Biomass energy comprises plant resources (like wood, residues, and farm waste) and animal waste, which can be transformed into gaseous fuel or liquid biofuels or burned directly as solid fuel (Umar and Abubakar 2014). It is burning wood and animal dung for heating and cooking accounts for 11% of global energy consumption and 30% of energy consumption in poor nations. When we burn biomass as fuel, carbon, sulphur, and nitrogen-based pollutants enter the atmosphere during the combustion processes. In some cases, this can result in air pollution and pollution levels that are higher than those produced by burning coal or natural gas, which are more common fuel sources.

Biofuel is considered a clean and promising renewable-energy source in the present day compared to other renewables. As a result of the limited ability to produce anything other than electricity, nearly all other renewables, including solar, wind, hydro, and nuclear power, cannot compete with petroleum on an equal basis (Arutyunov and Lisichkin 2017). There are other difficulties involved with electricity that affects the efficiency of some energy sources when it is used for long-distance transmission and switching to different sources of energy. Furthermore, compared to other energy alternatives, biofuels can be used with current infrastructure and don't need much technical advancements. Consequently, both developed, and developing nations have focused on expanding their bioenergy markets and developing global biofuel strategies. Biofuels are basically combustible biomass-derived fuels from organic materials like plant matter, wood, and animal wastes using microorganisms and different chemical processes. They can be gaseous, liquid, or solid, but as liquid biofuels are more readily integrated with modern technologies, they have the most potential to aid in the ecarbonisation of the transportation sector (Guo et al. 2015). The majority of liquid biofuel produced on earth, over 80%, is ethanol, making it the most widely used biofuel (Lee 2013). The usage of ethanol as a biofuel has increased significantly globally in recent years, with production expected to reach over 135 billion littles by 2024, where the USA accounting for 42% and Brazil accounting for 31%., respectively, making the most significant contributions (OECD/FAO 2015).

10.2 The Need for Plant-Based Biofuels

Renewable energy sources offer a significant number of potentials and benefits. They can enhance the quality of the environment since they generate little to no greenhouse gase (GHG) emissions. They are virtually ubiquitous on Earth, which makes them a more reliable energy source than fossil fuels while also distributing energy fairly, addressing difficulties with energy security, and tackling energy poverty.

The primary drawbacks of renewables like solar, wind, hydro, and nuclear power include their increased initial expenses, which can be excessive and serve as a disincentive for users, also the expense of storage systems, which is also fairly high. Additionally, as these energy sources are weather-dependent, prolonged periods of uncertain weather may result in an energy shortage (Owusu and Asumadu-Sarkodie 2016). In the case of wind energy, the primary adverse effects are on wildlife, both directly and indirectly, particularly migrating birds (despite the fact that it is thought to be the energy source most friendly to animals) (Saidur et al. 2011). Other adverse effects include aesthetic unattractiveness, noise, and communications signal interference (Mohtasham 2015). Additionally, even though geothermal energy is regarded as renewable, its use has a number of drawbacks, chief among them the release of specific GHGs into the environment.

Considering the present demand, biofuel is undoubtedly a promising renewable energy source in the transportation industry when compared to other renewable energies since it is sustainable, readily available, and useful for transportation, heating, and energy (Bildirici 2013). Because of this potential and its widespread use, biomass energy accounts for over 75% of all renewable energy sources worldwide (IRENA 2019). Additionally, it is carbon–neutral, which means it does not contribute any additional CO_2 to the climate (Abbasi and Abbasi 2010). In addition to its environmental advantages, the production of bioenergy offers social benefits as well because it can increase rural income and employment while also helping to alleviate poverty in emerging nations (Bildirici 2013).

10.3 Energy Crops for Biofuel Production

Traditionally, only vegetable crops were used to make biofuels, but due to concerns about food security, their usage in bioenergy was questioned. Based on specific traits like oil production, oil quality, and mitigating climate change, bioenergy crops are used. Traditional bioenergy crops may be grown for food and fodder, which would also help to slow down the effects of global warming (Singh 2008). They are primarily divided into five classes, including halophytes, specialized energy crops, first-, second-, and third-generation crops.

First-generation bioenergy crops are the regional or everyday food supplies from which biofuel production commenced. The first-generation bioenergy crops that were used to produce ethanol are sweet sorghum, corn, oil crops, sugarcane, and oil palm (Lobell et al. 2008). However, due to their higher production costs (Wang and Yan 2008), these crops are only partially capable of replacing petroleum-based goods (Chhetri et al. 2008; Carroll and Somerville 2008; Lorenz et al. 2009).

10.3.1 Sweet Sorghum

Sweet sorghum (*Sorghum bicolor* L.) is a cultivar of sorghum grass whose stalk has high sugar content. In order to produce more biomass, it stores a lot of sugars in stalks. Agronomic traits like the capacity to thrive in unfavorable environmental circumstances, fewer input requirements, cheap growing costs, and C4 photosynthesis make it an ideal feedstock for biofuel production. This plant has more easily fermentable soluble sugars in the form of glucose, fructose, and sucrose than other lignocellulosic biomass plant like thatchgrass and Chinese silvergrass (Regassa et al. 2014). In breeding programs, sweet sorghum was also used as a model crop to understand the genetic makeup of other bioenergy crops, like switchgrass (thatch grass), miscanthus (Chinese silvergrass), maize, and sugar crops (Paterson et al. 2009). A greater presence of metabolizing-sugar enzymes was seen as the stem developed due to the presence of high sugars in it (Qazi et al. 2012). During droughts, plant stores more sugars in the stems (Thomas and Howarth 2000). To boost crop yield, sorghum and sweet sorghum crops could potentially be crossed, and desired features could be found via genetic mapping (Okada et al. 2010).

10.3.2 Maize

Maize (*Zea mays* L) is a crucial crop as feedstock for bioethanol production due to its high grain production and rapid grain starch accumulation (Mabee et al. 2011). It is a preferred crop for bioconversion due to its high proportion of volatile components and simple transformation process. In the USA and other nations, maize is a common crop for ethanol production. However, the primary usage of maize as a dietary staple in several countries is its principal drawback as a feedstock. The development of bioenergy fuels from maize may raise food costs globally, resulting in famine and poverty. In order to address this problem, a sweet corn variety of maize was developed through recessive-mutation in the genes of maize that induce sucrose to starch conversion in the endosperm. Sweet corn hybrids with multiple uses and high photosynthetic efficiency may help farmers generate energy without harming the environment or the food supply (Takamizawa et al. 2010).

10.3.3 Oil Crops

Oil crops include rapeseeds, coconut, palm, hemp, castor, safflower, linseed, field mustard, sunflower, groundnut, and olive. Vegetable oils might either be used directly as heating fuels or processed to produce transport biofuels (Sims et al. 2006).

10.3.4 Sugarcane

Sugarcane (*Saccharum officinarum* L.) has enormous potential as a significant global feedstock for biofuel generation. Due to its outstanding ability to produce biomass, high carbohydrate content, and a favourable energy input/output ratio, it is regarded as one of the most significant possibilities for creating biofuels today. Basically, it is a perennial grass that can be cultivated in both warm temperate and tropical climates all over the year. As a result, this feedstock is perpetually available at a price that is lower than that of other energy crops (Yuan et al. 2008). This crop is mainly cultivated to extract juice from its stem that is utilized to make sugar. A significant portion of sugarcane juice is sucrose, a raw material used to make biofuel. There are numerous breeding efforts underway to enhance sucrose yield, sugarcane germplasm, and cellulosic biomass. Molasses, a byproduct of the sugar industry, is used to make commercial bioethanol.

Second-generation bioenergy crops that are perennial fodder plants include thatchgrass, canary grass, lucerne, Bermuda grass, and elephant grass (Sanderson and Adler 2008; Oliver et al. 2009). This generation of bioenergy uses the moderate strategy of using crop residue as feedstock. These plants use less energy than first-generation bioenergy crops because they produce biofuel from cellulosic biomass. Second-generation bioenergy crop-derived fuel is pure, has a less productional cost, avoids numerous environmental issues, and is an unadulterated fuel that is generated from hydrocarbons (Oliver et al. 2009). Here, lingo-cellulosic agricultural wastes are Thermochemically or biochemically converted into biofuels (Petersen 2008). Annual cereal crops and perennial biomass crops mainly constitute second-generation biofuel (Adler et al. 2007).

10.3.5 Thatchgrass

Thatchgrass (switchgrass) (*Panicum virgatum* L.) is a herbaceous perennial (C4) grass that has excellent efficiency for large-scale biofuel production. It is an ecologically sound crop for marginal and degraded lands as it requires less resources to grow (McLaughlin et al. 2006; Vogel and Mitchell 2008). Meanwhile, it needs around two-years to fully develop (McLaughlin et al. 2006). Recently scientists have given little attention to developing this crop, and numerous breeding projects have been conducted (Bouton 2007). Therefore a majority of thatchgrass varieties have germplasms that are quite similar to native genomes. There are only a few thatchgrass variants that cannot be distinguished from natural populations based on genetic makeup. Hence thatchgrass offers enormous genetic improvement potential for effective biomass production (Casler et al. 2007).

10.3.6 Miscanthus

Miscanthus is a warm-season large Asian grass, and the genus includes about 20 species of tall, attractive perennial grasses (Heaton et al. 2010). Because of its shape, the plant cannot be used as a feed crop. In Europe, the plant serves as a role model for herbaceous biomass feedstock. Basically, it is a C4 plant that has a low compensation point and low rate of photorespiration, making it an attractive feedstock for bioethanol production. Due to its advantages, such as quick response to inputs, disease resistance, and longer life of about 10–15 years, this plant is recognized as a dedicated crop (Villaverde et al. 2010). It has a 33% higher biomass output compared to thatchgrass (E. Heaton et al. 2004). *M. giganteus* L. is an excellent example of the Miscanthus genus; it uses 87% less space than prairie species to generate the same amount of biomass (Heaton et al. 2010). A few negatives to miscanthus crop production include the need for more energy during greenhouse propagation, longer period to propagate of about three years for rhizome-cutting and longer irrigation durations.

10.3.7 Lucern

Lucerne (*Medicago sativa* L.) is one of the earliest fodder crops that are native to North-America (Russelle 2001). These plants have fibrous stems that burn to produce electricity during the gasification process. Lucern is a good-quality animal feed and a promising feedstock for the manufacture of biofuels (DeLong et al. 1995). High levels of protein are present in the leaves (Lamb et al. 2003), and more significant levels of polysaccharides and lignin are present in stem cell walls which results in higher dry matter and potential bioethanol yield (Lamb et al. 2007).

10.3.8 Reed Canary Grass

Reed canary grass (*Phalaris arundinacea* L.) is a tall, perennial (C3) bunchgrass that is an excellent internal nitrogen recycler from the shoot to the roots. This plant is one of the highest-producing cold-season crops and is regarded as the most promising for the generation of biofuels. It thrives incredibly in areas with inadequate drainage and can endure prolonged flooding (Merigliano and Lesica 1998). Since grass produces a lot more biomass, it has the potential to generate a respectable amount of biofuel (Tahir et al. 2011).

10.3.9 Other Plants

Due to the benefits, certain plants are there like Napier grass and Bermunda grass, also contribute to bioenergy. Napier grass (*Cenchrus purpureus* L.) is a tall, perennial, tropical grass that is an excellent bioenergy crop since it is simple to establish, has drought tolerance, increased biomass yield per acre, and low lignin content (Yasuda et al. 2015). The grass has a pleasant flavour and is nourishing (Schmer et al. 2008). Its biomass has more volatile compounds and carbon content and is low in ash, sulphur, and nitrogen content (Mohammed et al. 2015). Bermuda grass (*Cynodon dactylon Pers.*) is another species utilized in bioenergy. Bermuda grass is a highly variable, hardy, C4 perennial grass with a limited lifespan that is mainly used as warm-season feed in tropical and subtropical regions. In addition to its outstanding biomass yield and widespread availability, Bermuda grass also has a high carbohydrate content that is advantageous for the economically viable generation of cellulosic ethanol. It acts as a soil binder in sand dams along riverbanks and seashores due to its pioneering characteristics and salinity tolerance. In irrigated lands, it is a worthwhile crop (Grassland 2011). Other potential perennial grass feedstocks include prairie cordgrass and eastern gama grass (*Tripsacum dactyloid* L.) (Springer and Dewald 2004).

The creation of third-generation bioenergy crops aims to provide a clean, renewable energy source that could slow down the effects of global warming (Bush and Leach 2007; Ehrlich and Pringle 2008). Mainly crops like boreal forest plants, CAM plants, eucalyptus, agave, and microalgae are included in this generation. The third-generation bioenergy crops In order to be a reliable biofuel source, cellulolytic bacteria must have efficient metabolisms during the fuel-conversion process. CAM and boreal plants can be used as cellulosic biomass-feed for direct fermentation (Patil et al. 2008; Schenk et al. 2008), where as Eucalyptus can be utilised to produce biofuels by thermo-conversion process (Carere et al. 2008). Microalgae and CAM plants can be suitable feedstock for the manufacture of biofuel. When feedstock undergoes an aerobic system, cellulose decomposes to yield water and carbon dioxide, while, in anaerobic systems, cellulose breaks down into CH_4 and H_2. Biofuel yields are being improved by newer approaches like metabolic engineering, system biology, biodiversity studies, and genomics.

10.3.9.1 Boreal Forest Plants

Boreal forest plants include perennial grasses like Timothy-grass (*Phleum pratense* L.) and canary-grass (*Phalari arundinacea* L.). Perennial grasses are the primary sources of plant biomass in boreal environments. Boreal plants are utilized to produce

CH_4 and are easily grown, harvested, and stored. The majority of phytopathogenic illnesses, drought, and frost don't harm the plants. Boreal plants can thrive in nutrient-poor soil and can tolerate harsh winters (Finckh 2008). Barbary fig (*Opuntia ficus-indica*), Sisal (*Agave sisalana* L.), Pineapple (*Ananas comosus* L.), and Blue agave (*Agave tequilana* L.) are a few boreal plants that are frequently used for the generation of bioenergy (Lehtomäki et al. 2008).

10.3.9.2 CAM Plants

CAM plants adapt well to photosynthesis in arid conditions. This pathway helps the plants in CO_2 absorption at night. These plants improve their efficiency of water and carbon absorption in arid regions. Since they can tolerate drought and utilize water up to three times more effectively than C3 and C4 plants, that makes it a potential bioenergy crop (Fraiture et al. 2008). Both solid and liquid biofuels are produced by these plants, making them potential versatile bioenergy crops (Grammelis et al. 2008; Borland et al. 2009).

10.3.9.3 Eucalyptus

Eucalyptus (*Eucalyptus sp.* L.) is a fast-growing, widely cultivating hardwood timber species in the world that is native to Australia. It has a huge genetic resource base and grows more quickly with continuous growth. The plant is tolerant of acidic soils, low fertility, insects, fire, drought, and other adverse environmental factors. This plant is widely cultivated because of its higher yield (70 m3/ha/year) and quicker development in tropical countries. The rotation cycle of the plant is just five years long. Around the world, 80% of plantations are made up of just four species—*E. urophylla*, *E. camaldulensis*, *E. grandis*, and *E. globulus*—and their hybrids. Out of them, only *E. globulus* is utilized for breeding purposes because of its wider adaptability, and it can develop more quickly than the other four species. Eucalyptus oil obtained from plant materials through thermo-conversion has enormous potential for use in the production of biofuels or bioenergy (Rockwood et al. 2008; Wang and Yan 2008).

10.3.9.4 Agave

Agave (*Agave* sp.) is a genus of monocots native to hot and dry areas of America and Carebian. The agave plant has a CAM pathway for photosynthesis; the plant opens its stomata at night to absorb CO_2 and lose less water through transpiration. This plant has thick leaves with sharp tips and is well-adapted to dry environments. Its nectar can be used in place of sugar in cooking. Furthermore, it is used in the production of alcoholic beverages, textiles, and sweeteners. Since it requires less water, thrives on swamps, and doesn't compete with feedstocks used for food crops, agave is a possible feedstock for biofuels (Escamilla-Trevio 2011).

10.3.9.5 Microalgae

Microalgae are photosynthetically efficient and excellent feedstock for the production of biofuels such as bio-hydrogen, bio-methane, bio-ethanol, and bio-diesel (Ahmad et al. 2011). They significantly reduce greenhouse gas emissions through carbon sequestration by absorbing carbon dioxide generated by plants and producing a tremendous amount of biomass in a short period of time (Schenk et al. 2008). Microalgal biofuels are less harmful to the ecosystem and the global food supply than usual biofuel-producing crops. (Patil et al. 2008; Schenk et al. 2008; Tilman et al. 2009). Due to their effectiveness in the transformation of photons into photosynthates, they have great promise to reduce the effects of climatic change (Patil et al. 2008). Furthermore, they can be conveniently collected at any time of the year (Williams et al. 2007). They are capable of producing highly biodegradable and non-toxic biofuels. The output of biofuel is being increased through a number of programs that genetically modify strains to boost productivity. It is more environmentally friendly than other conventional energy crops because microalgal fuel can be converted into biofuels more quickly.

10.3.9.6 Dedicated Energy Crops

These crops mainly include non-foody herbaceous perennials and woody plant types, which can be grown on marginal lands, specifically to provide biomass. They need lesser physical, chemical, and biological processes to generate biofuels. These plants are regarded as eco-friendly and may aid in reducing environmental alterations (Petersen 2008; Taherzadeh and Karimi 2008). They can address a range of environmental issues by reducing salt, storing carbon, strengthening biodiversity, and enhancing the quality of the soil and water (Ehrlich and Pringle 2008; Lal 2008). Cellulosic-biomass plants like cottonwood, sallows, eucalyptus, and silver birch, perennial grasses like canary-grass, giant reed, and Napiergrass, non-food oil crops like castor oil plant, and oil plants like *Sapium sebiferum, Pistacia chinensis, Vernicia fordii* and *Jatropha curcas* are a few of the dedicated plants specifically cultivated for bioenergy. Due to their shorter life cycles, these crops can be harvested multiple times each year (Boe and Lee 2007; Mathur 2008). Short rotation coppice (SRC) is considered the most vast potential dedicated crop for future global bioenergy supply (Rae et al. 2009). USA and Sweden are leading nations in the extensive cultivation of dedicated energy crops (Mola-Yudego and González-Olabarria 2010).

10.3.9.7 Halophytes

Halophytes are particular plants that thrive in sparsely populated, salty, and marshy environments. These typically live in estuaries, mangrove wetlands, and coastal areas (Glenn et al. 1999). Higher salt concentrations are better for these plants' growth and reproduction (Ventura et al. 2015). By aiding in the sequestration of carbon, the restoration of damaged land, and the stabilization of ecosystems by filling ecological niches, halophytes contribute to the slowing of global climate change. The adjacent plants and fauna are also protected by them from the elements and diseases (Jaradat 2010). Frost-sensitive Eucalyptus species and frost-tolerant Populus species are the finest genetic resources for biomass development in saline environments (Rockwood et al. 2008). In salt-degraded locations, halophytes could also be phytoremediate soils contaminated with heavy metals (Hasanuzzaman et al. 2014; Panta et al. 2014). Halophytes from the genus *Tamarix*, *Prosopis*, *Rhizophora*, *Casuarina*, *Eucalyptus*, and *Melaleuca* are frequently used in the production of biofuel. The ability of perennial halophyte (*Kosteletzkya pentacarpos*) seeds to produce biodiesel has been demonstrated (Moser et al. 2013). Halophytes have higher biofuel conversion efficiencies due to higher levels of secondary metabolites (Hastilestari et al. 2013).

10.3.9.8 Comparative Analysis of Energy Crops vs. Biowastes and Algae

Energy crops and biowaste can both be utilized as plant-based feedstock to produce biofuel. By using various biotechnologies, these can be transformed into fuels like biomethane, biohydrogen, bioethanol, and biodiesel. Algae, on the other hand, make up a hugely varied collection of primarily aquatic photosynthetic creatures and are responsible for roughly 50% of all photosynthesis on Earth. Compared to other plant-based feedstock, algae has more advantages in terms of producing biofuel and being environmentally friendly. Microalgae proliferate (e.g., 1–3 doublings/day) and produce vast amounts of neutral lipids/oil (20–50% of dry cell weight, or DCW). Microalgae cultures' oil yield per area could be much higher than the best oilseed crop yields [5000–20,000 L oil/acre/yr (Khan and el Dessouky 2009)]. Microalgae can grow in non-arable soil and marginal areas, benefiting wastewater bioremediation by utilizing nitrogen and phosphorus from sources like agricultural run-off and industrial and municipal wastewater. They also capture CO_2 from flue gases, reducing greenhouse gas emissions. According to Rodolfi et al. (2009), 1 kg of algal biomass requires roughly 1.8 kg of CO_2. They don't require pesticides and herbicides since they produce value-added co-products or byproducts, such as biopolymers, proteins, polysaccharides, pigments, animal feed, and fertilizer. Microalgae make more yearly biomass production per unit of area when grown in appropriate culture containers (photobioreactors) all year round.

Although microalgae constitute a superior source of feedstock for the manufacture of biofuels, there are numerous technical and topographical difficulties with algal biomass. In ideal development conditions, algae typically generate 1 to 7 g/L/d of dry matter (Chen et al. 2011). It suggests that considerable amounts of water are needed on an industrial scale, which poses a serious issue for the coldest nations where the average annual temperature is below zero. The high water content makes it challenging to separate lipids from the algal biomass, which first needs to be dewatered by centrifugation or filtration. Transesterification or hydrogenolysis can be used to convert algal lipids to produce kerosene-grade alkane that can be utilized as a drop in aviation fuel (Tran et al. 2010).

10.4 Types of Biofuels and Conversation Technologies

There are three categories of biofuels: first-, second-, and third-generation biofuels based on their biomass sources, limitations as a renewable energy source, and technological advances. First-generation biofuel involves food sources which are its main disadvantage. Second-generation biofuels are made from biomass that isn't used to grow food, but they still compete with it for land. Third-generation biofuels are considered the most efficient alternate option for fuel because they are not a food source. There are still significant challenges in making them economically viable.

10.4.1 First-Generation Biofuel

First-generation biofuels, made from sugar, starch, or vegetable oil, utilize advanced technologies and processes like fermentation, distillation, and transesterification. Traditionally, these procedures have been used for a variety of purposes, including producing alcohol. When sugars and starches undergo fermentation, they result in the production of ethanol, butanol, and propanol in lesser amounts. While pure ethanol (E100) can be used as an automobile fuel, it is more frequently added to petrol to raise octane levels and reduce emissions from moving vehicles. Despite only possessing a third of the energy density of petrol, it is presently utilized as an additive in several nations, including the US. Ethanol has the advantage of burning cleaner than petrol, which results in fewer greenhouse gases being produced. Another first-generation biofuel, called biodiesel, is made through the transesterification of plant or animal fat. This method includes exposing oils to alcohol, like methanol, while a catalyst is present. The main product of the reaction is separated from any byproducts of the reaction during the distillation phase. Although it can be used as a fuel for cars in its purest form (B100), it is typically added to diesel to lower emissions of particulate matter, carbon monoxide, and hydrocarbons from diesel-powered automobiles (Morales Bayetero et al. 2022).

10.4.1.1 Conversion Technologies

Sucrose to Ethanol

Sucrose, a disaccharide made up of glucose and fructose, is the most typical disaccharide utilized in the synthesis of bioethanol. According to Licht (2006), 48% of the gasoline-ethanol produced worldwide in 2006 came from sucrose. Using commercial yeast like Saccharomyces cerevisiae, sugar is fermented. Enzymatic hydrolysis of simple sugars is followed by fermentation in the chemical reaction. First, an enzyme found in yeast called invertase catalyzes the breakdown of sucrose into glucose and fructose. Then, a different enzyme (zymase), which is also present in yeast, turns the glucose and fructose into ethanol and carbon dioxide (CO_2)0.511 kg of ethanol can be produced theoretically from one tonne of hexose (glucose or fructose). However, the actual fermentation efficiency is only about 92% of this yield. The primary sucrose feedstocks used in the bioethanol business are sugarcane, molasses, sweet sorghum, and sugar beet. Within 24 to 72 h of harvesting, raw materials are processed, and sugar is extracted using specialized rollers to separate the juice. Brazil produces the most sugarcane juice in the world, which accounts for a sizeable portion of the fuel ethanol produced globally. In 2022, Brazil had 31.66 billion liters of gasoline-ethanol, 1.55 billion of which were exported. India, which ranks second in the world behind Brazil in terms of sugarcane production, is another potential significant producer of ethanol from sugarcane. However, only about 302 crore litters of bioethanol were generated in India in 2020–21, mostly from sugarcane molasses. A potential major producer of ethanol made from sugar beetroot juice is the European Union (EU). Sugar beets currently play a lower role in the production of ethanol in the EU than wheat, but new incentives for energy crops offered by the EU might significantly increase its future market share.

Starch to Ethanol

The alpha-glucose polymer is first hydrolyzed with the glucoamylase enzyme in order to convert starch to ethanol. Dextrose, an isomer of glucose known as D-glucose, is the resultant sugar. To generate anhydrous ethanol, enzymatic hydrolysis is followed by fermentation, distillation, and dehydration. Starch is mostly provided by grains (corn, wheat, or barley) in the fuel bioethanol sector. Corn, the primary feedstock used in the global starch-to-ethanol business, has 60 to 70% starch. In dry or wet mills, ethanol can be converted. Dry milling is the process of crushing grain into a powder, which then gets hydrolyzed. When the hydrolysate is combined with syrup, the sugar in the hydrolysate is converted to ethanol, and the residual flow, which contains fiber, oil, and protein, is converted into distillers grains (DG). The industry of grain-to-ethanol is dominated by dry mills. However, at a number of major facilities, the mills are comparable to bioreserves in that the grains are wet milled to separate the various components, namely starch, protein, fiber, and germ, before converting these intermediates into end byproducts. The largest ethanol producer

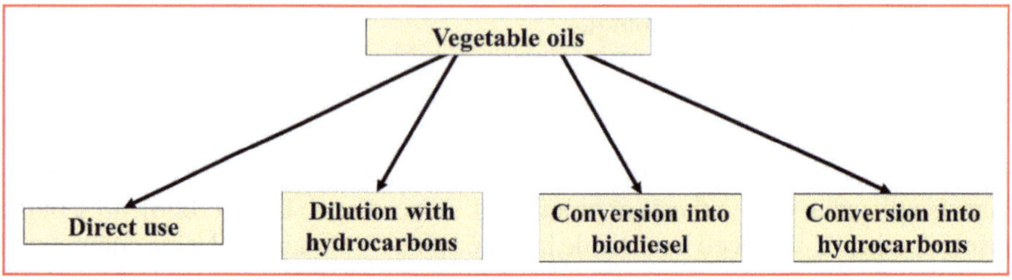

Fig. 10.1 Vegetable oils and their use as alternative biofuels

in the world from grains is the United States. From 8 billion litters in 2002 to 15 billion litters in 2005, it produced a lot more gasoline ethanol recently. In the United States, ethanol manufacturers used corn to produce almost 93% of the 18.5 billion liters of bioethanol that could be made in 2006. The European Union, where wheat is the predominant feedstock, is another important region that produces ethanol from grains. China and Canada are also producers. South Africa has started a large-scale program to convert maize into ethanol.

Vegetable Oil to Biodiesel

Biodiesel is a type of diesel made from reclaimed cooking fats, recycled vegetable oils, and animal fats consisting of mono-alkyl esters of long-chain fatty acids. Diesel engines may run on non-edible oils like Jatropha or kanuga seed as well as edible vegetable oils, including soybean, rape/canola, palm, sunflower, coconut, camelina, and used frying oil. The viscosities of vegetable oils used as alternative engine fuels are 10 to 20 times greater than those of petroleum-based diesel fuels. The high fuel viscosity that happens in compression ignition engines is the biggest drawback to using pure vegetable oils as diesel engine fuel.

To transform vegetable oils and fats into diesel fuel, a number of processes can be utilized, such as pyrolysis, dilution with hydrocarbons, and emulsification. The use of vegetable oils as alternative fuels to petroleum is depicted in Fig. 10.1 below (Demirbaş 2002).

It is common practice to convert fats and oils into biodiesel by transesterifying triglycerides with methanol (methanolysis) to create methyl esters of straight-chain fatty acids. To lower the viscosity of oil, the transesterification procedure is used, where ethanol is preferable to methanol due to its renewability, agricultural origin, and less harmful to the environment (Demirbaş 2002). Biodiesel has characteristics that are similar to those of diesel fuels. The ISO standards were used to characterize the biodiesel by measuring its viscosity, density, cetane number, cloud and pour points, features of distillation, ash and combustion points, and higher heating value (HHV).

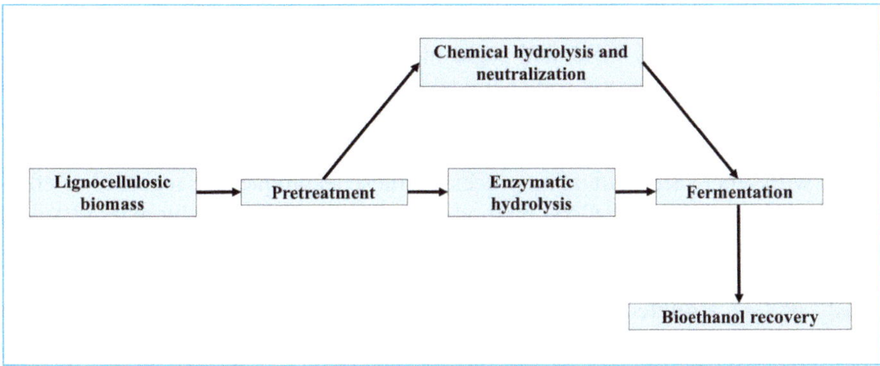

Fig. 10.2 Lignocellulosic biomass conversion into ethanol

10.4.2 Second-Generation Biofuel

A potential solution is provided by second-generation bioethanol, which makes use of the abundant non-edible lignocellulosic biomass found in agricultural wastes such as crop stover, straws, sugarmill byproducts, and wood. With the help of this generation's technology, cellulose and lignin may be separated from a plant so that cellulose can be fermented into alcohol. Since biomass refers to any source of organic carbon, a variety of biomasses can be used to produce these biofuels. This can quickly regenerate as a result of the carbon cycle. Second-generation biofuel technologies have been developed as a result of the substantial limitations in first-generation biofuel production. These problems can be fixed by allowing second-generation biofuels to produce a considerable proportion of ethanol sustainably, cheaply, and with better environmental benefits.

10.4.2.1 Conversion Technologies

Lignocellulosic Biomass to Ethanol

The main components of lignocellulosic biomass are cellulose, hemicellulose, and lignin. In order to make bioethanol, a multi-step process involving pre-treatment, saccharification, and detoxification is necessary due to the complex structure of lignocellulose (Fig. 10.2).

The lignocellulosic biomass's pretreatment promotes the depolymerization and removal of the lignin outer layer. This structural modification improves biomass surface area, exposes cellulose and hemicellulose, and is essential for the most effective hydrolysis.

The process of saccharification involves hydrolyzing cellulose and hemicellulose polymers to release their individual fermentable reducing sugars. Hexose sugars like glucose, galactose, and mannose, as well as pentose sugars like xylose and arabinose, are the most frequently generated sugar monomers. Saccharification is commonly accomplished via enzymatic and chemical processes.

Chemical procedures involve two methods called concentrated acid hydrolysis (CAH) and diluted acid hydrolysis (DAH). Since they can be employed in combination with other saccharification processes, these techniques are regarded as effective pre-treatments. In the CAH method, high-concentration dry biomass (10–40% substrate) is treated with concentrated acids, such as H_2SO_4 and HCL, in a batch process at temperatures below 160 °C. The DAH method, in contrast, involves continuously adding a diluted acid (4% H_2SO_4 solution) to low concentrations of dry biomass (5–10% substrate). In comparison to these two processes, CAH is thought to be more efficient since it encourages complete cellulose and hemicellulose hydrolysis and produces a large amount of fermentable sugars. Concentrated acids must be recovered and recycled for cost-effectiveness because of the harm they do to reactors due to their corrosive nature. The development of DAH is now more economically possible because higher temperatures have been shown to be beneficial for DAH, with glucose yields of about 70% delivered at 260 °C. However, higher temperatures result in more endproducts, like furfural, a fermentation inhibitor. As a result, DAH is usually carried out at temperatures of around 220 °C, with a 50–60% yield of fermentable sugar.

Enzymatic methods for saccharifying lignocellulose include fewer corrosion problems, work at temperatures between 40 and 50 °C, and are more environmentally friendly. In this approach, Endoglucanases, exoglucanases, cellobiohydrolases, and -glucosidases, also referred to as cellulosomes, are the primary enzymes that catalyze the hydrolysis of cellulose to glucose through the cleavage of -1,4 glucosidic linkages. In the end, Cellulases and hemicellulases convert cellulose and hemicellulose into hexose and pentose sugars.

Detoxification is the process of inhibiting the byproducts formed through the use of acid during pre-treatment and saccharification of lignocellulose. Inhibitors are classified into three types: weak acids, furans, and phenolic chemicals, such as acetic acid, furfural, and vanillin. The methods utilized for pre-treatment and saccharification have a significant impact on the types and amounts of inhibitory compounds contained in the hydrolysate. This process is employed to enable maximum potential bioethanol yields.

Ethanol Recovery

The ethanol concentration in the water and ethanol-containing broth that is recovered after fermentation is only 5 to 12 weight %; therefore, product purification is a crucial step to raise the ethanol concentration to a weight % that is suitable for use as a biofuel. Distillation can be used as a refinement method since the boiling temperatures of water (100 °C) and ethanol (78.37 °C) are different. This is because when the fermented broth is heated in a distillation column, the components take on their gaseous forms. A straightforward traditional distillation is insufficient to sufficiently separate ethanol and water because they combine to generate an azeotropic solution that causes co-distillation at 95.6% by weight of ethanol at 78.15 °C. Therefore, distillation, rectification, and dehydration are the three steps needed for an adequate ethanol purification procedure.

10.4.3 Third-Generation Biofuels

Third-generation biofuels get their energy from carefully bred plants like algae (Brennan and Owende 2010). To obtain the oil from these algae, they are cultured and harvested. The lipid content of the microorganisms is frequently a limiting component in the production of biofuels from algae. Typically, species like Chlorella are chosen for cultivation for their high lipid content (60–70%; Liang et al. 2009) and increased productivity (7.4 g/L/d for Chlorella protothecoides; Chen and Kuo 2011). The oil can subsequently be refined into alternative fuels to substitute petroleum-based fuels or transformed into biodiesel using a method similar to that used for first-generation biofuels. Third-generation biofuels have a higher energy density per harvested area compared to other-generation biofuels. Algae are advantageous because they can flourish in environments that are unsuitable for first- and second-generation crops, reducing pressure on available water and arable land, which makes them highly efficient, renewable, and affordable energy sources. It can be grown in sewage, wastewater, and saltwater from the sea or salt lakes. This would eliminate the requirement for using water that would otherwise be used for human consumption.

10.5 Recent Techniques and New Varieties of Crop Sources

It is standard practise to produce plants for food and livestock. Plant types with the necessary phenotypic, biochemical, and physical traits have been developed with the help of traditional breeding procedures and genetic modification (Lee 1998; Baenziger et al. 2006). Such initiatives primarily aim to increase agricultural quality and productivity. Additionally, food crops could be genetically altered to increase the production of starch and the C: N ratio in order to modify them for bioenergy production. However, the National Renewable Energy Laboratory (NREL) of the US Department of Energy has placed a focus on using lignin bioengineering to create biomass feedstocks with lower lignin contents (Larson and Initiative 2008). In general, this is because high lignin concentrations hinder the fermentation of ethanol and lower the amount of sugar that is produced during saccharification and hydrolysis. The presence of polysaccharides in plant cell walls and their concentration in lignocellulosic-biomass, such as mixed-linkage glucans, callose, cellulose, and galactan, are significant as they increase the yields of ethanol and other biofuels after processing. Transgenic production of enzymes that alter the polysaccharides in the cell wall before they reach full maturity may improve the quality of biomass

(Prasad et al. 2007; Brandon and Scheller 2020). Enhancing the ability of cellulolytic-enzymes to assist hydrolysis through genetic engineering is equally essential. After enzymatic saccharification, it was observed that the OsAT10 gene increased total sugar output by almost 40% in switchgrass lines and transgenic rice (de O Buanafina 2009; Bartley et al. 2013; Brandon and Scheller 2020). Increased levels of the Aspergillus aculeatus xyloglucanase gene (AaXEG2) in Populusalba resulted in enhanced cellulose deposition and growth, and enzymatic hydrolysis resulted in up to 81% more glucose being generated (Park et al. 2004; Kaida et al. 2009).

Additionally, the generation of biofuels from various bioresources employing a variety of cutting-edge biological techniques is expanding rapidly. Waste disposal difficulties could be resolved and environmental burdens reduced if biomass wastes and crop leftovers are used in biofuel production (Lee et al. 2019; Prasad, Kumar, et al. 2020a, b; Prasad, Singh et al. 2020a, b). Due to their environmental friendliness and carbon neutrality, biomass material derived from various plants and microorganisms has recently attracted increased scientific attention. This has led to more profound research into modern technologies for producing biofuels as a form of energy.

The various processes like mechanical, thermo-chemical, and biochemical conversion routes can be utilized to turn biomass into biofuels.

The mechanical use of biomass for the effective and affordable generation of bioenergy is significantly hampered by low biomass bulk density (BD) (Rentizelas et al. 2009; Lei et al. 2013; Vaccari et al. 2017). In this case, briquetting technology, which mechanically compresses or densifies biomass into solid-biofuel, could be advantageous (Lei et al. 2013). The technique involves compressing biomass wastes under high pressure to create thick briquettes (12–18% moisture) and pellets (15–30% moisture). In addition to these benefits, the automatic feeding of briquettes and pellets into boiler systems has produced excellent results. However, other raw materials are currently being looked for due to a lack of timber reserves. In order to produce energy and premium fuels of the highest quality, wood pellets, and briquettes are utilized in co-firing in nations such as Europe, North America, and Asia.

There are numerous thermo-chemical methods that are utilized to produce energy from lignocellulosic biomass, as well as heat, syngas, oxygenated bio-oil (liquid biofuels), charcoal (solid), and chemicals (Lee et al. 2019). This method mainly involves five energy production pathways such as combustion, torrefaction, pyrolysis, liquefaction, and gasification based on temperature, pressure, and heating (Demirbas, 2009).

Combustion is a known and developed method of using biomass (Kumar et al. 2022). In this process, biomass is burned at a high temperature, between 800 and 1000 °C, with enough oxygen, and the temperature may even go above 1400 °C at some points during the process (Demirbas 2009). Combustion is an exothermic-redox reaction in which energy is emitted as heat along with a product as the reaction's end product. Burning biomass accounts for more than 90% of energy use (Brown 2019).

The Pyrolysis technology is the most common method, which involves heating the biomass at temperatures between 350 and 550 °C, with the potential to reach 700 °C, to break it down into liquid fuels (py-oil/bio-oil), solid fuel (charcoal), and gases (syngas/fuel gas). This process takes place either in a vacuum or without air or oxygen. There are three types of pyrolysis: slow, quick, and flash. Pyrolysis can be carried out under various operating circumstances.

Gasification is a thermal conversion process where biomass is transformed into a gaseous product (such as syngas) at 800–1000 C through an endothermic reaction while being partially oxidized with oxygen or steam (Parmigiani et al. 2014; Vitali et al. 2013; Hirano et al. 1998). The gasification process begins with a reaction between biomass, steam, and oxygen-enriched air, which thermochemically degrades the biomass into methane, hydrogen, carbon monoxide, carbon dioxide, ethene, and ethane, as well as water and tar.

In the past, liquefaction has been employed by many countries. Biomass can be thermochemically converted into liquid biofuel at temperatures between 250 and 350 °C at high hydrogen partial pressures, usually between 100 and 200 bar. Bio-oil, a byproduct of biomass liquefaction, is a higher-quality fuel than py-oil (Gollakota et al. 2018; Liu et al. 2017). In this method, the pre-drying phase is eliminated, which makes it a perfect strategy for high moisture-containing feedstocks (Liu et al. 2017; Ruiz et al. 2013).

The primary biochemical processes used to convert biomass into biofuel are anaerobic digestion, fermentation, esterification, and photofermentation. The use of yeast, fungus, and bacteria in biochemical conversion is common. When it comes to converting the structural biopolymers in the biomass into gaseous or liquid biofuels like ethanol, biogas, and hydrogen, the enzymes produced by these bacteria are crucial. The production of biodiesel involves the well-known biochemical process of transesterification (Badruddin et al. 2018).

10.6 Environmental Impact of Plant-Based Biofuels

The fact that petroleum-based fuels are the leading cause of environmental issues such as air pollution, global warming, and acid rain which has a negative effect on human health and ecological balance. Compared to conventional diesel, biodiesel offers a number of outstanding advantages, such as it reduces the lifecycle of greenhouse gases by an average of 74%, hydrocarbon emissions by 67%, and generates 4.5 units of energy for every unit of fossil energy consumed over its lifetime, and it also has certain environmentally favourable features such as being non-hazardous, biodegradable, and safer for breathing (Pradhan et al. 2009). Biodiesel is an environmentally friendly and reliable biofuel that depends heavily on its sources (soybean, rapeseed, or animal fats) for characteristics such as oxygen content, cetane number, viscosity, density, and heat value (Wu et al. 2009). Moreover, these properties have a significant role in Engine performance and emissions of biodiesels. Highly oxygenated fuels like biodiesel can enhance combustion efficiency while lowering emissions of polycyclic aromatic hydrocarbons (PAHs), sulphur dioxide (SO_2), carbon dioxide (CO_2), carbon monoxide (CO) and unburned hydrocarbons (HCs). However, fuel usage for brakes marginally rises (Lin et al. 2006). It has made a considerable impact on carbon emission reductions in the transport sector, which is responsible for 25%

of the EU's overall GHG emissions. The European Commission highlighted that the use of renewable energy in transportation resulted in a reduction of over 35 million tonnes of carbon emissions in its Renewables Progress Report, which was published in June 2015. More specifically, the paper acknowledges that the usage of biofuels was responsible for the majority of these savings.

In addition, the sustainability standards established by European law ensure that biofuels will eventually adhere to even more demanding standards (50% less than fossil fuels by 2018), releasing 35% less CO_2 than fossil fuels now. Biodiesel is the biofuel that has the most significant impact on reducing emissions overall. In fact, the CARB emphasized that compared to fossil fuels, biodiesel reduces carbon emissions by an average of 50% to 81%, making it the most environmentally friendly liquid fuel. Biodiesel should be at the forefront of efforts to decarbonize global transportation because it directly reduces carbon emissions. Because of growing environmental awareness and increasing diesel costs, biodiesel is becoming more and more popular as a sustainable source of diesel fuel. Due to its clean emission characteristics, it is a consumer choice that is friendly to the environment and already makes up a sizeable part of the world's fuel market. The need to reduce the emission of greenhouse gasses, which is the main problem facing the globe today, is one of the elements influencing the development of biodiesel fuels in many nations. The shortage of sources for petroleum diesel also contributes to the global growth and production of biodiesel fuel.

10.7 Conclusion

Plants use carbon dioxide for their growth, which is released through biomass combustion. Hence, the use of such biomass for energy production won't result in any additional emissions of the gas since the quantity released during use was already fixed during plant development. The utilization of plant-based feedstock could make it simpler to employ this alternative renewable energy source. Using current engine technologies, the commercial generation of biofuels may help us become less dependent on fossil fuels in the transportation sector. Plant-based feedstocks can play a significant role in the production of biofuels such as ethanol and biodiesel, which can help to strengthen rural economies, offer efficient energy supply, and concentrate on the utilization of areas that have been devastated by environmental factors. Deforestation, biodiversity degradation, grassland conversion, and other problems can be resolved with bioenergy crops, which can also be cultivated on farms to provide an alternative source of wood. The cultivation of bioenergy crops has a significant potential to protect the natural forest. Such environment-related issues could be solved only by extensive plantation of bioenergy crops. Since they can control the dynamics of soil–water and nutrient levels, thus it is critical to consider how much water bioenergy crops need before field planting.

References

Abbasi T, Abbasi SA (2010) Biomass energy and the environmental impacts associated with its production and utilization. Renew Sustain Energy Rev 14(3):919–937. https://doi.org/10.1016/j.rser.2009.11.006

Adler PR, Del Grosso SJ, Parton WJ (2007) Life-cycle assessment of net greenhouse-gas flux for bioenergy cropping systems. Ecol Appl: Publ Ecol Soc Am 17(3):675–691. https://doi.org/10.1890/05-2018

Ahmad AL, Yasin NHM, Derek CJC, Lim JK (2011) Microalgae as a sustainable energy source for biodiesel production: A review. Renew Sustain Energy Rev 15(1):584–593. https://doi.org/10.1016/j.rser.2010.09.018

Ahuja D, Tatsutani M (2009) Sustainable energy for developing countries. S.A.P.I.EN.S. Surveys and Perspectives Integrating Environment and Society, 2.1, Article 2.1. https://journals.openedition.org/sapiens/823

Akinbami J-FK (2001) Renewable energy resources and technologies in Nigeria—Present situation, future prospects and policy framework. Mitig Adapt Strat Glob Change 6(2):155–182. https://doi.org/10.1023/A:1011387516838

Ali S, Anwar S, Nasreen S (2017, May 9) Renewable and Non-Renewable energy and its impact on environmental quality in south asian countries. Forman J Econ Stud (13):177–194

Arutyunov VS, Lisichkin GV (2017) Energy resources of the 21st century: Problems and forecasts. Can renewable energy sources replace fossil fuels†. Russ Chem Rev, 86(8), 777. https://doi.org/10.1070/RCR4723

Baenziger PS, Russell WK, Graef GL, Campbell BT (2006) Improving Lives: 50 Years of Crop Breeding, Genetics, and Cytology (C-1). Crop Sci 46(5):2230–2244. https://doi.org/10.2135/cropsci2005.11.0404gas

Bartley LE, Peck ML, Kim SR, Ebert B, Manisseri C, Chiniquy DM, Sykes R, Gao L, Rautengarten C, Vega-Sánchez ME, Benke PI, Canlas PE, Cao P, Brewer S, Lin F, Smith WL, Zhang X, Keasling JD, Jentoff RE, Ronald PC (2013) Overexpression of a BAHD acyltransferase, OsAt10, alters rice cell wall hydroxycinnamic acid content and saccharification. Plant Physiol 161(4):1615–1633. https://doi.org/10.1104/pp.112.208694

Bildirici ME (2013) Economic growth and biomass energy. Biomass Bioenerg 50:19–24. https://doi.org/10.1016/j.biombioe.2012.09.055

Boe A, Lee DK (2007) Genetic variation for biomass production in prairie cordgrass and switchgrass. Crop Sci 47(3):929–934. https://doi.org/10.2135/cropsci2006.05.0323

Borland AM, Griffiths H, Hartwell J, Smith JAC (2009) Exploiting the potential of plants with crassulacean acid metabolism for bioenergy production on marginal lands. J Exp Bot 60(10):2879–2896. https://doi.org/10.1093/jxb/erp118

Bouton JH (2007) Molecular breeding of switchgrass for use as a biofuel crop. Curr Opin Genet Dev 17(6):553–558. https://doi.org/10.1016/j.gde.2007.08.012

Brandon AG, Scheller HV (2020) Engineering of bioenergy crops: dominant genetic approaches to improve polysaccharide properties and composition in biomass. Front Plant Sci 11:282. https://doi.org/10.3389/fpls.2020.00282

Brennan L, Owende P (2010) Biofuels from microalgae—A review of technologies for production, processing, and extractions of biofuels and co-products. Renew Sustain Energy Rev 14(2):557–577. https://doi.org/10.1016/j.rser.2009.10.009

Brown RC (2019) Thermochemical processing of biomass: conversion into fuels. John Wiley & Sons, Chemicals and Power

Burns DA, Aherne J, Gay DA, Lehmann CMB (2016) Acid rain and its environmental effects: Recent scientific advances. Atmos Environ 146:1–4. https://doi.org/10.1016/j.atmosenv.2016.10.019

Bush DR, Leach JE (2007) Translational genomics for bioenergy production: there's room for more than one model. Plant Cell 19(10):2971–2973. https://doi.org/10.1105/tpc.107.191040

Carere CR, Sparling R, Cicek N, Levin DB (2008) Third generation biofuels via direct cellulose fermentation. Int J Mol Sci 9(7):1342–1360. https://doi.org/10.3390/ijms9071342

Carroll A, Somerville C (2008) Cellulosic biofuels. Annu Rev Plant Biol 60:165–182. https://doi.org/10.1146/annurev.arplant.043008.092125

Casler MD, Stendal CA, Kapich L, Vogel KP (2007) Genetic diversity, plant adaptation regions, and gene pools for switchgrass. Crop Sci 47(6):2261–2273. https://doi.org/10.2135/cropsci2006.12.0797

Chen C-Y, Yeh K-L, Aisyah R, Lee D-J, Chang J-S (2011) Cultivation, photobioreactor design and harvesting of microalgae for biodiesel production: A critical review. Biores Technol 102(1):71–81. https://doi.org/10.1016/j.biortech.2010.06.159

Chen W-H, Kuo P-C (2011) Isothermal torrefaction kinetics of hemicellulose, cellulose, lignin and xylan using thermogravimetric analysis. Energy 36(11):6451–6460. https://doi.org/10.1016/j.energy.2011.09.022

Chhetri AB, Tango MS, Budge SM, Watts KC, Islam MR (2008) Non-edible plant oils as new sources for biodiesel production. Int J Mol Sci 9(2):169–180. https://doi.org/10.3390/ijms9020169

de O Buanafina, M. M. (2009) Feruloylation in grasses: Current and future perspectives. Mol Plant 2(5):861–872. https://doi.org/10.1093/mp/ssp067

de Fraiture C, Giordano M, Liao Y (2008). Biofuels and Implications for Agricultural Water Use: Blue Impacts of Green Energy. https://doi.org/10.2166/wp.2008.054

DeLong MM, Swanberg DR, Oelke EA (1995). Sustainable biomass energy production and rural economic development using alfalfa as feedstock (NREL/CP-200–8098; CONF-9508104-). National Renewable Energy Lab. (NREL), Golden, CO (United States). https://www.osti.gov/biblio/115379

Demirbaş A (2002) Biodiesel from vegetable oils via transesterification in supercritical methanol. Energy Convers Manage 43(17):2349–2356. https://doi.org/10.1016/S0196-8904(01)00170-4

Demirbas A (2009) Progress and recent trends in biodiesel fuels. Energy Convers Manage 50(1):14–34. https://doi.org/10.1016/j.enconman.2008.09.001

Ehrlich PR, Pringle RM (2008) Where does biodiversity go from here? A grim business-as-usual forecast and a hopeful portfolio of partial solutions. Proceedings of the National Academy of Sciences, 105(supplement_1), 11579–11586. https://doi.org/10.1073/pnas.0801911105

Escamilla-Treviño L (2011) Potential of plants from the genus agave as bioenergy crops. BioEnergy Research, 5. https://doi.org/10.1007/s12155-011-9159-x

Finckh MR (2008) Integration of breeding and technology into diversification strategies for disease control in modern agriculture. Eur J Plant Pathol 121(3):399–409. https://doi.org/10.1007/s10658-008-9273-6

Glenn E, Brown J, Blumwald E (1999) Salt Tolerance and crop potential of halophytes. Critical Reviews in Plant Sciences—Crit Rev Plant Sci 18:227–255. https://doi.org/10.1080/07352689991309207

Gollakota ARK, Kishore N, Gu S (2018) A review on hydrothermal liquefaction of biomass. Renew Sustain Energy Rev 81:1378–1392. https://doi.org/10.1016/j.rser.2017.05.178

Grammelis P, Malliopoulou A, Basinas P, Danalatos NG (2008) Cultivation and characterization of *Cynara Cardunculus* for solid biofuels production in the Mediterranean region. Int J Mol Sci 9(7):1241–1258. https://doi.org/10.3390/ijms9071241

Grover S, Rajeshwari. (2009) Global warming and its impact on skin disorders. Indian J Dermatol Venereol Leprol 75(4):337. https://doi.org/10.4103/0378-6323.53127

Guo M, Song W, Buhain J (2015) Bioenergy and biofuels: History, status, and perspective. Renew Sustain Energy Rev 42:712–725. https://doi.org/10.1016/j.rser.2014.10.013

Hasanuzzaman M, Nahar K, Alam MM, Bhowmik PC, Hossain MA, Rahman MM, Prasad MNV, Ozturk M, Fujita M (2014) Potential use of halophytes to remediate saline soils. Biomed Res Int 2014:589341. https://doi.org/10.1155/2014/589341

Hastilestari BR, Mudersbach M, Tomala F, Vogt H, Biskupek-Korell B, Van Damme P, Guretzki S, Papenbrock J (2013) *Euphorbia tirucalli* L.-comprehensive characterization of a drought tolerant plant with a potential as biofuel source. PloS One 8(5):e63501. https://doi.org/10.1371/journal.pone.0063501

Heaton EA, Dohleman FG, Miguez AF, Juvik JA, Lozovaya V, Widholm J, Zabotina OA, McIsaac GF, David MB, Voigt TB, Boersma NN, Long SP (2010) *Miscanthus*. 56:75–137. https://doi.org/10.1016/B978-0-12-381518-7.00003-0

Heaton E, Voigt T, Long SP (2004) A quantitative review comparing the yields of two candidate C4 perennial biomass crops in relation to nitrogen, temperature and water. Biomass Bioenerg 27(1):21–30. https://doi.org/10.1016/j.biombioe.2003.10.005

Hirano A, Hon-Nami K, Kunito S, Hada M, Ogushi Y (1998) Temperature effect on continuous gasification of microalgal biomass: Theoretical yield of methanol production and its energy balance. Catal Today 45(1):399–404. https://doi.org/10.1016/S0920-5861(98)00275-2

Holechek J, Valdez R, Geli H, Sawalhah M (2022) A Global assessment: can renewable energy replace fossil fuels by 2050? Sustainability, 14. https://doi.org/10.3390/su14084792

Hwang J-H, Kabra AN, Ji M-K, Choi J, El-Dalatony MM, Jeon B-H (2016) Enhancement of continuous fermentative bioethanol production using combined treatment of mixed microalgal biomass. Algal Res 17:14–20. https://doi.org/10.1016/j.algal.2016.03.029

IRENA (2019) Renewable energy statistics 2019. The International Renewable Energy Agency, Abu Dhabi. https://www.irena.org/-/media/Files/IRENA/Agency/Publication/2019/Jul/IRENA_Renewable_energy_statistics_2019.pdf

Inc, AGRR (n.d.). Alberta geothermal resource recovery Inc. Alberta geothermal resource recovery Inc. Retrieved July 3, 2023, from https://albertageothermal.ca/home

Jaradat, A. (2010). Genetic resources of energy crops: biological systems to combat climate change. Australian journal of crop science. https://www.semanticscholar.org/paper/Genetic-Resources-of-Energy-Crops%3A-Biological-to-Jaradat/0c8dda344f7c722372531b220165a10e8b53b0a6

Kaida R, Kaku T, Baba K, Oyadomari M, Watanabe T, Nishida K, Kanaya T, Shani Z, Shoseyov O, Hayashi T (2009) Loosening xyloglucan accelerates the enzymatic degradation of cellulose in wood. Mol Plant 2(5):904–909. https://doi.org/10.1093/mp/ssp060

Khan NA, el Dessouky H (2009) Prospect of biodiesel in Pakistan. Renew Sustain Energy Rev 13(6):1576–1583. https://doi.org/10.1016/j.rser.2008.09.016

Kumar A, Singh E, Mishra R, Lo S-L, Kumar S (2022) A green approach towards sorption of CO_2 on waste derived biochar. Environ Res 214:113954. https://doi.org/10.1016/j.envres.2022.113954

Lal R (2008). Carbon sequestration. Philosophical Transactions of the Royal Society of London. Series B, Biological Sciences, 363(1492), 815–830. https://doi.org/10.1098/rstb.2007.2185

Lamb JAFS, Jung HJG, Sheaffer CC, Samac DA (2007) Alfalfa leaf protein and stem cell wall polysaccharide yields under hay and biomass management systems. Crop Sci 47(4):1407–1415. https://doi.org/10.2135/cropsci2006.10.0665

Lamb JFS, Sheaffer CC, Samac DA (2003) Population density and harvest maturity effects on leaf and stem yield in Alfalfa. Agron J 95(3):635–641. https://doi.org/10.2134/agronj2003.6350

Larson ED, Initiative UB (2008). Biofuel production technologies: Status, prospects and implications for trade and development. https://digitallibrary.un.org/record/1315430

Lee JW (Ed). (2013). Advanced biofuels and bioproducts. Springer. https://doi.org/10.1007/978-1-4614-3348-4

Lee M (1998) Genome projects and gene pools: New germplasm for plant breeding? Proc Natl Acad Sci USA 95(5):2001–2004. https://doi.org/10.1073/pnas.95.5.2001

Lee SY, Sankaran R, Chew KW, Tan CH, Krishnamoorthy R, Chu D.-T, Show P-L (2019). Waste to bioenergy: A review on the recent conversion technologies. BMC Energy, 1(1), Article 1. https://doi.org/10.1186/s42500-019-0004-7

Lehtomäki A, Viinikainen TA, Rintala JA (2008) Screening boreal energy crops and crop residues for methane biofuel production. Biomass Bioenerg 32(6):541–550. https://doi.org/10.1016/j.biombioe.2007.11.013

Lei T, Wang Z, Li Z, Xu J, He X, Zhu J (2013) A biomass briquetting fuel machine and its large-scale operation system. J Renew Sustain Energy 5(1):013107. https://doi.org/10.1063/1.4781089

Liang Y, Sarkany N, Cui Y (2009) Biomass and lipid productivities of Chlorella vulgaris under autotrophic, heterotrophic and mixotrophic growth conditions. Biotech Lett 31(7):1043–1049. https://doi.org/10.1007/s10529-009-9975-7

Licht F (2006). World ethanol markets: The outlook to 2015. https://www.semanticscholar.org/paper/World-ethanol-markets-%3A-the-outlook-to-2015-Licht/10bd09e581a87ac4159cf40c1af2747fedcc5032

Lin Y, Lee W, Wu T, Wang C (2006) Comparison of PAH and regulated harmful matter emissions from biodiesel blends and paraffinic fuel blends on engine accumulated mileage test. Fuel 85(17–18):2516–2523. https://doi.org/10.1016/j.fuel.2006.04.023

Liqun L, Zhixin W (2009) A variable voltage MPPT control method for photovoltaic generation system. WSEAS transactions on circuits and systems archive. https://www.semanticscholar.org/paper/A-variable-voltage-MPPT-control-method-for-system-Liqun-Zhixin/09457a3680d1eb4dc9c88a4a705213bc882c1c6c

Liu H, Ma M, Xie X (2017) New materials from solid residues for investigation the mechanism of biomass hydrothermal liquefaction. Ind Crops Prod 108:63–71. https://doi.org/10.1016/j.indcrop.2017.06.026

Lobell DB, Burke MB, Tebaldi C, Mastrandrea MD, Falcon WP, Naylor RL (2008). Prioritizing climate change adaptation needs for food security in 2030. Science (New York, N.Y.), 319(5863), 607–610. https://doi.org/10.1126/science.1152339

Lorenz AJ, Coors JG, Leon N, Wolfrum EJ, Hames BR, Sluiter AD, Weimer PJ (2009) Characterization, genetic variation, and combining ability of maize traits relevant to the production of cellulosic ethanol. Crop Sci 49(1):85–98. https://doi.org/10.2135/cropsci2008.06.0306

Mabee WE, McFarlane PN, Saddler JN (2011) Biomass availability for lignocellulosic ethanol production. Biomass Bioenerg 35(11):4519–4529. https://doi.org/10.1016/j.biombioe.2011.06.026

Mathur J, R., R. (2008) Ranade, S. A., Srivastava, A. P., Rana, T. S., Srivastava, J. and Tuli, R. 2008. Easy assessment of diversity in *Jatropha curcas* L. plants using two Single Primer Amplification Reaction (SPAR) methods. Biomass Bioenerg 32:533–540

McLaughlin SB, Kiniry JR, Taliaferro CM, De La Torre Ugarte D (2006) Projecting yield and utilization potential of switchgrass as an energy Crop. In Advances in Agronomy (90, pp 267–297). Academic Press. https://doi.org/10.1016/S0065-2113(06)90007-8

Merigliano MF, Lesica P (1998) The native status of reed canarygrass (Phalaris arundinacea L.) in the inland northwest, USA. Natural Areas Journal. https://www.semanticscholar.org/paper/The-native-status-of-reed-canarygrass-(Phalaris-L.)-Merigliano-Lesica/22265fe5eda6cae8336393027b7f5294984caa80

Mirza UK, Ahmad N, Majeed T (2008) An overview of biomass energy utilization in Pakistan. Renew Sustain Energy Rev 12(7):1988–1996

Mohajan HK (2018). Acid rain is a local environment pollution but global concern.

Mohammed IY, Abakr YA, Kazi K, F., Yusup, S., Alshareef, I., & Chin, S. A. (2015) Comprehensive characterization of Napier grass as a feedstock for thermochemical conversion. Energies 8(5):3403–3417. https://doi.org/10.3390/en8053403

Mohtasham J (2015) Review Article-Renewable energies. Energy Procedia 74:1289–1297. https://doi.org/10.1016/j.egypro.2015.07.774

Mola-Yudego B, González-Olabarria JR (2010) Mapping the expansion and distribution of willow plantations for bioenergy in Sweden: Lessons to be learned about the spread of energy crops. Biomass Bioenerg 34(4):442–448. https://doi.org/10.1016/j.biombioe.2009.12.008

Morales Bayetero C, Mafla Yépez C, Benavides Cevallos I, Hernández Rueda E (2022) Effect of the use of additives in biodiesel blends on the performance and opacity of a diesel engine. Materials Today: Proceedings 49:93–99. https://doi.org/10.1016/j.matpr.2021.07.478

Moser BR, Dien BS, Seliskar DM, Gallagher JL (2013) Seashore mallow (*Kosteletzkya pentacarpos*) as a salt-tolerant feedstock for production of biodiesel and ethanol. Renew Energy 50:833–839. https://doi.org/10.1016/j.renene.2012.08.016

OECD/FAO (2015) OECD-FAO agricultural outlook 2015. OECD Publishing, Paris. https://doi.org/10.1787/agr_outlook-2015-en

Okada M, Lanzatella C, Saha MC, Bouton J, Wu R, Tobias CM (2010) Complete switchgrass genetic maps reveal subgenome collinearity. Preferential Pairing and Multilocus Interactions. Genetics 185(3):745–760. https://doi.org/10.1534/genetics.110.113910

Oliver RJ, Finch JW, Taylor G (2009) Second generation bioenergy crops and climate change: A review of the effects of elevated atmospheric CO_2 and drought on water use and the implications for yield. GCB Bioenergy 1(2):97–114. https://doi.org/10.1111/j.1757-1707.2009.01011.x

Owusu PA, Asumadu-Sarkodie S (2016) A review of renewable energy sources, sustainability issues and climate change mitigation. Cogent Engineering 3(1):1167990. https://doi.org/10.1080/23311916.2016.1167990

Panta S, Flowers T, Lane P, Doyle R, Haros G, Shabala S (2014) Halophyte agriculture: Success stories. Environ Exp Bot 107:71–83. https://doi.org/10.1016/j.envexpbot.2014.05.006

Park S-I, Daeschel MA, Zhao Y (2004) Functional properties of antimicrobial Lysozyme-Chitosan composite films. J Food Sci 69(8):M215–M221. https://doi.org/10.1111/j.1365-2621.2004.tb09890.x

Parmigiani SP, Vitali F, Lezzi AM, Vaccari M (2014) Design and performance assessment of a rice husk fueled stove for household cooking in a typical sub-Saharan setting. Energy Sustain Dev 23:15–24. https://doi.org/10.1016/j.esd.2014.01.003

Paterson AH, Bowers JE, Bruggmann R, Dubchak I, Grimwood J, Gundlach H, Haberer G, Hellsten U, Mitros T, Poliakov A, Schmutz J, Spannagl M, Tang H, Wang X, Wicker T, Bharti AK, Chapman J, Feltus FA, Gowik U, Rokhsar DS (2009) The Sorghum bicolor genome and the diversification of grasses. Nature 457(7229):551–556. https://doi.org/10.1038/nature07723

Patil V, Tran K-Q, Giselrød HR (2008) Towards sustainable production of biofuels from microalgae. Int J Mol Sci 9(7):1188–1195. https://doi.org/10.3390/ijms9071188

Perera F (2017) Pollution from Fossil-Fuel combustion is the leading environmental threat to global pediatric health and equity: solutions exist. Int J Environ Res Public Health 15:16. https://doi.org/10.3390/ijerph15010016

Petersen J-E (2008). Energy production with agricultural biomass: environmental implications and analytical challenges (SSRN Scholarly Paper No. 1289122). https://papers.ssrn.com/abstract=1289122

Pradhan A, Shrestha D, McAloon A, Yee W, Haas M, Duffield J, Shapouri H (2009). Energy Life-Cycle assessment of soybean biodiesel. In Trans. ASABE (Vol. 54)

Prasad S, Kumar S, Yadav KK, Choudhry J, Kamyab H, Bach Q-V, Sheetal KR, Kannojiya S, Gupta N (2020a) Screening and evaluation of cellulytic fungal strains for saccharification and bioethanol production from rice residue. Energy 190:116422

Prasad S, Singh A, Joshi HC (2007) Ethanol as an alternative fuel from agricultural, industrial and urban residues. Resour Conserv Recycl 50(1):1–39. https://doi.org/10.1016/j.resconrec.2006.05.007

Prasad S, Singh A, Korres NE, Rathore D, Sevda S, Pant D (2020b) Sustainable utilization of crop residues for energy generation: A life cycle assessment (LCA) perspective. Biores Technol 303:122964

Qazi HA, Paranjpe S, Bhargava S (2012) Stem sugar accumulation in sweet sorghum—Activity and expression of sucrose metabolizing enzymes and sucrose transporters. J Plant Physiol 169(6):605–613. https://doi.org/10.1016/j.jplph.2012.01.005

Rae AM, Street NR, Robinson KM, Harris N, Taylor G (2009) Five QTL hotspots for yield in short rotation coppice bioenergy poplar: The Poplar Biomass Loci. BMC Plant Biol 9:23. https://doi.org/10.1186/1471-2229-9-23

Regassa TH, Wortmann CS (2014) Sweet sorghum as a bioenergy crop: Literature review. Biomass Bioenergy 64:348–355. https://doi.org/10.1016/j.biombioe.2014.03.052

Rentizelas AA, Tolis AJ, Tatsiopoulos IP (2009) Logistics issues of biomass: The storage problem and the multi-biomass supply chain. Renew Sustain Energy Rev 13(4):887–894. https://doi.org/10.1016/j.rser.2008.01.003

Rockwood DL, Rudie AW, Ralph SA, Zhu JY, Winandy JE (2008) Energy product options for eucalyptus species grown as short rotation woody crops. Int J Mol Sci 9(8):1361–1378. https://doi.org/10.3390/ijms9081361

Rodolfi L, Chini Zittelli G, Bassi N, Padovani G, Biondi N, Bonini G, Tredici MR (2009) Microalgae for oil: Strain selection, induction of lipid synthesis and outdoor mass cultivation in a low-cost photobioreactor. Biotechnol Bioeng 102(1):100–112. https://doi.org/10.1002/bit.22033

Ruiz HA, Rodríguez-Jasso RM, Fernandes BD, Vicente AA, Teixeira JA (2013) Hydrothermal processing, as an alternative for upgrading agriculture residues and marine biomass according to the biorefinery concept: A review. Renew Sustain Energy Rev 21:35–51. https://doi.org/10.1016/j.rser.2012.11.069

Russelle MP (2001) Alfalfa: After an 8,000-year journey, the "Queen of Forages" stands poised to enjoy renewed popularity. Am Sci 89(3):252–261

Saidur R, Rahim NA, Islam MR, Solangi KH (2011) Environmental impact of wind energy. Renew Sustain Energy Rev 15(5):2423–2430. https://doi.org/10.1016/j.rser.2011.02.024

Sanderson MA, Adler PR (2008) Perennial forages as second generation bioenergy crops. Int J Mol Sci 9(5):768–788. https://doi.org/10.3390/ijms9050768

Schenk PM, Thomas-Hall SR, Stephens E, Marx UC, Mussgnug JH, Posten C, Kruse O, Hankamer B (2008) Second generation biofuels: High-Efficiency microalgae for biodiesel production. BioEnergy Research 1(1):20–43. https://doi.org/10.1007/s12155-008-9008-8

Schmer MR, Vogel KP, Mitchell RB, Perrin RK (2008) Net energy of cellulosic ethanol from switchgrass. Proc Natl Acad Sci USA 105(2):464–469. https://doi.org/10.1073/pnas.0704767105

Sims REH, Hastings AFSJ, Schlamadinger B, Taylor G, Smith P (2006) Energy crops: Current status and future prospects. Glob Change Biol 12:2054–2076. https://doi.org/10.1111/j.1365-2486.2006.01163.x

Singh B, Agrawal M (2004) Impact of simulated acid rain on growth and yield of two cultivars of wheat. Water Air Soil Pollut 152(1–4):71–80. https://doi.org/10.1023/B:WATE.0000015331.02874.df

Singh PP (2008) Exploring biodiversity and climate change benefits of community-based forest management. Glob Environ Chang 18(3):468–478. https://doi.org/10.1016/j.gloenvcha.2008.04.006

Springer TL, Dewald CL (2004). Eastern gamagrass and other tripsacum species. In Warm-Season (C4) Grasses (pp. 955–973). John Wiley & Sons, Ltd. https://doi.org/10.2134/agronmonogr45.c29

Taherzadeh MJ, Karimi K (2008) Pretreatment of lignocellulosic wastes to improve ethanol and biogas production: A review. Int J Mol Sci 9(9):1621–1651. https://doi.org/10.3390/ijms9091621

Tahir MHN, Casler MD, Moore KJ, Brummer EC (2011) Biomass yield and quality of reed canarygrass under five harvest management systems for bioenergy production. BioEnergy Research 4(2):111–119. https://doi.org/10.1007/s12155-010-9105-3

Takamizawa K, Anderson W, Singh HP (2010) Ethanol from lignocellulosic crops. (B. P. Singh, Ed.; 1st ed., pp. 104–139). CABI. https://doi.org/10.1079/9781845936167.0104

Thomas H, Howarth CJ (2000) Five ways to stay green. J Exp Bot, *51 Spec No*, 329–337. https://doi.org/10.1093/jexbot/51.suppl_1.329

Tilman D, Socolow R, Foley JA, Hill J, Larson ED, Lynd L, Pacala SW, Reilly J, Searchinger T, Somerville C, Williams R (2009) Beneficial biofuels—The food, energy, and environment trilemma. Science 325(5938):270–271. https://doi.org/10.1126/science.1177970

TM, YK, Badruddin I, Ankalgi R, Badarudin A, Hungund B, Ankalgi F (2018). Biodiesel production by direct transesterification process via sequential use of Acid–Base Catalysis. Arab J Sci Eng, 43. https://doi.org/10.1007/s13369-018-3078-5

Tran NH, Bartlett JR, Kannangara GSK, Milev AS, Volk H, Wilson MA (2010) Catalytic upgrading of biorefinery oil from micro-algae. Fuel 89(2):265–274. https://doi.org/10.1016/j.fuel.2009.08.015

Umar D, Abubakar M (2014) Effects of energy utilization on the environment. Discovery 25:25–93

Vaccari M, Vitali F, Tudor T (2017) Multi-criteria assessment of the appropriateness of a cooking technology: A case study of the Logone Valley. Energy Policy 109:66–75. https://doi.org/10.1016/j.enpol.2017.06.052

Ventura Y, Eshel A, Pasternak D, Sagi M (2015) The development of halophyte-based agriculture: Past and present. Ann Bot 115(3):529–540. https://doi.org/10.1093/aob/mcu173

Villaverde JJ, De Vega A, Ligero P, Freire CSR, Neto CP, Silvestre AJD (2010) *Miscanthus* x *giganteus* bark organosolv fractionation: Fate of lipophilic components and formation of valuable phenolic byproducts. J Agric Food Chem 58(14):8279–8285. https://doi.org/10.1021/jf101174x

Vitali F, Parmigiani S, Vaccari M, Collivignarelli C (2013) Agricultural waste as household fuel: Techno-economic assessment of a new rice-husk cookstove for developing countries. Waste Management (New York, N.Y.), 33(12), 2762–2770. https://doi.org/10.1016/j.wasman.2013.08.026

Vogel KP, Mitchell RB (2008) Heterosis in switchgrass: biomass yield in swards. Crop Sci 48(6):2159–2164. https://doi.org/10.2135/cropsci2008.02.0117

Voloshin R, Kreslavskii V, Zharmukhamedov S, Bedbenov V, Ramakrishna S, Allakhverdiev S (2015). Photoelectrochemical cells based on photosynthetic systems: A review. Biofuel Res J, 2, 227–235. https://doi.org/10.18331/BRJ2015.2.2.4

Wang Y, Yan L (2008) CFD studies on biomass thermochemical conversion. Int J Mol Sci 9(6):1108–1130. https://doi.org/10.3390/ijms9061108

Williams C, Black I, Biswas T, Heading S (2007) Pathways to prosperity: Second generation biomass crops for biofuels using saline lands and wastewater. Agric Sci. https://www.semanticscholar.org/paper/Pathways-to-prosperity%3A-second-generation-biomass-Williams-Black/b3b76fe4832cadafaaac64d933f816b2dccdec31

Wu F, Wang J, Chen W, Shuai S (2009) A study on emission performance of a diesel engine fueled with five typical methyl ester biodiesels. Atmos Environ 43(7):1481–1485. https://doi.org/10.1016/j.atmosenv.2008.12.007

Xu H, Zhang J, Ouyang Y, Lin L, Quan G, Zhao B, Yu J (2015) Effects of simulated acid rain on microbial characteristics in a lateritic red soil. Environ Sci Pollut Res Int 22(22):18260–18266. https://doi.org/10.1007/s11356-015-5066-6

Yasuda M, Ishii Y, Ohta K (2015) Napier grass (*Pennisetum purpureum* Schumach) as raw material for bioethanol production: Pretreatment, saccharification, and fermentation. Biotechnol Bioprocess Eng 19:943–950. https://doi.org/10.1007/s12257-014-0465-y

Yuan JS, Tiller KH, Al-Ahmad H, Stewart NR, Stewart CN (2008) Plants to power: Bioenergy to fuel the future. Trends Plant Sci 13(8):421–429. https://doi.org/10.1016/j.tplants.2008.06.001

Chapter 11
Microbes-Induced Enhanced Biofuel Production

K. S. Shreenidhi, B. Vijaya Geetha, S. Priyanga, B. K. Raghunandhan, and K. Rooshini

Abstract The use of fossil fuels since its arrival is ceaseless, although its pernicious nature towards the living world is well-known. The Incessant climate crisis has enforced humankind to look for an alternative source of energy as a consequence of harmful carbon emissions. In the Next 30 years, the expected human population might increase up to 8 or even 10 billion thereupon without facing the demands of sustainable energy needs. Despite this, depending majorly on just fossil fuel, being a non-renewable source of energy is now a situation to be resolved. Biofuel act as a viable substitute for the reason that it is a renewable source of energy that provides energy security helps in the reduction of greenhouse gas emission and builds economic development. Innovative approaches such as Incorporating microbes will intensify biofuel production as they are efficiently accessible, and the enzymes within the microbes by their remarkable nature enhance the production. Therefore, this book chapter is written as an overview of microbe's efficacy in biofuel production.

11.1 Introduction

Our dependency on conventional fuels can be reduced by using alternative energy sources like biofuels, which can be produced using live organisms. Biodiesel has been more well-liked as a fossil fuel substitute globally in recent decades. It has the ability to displace current petroleum-based diesel fuels and produces renewable biological materials. Transesterification of various animal-based fats and vegetable oils, typically with ethanol or methanol, yields biodiesel fuels. The two most common seed oils from which they can be produced are rapeseed and soybean. The natural characteristics of the raw materials used to make biodiesel dictate its quality. In addition to being a key fuel for rural economies, biodiesel production generates little hazardous waste and primarily biodegradable input and output materials, making it sustainable and renewable. Engine performance can be increased by using biodiesel

K. S. Shreenidhi · B. Vijaya Geetha (✉) · S. Priyanga · B. K. Raghunandhan · K. Rooshini
Department of Biotechnology, Rajalakshmi Engineering College (Autonomous), Affiliated to Anna University, Thandalam, Chennai 602105, Tamil Nadu, India
e-mail: vijayageetha.b@rajalakshmi.edu.in

M. Shah and D. Deka (eds.), *Emerging Sustainable Technologies for Biofuel Production*, Environmental Science and Engineering, https://doi.org/10.1007/978-3-031-52167-6_11

which is good for the environment (Rodionova et al. 2017). It is composed of non-toxic compounds, unlike petrochemicals, and does not burn off sulphur and nitrogen-rich flue gas. A simple two-step process is used to make biodiesel.

Removing lipids or oils out of plant or animal tissues is the first step. The lipid fraction is trans-esterified with alcohols in the presence of catalysts in the second stage to create biodiesel. Biodiesel can be used as a "drop-in" fuel in internal combustion engines that are already in place. Its properties are similar to those of crude oil-derived diesel in terms of chain length, viscosity, and energy density (Tanvir et al. 2021). Vegetable oil esters can burn more quickly than hydrogen-based diesel because they have 10–11% more oxygen.

Additionally, compared to ordinary diesel, biodiesel has a somewhat decreased volumetric heat capacity. Transesterification reduces triglycerides' viscosity by a factor of roughly 8, reduces their molecular weight to one-third of triglycerides, and just slightly increases their volatility. The fact that biodiesel "recycles carbon dioxide" is another intriguing feature of the fuel production. Even bio alcohol has been used for many years as a fuel source using traditional methods. Agricultural wastes, sugarcane, rice straw, and lignocellulosic biomass are just a few of the various materials that can be utilized to produce bioethanol (Rodionova et al. 2017). Bioethanol is typically produced using feedstocks like bagasse (the dry, pulpy residue of sugarcane stems left behind after the juice extraction) and sucrose from sugarcane, beetroot, sugar, corn, sorghum, or wheat.

Algae can be used to produce biohydrogen in addition to producing bioethanol. Algae can serve as a source of feedstock for the production of biofuels since they contain roughly 50% of the lipids required to manufacture biodiesel and the other components, such as sugar and proteins, required to make bioethanol. One of the most important areas of biotechnology nowadays is the use of microbes and the byproducts of their metabolic processes by humans. The ability of using a variety of substrates for viable goods and by-products is crucial because biotechnology uses microorganisms to transform some chemicals into others with higher added value, enabling the wise and equitable use of natural resources. It is now commonly accepted that continuing to use fuels made from petroleum is not sustainable due to dwindling supplies and their part in the environment's carbon dioxide buildup. For long-term economic and environmental viability, renewable, carbon-free transportation fuels are crucial (Razzak et al. 2013). The significance of microbes in generating biofuel is further covered in this book chapter. The characteristics and roles of these bacteria. The contribution of bacteria to society, their effects, and the difficulties in creating biofuels. Also, it informs us of the current situation in the modern world as well as prospects.

11.2 Microbes and Biofuel

Biofuel is produced using waste produced by living things. Ethanol is a widely produced biofuel that is obtained from plants, just like the ethanol used to make alcohol. Sugar beetroot, maize, and sugar cane are some of the main plants used in the production of ethanol due to the quantity of sugar, which can be easily fermented by microorganisms like yeast (Saccharomyces cerevisiae). To create biofuel, a variety of microorganisms are used, including entire cells of bacteria, fungi, yeast, and algae. Transesterification, anaerobic fermentation, and aerobic fermentation are a few of the processes that make up this process (Kumari et al. 2022).

Pre-treatment of different lignocellulosic biomass, oily biomass, and starchy biomass with complex structures using a variety of microbes and enzymes to create a biofuel that is both environmentally and commercially viable (Xuan et al. 2009). Research is now being done on several methods for using microbes to make different biofuels (such as alcohols, hydrogen, biodiesel, and biogas) from different starting sources (Biofuels 2013). The process of extracting sugars from agricultural feedstock or of converting starch, especially from edible grains, into sugars is presently used to produce the bulk of biofuels. The sugars from both sources are then fermented by yeast to produce ethanol. Due to the recent increases in oil prices and the expanding global energy demand from emerging countries, research related to biofuel is currently a topic of great interest (Graboski and McCormick 1998). Because, of developments in omics-based research, our understanding of the variety and metabolic capacity of methanotrophs living in different habitats have significantly grown (Borowitzka 2013).

Currently of interest are new methods for methane conversion by methane producers and their huge potential for utilising methane to prevent climate change. Production of high-value bioproducts and biomaterials, including organic acids, single-cell proteins, biopolymers, single-cell proteins, biodiesel, surface layers, and biofuels, is also of interest. Numerous scientific disciplines have a particular focus on microorganisms. Most regions on Earth have enough of them, and their utilisation is sparking a tiny but important technological revolution. Microbes use lignocellulose, which is a component of the plant cell wall and is made of cellulose, hemicellulose, and lignin, to create ethanol for use in biofuels. An explanation of the lignin and the dissociation of lignin-carbohydrate bonds in the lignocellulosic matrix, as well as the molecular targets, pre-treatments, and feedstocks for biofuels. Only bioethanol is currently commercially generated by microorganisms, although plant biomass provides a plentiful and sustainable supply of energy-dense carbohydrates that can be effectively turned into biofuels (Saba et al. 2017).

Even though it is frequently produced, biomethane is not yet utilized in transportation. The sugars from both sources are then fermented to create ethanol using yeast. As cellulose is the most common polymer on Earth, second-generation biofuels (SGBs) are cellulosic ethanol products that employ it as their primary substrate. It is possible to successfully convert cellulose into cellobiose and then glucose monomers for fermentation when given a pure cellulose substrate. Without a doubt, the issues with global warming and the exhaustion of fossil fuels have renewed interest in other industrial fuel sources.

The need for sustainable fuels has sparked research into the biological conversion of lignocellulosic biomass feedstock into biofuels including bioethanol, biodiesel, and biohydrogen. All molecules with increased energy that are produced primarily biologically or through chemical synthesis from the biomass of once-living creatures are collectively referred to as biofuels (Rodionova et al. 2017). The main sources of biofuel are now assumed to be ethanol, alcohols, lipids, carbohydrates, cellulose, biodiesel, triglycerides, fatty acids and the biomass of living creatures. Biofuels are substances produced biologically or from the biomass of energetically dense living organisms such as plants, microalgae, and bacteria. Despite the fact that the concept of using microalgae as a fuel source is not new, it is currently gaining a lot of attention as a result of the rising price of petroleum and, more importantly, the growing concern over global warming brought on by the use of fossil fuels. In addition to oils, microalgal biomass offers a rich source of proteins, carbohydrates, and other nutrients. As a result, leftover biomass from the biodiesel production process may be given to animals. Anaerobic digestion of some of the leftover biomass might result in the creation of methane, which will be used to power the plant that produces microalgal biomass by being turned into electricity. To help defray the expense of creating biodiesel, extra power may be sold. It is technically possible to employ microalgal biomass to directly create methane through anaerobic digestion, but the amount produced would be insufficient to compete with the enormous number of easily accessible, inexpensive organic substrates. There is a chance that the algae biomass left over after the oil has been collected can be used to produce methane. A microalgal biorefinery can create biogas, biodiesel, animal feed, and power all at once (Chisti 2007).

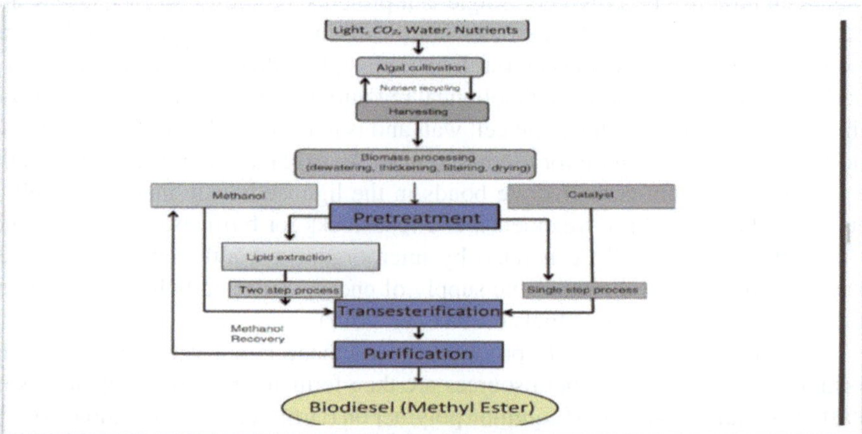

Roy and Das (2015)

Depending on the microalgae employed, additional high-value compounds might be recovered. To create biofuel, a variety of microorganisms are used, including entire cells of bacteria, fungi, yeast, and algae. This process involves several steps, such as transesterification and anaerobic and aerobic fermentation (Antolín 2002).

The different ways that microbes and enzymes can be utilized to pre-treat a variety of lignocellulosic biomass, starchy biomass, and oily biomass with complicated structures to create a biofuel that is both environmentally and commercially viable. Oil-crop biodiesel has the ability to sustainable and carbon-neutrally replace petroleum-based fuels. Unfortunately, even a small amount of biodiesel made from plants used to make animal fat and cooking oil cannot yet successfully supply the need for transportation.

11.3 Microbial Traits in Biofuel Production

A rich amount of both protein and carbs can be found in ordinary microalgae. For instance, a Scenedesmus sp. ecosystem that was utilized to treat wastewater produced 42–53% carbs and 43–55% proteins, but only 3% lipids. Like this, protein-rich nutrients were primarily beneficial to Spirulina species that were growing in digested garbage (Colla et al. 2004). A low lipid concentration was also evident in the distribution of macromolecules in certain other microalgae utilized to produce biofuels, such as Dunaliella or Chlamydomonas (Melis et al. 2000).

By allocating carbon largely as lipids and proteins, the main study core that was focused on Nanochloropsis sp. displayed a distinct macromolecular distribution. The enzyme that breaks down cellulose is called cellulase. Researchers have been examining the origins of this enzyme in diverse microbial species and environments. Two examples of these odd ecosystems are the termite stomach and volcanic soil. Another potential treatment is algae. Through photosynthesis, they turn carbon dioxide into sugar, which they use to create lipids. In small-scale laboratory bioreactors, scientists are turning lipids into biodiesel and algal carbohydrates into bioethanol (Kurano et al. 1995). The biofuel algal production could be a significant contributor to the mix if it can be ramped up to industrial levels.

Most photosynthetic organisms, including vascular land plants, micro- and macroalgae, and photosynthetic microbes, are used to produce biofuels. The primary byproducts of biofuel might be solid, liquid, or gaseous. These goods can undergo additional physical, thermochemical, and biological transformations. Primary and secondary biofuels are the two categories into which biofuels can be divided. The bulk of biofuels are produced by burning dry animal dung and woody or cellulosic plant material. All three of the secondary biofuel generations are derived inadvertently from plant and animal sources. The first generation of biofuels consists of ethanol made from starch-rich foods and biodiesel made from rendered animal fats like frying grease. The second generation consists of biodiesel and bioethanol, which are produced from cellulosic biomass and oil-rich plant seeds like soybean or jatropha, respectively (Timilsina and Cheng 2010). The world's energy needs may most likely be met by the third generation of biofuels, which are produced from cyanobacteria, microalgae, and other microbes. A method for growing cyanobacteria or microalgae that make biofuel is the growth of biofilms.

For a very long time, the most well-known source of biofuels was plant biomass. Algal green biomass is currently being examined as a potential source for the creation of biofuels, according to growing evidence. Algae and plants are the only life forms that can engage in photosynthetic activities. The method of utilizing solar energy to produce sugars from atmospheric carbon dioxide is referred to as photosynthesis. Photosynthesis is the process of harnessing solar energy to convert atmospheric carbon dioxide into carbohydrates. As photosynthesis is the only mechanism for producing plant and algal biomass, which serves as the foundation for the production of biofuels, it is intelligible that plants and algae are the only organisms that can fix carbon in nature. Sugar serves as the main molecular building block for the bioethanol and bio methanol production. The potential of diverse species of algae to produce biofuels holds the greatest promise for the production of alternative source of energy. The fact that various bacteria create varied amounts of hydrogen and biofuels should be emphasized. Genetic studies show that Caldicellulosiruptor, Pyrococcus, Thermococcus, and Thermotoga species produce more hydrogen and less ethanol. This discovery emphasises the importance of gene control in bacterial cell metabolism and the relationship between these processes and biofuel production. Recently, the importance of bacterial co-cultures in the generation of biofuels has increased. Ethanol has been successfully produced by co-cultivating Thermoanaerobacter species with cellulolytic bacteria. Saccharomyces cerevisiae, a well-known bacterium, has been used as a model organism for the efficient production of ethanol and fat through fermentative processes (Bro et al. 2006).

Growing photosynthetic bacteria, which generate molecular hydrogen, is one of the most promising techniques for creating renewable energy [28e30]. The process just requires sunlight, water, and a little quantity of macro- and micronutrients to function at room temperature. It is envisaged that future efforts will lead to the creation of an environmentally friendly platform for the commercial generation of renewable energy because photosynthesis produces hydrogen without generating greenhouse gases or other damaging pollutants. The biomass of plants and algae may be increased by using atmospheric CO_2 on a large scale through photosynthesis. Because of this dependence on photosynthesis, biofuels, or fuels made from biomass, may hold the key to efficiently and sustainably supplying the world's energy demands. Chlorophyll is a key component of the photosynthetic pigments that absorb light in both plants and microbes. In the stage without light, sugars are produced using the energy and electrons from ATP and NADPH, respectively. The two main processes for creating biohydrogen are listed below. Using photosynthesis' ability to create biomass, the first technique, often known as the indirect process, does so. To create biohydrogen, the gathered biomass particularly the carbohydrates is then further fermented and/or photo fermented (Singh and Nigam 2014).

This approach uses two discrete phases that can be separated either by a geographically occurring event in two different bioreactors or by a shift in the timing of fermentation and photosynthetic activity. Direct or indirect water bio-photolysis is used in the second approach, which will take up the most of this section, to split water into hydrogen and oxygen. The green alga Scenedesmus obliques's H_2 evolution during the light-bio photolysis process has been the subject of applied R&D for many years. The green alga Scenedesmus obliques's H_2 evolution during the light-bio photolysis process has been the subject of applied R&D for many years. The filamentous cyanobacterium Anabaena cylindrica has been seen to exhibit simultaneous H_2 and O_2 emission when exposed to an argon (Ar) atmosphere.

According to a recent study, particular modifications to the central carbon metabolism, such as overexpressing isocitrate dehydrogenase and deleting glutamate synthase, can greatly increase the quantity of ethylene produced by E. coli. Synechococcus sp. PCC 7002 is demonstrated to generate alkanes by heterologously producing an acyleacyl carrier protein reductase and an aldehyde decarbonylase. For microalgae to produce biofuel, an ideal environment must have the right balance of light, heat, nutrients, salinity, and p^H. A lower yield of biomass is typically the result of decreased environmental stress, even though it can boost the production of biofuel. It is suggested that the manufacture of biofuels can be combined with the treatment of municipal wastewater that is high in nitrogen and flue gas that is rich in CO_2. It is recommended that the production of biofuels can be combined with the treatment of municipal wastewater that is high in nitrogen and flue gas that is rich in CO_2. To maximise the production of biofuel, various genetic engineering approaches must be coupled to produce novel strains with commercial potential. In order to overcome the energy crisis, international cooperation is essential for incorporating biofuels into our current energy system. These joint initiatives will support the advancement of culture methodologies and technological innovation for the production of biofuels through fundamental mechanistic research, the construction of cell growth facilities, the genetic engineering of particular algal strains, and the optimisation of biofuel production conditions.

Understanding the intricate principles of photosynthesis for the generation of biofuel under diverse situations is the greatest approach to enhance the biomass of phototrophic organisms. Microalgae utilise light energy to convert carbon dioxide into organic molecules more efficiently than higher plants, making them a better source for the production of biofuel. Microalgae like C. protothecoides may contain up to 55% fat when cultivated heterotrophically with little nitrogen. The green colonial microalga B. braunii 765 may produce biodiesel, hydrocarbons, and biocrude oil at temperatures under 25 C. Chlorella minutissima produces more lipids at a temperature of 25 C when grown in a basic media. Because it contains C16 and C18 lipids, the species of C. minutissima UTEX2341 is also a source of biodiesel. Complex biofuels are made with the help of microalgae.

Following that, it goes through several procedures, including cell disruption, harvesting, drying, and fermentation distillation for the generation of bioethanol and biodiesel, respectively. The five primary families of microalgae are diatoms, blue-green algae (Cyanobacteria), red algae, green algae, and brown algae. Maybe the most common among them for the generation of biofuels is cyanobacteria. Some types of microalgae use the transesterification process to convert lipids into biodiesel (Goh et al. 2019). The density and viscosity of microalgal biodiesel are comparable to those of petroleum diesel. An archaeon called Sulfulobus solfatarticus resides in the crater lakes close to Mount Vesuvius. Recently, experiments with genetic manipulation have been conducted to increase this microbe's capacity to manufacture the required enzymes. Any soil can contain the fungus Trichoderma reesei. It feeds by releasing a significant amount of cellulase. This fungus, known as "jungle rot," was first found during World War II and destroyed the cellulose in US soldiers' tents and clothing. The fungus has been genetically altered by a Canadian company to create more cellulase and convert straw into glucose, which can be converted into ethanol. 75% of the effective conversion.

Depolymerization of polysaccharides, which is accomplished by a specific enzyme, is required by the majority of processes for turning lignocellulosic biomass into biofuels. Microalgal cultivation techniques for direct energy conversion to create biofuels have been proposed as potential. The use of contemporary methods to produce microbial biofuels is widely studied and recognised. The growing of microalgae or cyanobacteria in biofilms is one novel platform of biomass production methods that are finally being researched for the biofuel processing pathway. The growing of microalgae or cyanobacteria in biofilms is one novel platform of biomass production methods that are finally being researched for the biofuel processing pathway.

The only living things that can perform photosynthetic activities are algae and plants. Utilizing solar energy, photosynthesis transforms atmospheric carbon dioxide into carbohydrates. As photosynthesis is the only mechanism for producing plant and algal biomass, which serves as the foundation to produce biofuels, it is clear that plants and green algae are the only organisms that can fix carbon in nature. Sugar serves as the main molecular building block to produce bioethanol and bio methanol. When the bagasse leftover from sugarcane is used, a significant amount of bioethanol is produced. The growing of microalgae or cyanobacteria in biofilms is one novel platform of biomass production methods that are finally being researched for the biofuel processing pathway. The only living things that can perform photosynthetic activities are algae and plants. Utilizing solar energy photosynthesis transforms atmospheric carbon dioxide into carbohydrates. As photosynthesis is the only mechanism for producing plant and algal biomass, which serves as the foundation for the biofuel production, it is clear that plants and green algae are the only organisms that can fix carbon in nature. Sugar serves as the main molecular building block for the production of bioethanol and bio methanol.

A sizable amount of bioethanol is created when the bagasse residue of sugarcane is utilized. Chemicals are required for the growth of some types of microorganisms, in contrast to higher plants that require pesticides and herbicides for increased productivity. The abundance of proteins and carbs in microalgae, which may be processed to produce methanol and alcohol fuel, is another characteristic of these organisms. Algae have recently gotten more attention due to their ability to absorb CO_2 and their potential to be used as a renewable energy source (Cheng et al. 2006). The following reasons explain why algae cultivation is becoming more and more beneficial: On non-agricultural land, algae can be grown using wastewater and provide a high yield per unit of the area exposed to light. (ii) The combustion of fossil fuels and wastewater can provide algae with the CO_2 and nutrients they need to grow. Algae can range in complexity from basic multicellular giant bladder kelp species to single-cell microalgae species. There are numerous families of photosynthetic aquatic eukaryotes (Richmond et al. 1986). Microalgae are tiny creatures that frequently develop in water when suspended, and they derive their energy from the same mechanism of photosynthesis as larger plants. Green algae do not need a vascular system to carry nutrients, in contrast to higher plants. Every cell has photoautotrophic metabolism, which allows it to quickly take up dissolved nutrients (Costa and de Morais 2011). These eukaryotic organisms perform the role of solar-powered cell factories that can transform carbon dioxide into beneficial bioactive compounds, biofuels, and chemical feedstocks for animal feed (de la Noüe and Bassères 1989).

To create biomass and store carbon, microalgae can be genetically altered and grown on a huge scale. Microalgae cultures are made up of a single strain that has been carefully chosen to yield the required output and the best possible culture outcome. It is also crucial envisioned process flow diagram for producing biofuels and reducing CO_2 while treating algal wastewater (Fig. 11.1).

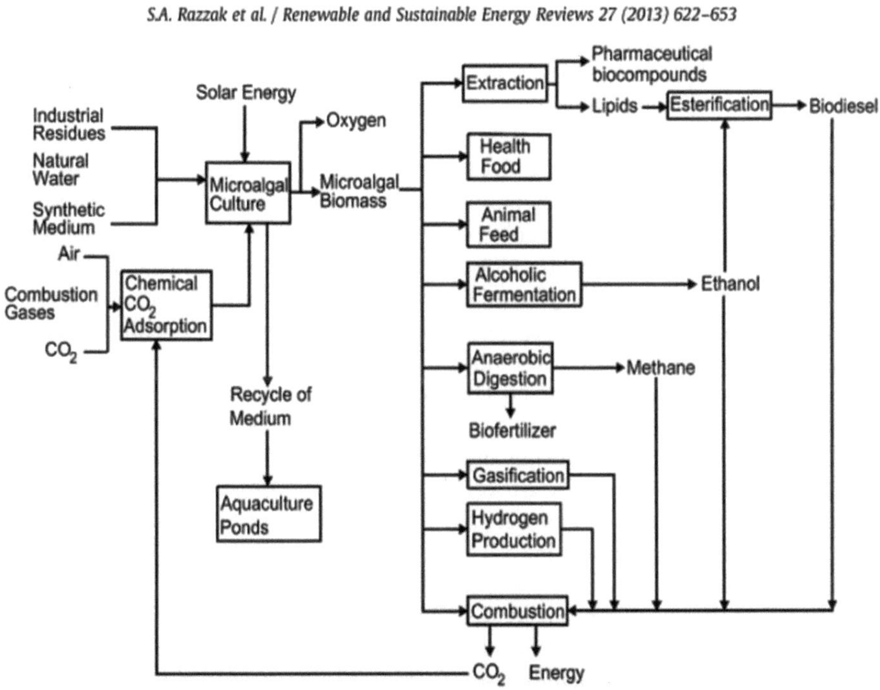

Fig. 11.1 Simplified process flow diagram envisioned for algae wastewater treatment with CO_2 mitigation and biofuel production (adapted from Chiu et al. 2008; Razzak et al. 2013)

11.4 Roles of Microbes in Biofuel Production

It is now well known that human activity increases the release of fossilised CO_2, which considerably contributes to the greenhouse effect. Ethanol was added to gasoline as an anti-knocking additive for internal combustion engines between 1925 and 1945. Because of the ethanol in the fuel, more piston compression was possible, which improved engine performance. These fuels, including ETBE, which is largely derived from bioethanol, account for more than 90% of the worldwide biofuel sector. All biofuels must have chemical and physical features that suit engine requirements, such as stability and predictable burning at high pressures, as well as transportation demands, such as safety and energy density. Liquid biofuels must be pumpable at all temperatures and must always remain liquid. Other factors to consider for liquid biofuels include a high heat of combustion value to reduce energy losses and expenditures during transportation, as well as storage stability (Lee et al. 2008).

Certain longer-chain alcohols, such as butanol, have a high enough heat of combustion to be employed in high thrust-to-weight applications like airplanes. Certain longer-chain alcohols, such butanol, have a high enough heat of combustion to be employed in high thrust-to-weight applications like aircraft. The ignition temperature and vapour pressure are essential parameters in safe and ecologically sustainable storage. Alcohol-based biofuels such as methanol, ethanol, and nbutanol can be used as oxygenators, liquefiers, and anti-knocking agents in current fossil fuels, as well as extending or replacing the supply of gasoline and diesel (Kumar and Saravanan 2016).

However, not all energy-dense byproducts of microbial fermentations are suitable for use as fuels because they may induce corrosion or swelling with particular materials, or have other undesirable properties. Biofuels derived from renewable biomass have the best potential for CO_2 neutral production of any sustainable energy source (Ghirardi 2000). Get rights and content. They are straightforward to introduce progressively as a substitute for fossil fuels by adopting procedures such as blending. The use of consolidated bioprocessing (CBP) microorganisms is the most successful way for producing low-cost biofuels. Despite this, because lignocelluloses are generated in such a vast variety of environments, they are inherently diversified.

As a result, it is critical to design and integrate effective microorganisms that can grow on a variety of biomasses, as well as pretreatment processes tailored to each application (Hatakka 1983). Fatty acid synthesis appears to be an active subject of biofuel technology research in industry. At least one company is currently capable of producing biodiesel from microbes, according to press reports. According to the literature, E. coli can overproduce free fatty acids, which can then be esterified in vivo with minor alcohols. Based of these results, we have reason to believe that we will be able to directly manufacture fuels from fatty acids using E. coli in the future. Microorganisms are key participants in the production of biofuel. The genetic modification of microbial strains, as well as the optimization of fermentation settings, have enhanced biofuel output. There is little question that the challenges of global warming and fossil fuel depletion have sparked interest in alternate commercial fuel sources.

The growing need for renewable fuels has motivated studies into the biological breakdown of lignocellulosic biomass feedstock to create biofuels such as bioethanol, biodiesel, and biohydrogen. Currently, starch and sugar-rich plant material, such as grain crops and sugarcanes, is the most common source of bioalcohols; however, focus has recently switched to perennial grasses such as switchgrass and miscanthus (Kapdan and Kargi 2006). Despite the fact that these crops do not compete with food consumption, the fermentation and distillation processes necessitate the transformation of cellulose biomass into sugars. Bioethanol is the most frequent bioalcohol, with biopropanol and biobutanol being less common. Microorganisms digesting a carbohydrate-rich substrate make these alcohols. The efficacy of a plant's photosynthetic activity, as well as the cultivation environment, determine the amount of plant biomass. There are three theoretical ways to utilise biofuel energy in engines. Engines must be modified to work with current fuels, or biofuel must be produced to closely resemble the fundamental features of traditional fossil fuels. The third approach is to employ a mix of conventional and biofuels. This typically results in a beneficial change of the fuel's qualities, such as an enhanced octane rating, without the use of hazardous chemicals. To minimize carbon dioxide emissions, bioethanol, for example, can be used in place of gasoline. Because oxygen exists in its molecular form, combustion may occur at low temperatures while emitting less hazardous gases including carbon monoxide (CO), nitric oxide (NO), and volatile organic compounds (Sakai et al. 1995). A diverse group of microalgae that use CO_2 and nutrients from wastewater to create high-quality biofuel. It's remarkable that unicellular microalgae may thrive in nutrient-rich waste water, so aiding in water treatment (Babel et al. 2008). In addition, there is significant interest in employing microalgae to clean dairy farm animal effluents. This is because uncontrolled nitrogen emissions from dairy manure have the potential to harm aquatic ecosystems and taint key groundwater supplies.

Traditional aerobic treatment methods, such as activated sludge, require a high-energy oxygen supply. Furthermore, these procedures result in inadequate recycling of essential nutrients found in dairy effluent. Chynoweth and colleagues (1999). Anaerobic digestion is routinely used to stabilize organic compounds in wastewater in a variety of forms and sizes. These measures can aid with pathogen and organic material reduction. They can also produce biogas, which can be used as fuel (Converti et al. 2009). As previously stated, the algal family of photosynthetic eukaryotes is varied. These species are mostly found in water. Several types of algae store oil and lipids, which are denser than water. This lesser density aids their capacity to float in ponds and lagoons. Algae are present in around 36,000 different species in the natural habitat. Red algae, green algae, brown algae, diatoms, blue-green algae (prokaryotes), and dinoflagellates are the different types of algae.

Many of these algae have been employed for a number of useful applications, including animal feed, fertilizer, pharmaceutical treatment manufacturing, pollution reduction, water treatment, colors, agar creation, and bioenergy generation. As a result, the advantage of microalgal anaerobic digestion implies a less complex production process while still producing an energy form and a mineralized digestate rich in nutrients. Biofuel is one solution to this challenge, and microbes are becoming increasingly important in the development of this alternative fuel source. Extensive research has been undertaken over the last several decades on the conversion of lignocellulosic resources into biofuels, most notably ethanol, lipids as biofuel precursors, and hydrogen. There are plentiful and renewable non-food lignocellulosic resources accessible around the world that might boost our energy independence while reducing our environmental effect. Various studies are being conducted to:—Improve lignocellulose's capacity to release carbs by combining physical and physicochemical pre-treatments. By combining pretreatments and hydrolysis, we may increase yields at large solids loadings and eventually attain economic feasibility for lignocellulose energy molecule production. Increase and expedite sugar monomer conversion into end products while also establishing integrated SSF bioprocesses and improving enzyme activity for separate hydrolysis and fermentation processes.—Improve fermentation performance by using nutritionally complete liquid and gaseous media, optimizing mass transfer with new bioreactor configurations, and developing engineered strains to better understand the metabolic pathways involved in biofuel synthesis so that overproducing engineered strains with increased inhibitor resistance can be developed.—Create a cleanup system before purifying energy molecules to eliminate any inhibitors from the cultured broth as well as the substrate (a liquid or gas) (Sun and Cheng 2002).

Microalgae are solar-powered cell factories capable of converting CO_2 into high-value bioactives, meals, feeds, and biofuels. With the aid of sunshine and photosynthesis, microalgae can collect CO_2 from a number of sources and convert it into biomass. Furthermore, the microalgae generated can be further processed to create biofuels. Microalgae may develop quite quickly. Neutral lipids, which are generated and stored in substantial quantities in cytosolic lipid bodies, can account for 20–50% of the dry weight of microalgae. As a result, selecting the appropriate algal strains is critical for building a profitable algae-based biofuel market (Table 11.1).

Most algae, according to this table, contain significant quantities of proteins, typically in the form of enzymatic or crude proteins. These proteins, mostly amino acids, offer high-quality nutrients present in both human and animal diets. Algae development requires water, nutrients, carbon dioxide, and adequate temperature management. Significant amounts of nutrients in wastewater are necessary for algae growth. As seen in the table, the two most researched algae species are Chlorella vulgaris and Phormididium laminosum. Anabaena doliolum, Chlorella emersonii, Chlamydomonas reinhardtii, Scenedesmus bijugatus, and Scenedesmus obliquus are among the other species investigated (Razzak et al. 2013). Microalgae appear to be the sole source of sustainable biodiesel capable of supplying world demand for transportation fuels, as seen below. Microalgae, like plants, create oils from sunlight, but they do it more effectively than agricultural plants. Several microalgae generate oil at rates much beyond the finest oil crops. Methods are being considered for producing microalgal biodiesel competitively competitive with petroleum fuel. These photosynthesis-based bacteria can be used as nitrogen-fixing biofertilizers as well as for bioremediation. This article discusses the possible use of microalgae as

Table 11.1 General composition of different Microalgae in percentage of dry weight basis (adapted from Mata et al. 2010; Trivedi et al. 2015)

Microalgae species	Protein	Carbohydrates	Lipids
Anabaena cylindrica	43–56	25–30	4–7
Aphanizomenon flos-aquae	62	23	3
Chlamydomonas rheinhardii	48	17	21
Chlorella pyrenoidosa	57	26	2
Chlorella vulgaris	51–58	12–17	14–22
Dunaliella salina	57	32	6
Euglena gracilis	39–61	14–18	14–20
Porphyridium cruentum	28–39	40–57	9–14
Scenedesmus obliquus	50–56	10–17	12–14
Spirogyra sp.	6–20	33–64	11–21
Arthrospira maxima	60–71	13–16	6–7
Spirulina platensis	46–63	8–14	4–9
Synechococcus sp.	63	15	11

a biodiesel source. Microalgae may be utilized to make a range of environmentally friendly biofuels (Patil et al. 2008). Biodiesel made from microalgal oil, photobiohydrogen, and methane generated by anaerobic digestion of algal biomass are among them. Transportation accounts for approximately 28% of the energy available for consumption in the EU25 nations, with road transport accounting for more than 80% of this. 2007. Eurostat. Transportation, the world's fastest-growing industry, accounts for approximately 27% of total primary energy consumption. As a result, lowering greenhouse gas emissions from transportation fuels is an appealing goal. Algae are important bioremediation agents.

They are already used in a number of wastewater treatment plants. However, the potential of algae in wastewater cleanup is significantly greater than its current function. Microalgae have greater dry weight nitrogen and phosphorus contents of 10–11% and 11%, respectively. These nitrogen and phosphorus levels are several orders of magnitude greater than those seen in plants. Despite their inability to compete financially with ion exchange resins, microalgae have been widely utilized to remove heavy metals from wastewater. Because of their capacity to absorb both CO_2 and NOx while growing, microalgae have grown in popularity. Microalgae, as previously noted, may yield potentially important biomass, which can be utilized as a biodiesel feedstock, a slow-release fertilizer, and an ingredient in animal feed. These photosynthesis-based bacteria can be used as nitrogen-fixing biofertilizers as well as for bioremediation. This article discusses the possible use of microalgae as a biodiesel source. Microalgae may be utilized to make a range of environmentally friendly biofuels. They include biodiesel made from microalgal oil, photobiohydrogen, and methane generated by anaerobic digestion of algal biomass.

11.5 Welfare to the Society

In this chapter, we look at how to employ functional genomics and machine learning technologies to make advanced biofuels and the possible cost savings, as well as difficulties with substrate and product toxicity in host microorganisms and ways for engineering tolerance. Methods for enhancing the conversion yield of microalgae organic matter into methane include biomass pretreatment, substrate co-digestion, and macromolecular composition management. More research is needed in this area before microalgae biofuels may be commercialized. Some industries provide numerous economic and environmental benefits, including as reduced use of fossil fuels, lower greenhouse gas (GHG) emissions, manufacture of low-cost raw materials, and synthesis of substrates required by many microorganisms. The ultimate result, energy from biofuels, is a convincing rationale for lawfully dumping feedstocks.

Aiming for a Better Future Climate change, the end of oil production, population growth, and land use are all grave challenges for humanity. Mitigation of such issues is possible because to cutting-edge approaches such as the use of microorganisms to make better biofuels. These solutions will require much more research before they can be given on an industrial scale, but as we progress, they are becoming more accessible. Concerns about climate change have compelled us to re-examine our transportation infrastructure. One potential substitute for carbon-polluting fossil fuels is biofuels produced by engineered microbes that utilize a sustainable carbon source. The quantity of advanced biofuels that may be used in current transportation infrastructure is not limited, but prices have kept these fuels from being widely adopted. Proteins for animal feed, for example, contribute to reduce climate change by decreasing the energy and greenhouse gas emissions necessary to create the feed. Biofuel byproducts, such as proteins for animal feed, contribute to climate change mitigation by lowering the amount of energy and greenhouse gas emissions necessary to produce the feed using traditional techniques. Biofuels have a substantially lesser influence on climate change than fossil fuels. Using biodiesel instead of diesel decreases carbon monoxide and smog-causing particles by around 50%, according to research (Lardon et al. 2009). In addition, biodiesel decreases hydrocarbon emissions by 75–90%. It also eliminates all sulfur emissions. Another concern is that of waste. This is tempered by the fact that the plant material used, such as straw, is inedible. As a consequence, using a microbial system to generate biofuel is less wasteful, moral, and cost-effective. It also cuts greenhouse gas emissions significantly.

Biofuels have a substantially lesser influence on climate change than fossil fuels. Using biodiesel instead of diesel decreases carbon monoxide and smog-causing particles by around 50%, according to research. In addition, biodiesel decreases hydrocarbon emissions by 75–90%. It also completely removes all sulfur emissions. Because of the recent significant increase in biofuel production, biofuel now meets more than 3% of the world's transport fuel requirement. Many people are concerned as a result of this rapid expansion (Tippmann et al. 2013). Food costs have risen dramatically, raising concerns about global food security. A key benefit of biofuels in preventing climate change has also been put into question. When ethanol and biodiesel factory gate costs were compared to fossil-based petrol and diesel refinery gate prices, ethanol, and biodiesel were about more expensive per unit of energy. Biofuel production costs must be greatly decreased for it to be competitive. Future technological developments have the highest likelihood of success. The subject of contemporary production methods and techniques has been thoroughly discussed above. A global effort to develop renewable energy sources is essential to prevent the detrimental effects of global climate change and meet future energy demands.

Biofuels are falling into this scenario as a sustainable energy source sourced from biomass. A sustainable biofuel production platform might help to better manage massive volumes of lignocellulosic waste biomass and lessen the negative consequences of environmental pollution. Traditional biomass is plant waste-derived biomass that has been used for cooking and heating since prehistoric times. Traditional biomass is plant waste-derived biomass that has been used for cooking and heating since prehistoric times. The development of ecologically safe biological pre-treatments for the manufacture of bioethanol can only profit from bioprospecting and maybe genetic modification of naturally isolated microorganisms.

Microbial pre-treatments are effective under the correct conditions and do not require a lot of energy, hazardous chemicals, or sophisticated processing. Scaling proven techniques from the laboratory to the industrial level is challenging. Biofuels derived from renewable biomass have the best potential for CO_2-neutral generation of any sustainable energy source. They are simple to deploy gradually as a replacement for fossil fuels through techniques such as blending (Murakami and Ikenouchi 1997).

Biofuels from biomass can be created using either thermochemical processes (such as BTL) or microbial fermentation. Both appear to be making equal use of the property. Novel and mostly microbially generated biofuels will attain industrial maturity faster. Biofuels of the future will be created by a mix of microbiological, chemical, and technological processes, as well as biotechnology. These processes do not even begin with the addition of bioethanol to gasoline. The biomass from the entire plant must be utilized to produce additional ethanol. This will also contribute to the resolution of the dilemma between food production and environmental conservation. Microbial biofuels have several potential applications in process operations such as pre-treatment, fermentation, substrate separation, energy coupling, and so on. Improved biofuel production will need biological research into energy plant development, enzymatic hydrolysis, specialized fermentation strains, and waste management. In contrast to the limited supply, geopolitical instability, and catastrophic global ramifications of fossil fuel energy, it is becoming clear that biofuels may be a sustainable source of renewable energy.

Biofuel production has increased to support bioethanol and biodiesel production, with market shifts resulting in increased production and consumption of biobutanol, biosyngas, and biohydrogen, among others. There are two phases of photosynthesis: light-dependent and light-independent. Light energy is acquired, processed into a charge separation, and eventually converted into the generation of ATP and NADPH during the light-dependent stage. Because sunlight gives energy and CO_2 from the atmosphere supplies carbon, employing photosynthetic organisms as a biofuel source is both economical and possible. As a result, a comprehensive strategy for commercial-scale CO_2 collection, wastewater treatment, and biofuel generation employing microalgae growth is clearly necessary. This method demands the efficient integration of several process stages within the framework of a single industrial site. This method requires the seamless integration of many process stages within the framework of a single industrial site. As a result, farming microalgae provides three key advantages: CO_2 emissions from fossil-fuel-powered power plants are absorbed, wastewater is treated, and renewable energy is generated. CO_2 serves as a carbon supply for microalgae and other creatures that develop during photosynthesis. There will be no growth if it does not exist. A scarcity of CO_2 supply typically hurts productivity.

Based on the average chemical composition of microalgae biomass, producing 1t of biomass takes around 1.8t of CO_2. Microalgae can create proteins, fatty acids, and nutritional supplements for both humans and animals. Microalgae may also create proteins, fatty acids, and nutritional supplements for people and animals. Furthermore, because microalgal lipids are chemically similar to those present in ordinary vegetable oils, they are great candidates for biodiesel manufacturing. Microalgae-based diesel, as opposed to biodiesel derived from fossil fuels, is renewable, biodegradable, and produces less pollution. As a result, decreasing atmospheric CO_2 levels through microalgal photosynthesis is thought to be both safe and advantageous to human ecology. Nonetheless, given the variety of molecules found in petroleum-based fuels and the massive amounts of renewable fuels necessary to power transportation, all of the solutions have the potential to be implemented. Economic feasibility, net energy gain, and carbon neutrality will all be important considerations. The collection and concentration of microalgal biomass from growing systems significantly increases the process's total operating expenses.

As a result, enhanced commercial sustainability of the microalgae biofuels sector necessitates the development of more effective and cost-effective harvesting technologies. Biodiesel has a lot of promise, but it does not represent a significant threat to petroleum-based fuels due to the high cost and scarcity of organic oils. Algae growth tactics that are innovative and different may enable biodiesel production to reach the price and scale of production necessary to compete with, and maybe replace, petroleum. More energy is required to improve the quality of life for the world's growing population. Biofuels might be one source of energy to help the world fulfill its demands.

Biofuels might be one source of energy for the planet. For many years, fossil fuels were the major energy source. Although biohydrogen may directly power fuel cells or internal combustion engines, the only result in both cases is water. A global effort to develop renewable energy sources is essential to prevent the detrimental effects of global climate change and meet future energy demands. Being a sustainable energy source derived from biomass, biofuels fit into this scenario. A sustainable biofuel production platform might help to better manage massive volumes of lignocellulosic waste biomass and lessen pollution's adverse impacts on the environment. Traditional biomass is biomass formed from plant waste that has been used for cooking and heating since prehistoric times.

11.6 Impact on Society

In response to the global increase of the use of biofuels as replacement transportation fuels, advanced chemical, biochemical, and thermochemical biofuel production technologies are rapidly being developed. The goal of this research and development is to increase the output and overall efficacy of biofuel-producing facilities, as well as the quality and environmental impact of biofuel. The Stern Review on the Economics of Climate Change stressed the need of limiting global warming economically. It is now well acknowledged that human activity causes an increased release of fossilized CO_2, which contributes significantly to the greenhouse effect. Over 28% of the energy available for consumption in the EU25 nations is accounted for by transportation, with road transport accounting for more than 80% of this. Eurostat, 2007. Transportation, the world's fastest-growing industry, consumes around 27% of all primary energy. As a result, lowering greenhouse gas emissions from transportation fuels is an appealing aim. Biogas is produced as a result of the anaerobic digestion of organic waste. Although the biological process of anaerobic digestion is well understood, some organic substrates may exhibit bottlenecks. However, because of the macromolecular dispersion and cell walls of microalgae, anaerobic digestion effectiveness varies by species and is difficult to estimate. Proteins, on the other hand, produce ammonia gas, which can limit methanogen activity during anaerobic digestion. It has also been found that biomass pretreatment and co-substrate digestion can boost methane output (Minale and Worku 2013). The comparison ultimately addresses the purported analogies between the activated sludge and microalgal biomass as an anaerobic digestion substrate.

In truth, anaerobic digestion is a simpler technique of generating energy since it does not need intense concentration, drying, or oil extraction. Because of the high sodium concentration in the digesting media, salty microalgae create an additional bottleneck. At this point in biofuel development, it is too early to determine whether the process of producing biofuels from algal biomass is best. Anaerobic digestion, however, appears to be the simplest of all of them; also, it can function as a significant energy source when paired with other biofuel generation (González-Fernández et al. 2011).

This complex feedstock may be fermented by a variety of biotechnologically interesting microorganisms, addressing the limitations associated with municipal solid waste and agricultural feedstocks for enzymatic hydrolysis. The cost of raw materials used in the production of biofuels has a considerable influence on its profitability. The cost of raw materials accounts for 60–90% of the entire cost of manufacturing of first-generation biofuels. Different biofuels are more or less cost-competitive with their petroleum-based counterparts, depending on the feedstock used and the country.

Due to its tremendous potential for global warming, methane concentrations in the environment have a significant impact on the climate of the Earth. There are economical and environmentally friendly options from methanotroph bacteria or bioconversion options from methanotrophs to prevent climate change. In addition to the fact that the sources of these fossil fuels are running out, it has been discovered that using fossil fuels to meet this energy demand significantly increases greenhouse gas (GHG) emissions. Its numerous detrimental effects, including climate change, glacier retreat, sea level rise, biodiversity loss, and others have prompted efforts to develop alternative, renewable, sustainable, effective, and affordable energy sources that emit fewer greenhouse gases. Although biomethane is produced in large quantities, it is not yet employed in transportation. Many industries are thinking about employing biobutanol, which might potentially replace petrol, diesel, and kerosene while also contributing to partially renewable energy. Fuels are required since contemporary civilization relies significantly on machinery to do even the most basic functions.

The development of new fuels to work with existing transportation infrastructure is undoubtedly a step in the right direction toward usability and economic feasibility (McGinn et al. 2011). It is obvious that the development of biological pre-treatments for the production of bioethanol that is ecologically safe can only benefit from bioprospecting and maybe genetic manipulation of naturally isolated microorganisms. Microbial pre-treatments are efficient in the right circumstances and don't need a lot of energy, dangerous chemicals, or special processing. Scaling-up established procedures from the laboratory to the industrial level presents difficulties. Nonetheless, substantial research is being conducted on a global basis to enhance biofuel output while simultaneously cutting production costs for sustainable sectors. The genetic modification of microbial strains, as well as the optimization of fermentation settings, have enhanced biofuel output. One of the biggest problems is the idea of improving production by different methods, but the business has benefited from a combination of engineering methods, fermentation technology, enzyme engineering, and metabolic engineering. It is obvious that technological advancements are necessary to make biofuels affordable and sustainable.

The world's initial energy consumption in 1980 was 302EJ/year (1 exajoule = 1018 J), but has since increased to 420EJ/year, 491EJ/year, and is expected to reach 826EJ/year by 2050. Avoid fossil fuels in favor of alternative, clean, renewable, and sustainable energy, including solar, hydro, wind, and bioenergy, due to the rapidly increasing energy needs and capacity of the world right now. Bioenergy (particularly biofuels) production from lignocellulosic biomass has lately attracted international interest, as biofuels are expected to be carbon–neutral and equivalent to the transportation of gasoline-based commodities in terms of energy content.

Various forms of biofuels have also been increased to assist bioethanol and biodiesel production, and commercial advancement has led to the production and use of new biofuels such as biobutanol, biosyngas, and hydrogen, among others. To convert biomass straight into diesel fuel, the bacteria do not need to undergo any additional chemical modifications. Recovery of biodiesel fuels is substantially less expensive than producing ethanol, which is now a common fuel. Environmental perturbations have a direct influence on the intracellular distribution of carbon flow, which decides which macromolecule is produced. Furthermore, creating specific operating circumstances might lead to the accumulation of a target macromolecule. Some sectors gain greatly from trade and commerce, such as reducing the use of fossil fuels, lowering carbon greenhouse gas (GHG) emissions, manufacturing affordable items, and building substrates that many creatures require. Energy from biofuels (final products) is the main reason for making food legal.

Through genetic and metabolic engineering, there is a huge need to enhance algae biology in order to produce low-cost microalgal biodiesel. The use of biorefineries and advances in photobioreactor engineering will further reduce production costs. Because their production costs are substantially greater than their operating costs, tubular photobioreactors are anticipated to be employed to produce the majority of the microalgae biomass necessary to produce biodiesel. Photobioreactors provide a controlled environment that can meet the unique needs of high microalgae, allowing considerable volumes of fuel to be generated each year. The industry for microalgal biodiesel is growing. The cost of producing microalgal biodiesel can be decreased by using biorefining-based production methods, enhancing photobioreactor engineering, and increasing microalgal ability through genetic breeding (Pulz 2001).

There are several types of biofuels available today. In the market now, only bioethanol and biodiesel are produced as fuels. These oils, including ETBE, which is partially generated from bioethanol, account for more than 90% of the worldwide biofuels industry. All biofuels must have particular chemical and physical features that suit engine demands, such as heat stability and predictability, as well as transportation needs, such as safety and speed. Microalgae is the only biodiesel source capable of totally replacing fossil fuels. Unlike other oily plants, microalgae grow fast and create a lot of oil.

Microalgae biomass often tripled within 24 h. Biomass may quadruple in 3.5 h during exponential development. Microalgae oil concentration can exceed 80% by weight of dry biomass. Fat content ranging from 20 to 50% is fairly uncommon. Fat content ranging from 20 to 50% is fairly uncommon (Ribeiro and Pereira da Silva 2015). The quantity of oil generated per unit volume of microalgal media per day, or oil production rate, is determined by algae growth rate and biomass oil content. Microalgae with a high oil output are ideal for biodiesel production. Microalgal biodiesel is theoretically achievable, as demonstrated by this example. It is the only renewable biodiesel capable of transforming gasoline into liquid fuel. To compete with gasoline-based diesel, the economy of producing microalgae biodiesel needs to be improved, and this appears to be happening.

11.7 Constraints

Biofuels are not flawless, though. Making usable ethanol from plant material is expensive. There are ethical issues as well. Fuel generation requires extensive land use, which limits the amount of land that can be utilized for food cultivation. This is a concern in developing nations with big populations to feed. In Brazil, there is a very real worry that rainforests may be removed to produce sugar cane to meet this expanding demand. The amount of acreage required also threatens important ecosystems. Second-generation biofuels (SGBs) are cellulosic ethanol products that use cellulose as their main substrate because it is the most prevalent polymer on Earth. Given a pure cellulose substrate, cellulose may be successfully broken down into cellobiose and then into glucose monomers for fermentation. However, it has been extremely difficult to come up with a straightforward, cost-effective method to disassemble the lignocellulosic matrix that contains the cellulose in typical SGB feedstocks including corn stover, bagasse, wheat straw, and grasses.

The majority of processes for turning lignocellulosic biomass into biofuels entail depolymerizing polysaccharides, which is accomplished by a specific enzyme. However, a commercially viable lignocellulose-based biofuel business cannot grow quickly due to the huge price of enzymes (Cheng and He 2014). For a long time, fossil fuels were the primary source of energy. However, this use is unsustainable and results in problems with the environment because of the burning of fossil fuels. In light of this difficulty, using ecologically friendly biofuels instead of fossil fuels could be possible. While there exist technologies for manufacturing sustainable biohydrogen, they currently have low yields and efficiency, posing challenges to the sector. Light-driven hydrogen production by microalgae has significant limitations since molecular oxygen is generated during photosynthesis and can permanently block hydrogenases. Furthermore, the cost of creating biohydrogen is not comparable with that of fossil fuels, thus it cannot fully replace them. Accordingly, the primary limiting factors that affect hydrogen photoproduction include the different efficiency of light utilization by phototrophs under varying solar light intensities, the high ambient oxygen level that harms the enzymes and pathways in cells that produce hydrogen, as well as the low rate of photosynthetic CO_2 assimilation necessary for an efficient accumulation of cell biomass and its further conversion to biohydrogen.

It's crucial to be able to strike a balance between biomass growth and biofuel production. Currently, biofuel struggles and costs too much to economically compete with fossil fuels. The natural strains' low product output makes it necessary to develop and improve them by utilizing metabolic engineering and genetic engineering procedures. Alcohol-based biofuels such as methanol, ethanol, and nbutanol are valuable as oxygenators, liquefiers, and anti-knocking agents in the manufacturing of current fossil fuels, in addition to delivering additional petrol and diesel or even replacing them. Biomass generation from microalgae in general produces microalgal biomass that costs more than cultivating crops. Light, CO_2, water, and inorganic salts are necessary for photosynthetic growth. The temperature must normally stay between 20 and 30 °C. Despite daily and yearly changes in light levels, the synthesis of biodiesel must rely on readily available sunlight to keep costs to a minimum (McGinn et al. 2011). The inorganic components of the algal cell must come from the growth media. Iron, nitrogen, phosphorus, and, in some situations, silicon is among the essential elements. The molecular composition of microalgal biomass should be utilized to estimate the lowest nutritional requirements.

However, not all energy-dense byproducts of microbial fermentations may be used as fuels because they may promote corrosion or swelling with particular materials or have other undesirable qualities. Because phosphates added to food mix with metal ions, not all of the nutritional value of the extra nutrients is available, hence nutrients like phosphorus must be provided in high quantities (Abdelaziz et al. 2013). Seawater is regularly treated with commercial fertilizers such as nitrate and phosphate, as well as a few other micronutrients, for the cultivation of marine microalgae (Molina Grima et al. 2003). The cost of growth media is often low. According to Cerón-García et al. (2013), The carbon content of microalgal biomass is over 50% by dry weight. Carbon dioxide is generally the source of all of this carbon. The amount of carbon dioxide fixed by creating 100 t of algal biomass is about 183 t. Carbon dioxide must be delivered continuously throughout the day. Carbon dioxide loss and pH fluctuations are decreased by feeding that is controlled in response to pH sensor signals. Some of the carbon dioxide released by fossil-fuel-burning power plants might potentially be utilized in the production of biodiesel. It's frequently inexpensive or even free to obtain this carbon dioxide. Since all the energy required to cultivate and process the algae would be obtained from the biodiesel and methane created by anaerobic digestion of the biomass waste products left over after the oils were extracted, microalgal biodiesel would ideally be carbon neutral (Singh and Nigam 2014). Even though microalgal biodiesel has the capacity to be carbon neutral, utilizing fossil fuels will still result in the accumulation of carbon dioxide, therefore there will be no net reduction in that quantity.

Degradation of lignin, pectin hydrolysis, hemicellulose depolymerization, and cellulose hydrolysis are the various steps that make up the complex process of depolymerizing lignocellulosic biomass into simple sugars. This process starts with reducing the structural complexity of lignocelluloses. Recent techno-economic evaluations on the manufacture of bioethanol from lignocellulose indicate that feedstock, enzyme, and pre-treatment costs would make up roughly two-thirds of the overall production costs, with enzyme costs being the highest (Himmel et al. 2007). The creation of sturdy cellulolytic bacteria could get around this financial restriction. Reduced expression of heterologous cellulases in CBP bacteria may be a result of effective pre-treatment. The activities of CBP bacteria, including cellulase secretion, cellulolytic capacity, and fermentation, as well as those of microorganisms utilized in SSF and SHF, should not be impacted by these upstream pre-treatment processes (Zhao et al. 2012). Most chemical pre-treatment processes result in liquid hydrolysates that are made up of a variety of soluble sugars, degradation by-products, and soluble lignin. If harsh reaction conditions are used for the pre-treatment, it may result in the production of significant amounts of unwanted inhibitory chemicals and degradation by-products.

For instance, it is well known that lignin and the products of its degradation are poisonous to microorganisms and interact irreversibly with the enzymes responsible for breaking down cellulose. Furthermore, at high temperatures, the pentose fraction (xylose and arabinose) quickly degrades to furfural, whereas the hexose fraction (glucose, galactose, and mannose) degrades to hydroxymethylfurfural (HMF), both of which are recognized fermentative microbe inhibitors. Pre-treated hydrolysates must be detoxified before fermentation since they include inhibitors such as phenolics, heavy metals, HMF, furfurals, weak acids, and tannins (Gaffron and Rubin 1942). The addition of a detoxification stage to the chemical pre-treatment process dramatically raises the cost of CBP. (But is necessary for CBP). The consequence of sugar from different pre-treatment methods and how it was handled. The following characteristics could make up an ideal CBP pre-treatment: High enzymatic hydrolysis rate, little or no toxic chemicals that inhibit CBP bacteria, little degradation of carbohydrate components, and low capital and operating costs are the expected outcomes, even though no pre-treatment method proposed thus far meets all of the aforementioned criteria. For the bioprocessing of lignocellulosic biomass to make biofuels, pre-treatment procedures, and synthetic CBP microbes must be correctly included in the entire bioconversion scheme, which is still being researched. During pre-treatment with sulfuric acid and water (including diluted sulfuric acid, steam explosion, and hot-water pre-treatment), hemicellulose and lignin are solubilized. Almost all processed solids are made of cellulose. It is a potential method for converting cellulose-rich material into biofuels utilizing CBP. The hemicellulosic part would be rendered useless during this pre-treatment process, lowering yield. When alkaline pre-treatment procedures such as AFEX, ARP, or lime pre-treatment are used, about half or the majority of the hemicellulose is retained with the solids. Because of the simultaneous bioconversion of hemicellulose-rich and cellulose-rich materials, alkaline pre-treatment is a potential alternative for CBP.

The microorganism utilized in the procedure is critical to LCB synthesis success. Biodiesel, bio-gasoline, and aviation fuel might be made by genetically modifying natural cellulolytic microorganisms, or lignocellulosic biomass could be broken down by industrial biofuel-producing microbes. The only simple secondary metabolites that native cellulolytic organisms can produce are ethanol and hydrogen gas. They are also resistant to genetic modification, and little is known about their physiology. Recent studies, however, point to the possibility of manipulating creatures like Clostridium and Thermobifida spp. On the other hand, industrial microorganisms are typically incapable of cellulolysis (Younesi et al. 2005). The majority of industrial bacteria do not metabolize other sugars in the presence of glucose due to carbon catabolite suppression (CCR), which is particularly problematic for their utilization with various substrates such as lignocellulosic biomass. To bypass this constraint, Escherichia coli and yeast were modified to co-metabolize a range of sugar combinations, such as glucose, xylose, cellobiose, galactose, or mannose (Parisutham et al. 2014). Ideal microorganisms for CBP with lignocellulosic substrate should be able to produce saccharifying enzymes for fast lignocellulose depolymerization, simultaneous utilization of numerous sugars, and resistance to toxic lignin molecules as well as the microbe's end metabolites. The production of energetic molecules in terms of titer is limited by low carbon concentrations in liquid substrates, low CO and H_2 solubilities in gaseous substrates, and the presence of inhibitors in lignocellulosic hydrolysates or syngas that inhibit microbial performance (Parisutham et al. 2014).

The most effective combination of pre-treatment, biological, and downstream processes with overproducing microbes for the best productivity, titers, and yields is currently a serious problem. Terpenes and butanol or other alcohols are the most common energetic molecules. Additionally, as the need for energy and chemical uses grows, so does the price of lignocellulosic resources. Also, the cost of the extraction procedure when chemicals are deposited intracellularly has a big impact on the economic assessment. In conclusion, the refinery process can successfully integrate high-value chemical and energetic applications. The economic and environmental feasibility of the lignocellulosic biofuel pathway will also be improved by optimizing by-product valorization and recycling, independent of changes in the price of fossil fuels.

11.8 Biofuels in the Contemporary World

Biofuel research is currently a hot topic due to rising worldwide energy demand from emerging nations and recent spikes in global oil prices. Biofuel presents opportunities for developed nations to achieve their emission reduction targets. Several ways are being investigated for the use of microbes in the manufacture of various biofuels (e.g., alcohols, hydrogen, biodiesel, and biogas) from various starting materials. As opposed to burning fossil fuels, burning biofuels (such as alcohol or hydrogen) results in significantly fewer (or no) carbon emissions into the atmosphere. For this reason, people view biofuels as being more environmentally friendly energy sources.

Biomass fuels have been used throughout human history. The term "biofuel" describes plant- and animal-based raw materials, such as molasses, rice husks, maize, and wood waste, that are transformed into fuel. Most of these are plant-derived alcohols produced by fermentation of carbohydrates or starches. Prior to the discovery of cheaper fossil fuel sources, they were used in many non-electrical applications such as solvents, oils, detergents or chemicals for the chemical industry. The use of bioalcohols as solvents or chemicals is once again discussed in light of the current rise in oil prices and conflicts in oil producing countries (Antoni et al. 2007).

Currently, 10% of the world's primary energy needs are met by biomass. In the face of rising crude oil prices, resource depletion, conflicts in developing countries and environmental problems, only biomass energy can replace conventional energy sources. It improves performance and smart use when hungry. Currently, on a commercial scale, only bioethanol is produced from plant biomass, an abundant and consistently energy rich carbohydrate that microbes can effectively convert to biofuels (Zhang et al. 2014).

Microorganisms have a crucial role in the production of biofuel. However, the natural strains' low product output makes it necessary to develop and improve them utilizing metabolic engineering and genetic engineering procedures. Therefore, any obstacles to using lignocellulosic biomass as a renewable fermentation feedstock may be removed by implementing these pathways in commercial fermentation hosts. By introducing helpful genes into the genome or removing obstructive ones, metabolic engineers must use cutting-edge technologies, such as omic technologies and the CRISPER/Cas9 system, to design and create novel strains of microbes with enhanced ability to produce biofuel from a variety of substrates (Tiedje et al. 2022).

The concerns about global warming and the depletion of fossil fuels have, without a doubt, rekindled interest in commercial alternative fuel sources. Research into the biological breakdown of lignocellulosic biomass feedstock to produce biofuels like bioethanol, biodiesel, and biohydrogen has been sparked by the need for renewable fuels. The profitability of biofuels is significantly impacted by the price of the raw materials required to produce them. In first-generation biofuels, the cost of raw materials makes up roughly 60–90% of total production. The cost of biofuels relative to petroleum equivalents varies with age (Khalid et al. 2022).

Nonetheless, major global research is being conducted to boost biofuel output while simultaneously cutting production costs in sustainable locations. The genetic modification of microbial strains and the improvement of fermentation settings have enhanced biofuel output. One of the primary obstacles is the strategy for maximizing yield from several processes; by combining process engineering, fermentation technology, enzyme engineering, and metabolic engineering, the industry has greatly benefited. Many biofuels are nearing economic viability; however, commercialization will demand additional subsidies until viability is achieved.

11.9 Future Prospects

Future research on biofuel production should focus on finding new species that produce fuel, enhancing and improving culture conditions, genetically modifying species that produce fuel, and thoroughly knowing biofuel production principles and practical ways for mass microorganism culture (Rodionova et al. 2017). Even though the demand for biofuels cannot be fully satisfied, they could have a big impact on the transportation industry. Based on current developments and improvements in the industry, it is claimed that greater scale biofuel production is essential and feasible to fulfill our future energy demands. Biotechnology-based fuel production is still a topic of great interest (Gavrilescu and Chisti 2005).

Other biofuels, except biohydrogen, are already being utilized commercially or in experimental studies. Despite a significant study on the bioconversion of lignocellulosic biomass to sugars for fermentation to ethanol, the problem remains unsolved (Aden et al. 2002). Bioconversion of starch to sugars for bioethanol production is becoming increasingly efficient and popular. Similarly, to that, gasoline-ethanol made from byproducts of the production of cane and beetroot sugar has been around for a while. Another popular strategy is the production of methane through the anaerobic digestion of organic waste. Traditional bioethanol production has already undergone a significant transformation thanks to modern biotechnology Gavrilescu et al. (2011). Using genetically modified corn as an example, higher yielding corn lowers the cost of primary feedstock; corn starch in genetically modified corn tends to be more suitable to enzymatic biological conversion to sugars than natural corn starch; microbial enzymes have been engineered for improved endurance and their capacity to rapidly convert starch to fermentable sugars; and microorganisms have been engineered to withstand greater concentrations of toxic ethanol and achieve rapid fermentation. Bioethanol will become more economically viable in the future thanks to these and other advancements (Kim et al. 2008). Anaerobic digestion methods are aiming for similar improvements. When unblended petrol is burned, hazardous pollutants such as volatile organic compounds, nitrous oxides, benzene, and particulates are produced, which can be eliminated by blending petrol with bioethanol (Gavrilescu and Chisti 2005).

Biodiesel emits significantly less pollution than petroleum diesel. The concept of converting biomass into electricity is enticing. Even though the annual land output of biomass is over five times the quantity of energy consumed globally, biomass presently accounts for less than 1% of commercial energy. Anaerobic digestion may transform organic waste from farms and landfills into combustible biogas, which contains roughly 45% carbon dioxide and 55% methane. Plant, animal, and microbial oils can be used to make liquid hydrocarbon fuels, using biotechnology to turn biomass into industrial chemicals. (Gavrilescu and Chisti 2005). The chemical sector uses hydrocarbon feedstocks worth close to US$24 billion per year.

Hydrocarbon purchases account for the vast bulk of the industry's raw material costs. As high-quality fossil fuel resources deplete, any hydrocarbon feedstocks that cannot be replaced must be obtained from alternate renewable sources. These resources are made up of plant, animal, and microbial materials that biodegrade. Chemical production will undergo a significant restructuring to move away from fossil hydrocarbons as feedstocks and towards plant-based matter to enable sustainable production. The feedstocks would come from regional agricultural produce. Because of the local availability of feedstock, fewer energy requirements for processing, less waste disposal, and efficient manufacturing, small production facilities near markets would become economically viable, particularly for high-value commodities. It is feasible to make net reductions in greenhouse gas emissions without reducing living standards. Until the 1930s, the majority of bulk chemicals were produced by chemical and fermentation processes using biomass such as maize, potatoes, wood, and plant oils. Traditional biomanufacturing's scope and capabilities are being substantially expanded by modern biotechnology. From primary resources, a wide range of industrial and consumer goods are currently available. Fatty acid synthesis appears to be an active area of biofuel technology study in the industry. According to press reports, at least one company can currently produce biodiesel from microorganisms. According to the literature, free fatty acids can be overproduced in E. coli and can be esterified in vivo with minor alcohols. Because of these successes, we have reason to expect that in the future we will be able to directly generate fuels made from fatty acids utilizing E. coli. No cutting-edge biofuel production technology has yet proven to be superior. Major contributors to the creation of biofuel are microorganisms.

Fossil fuels, such as coal and oil, have been essential to human history in the modern era because they provided a plentiful source of energy that drove the majority of social advancement and industrialization. It is generally accepted that these fuels cannot sustain the long-term growth of humanity, even if they remain the primary source of energy for the industrialised nations of the world. Oil is in high demand, and as new nations form, this need will only increase. As fewer and fewer new oil sources are discovered, many analysts predict that the rate of global oil production has already peaked and will only start to decline. Due to rising demand and dwindling supply, oil and other fossil fuels will eventually become economically unviable. The fact that using fossil fuels is not environmentally feasible is one of the problems with doing so. Massive amounts of carbon dioxide are released into the atmosphere when fossil fuels are burned, harming the biosphere by causing global warming. There is a strong incentive to investigate the development of renewable energy sources, especially microbial biofuels, for all of the reasons listed above.

It is incredible the variety of useful substances that microbes can use and generate. There are also many well-defined genetically manipulatable microbial systems and future scientific advancements will only make it easier to work with these systems. Microbial biofuels are an interesting study subject because of the possibility of "fine-tuning" biofuel-producing bacteria and genetically altering species to successfully use resources and wastes that would otherwise be worthless. Microbial biofuels have the potential to address both the economic and environmental issues associated with fossil fuels, even though there is presently no biofuel alternative that accomplishes so (The Next 25 Years 2021).

A persistent problem with the production of biofuels is that the biofuel itself is frequently harmful to cells. The most popular biofuel, ethanol, and others fit this description. Additionally, the route that creates the biofuel is frequently downregulated by the accumulation of the fuel within the organism itself, lowering production efficiency. Dunlop et al. modified E. coli and inserted numerous unique efflux pumps from various microbe species in an effort to solve this problem. They wanted to know if the pumps would improve the organism's capacity to withstand the stress brought on by the production of biofuel. To achieve this, researchers compiled a database of all bacterial protein pumps from the hydrophobe/amphiphile efflux (HAE1) family known to confer solvent resistance in gram-negative bacteria.

The scientists next performed a bioinformatic study, looking for proteins with substrate-binding area sequences that matched those of TtgB, a well-researched efflux pump recognized for its ability to export solvents from the cytoplasm. They used one of each of these plasmids to convert E. coli after cloning the operon of each discovered efflux pump into a plasmid vector. To find practical efflux pumps, the authors ran a competitive experiment. With the hope that the population would eventually be dominated by E. coli that expressed efflux pumps that conferred resistance to that biofuel, they did this by combining all E. coli strains into a single culture that also included a sizable amount of the biofuel. After maintaining and dilution each culture for four days, they measured relative plasmid concentrations and found that some efflux pumps can confer resistance to biofuels and that the pump also improves the efficiency of biofuel production in a limonene production strain.

11.10 Conclusion

This book chapter highlights the necessity for alternate energy sources owing to the negative environmental effects of fossil fuels. Biofuels are mentioned as a feasible option since they are renewable, minimize greenhouse gas emissions, and encourage economic development. The employment of microbes in biofuel production is highlighted as a novel method of increasing output. Because it is formed of non-toxic components and produces minimal hazardous waste, biodiesel is being proposed as a popular option for traditional diesel fuels. Bioethanol is also cited as a potential biofuel source, and algae is identified as a prospective feedstock source. The usage of fossil fuels has become a serious issue in recent decades, as the climate catastrophe has compelled humanity to seek alternate energy sources. Biofuels are a renewable energy source that increases energy security, reduces greenhouse gas emissions, and promotes economic development. Incorporating microorganisms into biofuel production is a potential alternative because they are easily available and contain enzymes that boost production. Biodiesel has the potential to replace traditional petroleum-based diesel fuels while also producing sustainable biological materials. It is comparable to diesel derived from crude oil in terms of chain length, viscosity, and energy density. Microorganisms are used in biotechnology to convert some compounds into others with higher added value, allowing for the sensible and equitable use of natural resources. It is now widely understood that continuing to utilize petroleum-based fuels is unsustainable due to declining supplies and their contribution to carbon dioxide accumulation in the environment.

References

Abdelaziz AE, Leite GB, Hallenbeck PC (2013) Addressing the challenges for sustainable production of algal biofuels: II. Harvesting and conversion to biofuels. Environ Technol 34(13–14):1807–1836. https://doi.org/10.1080/09593330.2013.831487

Aden A, Ruth M, Ibsen K, Jechura J, Neeves K, Sheehan J, Wallace B, Montague L, Slayton A, Lukas J (2002) Lignocellulosic biomass to ethanol process design and economics utilizing co-current dilute acid prehydrolysis and enzymatic hydrolysis for corn stover. https://doi.org/10.2172/15001119

Antolín G (2002) Optimisation of biodiesel production by sunflower oil transesterification. Bioresour Technol 83(2):111–114. https://doi.org/10.1016/s0960-8524(01)00200-0

Antoni D, Zverlov VV, Schwarz WH (2007) Biofuels from microbes. Appl Microbiol Biotechnol 77(1):23–35. https://doi.org/10.1007/s00253-007-1163-x

Babel S, Sae-Tang J, Pecharaply A (2008) Anaerobic co-digestion of sewage and brewery sludge for biogas production and land application. Int J Environ Sci Technol 6(1):131–140. https://doi.org/10.1007/bf03326067

Biofuels production (2013). Wiley

Borowitzka MA (2013) High-value products from microalgae-their development and commercialisation. J Appl Phycol 25:743–756. https://doi.org/10.1007/s10811-013-9983-9

Bro C, Regenberg B, Forster J, Nielsen J (2006) In silico aided metabolic engineering of Saccharomyces cerevisiae for improved bioethanol production. Metab Eng 8(2):102–111. https://doi.org/10.1016/j.ymben.2005.09.007

Cerón-García M, Macías-Sánchez M, Sánchez-Mirón A, García-Camacho F, Molina-Grima E (2013) A process for biodiesel production involving the heterotrophic fermentation of Chlorella protothecoides with glycerol as the carbon source. Appl Energy 103:341–349. https://doi.org/10.1016/j.apenergy.2012.09.054

Cheng L, Zhang L, Chen H, Gao C (2006) Carbon dioxide removal from air by microalgae cultured in a membrane-photobioreactor. Sep Purif Technol 50(3):324–329. https://doi.org/10.1016/j.seppur.2005.12.006

Cheng D, He Q (2014) Assessment of environmental stresses for enhanced microalgal biofuel production—an overview. Front Energy Res 2. https://doi.org/10.3389/fenrg.2014.00026

Chisti Y (2007) Biodiesel from microalgae. Biotechnol Adv 25(3):294–306. https://doi.org/10.1016/j.biotechadv.2007.02.001

Chiu SY, Kao CY, Chen CH, Kuan TC, Ong SC, Lin CS (2008) Reduction of CO_2 by a high-density culture of Chlorella sp. in a semicontinuous photobioreactor. Bioresour Technol 99(9):3389–3396. https://doi.org/10.1016/j.biortech.2007.08.013

Chynoweth DP, Wilkie AC, Owens JM (1999) Anaerobic treatment of piggery slurry-review. Asian-Australas J Anim Sci 12(4):607–628. https://doi.org/10.5713/ajas.1999.607

Colla LM, Bertolin TE, Costa JAV (2004) Fatty acids profile of spirulina platensis grown under different temperatures and nitrogen concentrations. Zeitschrift Für Naturforschung C 59(1–2):55–59. https://doi.org/10.1515/znc-2004-1-212

Converti A, Oliveira R, Torres B, Lodi A, Zilli M (2009) Biogas production and valorization by means of a two-step biological process. Bioresour Technol 100(23):5771–5776. https://doi.org/10.1016/j.biortech.2009.05.072

Costa JAV, de Morais MG (2011) The role of biochemical engineering in the production of biofuels from microalgae. Bioresour Technol 102(1):2–9. https://doi.org/10.1016/j.biortech.2010.06.014

de la Noüe J, Bassères A (1989) Biotreatment of anaerobically digested swine manure with microalgae. Biol Wastes 29(1):17–31. https://doi.org/10.1016/0269-7483(89)90100-6

Gaffron H, Rubin J (1942) Fermentative and photochemical production of hydrogen in algae. J Gen Physiol 26(2):219–240. https://doi.org/10.1085/jgp.26.2.219

Gavrilescu M, Chisti Y (2005) Biotechnology—a sustainable alternative for chemical industry. Biotechnol Adv 23(7–8):471–499. https://doi.org/10.1016/j.biotechadv.2005.03.004

Gavrilescu M, Fortuna ME, Simion IM (2011) Sustainability in environmental remediation. Environ Eng Manag J 10(12):1987–1996. https://doi.org/10.30638/eemj.2011.264

Ghirardi M (2000) Microalgae: a green source of renewable H_2. Trends Biotechnol 18(12):506–511. https://doi.org/10.1016/s0167-7799(00)01511-0

Goh BHH, Ong HC, Cheah MY, Chen WH, Yu KL, Mahlia TMI (2019) Sustainability of direct biodiesel synthesis from microalgae biomass: a critical review. Renew Sustain Energy Rev 107:59–74. https://doi.org/10.1016/j.rser.2019.02.012

González-Fernández C, Sialve B, Bernet N, Steyer JP (2011) Impact of microalgae characteristics on their conversion to biofuel. Part I: focus on cultivation and biofuel production. Biofuels Bioprod Biorefining 6(1):105–113. https://doi.org/10.1002/bbb.338

Graboski MS, McCormick RL (1998) Combustion of fat and vegetable oil derived fuels in diesel engines. Prog Energy Combust Sci 24(2):125–164. https://doi.org/10.1016/s0360-1285(97)00034-8

Hatakka AI (1983) Pretreatment of wheat straw by white-rot fungi for enzymatic saccharification of cellulose. Appl Microbiol Biotechnol 18:350–357. https://doi.org/10.1007/BF00504744

Himmel ME, Ding SY, Johnson DK, Adney WS, Nimlos MR, Brady JW, Foust TD (2007) Biomass recalcitrance: engineering plants and enzymes for biofuels production. Science 315(5813):804–807. https://doi.org/10.1126/science.1137016

Kapdan IK, Kargi F (2006) Bio-hydrogen production from waste materials. Enzyme Microb Technol 38(5):569–582. https://doi.org/10.1016/j.enzmictec.2005.09.015

Khalid Z, Alam SN, Guldhe A, Singh B (2022) Novel feedstocks for biofuels: current scenario and recent advancements. In: Guldhe A, Singh B (eds) Novel feedstocks for biofuels production. Clean energy production technologies. Springer, Singapore. https://doi.org/10.1007/978-981-19-3582-4_2

Kim TH, Taylor F, Hicks KB (2008) Bioethanol production from barley hull using SAA (soaking in aqueous ammonia) pretreatment. Biores Technol 99(13):5694–5702. https://doi.org/10.1016/j.biortech.2007.10.055

Kumar BR, Saravanan S (2016) Use of higher alcohol biofuels in diesel engines: a review. Renew Sustain Energy Rev 60:84–115. https://doi.org/10.1016/j.rser.2016.01.085

Kumari A, Kundu PK, Gupta MM, Bala K, Chandra S, Dutta R, Das A (2022) Role of Microorganisms in Production of Biofuels. Clean Energy Prod Technol 65–116. https://doi.org/10.1007/978-981-19-3582-4_4

Kurano N, Ikemoto H, Miyashita H, Hasegawa T, Hata H, Miyachi S (1995) Fixation and utilization of carbon dioxide by microalgal photosynthesis. Energy Convers Manage 36(6–9):689–692. https://doi.org/10.1016/0196-8904(95)00099-y

Lardon L, Hélias A, Sialve B, Steyer JP, Bernard O (2009) Life-cycle assessment of biodiesel production from microalgae. Environ Sci Technol 43(17):6475–6481. https://doi.org/10.1021/es900705j

Lee SK, Chou H, Ham TS, Lee TS, Keasling JD (2008) Metabolic engineering of microorganisms for biofuels production: from bugs to synthetic biology to fuels. Curr Opin Biotechnol 19(6):556–563. https://doi.org/10.1016/j.copbio.2008.10.014

Mata TM, Martins AA, Caetano NS (2010) Microalgae for biodiesel production and other applications: a review. Renew Sustain Energy Rev 14(1):217–232. https://doi.org/10.1016/j.rser.2009.07.020

McGinn PJ, Dickinson KE, Bhatti S, Frigon JC, Guiot SR, O'Leary SJB (2011) Integration of microalgae cultivation with industrial waste remediation for biofuel and bioenergy production: opportunities and limitations. Photosynth Res 109(1–3):231–247. https://doi.org/10.1007/s11120-011-9638-0

Melis A, Zhang L, Forestier M, Ghirardi ML, Seibert M (2000) Sustained photobiological hydrogen gas production upon reversible inactivation of oxygen evolution in the green alga Chlamydomonas Reinhardtii. Plant Physiol 122(1):127–136. https://doi.org/10.1104/pp.122.1.127

Minale M, Worku T (2013) Anaerobic co-digestion of sanitary wastewater and kitchen solid waste for biogas and fertilizer production under ambient temperature: waste generated from condominium house. Int J Environ Sci Technol 11(2):509–516. https://doi.org/10.1007/s13762-013-0255-7

Molina Grima E, Belarbi EH, Acién Fernández F, Robles Medina A, Chisti Y (2003) Recovery of microalgal biomass and metabolites: process options and economics. Biotechnol Adv 20(7–8):491–515. https://doi.org/10.1016/s0734-9750(02)00050-2

Murakami M, Ikenouchi M (1997) The biological CO_2 fixation and utilization project by rite (2)—screening and breeding of microalgae with high capability in fixing CO_2—. Energy Convers Manage 38:S493–S497. https://doi.org/10.1016/s0196-8904(96)00316-0

Parisutham V, Kim TH, Lee SK (2014) Feasibilities of consolidated bioprocessing microbes: from pretreatment to biofuel production. Biores Technol 161:431–440. https://doi.org/10.1016/j.biortech.2014.03.114

Patil V, Tran KQ, Giselrød HR (2008) Towards sustainable production of biofuels from microalgae. Int J Mol Sci 9(7):1188–1195. https://doi.org/10.3390/ijms9071188

Pulz O (2001) Photobioreactors: production systems for phototrophic microorganisms. Appl Microbiol Biotechnol 57(3):287–293. https://doi.org/10.1007/s002530100702

Razzak SA, Hossain MM, Lucky RA, Bassi AS, de Lasa H (2013) Integrated CO_2 capture, wastewater treatment and biofuel production by microalgae culturing—a review. Renew Sustain Energy Rev 27:622–653. https://doi.org/10.1016/j.rser.2013.05.063

Ribeiro LA, Pereira da Silva P (2015) Qualitative Delphi approach of advanced algae biofuels. Manag Environ Qual Int J 26(6):852–871. https://doi.org/10.1108/meq-03-2014-0046

Richmond A (ed) (1986) Handbook of microalgal mass culture (1986) (1st ed.). CRC Press. https://doi.org/10.1201/9780203712405

Rodionova M, Poudyal R, Tiwari I, Voloshin R, Zharmukhamedov S, Nam H, Zayadan B, Bruce B, Hou H, Allakhverdiev S (2017) Biofuel production: challenges and opportunities. Int J Hydrog Energy 42(12):8450–8461. https://doi.org/10.1016/j.ijhydene.2016.11.125

Roy S, Das D (2015) Liquid fuels production from algal biomass. Algal Biorefinery Integr Approach 277–296. https://doi.org/10.1007/978-3-319-22813-6_13

Saba B, Christy AD, Yu Z, Co AC (2017) Sustainable power generation from bacterio-algal microbial fuel cells (MFCs): an overview. Renew Sustain Energy Rev 73:75–84. https://doi.org/10.1016/j.rser.2017.01.115

Sakai N, Sakamoto Y, Kishimoto N, Chihara M, Karube I (1995) Chlorella strains from hot springs tolerant to high temperature and high CO_2. Energy Convers Manage 36(6–9):693–696. https://doi.org/10.1016/0196-8904(95)00100-r

Singh A, Nigam P (2014). Microbial biofuels production. Microb Biotechnol 155–168. https://doi.org/10.1201/b17587-8

Sun Y, Cheng J (2002) Hydrolysis of lignocellulosic materials for ethanol production: a review. Biores Technol 83(1):1–11. https://doi.org/10.1016/s0960-8524(01)00212-7

Tanvir RU, Zhang J, Canter T, Chen D, Lu J, Hu Z (2021) Harnessing solar energy using phototrophic microorganisms: a sustainable pathway to bioenergy, biomaterials, and environmental solutions. Renew Sustain Energy Rev 146:111181. https://doi.org/10.1016/j.rser.2021.111181

The next 25 years (2021) Nat Biotechnol 39(3):249–249. https://doi.org/10.1038/s41587-021-00872-0

Tiedje JM, Bruns MA, Casadevall A, Criddle CS, Eloe-Fadrosh E, Karl DM, Nguyen NK, Zhou J (2022) Microbes and climate change: a research prospectus for the future. MBio 13(3). https://doi.org/10.1128/mbio.00800-22

Timilsina GR, Cheng JJ (2010) Advanced biofuel technologies: status and barriers. Policy research working papers. https://doi.org/10.1596/1813-9450-5411

Tippmann S, Chen Y, Siewers V, Nielsen J (2013) From flavors and pharmaceuticals to advanced biofuels: production of isoprenoids in Saccharomyces cerevisiae. Biotechnol J 8(12):1435–1444. https://doi.org/10.1002/biot.201300028

Trivedi J, Aila M, Bangwal D, Kaul S, Garg M (2015) Algae based biorefinery—how to make sense? Renew Sustain Energy Rev 47:295–307. https://doi.org/10.1016/j.rser.2015.03.052

Xuan J, Leung MK, Leung DY, Ni M (2009) A review of biomass-derived fuel processors for fuel cell systems. Renew Sustain Energy Rev 13(6–7):1301–1313. https://doi.org/10.1016/j.rser.2008.09.027

Younesi H, Najafpour G, Mohamed AR (2005) Ethanol and acetate production from synthesis gas via fermentation processes using anaerobic bacterium Clostridium Ljungdahlii. Biochem Eng J 27(2):110–119. https://doi.org/10.1016/j.bej.2005.08.015

Zhang X, Rong J, Chen H, He C, Wang Q (2014) Current status and outlook in the application of microalgae in biodiesel production and environmental protection. Front Energy Res 2. https://doi.org/10.3389/fenrg.2014.00032

Zhao X, Zhang L, Liu D (2012) Biomass recalcitrance. Part II: fundamentals of different pretreatments to increase the enzymatic digestibility of lignocellulose. Biofuels Bioprod Biorefining 6(5):561–579. https://doi.org/10.1002/bbb.1350

Chapter 12
Microorganisms: An Efficient Enhancer of Biofuel Production

Sriparna De, Dipankar Das, Aishwarya Saha, Debayan Chakraborty, and Dipankar Chattopadhyay

Abstract The microbial fuel cell (MFC) technology has annexed recognition within the scientific community for the organic waste derived electricity production assisted by microbially catalyzed anodic, and microbial/enzymatic/abiotic cathodic electrochemical reactions. MFC also captured utmost attention and in demand stemming from the continuous rise in human population and world globalization. The continued burning of such non-renewable energy sources not only depletes fossil fuel reservoirs but also contributes to increase in atmospheric CO_2 levels which is a great threat to the environment i.e. global warming. To, reduce the exponential oversaturation of carbon dioxide in the ecosystem and sustainable energy production noteworthy technological advances for biofuel production has offered excellent opportunities in lipid accumulation and bioconversion. Biofuels of different generations can contribute synergistically to intend the feasibility of the microbial induced pathways in fulfilling energy demand. The chapter portrays, initially on the primitive stage, nutritional aspects for inducing biomass yielding from various microorganism feedstock. Also, this chapter highlights the pretreatment of biomass, the increased production of bio-fuel (biodiesel, bioalcohol, and biogas) using advanced microbial systems. Microbial Biomass production in synergism with nutrient optimization and supplementation have summarized in pertinent to various perspectives. More significantly, the economic perspectives of oleaginous microorganism cultivations are also reviewed along with its possible scheme of utilizing alternative and non-conventional nutrient sources for possible technology scale-ups and commercialization.

Keywords Biomass · Biodiesel · Fuel cell · Biogas · Fermentation · Microalgae · Microorganism

---

S. De (✉) · D. Das · A. Saha · D. Chakraborty
Department of Allied Health Sciences, Brainware University, Kolkata 700125, India
e-mail: sriparna.de2@gmail.com

D. Chattopadhyay
Department of Polymer Science and Technology, University of Calcutta, Kolkata 700009, India

M. Shah and D. Deka (eds.), *Emerging Sustainable Technologies for Biofuel Production*, Environmental Science and Engineering, https://doi.org/10.1007/978-3-031-52167-6_12

12.1 Historical Background

The pioneer in the history of electrochemistry and bioelectricity is a physician and professor from the University of Bologna named Luigi Galvani. He was the first to discover that the muscles of a dead frog's legs twitch when it is struck by an electric spark. He described the activating force that activated muscles of the frog's legs being generated by an electrical fluid that is carried to the muscles by the nerves which he coined as "animal electricity". There has been remarkable development in the field of electrochemical systems for power generation or storage of energy during a few of the beginning decades of the nineteenth century. One of the most important inventions happened in 1859 of the lead acid battery by Gaston Plante, the French physicist. But the first H_2/O_2 acid fuel cell was discovered by William Grove who happens to be considered the 'father of fuel cells (Santoro et al. 2017). Subsequently in the following years, several advancements in Fuel Cells (FCs) were achieved in which various national agencies, NASA and vehicle manufactures play a part. Among them, the Biological Fuel Cells (BFC) are very effective as they are devices that transform chemical energy to electrical energy via electrochemical reactions which consist of biochemical pathways. One of the major types of BFC is Microbial Fuel Cell which uses microorganisms to directly convert chemical energy to electrical energy (Minteer et al. 2007). It is an energy enriched chemical compound which is derived from living biomass of microbial cells such as microalgae, bacteria. As the global population is increasing in regular fashion and it demands more energy supplies to fulfill daily needs, it should increase quality of living. Fossil fuel that is used as the main source of energy produces gases responsible for global warming. As a result biofuel comes forward as a promising source of energy. It is renewable as well as an environmentally friendly source of energy. The advantage of using a Microbial Fuel Cell (MFC) is that it uses a whole microbial cell for biochemical oxidation of fuels which allows multiple enzymatic reactions to occur but without any enzyme isolation. During the past few years, the prospect of using MFCs for large scale wastewater treatment has increased significantly (Santoro et al. 2017; Minteer et al. 2007; Bhatia et al. 2017a).

12.2 Types of Biofuel

Major two categories of biofuel are primary biofuel and secondary biofuel (Fig. 12.1). If biofuels are produced from wood pellets, crop residues of sugarcane, corn and other crops, animal waste and used in unprocessed form, it is termed as primary biofuel. In the case of secondary biofuel, it is directly produced from biomass of different organisms, especially microorganisms. Secondary biofuel further can be divided into three categories, First generation, second generation and third generation biofuel. First generation biofuels are generated from edible crops like sugarcane (sugar based crop), wheat (starch based crop), sunflower (oil based crop). Production

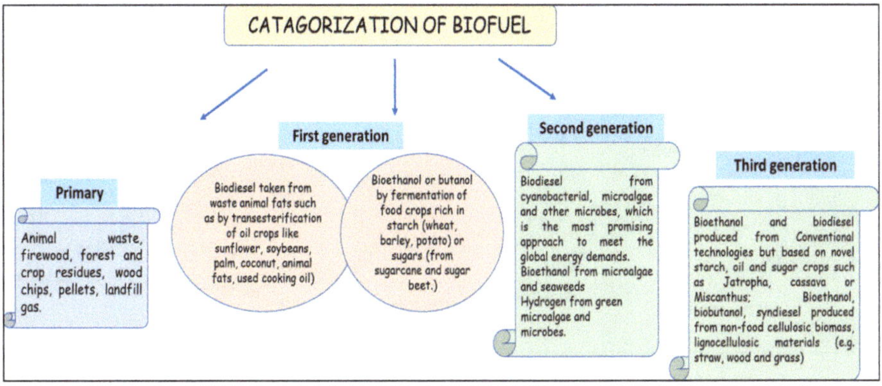

Fig. 12.1 Categorization of biofuel

of bioethanol from such crops is under the first generation of biofuel (Abdelaziz et al. 2013). Secondary generation of biofuels are produced from lignocellulosic biomass of straw, grass via conventional technologies. Third generation of biofuels are the most advanced form of biofuel which are produced from biomass of microorganisms such as microalgae. Third generation biofuel gives the solution against energy insecurity and climate related issues (Rafa et al. 2021; Rodionova et al. 2017).

12.2.1 Microbial Fuel Cells (MFCs) System

MFCs system is used as a tool/device for processing of waste matter in different sewages/wastewater resources. MFC is operated by microbial mediated catalysis process in anode surface wherein bioelectric power generated in cathode surface to external circuit system (Fig. 12.2). In this connection MFC units are stacked configurations, the greater amount of power output (2×10^2 W.m^{-2}) via using a hexacyanoferrate cathode system at hourly periods (Razeghifard 2013). Application of power management systems (PMS) can enable step-up voltage conversion processes by displacing the unstable stacking arrangement of the MFCs system, which can be advantageous for practical uses of MFCs technology. This approach can be tuned and operated in power management systems (PMS). MFCs system comprised of anode and cathode compartments and is specifically intricate in the degradation of organic waste substrates via bio-electric power generator. To tune the electrocatalytic property as well as efficiency of MFCs the designing of needs to be altered in terms of technological advancement and membrane constituents (Tran et al. 2010). The fundamental reason for developing this technology is the possibility to replace the current expensive wastewater treatment systems with one that can be self-sustaining or even produce a net amount of energy that is positive while removing contaminants.

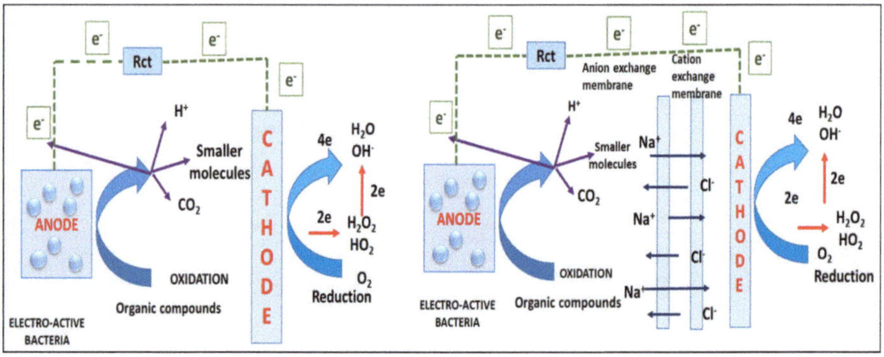

Fig. 12.2 Schematic diagram of **a** microbial fuel cell and **b** general microbial electrosynthesis cell

MEC needs an external source of electricity for electrolysis to make hydrogen at the cathode, however this external energy source is just a minor quantity needed because the majority of the energy is chemical energy derived from substrates oxidized at the anode. Therefore, using bioelectrocatalysis (BES) in conjunction with other low energy power sources, hydrogen can be created with a low energy usage. Since hydrogen is a precious gas that is created and is essential for the developing hydrogen-energy economy. Most recently, The development of the microbial desalination cell (MDC), which has the potential to clean wastewater, produce power, and desalinate water all at once, has proved successful. Recent studies and work reports on the major arrangements utilized in microbial desalination cells have made use of air-breathing arrangements as well as biocathodes and osmotic membranes (Kruse et al. 2005). BESs have recently been introduced as microbial electro-synthesis devices that, in conjunction with this study, enable the creation of valuable chemicals from CO_2 or other molecules, including gas transformation or reduction, through the use of particular bacteria or operating conditions. Because renewable energy could be used in this relatively new way even when it is not connected to the primary power lines for distribution, it is of great importance. Both direct and indirect methods can be used to deliver electrons from an electrode to bacteria trapped inside biofilms. Some bacteria, including Pseudomonas aeruginosa, Shewanella oneidensis, and Pseudomonas alcaliphila, have the ability to make their own redox mediators. Similar to pyocianine in P. aeruginosa, these mediators are in charge of the electrochemical activity of the bacteria. The synthesis of a quinone mediator by S. oneidensis increases a microbial fuel cell's (MFC) power density. The oxidation of waste products from bacterial metabolism, such as hydrogen produced by fermentative bacteria and then oxidized at the anode surface, can also result in indirect electron transfer. When electroactive bacteria come into direct touch with the anode surface, their outer membranes transfer electrons directly. Cytochrome C membrane proteins assist the transfer of electrons from the cell to the electrode. This direct transfer process has been shown in tests with Geobacter sulfurreducens mutants, where it was made obvious that cytochrome c is involved in electron transfer (Prince and Kheshgi 2005).

12.2.2 Microalgae as Source of Biofuel

As discussed earlier, to fulfil the need of energy and at the same time to eliminate pollution status of energy sources, microorganisms are the best recognised source. Microalgal biofilm cultivation can be a way for direct energy conversion to produce biofuels (Fig. 12.3). Algae contain green pigment, photosynthetic in nature and act as raw material for biomass production. Cultivation of microalgae is feasible as it needs very little space for growth and can increase the number of cells with a high exponential rate as they have a growth time of 24 h. Photosynthetic microorganisms like algae and cyanobacteria can act as the cheapest source of biofuel as atmospheric carbon-di-oxide acts as a source of carbon and sunlight serves as an energy source to make sugar. For production of bioethanol, biomethanol sugar is a basic molecule that is provided by algae through photosynthesis. During oxygenic photosynthesis, molecular hydrogen is generated which indicates the production of biohydrogen as microbial biofuel. Depending on species of algae or other micro-organisms major biofuels like biodiesel, fatty acid, bioethanol, bioethanol can be produced and even biomass (in terms of carbohydrate, protein, lipid) of such microorganisms act as biofuel itself (Rodionova et al. 2017) *Spirulina maxima* has maximum 70% w/w of protein in biomass, *Porphyridium cruentum* has maximum 60% w/w of carbohydrate in biomass, *Schizochytrium sp.* has a maximum 80% w/w of lipids in biomass (in case of dry weight). Thus they are promising sources of microbial biofuel to fulfill ever rising energy demand of society (Razeghifard 2013) Production of microalgae based biofuel is more than that of green plant produced biofuel due to presence of special genes responsible for bioenergy production. Cyanobacteria *Synechococcus elongatus* PCC 7942 contain ketoacid decarboxylase gene and can produce isobutyraldehyde, butanol. Crude oil can be derived from green algae *Botryococcus braunii* and *Chlorella protothecoides* as it has high levels of terpenoid hydrocarbon and glyceryl lipid. In the current scenario by the fermentative pathway *Saccharomyces cerevisiae* is considered as a model microorganism for production of bioethanol (Tran et al. 2010).

12.3 Biohydrogen Production: Microbial Biosynthesis

Molecular hydrogen production by photosynthetic microorganisms is the most important way to get such biofuel. Photosynthetic process is involved in emission of zero percent of greenhouse gas and due to this it is eco-friendly to generate biofuel by utilizing photosynthetic pathways. With expense of energy, molecular

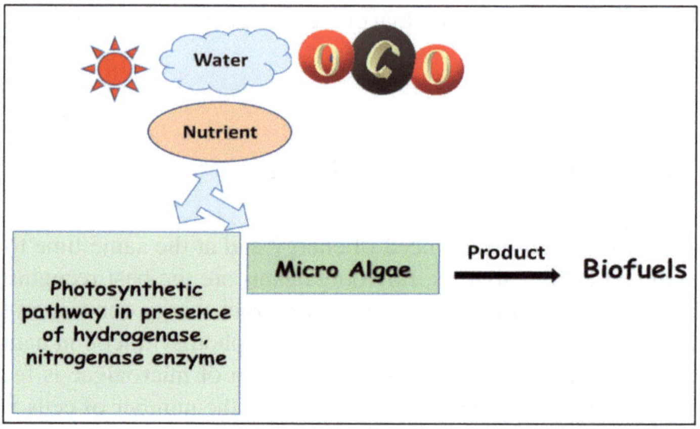

Fig. 12.3 Microalgae as source of biofuel

hydrogen can be liberated by photosynthetic bacteria and with loss of energy photosynthetic capacity of microorganisms such as microalgae can be decreased (Kruse et al. 2005; Prince and Kheshgi 2005). However, inadequate yield and low efficiencies turn out to be the main challenges. Production of hydrogen by phototrophic microalgae has a problem as the production of oxygen can inhibit the action of hydrogenases irreversibly. On top of that, production of biohydrogen is not economically viable in comparison to hydrogen production using fossil fuels. Also, during photosynthetic hydrogen production the rate of carbon assimilation is lower than the ideal requirement for the accumulation of cell biomass (Nikolaidis and Poullikkas 2017). Hydrogen can be produced from many prokaryotic as well as eukaryotic organisms mainly including cyanobacteria, photosynthetic bacteria and green algae. Unlike photosynthetic bacteria, cyanobacteria and green algae are capable of photolysis of water; the end products of which are converted to energy rich compounds like ATP and NADPH which are further utilized for biomass accumulation (Ghirardi et al. 2009).

12.3.1 Use of Microbes in Third Generation Biofuel Production

Utilization of microbes is a sustainable way for the production of biofuels like biodiesel, bio alcohol and biogas. Biodiesel production: Biodiesel can be produced by biological elements to substitute petroleum products which emits pollutants and greenhouse gases as well. Transesterification of various animal fats and vegetable oils, typically with methanol or ethanol, produces biodiesel fuels. Nowadays, use of biodiesel not only reduces environmental pollution but also can give rural economies access to affordable, local fuel. In addition to that, by-products of the biodiesel production pathway are mostly less toxic and biodegradable in nature as it does not release sulphur or nitrogen containing flue gases.

12.3.2 Biodisel Production

Production of biodiesel is easy and it can be achieved by a simple two-step process. First, extraction of animal fat or plant oil specially from rapeseed and soybean oil (Fig. 12.4). In order to produce biodiesel, the lipid fraction is then transesterified with alcohols in the presence of catalysts in the second stage. Recently, plant *Jatropha curcas* has been identified as an important source to obtain biodiesel. Microbes are able to utilize lignocellulose derived sugars, use them as carbon source and ferment to accumulate Triacylglycerol (TAG). On further transesterification of TAGs with alcohols, biodiesel is produced. Bacteria which can accumulate large amounts of TAG are *Acinetobacter sp., Mycobacterium sp.,Streptomyces sp.* It is observed that few bacterial strains like *Rhodococcus opacus* able to accumulate 87% of TAG as per their dry weight (Shuobo Shi et al.). The advantage of using microbes in the production of biodiesel is that all the required components namely TAG, alcohol and the enzyme for transesterification (lipase) are all produced by the microbe itself. Bhatia and co-workers used hot water treated oil palm hydrolysate as a carbon source in a process which ultimately resulted in FAME production (Bhatia et al. 2017a). For the microbes who cannot produce raw materials and the ones which are non-oleaginous, bioengineering is used to improve their potential in utilizing carbon sources. *E coli*, being a non-oleaginous microbe is bioengineered by Bokinsky and co-workers so as to secrete different hydrolase enzymes like B-glucosidase, Xylanase, cellulase which ultimately resulted in utilization of switchgrass as carbon source and production of biodiesel FAME (Bokinsky et al. 2011). Other co culturing techniques can be useful in downstream processes. Co culture of fungi *Aspergillus fumigates* with algae makes the harvesting process easier by enhancing flocculation efficiency of the later.

12.3.3 Bio-Alcohol Production

Fermentation is the process by which most alcohols are produced. Butanol, Isobutanol, Propanol are all products of fermentation. For example, **Butanol** is a renewable fuel and can be used as a replacement for gasoline. For fermentation of Clostridium sp., Komonkiat and coworkers used oil palm trunk sap as raw material which resulted in butanol production (Bhatia et al. 2017b). Even a Co-culture technique can be used to produce butanol. One such experiment was performed by Abd-Alla and coworkers using date palm as carbon source. The experiment was done by co-culturing *Bacillus*

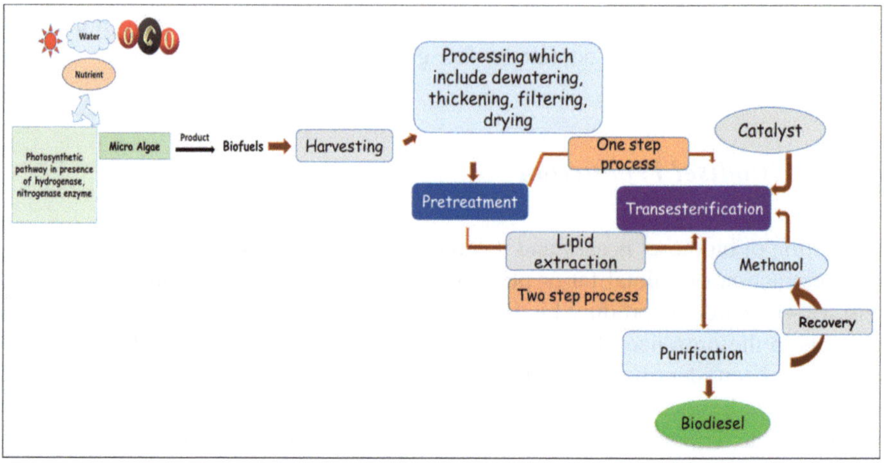

Fig. 12.4 Schematic representation of Biodisel production pathway

subtilis, which hydrolyze biomass and scavenge oxygen to create an anaerobic environment which is again used for the growth and fermentation of *Clostridium acetobutylicum.* This co-culture technique finally produced 14.9 g/l butanol (Komonkiat and Cheirsilp 2013).

12.3.4 Ethanol Production

For the synthesis of bioalcohol, both naturally occurring and artificially created microorganisms have been documented. When *Miscanthus sacchariflorus* was treated with alkali, Kang and colleagues observed that *Saccharomyces cerevisiae* CHY1011 produced 69.2 g/L of ethanol (Abd-Alla and El-Enany 2012).It has been reported that surfactant is used to speed up the hydrolysis of biomass. Tsuji and colleagues observed 7.2 g/L ethanol production using Mrakia blollopis after applying Tween 80 to accelerate the hydrolysis of cellulose biomass (Kang et al. 2015). Liu and colleagues modified *Saccharomyces cerevisiae* with cell surface expression of cellulase to expand the substrate utilisation range, and they obtained 18 g/L ethanol production from rice straw. For the purpose of producing ethanol from cellulose, Zuroff and colleagues reported a co-culture of the microbes *Clostridium phytofermentans* and *Saccharomyces cerevisiae* cdt-1 (Tsuji et al. 2013).

2,3-butanediol: A direct liquid fuel or a starting point for the synthesis of additional compounds, 2,3-butanediol has two possible uses. It is made by the use of a fermentation process that goes through an intermediate of acetolactate and acetoin. It has been stated that the manufacture of 2,3-butanediol can use a variety of pure carbon sources and other sustainable feedstocks. The group of microorganisms responsible for 2,3-butanediol fermentation that has received the most research includes Enterobacter and Klebsiella. Utilising Enterobacter cloacae, Zhang and colleagues demonstrated 81.4 g/L 2,3-butanediol synthesis from xylose obtained from corncobs (Zuroff et al. 2013). For the manufacture of 2,3-butanediol, both simultaneous saccharification and fermentation (SSF) and separate hydrolysis and fermentation (SHF) procedures can be used. When Klebsiella pneumonia was fermented using Jerusalem artichoke tubers as the carbon source, Sun and colleagues found 81.59 and 91.63 g/L 2,3-butanediol production, respectively (Zhang et al. 2016).

Isobutanol: High energy fuel isobutanol is less corrosive and volatile than ethanol. Compared to ethanol, isobutanol has a lower solubility in water (8.5%), but it has a 30% higher energy content. More isobutanol can be blended with petrol because of the decreased O_2 level. In the industrial setting, propylene is carbonylated to create isobutanol. The production of the amino acid valine, which is present in the majority of microorganisms, shares intermediates with the biosynthesis of isobutanol (Sun et al. 2009).

However, are able to naturally create isobutanol. It is possible to build isobutanol production pathways in bacteria using metabolic engineering. The synthesis of isobutanol at a high temperature (50 °C) using the thermophilic *Geobacillus thermoglucosidasius* was reported by Lin and coworkers (Atsumi et al. 2008). Fermentation at higher temperatures (>40 °C) is beneficial because it lowers enzyme loading, reduces the risk of contamination, and reduces the cost of product separation (Lin et al. 2014). Cellobiose was converted by engineered G. thermoglucosidasius to 0.6 g/L isobutanol (Komonkiat and Cheirsilp 2013). A synthetic microbial consortium of bacteria and fungi was described by Minty and co-workers for the manufacture of isobutanol from lignocelluloses. *Trichoderma reesei* strains release cellulase enzymes to break down lignocellulosic biomass into free sugars, which are then processed into 1.9 g/L isobutanol by modified E. coli (Lynd 1989).

12.4 Biogas Production

Biogas, a highly effective source of renewable energy, can serve as an alternative for fossil fuels. Biohydrogen & methane can be produced through microbial fermentation techniques. Hydrogen, conventionally produced from fossil fuels, has both advantages and disadvantages. Though it has the highest energy content on a mass basis and no detrimental effect to the environment during use, one major limitation is the storage and distribution. Biohydrogen production is also more eco-friendly as compared to methods like pyrolysis, photolysis and electrolysis. The most common methods used for biohydrogen production include dark fermentation and photofermentation. Dark fermentation involves the carbohydrate fermentation by strict or facultative anaerobes resulting in the production of Volatile fatty acids (VFA) and alcohol as by-products, whereas photofermentation utilizes light as the chief source of energy (Minty et al. 2013). Organic acids like acetic and propionic acids are found to have important roles in biohydrogen production. However, despite continuous efforts, production of biohydrogen at an industrial scale is still not feasible. In order to overcome this and make the production economically viable, utilization of lignocellulose can serve as an important way. Fermentation using sugarcane bagasse as raw material by *Clostridium butyricum* has reported hydrogen production. Further experiments have shown that an increase in temperature can increase the production manifold. To increase hydrogen production, genetic engineering of *Caldicellulosiruptor bescii* was done by Cha & co-workers by deletion of lactate dehydrogenase gene which showed 21–24% increase in hydrogen production. Bioaugmentation in dark augmentative hydrogen production can be utilized. However, its efficacy depends on the ability of the chosen organism to grow in the respective culture media (Xie et al. 2015).

Biomethane: a suitable fuel for internal combustion engines, can be produced during anaerobic fermentation of organic materials. Having a higher octane rating than gasoline it also produces less carbon dioxide as compared to other conventional fossil fuels. Production of methane is a complex process involving hydrolysis, acidogenesis, acitogenesis, and methanation. During hydrolysis, hydrolase enzymes secreted by obligate and facultative anaerobes break down complex materials like cellulose and proteins to simple materials. Free sugars produced in this step are then utilized by other bacteria to produce organic acids. Several other methods like acid, alkali, thermal & ultrasonic treatments can be utilized. The amount of methane produced depends on the type of feedstock used. Corn stover, wheat straw, yard waste can be used. Banana stem can also be an efficient substrate for methane production due to its high lignin content (Dan et al. 2016).

Most microorganisms are incapable of utilising lignocellulose, which is an abundantly available carbon source and can be engineered to produce biofuel. Several strategies may be employed to counter the lignocellulose inhibitor effect. One of the major inhibitors include Furfural, which inhibits microbial growth by affecting sulfate assimilation. *E.coli* tolerance for furan can be increased by addition of cysteine and the expression of pntAB gene. In case of *S.cerevisiae*, tolerance can be increased by overexpression of G6P dehydrogenase. Various new target genes like thyA or groESL have been identified in *E.coli* which can affect furfural tolerance (Chandra et al. 2012).

12.5 Recent Approaches to Enhance Biofuel Production

The production of biohydrogen can be one of the major sources of biofuel. It is governed by several biotic & abiotic factors. While dark fermentation seems to be the most effective method, the major limitation lies in the inability to utilize the materials at their full potential. The formation of substances like acids and alcohol serves to be the principal reason for getting lower hydrogen yields. The biggest obstacle to be overcome in the biofuel sector is the technology scale-up and commercialization of third generation biofuel feedstock, despite the fact that these materials show promising potential for reducing reliance on fossil fuels for the production of energy. Third generation biofuels derived from oleaginous microorganisms; namely microalgae, activated sludge (bacteria), yeast and fungi are found to be sustainable considering the economic & environmental impact. Recently, enormous efforts to translate laboratory-scale technology into commercial operations had been undertaken through evaluating life cycle assessments and improving various oleaginous microbe growing methods. With a thorough awareness of the TEA's nutritional requirements, oleaginous bacteria are expected to be able to produce a precise corrective action that will increase profitability. Risk maximization in investments-reducing, absorbing the severe effects of scale-up production. By analyzing life cycle assessments and improving several cultivating techniques for oleaginous microorganisms, laboratory scale technology has been transformed into commercialized operations. There are now numerous investigations and developments into the use of nutrient-rich wastewaters as alternatives to water and fertilizer resources while growing microalgae. This interdisciplinary method in conjunction with wastewater treatment, where the microalgae biomass would absorb the eliminated nutrients (Chandra et al. 2012; Quiroz Arita et al. 2015; Gouveia et al. 2016). Utilizing these waste streams would come with high risks of contamination from potential pathogens, chemical compounds, and heavy metals that could be harmful to microalgae cells, even though municipal and agricultural wastewaters do contain high amounts of nutrients (nitrogen and phosphorus) required for microalgae growth. Despite the fact that municipal and agricultural wastewaters contain significant amounts of the nutrients (nitrogen and phosphorus) needed for microalgal growth, using these waste streams would come with a high risk of contamination from potential pathogens, chemical compounds, and heavy metals that could be harmful to microalgae cells.

Therefore, research into the ecology and pathophysiology of microalgae growing ponds is crucial to lowering the dangers of contamination for future commercialization, risk mitigation, and bioremediation tactics harmful to microalgae cells. In comparison to municipal primary sludge (8–17% w/w of dry weight), mixed wastewater sludge (16–25% w/w of dry weight), and paper and pulp wastewater sludge (28–33% w/w of dry weight), Zhang et al. reported that the use of municipal secondary sludge had resulted in the highest lipid accumulation, ranging from 30 to 38% w/w of dry weight (Fenton and Ó'hUallacháin 2012). The average obtainable lipid content from Chlorella protothecoides cultured with supplemented waste activated sludge hydrolyzate was 21.5u 1.44%, according to another study on using waste activated sludge in the form of hydrolyzate as a carbon source for growing the heterotrophic microalgae (Zhang et al. 2014).

12.5.1 *Tailoring Metabolic Engineering for Biofuel Production*

Despite the fact that metabolic engineering is a critical enabler technology for transforming microbes into efficient cell factories. Synthetic biologists and metabolic engineers rely on a wide range of strategies for exploring the pathway's limiting elements. Such metabolic systems techniques are helpful in the production of biofuels. Among the new approaches are high-throughput sequencing for genome decipherment, sensitive transcriptomics, proteomics, and metabolomics technologies, and breakthroughs in fluxomics and computational tools. Engineered microbial cells with surface-displayed bioactive proteins have the potential to be effective biocatalysts for the generation of biofuels and other goods. S. cerevisiae was the first yeast to acquire surface arming. Following that, other yeasts, including Pichia pastoris and Yarrowia lipolytica, were transformed using comparable display systems (Wen et al. 2013; Wang et al. 2008). Some clostridia manufacture cellulosomes, a combination of scaffolding proteins and several enzymes required for cellulose and hemicellulose hydrolysis. Tsai et al. genetically modified yeast to produce a mini-cellulosome. This cellulosome's scaffolding proteins were made up of three cohesin domains obtained from Clostridium thermocellum, Clostridium cellulolyticum, and Ruminococcus flavefaciens (Yuzbasheva et al. 2011; Hyeon et al. 2014).

In general, microalgae have received the utmost attention as oil producers for fuels. The marine diatom Thalassiosira pseudonana was modified to boost lipid production without impacting growth by knocking out a multifunctional lipase/phospholipase/acyltransferase. During exponential growth, the resulting strains possessed up to three times the lipid content of the wildtype. The marine diatom Phaeodactylum tricornutum's lipid composition has also been altered through metabolic engineering (Tsai et al. 2009). Microalgae can be divided into five main groups e.g. blue-green algae (Cyanobacteria), green algae, Diatoms, Red algae, and Brown algae (Radakovits et al. 2011). Cyanobacteria are most likely the most dominant for biofuel production. Many microalgae species may transesterify lipid to produce biodiesel. In terms of viscosity and density, biodiesel made from microalgae is comparable to petroleum diesel (Bhatt et al. 2014). These bacteria require the use of chemicals, whereas higher plants require herbicides and insecticides for increased productivity. Another characteristic of microalgae is the quantity of proteins and carbohydrates that can be processed to produce methanol and alcohol fuel (Schenk et al. 2008).

12.5.2 *Electroactive (EA) Biofilms: Microbial Electron Transfer*

The tendency of many bacteria to cluster together and form biofilms make them more resilient to environmental stress, shear forces and even antibiotics. The process of biofilm formation starts with reversible attachment by electrostatic and hydrodynamic interactions, followed by Van der Waals interaction which is an irreversible attachment between cells and the surface. Presence of Pilli, flagella, adhesins and Extracellular polymeric substance (EPS) facilitate biofilm formation and play a significant role in the adhesion to the surfaces. In bio electrochemical systems, increased electroactivity results in more efficiency in formation of products (McGinn et al. 2011). Electroactive bacteria are found in microbially-rich environments like sewage-sludge, domestic effluents, industrial effluents and also in soils, seawater, sediment like natural ecosystems. Bacterial strains like *Rhodoferax ferrireducens* have the electro-catalytic ability to exchange electrons with electrodes (Conners, et al. 2022).

In 2004, Rabaey et al. conducted the first study on the EA biofilm transplanting approach. The process involved multiple subsequent EA biofilm transplantation operations in which the biofilm on the MFC's anode was scraped off and utilized to inoculate a new MFC (Niessen et al. 2005). It has proven possible to create EA biofilms in batch, fed-batch, or continuous-flow modes, with fed-batch being the most used method so far. When the electrical output falls below a baseline, fed-batch is frequently carried out by replacing the entire solution or a portion of the solution or by adding a fresh dose of substrate. In the initial batches, it may be essential to introduce an inoculum repeatedly; following that, only new medium is supplied in subsequent replacements or replenishments. Higher performance has occasionally been achieved using a different method that prevents current decline by keeping substrate concentration above a predetermined threshold. The second MFC's coulombic efficiency had grown from 4 to 81%, and its power density had increased from 0.6 W m2 to 4.3 W m2. A biofilm gathered below a movable metal wharf in a harbor was exploited by Erable et al. While the inoculum gathered from nearby sediments produced current densities, this wild biofilm examined under chronoamperometry at 0.1 V versus SCE on a graphite electrode and with acetate as a fuel had obtained a power density of 2.5 A m^2. The Schroeder group has created what are referred to be secondary EA biofilms using the whole primary microbial anode created from wastewater to inoculate a new reactor, which falls under the same category. In a sterilized synthetic medium with only the primary bioanode injected and kept at a fixed potential, secondary EA biofilm was formed. Results from this process were quite reproducible (Rabaey et al. 2004; Liu et al. 2010).

The most widely used electrodes have all the qualities described above and are made of carbonaceous and metallic materials. To enhance interactions between the EA biofilm and the material surface, the materials utilized as anode electrodes must possess a number of specific properties (Fig. 12.5). According to a recent review, the most crucial properties are electrical conductivity, corrosion resistance, high mechanical strength, developed surface area, biocompatibility, environmental friendliness, and low cost. Among carbonaceous material, Carbon cloth ensures a large surface area and a comparatively high porosity while also displaying strong electrical conductivity, flexibility, and mechanical strength when constructing more intricate 3D structures. Carbon brush, Carbon rods, Carbon veil, Carbon paper all possess versatile characteristic features and utilized as anode material. surface morphology is crucial

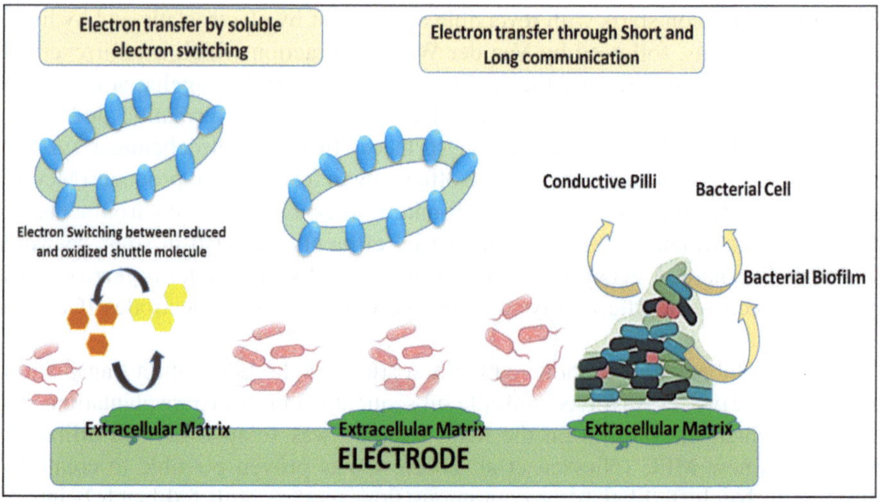

Fig. 12.5 Mechanistic insight on the interaction of Electrode and Biofilm for microbial electron transfer

to the development of the biofilm and the current generated. In order to expand the available surface area and improve the bio-interface between bacteria and electrode. When transitioning from a flat surface to a more complex 3-D surface, limitations brought on by diffusion transport phenomena of both products and reactants as well as relative pH gradients are also possible (Blanchet et al. 2014; Chen et al. 2011; Patil et al. 2011; Babauta et al. 2013). The electron transfer between individual electro-active bacteria in EA biofilm are of two types illustrated below.

12.6 Current Production Status and Future Perspectives

Metabolic engineering approaches are being used to produce biofuels and biofuel feedstocks in a wide range of microorganisms, including bacteria, yeasts, cyanobacteria, and microalgae. Commercialization of biofuels will eventually entail microbial cell factories that outperform wildtypes and some of the modified strains already available. Microorganisms will be able to use substrates that they could not previously use utilizing surface display technology. This skill will augment the economics of some biofuel production. Third generation biofuels generated from oleaginous microorganisms are currently recognised and accepted to be sustainable in terms of both environmental and global demand issues (George et al. 2014; Babauta et al. 2012). Third generation biofuels are becoming more appealing due to their readily available and accessible feedstock, which addresses the shortcomings of their predecessors. In general, all oleaginous microorganisms will grow a high biomass with a high lipid output in favour of carbon source replenishment. Because most microalgae may use either or both organic and inorganic carbon sources for growth, microalgae grown in the mixotrophic mode produce more biomass and lipids than autotrophic and heterotrophic modes. Nonetheless, exposing oleaginous microbes to these waste nutrient sources may result in contamination of collected biomass, hindering downstream operations. In this regard, preparation of obtained waste nutrient sources before to use as culture mediums for oleaginous bacteria is thought required in order to avoid untoward incidences that cost more on application (Liebeke et al. 2011; Renslow et al. 2013). Comprehensive studies of photosynthesis mechanisms for biofuel production under a variety of situations are critical because they provide particular information on the optimisation of phototrophic organism biomass growth. Although biofuels alone could not satisfy world demand, they possess an essential role in transportation. Based on existing and projected biofuel production developments, we believe that large-scale biofuel production is both necessary and viable in order to meet our future energy needs.

References

Abd-Alla MH, El-Enany A (2012) Production of acetone-butanol-ethanol from spoilage date palm (Phoenix dactylifera L.) fruits by mixed culture of Clostridium acetobutylicum and Bacillus subtilis. Biomass Bioenerg 42. https://doi.org/10.1016/j.biombioe.2012.03.006

Abdelaziz AE, Leite GB, Hallenbeck PC (2013) Addressing the challenges for sustainable production of algal biofuels: II. Harvesting and conversion to biofuels. Environ Technol 34(13–16):1807–1836. https://doi.org/10.1080/09593330.2013.831487

Atsumi S, Hanai T, Liao JC (2008) Non-fermentative pathways for synthesis of branched-chain higher alcohols as biofuels. Nature 451. https://doi.org/10.1038/nature06450

Babauta JT, Nguyen HD, Harrington TD, Renslow R, Beyenal H (2012) Biotechnol Bioeng 109:2651–2662

Babauta JT, Nguyen HD, Istanbullu O, Beyenal H (2013) Chemsuschem 6:1252–1261

Bhatia SK, Kim J, Song H-S, Kim HJ, Jeon J-M, Sathiyanarayanan G et al (2017a) Microbial biodiesel production from oil palm biomass hydrolysate using marine Rhodococcus sp. YHY01. Bioresour Technol 233:99–109. https://doi.org/10.1016/j.biortech.2017.02.061

Bhatia SK et al (2017b) Energy Convers Manage 148:1142–1156

Bhatt NC, Panwar A, Bisht TS, Tamta S (2014) Coupling of algal biofuel production with wastewater. Sci World J 10:210504. https://doi.org/10.1155/2014/210504

Blanchet E, Pecastaings S, Erable B, Roques C, Berge A (2014) Bioresour Technol 173:224–230

Bokinsky G, Peralta-Yahya PP, George A, Holmes BM, Steen EJ, Dietrich J et al (2011) Synthesis of three advanced biofuels from ionic liquid-pretreated switchgrass using engineered Escherichia coli. Proc Natl Acad Sci USA 108:19949–19954. https://doi.org/10.1073/pnas.1106958108

Chandra R, Takeuchi H, Hasegawa T (2012) Methane production from lignocellulosic agricultural crop wastes: a review in context to second generation of biofuel production. Renew Sust Energy Rev 16:1462–1476. https://doi.org/10.1016/j.rser.2011.11.035

Chen S, Hou H, Harnisch F, Patil SA, Carmona-Martinez AA, Agarwal S, Zhang Y, Sinha-Ray S, Yarin AL, Greiner A, Schroder U (2011) Energy Environ Sci 4:1417–1421

Conners EM et al (2022) Electroactive biofilms: how microbial electron transfer enables bioelectrochemical applications. J Ind Microbiol Biotechnol 49 (4):kuac012. https://doi.org/10.1093/jimb/kuac012

Dan J, Fang Z, Chin SX, Tian XF, Su TC (2016) Biohydrogen production from hydrolysates of selected tropical biomass wastes with Clostridium Butyricum. Sci Rep 6:27205. https://doi.org/10.1038/srep27205

Fenton O, Ó'hUallacháin D (2012) Agricultural nutrient surpluses as potential input sources to grow third generation biomass (microalgae): a review. Algal Res 1:49–56

George KW, Chen A, Jain A, Batth TS, Baidoo EE, Wang G, Adams PD, Petzold CJ, Keasling JD, Lee TS (2014) Correlation analysis of targeted proteins and metabolites to assess and engineer microbial isopentenol production. Biotechnol Bioeng 111:1648–1658

Ghirardi ML, Dubini A, Yu J, Maness P-C (2009) Photobiological hydrogen-producing systems. Chem Soc Rev 38:52–61

Gouveia L, Graça S, Sousa C, Ambrosano L, Ribeiro B, Botrel EP et al (2016) Microalgae biomass production using wastewater: treatment and costs: scale-up considerations. Algal Res 16:167–176

Hyeon JE, Kang DH, Han SO (2014) Signal amplification by a self-assembled biosensor system designed on the principle of dockerin–cohesin interactions in a cellulosome complex. Analyst 139:4790–4793

Kang KE, Chung DP, Kim Y, Chung B-W, Choi GW (2015) High-titer ethanol production from simultaneous saccharification and fermentation using a continuous feeding system. Fuel 145:18–24. https://doi.org/10.1016/j.fuel.2014.12.052

Komonkiat I, Cheirsilp B (2013) Felled oil palm trunk as a renewable source for biobutanol production by Clostridium spp. Bioresour Technol 146:200–207. https://doi.org/10.1016/j.biortech.2013.07.067

Kruse O, Rupprecht J, Mussgnug JH, Dismukes GC, Hankamer B (2005) Photosynthesis: a blueprint for solar energy capture and biohydrogen production technologies. Photochem Photobiol Sci 04:957–970

Liebeke M, Dorries K, Zuhlke D, Bernhardt J, Fuchs S, Pane-Farre J, Engelmann S, Volker U, Bode R, Dandekar T, Lindequist U, Hecker M, Lalk MA (2011) Metabolomics and proteomics study of the adaptation of Staphylococcus aureus to glucose starvation. Mol Biosyst 7:1241–1253

Lin PP, Rabe KS, Takasumi JL, Kadisch M, Arnold FH, Liao JC (2014) Isobutanol production at elevated temperatures in thermophilic Geobacillus thermoglucosidasius. Metab Eng 24:1–8. https://doi.org/10.1016/j.ymben.2014.03.006

Liu Y, Harnisch F, Fricke K, Schroder U, Climent V, Feliu JM (2010) Biosens Bioelectron 25:2167–2171

Lynd LR (1989) Production of ethanol from lignocellulosic materials using thermophilic bacteria: critical evaluation of potential and review. Lignocell Mater 24:1–52. https://doi.org/10.1007/BFb0007858.(Springer,Heidelberg)

McGinn PJ, Dickinson KE, Bhatti S, Frigon J, Guiot SR, O'Leary SJ (2011) Integration of microalgae cultivation with industrial waste remediation for biofuel and bioenergy production: opportunities and limitations. Photosynth Res 109:231e47

Minteer SD, Liaw BY, Cooney MJ (2007) Enzyme-based biofuel cells. Curr Opin Biotechnol 18(3):228–234. https://doi.org/10.1016/j.copbio.2007.03.007

Minty J, Singer M, Scholz S, Bae C, Ahn J, Foster C et al (2013) Design and characterization of synthetic fungal-bacterial consortia for direct production of isobutanol from cellulosic biomass. Proc Natl Acad Sci 110. https://doi.org/10.1073/pnas.1218447110

Niessen J, Schröder U, Harnisch F, Scholz F (2005) Gaining electricity from in situ oxidation of hydrogen produced by fermentative cellulose degradation. Lett Appl Microbiol 41(3):286–290. https://doi.org/10.1111/j.1472-765X.2005.01742.x

Nikolaidis P, Poullikkas A (2017) A comparative overview of hydrogen production processes. Renew Sust Energy Rev 67:597–611. https://doi.org/10.1016/j.rser.2016.09.044

Patil SA, Harnisch F, Koch C, Hübschmann T, Fetzer I, Carmona-Martinez AA, Müller S, Schroder U (2011) Bioresour Technol 102:9683–9690

Prince RC, Kheshgi HS (2005) The photobiological production of hydrogen: potential efficiency and effectiveness as a renewable fuel. Crit Rev Microbiol 31:19–31

Quiroz Arita CE, Peebles C, Bradley TH (2015) Scalability of combining microalgae-based biofuels with wastewater facilities: a review. Algal Res 9:160–169

Rabaey K, Boon N, Siciliano SD, Verhaege M, Verstraete W (2004) Appl Environ Microbiol 70:5373–5382

Radakovits R, Eduafo PM, Posewitz MC (2011) Genetic engineering of fatty acid chain length in Phaeodactylum tricornutum. Metab Eng 13:89–95

Rafa N, Ahmed SF, Badruddin IA, Mofijur M, Kamangar S (2021) Front Energy Res, Sec Bioenergy Biofuels 9. https://doi.org/10.3389/fenrg.2021.749968

Razeghifard R (2013) Algal biofuels. Photosynth Res. 117(1–3):207–219. https://doi.org/10.1007/s11120-013-9828-z. (Epub 2013 Apr 21, PMID: 23605290)

Renslow RS, Babauta JT, Majors PD, Beyenal H (2013) Energy Environ Sci 6:595–607

Rodionova MV, Poudyal RS, Tiwari I, Voloshin RA, Zharmukhamedov SK, Nam HG, Zayadan BK, Bruce BD, Hou HJM, Allakhverdiev SI (2017) Biofuel production: challenges and opportunities. Int J Hydrogen Energy 42(12):8450–8461. https://doi.org/10.1016/j.ijhydene.2016.11.125

Santoro C, Arbizzani C, Erable B, Ieropoulos I (2017) Microbial fuel cells: from fundamentals to applications. A Rev J Power Sources 356:225–244. https://doi.org/10.1016/j.jpowsour.2017.03.109

Schenk PM, Thomas-Hall SR, Stephens E, Marx UC, Mussgnug JH, Posten C et al (2008) Second generation biofuels: high efficiency microalgae for biodiesel production. BioEnergy Res 01:20–43

Sun LH, Wang XD, Dai JY, Xiu ZL (2009) Microbial production of 2,3-butanediol from Jerusalem artichoke tubers by Klebsiella pneumoniae. Appl Microbiol Biotechnol 82:847–852. https://doi.org/10.1007/s00253-008-1823-5

Tran NH, Bartlett JR, Kannangara GSK, Milev AS, Volk H, Wilson MA (2010) Catalytic upgrading of biorefinery oil from micro-algae. Fuel 189:265–274

Tsai SL, Oh J, Singh S, Chen R, Chen W (2009) Functional assembly of minicellulosomes on the Saccharomyces cerevisiae cell surface for cellulose hydrolysis and ethanol production. Appl Environ Microbiol 75:6087–6093

Tsuji M, Goshima T, Matsushika A, Kudoh S, Hoshino T (2013) Direct ethanol fermentation from lignocellulosic biomass by Antarctic basidiomycetous yeast Mrakia blollopis under a low temperature condition. Cryobiology 67:241–243. https://doi.org/10.1016/j.cryobiol.2013.06.003

Wang Q, Li L, Chen M, Qi Q, Wang PG (2008) Construction of a novel Pichia pastoris cell-surface display system based on the cell wall protein Pir1. Curr Microbiol 56:352–357

Wen Q, Chen Z, Li P, Duan R, Ren N (2013) Lipid production for biofuels from hydrolyzate of waste activated sludge by heterotrophic Chlorella protothecoides. Bioresour Technol 143:695–698

Xie G-J, Liu B-F, Wang R-Q, Ding J, Ren H-Y, Zhou X et al (2015) Bioaggregate of photo-fermentative bacteria for enhancing continuous hydrogen production in a sequencing batch photobioreactor. Sci Rep 5:16174. https://doi.org/10.1038/srep16174

Yuzbasheva EY, Yuzbashev TV, Laptev IA, Konstantinova TK, Sineoky SP (2011) Efficient cell surface display of Lip2 lipase using C-domains of glycosylphosphatidylinositol-anchored cell wall proteins of Yarrowia lipolytica. Appl Microbiol Biotechnol 91:645–654

Zhang C, Li W, Wang D, Guo X, Ma L, Xiao D (2016) Production of 2,3-butanediol by Enterobacter cloacae from corncob-derived xylose. Turkish J Biol 40:856–865. https://doi.org/10.3906/biy-1506-66

Zhang X, Yan S, Tyagi RD, Surampalli RY, Valéro JR (2014) Wastewater sludge as raw material for microbial oils production. Appl Energy 135:192–201

Zuroff TR, Xiques SB, Curtis WR (2013) Consortia-mediated bioprocessing of cellulose to ethanol with a symbiotic Clostridium phytofermentans/yeast co-culture. Biotechnol Biofuels 6. https://doi.org/10.1186/1754-6834-6-59

Chapter 13
Utility of Endophytes for the Enhancement of Biofuel Production

Susmita Paul, Birson Ingti, and S. R. Joshi

Abstract The need for biofuel is growing with the tenacity of reducing green-house gaseous emission leading to deviations in climate and global warming effect. For worldwide constancy, commercial affluence, and quality of life a clean energy sources like microbial fuels could be an alternative substitute to that of fossil fuels. These biofuels are safe, renewable and reduce green-house gas emissions. Microorganisms in such platform is useful as they do not involve land for cultivation, do not strive for food production and these can be cultured easily under laboratory conditions. Altogether, plants in nature hold endophytes and it is essential to comprehend their biodiversity and role in the host plants. These endophytic microbes' mainly fungal endophytes are considered, to be a treasure of structurally and biologically exceptional natural products. Most of these compounds are those formed by their corresponding host plants as well and henceforth for the construction of equivalent compounds, endophytes from the host plants are valuable. Endophytic fungi possess a suitable lipid matrix at high concentrations and volatile organic compounds containing resemblance with conventional diesel fuel that make them promising sources for next generation biofuels.

Keywords Biofuel · Lipid matrix · Microbial fuel · Endophytes

S. Paul · S. R. Joshi (✉)
Microbiology Laboratory, Department of Biotechnology and Bioinformatics, North-Eastern Hill University, Shillong, Meghalaya 793022, India
e-mail: srjoshi2006@yahoo.co.in

S. Paul
e-mail: spaul@rgu.ac

S. Paul · B. Ingti
Royal School of Biosciences, The Assam Royal Global University, Guwahati 781035, India
e-mail: bingti@rgu.ac

M. Shah and D. Deka (eds.), *Emerging Sustainable Technologies for Biofuel Production*, Environmental Science and Engineering, https://doi.org/10.1007/978-3-031-52167-6_13

13.1 Introduction

Globally fossil fuel (coal, diesel, gasoline, oil natural gas etc.) were drastically used for the generation of energy and accounts for more than 80% of the total primary energy supply. However, burning of these fuels generate highly toxic gases such as sulfur dioxide, nitrogen oxide etc. that adversely affects human health and, also emit greenhouse gases (CO, CO_2) responsible for climate change (Perera 2017) Considering the tremendous rise in global population, the only way to ensure the continued existence of life on our planet in the modern day is to transform society into a pleasant, enjoyable place to live. Therefore, biofuels have now been proposed as a potential alternative for fossil fuels in order to achieve sustainability goals because they are a safe, renewable, and possibly useful liquid that can help to cut greenhouse gas emissions. Biofuels are renewable energy that can be produced from animal fats, edible vegetable oil, organic wastes, etc. through the process of transesterification. According to the biomass sources used in its production, biofuel is primarily divided into four types: first, second, third, and fourth generation. First-generation biofuels were made from plant oil, animal fats, etc.; whereas second-generation biofuels were made from non-food feedstocks such organic wastes, etc. The third generation of biofuels, however, is made from microbial cells (oleaginous microorganisms), and lately the fourth generation, which makes use of genetically modified feedstock (Singh et al. 2020; Hossain and Mahlia 2020; Koreti et al. 2022). The generation of biofuels from microbial lipid, such as endophytic microbes, has recently attracted increasing attention on a global scale due to the possibility to manufacture numerous value-added products (such citric acid, succinic acid, butanol, ethanol, etc.). Endophytes belonging to different taxa such as bacteria, fungi etc. are hidden companions of plants living mutually beneficial life inside the plants and known to posse's higher lipid content (Carsanba et al. 2018; Chilakamarry et al. 2022; Abdalla and Matasyoh 2014).

In this chapter, we therefore attempt to summarise the potential of endophytes that can be employed for biofuel production, with a focus on how endophytes can improve biofuel production.

13.2 Biofuels and Its Potential Sources

Due to diminishing oil supplies, energy security concerns, and growing environmental worries caused by rising climate change and greenhouse gas emissions, biofuels have gained major relevance in recent years. Biofuels are any fuel created and derived from organic material, such as plants and their by-products, agricultural crops and by-products that can serve as a suitable replacement for fuel derived from petroleum (Demirbas et al. 2016; Mahapatra et al. 2021). It contains high level of oxygen ranging from 10 to 45% compared to petroleum products, and significantly differs from other petroleum feedstocks. They are divided into three categories based on the type of biomass used in their production: solid, liquid, and gaseous biofuels. Many microorganisms can act as potential sources for biofuels production. Some of these microbes are discussed as follows.

13.2.1 Oleaginous Microbes as Potential Source of Biofuel

The term "oleaginous microorganisms" (OMs) refers to any microorganism that possess higher lipid content (more than 20%), such as *Arthrobacter, Bacillus, Streptomyces, Chlorella pyrenoidosa, Rhodococcus, Cryptococcus curvatus* etc. (Meng et al. 2009; Maa et al. 2018; Chintagunta et al. 2021). These organisms can acquire lipid at a rate greater than 20% of their biomass, and may even reach up to 70% or more, when under stress conditions, such as rich carbon but low nitrogen sources. The lipid composition is similar to vegetative oils, and transesterification can be used to produce microbial oil from a range of renewable resources. In 1985, the first commercially available batch of microbial oil was produced utilising the fungus *Mucor circinelloides*. However, there has not been much study on its possible uses in biofuel production. The three main groups of microorganisms that can accumulate substantial levels of lipids in their cellular counterparts and be utilised to make biodiesel are microalgae, bacteria, and fungi (Ma et al. 2008; Rossi et al. 2011; Cho and Park 2018).

Oleaginous microalgae are regarded as reliable raw materials for the synthesis of sustainable biofuels as it has the potential to produce enough feedstock for oil production on a worldwide scale. Since biodiesel yields from this microbial oil are much higher than those from plant feedstock, research into microalgae-based biodiesel is currently gaining relevance. Microalgae normally have a lipid content of 20–50% of their dry weight, but under certain circumstances, it can exceed 90% (Chisti 2007; Papanikolaou and Aggelis 2011; Zuccaro et al. 2020). This organism transforms CO_2 into sugars in the chloroplast, which are then further processed to produce acetyl-CoA, a precursor to the production of fatty acids and employs the de novo pathway to manufacture lipids. Light energy was first used to produce microalgae for the purpose of making biodiesel. However, due to its numerous advantages, including affordability, simplicity of cultivation, and low maintenance, it has been switched to a heterotrophic form of cultivation (Perez-Garcia et al. 2011; Guldhe et al. 2017). Because this heterotrophic method of cultivation often uses the pricy carbon source glucose, microalgae must be grown utilising other renewable sources to cut costs (Mohamed et al. 2011; Chen and Jiang 2017).

Apart from algae a few bacterial species, including *Rhodococcus, Streptomyces,* and *Mycobacterium,* are also known to accumulate significant amounts of lipids. Although their use is less common than that of microalgae and yeast for the production of biodiesel, these oleaginous bacteria are known to be a possible source of triacylglycerols (TAGs). Other significant oleaginous bacterial genera include *Arthrobacter* species, *Gordonia* sp., *Acinetobacter* sp. etc. However, due to its increased lipid content, the *Rhodococcus* strain (70%) has received the most attention out of all the oleaginous bacteria that have been identified to date. Many of these bacteria generate a sizable quantity of lipids from CO_2, which can then be utilised to make biodiesel and reduce CO_2 emissions. Many of these bacteria can also produce neutral lipids called polyhydroxyl alkanoic acids, which are produced as tiny droplets in the bacterial cytoplasm and function as intracellular carbon and energy storage molecules. These bacteria offer a few characteristics that make them an excellent feedstock for the production of biodiesel, including a rapid growth rate, ease of cultivation, and the ability to grow on a range of substrates, the potential for genetic modification, and the capacity to collect carbon dioxide. As a result, it could soon serve as a substantial resource to produce biofuel (Cho and Park 2018; Zuccaro et al. 2020; Koreti et al. 2022).

Additionally, yeasts and other fungi are also extensively used in the production of lipids. About 30 yeast species have been identified as oleaginous and used in the production of biofuels, including those from the genera *Rhodosporidium, Candida, Lipomyces, Cryptococcus, Rhodotorula, Yarrowia, and Trichosporon* (Luis et al. 2016). They may gather lipids up to 80% of their dry biomass when nutrients are scarce. Additionally, other non-oleaginous yeast species such as *Saccharomyces cerevisiae,* were also used and genetically modified to produce lipids. Apart from yeast, oleaginous filamentous fungi are potential microbes for the manufacture of biodiesel and have a number of advantages, such as the presence of linolenic acid (GLA) (Subramanian et al. 2010; Subhash and Mohan 2011). These fungi have the capacity to store up to 80% of their biomass as lipid, and their lipids typically include more unsaturated fatty acids (Economou et al. 2012; Papanikolaou and Aggelis 2019). Therefore, oleaginous fungi were frequently utilised for the generation of biodiesel due to their high lipid content and capacity to thrive on renewable carbon sources such as molasses, wastewater, sewage sludge, agricultural waste, etc. An oleaginous fungus called *Cunninghamella echinulate,* for instance, produced 46.6% of the total lipids when cultivated on orange peel and glucose. Simillarly when cultivated on glucose and potassium nitrate, another fungus called *Mortierellaalpina* produced a significant amount of lipids (31.1%) with a content of arachidonic acid that was 60.4%.

13.2.2 Endophytes as Potential Source of Biofuels

Endophytes are microbes (Table 13.1) that reside latently or actively inside the plant tissue (Fig. 13.1), colonizing locally or systematically, without causing obvious harm to the plant. In some situations, endophytes can even improve plant development and lessen illness symptoms brought on by various plant pathogens. De Bary initially used the term "endophyte" in *Lolium temulentum,* which was first characterized in 1866. Endophytes stay in close proximity to their hosts for at least a portion of their life cycle, and during this time, none of the parties involved are clearly harmed. Endophytes receive shelter and nutrition from host plants while supplying/ facilitating nutrient uptake and safeguarding the plant from biotic and abiotic stresses and pests. These endophytes are known to synthesize many secondary metabolites such as alkaloids, chinones, benzo-pyranones, phenolic acids, flavonoids, steroids, quinones, terpenoids, xanthones, tetralones and other bioactive secondary metabolites with particular structures. These bioactive metabolites can be used as antibiotics, antiparasitics, antioxidants, anticancer agents, and many more (Stone et al. 2000; De Bary 1866; Arora and Ramawat 2017).

13.3 Classification of Endophytes

Many endophytes reported so far are fungi, bacteria or actinomycetes). Endophytes are primarily categorized based on the microorganism as endophytic fungus, endophytic bacterium, and endophytic actinomyces (Strobel et al. 2004).

13.3.1 Endophytic Fungi

An endophytic fungus can be found in the tissues of plants, such as their stems, leaves, and roots, in the mycelial form. Researchers have found that they are connected to the above-ground tissue of liverwort, hornwort, mosses, lycophytes, equisetopsids, fern, and seed plants everywhere from the arctic to the tropics, from agricultural areas to the most diverse tropical forests (Arnold 2007). These endophytic fungus associations vary from that of the mycorrhizae as they lack a localized interface of specialized hyphae, synchronized plant-fungus development, and nutrient transfer benefits for the plants. Endophytes may indirectly benefit plants by enhancing their resilience to herbivores, pathogens, stress, or other undiscovered mechanisms (Schulz and Boyle 2005). Additionally, endophytic fungi are present in marine plants. One example is the marine obligate fungus *Ascochyta salicorniae,* which is found on green algae (Strobel et al. 1997; Osterhage et al. 2000). Endophytic fungi are frequently used to protect plants from disease and insect pests as well as to promote plant growth.

13.3.2 Endophytic Bacteria

Endophytic bacteria on a certain plant host are divided into multiple genera and species rather than being categorized as a single species. They typically reside in the plant's vascular tissue and intercellular space. More than 129 endophytic bacteria were isolated from different plant species which includes both Gram-negative and Gram-positive species representing over 54 genera such as *Acidovorax, Acinetobacter Actinomyces, Aeromonas, Afipia, Agrobacterium, Agromonas, Alcaligenes, Alcanivorax, Allorhizobium, Alteromonas, Aminobacter, Aquaspirillum* etc.

Table 13.1 List of common endophytes found in different plant species (Arora and Ramawat 2017)

Sl. No.	Endophytes	Types of endophytes	Plant host
1	*Acremonium* sp.	Fungal endophytes	*Taxus chinensis, Huperzia serrata*
2	*Aspergillus* sp.	Fungal endophytes	*Datura stramonium, Moringa olifera, Prosopis chilensis*
3	*Cladosporium* sp. *C. herbarum*	Fungal endophytes	*Opuntia ficus indica, Cinnamomum camphora, Lycopersicum esculentum* Mill., *Triticum aestivum*
4	*Colletotrichum* sp. *C. gloeosporiodes*	Fungal endophytes	*Triticum aestivum*, Citrus plants *Cinnamomum camphora, Pasania edulis, Ginkgo biloba* L., *Tectona grandis and Samanea saman Huperzia serrata, Cinnamomum camphora, Lycopersicum esculentum* Mill
5	*Curvularia* sp.	Fungal endophytes	*Datura stramonium, Moringa olifera*
6	*Penicillium* sp.	Fungal endophytes	*Lycopersicum esculentum* Mill., *Huperzia serrata*
7	*Phyllosticta* sp.	Fungal endophytes	*Citrus* sp., *Pasania edulis, Coffea Arabica, Quercus variabilis, Centella asiatica, Panax quinquefolium, Ginkgo biloba* L.
8	*Phomopsis* sp.	Fungal endophytes	*Pasania edulis, Ginkgo biloba* L. *Tectona grandis and Samanea saman, Taxus chinensis*
9	*Stemphylium globuliferum*	Fungal endophytes	*Avicennia marina*
10	*Bacillus megatarium*	Bacterial endophytes	*Medicago satavia*
11	*B. thuringiensis, B. subtilis subsp subtilis*	Bacterial endophytes	*Musa* sp.
12	*Burkholderia cepacia*	Bacterial endophytes	*Lupinus luteus*
13	*Enterobacter asburiae*	Bacterial endophytes	*Ipomoea batatas*
14	*Citrobacter*	Bacterial endophytes	*Musa* sp.
15	*Microbacterium* sp.	Bacterial endophytes	*Pogonatherum paniceum*
16	*Pseudomonas saponiphilia*	Bacterial endophytes	*Dendrobium candidum*
17	*Rhizobium radiobacter*	Bacterial endophytes	*Daucus carota*

(continued)

Table 13.1 (continued)

Sl. No.	Endophytes	Types of endophytes	Plant host
18	*Staphylococcus saprophyticus*	Bacterial endophytes	*Carrot*
19	*Micrococcus*	Bacterial endophytes	*Oryza sativa*

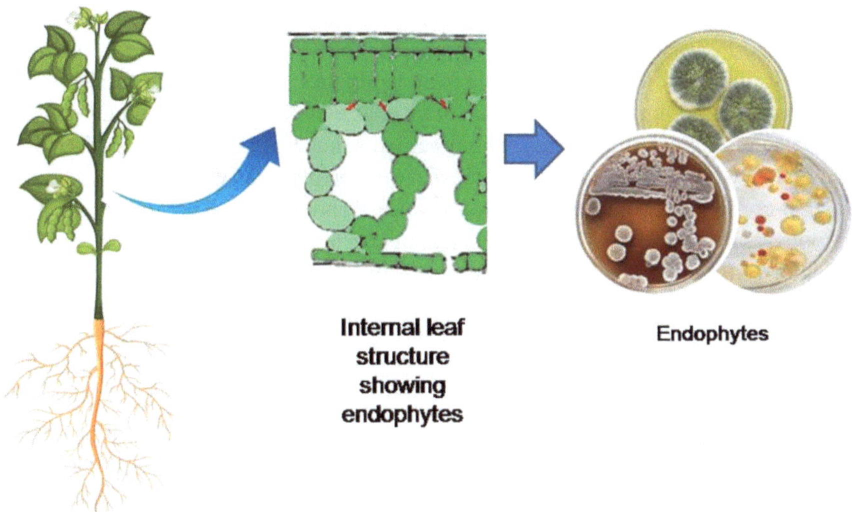

Fig. 13.1 Host plant showing endophytes in the plant tissue

However, the dominant bacterial taxa are members of the Enterobacteriaceae and the erstwhile *Pseudomonas* group (Lodewyckx et al. 2002; Gaiero et al. 2013).

13.3.3 Endophytic Actinomyces

Actinomycetes and other plant-associated bacteria operate as growth regulators. The most endophytic actinomycete that has been isolated is *Streptomyces*, which frequently colonizes the roots, stems, and leaves of plants (Wang and Paul 2008). Medicinal plants are the primary source of isolating endophytic actinomyces, which can produce secondary metabolites of substantial importance. The majority of endophytic actinomycetes latently reside in the tissues of asymptomatic plants without having any negative consequences on the host. It has been demonstrated that endophytic actinomycetes help plants grow and develop while also reducing disease symptoms and other environmental challenges. The effectiveness of endophytes in farm systems was assessed by Stoyke and Currah in 1991 and concluded that the endophyte defends the plant against many insect pests by generating alkaloids, some of which are poisonous to grazing animals (Hasegawa et al. 2006).

13.4 Production of Biofuels

One of the most promising replacements amongst the renewable resources also being eco-friendly energy source is biodiesel. Perspective to this, microorganisms have the potential to be used in biofuel production. Oleaginous endophytic microorganisms such as fungi, particularly are capable in accumulating high amount of lipids in their biomass and are appropriate for fermentation at a large-scale, principally attractive for biotechnological approaches. The organisms that are found inside surface sterilized plant tissues or drawn out from internal portion of plants which have no apparent adverse consequence on the host plants are the endophytes. The commonly found endophytes inside the tissues of the host plants are recognized to mimic physiological behavior of the host counting production of metabolites that comprises production of primary as well as secondary metabolites. Production oil and fat are amongst such physiological resemblances. As a result of depletion of nitrogen source or other nutrients, growth is limited sometimes and microorganisms like yeast, fungi and algae are then set in motion to amass lipid as a storage material. The evaluation of the effect of different media constituents in enhancing growth and lipid production by fungal isolates, may also be carried out (Fig. 13.2). Lately, several endophytic fungi have been known to produce volatile organic compounds (VOCs) having properties comparable to fossil fuels, called myco-diesel (Fig. 13.2), while growing on lignocellulosic plant and agricultural residues. Endophytes are plant symbionts, this statement suggests that some may be, are capable of producing lignocellulolytic enzymes, making them capable of both deconstructing lignocellulose and converting it into myco-diesel.

It has been observed that, in the current years, biodiesel production is growing both in developed along with the developing countries, leading in a few cases to thoughtful adverse impacts on ecosystems and, leading to conditions of possible conflict of fuel versus food security owing to use of land for planting of biofuel plants (Ajanovic 2011). One approach which is likely to overcome these imperative but often contradictory issues is to look for possible biodiesel precursors in organisms other than plants. There are certain advantages of microorganisms over plants and animals that can gather high levels of lipids because they do not necessitate land for cultivation and do not compete in food production. Subsequently, microorganisms for instance endophytes are found to express similarity with respect to physiological behavior of the host counting production of metabolites (Bhagobaty and Joshi 2011a), fungal endophytes were preferred for the current study. Moreover, they can be shaped in large scale fermentation developments, which make them a potentially more viable source for biofuel production (Janßen and Steinbüchel 2014).

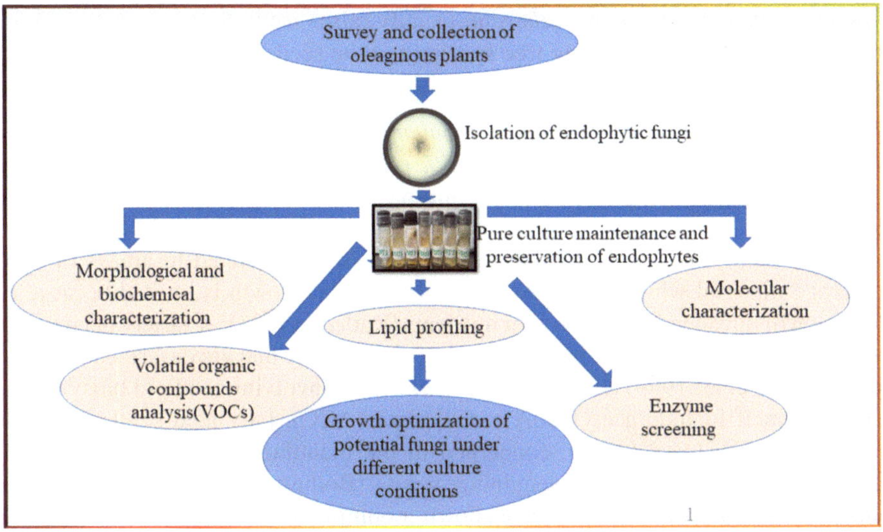

Fig. 13.2 Schematic representation of biofuel production

There are a few plants having history of being characterized as biodiesel plants. These were selected for isolation of endophytic fungi. Following are a few such plants including *Jatropha curcas L: Jatropha curcas* has been getting great consideration due to its enormous role in production of biodiesel as an environment friendly fuel, biodegradable, renewable andnon-hazardous in nature compared to petro-diesel, excluding some oncogenic compounds found in oil cake (Gmünder et al. 2012).

Pongamia pinnata: Oil of *Pongamia pinnata* is considered as a non-edible oil based in India with the oil content of the kernel observed to be about 30–40% (Halder et al. 2014).

Mesua ferrea L: Seed oil of *Mesua ferrea* is observed to be the potential feedstocks for biofuel production. This classic tree is discovered to grow naturally in north-eastern part of the India. An ordinary 15–20-years-old tree was observed to harvest in an average of 30 kg of seeds. In addition, it is assessed that 5500 tons of *Mesua ferrea* seeds are yearly available from Lakhimpur and sibsagar districts of Assam (Bora et al. 2014).

Sapindus mukorossi: *Sapindus mukorossi* a deciduous tree that is extensively full-grown from upper reaches of Indo-Gangetic plains, Shivaliks and sub-Himalayan tracts, at altitudes ranging from 200 to 1500 m (Upadhyay and Singh 2012). The oil contented in the kernel of the soapnut seeds was observed to be about 42.7 weight (%) percent (Chen et al. 2012).

Terminalia belerica: Seeds of *Terminalia belerica* are observed as a substitute bio-resource for production of biodiesel. The oil production of *T. belerica* seed is around 31% (dry weight basis) (Sarin 2012).

Cascabela thevetia: *Cascabela thevetia* usually raises naturally in north-east India, is a non-edible oil-seed bearing tree species and the seeds of *C. thevetia* are known for biodiesel production (Sut et al. 2016).

Ricinus communis: Seeds of *Ricinus communis* are recognized as possible source of biodiesel which contain around 30–35% oil that can be administered to yield a high-quality biodiesel fuel operational in typical diesel engines (Okechukwu et al. 2015).

Yeast, fungi, and algae, all these microorganisms are stated to amass lipid as a storage material while development in terms of the growth is limited because of depletion of nitrogen source, or other nutrients (Ratledge 1991). A variety of enzymes potentially comprised in the biochemistry of oleaginicity that are described amongst fungi (Wynn et al. 1999) (Fig. 13.1). Such enzymes involving the lipid biosynthetic pathway itself include acetyl-CoA carboxylase (ACC), fatty acid synthase (FAS) and ATP citrate lyase (ACL), concerned in the manufacture of cytosolic acetyl-CoA for lipid production in oleaginous yeasts (Botham and Ratledge 1979). As anover-alloutline, a few results have shown mounting attention in bio-fuels, particularly from microbial sources. Exploration in this regard is on for other microbes that have the genetic machinery for lipid production (Schirmer et al. 2010). Biofuel has been shown to be a very attractive substitute to petroleum derivatives, and researchers have been devoting substantial effort to gain ample technological expertise for operative biodiesel production (Demirbas 2009). There are reports presented on plants having potential for biodiesel generation, while some studies have been made on microbes related with such plants (Sivaramakrishnan and Ravikumar 2011). Amongst the heterotrophic microorganisms, oleaginous fungi are progressively been conveyed as good triacylglycerol producers (Patel et al. 2018).

There are several tailbacks related with the first- as well as the second-generation biofuels as for example greenhouse gas (GHG) discharges, contamination in air, biodiversity, endowments, etc. The second-age biofuel structures incline to eliminate supportive feedstocks from the woody or fibrous biomass. The creation of biofuels from fungi has a wider number of target sites than the first- and second-era biofuels since the biofuels are directly distributed from horticulture leftovers and other home trash like grounds. While biofuels created from a little amount of biomass on arable land or in waste areas could provide an improved and more significant benefit to the environment, ethanol produced from biomass is sustainable and intentionally strengthens the economy and ecology. The constraint, though, comes from plain land usage, which has an impact on eutrophication and greenhouse dioxide emissions. Because it is a sustainable energy source, third generation biodiesel is one of the samples being used to identify practical solutions to deal with the issues with fossil-based fuels. They can be made from a variety of sources, such as mushrooms, algae, and other things, and they are a substitute fuel that does not hurt the environment. The ability of fungus to collect intracellular lipids makes it possible to produce biodiesel. Advanced methods are certainly desirable for the quick commercialization of fungi-derived biofuels (Raven et al. 2019).

13.5 Biofuel Production Potential Worldwide

In the past few years, biofuel is coming into view as an effective fuel substitute to overcome the shortage of fuel (Strobel 2003). Although, biofuels can be utilized in their pure form, they are generally used as a blend to advance efficiency of other fuels, for example, ethanol is mixed with gasoline, and biodiesel is blended with petroleum diesel. Nearly all gasoline we pump now a days is E10, that is 10% ethanol and 90% gasoline (Sovacool and Brown 2007). About 90% biodiesel is derivative of plants (Kywe and Oo 2009).

There are several plants such as *Cascabela thevetia, Terminalia belerica Pongamia pinnata, Sapindus mukorossi, Mesua ferrea, and Ricinus communis* which are being studied for search of their prospective as biodiesel plants. Nevertheless, utilizing such energy crops for bio-fuels necessitates land for cultivation and hence to overcome such issues, possible sources for bio-fuels are being patterned in organisms other than plants. Endophytic fungi may perhaps be the probable source here as they live in internal part of the plant without producing any apparent undesirable consequence on the host. As an alternative, they protect the host plant from pests and certain ailments (Kumar et al. 2011).

The endophytes explored in a study have been described to yield numerous VOCs with potential use as myco-diesel, grounded on high energy density and other vital chemical properties to be used as a fuel. GC-MS analysis identified a wide spectrum of myco-diesel VOCs, principally monoterpenes or sesquiterpenes, that comprises high energy density and near-zero oxygen content, and are therefore considered good potential drop-in fuel compounds. The capacity to both deconstructs plant cell walls and to yield myco-diesel show that *Hypoxylon* and *Daldinia* were described to be promising new genera to discover and its consolidated bioprocessing (Wu et al. 2017).

As it is known that endophytic microorganisms are predominant in the tissues of the host plants and are identified to mimic physiological behavior of the host counting production of metabolites (Bhagobaty and Joshi 2011b). These comprises synthesis of primary and secondary metabolites, and oil and fat production are among such physiological similarities. But all microorganisms cannot be anticipated as sufficient sources of oils and fats. Though, microorganisms perpetually hold lipids for the vital presentation of membranes and membranous structures. Organisms which can accumulate about 20% or more of their biomass as lipid have been selected as—"oleaginous" in keeping with oil-bearing plants that are likewise named. Of approximately 600 yeast species, around 25 or so can accumulate more than 20% lipid and of the 60,000 fungal species fewer than 50 accumulate more than around 25% lipid (Akpinar-Bayizit 2014). Conferring to the phase of development of fungi, age and state under which fungus is cultured, proportions of all the lipid elements may vary. Maximum fungal species contain, in order of plentifulness, oleic acid (C18), palmitic acid (C16) and linoleic acid (C18) as the key acids, with stearic acid (C18), linolenic acid (C18) and palmitoleic acid (C16) as the trivial ones (Moser 2010).

Lipids synthesized from filamentous fungi show substantial capacity for production of biofuel. As it has been observed that fungi yield diverse mixtures of gas-phase, carbon-based composites called volatile organic compounds (VOCs) and due to their small size, they are able to diffuse through the atmosphere and soils. Volatile organic compounds (VOCs) are carbon-based solids and liquids that readily enter the gas phase by vaporizing at 0.01 kPa at a temperature of about 20 °C. Nearly all are lipid soluble and consequently, have low water solubility. Meanwhileit is observed that VOCs can disperse across the atmosphere and soil, they are ideal "infochemicals" (Morath et al. 2012). Around 250 VOCs have been documented from fungi where they arise as mixtures of simple hydrocarbons, heterocycles, aldehydes, ketones, alcohols, phenols, thioalcohols, thioesters, and their byproducts, counting, amongst others, benzene derivatives and cyclo-hexanes. The structure of VOCs shaped by microorganisms varies with the microbial species and its growth environment. Some microorganisms yield numerous VOCs when grown under microaerophilic conditions, the utmost interesting of which are the acetyl esters of alkanes counting those of hexane, heptane, octane and decane (Strobel et al. 2008). These alkanes make up the backbone constituents of diesel fuel and as such make these organisms of attention as myco-diesel producers. Terpenoids too, signify a class of metabolites with over 40,000 identified structures, that are shaped as VOCs from various fungal species (Wawrzyn et al. 2012) and this chemical diversity is produced by terpene synthase enzymes. This compound is a commercially imperative component of oil utilized in pharmaceutical applications and has been discovered as a possible biofuel additive (Shaw et al. 2015).

13.5.1 Enzymes and Their Role in Biofuel Production Potential

In a recent study of endophytic fungi which was reported to be a prospective source of producers of industrial enzymes, it was stated that Endophytic fungus produce a wide range of secondary metabolites and enzymes that illustrate several biological processes. Their latest research has demonstrated the potential use of endophytic fungus-produced enzymes in the food, pharmaceutical, agricultural, and biofuel industries. Additionally, the distinctive qualities of these enzymes were covered. It was also claimed that endophytic fungi that produce ligninolytic enzymes may be able to be used in lignocellulosic biorefineries for biotechnological reasons. The analysis described above outlined the issues, potential outcomes, and global market for industrially important enzymes. Nevertheless, the commercialization of enzymes from fungal endophytes for industrial purposes is yet to be discovered. It was proposed through the study that endophytic fungi can yield numerous enzymes and may turn out to be a fresh source for upscaling the synthesis of enzymes of industrial use (Bhadra et al. 2022). In one more current study of endophytes by Mattoo and Nonzom (2022), a new line on lignin valorization was described. It was stated that due to the structural complexity and variability of lignin, which hinders the synthesis of necessary and precise results, lignin valorization is currently facing complicated challenges. While increased reaction rates and shorter time requirements are potential benefits of thermochemical processes, their high energy requirements and excessive chemical use degrade environmental quality. Lignin needs to be valued by turning it into biofuels and chemicals using environmentally acceptable methods, focusing on the aromatic reserve. The greatest number of drawbacks of thermochemical approaches may be solved by using biological agents in the valorization of lignin. However, we are still having trouble separating the numerous mechanisms that microorganisms use to break down lignin. Microbes from various bases must be investigated for their delignification effectiveness in order to perform competent valorization. Due to their remarkable ability to convert lignin into valuable chemicals, endophytes are seen as a novel and potential source for the valorization of lignin. Their ability to impersonate essential host metabolites is one of the distinctive characteristics of endophytes. As a result, plants and vegetables are used as the primary sources for the manufacture of biofuel, and bacteria that are endophytic with these plants imitate their hosts by producing lipids. Our reliance on these biodiesel plants has decreased as a result of substituting to the production of endophytic lipid, aiding in their preservation. There is much to learn about endophytes' translational capacities because they have only been sparingly investigated for their ability to degrade lignin. It has been determined that these organisms can degrade lignin more quickly than white rot fungus. As a result, many pathways and the processes underlying this requirement for potency must be exposed (Tuomela 2002).

A few fungal isolates have been described to hold consolidated bioprocessing (CBP) features, recognized to translate cellulose to ethanol. Such fungi were similarly recognized to yield a substantial quantity of byproducts, signifying that additional progress and search need to be started to categorize and progress fungal CBP systems. In a study (Wu et al. 2017), four endophytes a few strains of *Hypoxylon* sp. were selected and evaluated for their CBP potential. Analysis of few strains 'genomes were found to indicate that these endophytes have a rich reservoir of biomass-deconstructing carbohydrate-active enzymes (CAZys), which include enzymes active on both polysaccharides and lignin, as well as terpene synthases (TPSs), enzymes that may produce fuel-like molecules, suggesting that they do indeed have CBP potential (Wu et al. 2017). Similarly, The pentose phosphate pathway (PPP) enzyme glucose-6-phosphate dehydrogenase (G-6-PD) was shown to be the NADPH producers reacting to lipogenesis in the oleaginous microbes (Chen et al. 2015).

13.6 Conclusion

In this chapter range of endophytic fungi related with oleaginous plants found in are discussed. The distribution of fungal endophytes indifferent oleaginous plants would be helpful in discovering multiplicity in colonization by endophytic fungi. The evidence on fungi endophytic to these plants is sparse as far as their isolation and diversity studies are concerned. Primary information about fungal resources of oleaginous plants drawn from studies divulges that their delivery is found to be affected by hosts and tissues.

The progression of molecular techniques in characterization of filamentous fungi can be utilized for a deeper determination of fungal distinctiveness at and below the genus level, henceforth the identification can be established through molecular techniques. The concept of fatty acid assessment for the fungal mycelium can be applied in works associated to microbial fuel production. Few fatty acids, which are the chief components of microbial biodiesel, can be examined for the identified positive isolates that demonstrates the compounds which will have designated hydrocarbons are accountable for biodiesel production. Biodiesel is a mixture of esters of long chain fatty acids derived from a renewable lipid feedstock.

Profiling of enzymes is vital in progress of techniques that use lignocellulosic resources. In production of ethanol, sugarcane bagasse is subjected to pretreatment; hydrolysis and fermentation which is considered substantial source of raw material to produce renewable fuels. For biodiesel production by enzymatic catalyzed processes, use of microbes is more environmentally responsive and needs less energy when associated to its production by conventional alkaline catalyzed procedures, for example, the use of enzymes for the hydrolysis of cellulose to yield fermentable sugars for bio-ethanol manufacture.

Numerous fungi have been described to hold CBP characteristics, by converting cellulose to ethanol and a substantial figure of by-products were found to be shaped by them, offering scope for additional development and investigation to classify and progress fungal CBP systems. In the chapter, it was discussed that the endophytes discovered were found to yield several VOCs, grounded on high energy density and other chemical properties crucial for usage as a fuel.

These endophytes are found to be generating a comprehensive array of volatile organic compounds (myco-diesel), in consequence supporting their potential as CBP organisms. The complete conclusion is that the endophytic fungi isolated from the selected oleaginous plants may show distinction in diversity that can then be assessed for several bioactivities.

References

Abdalla MA, Matasyoh JC (2014) Endophytes as producers of peptides: an overview about the recently discovered peptides from endophytic microbes. Nat Prod Bioprospect 4:257–270

Ajanovic A (2011) Biofuels versus food production: does biofuels production increase food prices? Energy 36:2070–2076. https://doi.org/10.1016/j.energy.2010.05.019

Akpinar-Bayizit A (2014) Fungal lipids: the biochemistry of lipid accumulation. Int J Chem Eng Appl 5:409–414. https://doi.org/10.7763/IJCEA.2014.V5.419

Arnold AE (2007) Understanding the diversity of foliar endophytic fungi: progress, challenges, and frontiers. Fungal Biol Rev 21:51–66

Arora J, Ramawat K (2017) An introduction to endophytes. https://doi.org/10.1007/978-3-319-665 41-2

Bhadra F, Gupta A, Vasundhara M, Reddy MS (2022) Endophytic fungi: a potential source of industrial enzyme producers. 3 Biotech 12:86. https://doi.org/10.1007/s13205-022-03145-y

Bhagobaty DR, Joshi SR (2011a) Metabolite profiling of endophytic fungal isolates of five ethno-pharmacologically important plants of Meghalaya, India

Bhagobaty RK, Joshi SR (2011b) Fungal endophytes of five medicinal plants prevalent in the traditionally preserved 'Sacred forests' of Meghalaya, India. Forest Sci Technol 7:151–154. https://doi.org/10.1080/21580103.2011.621381

Bora DK, Saha R, Saikia M et al (2014) Biofuel production from Mesua ferrea L. seed oil. Int J Gen Microbiol 114:361–375

Botham PA, Ratledge C (1979) A Biochemical explanation for lipid accumulation in Candida 107 and other oleaginous micro-organisms

Carsanba E, Papanikolaou S, Erten H (2018) Production of oils and fats by oleaginous microorganisms with an emphasis given to the potential of the nonconventional yeast Yarrowia lipolytica. Crit Rev Biotechnol 38:1–14

Chen HH, Jiang JG (2017) Lipid accumulation mechanisms in auto- and heterotrophic microalgae. J Agric Food Chem 65:8099–8110

Chen YH, Tang TC, Chiang TH et al (2012) A complementary biodiesel blend from soapnut oil and free fatty acids. Energies 5:3137–3148. https://doi.org/10.3390/en5083137

Chen H, Hao G, Wang L et al (2015) Identification of a critical determinant that enables efficient fatty acid synthesis in oleaginous fungi. Sci Rep 5:11247. https://doi.org/10.1038/srep11247

Chilakamarry CR, Sakinah AM, Zularisam AW (2022) Opportunities of biodiesel industry waste conversion into value-added products. Mater Today Proc 57:1014–1020

Chintagunta AD, Zuccaro G, Kumar M, Kumar SPJ, Garlapati VK, Postemsky PD, Kumar NSS, Chandel AK, Simal-Gandara J (2021) Biodiesel production from lignocellulosic biomass using oleaginous microbes: prospects for integrated biofuel production. Front Microbiol 12:658284

Chisti Y (2007) Biodiesel from microalgae. Bitechnol Adv 25:294–306

Cho HU, Park JM (2018) Biodiesel production by various oleaginous microorganisms from organic wastes. Bioresour Technol 256:502–508

De Bary A (1866) Morphologie und physiologie pilze, flechten, und myxomyceten, vol 2. Hofmeister's Handbook of Physiological Botany, Leipzig

Demirbas A (2009) Progress and recent trends in biodiesel fuels. Energy Convers Manag 50:14–34. https://doi.org/10.1016/j.enconman.2008.09.001

Demirbas A, Bafail A, Ahmad W, Sheikh M (2016) Biodiesel production from non-edible plant oils. Energy Explor Exploit 34(2):290–318. https://doi.org/10.1177/0144598716630166

Economou CN, Makri A, Aggelis G, Pavlou S, Vayenas DV (2012) Semi-solid state fermentation of sweet sorghum for the biotechnological production of single cell oil. Bioresour Technol 101:1385–1388

Gaiero JR, McCall CA, Thompson KA et al (2013) Inside the root microbiome: bacterial root endophytes and plant growth promotion. Am J Bot 100:1738–1750

Gmünder S, Singh R, Pfister S et al (2012) Environmental impacts of Jatropha curcas biodiesel in India. J Biomed Biotechnol 2012. https://doi.org/10.1155/2012/623070

Guldhe A, Ansari FA, Singh P, Bux F (2017) Heterotrophic cultivation of microalgae using aquaculture wastewater: a biorefinery concept for biomass production and nutrient remediation. Ecol Eng 99:47–53

Halder PK, Paul N, Beg MRA (2014) Prospect of Pongamia pinnata (Karanja) in Bangladesh: a sustainable source of liquid fuel. J Renew Energy 2014:1–29. https://doi.org/10.1155/2014/647324

Hasegawa S, Meguro A, Shimizu M et al (2006) Endophytic actinomycetes and their interactions with host plants. Actinomycetologica 20:72–81

Hossain N, Mahlia TMI (2020) Progress in physicochemical parameters of microalgae cultivation for biofuel production. Crit Rev Biotechnol 39:835–859

Janßen HJ, Steinbüchel A (2014) Fatty acid synthesis in Escherichia coli and its applications towards the production of fatty acid based biofuels. Biotechnol Biofuels 7:7. https://doi.org/10.1186/1754-6834-7-7

Koreti D, Kosre A, Jadhav SK, Chandrawanshi NK (2022a) A comprehensive review on oleaginous bacteria: an alternative source for biodiesel production. Bioresour Bioprocess 9(1):1–19

Kumar V, Sahai V, Bisaria VS (2011) High-density spore production of Piriformospora indica, a plant growth-promoting endophyte, by optimization of nutritional and cultural parameters. Bioresour Technol 102:3169–3175. https://doi.org/10.1016/j.biortech.2010.10.116

Kywe T, Oo M (2009) Production of biodiesel from Jatropha oil (Jatropha curcas) in pilot plant. Proceeding World Acad Sci 38:481–487

Lodewyckx C, Vangronsveld J, Porteous F, Moore E, Taghavi S, Mezgeay M, Van der LD (2002) Endophytic bacteria and their potential applications. Crit Rev Plant Sci 21:583–606. https://doi.org/10.1080/0735-260291044377

Luis A, Garay LA, Sitepu IR, Cajka T, Chandra I, Shi S (2016) Eighteen new oleaginous yeast species. J Ind Microbiol Biotechnol 43:887–900

Ma B, Li Q, Niwa T, Oura T, Du W, Liu D, Susumu K (2008) Improvement of lipid accumulation in an oleaginous yeast. J Biotechnol 136:S415–S415

Maa X, Gaoa Z, Wang Q, Liu Y (2018) Biodiesels from microbial oils: opportunity and challenges. Bioresource Technol 263:631–641

Mahapatra S, Kumar D, Singh B, Kumar PS (2021) Biofuels and their sources of production: a review on cleaner sustainable alternative against conventional fuel, in the framework of the food and energy nexus. Energy Nexus 4:100036. https://doi.org/10.1016/j.nexus.2021.100036

Mattoo AJ, Nonzom S (2022) Endophytes in lignin valorization: a novel approach. Front Bioeng Biotechnol 10:1–30. https://doi.org/10.3389/fbioe.2022.895414

Meng X, Yang J, Xu X, Zhang L, Nie Q, Xian M (2009) Biodiesel production from oleaginous microorganisms. Renew Energy 34:1–5

Mohamed MS, Lai ZW, Ariff AB (2011) Heterotrophic cultivation of microalgae for production of biodiesel. Recent Pat Biotechnol 5(2):95–107

Morath SU, Hung R, Bennett JW (2012) Fungal volatile organic compounds: a review with emphasis on their biotechnological potential. Fungal Biol Rev 26:73–83. https://doi.org/10.1016/j.fbr.2012.07.001

Moser B (2010) Biodiesel production, properties, and feedstocks

Okechukwu RI, Iwuchukwu AC, Anuforo HU (2015) Production and characterization of biodiesel from ricinus communis seeds. Res J Chem Sci Res J Chem Sci 5:2231–2606

Osterhage C, Höller U, König GM, Wright AD (2000) Ascosalipyrrolidinone A, an antimicrobial alkaloid, from the obligate marine fungus *Ascochyta salicorniae*. J Org Chem 65:6412–6417

Papanikolaou S, Aggelis G (2019) Sources of microbial oils with emphasis to Mortierella (Umbelopsis) isabellina fungus. World J Microbiol Biotechnol 35:63

Papanikolaou S, Aggelis G (2011) Lipids of oleaginous yeasts. Part I: biochemistry of single cell oil production. Eur J Lipid Sci Technol 113(8):1031–1051

Patel A, Mikes F, Matsakas L (2018) An overview of current pretreatment methods used to improve lipid extraction from oleaginous microorganisms. Molecules 23. https://doi.org/10.3390/molecules23071562

Perera F (2017) Pollution from fossil-fuel combustion is the leading environmental threat to global pediatric health and equity: solutions exist. Int J Environ Res Public Health 15(1):16
Perez-Garcia O, Escalante FME, de-Bashan LE, Bashan Y (2011) Heterotrophic cultures of microalgae: metabolism and potential products. Water Res 45:11–36
Ratledge C (1991) Microorganisms for lipids. Acta Biotechnol 11:429–438. https://doi.org/10.1002/abio.370110506
Raven S, Francis A, Srivastava C, et al (2019) Fungal biofuels: innovative approaches. In: Yadav AN, Singh S, Mishra S, Gupta A (eds) Recent advancement in white biotechnology through fungi: volume 2: perspective for value-added products and environments, pp 385–405. Springer International Publishing, Cham
Rossi M, Amaretti A, Raimondi S, Leonardi A (2011) Getting lipids for biodiesel production from oleaginous fungi, vol 1, pp 72–74. InTech Publishers
Sarin A (2012) Biodiesel. Major resources for biodiesel production. In: Biodiesel production and properties, pp 168–196. The Royal Society of Chemistry Publisher, Cambridge, UK
Schirmer A, Rude MA, Li X et al (2010) Microbial biosynthesis of alkanes. Science 329(80-) :559–562
Schulz B, Boyle C (2005) The endophytic continuum: review. Mycol Res 109:661–698
Shaw JJ, Spakowicz DJ, Dalal RS et al (2015) Biosynthesis and genomic analysis of medium-chain hydrocarbon production by the endophytic fungal isolate Nigrograna mackinonnii E5202H. Appl Microbiol Biotechnol 99:3715–3728. https://doi.org/10.1007/s00253-014-6206-5
Singh D, Sharma D, Soni SL, Sharma S, Sharma PK, Jhalani A (2020) A review on feedstocks, production processes, and yield for different generations of biodiesel. Fuel 262:116553
Sivaramakrishnan K, Ravikumar P (2011) Determination of higher heating value of biodiesels. Int J Eng Sci Technol 3:7981–7987
Sovacool BK, Brown MA (2007) Energy and American society-thirteen myths. Springer, Netherlands
Stone JK, Bacon CW, White JF (2000) An overview of endophytic microbes: endophytism defined. In: Bacon CW, White JF (eds) Microbial endophytes. M Dekker Inc., New York, pp 3–5
Strobel GDB (2003) Bioprospecting for microbial endophytes and their natural product. Microbiol Mol Biol Rev 67:491–502. https://doi.org/10.1128/MMBR.67.4.491
Strobel GA, Hess WM, Li JY (1997) Pestalotiopsis guepinii, a taxol-producing endophyte of the wollemi pine, Wollemia nobilis. Aust J Bot 45:1073–1082
Strobel G, Daisy B, Castillo U, Harper J (2004) Natural products from endophytic microorganisms. J Nat Prod 67:257–268
Strobel GA, Spang S, Kluck K et al (2008) Synergism among volatile organic compounds resulting in increased antibiosis in Oidium sp. FEMS Microbiol Lett 283:140–145
Subhash GV, Mohan SV (2011) Bioresource technology biodiesel production from isolated oleaginous fungi Aspergillus sp. using corncob waste liquor as a substrate. Bioresour Technol 102:9286–9290
Subramanian R, Dufreche S, Zappi M, Bajpai R (2010) Microbial lipids from renewable resources: production and characterization. J Ind Microbiol Biotechnol 37:1271–1287
Sut D, Chutia RS, Bordoloi N et al (2016) Complete utilization of non-edible oil seeds of Cascabela thevetia through a cascade of approaches for biofuel and by-products. Bioresour Technol 213:111–120. https://doi.org/10.1016/j.biortech.2016.02.066
Tuomela M (2002) Degradation of lignin and other 14 C-labelled compounds in compost and soil with an emphasis on white-rot fungi
Upadhyay A, Singh DK (2012) Pharmacological effects of Sapindus mukorossi. Rev Inst Med Trop Sao Paulo 54:273–280. https://doi.org/10.1590/S0036-46652012000500007
Wang Y, Lie P H (2008) The summarize about recent research process on Gramineae endophyte symbiosis. Biotechnol Bull 3:33–38. (In Chinese)
Wawrzyn G, E Bloch S, Schmidt-Dannert C (2012) Discovery and characterization of terpenoid biosynthetic pathways of fungi

Wu W, Davis RW, Tran-Gyamfi MB et al (2017) Characterization of four endophytic fungi as potential consolidated bioprocessing hosts for conversion of lignocellulose into advanced biofuels. Appl Microbiol Biotechnol 101:2603–2618. https://doi.org/10.1007/s00253-017-8091-1

Wynn JP, Hamidt A, Ratledge C (1999) Lipid accumulation in filamentous fungi. Microbiology 145:1911–1917

Zuccaro G, Pirozzi D, Yousuf A (2020) Lignocellulosic biomass to biodiesel. Lignocellulosic biomass to liquid biofuels. Academic Press, Cambridge, MA, pp 127–167

Chapter 14
Recent Advancements in Microalgae-Based Biofuel Production

Anil Kumar Singh

Abstract Biofuels are fuels derived from biological materials or renewable carbon sources. Microalgae have emerged as a sustainable and renewable feedstock for third generation biofuels. Microalgae are considered ideal for biofuel production owing to several advantages such as higher lipid accumulation, rapid growth rate, higher photosynthetic yield, opportunity for wastewater treatment and carbon dioxide sequestration. Unlike plant based biomass, microalgae do not impart much pressure on agriculture and forest ecosystem. Microalgae are suitable for production of biofuels like bioethanol, biomethane, biohydrogen, biodiesel, bio-oil and biogas. The present chapter provides contemporary knowledge of microalgae-based biofuel production including microalgae cultivation, harvesting and biofuel extraction technologies. Advance strategies for improving algal strain and maximizing the cost affordability of biofuel production have also been discussed elaborately in the chapter.

Keywords Sustainability · Renewable · Triacylglycerol · Genetic engineering · CRISPR/Cas9

14.1 Introduction

Energy is the fundamental requirement of human society. Classically, energy is described as an ability to do work. Human may harness energy from diverse resources ranging from renewable energy resources to non-renewable energy resources. Global energy requirement has increased continuously with increase in human population and advancement in technologies. According to International Energy Outlook 2017 from the US Energy Information Administration the global energy consumption is going to increase from 575 quads in 2015 to 736 quads in 2040 (IEO 2017). Large portion (~80%) of energy requirement is fulfilled by non-renewable sources particularly petroleum based fuel such as coal, oil and natural gas. Petroleum based fuels have driven economic growth and global development for decades. Use of

A. K. Singh (✉)
Department of Agriculture, Sant Baba Bhag Singh University, Khiala, Punjab 144030, India
e-mail: anilimtech@gmail.com

M. Shah and D. Deka (eds.), *Emerging Sustainable Technologies for Biofuel Production*, Environmental Science and Engineering, https://doi.org/10.1007/978-3-031-52167-6_14

petroleum based fuels releases stored carbon and greenhouse gases in atmosphere, causing global warming and pollution (Hasan et al. 2023).

Petroleum based fuels are formed over millions of years from plant and animal based biomass. Decomposition of biomass under the high compression and heat result in formation of fossil fuels. Petroleum based fuels are non-renewable in nature. According to an estimate, at current consumption rate the global petroleum reservoir will be exhausted within 50 years (Mat Aron et al. 2020). Thus, limited amount of fossil fuels, increase in the price of petroleum based fuels, global warming and pollution has attracted global attention towards renewable energy sources such as solar, wind, tidal, and biomass (Jeswani et al. 2020; Behera et al. 2015). Researcher and policy makers are of view that use of renewable energy source will not only reduce the global dependence on petroleum-based fuels but will also help in mitigating global warming and environmental pollution (Hasan et al. 2023).

Biomass obtained from animals, plants and microorganisms are an interesting renewable material for generating energy. Carbohydrate and lipids present in the biomass can be converted to different energy forms such as heat, electricity, biogas, bio-oil and biofuel. Energy derived from biomass has several advantages such as sustainability, secure supply, decrease in greenhouse gas emission, reduced adverse environmental impact and development of local region by supporting agriculture. Efforts are being made by different governments to harness the benefits of biomass derived biofuel by supporting the research and production of biofuel (Khan et al. 2023; Hasan et al. 2023).

14.2 Biofuels

Biofuels are fuels produced from biological materials or renewable carbon sources. Fermentation of biological materials can produce biofuels such as biodiesel, bioethanol, biohydrogen, bio-oil and biogas (Behera et al. 2015). Idea of biofuels is known since ancient times. Ethanol produced from plant material was used as lamp oil and cooking in ancient period. In early nineteenth century, plant based ethanol was used as fuel in internal combustion engine. Petroleum based fuel replaced ethanol by the end of nineteenth century. Ethanol as source of fuel was again introduced during Arab oil embargo in the 1970s to cope up with the rising price of petroleum based fuels (Jeswani et al. 2020). Advantages associate with biofuels has attracted lot of global attention. According to IEA the global demand for biofuels are likely to increase by nearly 28% during the period of 2021–2026. Biofuels such as bioethanol and biodiesel would be the major contributor towards the forecasted global demand (Hasan et al. 2023).

Depending upon the biomass utilized for biofuel production, biofuels has been categorized as first, second and third generation biofuels. First-generation biofuels are produced from food or animal feed crops such as wheat, rice potato, sugarcane, sugar beet and barley. Biomass obtained from crop plants such as sugar beet and sugarcane are suitable for producing bioethanol whereas crop plants such as rapeseed, soybeans and sunflowers are ideal feedstock for biodiesel production. First-generation biofuels are produced through well-established technologies and process such as fermentation, transesterification and distillation and hence also referred to as conventional biofuels. Major problem with first generation biofuel is that feedstock competes with food crops for arable land, fertilizer and water (Mat Aron et al. 2020).

Second-generation biofuels came to existence so as to overcome the drawbacks of first-generation biofuels. Second-generation biofuels are produced from non-food feed stocks, such as energy crops (Switchgrass, Jatropha, Elephant grass), agriculture residues, forest residues and waste materials. Feedstock for second-generation biofuels does not compete with the food crops. Another advantage associated with second-generation biofuels is that they emit less greenhouses gases as compared to the first-generation biofuels. Most of the feedstock for second-generation biofuels are waste biomass and need not be cultivated. Moreover, feedstock for second-generation biofuels has short life span and thus allows continues availability of biomass for producing biofuels. Feedstock used in second-generation biofuels needs to be pre-treated before it become suitable for biofuel production. Set-up required for producing second-generation biofuels are relatively expensive as compared to the first-generation biofuels (Mat Aron et al. 2020).

The third-generation biofuels are produced from algae through techniques or pathways that are not well established and generally requires more research. The production techniques of second and third generation biofuel have potential for large scale production and require more effort in form of research and development. Hence, researchers prefer designating second and third generation biofuel as advance biofuels. According to Mat Aron et al. (2020), the third-generation biofuel has tremendous growth potential as the biofuel market is expanding day by day. As evident form Fig. 14.1 the annual scientific publication with keywords biofuel and algae has increased continuously after 2003. Research publication pertaining to algae based biofuel started to grow rapidly in 2005 and beyond. The number of articles on the algae based biofuel climbed steadily in the last 18 years. In 2020, 2021 and 2022, the number of scientific publication was 1836, 2193 and 2738, respectively. The reason behind the significant increase in algae based biofuel research is primarily due to initiatives taken up by many countries to support renewable energy programs.

Past decades have also seen considerable increase in number of companies investing and making effort to commercialize algal biofuels. Some of the prominent companies attempting to commercialize algal fuels are Green Fuel Technologies (USA; www.greenfuelonline.com), IHI Corporation (Japan; www.ihi.co.jp), Neste Oil (Finland; www.nesteoil.com), Oil Fox (Argentina; www.oilfox.com.ar), Pond Biofuels (Canada; www.pondbiofuels.com), Algae Link (Netherlands; www.algaelink.com), ALG Western Oil (South Africa; www.algbf.co.za), Alga Fuel (Portugal; www.a4f.pt), BP (England; www.bp.com), BR Team (Iran; http://brteam.ir), Eni (Italy; www.eni.com), Euglena (Japan; http://euglena.jp/en/) and Greon (Bulgaria; www.greon.eu).

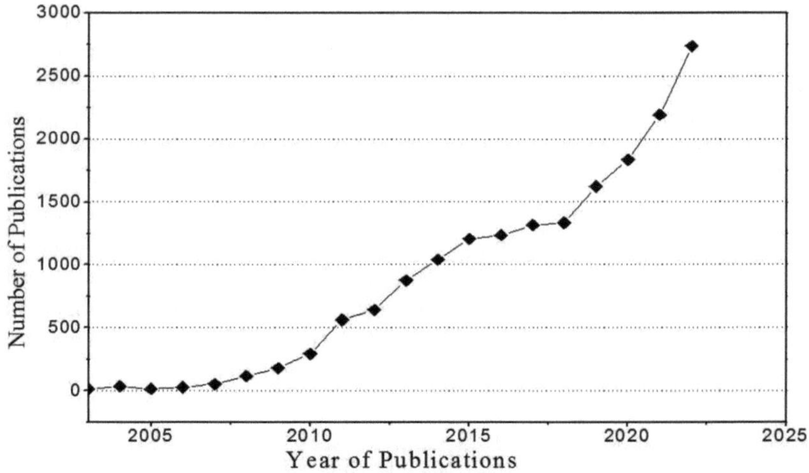

Fig. 14.1 Annual number of scientific documents published with Algae and Biofuels during the period of 2003–2022

14.3 Algae as a Source of Biofuels

Algae are a diverse group of autotrophic or heterotrophic organisms including both unicellular (such as Chlorella, Chlamydomonas, Spirogyra, and Euglena) and multicellular forms (such as Giant Kelp). The autotrophic algae carry photosynthesis and require only inorganic compounds such as carbon dioxide, salts, and a light energy for their growth. The heterotrophs algae are non-photosynthetic and require external organic compounds as nutrient and energy sources (Khoo et al. 2023). Depending upon the size, algae have been grouped into microalgae and macroalgae. Study by Xu et al. (2018) exhibited that microalgae feedstock are better than macroalgae feedstock for biofuel generation. Microalgae are very small in sizes, grow at much faster rate in ideal condition and possess high lipid content. More than 40% of the global carbon dioxide is fixed by the microalgae. Algae are important feedstock for third-generation biofuels. Fast growth and high lipid content makes microalgae an ideal feedstock for biofuel production. Use of algae for biofuel generation avoids the issue of competition for food and land use because microalgae can be grown on non-arable land and water bodies (Neag et al. 2023). Figure 14.2 summarizes the advantages of producing biofuel from microalgae.

Biofuel generation from microalgae is energy-intensive process and often regarded as economical unviable for large scale production. Growing and harvesting algae is relatively costlier than the conventional crops. According to Behera et al. (2015), nearly 20–30% of the total biofuel production cost is taken up by harvesting.

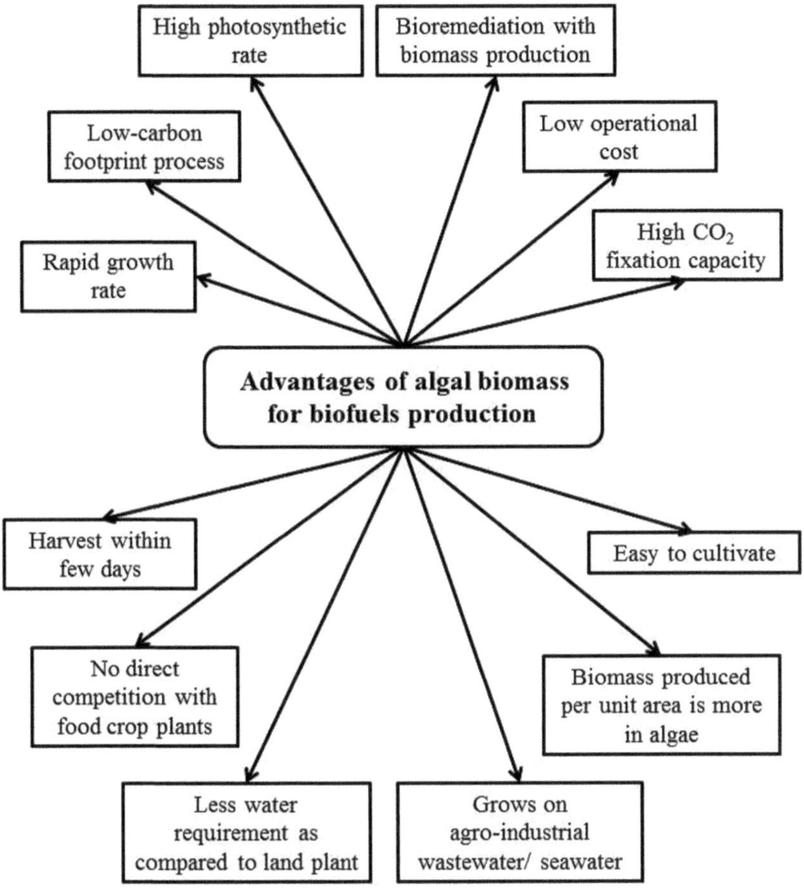

Fig. 14.2 Advantages of producing biofuel from algae

14.4 Large-Scale Algae Cultivation

Economical large-scale cultivation of algae is prerequisite for generating cost-effective biofuels. Algal species such as *Scendesmus obliquus*, *Ulva lactuca*, *U. pinnatifide*, *Gracilaria amansii*, *Chlorella sorokiniana* and *Monoraphidium miutum* have demonstrated potential to serve as good source of biomass for producing biofuels (Khoo et al. 2023; Neag et al. 2023). Large amount of algal biomass is required to compete with other feedstock used for producing biofuels. Understanding the algal growth requirement is essential for developing economical large-scale cultivation technology. Algal biofuel production can be made sustainable, feasible and economically viable by developing efficient culturing technologies for obtaining large scale production of biomass.

Algae can be cultivated by different methods and under different conditions. Growth media composition may vary for different algal species. However some basic factors such as light, pH, nutrients, temperature, organic or inorganic carbon source, nitrogen, phosphorus, and iron are essential for cultivating any algal species. The macronutrients (namely Na, Mg, Ca, and K) and micronutrients (such as Mo, Mn, B, Co, Fe, and Zn) are also essential for cultivating algae. Presence of few other trace elements such as Cu, Mo and Ni are documented to increase both biomass production and lipid content of algal cells. Usually the wastewater generated from food and agriculture sectors contains these micronutrients and trace elements (Mahmood et al. 2023).

Large scale cultivation of algae is usually performed in photobioreactors and open raceway ponds. Cultivation of photoautotrophic algae requires regular supply of CO_2. Diffusion of atmospheric CO_2 in large scale growth medium is restricted and thus, hindering the proper growth of algae. Use of bioreactor for culturing algae reduces the chance of contamination. Bioreactors are design according to the species and requirement. Continuous bioreactors or batch culture bioreactors are suitable for culturing the algae in controlled environment. Bioreactors require small area for operation. It also provides opportunity to fully exploit light and nutrition availability due to the use of artificial sources. Various types of photobioreactors such as Tubular Photobioreactor, Flat Plate Photobioreactor and Plastic Bag Photobioreactor have been designed for algae production (Benner et al. 2022).

Tubular photobioreactors can be vertical, horizontal, and helical form tubular reactors. It is widely used for industrial purposes or large scale production due to easy scale up process. In a tubular photobioreactor, the algae and growth media are continuously circulated through the tubes between reservoir and reactor using an airlift or mechanical pump. Tubular photobioreactors have been successfully employed for growing wide range of algae such as Arthrospira, Porphyridium, Chlorella, Dunaliella, Haematococcus, Tetraselmis, and Phaeodactylum (Abdel-Raouf et al. 2012). Algae growing in tubular photobioreactors often encounter adverse pH, CO_2 gradients and photoinhibition (Benner et al. 2022).

Flat-plate photobioreactors offers large surface area for algal growth. Large surface area allows high degree of solar light exposure and lower accumulation of dissolved oxygen. Convenient modular design of flat-plate photobioreactors helps in large-scale production of algae both under indoor and outdoor facilities. The major disadvantage with flat plate photobioreactors are difficulty in controlling the temperature, algal biofilms formation on the plates and the hydrodynamic stress (Benner et al. 2022).

Plastic Bag Photobioreactors are large bags fitted with aerators. These photobioreactors are hung vertically or placed in a metal or plastic cage for support and exposed to direct sunlight. Here the algae cultures are continuously mixed with help of diffused air pump. Algae growing in plastic bag photobioreactors often fail due to poor mixing. This kind of photobioreactor also requires frequent maintenance (Benner et al. 2022).

Open culture systems are relatively less expensive, but prone to contamination and predation. Open culture systems are also known as high rate algal ponds. These algal culture systems are cost-effective and require less energy as compared to photobioreactors. Open raceway ponds have depths of 20–50 cm to ensure ample penetration of sunlight for enhanced biomass production. However, the productivity of open raceway ponds is considerably affected by several environmental factors, such as a large area requirement, the influence of variable weather conditions, water loss due to evaporation, poor mixing efficiency, inefficient use of CO_2, shading effect and inefficient light utilization. Open raceway ponds have been reported to produce biomass of up to 60–100 mg dry weight/L/d. Algal species such as *Chlorella* spp., *Arthrospira platensis*, *Hematococcus* spp., and *D. salina* has been successfully cultivated for industrial purpose in open raceway ponds (Kratzer and Murkovic 2021).

Interestingly, algae can also be cultured on waste water generated from agriculture, industries and sewage. This serves as dual benefits of water treatment and algal biomass production (Ali et al. 2022). Sea water is another good alternative for cultivating algae. Sea water contain inorganic nutrient essential for algal growth and thus reduces the cost of biomass production. Algae growing in sea water have good percentage of stored lipids (Gao et al. 2022).

Improving algae cultivation is a multivariable trial process that must not only include strain selection but also optimization of culture condition. In general there are some crucial parameters namely light, temperature, nutrients, agitation or mixing of growth medium, pH of growth medium, and mixotrophic cultivation that must be considered for economical large-scale production (Mahmood et al. 2023).

14.4.1 Light

Light is one of the important limiting factors in algae cultivation. Light intensity and duration has direct influence on photosynthesis, which in turn influences biochemical composition and yield of algal biomass. Optimum light intensity and duration is essential for achieving maximum algal growth and biomass. Algae demonstrate inappropriate growth at very low and high light intensity. Net zero algal growth is seen at compensation point. Compensation point is the irradiance at which photosynthetic CO_2 uptake is equal to respiratory CO_2 release. A specific duration of light and dark periods is also essential for algal photosynthesis. Algae use light for synthesizing ATP and NADPH which in turn is utilized in the dark reaction of photosynthesis to produce organic compounds. Same algal species grown under different light duration and intensity have different growth rate and biomass yield. Literatures suggest that 16 h of light period and 18 h of dark period is ideal for most of the algal species. Uniform light within the bioreactor is necessary for good algal growth. Self-shading effect of algae within the bioreactor is well known to adversely influence growth rate and biomass yield. Light source such as LED lights and fluorescent tubes are considered ideal for illuminating photobiorectors (Nzayisenga et al. 2020).

Nzayisenga et al. (2020) reported that by increasing the light intensity both biomass and fatty acid contents of algae namely *Desmodesmus* sp. and *S. obliquus* increased significantly. Both algae strain was grown in BG11 medium at 25 °C with a photoperiod of 16 h light:8 h dark, and shaking at 150 rpm. Algae were exposed to three different light intensity 50, 150 and 300 μE m^{-2} s^{-1}. Algal biomass obtained from higher light intensity has lower amounts of 18:3, and higher contents of 18:1 fatty acids. Biodiesel having high content of polyunsaturated fatty acids such as 18:3 is prone to oxidation-dependent degradation and not ideal for biofuel production. Contrary high monounsaturated fatty acids such as 18:1 content are not susceptible to oxidation and reduce solidification temperature. Overall study shows that optimizing the light intensity can improve the quality of microalgae-derived biodiesel (Nzayisenga et al. 2020).

Algae exposed to high irradiance for longer period often experiences phenomena known as photoinhibition. It causes reduction in photosynthetic capacity of algae and reduces overall biomass yield. Photoinhibition can be prevented by increasing the light intensity or by thoroughly mixing the culture continuously. Cazzaniga et al. (2022) demonstrated that by tailoring photosynthetic pigment composition and addition of ketocarotenoids can improve photosynthetic efficiency of *Chlamydomonas reinhardtii*. Astaxanthin accumulated in engineered *C. reinhardtii* and improved growth under high light irradiance by reducing photoinhibition (Cazzaniga et al. 2022).

14.4.2 Temperature

Like any living organisms temperature has crucial influence on algal growth rate. Each algal species has optimum growth temperature. Temperature influences algal growth by influencing biochemical processes and photosynthesis. Deviation from optimum growth temperature decreases biomass yield. Literatures suggest that most of the algal species has optimum growth temperature in the range of 20–30 °C. However, thermophile algae such as *Anacystis nidulans* and *Chaetoceros* have higher optimum growth temperature. The advantage of using thermophile algae is that it minimizes the chance of contamination. In some of the thermophile algae such as *Ochromonas danica* and *Nannocloropsis salina*, lipid content increases with increase in temperature (Barnard et al. 2010; Saini et al. 2021).

Temperature can be easily controlled in bioreactors but it is difficult in open pond culture. Careful monitoring of temperature is essential in open pond culture. Exposure to high temperature reduces photosynthesis by inactivating photosynthetic proteins whereas low temperature adversely affects photosynthesis by reducing the carbon assimilation process. The decline in photosynthesis reduces growth rate and overall biomass yield (Saini et al. 2021).

14.4.3 Nutrients

Nutritional requirement for most optimal growth is different for different algal species however the basic requirement remains same for all the species. Rapid growth of algae is essential for commercial production of biomass. Fast algal growth can be achieved by providing proper nutrients and growth condition. Nitrogen, phosphorus, and carbon constitute the backbone of microalgae and are included in macronutrients required for algal growth. Some marine microalgae species such as *Nannochloropsis oculata* also require silicon as a macronutrient (Sukarni et al. 2014). Amount of available nitrogen and phosphors directly influence cell growth and size. Nitrogen limitation reduces algal growth rate and overall biomass yield. Nitrogen limitation also alters biochemical composition by increasing carbohydrate and lipid production. Usually Urea is used as a cost-effective source of inorganic nitrogen. The micronutrients Mo, K, Co, Fe, Mg, Mn, B, and Zn are required in minute quantity but are essential for good microalgae growth.

Carbon can be supplied to algae culture media in organic forms, such as glycerol or acetates, or inorganic form as CO_2. Use of environmental CO_2 in algae culture media not only reduces production cost but also mitigate CO_2. In recent past, algae particularly microalgae have emerged as promising option for treating wastewater generated from household, pharmaceutical industries and agro-industries. Wastewater usually contains important nutrients (such as organic carbon, phosphorus, nitrogen, and potassium) which may support growth and bioremediation. Algae can also be cultivated on municipal, industrial, agro-industrial, and livestock wastewaters. Algae utilize nutrient (such as nitrates and phosphates) present in wastewater for growth and thus reduce production cost as well as help in reclaiming wastewater (Deng et al. 2022). However, wastewater may contain some hostile contaminates such as heavy metals, pesticides, organic and inorganic toxins that may hinder algal growth. In such scenario, algae with special attributes may be utilized for biomass generation and wastewater reclamation. Algal species such as *Euglena*, Oscillatoria, Chlamydomonas, Scenedesmus, Chlorella, Nitzschia, Navicula, and Stigeoclonium have been reported as genera resistant to organic pollutants (Palmer 1969).

14.4.4 Agitation or Mixing

Agitation or mixing of culture media results in uniform distribution of light, nutrients, air, and CO_2. It also prevents settling down and aggregation of algal biomass. Absence of proper mixing reduces overall biomass yield even if nutrients are supplied appropriately (Mahmood et al. 2023).

14.4.5 PH of Growth Media

The pH of the culture media is another vital factor influencing microalgae growth. Different algal species favor different pH for growth. Most of the algal species favor very narrow pH range for growth. Very few algae like *C. vulgaris* can tolerate and grow under broad range of pH. Literature suggests that pH range from 6 to 8 is most suitable pH for algal growth. Study has shown that *C. vulgaris* can grow in broad pH range with the maximum growth rate and biomass production at pH 9–10 (Daliry et al. 2017).

14.4.6 Mixotrophic Cultivation

Large scale production of algae can be achieved through autotrophic cultivation, heterotrophic cultivation or mixotrophic cultivation. Under autotrophic cultivation, energy from light is utilized to assimilate carbon. In the heterotrophic cultivation, organic carbon sources are metabolized to generate energy. In mixotrophic cultivation combination of autotrophic and heterotrophic models are utilized for culturing algae. Under mixotrophic cultivation, algae have luxury to fix inorganic carbon by photosynthesis as well as utilize organic carbon source (such as glucose, glycerol, and acetate) supplied in the culture media. Algae growing under mixotrophic cultures condition exhibits faster growth as both autotrophic and heterotrophic pathways are utilized for growth (Wang et al. 2014).

In one of the interesting study Yun et al. (2021) evaluated biomass productivity, and compositional changes in Chlorella growing under photoautotrophic, mixotrophic, and heterotrophic cultivation conditions. The biomass yield of Chlorella KNUA104, Chlorella KNUA114, and Chlorella KNUA122 was observed to be 0.35 $\pm$ 0.04 g/L/d, 0.40 $\pm$ 0.08 g/L/d and 0.38 $\pm$ 0.05 g/L/d under photoautotrophic culture condition, respectively. The biomass yield under mixotrophic culture condition was observed to be 2.99 $\pm$ 0.10 g/L/d, 5.18 $\pm$ 0.81 g/L/d and 5.07 $\pm$ 0.22 g/L/d for Chlorella KNUA104, Chlorella KNUA114, and Chlorella KNUA122, respectively. Under heterotrophic conditions the biomass yield for Chlorella KNUA104, Chlorella KNUA114, and Chlorella KNUA122 was observed to be 1.72 $\pm$ 0.26 g/L/d, 4.26 $\pm$ 0.27 g/L/d and 4.32 $\pm$ 0.32 g/L/d, respectively. Chlorella growing under mixotrophic condition demonstrated enhanced lipid (68.80 $\pm$ 0.54%) and pigment (0.33 $\pm$ 0.01 g/L) content as compared to heterotrophic and photoautotrophic culture condition. Saturated fatty acids content also increased under mixotrophic cultivation with 29.76 $\pm$ 1.31%, 37.01 $\pm$ 0.98% and 33.37 $\pm$ 0.17% for strain KNUA104, KNUA114 and KNUA122, respectively (Yun et al. 2021).

14.5 Conversion Techniques for Biofuel Production from Algal Biomass

The algal biomass can be converted into several types of biofuels such as bioethanol, biodiesel, biogas, biohydrogen and bio-oil. Conversion of algae biomass to biofuel can be achieved through any of the three methods namely biochemical, chemical, and thermochemical conversion (Shokravi et al. 2021). Biochemical methods of conversion produce liquid or gaseous fuels through fermentation or anaerobic respiration. Thermochemical conversion includes methods such as combustion, gasification and pyrolysis. Combustion converts algae biomass to heat, Carbon dioxide and water in the presence of oxidant. According to Osman (2020) biomass combustion is not the ideal strategy for utilizing biomass as it does not allow maximum energy recovery and causes severe environmental damage. Gasification of algae biomass involves heating process to produce mixture of carbon monoxide and hydrogen in presence of partial oxidation. Pyrolysis is a thermal heating of biomass in the absence of oxygen to produce products like gases, pyrolytic oil and char (Khan et al. 2023).

14.5.1 Bioethanol

The algal biomass is considered more suitable for bioethanol production since algae do not have lignin and minuscule amount of hemicellulose. Absence of lignin allows easy conversion of algal carbohydrate to simple sugar as compared to lignocellulosic materials. Biomass obtained from algae is converted to bioethanol by fermentation process. Algae are considered better alternative to energy crops such as corn and soybean. Several algae such as *Chlorococcum* sp., *Prymnesium parvum*, *Gelidium amansii*, *Gracilaria* sp., *Laminaria* sp., *Sargassum* sp., and *Spirogyra* sp., have been used as feedstock for producing bioethanol. These algae use light, nutrients and carbon dioxide to produce polysaccharides. The stored polysaccharides can be extracted and converted to simple sugars through the process of hydrolysis. The simple sugar obtained after hydrolysis is fermented to produce bioethanol. Distillation helps in recovering bioethanol from fermentation mixture (Khan et al. 2023).

Algae species having good cellular starch concentration have better bioethanol yield whereas algae with high percentage of polysaccharides need to be pre-treated and scarification before fermentation. Single step bioethanol production can be carried out by use of amylase producing microorganisms. Enzyme amylase has ability to simultaneously bring about saccharification and fermentation. Pre-treatment of algae facilitates accessibility of biomass to enzymes for the release of simple sugars. Algae also contain certain polymers such as alginate, mannitol, and fucoidan that require additional pre-treatment and scarification before fermentation. Pre-treatment of algal biomass can be carried out by acid hydrolysis, enzymatic digestion and gamma radiation. Phwan et al. (2019) obtained 0.28 g ethanol/g microalgae upon pre-treating algal biomass with 5% (v/v) sulfuric acid. Pre-treating algal biomass with 5% (v/v) acetic acid gave 0.23 g ethanol/g microalgal biomass (Phwan et al. 2019).

Fermentation of algal biomass can be carried out by yeast, bacteria and fungi. These microorganisms convert simple sugar under anaerobic condition to bioethanol. However, some algae have mannitol and anaerobic fermentation of mannitol is not feasible. Horn et al. (2000) has exhibited mannitol fermentation in the presence of oxygen by *Zymobacter palmae*. Most commonly *Saccharomyces cerevisiae* and *Zymomonas mobilis* are employed for generating bioethanol from algal biomass. *Z. mobilis* is a facultative anaerobic Gram-negative bacterium widely used for fermenting biomass to ethanol. Several factors (such as inoculum size, agitation rate, temperature, pH, strain, sugar concentration and fermentation time) are known to influence bioethanol production during fermentation of algal biomass (Selim et al. 2018).

Recently, Condor et al. (2022) demonstrated that microalgae biomass can be fermented at high-solids loading (200 g/L) to improve the economic viability of microalgae bioethanol production. Microalgae biomass was produced by cultivating *Chlorella vulgaris* FSP-E with high carbohydrate content. The fermentation of microalgal biomass hydrolysate was performed by *Saccharomyces cerevisiae* FAY-1 to produce bioethanol. The final titer suggested ethanol concentration to be 43.9 g/L. The maximum productivity of 7.387 g/L/h was achieved in this study thus strongly indicating the economic viability of the high-solids loading fermentation (Condor et al. 2022).

14.5.2 Biodiesel

Biodiesel is a mixture of monoalkyl esters of long chain fatty acids produced from renewable lipid containing biomass. It is an ideal candidate to replace petroleum based diesel oil. Algal biomass produces higher biodiesel as compared to energy crop plant thus making it economical. The mass of oil produced per unit volume of algal broth per day depend on growth rate and biomass. The algae such as *Kirchneriella lunaris*, *Ankistrodesmus fusiformis*, *Chlamydocapsa bacillus*, and *Ankistro desmusfalcatus* have good content of polyunsaturated fatty acid methyl esters and preferred for biodiesel production. Oil content of these algae is very high and often reaches up to 80% by weight of its dry weight. According to an estimate nearly 5,000–15,000 gal of biodiesel can be produced from algal biomass per acre per year. Biodiesel produced from algae needs to comply on the physical and chemical properties set by International organization such as International Biodiesel Standard for Vehicles (EN14214) and American Society for Testing and Materials (ASTM) for its acceptance as substitute to fossil fuels. Presence of higher degree of polyunsaturated fatty acids in algal oil makes it more prone to oxidation and thus making prolong storage difficult and limit its extensive utilization (Khan et al. 2023).

Biodiesel production from algae has several steps. Initially, algae are cultured and algal biomass is harvested, dried and oil is extracted from dried biomass. The extracted oil is converted to diesel by process of transesterification. Harvesting of macroalgae is easy as it can be performed by simple nets. Conventional process such as filtration, flocculation, foam fractionation, sedimentation and ultrasonic separation is used for harvesting microscopic algae. Proper drying of biomass is essential for efficient recovery of oil by mechanical and solvent extraction. Post-harvest storage and self-life of harvested algal biomass is significantly increased in properly dried algae. Drying methods like spray-drying, drum-drying, freeze drying or lyophilization, and sun-drying have been employed for drying of harvested algal biomass. Simple sun-drying is often considered not ideal for drying of algae as it contains large percentage of water. Sun-drying with of aid of especially designed solar drying device is better than simple sun-drying (Neag et al. 2023).

Drying temperature has significant effect on overall yield of lipids and triglycerides. The extraction of lipids from algae biomass is energy intensive and expansive process. Mechanical extraction of oil from biomass can be achieved either by expellers or mechanical press. Mechanical force break cell and allows oil to leach out. Mechanical extraction allows nearly 75% recovery of oil from algal biomass (Neag et al. 2023).

Solvent extraction of oil is more efficient than mechanical extraction methods. Solvent extraction methods allow more than 99% recovery leaving only 0.5–0.7% residual oil in the algal biomass. The nature of solvent used for extraction has important influence on the biodiesel yield. Organic solvent like chloroform, hexane, cyclohexane, acetone, and benzene has been used either solely or in form of mixture. Solvent breakdowns cell and releases oil in surrounding. The oil gets easily dissolved in the solvent. Further, solvent is recovered and oil separated from solvent. Solvent can be recycled for next extraction after the recovery of oil. Hexane and mixture of chloroform/methanol in the ratio 2:1 (v/v) has been most favored solvent for extracting lipid form algal biomass. Selecting right solvent system is important for the maximum and economical extraction of oil from algal biomass (Neag et al. 2023; Khan et al. 2023).

Transesterification is reaction between triglycerides or fatty acids and alcohol (Fig. 14.3). Alcohol such as ethanol, butanol, methanol, propanol, and amyl alcohol has been successfully employed for transesterification process. Alcohol such as ethanol and methanol are preferred for transesterification reaction due to economical, physical and chemical advantages. Inorganic catalyst or lipase enzymes are added to reaction. Biodiesel are periodically or continuously remove from rector to drive the reaction forward. Acid catalyzed transesterification reaction is very slow as compared to alkali catalyzed transesterification reaction. Sodium and potassium hydroxides are most commonly used alkali catalysts (Khan et al. 2023). Singh et al. (2006) has demonstrated that sodium methoxide has better catalytic activity in comparison to sodium hydroxide.

Lipases (triacylglycerol hydrolases E.C.3.1.1.3) catalyzed reaction is often considered better than inorganic catalyzed reaction. Lipases have several advantages such as versatility, substrate selectivity, regioselectivity, enantioselectivity and

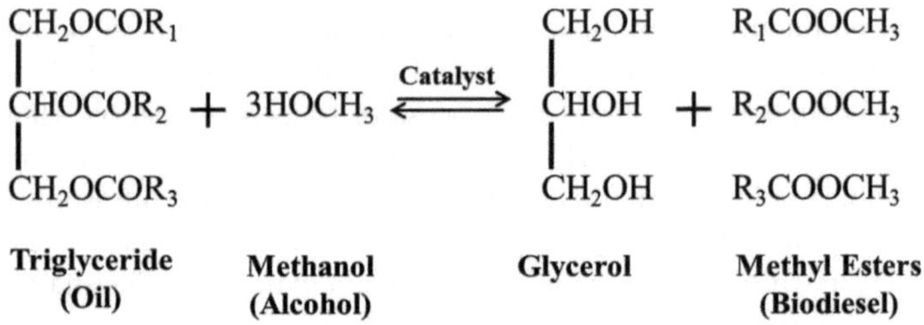

Fig. 14.3 Transesterification of oil to biodiesel. (R1, R2, and R3 are hydrocarbon)

high catalytic activity at ambient temperature and pressure. Some lipases have ability to hydrolyze ester bonds at secondary positions, while some other group of lipases hydrolyzes both primary and secondary esters. There are other group of lipases that demonstrate fatty acid selectivity by cleaving ester bond at particular type of fatty acids. Huang et al. (2015) constructed recombinant *Rhizomucor miehei* lipase and expressed in *Pichia pastoris*. The Lipase GH2 was able to convert microalgae oil mixed with methanol or ethanol to biodiesel in an n-hexane solvent system. Both fatty acid methyl ester (FAME) and fatty acid ethyl ester (FAEE) were produced by lipase GH2. Lipase with low temperature activity is believed to provide energy saving (Huang et al. 2015).

Extractive transesterification includes several steps (such as drying, cell disruption, oils extraction, transesterification, and biodiesel refining) to produce biodiesel. The high water content in algal biomass (more than 80%) causes problem in extractive transesterification by increasing the cost of whole process (Mota et al. 2022).

In situ transesterification skips the oil extraction step. This method uses alcohol as an extraction solvent and an esterification reagent. Alcohol enhances the porosity of the cell membrane and subsequently increases yields as compared to the conventional routes. Waste generated during in situ transesterification is comparatively less as compared to the conventional method. However, release of large volume of extraction solvent in environment contribute towards global warming and create problem for atmosphere (Mota et al. 2022).

14.5.3 Biogas Production

Anaerobic digestion of algal biomass for generating biogas has received a lot of attention in recent past. Presence of polysaccharides such as agar, alginate, carrageenam and mannitol with minute amount of cellulose and lignin makes algae suitable for biogas generation. Algae species like *Scenedesmus*, *Spirulina*, *Euglena*, and *Ulva* have been successfully employed for producing biogas. Seaweeds such as Chlorophyta, Rhodophyta and Phaeophyceae are excellent source of feedstock for producing biogas (Wang et al. 2021).

Anaerobic transformation of algal biomass involves basically four steps. The first step is to convert insoluble and high molecular organic compounds into soluble organic mass. Obligate anaerobes such as *Clostridia* and *Streptococci* releases enzymes that bring about afore stated conversion. In the second step, acidogenic bacteria release enzymes that convert soluble organic mass to alcohols as well as releases volatile fatty acids. In the third step acetogenic bacteria convert alcohols and volatile fatty acids to acetic acid and hydrogen. In the fourth step acetic acid and hydrogen are metabolized by methanogens to produce methane and carbon dioxide. Studies have shown that use of pretreatment like heat, ultrasound and microwave can enhance biogas production (Wang et al. 2021). *Ulva intestinalis* pretreated with combination of microwave and Fe_3O_4 nanoparticles gave higher biogas yield of 206 mL/g VS as compared to non-treated algal biomass (84 mL/g VS). Pretreatment of algal biomass with microwave and nanoparticles enhanced cell wall dissociation for biogas production (Nemr et al. 2021).

Biogas production from algal biomass has some concerns like requirement for large space, sophisticated infrastructure and energy for digesters. Another problem commonly encountered in algal biomass based biogas production is the generation of ammonia that inhibits growth of anaerobic microorganisms. The protein present in algae lowers carbon is to nitrogen ratio, thus reducing the biogas production. Sodium ions are also known to adversely affect the growth of anaerobic microorganisms. Researchers are of view that salt tolerant microorganisms for the anaerobic digestion of algal biomass can overcome the problems (Nemr et al. 2021; Saini et al. 2021).

14.5.4 Biohydrogen Production

Hydrogen can be utilized as gaseous fuel as well as source for generating electricity. The combustion of hydrogen produces water without generation of CO_2, thus establishing itself as most desirable renewable energy source. Unlike several energy sources, utilization of hydrogen as energy source does not add greenhouses gases to atmosphere. In the recent past hydrogen has attracted lots of interest as the most efficient fuel because it has the highest energy density and the best conversion efficiency among all fuels problems (Nemr et al. 2021; Saini et al. 2021).

In nature, living microorganisms can generate hydrogen by two different processes namely light-dependent process (like bio-photolysis and photo-fermentation) and the light-independent process. Microalgae and cyanobacteria generate hydrogen by bio-photolysis of water. Photosynthetic bacteria generate hydrogen by photo-fermentation. Anaerobic bacteria carry fermentation (also called dark fermentation) of organic molecules to generate hydrogen.

Enzyme hydrogenase is a proton-lowering enzyme is responsible for the splitting of water and production of hydrogen in algae (Li et al. 2022). Oxygen generated during photolysis of water deactivates hydrogenase enzyme, thus permitting only short duration reaction. Commercial utilization of hydrogenase enzyme for hydrogen production requires oxygen resistant enzyme. Researchers are striving to modify hydrogenase enzymes by random mutation or design new hydrogenase to make it more oxygen resistant (Li et al. 2022).

Additionally, use of dark fermentation of algal biomass with anaerobic bacteria is affordable means for bio-hydrogen generation. Algal biomass of *Chlorella vulgaris* and Scenedesmus sp. has been widely used for generation of bio-hydrogen generation by dark fermentation. Usually the hydrogen yield by dark fermentation of algal biomass ranges from 0.37 to 19 mL H_2/g VS (Wang et al. 2021; Wang and Yin 2018). Lunprom et al. (2019) reported that production of 2.87 mmol H_2/g of Chlorella sp. by dark fermentation with *Clostridium butyricum* under anaerobic conditions.

Pretreating of algal biomass often increases biohydrogen production under dark fermentation by bacteria. Main purpose of algal biomass pretreatment is to release the organic substance from cell and make it readily available for fermenting microorganism. Different researchers have used different methods namely physical (mechanical, heat and ultrasonic treatment), chemical (acid, base and ozone) and biological (enzymatic and microbiological treatment) methods pretreatment. Combination of aforementioned methods has also been employed to for pretreating the biomass.

14.5.5 Bio-Oil Production

Bio-oil is an alternative to liquid fossil fuel. It is synthesized from algal biomass under anaerobic condition at high temperature. The nature of feedstock and processing condition determine composition of bio-oil. Bio-oil productivity is influenced by several factors such as biomass composition, pyrolysis temperature, and vapor residence time. Use of crude bio-oil as fuel is restricted due to the presence of undesirable components such as water, oxygen content, unsaturated and phenolic moieties. Treating crude bio-oil is pre-requisite for improving the quality and making it suitable as a substitute for fossil fuel (Xu et al. 2021).

Bio-oils are dark brown viscous liquid with distinctive odor. Chemically bio-oil is a complex mixture of several hundreds of organic compounds. It primarily contains acids, alcohols, aldehydes, esters, ketones, phenols, and guaiacols. Xu et al. (2018) carried out hydrothermal liquefaction of eight algal biomasses at 350 °C for 1 h. Microalgae produced higher crude bio-oil yield than the macroalgae. Among all algae, Schizochytrium limacinum provided the maximum crude bio-oil yield of 54.42 wt% due to their high lipid content. Crude bio-oils were treated at 400 °C for 2 h with 10 wt% Ru/C using tetralin as the hydrogen donor for improving the quality of bio-oils (Xu et al. 2018).

14.6 Bioengineering Algae for Enhanced Biofuel Production

Algal is an ideal candidate for production of fourth generation biofuels. However, economic viability of algae-based biofuel has restricted its extensive commercialization. Successful commercialization of algae-based biofuel would not only require efficient downstream processing but also an ideal strain. Algae strain that have fast growth rate, good lipid synthesis pathway and high lipid accumulating capability is considered ideal for biofuel production. Conventional ways to improve growth rate and lipid productivity include culture conditions, chemical additives and applying stress (Shokravi et al. 2021; Varela et al. 2020). Hossain et al. (2019) highlighted the importance of nano-additives like nanoparticles for microalgae cultivation, harvesting and improved biofuel extraction.

In recent past researchers are looking up to more sophisticated and contemporary means to introduce desirable traits and improve algal strain for biofuel production. Engineering algae traits by modifying genetic constitution or metabolic properties of microalgae are seen as future of strain improvement. Advance genetic engineering tools have emerged at a greater pace however their utilization for microalgae strain improvement is slow. The recent availability of genome sequences of algae like *Scenedesmus obliquus* AS-6-11, *Chlorella variabilis* NC64A, *Chlorella vulgaris* UTEX 395, *Chlamydomonas reinhardtii* CC503, *Geitlerinema* sp. FC II, *Shewanella* SYT3, and other omics datasets have good potential to help in algae strain improvement program (Shokravi et al. 2021).

Gene editing or genome editing is type of genetic engineering approach to modify genes in living organism. In gene editing the DNA is inserted, deleted, modified or replaced in the genome of living organism. Earlier gene editing was random process. However, recent advancement has made gene editing highly precise. Modern gene editing technologies such as transcription activator-like effector nucleases (TALEN), zinc-finger nuclease (ZFN), and clustered regularly interspaced palindromic sequences (CRISPR) have been successfully used in diverse living organisms (Shokravi et al. 2021; Varela et al. 2020).

ZFNs are sequence-specific nucleases with a synthetic zinc finger DNA binding domain and a DNA cleavage domain. The zinc finger DNA binding domain can be altered to specifically introduce cleavage at targeted double stranded DNA. Greiner et al. (2017) used ZFN to disrupt photoreceptor genes in *Chlamydomonas reinhardtii.* TALENs are sequence-specific nucleases consisting of transcription activator-like effectors (TALE) and catalytic domain of the *Fok I* endonuclease. The DNA-binding domain of TALE monomers consist of central repeat domain (CRD) that determines DNA binding specificity. Kurita et al. (2020) targeted nitrate reductase gene, *NoNR*, and acyltransferase gene, *LPAT1* with Platinum transcription activator-like effector nucleases (TALENs) in *Nannochloropsis oceanica.*

CRISPR/Cas 9 gene editing system is most powerful method of site-specific genetic manipulation. It uses RNA-guided interaction with DNA for manipulating gene. Lakhawat et al. (2022) CRISPR-Cas9 modification of microalgae has demonstrated improved total lipid content, a prerequisite for biofuel production. Dhokane et al. (2020) targeted MAA7 gene using CRISPR-Cas9 ribonucleoproteins to demonstrate knock-out and knock-in through homology-dependent repair template at the target site.

Modifying genes responsible for triacylglycerol and fatty acid biosynthesis pathway are seen as most viable approach to increase lipid content in the algae. Fatty acid synthesis pathway is responsible for synthesizing the shorter and saturated fatty acids in microalgae. Overexpression of genes involved in triacylglycerol and fatty acid biosynthesis pathway or blocking competing pathways are important means of enhancing lipid synthesis and lipid accumulation (Shokravi et al. 2021). Gomma et al. (2015) reported that overexpression of ACCase in *Scenedesmus quadricauda* increased fatty acid accumulation by 1.6-fold. Chen et al. (2017) and Li et al. (2018) enhanced TAG biosynthesis pathway in Nannochloropsis oceanica, and Schizochytrium to achieve 36%, and 39.6% increase in total fatty acids, respectively. Increase in acetyl-CoA supply in Yarrowia lipolytica increased lipid accumulation by 60-fold (Blazeck et al. 2014).

14.7 Conclusions

Algae are diverse group of organism with lots of potential to serve as feedstock for production of third generation biofuel. Affordable biofuel production from algal biomass is major obstacle. Strain improvement by gene editing system particularly CRISPR/Cas 9 gene editing system can provide unprecedented results. Improving culture condition for continuous and rapid production of algae biomass by use of nano-additives has good potential to improve biofuel production. Research and technical development are need of hour to make algae based biofuel production competitive with fossil fuels.

References

Abdel-Raouf N, Al-Homaidan AA, Ibraheem IBM (2012) Microalgae and wastewater treatment. Saudi J Biol Sci 19(3):257–275

Ali SS, Mastropetros SG, Schagerl M, Sakarika M, Elsamahy T, El-Sheekh M, Sun J, Kornaros M (2022) Recent advances in wastewater microalgae-based biofuels production: A state-of-the-art review. Energy Rep 8:13253–13280

Barnard D, Casanueva A, Tuffin M, Cowan D (2010) Extremophiles in biofuel synthesis. Environ Technol 31(8–9):871–888

Behera S, Singh R, Arora R, Sharma NK, Shukla M, Kumar S (2015) Scope of algae as third generation biofuels. Front Bioeng Biotechnol. 11(2):90

Benner P, Meier L, Pfeffer A et al (2022) Lab-scale photobioreactor systems: principles, applications, and scalability. Bioprocess Biosyst Eng 45:791–813

Blazeck J, Hill A, Liu L, Knight R, Miller J, Pan A, Otoupal P, Alper HS (2014) Harnessing Yarrowia lipolytica lipogenesis to create a platform for lipid and biofuel production. Nat Commun 5:1–10

Cazzaniga S, Perozeni F, Baier T, Ballottari M (2022) Engineering astaxanthin accumulation reduces photoinhibition and increases biomass productivity under high light in Chlamydomonas reinhardtii. Biotechnol Biofuels Bioprod 15(1):77

Chen JW, Liu WJ, Hu DX, Wang X, Balamurugan S, Alimujiang A, Yang WD, Liu JS, Li HY (2017) Identification of a malonyl CoA-acyl carrier protein transacylase and its regulatory role in fatty acid biosynthesis in oleaginous microalga Nannochloropsis oceanica. Biotechnol Appl Biochem 64:620–626

Condor BE, de Luna MDG, Chang Y, Chen J, Leong YK, Chen P, Chen C, Lee D, Chang J (2022) Bioethanol production from microalgae biomass at high-solids loadings. Biores Technol 363:128002

Daliry S, Hallajisani A, Roshandeh JM, Nouri H, Golzary A (2017) Investigation of optimal condition for Chlorella vulgaris microalgae growth. Global J Environ Sci Manage. 3(2):217–230

Deng Z, Khatiwada JR, Fan L et al (2022) Algae for Wastewater Treatment and Production of Biofuels and Bioproducts. Int J Environ Res 16:34

Dhokane D, Bhadra B, Dasgupta S (2020) CRISPR based targeted genome editing of Chlamydomonas reinhardtii using programmed Cas9-gRNA ribonucleoprotein. Mol Biol Rep 47(11):8747–8755

El Nemr A, Hassaan MA, Elkatory MR, Ragab S, Pantaleo A (2021) Efficiency of Fe_3O_4 nanoparticles with different pretreatments for enhancing biogas yield of Macroalgae ulva intestinalis Linnaeus. Molecules 26(16):5105

Gao F, Zhang XL, Chang-Jian Zhu, Ke-Hao Huang, Qiang Liu (2022) High-efficiency biofuel production by mixing seawater and domestic sewage to culture freshwater microalgae. ChemEng J 443:136361

Gomma AE, Lee SK, Sun SM, Yang SH, Chung G (2015) Improvement in oil production by increasing malonyl-CoA and glycerol-3-phosphate pools in Scenedesmus quadricauda. Indian J Microbiol 55:447–455

Greiner A, Kelterborn S, Evers H, Kreimer G, Sizova I, Hegemann P (2017) Targeting of photoreceptor genes in Chlamydomonas reinhardtii via zinc-finger nucleases and CRISPR/Cas9. Plant Cell 29(10):2498–2518

Hasan M, Abedin MZ, Amin MB, Nekmahmud M, Oláh J (2023) Sustainable biofuel economy: A mapping through bibliometric research. J Environ Manage 336:117644

Horn S, Aasen I, Østgaard K (2000) Production of ethanol from mannitol by Zymobacter palmae. J Ind Microbiol Biotech 24:51–57

Hossain N, Mahlia TMI, Saidur R (2019) Latest development in microalgae-biofuel production with nano-additives. Biotechnol Biofuels 12:125

Huang J, Xia J, Jiang W, Li Y, Li J (2015) Biodiesel production from microalgae oil catalyzed by a recombinant lipase. Biores Technol 180:47–53

IEO (2017) https://www.eia.gov/outlooks/ieo/pdf/0484(2017).pdf

Jeswani HK, Chilvers A, Azapagic A (2020) Environmental sustainability of biofuels: A review. Proc Math Phys Eng Sci 476(2243):20200351

Khan S, Das P, Abdul Quadir M, Thaher MI, Mahata C, Sayadi S, Al-Jabri H (2023) Microalgal feedstock for biofuel production: Recent advances, challenges, and future perspective. Fermentation 9(3):281

Khoo KS, Ahmad I, Chew K W, Koji Iwamoto, Amit Bhatnagar, Pau Loke Show (2023) Enhanced microalgal lipid production for biofuel using different strategies including genetic modification of microalgae: A review. Prog Energy Combust Sci 96:101071

Kratzer R, Murkovic M (2021) Food ingredients and nutraceuticals from microalgae: Main product classes and biotechnological production. Foods 10:1626

Kurita T, Moroi K, Iwai M et al (2020) Efficient and multiplexable genome editing using Platinum TALENs in oleaginous microalga, Nannochloropsis oceanica NIES-2145. Genes Cells 25:695–702

Lakhawat SS, Malik N, Kumar V, Kumar S, Sharma PK (2022) Implications of CRISPR-Cas9 in developing next generation biofuel: A mini-review. Curr Protein Pept Sci 23(9):574–584
Li Z, Meng T, Ling X, Li J, Zheng C, Shi Y, Chen Z, Li Z, Li Q, Lu Y (2018) Overexpression of malonyl-CoA: ACP transacylase in Schizochytrium sp. to improve polyunsaturated fatty acid production. J Agric Food Chem 66:5382–5391
Li S, Li F, Zhu X, Liao Q, Jo-Shu Chang, Shih-Hsin Ho (2022) Biohydrogen production from microalgae for environmental sustainability. Chemosphere 291(1): 132717
Lunprom S, Phanduang O, Salakkam A, Liao Q, Reungsang A (2019) A sequential process of anaerobic solid-state fermentation followed by dark fermentation for bio-hydrogen production from Chlorella sp. Int J Hydrog Energy 44:3306–3316
Mahmood T, Hussain N, Shahbaz A et al (2023) Sustainable production of biofuels from the algae-derived biomass. Bioprocess Biosyst Eng 46:1077–1097
Mat Aron NS, Khoo KS, Chew KW, Show PL, Chen W-H, Nguyen THP (2020) Sustainability of the four generations of biofuels—A review. Int J Energy Res 44:9266–9282
Mota GF, de Sousa IG, de Oliveira ALB et al (2022) Biodiesel production from microalgae using lipase-based catalysts: Current challenges and prospects. Algal Res 62:102616
Neag E, Stupar Z, Maicaneanu SA, Roman C (2023) Advances in biodiesel production from microalgae. Energies 16(3):1129
Nzayisenga JC, Farge X, Groll SL et al (2020) Effects of light intensity on growth and lipid production in microalgae grown in wastewater. Biotechnol Biofuels 13:4
Osman AI (2020) Mass spectrometry study of lignocellulosic biomass combustion and pyrolysis with NOx removal. Renew Energy 146:484–496
Palmer CM (1969) A composite rating of algae tolerating organic pollution (2). J Phycol 5(1):78–82
Phwan CK, Chew KW, Sebayang AH et al (2019) Effects of acids pre-treatment on the microbial fermentation process for bioethanol production from microalgae. Biotechnol Biofuels 12:191
Saini N, Kaushik Pal, Sujata, Bansal Deepak, Sharma Mona (2021) Thermophilic algae: A new prospect towards environmental sustainability. J Clean Prod 324:129277
Selim KA, El-Ghwas DE, Easa SM, Abdelwahab Hassan MI (2018) Bioethanol a microbial biofuel metabolite; new insights of yeasts metabolic engineering. Fermentation 4:16
Shokravi H, Shokravi Z, Heidarrezaei M, Ong HC, Koloor SSR, Petrů M, Lau WJ, Ismail AF (2021) Fourth generation biofuel from genetically modified algal biomass: Challenges and future directions. Chemosphere 285:131535
Singh A, Thompson B, He J, Gerpen JV (2006) Process optimization of biodiesel production using alkaline catalysts. Appl Eng Agric 22(4):597–600
Sukarni, Sudjito, Hamidi N et al. (2014) Potential and properties of marine microalgae Nannochloropsis oculata as biomass fuel feedstock. Int J Energy Environ Eng 5:279–290
Varela Villarreal J, Burgués C, Rösch C (2020) Acceptability of genetically engineered algae biofuels in Europe: opinions of experts and stakeholders. Biotechnol Biofuels 13:92
Wang J, Yin Y (2018) Fermentative hydrogen production using pretreated microalgal biomass as feedstock. Microb Cell Fact 17:22
Wang J, Yang H, Wang F (2014) Mixotrophic cultivation of microalgae for biodiesel production: Status and prospects. Appl Biochem Biotechnol 172:3307–3329
Wang K, Khoo KS, Chew KW, Selvarajoo A, Chen W-H, Chang J-S, Show PL (2021) Microalgae: The future supply house of biohydrogen and biogas. Front Energy Res 9:660399
Xu YP, Duan PG, Wang F et al (2018) Liquid fuel generation from algal biomass via a two-step process: effect of feedstocks. Biotechnol Biofuels 11:83
Xu W, Ding K, Hu L (2021) A mini review on pyrolysis of natural algae for bio-fuel and chemicals. Processes 9:2042. https://doi.org/10.3390/pr9112042
Yun HS, Kim YS, Yoon HS (2021) Effect of different cultivation modes (photoautotrophic, mixotrophic, and heterotrophic) on the growth of Chlorella sp. and biocompositions. Front Bioeng Biotechnol 17(9):774143

Chapter 15
Fatty Acid Composition of Inedible Seeds as a Source of Bioenergy

Naga Mangalam Sudheep, Kalingar Subrahmanya Muralikrishna, and Kandikere Ramaiah Sridhar

Abstract Edible cultivated fruit trees and wild edible fruit trees have been considered important sources of human nutrition and health. Inedible fruits and their seeds, as non-timber forest products, are highly valuable natural sources for the production of bioenergy. Inedible seeds of seven tree species of the Western Ghats have been compared for fatty acid methyl esters (FAMEs) as energy sources. Seeds of *Albizia saman, Pongamia pinnata* and *Delonix regia* are found to be suitable for the production of biodiesel based on their FAMEs content, total fatty acids and fatty acids with carbon chain lengths 16–18. The use of FAMEs from these seeds could be improved using alternate methods of total lipid extraction, modification of the quality of FAMEs and assessment of their essential oils. Development of ecofriendly energy parks and agroforests with these tree species offers a renewable source of energy against fossil fuels to meet future energy demand.

Keywords Biodiesel · Energy parks · Conservation · FAMEs · Western Ghats

15.1 Introduction

There is a huge demand for bioenergy worldwide due to depletion of the conventional energy resources. The developed countries have initiated employing biomass using advanced technology to improve the renewable energy, whereas the developing countries are dependent on conventional bioresources (Joshi et al. 2016). India requires

N. M. Sudheep
Institute of Integrated Programmes and Research in Basic Sciences, Mahatma Gandhi University, Priyadarshini Hills, Kottayam, Kerala, India

K. S. Muralikrishna
Indian Council of Agricultural Research—Central Plantations Crops Research Institute, Kasaragod, Kerala, India

K. R. Sridhar (✉)
Department of Biosciences, Mangalore University, Mangalagangotri, Mangalore, Karnataka, India
e-mail: kandikeremanasa@gmail.com

M. Shah and D. Deka (eds.), *Emerging Sustainable Technologies for Biofuel Production*, Environmental Science and Engineering, https://doi.org/10.1007/978-3-031-52167-6_15

about 550–600 million tonnes of biomass per annum as a primary source of energy to cater to the needs of up to 70% of the population (http://mnre.gov.in/). Owing to such increased demand for energy, exploration of alternate or indigenous renewable feedstocks (e.g., bioindustrial wastes, agricultural residues and forest biproducts) to replace the petrochemical-based fuels. In principle, the use of edible oil seeds for the production of energy has been ruled out as they are required for human nutrition (Islam et al. 2018). Forest and plantation-based inedible oils as feedstock are in demand for bioenergy production similar to conventional seeds (e.g., Neem: *Azadiricta*; Physic Nut: *Jatrpopha*; Castor: *Ricinus*) (Chhetri et al. 2008; Ahmed et al. 2011; Prasad et al. 2012; Patel et al. 2013). Oil extracted from inedible seeds blended in a suitable proportion (10–20%) with diesel resulted in no change in engine performance (Oza et al. 2012).

The physical characteristics of fossil diesel and fatty acid methyl esters (FAMEs) (density, lower heating value, viscosity, cetane number and fuel equivalence) are almost similar, hence, a mixture of FAMEs is also referred to as biodiesel. The advantage of biodiesel is that it is non-toxic, biodegradable and renewable. Up to 77 plant species as sources of potential inedible seeds could be exploited as feedstock in India as they possess an oil content of about 30% and above (Azam et al. 2005). Advantages of these plants include their renewable source of energy, their cultivation in wastelands, their adaptation as agroforestry crops, their ability to withstand extreme climatic conditions and yield of seeds over a long period of time. In addition, FAMEs derived from inedible seeds have several medicinal applications. Thus, new cultivable energy crops should be harnessed for sustainable energy generation without using edible oils or encroaching on the land necessary for food production. According to Azam et al. (2005), the FAMEs of 26 plant species are suitable to use as biodiesel owing to their fulfiling the standards necessary for biodiesel (e.g., *Azadirachta, Calophyllum, Jatropha* and *Pongamia*). The present chapter aims to evaluate the FAMEs of seven inedible forest tree seeds obtained from the Western Ghats to forecast the baseline data and significance of inedible wild seeds as a source of bioenergy for future applications.

15.2 Source of Seeds

Seeds of seven tree species: *Adenanthera pavonina* L., *Albizia saman* (Jacq.) Merr., *Casia fistula* L., *Cassia Javanica* L., *Delonix regia* (Hook.) Raf., *Pongamia pinnata* (L.) Pierre and *Senna spectabilis* (DC.) H.S. Irwin & Barneby were collected from the Western Ghat region for evaluation of FAMEs (Fig. 15.1). Seeds were sampled from three plant species as replicates. They were sun-dried and whole seeds were milled (Wiley Mill, mesh 30) to extract the total lipids.

Fig. 15.1 Seeds of *Adenanthera pavonina* (**a**), *Albizia saman* (**b**), *Casia fistula* (**c**), *Cassia Javanica* (**d**), *Delonix regia* (**e**), *Pongamia pinnata* (**f**), *and Senna spectabilis* (**g**)

15.3 Extraction of Total Lipids

Flours of seeds (1 g; moisture, 5–6%) were transferred to thimbles were arranged in a Soxhlet extractor to extract total lipids using petroleum ether (200 ml) at 60–80 °C (AOAC 1999). The pace of condensation was fixed at 160 drops/min up to 7 h, the samples were cooled, transferred to a pre-weighed Petri dish, evaporated to dryness at laboratory temperature (28 ± 2 °C) and the total lipid quantity was determined gravimetrically.

15.4 Gas Chromatography

The esterified 100 μl lipid samples taken in ampules were diluted with HPLC grade n-hexane (900 μl). One μl as an aliquot was injected into the gas chromatograph (GC-2010, Shimadzu, Japan), equipped with an auto injector (AOI) as well as the capillary column (BPX-70). The latter was allowed to be conditioned for 10 h before use. The flame ionization detector (FID) detected the elutants and the signals were monitored using GC-Solutions software. The analytical specifications were followed based on Nareshkumar (2007) (Box 1). The quantity of FAMEs was determined using a standard mixture (C4-C24; Sigma, USA), which was run similar to the samples. The concentration as well as area of peaks of FAMEs were computed by the GC Postrun analysis software (GC–MS Solution version 4.52, Shimadzu Corporation).

Box 1. Details of analytical conditions of gas chromatography (modified from Nareshkumar 2007).

1. Auto-sampler settings: Injection sample volume, 1 μl; terminal air gap, nil; number of rinses with solvent during pre-run, 4; number of rinses with solvent during post-run, 6; number of rinses with sample, 5; washing volume, 8 μl; plunger suction and injection speed, high; syringe injection speed, low; injection port dwell time, 1 sec.
2. Injection port settings: Injection mode, split; temperature, 225 °C; carrier gas, N2/air; pressure, 114.9 kPa; column flow, 1.29 ml/min; linear velocity, 34 cm/sec; purge flow, 3ml/min; split ratio, 50.
3. Column oven settings: initial temperature, 100 °C.
4. Column oven temperature program: equilibrium time, 3 min; total program time, 30 min.
5. Temperature hold time: 100 °C, 1 min; 220 °C, 5 min; rate, 5 °C/min.
6. Column information: column, BPX-70; film thickness, 0.25 μm; inner diameter, 0.25 mm; column length, 30 m; column maximum temperature, 260 °C.
7. Detector settings: detector, FID, temperature, 280 °C; makeup gas, N2/air; makeup flow, 30 ml/min; H2 flow, 47ml/min; airflow, 400 ml/min; sampling rate, 40 ms; stop time, 30 min; delay time, nil).

15.5 Fatty Acid Methyl Esters

A total of 28 fatty acids (14 saturated and 14 unsaturated) were obtained from the total lipids of seven inedible seeds (Table 15.1). The seeds of *A. pavonina* consist of the highest quantity of palmitic acid followed by oleic, cis-10-heptadecenoic and stearic acids. The seeds of *A. saman* consist of the highest quantity of oleic acid, followed by behenic, linoleic, Cis-11-eicosadienoic, palmitic, stearic and tricosanoic acids. Seeds of *C. fistula* showed a relatively low profile of FAMEs with the highest quantity of oleic acid (307%), followed by palmitic acid (2%). Seeds of *C. javanica* possess the highest quantity of linoleic acid, followed by stearic and palmitic acids. Seeds of *D. regia* possess the highest quantity of linoleic acid, followed by palmitic, oleic and stearic acids. Seeds of *P. pinnata* possess the highest quantity of stearic acid followed by Cis-10-pentadecenoic, Cis-11,14,17-eicosadienoic, oleic, linoleic, arachidic, palmitic and lignoceric acids. Seeds of *S. spectabilis* consist of palmitic acid, followed by oleic, stearic, palmitic and linoleic acids.

Total FAMEs with carbon chain length 16–18 were highest in *P. pinnata*, followed by *A. saman* and *D. regia*, while they were least in *C. fistula* (Table 15.2). Seeds of *P. pinnata, A. saman* and *D. regia* seem to be well suited for biodiesel production.

Table 15.1 Fatty acid methyl esters (g/100 g oil) (n = 3, mean ± SD; BDL, below detectable level)

Fatty acid	*Adenanthera pavonina*	*Albizia saman*	*Cassia fistula*	*Cassia javanica*	*Delonix regia*	*Pongamia pinnata*	*Senna spectabilis*
Butyric acid (C4:0)	BDL	1.75 ± 0.02	BDL	BDL	BDL	5.48 ± 0.03	BDL
Caproic acid (C6:0)	11.91 ± 0.36	2.15 ± 0.03	BDL	BDL	BDL	6.97 ± 0.04	BDL
Caprylic acid (C8:0)	3.65 ± 0.11	2.98 ± 0.04	BDL	1.16 ± 0.04	BDL	BDL	7.632 ± 0.06
Capric acid (C10:0)	BDL	BDL	BDL	0.78 ± 0.02	BDL	BDL	BDL
Lauric acid (C12:0)	2.91 ± 0.08	4.69 ± 0.06	BDL	4.51 ± 0.16	BDL	BDL	14.78 ± 0.11
Myristic acid (C14:0)	8.17 ± 0.24	4.21 ± 0.06	BDL	2.09 ± 0.07	0.77 ± 0.01	6.61 ± 0.04	8.94 ± 0.07
Myristolic acid (C14:0)	BDL	4.64 ± 0.06	BDL	BDL	BDL	BDL	BDL
Cis-10- pentadecenoic acid (C15:1))	7.44 ± 0.22	2.92 ± 0.04	BDL	BDL	BDL	428.71 ± 2.67	BDL
Palmitic acid (C16:0)	137.90 ± 4.19	148.78 ± 2.14	1.98 ± 0.07	54.22 ± 1.93	80.67 ± 0.74	56.2 ± 0.35	342.12 ± 2.71
Palmitoleic acid (C16:1)	2.69 ± 0.08	17.76 ± 0.25	0.05 ± 0.002	2.15 ± 0.07	0.92 ± 0.01	BDL	43.61 ± 0.34
Heptadecenoic acid (C17:0)	BDL	2.35 ± 0.03	BDL	BDL	0.63 ± 0.01	BDL	BDL
Cis-10-heptadecenoic acid (C17:1)	74.50 ± 2.26	BDL	BDL	BDL	BDL	BDL	BDL
Stearic acid (C18:0)	64.12 ± 1.95	138.38 ± 1.99	BDL	10.89 ± 0.38	54.57 ± 0.5	2118.53 ± 13.21	45.59 ± 0.36
Oleic acid (C18:1)	107.41 ± 3.26	306.57 ± 4.41	3.73 ± 0.13	65.74 ± 2.34	56.96 ± 0.52	123.86 ± 0.77	123.83 ± 0.98
Linoleic acid (C18:2)	BDL	204.61 ± 2.94	0.88 ± 0.03	144.12 ± 5.14	BDL	BDL	40.14 ± 0.31
Linolenic acid (C18:3)	BDL	BDL	BDL	2.27 ± 0.08	378.02 ± 3.46	97.87 ± 0.61	BDL
Arachidic acid (C20:0)	30.72 ± 0.93	BDL	0.15 ± 0.01	3.96 ± 0.14	4.71 ± 0.04	68.73 ± 0.42	15.08 ± 0.11
Cis-11-eicosadienoic acid (C20:1)	8.91 ± 0.27	185.94 ± 2.67	BDL	0.93 ± 0.03	0.6 ± 0.01	16.35 ± 0.10	11.99 ± 0.09
Cis-11,14-eicosadienoic acid (C20:1)	BDL	1.71 ± 0.02	BDL	BDL	BDL	BDL	BDL
Cis-11,14,17-eicosadienoic acid (C20:3)	BDL	3.13 ± 0.04	BDL	0.51 ± 0.01	1.25 ± 0.01	135.76 ± 0.84	BDL

(continued)

Table 15.1 (continued)

Fatty acid	*Adenanthera pavonina*	*Albizia saman*	*Cassia fistula*	*Cassia javanica*	*Delonix regia*	*Pongamia pinnata*	*Senna spectabilis*
Arachidonic acid (C20:4)	BDL	1.29 ± 0.01	BDL	BDL	BDL	BDL	BDL
Behenic acid (C22:0)	38.24 ± 1.16	268.07 ± 3.85	0.05 ± 0.002	3.48 ± 0.12	BDL	BDL	16.73 ± 0.13
Erucic acid (C22:1)	3.08 ± 0.09	5.23 ± 0.07	BDL	BDL	1.72 ± 0.01	BDL	BDL
Cis-13,16-docosadeienoic acid (C22:2)	BDL	4.23 ± 0.06	BDL	BDL	2.58 ± 0.01	6.22 ± 0.03	BDL
Cis-4,7,10,13,16,19-docosahexaenoic acid (C22:6)	2.69 ± 0.08	BDL	BDL	0.97 ± 0.03	2.11 ± 0.02	5.59 ± 0.03	BDL
Tricosanoic acid (C23:0)	3.13 ± 0.09	42.54 ± 0.61	BDL	BDL	BDL	BDL	BDL
Lignoceric acid (C24:0)	BDL	7.38 ± 0.1	BDL	5.54 ± 0.19	1.19 ± 0.01	53.86 ± 0.03	13.36 ± 0.10
Nervonic acid (C24:1)	244.11 ± 7.43	BDL	0.06 ± 0.002	0.96 ± 0.03	BDL	BDL	BDL

Table 15.2 FAMEs with carbon chain lengths between 16 and 18 (g/100 g oil)

	Adenanthera pavonina	*Albizia saman*	*Cassia fistula*	*Cassia javanica*	*Delonix regia*	*Pongamia pinnata*	*Senna spectabilis*
Cis-10-pentadecenoic acid (C15:1)	7.44	2.92	0	0	0	428.71	0
Palmitoleic acid (C16:1)	2.69	17.76	0.05	2.15	0.92	0	43.61
Heptadecenoic acid (C17:0)	0	2.35	0	0	0.63	0	0
Cis-10-heptadecenoic acid (C17:1)	74.5	0	0	0	0	0	0
Stearic acid (C18:0)	64.12	138.38	0	10.89	54.57	2118.53	45.59
Oleic acid (C18:1)	107.41	306.57	3.73	65.74	56.96	123.86	123.83
Linoleic acid (C18:2)	0	204.61	0.88	144.12	0	0	40.14
Linolenic acid (C18:3)	0	0	0	2.27	378.02	97.87	0
Total	256.16	672.59	4.66	225.17	491.1	2768.97	253.17

15.6 Total Fatty Acids and Ratios

The total unsaturated fatty acids (TUFA) were highest in *A. saman,* followed by *P. pinnata* and *C. javanica* (Table 15.3). The total saturated fatty acids (TSFA) were highest in *S. spectabilis* seeds, followed by *P. pinnata, A. saman* and *A. pavonina.* The ratio of TUFA/TSFA was highest in *A. saman*, followed by *C. javanica* and *C. fistula.* Polyunsaturated fatty acids (PUFA) were the highest in *A. saman* followed by *P. pinnata* and *C. javanica.* Monounsaturated fatty acids (MUFA) were also highest in *A. saman* followed by *P. pinnata, A. pavonina* and *S. spectabilis* seeds. The ratio of PUFA/MUFA was highest in *A. saman, P. pinnata* and *S. spectabilis* seeds.

Table 15.3 Total fatty acids (g/100 g oil) and ratio of fatty acids

	Adenanthera pavonina	*Albizia saman*	*Cassia fistula*	*Cassia javanica*	*Delonix regia*	*Pongamia pinnata*	*Senna spectabilis*
TUFA	112.79	991.82	3.99	214.41	64.33	613.51	95.74
TSFA	300.68	345.49	2.18	86.62	144.9	576.91	806.34
TUFA/TSFA	0.38	2.87	1.84	2.48	0.44	1.06	0.12
PUFA	11.6	406.13	0.88	146.53	1.85	198.8	52.13
MUFA	354.6	582.02	6.18	66.70	56.96	443.12	123.83
PUFA/MUFA	0.033	0.70	0.14	2.20	0.032	0.45	0.42

15.7 Discussion

Among the seven inedible seeds investigated in our study, investigations have been carried out on *P. pinnata* as a potential source of biodiesel (e.g., Ahmed et al. 2011; Islam et al. 2018). The current study also advocates the use of FAMEs from *P. pinnata* seeds as a top candidate for biodiesel production.

The FAMEs, consisting of carbon chain lengths between 16 and 18, show boiling points between 330 and 357 °C, which meets the required specification necessary for biodiesel (360 °C) (Ahmed et al. 2011). In the current study, total FAMEs with carbon chain lengths between 16 and 18 were highest in *P. pinnata* followed by *A. saman* and *D. regia*. These seeds seem to be well suited for biodiesel production. In addition, linolenic acid concentration and acids possessing four double bonds in FAMEs should not be more than 12% and 1%, respectively (Pasto et al. 1992). In our study, these FAMEs were in limited quantity and almost all seeds were within the specified limit, thus qualifying them for the production of biodiesel with a few exceptions. The present study results suggest using seeds of *A. pavonina, S. spectabilis, P. pinnata* and *A. saman* as suitable feedstock for the production of biodiesel.

Methods of total lipid extraction selectively change the composition of FAMEs, thus advocating the use of suitable methods favourable for biodiesel production in inedible seeds. In *Canavalia* seeds, alterations in quantity of saturated fatty acids (UFA) and unsaturated fatty acids (SFA) in FAMEs were dependent on the method of total lipid extraction (hot extraction by Soxhlet with petrolium ether; cold extraction with a blend of chloroform–methanol-water) (AOAC 1999; Bligh and Dyer 1959; Supriya et al. 2012). Among these methods, cold extraction proved the best for *Canavalia* seeds. Cold extraction methods need to be performed for the inedible seeds in this study, which may give a better perspective than hot extraction.

Assessment of FAMES needs to be performed for several properties to fulfil the needs of biodiesel, especially cetane number, low temperature properties, viscosity and stability of the products. Viscosity is one of the important features of biodiesel and is dependent on the quantity and kind of fatty acids in the FAMEs. In addition to the present study, the determination of desired qualities of FAMEs obtained in inedible seeds to fulfil as biodiesel needs to be addressed (e.g., density, viscosity, surface tension, cetane number and flash points).

The determination of essential oils in inedible seeds has several advantages. Although essential oils are used in aroma industries, the byproducts of such industries are of immense value as they can be used in biodiesel production (Rahman et al. 2019). Essential oils possess similar qualities to diesel, especially the heating value and lower viscosity, which have potential to produce fuel. An appropriate blend of essential oils with diesel results in a reduction in smoke and carbon monoxide emissions (Fattah et al. 2013; Zare et al. 2016).

15.8 Conclusion

This study compared the composition of FAMEs in seven inedible seeds of the Western Ghats, where except for *C. fistula*, the rest of the six seeds showed a substantial amount of FAMEs. Based on the composition of FAMEs (qualitative and quantitative), total fatty acids and FAMEs with carbon chain lengths between 16 and 18, in order of preference, *P. pinnata, A. saman, S. spectabilis* and *Delonix regia* are suitable for the production of biodiesel. The latter three inedible seeds are the new source for the prospective production of bioenergy. However, alternate methods of extraction of FAMEs, improvement of the qualities of FAMEs suitable for production of biodiesel and assessment of the essential oils of these seeds are worth exploring. Being ecofriendly, these tree species deserve to be grown in plantations and agroforests as energy parks to meet the energy challenges of the future.

Acknowledgements The authors are grateful to Mangalore University and the Department of Biosciences for offering the laboratory facilities. We are thankful to Dr. S. Mahadevakumar, Scientist, Botanical Survey of India, Port Blair, for constructive suggestions and logistical support.

References

Ahmad M, Khan MA, Zafar M, Sultana S (2011) Biodiesel from non edible oil seeds: a renewable source of bioenergy, economic effects of biofuel production. In: Bernardes MADS (ed), InTech, pp 259–280

Association of Official Analytical Chemists (AOAC) (1999) In: Cunniff P (ed), Official Methods of Analysis, 16th Edition, AOAC International, Maryland, USA

Azam M, Waris A, Nahar NM (2005) Prospects and potential of fatty acid methyl esters of some non-traditional seed oils for use as bio-diesel in India. J Biomass Bioener 29:293–302

Bligh EG, Dyer WJ (1959) A rapid method of total lipid extraction and purification. Can J Biochem Physiol 37:911–917

Chhetri AB, Tango MS, Budge SM, Watts KC, Islam MR (2008) Non-edible plant oils as new sources for biodiesel production. Int J Mol Sci 9:169–180

Fattah IMR, Masjuki HH, Liaquat AM, Ramli R, Kalam MA, Roiazuddin VN (2013) Impact of various biodiesel fuels obtained from edible and non-edible oils on engine exhaust gas and noise emissions. Renew Sustain Energy Rev 18:552–567

Islam AKMA, Primandari SRP, Yaakob Z (2018) Non-edible vegetable oils as renewable resources for biodiesel production: South-East Asia perspective. In: Soneji JR (ed) Nageshwara-Rao M. Advances in biofuels and bioenergy, InTech, pp 201–214

Joshi P, Sharma N, Sarma PM (2016) Assessment of biomass potential and current status of bio-fuels and bioenergy production in india. Curr Biochem Eng 3:4–15

Nareshkumar S (2007) Capillary gas chromatography method for fatty acid analysis of coconut oil. J Plant Crops 35:23–27

Oza NP, Rahtod PP, Patel NK (2012) A Review of Recent Research on non-edible vegetable oil as fuel for CI Engine. J Irrig Drain Eng 3:84–86

Pasto D, Johnson C, Miller M (1992) Experiments and techniques in organic chemistry. Prentice Hall, New Jersey, USA

Patel NK, Nagar PS, Shah DSN (2013) Identification of non-edible seeds as potential feedstock for the production and application of bio-diesel. Ener Pow 3:67–78

Prasad S, Dhanya MS, Gupta Nm, Kumar A (1012) Biofuels from biomass: a sustainable alternative to energy and environment. Biochem Cell Arch 12:255–260

Rahman SMA, Van TC, Hossain FM, Jafari M, Dowell A, Isla MAQ, Nabi MN, Marchese AJ, Tryner J, Rainey T, Ristovski ZD, Brown RJ (2019) Fuel properties and emission characteristics of essential oil blends in a compression ignition engine. Fuel 238:440–453

Supriya P, Sridhar KR, Nareshkumar S, Ganesh S (2012) Impact of electron beam irradiation on fatty acid profile of *Canavalia* seeds. Food Bioproc Technol 5:1049–1060

Zare A, Nabi MN, Bodisco TA, Hossain FM, Rahman MM, Ristovski ZD, Brown RJ (2016) The effect of triacetin as a fuel additive to waste cooking biodiesel on engine performance and exhaust emissions. Fuel 182:640–649

Chapter 16
Molecular Mechanisms Behind the Plant/ Microbe Induced Biofuel Production: Systems Biology Approach

Jyoti K. Choudhari, J. Anandkumar, Jyotsna Choubey, and Biju P. Sahariah

Abstract Biofuel production by plants and microbial consortia often involves production or release of enzymes, which catalyze the transformation of simple compounds into more complex molecules. Such enzymes, which are mainly secreted by microorganisms, can be released into the environment and act on the substrates that are available in the root exudates, or on the cell wall of the plants. These enzymes are the main contributor for the conversion of molecules into biofuels by plants or microbes. The first step of application of microbes for biofuel production is identifying bacteria capable of producing the desired fuel. Microbial fuel cells can allow the identification of such bacteria. The cost and efficiency of biofuel production is determined by the capabilities of the microbe being used. Currently, some of the most promising bacteria being used in biofuel production are *Clostridia, Bacillus* and *E. coli*. Once the appropriate microbe has been identified, it is commonly genetically engineered to increase fuel production, i.e., incorporating genes that increase protein production or adding in genes related to metabolism pathways. Once the microbe has been engineered, the microbe is typically then harvested and converted into biofuels. This can be done through a fermentation process. The fermentation process typically involves nutrients and a carbon source. The microbe is then manipulated to convert the energy source into a desired product, in this case a biofuel. The Microbe-induced biofuel production is a method of producing biofuels utilizing microorganisms mostly algae and bacteria. This involves the cultivation of microorganisms in an environment where they can convert energy sources such as sunlight and carbon dioxide into oils that can be converted into biofuels. These microorganisms are usually grown in photo-bioreactors, which allow them to optimize their energy production. Utilizing microbes to produce biofuels has a number of advantages, such as being more efficient than traditional biofuel production methods and requiring little

J. K. Choudhari
Maulana Azad National Institute of Technology, Bhopal, Madhya Pradesh 462003, India

J. Anandkumar · B. P. Sahariah (✉)
National Institute of Technology Raipur, Raipur, Chhattisgarh 492010, India
e-mail: biju.sahariah@gmail.com

J. Choubey
Raipur Institute of Technology Raipur, Raipur, Chhattisgarh 492001, India

M. Shah and D. Deka (eds.), *Emerging Sustainable Technologies for Biofuel Production*, Environmental Science and Engineering, https://doi.org/10.1007/978-3-031-52167-6_16

external input. One limitation about microbe-induced biofuel production is the fact that it created biofuels that are still relatively expensive to produce. Present Chapter describe cellular and molecular mechanisms in microbial cell for biofuel production and predominant genetic material associated with the same. This will help finding more biofuel production efficient organisms beneficial for sustainable development.

Keywords Biofuel · Genetically engineered organisms · Enzyme production · Metabolism pathways

16.1 Introduction

Biomass, a renewable and sustainable source of energy, is derived from various organic materials, including microorganisms, animal waste and plants. The use of peanut oil for diesel first occurred around a hundred years ago; however, today these fuels are generally blended with petroleum-based fuels but can be used independently under specific conditions. For an engine to operate correctly with biofuel, it must comply with the performance criteria required for conventional engines and needs to work well in tandem with existing technology utilizing traditional petroleum-derived fuels. Consequently, there may be considerable physicochemical similarity between biofuels and their petroleum counterparts due to this overlap in characteristics. The term "biofuel" describes a fuel category that is produced through biomass conversion in a much shorter timeframe compared to fossil fuels, namely oil, which undergoes a significantly slower natural process. This biomass can originate from a variety of sources, including plants or biowaste produced as a result of agricultural, domestic, or industrial activities (Roberts and Smagala 2022). Biofuels are a promising substitute for fossil fuels, and investment in their commercial production started in the 1970s to mitigate carbon dioxide emissions. Biofuel has since grown exponentially, with over 100 billion liters produced in 2010 (Paul et al. 2019). The technology is categorized into various generations of development as depicted in Fig. 16.1.

Biofuels are usually classified as follows:

First generation—Conventional biofuels, derived mainly from food crops containing sugar, starch and vegetable oil, are the predominant type currently produced in commercial volumes (Lee and Lavoie, 2013).

Second generation—Advanced biofuels are produced through the utilization of non-edible vegetation and lignocellulosic refuse, whereby an enzymatic breakdown and fermentation process is necessary for manufacturing (Dahman et al. 2019).

Third generation—algae biofuels are sometimes referred to as third-generation biofuel and are manufactured from photosynthetic algae (Neto et al. 2019).

In contrast to alternative sources of sustainable energy, biomass can be transformed into liquid fuels known as "biofuels" for the purpose of fulfilling transportation fuel requirements. The primary biofuel technologies currently employed are ethanol and biodiesel, both belonging to the initial generation of such developments.

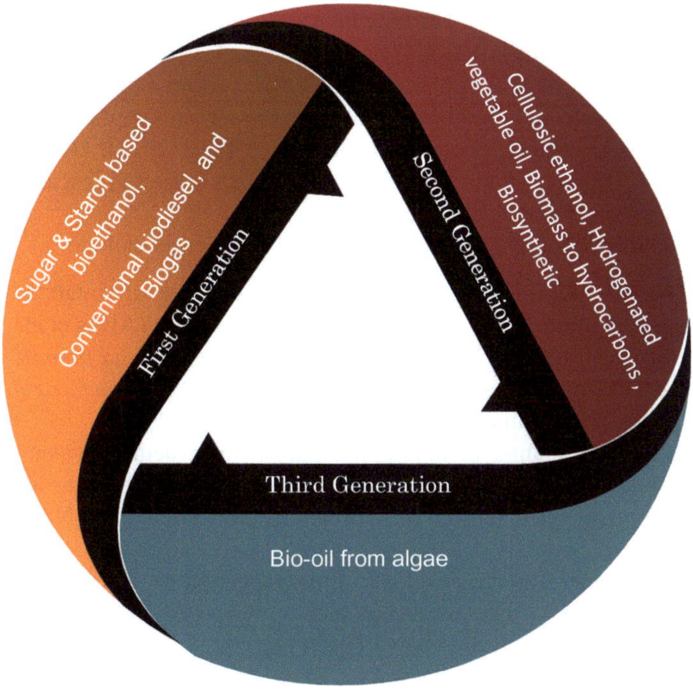

Fig. 16.1 Technological generations of biofuel technology

16.2 Sources of Biofuel Raw Materials

16.2.1 Plant-Based Biofuels

Biofuels from plants are the most productive source of renewable fuels. Starch, cellulose, hemicellulose, and oils can be used to produce ethanol and butanol (as gasoline additives) and long-chain hydrocarbons (for diesel additives or as jet fuel). The source of the energy absorbed by plants is the sun, which will be a constant source of energy for the next billion years. The carbon released when biofuels are burned is in a constant cycle and is not released from old, solid carbon sources as is the case with fossil oil and natural gas. Biodiesel can be produced from numerous oils and fats, including virgin vegetable oils such as canola, soybeans and camelina, used cooking oils (Hood 2016). Various plants can be used for biofuel production, including Jatropha, microalgae, and lignocellulosic biomass (Sharma et al. 2020). Jatropha is a plant that produces oily seeds that can be used to produce biodiesel. It is a drought-resistant plant and can be grown in arid regions, making it an ideal resource for biofuels production in areas with limited water resources. Microalgae are considered promising feedstock for the production of biofuels and other bioactive compounds (Chia et al. 2018). Lignocellulosic biomass, such as wood and agricultural residues, can be converted into biofuels through various processes, including fermentation and gasification (Sharma et al. 2020). In addition, genetic and metabolic techniques have been used to improve the potential of woody plants such as poplar and eucalyptus for biofuel production (Tang and Tang, 2014). Overall, there are several plant-based options for biofuel production, and ongoing research is focused on improving their efficiency and feasibility.

The most promising plant species for biofuel production include:

First-generation biofuel crops like sugarcane and maize, although their use may decline in the future due to concerns about food security (Alves et al. 2015).

Non-edible plant oils, such as those from Karanja (*Pongamia pinnata*), Jatropha (*Jatropha curcas*), *Leonurus artemisia* L. (LAL), and *Peganum harmala L.* (PHL) (Arefin et al. 2021; Banković-Ilić et al. 2012).

Floating aquatic plants, including Azolla, water hyacinth, water fern, water lettuce, and duckweed, which can be processed into biofuel through methods like transesterification, pyrolysis, hydrolysis, and torrefaction (Nwosu-Obieogu et al. 2016).

Dedicated perennial crops developed through genomics research and breeding programs, which can increase the efficiency of breeding programs and shorten the length of breeding cycles (Satya 2015).

These plant species offer renewable, environmentally friendly, and energy-efficient alternatives to fossil fuels for biofuel production.

16.2.2 Plant Oils to Biofuels

Plant oils are a great alternative to traditional biofuels such as ethanol. These fuels can be used for transportation and heating purposes, making them a viable option for the future of biofuel production (Vijayakumar et al. 2016). Since they don't compete with food crops like maize or soybeans for space on our plates or in our refrigerators, it is frequently generated from non-food crops. In contrast to corn-based ethanol, which needs a lot of water during cultivation and processing, they don't additionally require any more water during production. Plant oils may be transformed into fuel for vehicles using processes that do not need chemical processing, such as fermentation or distillation; they are already available in nature. Oils extracted from plants include a high concentration of fatty acids and can be utilised in place of animal fats. These oils are commonly utilised because of their high fatty acid content. They have a high energy density, making them suitable for use as biofuel. Hydrogenation and transesterification processes are used to transform plant oils into biodiesel. The two most common types of biofuels in use today are ethanol, a fuel produced by fermentation of sugars, and biodiesel, a fuel produced by transesterification of vegetable oils. Diesel fuel of high grade, usable in internal combustion engines, is the product of this procedure. When compared to other biofuels like soybean oil or maize oil, the key benefit of plant oils is their cheap cost. The main raw materials used in the production of biodiesel are rapeseed oil, canola oil, soybean oil, sunflower oil, tallow from cattle and sheep and poultry fat. There are various other sources of raw material: almond andiroba (*Carapa guianensis*), babassu (*Orbignia sp.*), barley, camelina (*Camelina sativa*), coconut, copra, cumaru (*Dipteryx odorata*), *Cynara cardunculus* (oil extracted from a plant native to North America), fish oil, groundnut (*Arachis hypogaea*), Jatropha curcas (a plant native to Africa but now grown in many regions), karanja (*Pongamia glabra*; a leguminous tree native to Asia), laurel (*Laurus nobilis or Laurus spp.*), *Lesquerella fendleri* (a shrub related to the laurel genus), *Madhuca indica*, microalgae (*Chlorella vulgaris*), oat, piqui (*Caryocar sp.*), poppy seed, rice, rubber seed, sesame, sorghum, tobacco seed, and wheat (Demirbas 2009; Vijayakumar et al. 2016). There are more than 350 oil-bearing crops that have been identified, of which only soybean, palm, sunflower, safflower, cottonseed, rapeseed and peanut oils are considered potential alternative fuels for diesel engines. Oil-bearing crops are among the most important and diverse crops cultivated worldwide. However, only soybean, palm, sunflower, safflower, cottonseed, rapeseed, and peanut oils are considered potential alternative fuels for diesel engines. These oils contain a variety of fatty acids that can be used to produce liquid fuels like diesel fuel or jet fuel. The use of oil-bearing crops as biofuels has been growing in recent years because they can be grown on marginal lands where other crops cannot grow. These also provide a sustainable resource that can help meet our energy needs while conserving natural resources such as water and land.

16.2.2.1 Non-Edible Plant Oils

Non-edible plant oils are a viable alternative to edible oils for biodiesel production, as they do not compete with food supply and can be more cost-effective. Some examples of non-edible plant oils used for biodiesel production include Karanja (*Pongamia pinnata*) oil, Jatropha (*Jatropha curcas*) oil, Leonurus artemisia L. (LAL) oil, and Peganum harmala L. (PHL) oil (Banković-Ilić et al. 2012; Chang et al. 2017; Sahu et al. 2011). These oils can be converted into biodiesel through processes such as transesterification, which involves the use of an alkali catalyst like KOH and methano (Sahu et al. 2011). The fuel properties of biodiesel produced from non-edible plant oils have been found to be within the range of ASTM specifications, making them suitable for use in internal combustion engines (Sahu et al. 2011).

16.2.2.2 Edible Plant Oils

Edible plant oils are derived from various plant sources and are commonly used for cooking and other purposes. Some examples of edible plant oils are olive oil, sunflower oil, and canola oil.. These oils can be beneficial to health, for example, by providing antioxidants and mitigating oxidative stress (Abu Zeid et al. 2022). Metabolic engineering has been used to alter the composition of edible plant oils to improve their nutritional value by increasing the content of desirable components or reducing undesirable components (Zhou et al. 2023). However, it should be noted that some plant oils are used for non-edible purposes, such as biodiesel production, which usually uses non-edible oils and used cooking oils as feed stocks (Banković-Ilić et al. 2012; Ilham et al. 2019). In fact, there are many plants with the potential to produce biofuels (Table 16.1). Some of these plants include sugarcane, corn, and other first-generation biofuel crops. However, the use of food crops for biofuel production may threaten food security, leading to increased interest in next-generation biofuel crops (Satya 2015). Floating aquatic plants such as Azolla, water hyacinth, water fern, water lettuce, and duckweed have also been considered for biofuel production due to their physicochemical properties and potential for energy production (Arefin et al. 2021). Perennial plants are another option for biofuel production as they can provide a renewable resource for bioenergy. Genomics research can help improve the efficiency of breeding programs and shorten breeding cycles for these specific perennial biofuel crops (Alves et al. 2015) (Table 16.2).

16.2.3 Microbes-Based Biofuels

Microbes have been extensively studied as a potential source for the production of biofuels. They can convert a variety of feedstocks such as agricultural waste, lignocellulosic biomass, and algae into biofuels such as ethanol, butanol, and biodiesel. The advantage of using microbes for biofuel production is that they can convert these feedstocks into fuel in a highly efficient manner, and they can also produce a wide range of biofuels that are suitable for different applications. One of the most promising microbial biofuels is bioethanol. This is produced by fermenting sugars extracted from biomass using microorganisms such as yeast or bacteria. The resulting ethanol can be used as a fuel additive or blended with gasoline to produce a biofuel that can be used in standard combustion engines. Another promising microbial biofuel is butanol. Butanol has a higher energy content than ethanol and can be used as a drop-in replacement for gasoline in existing engines. It is produced by fermenting sugars with bacteria such as Clostridium acetobutylicum. Biodiesel is another microbial biofuel that is produced by using algae. Algae can be grown in ponds or photobioreactors, and the oil extracted from the algae can be converted into biodiesel using a process called transesterification. *Corynebacterium glutamicum, autoethanogenu Synechocystis sp., S. elongates, E. coli, Saccharomyces cerevisiae, Klebsiella pneumoniae, C. cellulolyticum, Thermoanaerobacterium saccharolyticum, Enterobacter cloacae, Methylobacterium extorqens, C. cellulovorans, Caldicellulosiruptor, Bescii, Clostridium Tyrobutyricum, acetobutylicum,* and *thermocellum* are all microorganisms that have been studied for their potential to produce biofuels.

Corynebacterium glutamicum is a species of gram-positive bacteria that is commonly used in industrial biotechnology for the production of various compounds, including amino acids, organic acids, and biofuels (Sasaki et al. 2019). In terms of

Table 16.1 Edible plant oils for biofuel production

NCBI Taxonomy ID	Scientific name	Common Name	Plant type
13894	*Cocos nucifera* (L)	Coconut	Tree
4577	*Zea mays*	Corn	Herb
3633	*Gossypium*	Cottonseed	Herb
3708	*Brassica napus*	Canola (Rapeseed)	Herb
4146	*Olea europaea*	Olive	Herb
3818	*Arachis hypogaea*	Peanut (groundnut)	Herb
4222	*Carthamus tinctorius*	Safflower	Herb
4182	*Sesamum indicum*	Sesame	Herb
3847	*Glycine max*	Soybean	Herb
4232	*Helianthus annuus*	Sunflower	Herb
3755	*Prunus amygdalus*	Almond	Tree
171929	*Anacardium occidentale*	Cashew	Shrub
13451	*Corylus avellana*	Hazelnut	Tree
60698	*Macadamia integrifolia*	Macadamia	Tree
32201	*Carya illinoinensis*	Pecan	Tree
277990	*Amaranthus tuberculatus*	Amaranth	Shrub
36596	*Prunus armeniaca*	Apricot	Tree
85884	*Argania spinose*	Argan	Tree
4265	*Cynara cardunculus*	Artichoke	Herb
145700	*Attalea speciose*	Babassu	Tree
85223	*Laurus nobilis*	Bay laurel	Shrub
21024	*Fagus*	Beech nut	Tree
3735	*Moringa oleifera*	Ben	Tree
152440	*Shorea stenoptera*	Borneo tallow nut	Tree
20340	*Ceratonia silique*	Carob pod	Herb
290272	*Attalea cohune*	Cohune	Tree
4047	*Coriandrum sativum*	Coriander seed	Herb
90675	*Camelina sativa*	False flax	Herb
29760	*Vitis vinifera*	Grape seed	Herb
3483	*Cannabis sativa*	Hemp	Herb
193163	*Ceiba pentandra*	Kapok seed	Tree
1585253	*Lallemantia royleana*	Lallemantia	Herb
2708	*Citrus limon*	Lemon seed	Herb
48386	*Acrocomia aculeate*	Macauba fruit	Tree
42439	*Limnanthes alba*	Meadowfoam seed	Herb
3710	*Brassica nigra*	Mustard	Herb
455045	*Abelmoschus esculentus*	Okra	Herb

(continued)

Table 16.1 (continued)

NCBI Taxonomy ID	Scientific name	Common Name	Plant type
48386	*Perilla frutescens*	Perilla seed	Herb
480971	*Caryocar brasiliense*	Pequi	Tree
3337	*Pinus*	Pine nut	Tree
3469	*Papaver somniferum*	Poppy seed	Herb
3758	*Prunus domestica*	Prune kernel	Herb
63459	*Chenopodium quinoa*	Quinoa	Herb
4230	Guizotia abyssinica	Ramtil (Niger pea)	Herb
4530	*Oryza sativa*	Rice bran	Herb
139772	*Triadica sebifera*	Tallow	Herb
4442	*Camellia sinensis*	Tea (camellia)	Herb
92921	Silybum marianum	Thistle (seed)	Herb
4565	*Triticum aestivum*	Wheat germ	Herb

Table 16.2 Non-edible oil crops for biofuel production

NCBI Taxonomy ID	Scientific name	Common name	Plant type
180498	*Jatropha curcas*	Jatropha	Tree
317856	*Madhuca longifolia*	Mahua	Tree
123498	*Aleurites moluccanus*	Candlenut	Tree
3981	*Hevea brasiliensis*	Rubber	Tree
57655	*Sapindus mukorossi*	Soapnut	Tree
3999	*Simmondsia chinensis*	Jojoba	Shrub
4097	*Nicotiana tabacum*	Tobacco	Herb
124943	*Azadirachta indica*	Neem	Tree
56065	*Millettia pinnata*	Karanja	Tree
3988	*Ricinus communis*	Castor	Tree/shrub
158927	*Calophyllum inophyllum* L	Polanga	Tree
3633	*Gossypium*	Cotton	Tree
4222	*Carthamus tinctorius*	Kusum	Tree
52862	*Cascabela thevetia*	Yellow oleander	Tree
141546	*Cerbera odollam*	Sea mango	Tree
73154	*Vernicia fordii*	Tung	Tree
190903	*Brachychiton rupestris*	Bottle tree	Tree

biofuels, C. glutamicum has been studied for its ability to produce ethanol (Inui et al. 2004) and isobutanol (Smith et al. 2010). Ethanol is a commonly used biofuel that can be blended with gasoline or used as a standalone fuel (Balat and Balat 2009). Isobutanol is a promising biofuel that has a higher energy density than ethanol and can be used as a drop-in replacement for gasoline (Rahaju et al. 2022). According to the search results, Corynebacterium glutamicum has been engineered for the production of various biofuels, including 3-methyl-1-butanol (Zhang et al. 2019a, b, 2019a, b), 1,2-propanediol, and 1-propano (Siebert and Wendisch 2015). One study (Zhang et al., 2019a, b) chose *C. glutamicum* as a host bacterium for 3-methyl-1-butanol production because of its strong resistance to isobutanol. The study designed several strategies to engineer *C. glutamicum* mutants for 3-methyl-1-butanol production, including overexpression of the 2-ketoisocaproate decarboxylase gene and deletion of the leucine dehydrogenase gene. Another study (Siebert and Wendisch, 2015) engineered *C. glutamicum* for the production of 1,2-propanediol and 1-propanol by metabolic pathway engineering. The study overexpressed the glycerol dehydratase and 1,2-propanediol dehydrogenase genes and deleted the lactate dehydrogenase gene to redirect the carbon flux towards 1,2-propanediol and 1-propanol production. These findings suggest that *C. glutamicum* has potential as a biofuel producer and can be engineered for the production of various biofuels. *C. glutamicum* has several advantages as a biofuel producer, including its ability to use a variety of carbon sources, its tolerance to high concentrations of ethanol and other biofuels, and its ability to produce high yields of biofuels. However, there are still some challenges that need to be overcome in order to make *C. glutamicum* a commercially viable biofuel producer, such as improving its efficiency and scalability.

Similarly, *Clostridium autoethanogenum* is an acetogenic bacterium that has the inherent ability to produce acetate and ethanol while growing on industrial waste gases such as carbon monoxide and carbon dioxide by using the Wood-Ljungdahl pathway (WLP) to fix carbon (Breitkopf 2018). One study (Cotter et al. 2009) investigated the production of ethanol and acetate by *Clostridium ljungdahlii* and *C. autoethanogenum* using quiescent cells. Another study (Wichlacz 2018) aimed to investigate metabolic flux through *C. autoethanogenum* pathways using isotopically labelled compounds to be assessed by mass spectrometry and NMR. The study also developed inhibitors for the pathway enzymes to direct metabolic processes toward higher-value chemical compounds by 'turning off' other branches of the pathway. In another study (Ricci et al. 2021), a laboratory-scale screening campaign was developed with the aim of increasing the production of 2,3-butanediol using *C. autoethanogenum* grown on different gas substrates of primary application interest (CO_2–H_2, CO–CO_2, syngas) and on different media with different compositions in terms of trace metals, mineral elements, and vitamins.

Synechocystis sp. is a species of cyanobacteria that has been genetically engineered to produce biofuels, such as ethanol and butanol, from carbon dioxide and sunlight (Savakis and Hellingwerf 2015). This microorganism has the ability to carry out photosynthesis, which makes it an attractive candidate for producing biofuels sustainably (Dienst et al. 2014). *S. elongatus* is another species of cyanobacteria that has been studied for its potential to produce biofuels (Mashayekhi et al. 2017). This microorganism has been engineered to produce isobutanol, a biofuel that has a higher energy density than ethanol and can be used in gasoline blends. *S. elongatus* has the advantage of being able to grow in seawater, which could make it a sustainable source of biofuels for coastal communities (Lan and Liao 2011). Similarly, E. coli, a commonly used model organism in molecular biology (Cobo-Simón et al. 2023), has been engineered to produce biofuels such as ethanol and isobutanol (Blombach and Eikmanns 2011). These biofuels can be produced from renewable resources such as agricultural waste, which can reduce our dependence on fossil fuels and decrease greenhouse gas emissions. Researchers have engineered *E. coli* to produce biofuels by introducing new pathways that convert sugar into these fuels (Koppolu and Vasigala 2016). This involves introducing genes that encode enzymes that can catalyze the chemical reactions needed to convert sugar into biofuels. Scientists have also used directed evolution techniques to improve the efficiency of these pathways and increase the amount of biofuels produced. Klebsiella pneumoniae is mentioned in one of the provided as a potential host organism for the de novo production of 2-butanol, another butanol isomer, through metabolic engineering (Chen et al. 2015). In another study, *K. pneumoniae UT3* was used for fermentative ethanol production from *Shorea robusta* through simultaneous saccharification and fermentation (SSF) (Choudhary et al. 2018). Numerous studies point to Saccharomyces cerevisiae as a suitable host organism for the generation of biofuels because of its complex genetic toolbox, easy modification of metabolic pathways and strain characteristics, and well-established commercial fermentations (Du and Zhao 2011; Khan et al. 2023; Lopez and Anne 2015). Using metabolic engineering approaches, such as overexpressing enzymes involved in the generation of bioethanol and biobutanol, knocking out competing pathways, improving carbon flow, and tolerance, *S. cerevisiae* has been modified to produce biofuel from lignocellulosic biomass (Khan et al. 2023). Another development towards the creation of second-generation biofuels is the engineering of S. cerevisiae for C5 fermentation (Yusuf and Gaur 2017). Additionally, S. cerevisiae's fatty acid metabolic pathway has been modified to produce advanced biofuels (Tang et al. 2015). Several research mentions *C. cellulolyticum* as one of four Clostridium species described for possible biofuel and biochemical production in a consolidated bioprocessing (CBP) system. *C. cellulolyticum* and *C. phytofermentans* had higher ethanol yields averaged across all substrates, and *C. cellulolyticum* had higher acetic acid production, indicating its potential for ethanol and co-product production from lignocellulosic biomass (Williams et al. 2012). *Thermoanaerobacterium saccharolyticum,* a bacterium is enriched with high potential in biofuel production. It can ferment compounds of industrial interest, such as xylan and xylose, which are not easily metabolized by other industrial microorganisms (Fiamenghi et al. 2023). *T. saccharolyticum* has also been used in a consortium with *C. thermocellum* to investigate the dynamics of their interaction with respect to cellulosic and lignocellulosic substrate attachment, respective numbers, and extent of solid substrate hydrolysis and desired end product formation (Miscevic 2021). Additionally, research has been done on the pyruvate kinase isozyme from *T. saccharolyticum,* which has been used in metabolic engineering for improved ethanol production (Fenton et al. 2021). Overall, microbial biofuels offer a sustainable and renewable alternative to fossil fuels, and their development has the potential to reduce our dependence on non-renewable energy sources while mitigating climate change. However, further research and development are needed to make microbial biofuels commercially viable and competitive with traditional fossil fuels.

16.2.4 Biofuel Production Process

Biofuel production involves several process, including catalytic reactions, genetic engineering, and understanding the impact of policy changes on agriculture. Microwave technology is one of the green technologies that is widely applied in the field of biofuel production, and microwave-absorbing catalysts have been reviewed for their potential in catalytic reactions (Chia et al., 2022; El Sherbiny et al., 2010). Additionally, microwave-absorbing catalysts have been used in catalytic reactions of biofuels to improve the efficiency of the reaction and reduce the production costs (Chia et al. 2022) The red alga *Cyanidioschyzon merolae* has been considered an excellent model system to understand the fundamental mechanisms behind the accumulation of triacylglycerols (TAGs)/starch in microalgae, and genetic engineering of key genes and proteins has resulted in a drastic increment of the amount of TAGs and starch (Pancha et al. 2021). Plant cell walls, which are composed of lignocellulose, represent the bulk of plant dry mass and are a key strategy for biofuel production (Bartley and Ronald 2009). However, the recalcitrant nature of the plant cell wall is a major limitation for efficient conversion of plant biomass to biofuels (Dodd and Cann 2009). Microbes have evolved enzymatic strategies for hydrolyzing plant cell walls into their constituent sugars for subsequent fermentation to biofuels (Dodd and Cann 2009). Recent research has focused on understanding and harnessing the mechanisms by which plants build cell walls, as well as developing enzymes and microbes that facilitate sugar release and biofuel production (Bartley and Ronald 2009). In addition, metabolic engineering in *Saccharomyces cerevisiae* has been performed to enhance and sustain triacylglycerol (TAG) production, which is a potential biofuel (Hung et al. 2016).

16.2.4.1 Transesterification Process

Transesterification is a popular approach for biodiesel production that involves the reaction of oil and/or fat with an alcohol to produce fatty acid methyl esters (FAMEs) biodiesel (Zhang et al. 2019a,b). The transesterification reaction takes place in four steps, which involve mixing the alcohol with a catalyst, adding the oil or fat, heating the mixture, and separating the FAMEs biodiesel from the glycerol byproduct. The efficiency and yield of the transesterification reaction can be improved by optimizing the reaction conditions, such as the temperature, pressure, and catalyst concentration (Sivaprakasam and Saravanan 2007). Additionally, microwave heating has been used to intensify and optimize the transesterification reaction for biodiesel production, resulting in higher reaction rates and improved yields (Athar et al. 2020). Overall, transesterification is a widely used approach for biodiesel production, and ongoing research is focused on optimizing the reaction conditions and developing new approaches to improve the efficiency and feasibility of biodiesel production. Figure 16.1 shows the biodiesel process using transesterification. The alcohol, the catalyst, and the oil all contribute to the formation of a distinct ester. Extraction of crude glycerol from a mixture of methyl esters is the only way to obtain this form of glycerol. To extract the alcohol that is present in crude glycerol, an apparatus known as a heat-vacuum separator is applied. The end product is glycerol that has been neutralised, which is achieved by performing a water wash followed by an acid neutralisation process. In order to get rid of the alcohol, the mixed methyl esters that were left over after the process of transesterification need to be separated. After going through the first separator and having some of the water and biodiesel removed, they are put through a second separator that utilizes heat and a vacuum to accomplish the same purpose (Fig. 16.2).

16.2.4.2 Esterification Process

Esterification is a process used for the production of biodiesel from feedstocks with high free fatty acid (FFA) content. The process involves the reaction of FFA with an alcohol in the presence of an acidic catalyst to produce fatty acid alkyl esters (FAAEs) biodiesel (Ahmed and Huddersman 2022; Khan et al. 2021). The esterification process is different from the transesterification process, which is used for the production of biodiesel from triglycerides. The efficiency and yield of the esterification reaction can be improved by optimizing the reaction conditions, such as the temperature, pressure, and catalyst concentration (Athar et al. 2020). Additionally, microwave heating has been used to intensify and optimize the esterification reaction for biodiesel production, resulting in higher reaction rates and improved yields. Overall, esterification is a useful process for the production of biodiesel from feedstocks with high FFA content, and ongoing research is focused on optimizing the reaction conditions and developing new approaches to improve the efficiency and feasibility of biodiesel production (Fig. 16.3).

16.2.5 Genetic Resources for Biofuel Production

Genetic resources play an important role in biofuel production, particularly in the development of advanced biofuels from non-food feedstock sources (Torto-Alalibo et al. 2014). Advances in genetic engineering have been used to improve the potential of various organisms, including microorganisms and plants, for biofuel production (Rasala et al. 2013; Sticklen 2008). For example, genetic resource have been used to identify and optimize the genes associated with the synthesis of advanced biofuels,

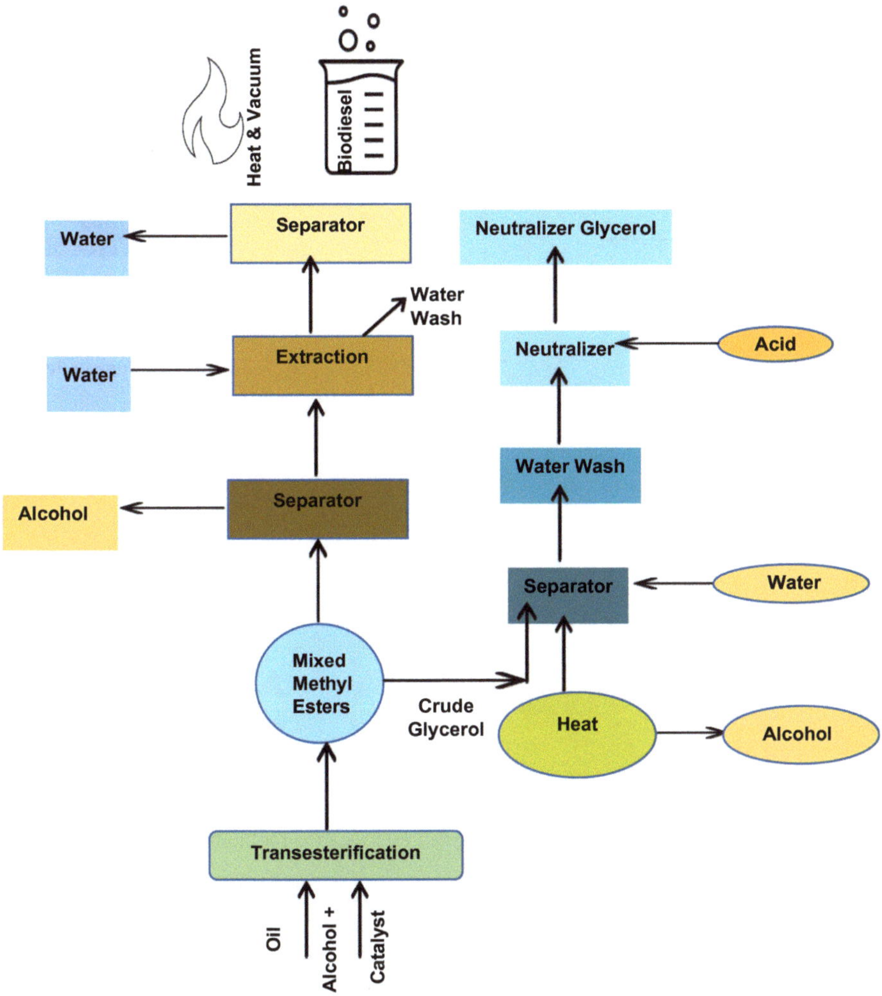

Fig. 16.2 Schematic of the biodiesel process using transesterification

such as alcohols, fatty acids, isoprenoids, and polyhydroxyalkanoates (Torto-Alalibo et al. 2014) Several genes and proteins have been studied for their potential use in biofuel production. Cell wall proteomics has allowed the identification of numerous cell wall proteins (CWPs) involved in biosynthesis and rearrangements, including glycosyl hydrolases and oxidoreductases, which are assumed to contribute to the remodelling of cell wall polysaccharides by enzymatic or non-enzymatic mechanisms (Calderan-Rodrigues et al. 2017). *Thermoanaerobacterium saccharolyticum* is an organism capable of naturally fermenting compounds of industrial interest, such as xylan and xylose. A deep evolutionary study of members of the firmicutes clade, including *T. saccharolyticum,* has been conducted to understand adaptations that may

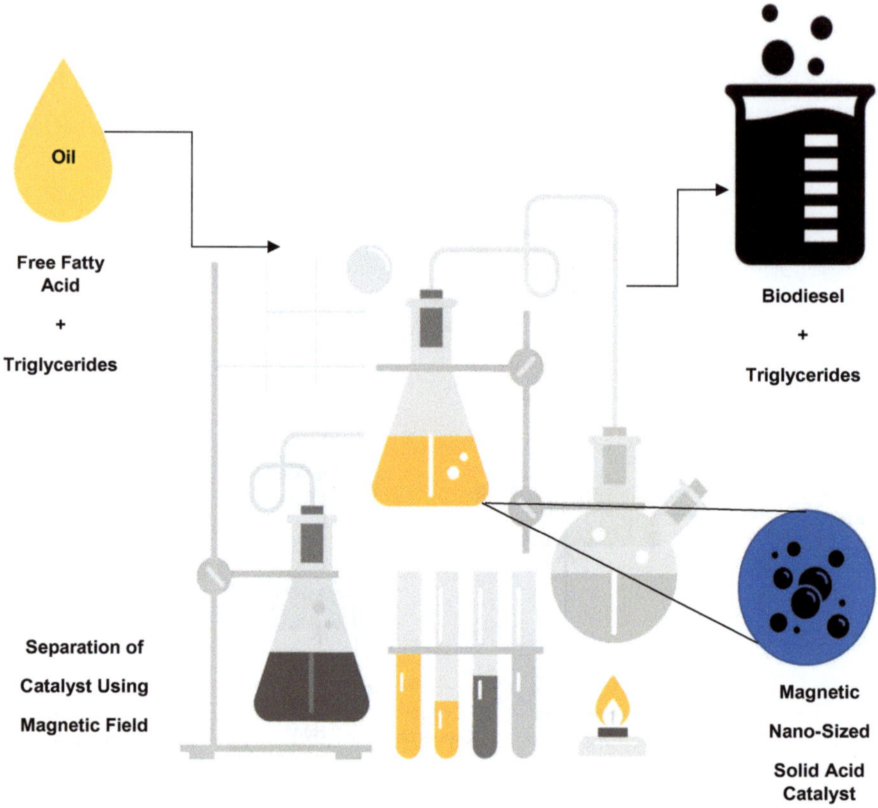

Fig. 16.3 Scheme of magnetic solid acid-catalyzed esterification for biodiesel production

be related to overall fermentation metabolism, especially for xylose fermentation (Fiamenghi et al. 2023). Additionally, *Yarrowia lipolytica,* a non-conventional yeast capable of high lipogenesis, has been studied for its potential to produce lipid-based biofuels and chemicals from renewable resources such as lignocellulosic biomass. The endogenous genes encoding the aldehyde dehydrogenase family proteins were overexpressed in *Y. lipolytica* to support the conversion of furfural to furoic acid, improving its furfural tolerance and increasing cell growth and lipid production (Kim et al. 2022) The NCBI Genes, Nucleotides, and BioMart-Ensembl are major genetic databases that can be used to identify and improve genes associated with the production of advanced biofuels.

NCBI Database

NCBI offers a range of resources and conducts research pertaining to biofuel production. A study was conducted to evaluate the bioproduction potential of certain strains of cyanobacteria, particularly those related to Cyanothece. This study presents a methodical approach for the integration and ranking of annotation data sourced from the Universal Protein Resource (Uniprot), NCBI Protein Clusters, and Rapid Annotations using Subsystems Technology (RAST) techniques. This research revealed that the five Cyanothece species exhibited the ability to perform nitrogen fixation and hydrogen production. However, Cyanothece PCC 7425 is the only species that cannot execute this function in an aerobic milieu. The present study underscores the prospective utilization of aquatic phototrophs, particularly cyanobacteria, as viable substitutes for terrestrial crops in the context of biofuel generation (Mueller et al. 2013.

BioCyc Database

The BioCyc database comprises a set of pathway/Genome Databases (PGDBs) that provide data pertaining to metabolic pathways and genome annotations for diverse organisms. The BioCyc database has been utilized in the realm of biofuel production to develop metabolic models and scrutinize gene expression data for diverse organisms, including the marine diatom Thalassiosira pseudonana. The investigation employed the BioCyc pathway tool from the Stanford Research Institute (SRI) and the Integrated Microbial Genomes/Expert Review (IMG/ER) tool from the Joint Genome Institute (JGI) to construct the initial comprehensive BioCyc pathway model for the entire genome of *T. pseudonana.* This model encompassed more than 300 metabolic pathways. The utilization of transcriptome data in conjunction with the pathway model facilitated the comprehension of the functionality of fatty acid and lipid pathways, along with the interrelationship of other pathways that impact the accrual of triacylglycerols (TAGs) under varying environmental conditions (Parker 2013).

Biomart-Ensemble Database

BioMart is a robust data-management system designed for query-based operations. It facilitates access to a diverse range of biological databases including Ensembl. Although the search findings did not explicitly reference the use of BioMart-Ensembl in biofuel research, it is noteworthy that BioMart-Ensembl holds potential as a valuable asset for researchers engaged in biofuel-related investigations. Ensembl is a comprehensive repository that encompasses genome annotations, gene sequences, and other genomic information for diverse organisms, including plants and microorganisms, which play crucial roles in biofuel production. The BioMart-Ensembl platform is a valuable tool for researchers seeking to access, retrieve, and analyze genomic data to identify candidate genes, investigate metabolic pathways, and comprehend the genetic underpinnings of biofuel production across diverse organisms. The aforementioned data hold significant importance in formulating strategies to augment biofuel production via genetic modification or other biotechnological methodologies (Wang et al. 2017).

BRENDA Enzyme Database

The BRENDA Enzyme Database is a comprehensive repository of functional data pertaining to enzymes, encompassing enzyme classification, substrates, products, and kinetic parameters. Although the search outcomes do not explicitly refer to the utilization of the BRENDA Enzyme Database in biofuel research, it can serve as a significant tool for scholars engaged in biofuel-related investigations (Schomburg et al., 2002). Enzymes are of paramount importance in the conversion of biomass into biofuels.

A comprehensive understanding of their properties and functions can facilitate the optimization of biofuel production processes. The BRENDA Enzyme Database is a valuable resource for researchers seeking comprehensive data on enzymes that participate in the breakdown of lignocellulosic biomass, the fermentation of sugars, and other pathways relevant to biofuel production. These data have the potential to facilitate the identification of enzyme targets for genetic manipulation or other biotechnological methodologies aimed at augmenting the efficacy and output of biofuel production (Wang et al. 2017).

16.2.6 Kyoto Encyclopedia of Genes and Genomes

The KEGG (Kyoto Encyclopedia of Genes and Genomes) database is a valuable resource for biofuel research, as it provides comprehensive information on metabolic pathways, enzymes, and gene functions related to various organisms involved in biofuel production. Researchers can use the KEGG database to study metabolic pathways and enzymes involved in the conversion of biomass to biofuels, such as lipid synthesis in microalgae and anaerobic fermentation processes (Misra et al. 2016). By analyzing the KEGG database, researchers can identify key enzymes and metabolic pathways that can be targeted for genetic engineering or other biotechnological approaches to enhance biofuel production efficiency and yield. Additionally, the KEGG database can be integrated with other bioinformatics databases and tools, such as BioCyc and The SEED, to provide a more comprehensive understanding of the metabolic networks involved in biofuel production (Milne et al. 2011).

Reactome Database

The Reactome database is a thorough and rigorously reviewed compilation of molecular reactions and biological pathways in humans as well as the corresponding molecular reactions in model organisms. The database is composed of expert-authored and well-documented information. The resources in question offer significant value to researchers in their exploration of biological pathways, molecular interactions, and cellular processes. Such insights can be beneficial across a range of disciplines, including biofuel research. The Reactome database enables users to explóre biological pathways, locate pathways that involve a specific molecule of interest, and utilize pathway and Expression analysis tools to investigate potential correlations between user-generated experimental data sets and Reactome pathways. Furthermore, the Species Comparison tool enables users to conduct a comparative analysis of the pathways between humans and model organisms. The Reactome database represents a significant resource for investigating the metabolic pathways and enzymes that participate in the conversion of biomass to biofuels, along with other biofuel production processes. Through the examination of the Reactome database, scholars have the ability to discern pivotal enzymes and metabolic pathways that may be subject to genetic manipulation or other biotechnological methods to optimize the efficacy and output of biofuel production. Upon identification of potential genes, researchers may employ techniques such as gene editing and metabolic engineering to enhance the expression of these genes, thereby augmenting the quantity and calibration of biofuels. Through precise manipulation of genes and metabolic pathways, researchers can mitigate the production of byproducts and waste, thereby enhancing the efficacy and ecological viability of biofuel manufacturing. Genetic resources such as NCBI genes, nucleotides, and biomart ensembles provide notable benefits to researchers involved in advanced biofuel production studies. By utilizing these resources, researchers can acquire a more profound understanding of the genetic foundation of biofuel synthesis and execute biofuel production procedures that are more efficient and ecologically sustainable (Table 16.3).

16.3 Current Challenges in Biofuel Production

Biofuel production is becoming a viable renewable energy source, but negative impacts from global climate change make it difficult to meet the projected increases in biofuels' share in the fuel market (Fink and Medved 2011). Technological and biological advancements are necessary for sustainable and economical biofuel production at the scales necessary to support the world's energy needs (M. 2013). There are several challenges in biofuel production from plants and microbes. One of the major challenges in the economic scaling-up of microalgae- and cyanobacteria-based biofuels production is the lack of a reliable preventing and controlling culture system of biological contamination, which can attack the cell growth or product accumulation causing crashing effects (Zhu et al. 2020). The hydrolysis of cellulose present in biomass to sugars and other organic compounds that can later be converted to fuels is hindered by many physicochemical structural and compositional factors (Sarrouh et al. 2009). Different pretreatment techniques can change the physical and chemical structure of the lignocellulosic biomass and improve their hydrolysis rates (Sarrouh et al. 2009). The production of bioethanol from molasses, a sugar industry waste, faces several challenges, including the type of substrate, the fermenting microorganisms, optimization of physicochemical parameters like pH, temperature, and the nutrient requirements of fermenting microbes for the conversion of sugar-rich substrate into bioethanol (Hashmi and Hameed 2020) Yeasts, such as Saccharomyces cerevisiae, are promising host organisms for biofuel production, but challenges such as limited production and product toxicity need to be overcome (Lopez and Anne 2015).

16.4 Systems Biology Contribute to Biofuel Production

Systems biology can play an important role in biofuel production by providing a comprehensive understanding of the metabolic pathways involved in biofuel production and identifying potential targets for genetic and metabolic engineering (Sharma et al. 2020). For example, genetic and metabolic engineering have been used to

Table 16.3 List of database resource used for the biofuel production

Name of database	Availability/link	Uses/function	Reference
DNA data bank of japan (DDBJ)	http://www.ddbj.nig.ac.jp/	The nucleotide sequence archive of a database	Tanizawa et al. (2023)
Ensembl	http://www.ensembl.org	The genome browser designed specifically for vertebrates facilitates comparative genomics, evolutionary analysis, investigation of sequence variation, and exploration of transcriptional data	Birney et al. (2004)
GeneCards	http://www.genecards.org	GeneCards is a comprehensive database that provides information on human genes	Safran et al. (2010)
KEGG	http://www.genome.jp/kegg	KEGG is a comprehensive database that provides information on biological pathways, genes, and genomes	Kanehisa and Goto (2000)
NCBI RefSeq	http://www.ncbi.nlm.nih.gov/refseq	NCBI RefSeq is a database of curated genomic, transcriptomic, and protein sequences maintained by the NCBI	O'Leary et al. (2016)
UCSC Genome Browser	http://genome.ucsc.edu	The UCSC Genome Browser is an online tool that allows researchers to visualize and analyze genome sequences from various organisms	Karolchik et al. (2009)
GenBank	https://www.ncbi.nlm.nih.gov/genbank/	GenBank is a public database that provides annotated DNA and RNA sequences, including complete genomes, for various organisms	Benson et al. (2013)
PANTHER	http://www.pantherdb.org	PANTHER is a comprehensive database that provides information on the evolutionary relationships, functions, and pathways of proteins across various organisms	Mi et al. (2005)
Protein Data Bank (PDB)	http://www.rcsb.org/PDB	The PDB is a public database that provides information on the three-dimensional structures of proteins, nucleic acids, and other biomolecules	Berman et al. (2000)
UniProt	http://www.uniprot.org	UniProt (Universal Protein Resource) is a comprehensive database that provides information on protein sequences and their functions	Apweiler et al. (2004)

(continued)

Table 16.3 (continued)

Name of database	Availability/link	Uses/function	Reference
Golm Metabolome Database (GMD)	http://gmd.mpimp-golm.mpg.de	The GMD is a comprehensive database that provides information on the metabolites (small molecules) found in various organisms, including plants, animals, and microorganisms	Kopka et al. (2005)
MetaCyc	http://metacyc.org	MetaCyc is a comprehensive database that provides information on metabolic pathways and enzymes for various organisms, including bacteria, plants, and humans	Karp et al. (2002)
Pathway Commons	http://www.pathwaycommons.org	Pathway Commons is a collaborative database that provides information on biological pathways and their interactions	Rodchenkov et al. (2020)
Pathway Interaction Database (PID)	http://pid.nci.nih.gov	The PID is a comprehensive database that provides information on biomolecular interactions and signaling pathways	Schaefer et al. (2009)
Reactome	http://www.reactome.org	Reactome is a comprehensive database that provides information on biological pathways and processes in various organisms, including humans	Fabregat et al. (2017)
UniPathway	http://www.grenoble.prabi.fr/obiwarehouse/unipathway	UniPathway is a comprehensive database that provides information on the metabolic pathways and enzymes for various organisms, including bacteria, plants, and humans	Morgat et al. (2012)
BioCarta	http://www.biocarta.com/	BioCarta is a database that provides information on biological pathways and processes in humans	Nishimura (2001)
MIPS CYGD	http://mips.gsf.de/genre/proj/yeast	MIPS CYGD (Munich Information Center for Protein Sequences Comprehensive Yeast Genome Database) is a comprehensive database that provides information on the genome, proteome, and metabolic pathways of the yeast Saccharomyces cerevisiae	Mewes et al. (2002)
SaccharomycesGenome Database (SGD)	http://www.yeastgenome.org/biocyc	The SGD is a comprehensive database that provides information on the genome, genes, and proteins of the yeast Saccharomyces cerevisiae	Cherry et al. (1998)

improve biofuels production from microalgae by editing the genomes of microalgae to improve their potential to produce biofuels and bio-based chemicals (Rasala et al. 2013). Additionally, synthetic biology has been used to improve biofuel production from yeast by engineering the metabolic pathways involved in biofuel production (Liu et al. 2022). Biofuels produced by microorganisms have similar properties to petroleum-based fuels and can be used as a substitute for them (Peralta-Yahya et al. 2012). However, producing these biofuels in yields high enough to be useful requires the engineering of the microorganism's metabolism. Synthetic biology and data-driven approaches can be used to optimize both the host and pathways to maximize fuel production. The interdependency among the systems involved in biofuel development, including biofuel refinery location, transportation infrastructure, agricultural production and markets, environment, and social communities, can be considered using a system of systems (SoS) Biofuel model (Housh et al. 2015). The optimal infrastructure development and land-use allocation for biofuel production in a region considering socioeconomic and water quality and quantity effects can be determined using this model. The metabolic responses of Escherichia coli induced by three major biofuel products, ethanol, butanol, and isobutanol, can be determined using global metabolomic and network analysis. Machine learning and probabilistic modeling techniques can be used to guide synthetic biology in a systematic fashion, without the need for a full mechanistic understanding. The use of whole corn germ, a by-product of corn grain, can be used in animal and human feeding, biofuel and feedstock production, or other systems (Nascimento et al. 2021) Building synthetic feedback control systems may yield insight into the roles that feedback can play in regulation since it can be introduced independently of native regulation, and alternative control architectures can be compared (Dunlop et al. 2010). An integrated approach involving metabolic engineering and systems biology would help in improving titer of advanced biofuels (Das et al. 2020). In the foreseeable future, formal integration of functional genomics and systems biology with synthetic biology and metabolic engineering will undoubtedly support the discovery, characterization, and engineering of new metabolic routes and more efficient microbial systems for the production of biofuels (Dellomonaco et al. 2010). The integration of activity-based protein profiling (ABPP) with molecular and systems biology approaches will provide new insights into the catalytic and regulatory mechanisms of functional enzymes and their synergistic effects in biofuel production. (Liu et al. 2015).Hence, the field of synthetic biology exhibits immense potential in expediting the advancement of microbial systems for the generation of sustainable fuels and chemicals (Picataggio 2009). Recent advancements in metabolic engineering, genome engineering, and synthetic biology have the potential to enable the production of an economically viable advanced biofuel. This could be achieved through the use of genetically modified or synthetic microbes possessing well-defined metabolism (Vinuselvi et al. 2011). Fermentation engineers can optimise large-scale biofuel production by employing rational pathway modifications, synthetic biology circuits, and bioreactor control algorithms based on metabolic analyses (Hollinshead et al. 2014). The potential for biofuels to serve as a reliable source of renewable energy is evident. However, the attainment of sustainable and cost-effective biofuel production on a global scale necessitates technological and biological advancements (M., 2013). This platform presents a route for expanding the repertoire of genetic elements available to synthetic biology and provides a starting point for efforts to engineer robust strains for biofuel generation (Sommer et al. 2010). The potential for systems biology to significantly enhance the development of cell factories has been widely acknowledged. As such, we suggest the term "Industrial Systems Biology" to describe the application of systems biology in the advancement of industrial biotechnology for sustainable chemical production (Papini et al. 2010). In general, the use of systems biology methodologies can provide significant insights into the intricate metabolic pathways implicated in the production of biofuels and can facilitate the identification of prospective targets for genetic and metabolic engineering with the aim of enhancing the efficiency and practicability of biofuel production.

16.5 Future Scope of Biofuel from the Plant and Microbes

Biofuels, composed of, derived from or produced by microbial cell factories from biological resources, are presently being recognized as drop-in fuels, and the usage of biofuels, such as biodiesel, bioethanol, and isoprene, is increasing (Zhu et al. 2020). Enzymatic biofuel cells, which employ biological molecules such as enzymes, organelles, or microorganisms to convert chemical energy into electricity, offer several advantages over traditional batteries and have promising applications in low-power devices (Aquino Neto and Andrade 2013). Diatoms are priority candidates for biofuel production due to their photosynthetic efficiency and high-lipid content, and their value-added molecule production can be modulated by spectral light control during growth (Duarte et al. 2021; Sittmann et al. 2020). Bacterially produced small molecules, such as diketopiperazines, can stimulate diatom growth and increase lipid content, offering new approaches to enhance diatom-based biofuel production (Sittmann et al. 2020). Future research in this area should focus on overcoming the challenges of biological contamination in mass cultivations of microalgae and cyanobacteria and developing new varieties of sugarcane that can sustain sugar and energy production under climate change scenarios (Bhaskaran and Nair 2014; Zhu et al. 2020). Several study shows strong evidence that a next-generation biofuel production facility utilizing genetically engineered strains of filamentous cyanobacteria could become economically feasible in the future if strains are developed with increased biofuel productivities (Johnson et al. 2016). Ultimately, these studies demonstrate a clear need to leverage integrated modeling platforms to advance microalgae biofuel systems as a whole, and specific recommendations are made for prioritizing research and development pathways to achieve economical biofuel production from microalgae. This contributes towards greater feasibility in microalgae cultivation for biofuel productions and as well towards environmental sustainability.

16.6 Conclusion

The utilization of plant–microbe interactions for the production of biofuels holds promise as a feasible substitute for the limited reserves of fossil fuels. Genetic engineering has the potential to enhance the production of biofuels through the development of organisms that exhibit improved tolerance to inhibitory compounds, increased biofuel yields, and efficient fermentation of industrially relevant compounds. *Thermoanaerobacterium saccharolyticum* has undergone genetic modifications to enhance its ethanol production capacity and improve its resistance to inhibitory substances present in pretreated hardwood. Microalgae have been the subject of research because of their capacity to generate biofuels and bio-derived chemicals. In this regard, genetic modification has been employed to manipulate the genomes of microalgae to enhance their potential to produce biofuels and bio-derived chemicals. In addition, the application of plant cell wall proteomics has facilitated the recognition of a multitude of CWPs that participate in the biosynthesis and rearrangement processes. These proteins include glycosyl hydrolases and oxidoreductases, which are believed to play a role in the modification of cell wall polysaccharides via enzymatic or non-enzymatic means. Renewable energy sources have the potential to decrease reliance on nonrenewable fossil fuels, thereby offering a sustainable energy solution. Furthermore, it has the potential to decrease greenhouse gas emissions and alleviate the impact of climate change. In addition, biofuel production has the potential to generate economic benefits, including the expansion of employment opportunities through the establishment and management of novel biorefineries. Biofuel crops have the potential to be cultivated on lands not suitable for conventional agriculture, such as reclaimed mine lands. This could serve as a viable option for alternative agricultural land-use and aid in mine reclamation. It is imperative to contemplate the plausible repercussions of land-use alteration caused by biofuels and devise tactics to mitigate its effects. In general, the production of biofuels induced by plants and microbes holds promise as a viable and ecologically sound means of generating energy while also making a positive contribution to economic expansion and land restoration initiatives. Nevertheless, additional investigation is required to ascertain the most promising genes and proteins involved in the production of biofuels. The implementation of policies aimed at augmenting the proportion of biofuels in transportation fuels can have implications for the agricultural sector, given that biofuel production is vital to the food industry (Gradziuk et al. 2021). Furthermore, the use of organic solvents in biofuel production may have detrimental effects on microorganism proliferation. However, only a limited number of microorganisms have developed the capacity to endure elevated levels of organic solvents, rendering them promising candidates for prospective industrial implementation (Xiang-yang 2013).

References

Abu Zeid IM, Al-Asmari KM, Altayb HN, Al-Attar AM, Qahl SH, Alomar MY (2022) Predominance of antioxidants in some edible plant oils in ameliorating oxidative stress and testicular toxicity induced by malathion. Life 12:350. https://doi.org/10.3390/life12030350

Ahmed R, Huddersman K (2022) Review of biodiesel production by the esterification of wastewater containing fats oils and grease (FOGs). J Ind Eng Chem 110:1–14. https://doi.org/10.1016/j.jiec.2022.02.045

Alves AA, Laviola BG, Formighieri EF, Carels N (2015) Perennial plants for biofuel production: bridging genomics and field research. Biotechnol J 10:505–507. https://doi.org/10.1002/biot.201400201

Apweiler R, Bairoch A, Wu CH, Barker WC, Boeckmann B, Ferro S, Gasteiger E, Huang H, Lopez R, Magrane M, Martin MJ, Natale DA, O'Donovan C, Redaschi N, Yeh L-SL (2004) UniProt: the universal protein knowledgebase. Nucleic Acids Res 32:D115-119. https://doi.org/10.1093/nar/gkh131

Aquino Neto S, Andrade ARD (2013) New energy sources: the enzymatic biofuel cell. J Braz Chem Soc. https://doi.org/10.5935/0103-5053.20130261

Arefin MA, Rashid F, Islam A (2021) A review of biofuel production from floating aquatic plants: an emerging source of bio-renewable energy. Biofuels Bioprod Biorefining 15:574–591. https://doi.org/10.1002/bbb.2180

Athar M, Zaidi S, Hassan SZ (2020) Intensification and optimization of biodiesel production using microwave-assisted acid-organo catalyzed transesterification process. Sci Rep 10:21239. https://doi.org/10.1038/s41598-020-77798-1

Balat M, Balat H (2009) Recent trends in global production and utilization of bio-ethanol fuel. Appl Energy 86:2273–2282. https://doi.org/10.1016/j.apenergy.2009.03.015

Banković-Ilić IB, Stamenković OS, Veljković VB (2012) Biodiesel production from non-edible plant oils. Renew Sustain Energy Rev 16:3621–3647. https://doi.org/10.1016/j.rser.2012.03.002

Bartley L, Ronald P (2009) Plant and microbial research seeks biofuel production from lignocellulose. Calif Agric 63:178–184

Benson DA, Cavanaugh M, Clark K, Karsch-Mizrachi I, Lipman DJ, Ostell J, Sayers EW (2013) GenBank. Nucleic Acids Res 41:D36-42. https://doi.org/10.1093/nar/gks1195

Berman HM, Westbrook J, Feng Z, Gilliland G, Bhat TN, Weissig H, Shindyalov IN, Bourne PE (2000) The protein data bank. Nucleic Acids Res 28:235–242. https://doi.org/10.1093/nar/28.1.235

Bhaskaran A, Nair NV (2014) Challenges and opportunities in sugarcane cultivation under climate change scenario. J Sugarcane Res 4:1–18

Biofuels: Their characteristics and analysis—ScienceDirect [WWW Document], n.d. URL https://www.sciencedirect.com/science/article/abs/pii/B978008102426300014X. Accessed 17 April 2023)

Birney E, Andrews TD, Bevan P, Caccamo M, Chen Y, Clarke L, Coates G, Cuff J, Curwen V, Cutts T, Down T, Eyras E, Fernandez-Suarez XM, Gane P, Gibbins B, Gilbert J, Hammond M, Hotz H-R, Iyer V, Jekosch K, Kahari A, Kasprzyk A, Keefe D, Keenan S, Lehvaslaiho H, McVicker G, Melsopp C, Meidl P, Mongin E, Pettett R, Potter S, Proctor G, Rae M, Searle S, Slater G, Smedley D, Smith J, Spooner W, Stabenau A, Stalker J, Storey R, Ureta-Vidal A, Woodwark KC, Cameron G, Durbin R, Cox A, Hubbard T, Clamp M (2004) An overview of ensembl. Genome Res. 14:925–928. https://doi.org/10.1101/gr.1860604

Blombach B, Eikmanns BJ (2011) Current knowledge on isobutanol production with *Escherichia coli*, *Bacillus subtilis* and *Corynebacterium glutamicum*. Bioeng Bugs 2:346–350. https://doi.org/10.4161/bbug.2.6.17845

Breitkopf R (2018) Understanding the C4 dicarboxylic acid metabolism in Clostridium autoethanogenum

Calderan-Rodrigues MJ, Fonseca JG, Labate CA, Jamet E (2017) Cell wall proteomics as a means to identify target genes to improve second-generation biofuel production. In: Jacob-Lopes E, Zepka LQ (Eds.). InTech. https://doi.org/10.5772/66189

Chang F, Zhou Q, Zhou L (2017) Fuel properties of biodiesel from nonedible herbaceous oil feedstocks: *Leonurus artemisia* L. and *Peganum harmala* L. Energy Sour Recovery Util Environ Eff 39:1879–1885. https://doi.org/10.1080/15567036.2017.1381783

Chen Z, Wu Y, Huang J, Liu D (2015) Metabolic engineering of *Klebsiella pneumoniae* for the de novo production of 2-butanol as a potential biofuel. Bioresour Technol 197:260–265. https://doi.org/10.1016/j.biortech.2015.08.086

Cherry JM, Adler C, Ball C, Chervitz SA, Dwight SS, Hester ET, Jia Y, Juvik G, Roe T, Schroeder M (1998) SGD: Saccharomyces genome database. Nucleic Acids Res 26:73–79

Chia SR, Chew KW, Show PL, Yap YJ, Ong HC, Ling TC, Chang J-S (2018) Analysis of economic and environmental aspects of microalgae biorefinery for biofuels production: a review. Biotechnol J 13:e1700618. https://doi.org/10.1002/biot.201700618

Chia SR, Nomanbhay S, Milano J, Chew KW, Tan C-H, Khoo KS (2022) Microwave-absorbing catalysts in catalytic reactions of biofuel production. Energies 15:7984. https://doi.org/10.3390/en15217984

Choubey J, Choudhari JK, Verma MK, Chaterjee T, Sahariah BP (2022) Systems biology aided functional analysis of microbes that have rich bioremediation potential for environmental pollutants. In: Microbial remediation of azo dyes with prokaryotes. CRC Press

Choudhari JK, Chatterjee T, Gupta S, Garcia-Garcia JG, Vera-González J (2021) Network biology approaches in ophthalmological diseases: a case study of glaucoma. In: Wolkenhauer O (Ed.), Systems medicine. Academic Press, Oxford, pp 190–202. https://doi.org/10.1016/B978-0-12-801238-3.11586-7

Choudhary A, Jadhav K, Tiwari K (2018) Simultaneous saccharification and *Shorea robusta* for bioethanol production

Cobo-Simón M, Hart R, Ochman H (2023) *Escherichia Coli*: what is and which are? Mol Biol Evol 40:msac273. https://doi.org/10.1093/molbev/msac273

Cotter JL, Chinn MS, Grunden AM (2009) Ethanol and acetate production by *Clostridium ljungdahlii* and *Clostridium autoethanogenum* using resting cells. Bioprocess Biosyst Eng 32:369–380

Dahman Y, Syed K, Begum S, Roy P, Mohtasebi B (2019) 14 - Biofuels: Their characteristics and analysis. Biomass, Biopolymer-Based Materials, and Bioenergy. Constr Biomed Ind Appl 277–325

Das M, Patra P, Ghosh A (2020) Metabolic engineering for enhancing microbial biosynthesis of advanced biofuels. Renew Sustain Energy Rev 119:109562. https://doi.org/10.1016/j.rser.2019.109562

Dellomonaco C, Fava F, Gonzalez R (2010) The path to next generation biofuels: successes and challenges in the era of synthetic biology. Microb Cell Factories 9:3. https://doi.org/10.1186/1475-2859-9-3

Demirbas A (2009) Progress and recent trends in biodiesel fuels. Energy Convers Manag 50:14–34. https://doi.org/10.1016/j.enconman.2008.09.001

Dienst D, Georg J, Abts T, Jakorew L, Kuchmina E, Börner T, Wilde A, Dühring U, Enke H, Hess WR (2014) Transcriptomic response to prolonged ethanol production in the *cyanobacterium synechocystis* sp. PCC6803. Biotechnol Biofuels 7:21. https://doi.org/10.1186/1754-6834-7-21

Dodd D, Cann IKO (2009) Enzymatic deconstruction of xylan for biofuel production. Glob Change Biol Bioenergy 1:2–17. https://doi.org/10.1111/j.1757-1707.2009.01004.x

Du J, Zhao H (2011) Engineering saccharomyces cerevisiae for biofuel production from lignocellulosic biomass

Duarte B, Feijão E, Goessling JW, Caçador I, Matos AR (2021) Pigment and fatty acid production under different light qualities in the diatom *Phaeodactylum tricornutum*. Appl Sci 11:2550. https://doi.org/10.3390/app11062550

Dunlop MJ, Keasling JD, Mukhopadhyay A (2010) A model for improving microbial biofuel production using a synthetic feedback loop. Syst Synth Biol 4:95–104. https://doi.org/10.1007/s11693-010-9052-5

El Sherbiny SA, Refaat AA, El Sheltawy ST (2010) Production of biodiesel using the microwave technique. J Adv Res 1:309–314. https://doi.org/10.1016/j.jare.2010.07.003

Fabregat A, Sidiropoulos K, Viteri G, Forner O, Marin-Garcia P, Arnau V, D'Eustachio P, Stein L, Hermjakob H (2017) Reactome pathway analysis: a high-performance in-memory approach. BMC Bioinformatics 18:142. https://doi.org/10.1186/s12859-017-1559-2

Fenton CA, Tang Q, Olson DG, Maloney MI, Bose JL, Lynd LR, Fenton AW (2021) Inhibition of pyruvate kinase from *Thermoanaerobacterium saccharolyticum* by IMP is independent of the extra-C domain. Front Microbiol 12:628308. https://doi.org/10.3389/fmicb.2021.628308

Fiamenghi MB, Prodonoff JS, Borelli G, Carazzolle MF, Pereira GAG, José J (2023). Comparative Genomics of Firmicutes Reveals Probable Adaptations for Xylose Fermentation in Thermoanaerobacterium Saccharolyticum. https://doi.org/10.1101/2023.03.08.531619

Fink R, Medved S (2011) Global perspectives on first generation liquid biofuel production. Turk J Agric For

Gradziuk P, Jończyk K, Gradziuk B, Wojciechowska A, Trocewicz A, Wysokiński M (2021) An economic assessment of the impact on agriculture of the proposed changes in EU biofuel policy mechanisms. Energies 14:6982. https://doi.org/10.3390/en14216982

Hashmi M, Hameed A (2020) Bioethanol production from sugarcane molasses an industrial waste product. Natl J Biol Sci 1:01–23. https://doi.org/10.37605/njbs.v1i2.1

Hollinshead W, He L, Tang YJ (2014) Biofuel production: an odyssey from metabolic engineering to fermentation scale-up. Front Microbiol 5:344. https://doi.org/10.3389/fmicb.2014.00344

Hood EE (2016) Plant-based biofuels. F1000Research 5, F1000 Faculty Rev-185. https://doi.org/10.12688/f1000research.7418.1

Housh M, Cai X, Ng TL, McIsaac GF, Ouyang Y, Khanna M, Sivapalan M, Jain AK, Eckhoff S, Gasteyer S, Al-Qadi I, Bai Y, Yaeger MA, Ma S, Song Y (2015) System of systems model for analysis of biofuel development. J Infrastruct Syst 21:04014050. https://doi.org/10.1061/(ASCE)IS.1943-555X.0000238

Hung C-H, Kanehara K, Nakamura Y (2016) In vivo reconstitution of algal triacylglycerol production in *Saccharomyces cerevisiae*. Front Microbiol 7. https://doi.org/10.3389/fmicb.2016.00070

Ilham Z, Haque Nimme F (2019) Institute of Biological Sciences, Faculty of Science, University of Malaya, 50603, Kuala Lumpur, Malaysia, 2019. Quantitative priority estimation model for evaluation of various non-edible plant oils as potential biodiesel feedstock. AIMS Agric Food 4:303–319. https://doi.org/10.3934/agrfood.2019.2.303

Inui M, Kawaguchi H, Murakami S, Vertès AA, Yukawa H (2004) Metabolic engineering of *Corynebacterium glutamicum* for fuel ethanol production under oxygen-deprivation conditions. J Mol Microbiol Biotechnol 8:243–254. https://doi.org/10.1159/000086705

Johnson TJ, Jahandideh A, Johnson MD, Fields KH, Richardson JW, Muthukumarappan K, Cao Y, Gu Z, Halfmann C, Zhou R, Gibbons WR (2016) Producing next-generation biofuels from filamentous cyanobacteria: an economic feasibility analysis. Algal Res 20:218–228. https://doi.org/10.1016/j.algal.2016.10.020

Kanehisa M, Goto S (2000) KEGG: Kyoto encyclopedia of genes and genomes. Nucleic Acids Res 28:27–30. https://doi.org/10.1093/nar/28.1.27

Karolchik D, Hinrichs AS, Kent WJ (2009) The UCSC genome browser. Curr Protoc Bioinform Ed. Board Andreas Baxevanis Al CHAPTER, Unit1.4. https://doi.org/10.1002/0471250953.bi0104s28

Karp PD, Riley M, Paley SM, Pellegrini-Toole A (2002) The metacyc database. Nucleic Acids Res 30:59–61

Khan Z, Javed F, Shamair Z, Hafeez A, Fazal T, Aslam A, Zimmerman WB, Rehman F (2021) Current developments in esterification reaction: a review on process and parameters. J Ind Eng Chem 103:80–101. https://doi.org/10.1016/j.jiec.2021.07.018

Khan S, Hussain A, Hussain MI, Aqeel SM, Basharat S, Hussain A, Al-Ansi W, Abdulqader A, Yao N (2023) Metabolic engineering of Saccharomyces cerevisiae for ethanol and butanol biofuel production. Int J Environ Agric Biotechnol 8:047–056. https://doi.org/10.22161/ijeab.82.5

Kim J, Son HF, Hwang S, Gong G, Ko JK, Um Y, Han SO, Lee S-M (2022) Improving lipid production of *Yarrowia lipolytica* by the aldehyde dehydrogenase-mediated furfural detoxification. Int J Mol Sci 23:4761. https://doi.org/10.3390/ijms23094761

Kopka J, Schauer N, Krueger S, Birkemeyer C, Usadel B, Bergmüller E, Dörmann P, Weckwerth W, Gibon Y, Stitt M, Willmitzer L, Fernie AR, Steinhauser D (2005) GMD@CSB.DB: the golm metabolome database. Bioinformatics 21:1635–1638. https://doi.org/10.1093/bioinformatics/bti236

Koppolu V, Vasigala VK (2016) Role of *Escherichia coli* in biofuel production. Microbiol Insights 9, MBI-S10878

Lan EI, Liao JC (2011) Metabolic engineering of cyanobacteria for 1-butanol production from carbon dioxide. Metab Eng 13:353–363. https://doi.org/10.1016/j.ymben.2011.04.004

Lee RA, Lavoie J-M (2013) From first- to third-generation biofuels: challenges of producing a commodity from a biomass of increasing complexity. Anim Front 3:6–11. https://doi.org/10.2527/af.2013-0010

Liang M, Zhou X, Xu C (2016) Systems biology in biofuel. Phys Sci Rev 1. https://doi.org/10.1515/psr-2016-0047

Liu Y, Fredrickson JK, Sadler NC, Nandhikonda P, Smith RD, Wright AT (2015) Advancing understanding of microbial bioenergy conversion processes by activity-based protein profiling. Biotechnol Biofuels 8:156. https://doi.org/10.1186/s13068-015-0343-7

Liu Z, Wang J, Nielsen J (2022) Yeast synthetic biology advances biofuel production. Curr Opin Microbiol 65:33–39. https://doi.org/10.1016/j.mib.2021.10.010

Lopez D, Anne S (2015) Improving Saccharomyces cerevisiae as a biofuel production organism

MA (2013) Metabolic engineering of hydrocarbon biosynthesis for biofuel production. In: Fang Z (Ed.), InTech. https://doi.org/10.5772/52050

Madhavan A, Jose AA, Binod P, Sindhu R, Sukumaran RK, Pandey A, Castro GE (2017) Synthetic biology and metabolic engineering approaches and its impact on non-conventional yeast and biofuel production. Front Energy Res 5

Mashayekhi M, Sarrafzadeh MH, Tavakoli O, Soltani N, Faramarzi MA (2017) Potential for biodiesel production and carbon capturing from *Synechococcus Elongatus*: an isolation and evaluation study. Biocatal Agric Biotechnol 9:230–235

Mewes HW, Frishman D, Güldener U, Mannhaupt G, Mayer K, Mokrejs M, Morgenstern B, Münsterkötter M, Rudd S, Weil B (2002) MIPS: a database for genomes and protein sequences. Nucleic Acids Res 30:31–34

Mi H, Lazareva-Ulitsky B, Loo R, Kejariwal A, Vandergriff J, Rabkin S, Guo N, Muruganujan A, Doremieux O, Campbell MJ (2005) The PANTHER database of protein families, subfamilies, functions and pathways. Nucleic Acids Res 33:D284–D288

Milne CB, Eddy JA, Raju R, Ardekani S, Kim P-J, Senger RS, Jin Y-S, Blaschek HP, Price ND (2011) Metabolic network reconstruction and genome-scale model of butanol-producing strain *Clostridium beijerinckii* NCIMB 8052. BMC Syst Biol 5:130. https://doi.org/10.1186/1752-0509-5-130

Miscevic D (2021) Interaction between *Clostridium thermocellum* ATCC 27405 and *Thermoanaerobacterium Saccharolyticum* DSM 7060 on cellulosic and lignocellulosic substrates: substrate attachment, concentration dynamics, and ethanol production. https://doi.org/10.32920/ryerson.14668251.v1

Misra N, Panda PK, Parida BK, Mishra BK (2016) DEMBF: a comprehensive database of enzymes of microalgal biofuel feedstock. PLoS ONE 11:e0146158. https://doi.org/10.1371/journal.pone.0146158

Morgat A, Coissac E, Coudert E, Axelsen KB, Keller G, Bairoch A, Bridge A, Bougueleret L, Xenarios I, Viari A (2012) UniPathway: a resource for the exploration and annotation of metabolic pathways. Nucleic Acids Res 40:D761–D769. https://doi.org/10.1093/nar/gkr1023

Mueller TJ, Berla BM, Pakrasi HB, Maranas CD (2013) Rapid construction of metabolic models for a family of Cyanobacteria using a multiple source annotation workflow. BMC Syst Biol 7:142. https://doi.org/10.1186/1752-0509-7-142

Nascimento C de O, Pina D dos S, Cirne LGA, Santos SA, Araújo MAL, Rodrigues TCG de C, Silva WP, Souza MNS, Alba HDR, Carvalho GGP de (2021) Effects of whole corn germ, a source of linoleic acid, on carcass characteristics and meat quality of feedlot lambs. Animals 11. https://doi.org/10.3390/ani11020267

Neto JM, Komesu A, da Silva Martins LH, Gonçalves VOO, de Oliveira JAR, Rai M (2019) Chapter 10—Third generation biofuels: an overview. In: Rai M, Ingle AP (Eds.), Sustainable bioenergy. Elsevier, pp. 283–298. https://doi.org/10.1016/B978-0-12-817654-2.00010-1

Nishimura DB (2001) Biotech Softw. Internet Rep 2:657

Nwosu-Obieogu K, Chiemenem L, Adekunle K (2016) Biodiesel production from plant seed oil—a review. Int J Sci Res Sci Engl Technol

O'Leary NA, Wright MW, Brister JR, Ciufo S, Haddad D, McVeigh R, Rajput B, Robbertse B, Smith-White B, Ako-Adjei D, Astashyn A, Badretdin A, Bao Y, Blinkova O, Brover V, Chetvernin V, Choi J, Cox E, Ermolaeva O, Farrell CM, Goldfarb T, Gupta T, Haft D, Hatcher E, Hlavina W, Joardar VS, Kodali VK, Li W, Maglott D, Masterson P, McGarvey KM, Murphy MR, O'Neill K, Pujar S, Rangwala SH, Rausch D, Riddick LD, Schoch C, Shkeda A, Storz SS, Sun H, Thibaud-Nissen F, Tolstoy I, Tully RE, Vatsan AR, Wallin C, Webb D, Wu W, Landrum MJ, Kimchi A, Tatusova T, DiCuccio M, Kitts P, Murphy TD, Pruitt KD (2016) Reference sequence (RefSeq) database at NCBI: current status, taxonomic expansion, and functional annotation. Nucleic Acids Res 44:D733-745. https://doi.org/10.1093/nar/gkv1189

Pancha I, Takaya K, Tanaka K, Imamura S (2021) The Unicellular Red Alga *Cyanidioschyzon merolae*, an excellent model organism for elucidating fundamental molecular mechanisms and their applications in biofuel production. Plants 10:1218. https://doi.org/10.3390/plants10061218

Papini M, Salazar M, Nielsen J (2010) Systems biology of industrial microorganisms. Adv Biochem Eng Biotechnol 120:51–99. https://doi.org/10.1007/10_2009_59

Parker K (2013) Metabolic network construction based on the genome of the marine diatom *Thalassiosira Pseudonana* and the analysis of genome-wide transcriptome data to investigate triacylglyceride accumulation. Masters Theses. https://doi.org/10.31979/etd.mved-nx78

Paul PEV, Sangeetha V, Deepika RG (2019) Chapter 9—emerging trends in the industrial production of chemical products by microorganisms, In Buddolla V (Ed.), Recent developments in applied microbiology and biochemistry. Academic Press, pp. 107–125. https://doi.org/10.1016/B978-0-12-816328-3.00009-X

Peralta-Yahya PP, Zhang F, del Cardayre SB, Keasling JD (2012) Microbial engineering for the production of advanced biofuels. Nature 488:320–328. https://doi.org/10.1038/nature11478

Picataggio S (2009) Potential impact of synthetic biology on the development of microbial systems for the production of renewable fuels and chemicals. Curr Opin Biotechnol 20:325–329. https://doi.org/10.1016/j.copbio.2009.04.003

Rahaju SMN, Veza I, Tamaldin N, Sule A, Opia AC, Abdulrahman MB, Djamari DW (2022) Acetone-butanol-ethanol as the next green biofuel-a review. Automot Exp 5:251–260

Rasala BA, Gimpel JA, Tran M, Hannon MJ, Miyake-Stoner SJ, Specht EA, Mayfield SP (2013). Genetic engineering to improve algal biofuels production. In: Borowitzka MA, Moheimani NR (Eds.), Algae for biofuels and energy, developments in applied phycology. Springer Netherlands, Dordrecht, pp 99–113. https://doi.org/10.1007/978-94-007-5479-9_6

Ricci L, Agostino V, Fino D, Re A (2021) Screening of gas substrate and medium effects on 2,3-butanediol production with *C. ljungdahlii* and *C. autoethanogenum* aided by improved autotrophic cultivation technique. Fermentation 7:264. https://doi.org/10.3390/fermentation7040264

Roberts LG, Smagala T (2022) Biofuels. In: Reference module in biomedical sciences. Elsevier. https://doi.org/10.1016/B978-0-12-824315-2.00332-8

Rodchenkov I, Babur O, Luna A, Aksoy BA, Wong JV, Fong D, Franz M, Siper MC, Cheung M, Wrana M, Mistry H, Mosier L, Dlin J, Wen Q, O'Callaghan C, Li W, Elder G, Smith PT,

Dallago C, Cerami E, Gross B, Dogrusoz U, Demir E, Bader GD, Sander C (2020) Pathway Commons 2019 Update: integration, analysis and exploration of pathway data. Nucleic Acids Res 48:D489–D497. https://doi.org/10.1093/nar/gkz946

Safran, M., Dalah, I., Alexander, J., Rosen, N., Iny Stein, T., Shmoish, M., Nativ, N., Bahir, I., Doniger, T., Krug, H., Sirota-Madi, A., Olender, T., Golan, Y., Stelzer, G., Harel, A., Lancet, D., 2010. GeneCards Version 3: the human gene integrator. Database J Biol Databases Curation 2010, baq020. https://doi.org/10.1093/database/baq020

Sahu G, Das LM, Sharma BK, Naik SN (2011) Pilot plant study on biodiesel production from Karanja and Jatropha oils. Asia-Pac J Chem Eng 6:38–43. https://doi.org/10.1002/apj.443

Sarrouh BF, Philippini RR, da Silva SS (2009) Lignocellulosic bioethanol production: perspectives and challenges. Int Rev Chem Eng 1:614–622

Sasaki Y, Eng T, Herbert RA, Trinh J, Chen Y, Rodriguez A, Gladden J, Simmons BA, Petzold CJ, Mukhopadhyay A (2019) Engineering *Corynebacterium glutamicum* to produce the biogasoline isopentenol from plant biomass hydrolysates. Biotechnol Biofuels 12:41. https://doi.org/10.1186/s13068-019-1381-3

Satya P (2015) Crop plants as renewable resources for biofuel production

Savakis P, Hellingwerf KJ (2015) Engineering cyanobacteria for direct biofuel production from CO_2. Curr Opin Biotechnol 33:8–14

Schaefer CF, Anthony K, Krupa S, Buchoff J, Day M, Hannay T, Buetow KH (2009) PID: the pathway interaction database. Nucleic Acids Res 37:D674–D679

Schomburg I, Chang A, Schomburg D (2002) BRENDA, enzyme data and metabolic information. Nucleic Acids Res 30:47–49

Sharma R, Joshi R, Kumar D (2020) Chapter 12—Present status and future prospect of genetic and metabolic engineering for biofuels production from lignocellulosic biomass. In: Kuila A, Sharma V (Eds.), Genetic and metabolic engineering for improved biofuel production from lignocellulosic biomass. Elsevier, pp. 171–192. https://doi.org/10.1016/B978-0-12-817953-6.00012-9

Siebert D, Wendisch VF (2015) Metabolic pathway engineering for production of 1,2-propanediol and 1-propanol by *Corynebacterium glutamicum*. Biotechnol Biofuels 8:91. https://doi.org/10.1186/s13068-015-0269-0

Sittmann J, Bae M, Mevers E, Li M, Quinn A, Sriram G, Clardy J, Liu Z (2020). Bacterially Produced Small Molecules Stimulate Diatom Growth. https://doi.org/10.1101/2020.11.02.365239

Sivaprakasam S, Saravanan CG (2007) Optimization of the transesterification process for biodiesel production and use of biodiesel in a compression ignition engine. Energy Fuels 21:2998–3003. https://doi.org/10.1021/ef060516p

Smith KM, Cho K-M, Liao JC (2010) Engineering *Corynebacterium glutamicum* for isobutanol production. Appl Microbiol Biotechnol 87:1045–1055. https://doi.org/10.1007/s00253-010-2522-6

Sommer MO, Church GM, Dantas G (2010) A functional metagenomic approach for expanding the synthetic biology toolbox for biomass conversion. Mol Syst Biol 6:360. https://doi.org/10.1038/msb.2010.16

Sticklen MB (2008) Plant genetic engineering for biofuel production: towards affordable cellulosic ethanol. Nat Rev Genet 9:433–443. https://doi.org/10.1038/nrg2336

Tang W, Tang AY (2014) Transgenic woody plants for biofuel. J for Res 25:225–236. https://doi.org/10.1007/s11676-014-0454-1

Tang X, Lee J, Chen WN (2015) Engineering the fatty acid metabolic pathway in *Saccharomyces cerevisiae* for advanced biofuel production. Metab Eng Commun 2:58–66. https://doi.org/10.1016/j.meteno.2015.06.005

Tanizawa Y, Fujisawa T, Kodama Y, Kosuge T, Mashima J, Tanjo T, Nakamura Y (2023) DNA Data Bank of Japan (DDBJ) update report 2022. Nucleic Acids Res 51:D101–D105. https://doi.org/10.1093/nar/gkac1083

Torto-Alalibo T, Purwantini E, Lomax J, Setubal JC, Mukhopadhyay B, Tyler BM (2014) Genetic resources for advanced biofuel production described with the Gene Ontology. Front Microbiol 5:528. https://doi.org/10.3389/fmicb.2014.00528

Vijayakumar C, Ramesh M, Murugesan A, Panneerselvam N, Subramaniam D, Bharathiraja M (2016) Biodiesel from plant seed oils as an alternate fuel for compression ignition engines-a review. Environ Sci Pollut Res Int 23:24711–24730. https://doi.org/10.1007/s11356-016-7754-2

Vinuselvi P, Park JM, Lee JM, Oh K, Ghim C-M, Lee SK (2011) Engineering Microorganisms for Biofuel Production. Biofuels 2:153–166. https://doi.org/10.4155/bfs.11.4

Wang Y, Li X, Guo K, Peng L, Wang Y (2017) Gene profiling of plant cell wall biosynthesis for genetic enhancing biomass enzymatic saccharification in cereal crops. J Plant Biochem Physiol 05. https://doi.org/10.4172/2329-9029.1000179

Wichlacz AT (2018) Development of probing strategies to investigate metabolic flux of biofuel production pathways in Clostridia

Williams K, Zheng Y, McGarvey J, Fan Z, Zhang R (2012) Ethanol and volatile fatty acid production from lignocellulose by *Clostridium cellulolyticum*. ISRN Biotechnol 2013:137835. https://doi.org/10.5402/2013/137835

Xiang-yang L (2013) Tolerance mechanisms of microorganisms to organic solvent in biofuel production. Chem Bioeng

Yusuf F, Gaur NA (2017) Engineering *Saccharomyces cerevisiae* for C5 Fermentation: a step towards second-generation biofuel production. In: Kalia VC, Saini AK (Eds.). Springer Singapore, Singapore, pp. 157–172. https://doi.org/10.1007/978-981-10-5511-9_8

Zhang Y, Zhang X, Xiao S, Qi W, Xu J, Yuan Z, Wang Z (2019b) Engineering *Corynebacterium glutamicum* Mutants for 3-Methyl-1-butanol Production. Biochem Genet 57:443–454. https://doi.org/10.1007/s10528-019-09906-4

Zhang X, Yan S, Tyagi RD, Surampalli RY (2019) Biodiesel Production by Transesterification 27–52. https://doi.org/10.1061/9780784415344.ch03

Zhou X-R, Liu Q, Singh S (2023) Engineering nutritionally improved edible plant oils. Annu Rev Food Sci Technol 14:247–269. https://doi.org/10.1146/annurev-food-052720-104852

Zhu Z, Jiang J, Fa Y (2020) Overcoming the Biological contamination in microalgae and cyanobacteria mass cultivations for photosynthetic biofuel production. Molecules 25:5220. https://doi.org/10.3390/molecules25225220

Chapter 17
Physiological Mechanisms Behind the Microbe-Induced Biofuel Production

Shruti Awasthi, Yasrib Qurishi, Deepa Sharma, and Naveen Dwivedi

Abstract The global economy depends on one of the important constituents of life i.e., energy Globalization and industrialization in the world have led to increases in petroleum-based fuel and global oil prices, henceforth biofuel research is considered a hot topic due to the increasing demand for global energy which presently seems to be the alternative source of energy for sustainable development considering the environmental aspects. The development of different methods for biofuel production using plants and microbes has gained considerable attention. Inexhaustible biological resources are available in the form of agricultural biomass and various biological wastes which can be transformed into liquid biofuels. However, the conversion process is very expensive and not worthwhile for large-scale production of biofuel for commercial use, therefore a lot of research needs to be done for the efficient, economical, effective and sustainable production method. The present chapter aims to discuss the microbial mechanism for the production of biofuel with the advancement in metabolic engineering for new organisms and to improve biofuel production. We also discuss the economic viability of various approaches used in biofuel production.

Keywords Globalization · Industrialization · Biofuel · Energy · Sustainable development

S. Awasthi
Department of Life Sciences, Garden City University, Bangalore, India

Y. Qurishi
School of Allied Healthcare and Sciences (SAHS), Jain University, Bangalore, India

D. Sharma
Department of Chemistry, SoBS, IIMT University, Meerut, India

N. Dwivedi (✉)
Department of Biotechnology Engineering, Chandigarh University, Mohali, Punjab (INDIA), India
e-mail: naveen.dwivedi@gmail.com

M. Shah and D. Deka (eds.), *Emerging Sustainable Technologies for Biofuel Production*, Environmental Science and Engineering, https://doi.org/10.1007/978-3-031-52167-6_17

17.1 Introduction

Industrial development and the global economy depend on one of the important constituents of life i.e., energy. Due to rapid globalization, industrialization, and the cumulation of greenhouse gases in the atmosphere, the available energy sources such as fossil fuels are getting depleted. In the current scenario, we can say that the total consumption of crude oil is near off 11.6 million tons per day as a result resources are rapidly depleting and may soon be vanished (Wall et al. 2008; Klasson et al. 1992) without any alternatives source to the crude oil the global economy will suffer and it will fall off as there will high shoot up of the oil prices as the demand will keep on increasing due to globalization (Outlaw et al. 2005) Therefore, we have to do research and find the alternative and economic energy sources that can be renewed and release lesser emission of harmful gases (Baez Vasquez et al. 2008; Nichols et al. 2008). So, we can think about biofuels based on the cellular composition using different plant or animal sources or even we can take the waste resources from plants, animals, or industry and then convert this biomass into fuels which will lead to sustainable forms of energy (Gulati et al. 1996).

Recent development in R&D resulted in many alternative sources for the production of biofuels among them biodiesel has got more attention as it is very similar to diesel in chemical structure as well as energy composition. One plus point is that there is no change needed in existing models of the engine as it is commercially blended with diesel and is being exported to many countries like Italy, Germany, and Malaysia. Production of biodiesel can be done by transesterification of fats and vegetable oils into fatty acid methyl esters or fatty acid ethyl esters in the presence of a catalyst by alcohol (Tilman et al. 2006). Biodiesel has an advantage over petroleum diesel, it is biodegradable, non-toxic, and reduces toxicants. Various traditional methods are therefor-biodiesel productions of these two common methods used in the industry are chemical catalysis and enzymatic catalysis (Raghu et al. 2006). For industrial production chemical catalysis is of prime importance whereas enzymatic catalysis is performed for mild reactions and as well as easy separation of the product. In the current scenario production of biodiesel faces a lot of problems like insufficient stable supply system of feed stock, performance are not consistent and challenging economics (Doi et al. 2008).

Biofuel production from microorganisms has a huge potential to overcome these problems but they do have some advantages like a short producing period, little labour required, an easy scale-up is possible, different venues, seasons, climatic change, and other factors (Filho et al. 1991) using microorganisms to the fullest for the production of biofuels or biodiesel.

17.2 Important Microorganisms for Biofuel Production

Various microorganisms are responsible to produce biofuel. Single-celled microorganisms not only act as catalysts but also synthesize biodiesel by using alcohol and fatty acid as a substrate.

17.2.1 Biofuel Production Using Bacteria

17.2.1.1 Clostridium

Clostridium sp. is one of the most explored species in the production of the biofuel industry. *Clostridium* has a direct pathway that follows the fermentation of carbohydrate, butanol, and ethanol which is biphasic and is produced at the solventogenic phase by exploiting the Embden–Meyerhof pathway which will grow in the acidogenic phase. The anaerobic bacterium *Clostridium ljungdahlii* has the capability to ferment sugar to produce ethanol under anaerobic conditions with the help of NAD^+dependent acetaldehyde & ethanol dehydrogenases (Kopke et al. 2010). If we culture *Clostridium ljungdahlii* in two different bioreactors each for the development of cells as well as ethanol production by using this method ethanol production is enhanced (Klasson et al. 1992).

17.2.1.2 Pseudomonas

Microbial fuel cells are extensively used for the isolation of several bacteria like *Pseudomonas* which has the capability of converting energy obtained from metabolic processes into electricity (Rabaey et al. 2005). Bacteria will use the redox mediators which can shuttle the electrons from bacteria and electron acceptors, which will produce a tremendous power output (Fig. 17.1).

17.2.1.3 *Escherichia coli*

Research has proved that this microorganism *Escherichia coli* is one of the finest model organisms which have been extensively used for the manufacturing of biofuels along with bioethanol and biodiesel. *Escherichia coli* has been modified genetically so that we can optimize good quality efficient biofuel. One of the best examples is the production of fatty acid methyl esters which is obtained by genetically modifying *Escherichia coli* in such a manner that will express the enzyme fatty acid methyltransferase we can even increase the production by deleting methionine regulator metJ (Ingram et al. 1987).

Fig. 17.1 Biofuel production using bacteria

17.2.1.4 *Zymomonas mobilis*

Microorganism uses various pathways for the production of biofuel but Entner–Doudoroff pathway is used by *Z. mobilis* as they need require less amount of ATP for the production of ethanol (Yang et al. 2016). *Z. mobilis*which uses hexose can be genetically modified to produce biofuel like ethanol from lignocellulose hydrolysateconsistinga mixture of glucose and xylose (Clarke et al. 2017).

17.2.1.5 Bacillus

Biofuel can be produced from various strains of *Bacillus Sp.*e.g.; strain *B. thuringiensis*can produce 2450 and 2300 ml/L of biohydrogen similarly *B. subtilis*can produce 1.55 and 1.03 ml/L of ethanol from molasses obtained from sugarcane (Gabra et al. 2019).

Biofuel can be produced from *B. subtilis* by genetically modifying they consist of transcriptional repressor codY which can regulate two monocistronic transcription units, i.e., ilvD and ybgE, and another operon will consist of seven genes that are involved in the biosynthesis of an amino acid (Shivers and Soneshein 2004).

17.2.2 Biofuel Production Using Algae

Research has proved that algae are well off in oil and fatty acid composition and they are present in triacylglycerol form which is the main source for the production of biofuel. Various methods employed for biofuel production from microalga are pyrolysis, gasification, liquefaction, etc. separate hydrocarbons in liquid form e.g., *Dunaliella tertiolecta* (Minowa et al. 1995).

Another method is hydrothermal liquefaction which is cost-effective as algae are well off in water content so they need less energy (Barreiro et al. 2013). In this method, algae are harvested in both small as well as large-scale industries after that they will separate proteins & fatty acids later, they are subjected to thermochemical producing crude biofuel (Alba et al. 2012) temperature needs to be from 250 to 375 c and the pressure needs to be maintained 10 to 20 MPa.

Algal biorefinery is another system that consists of two different kinds of feedstock of algal one will have lipid and the other feedstock will have algaenan which will produce biodiesel and biofuel (Alba et al. 2012) the products from these would be mixed to give the third type of energy system.

There are four types of pyrolysis methods slow, fast, catalytic, and microwave pyrolysis that are used to produce bio-oils and fuels from algae (Fig. 17.2).

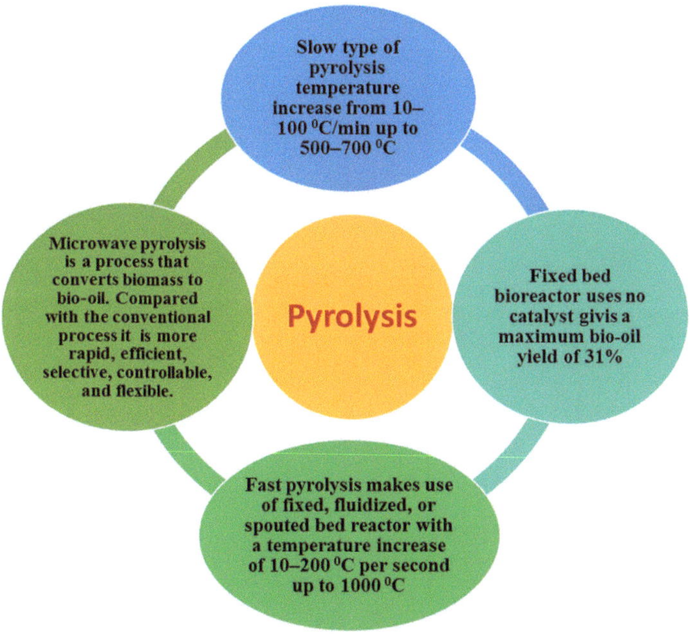

Fig. 17.2 Types of pyrolysis (Pan et al. 2010; Chen et al. 2015)

17.2.3 *Biofuel Production Using* Cyanobacteria

Microalgae or *cyanobacteria* organisms can be used for biodiesel production by growing them with light or without light in open or closed lakes or ponds by providing them with different sources of carbon. Harvesting will allow the biomass to get concentrated finally biomolecules can be extracted by using various techniques which will depend on the product to be produced e.g. biodiesel and gasoline (Blinová et al. 2015).

17.2.3.1 Synechocystis

Synechocystis sp. PCC6803 is a cyanobacterium that was genetically engineered by inserting an enzyme pyruvate decarboxylase isolated from *Z. mobilis* (Gao et al. 2012) similarly another enzyme alcohol dehydrogenase which was isolated from the *cyanobacterium* was introduced to overexpress in conjunction and they will interfere with the pathway of poly-β-hydroxybutyrate will enhance the manufacture of ethanol (Fig. 17.3).

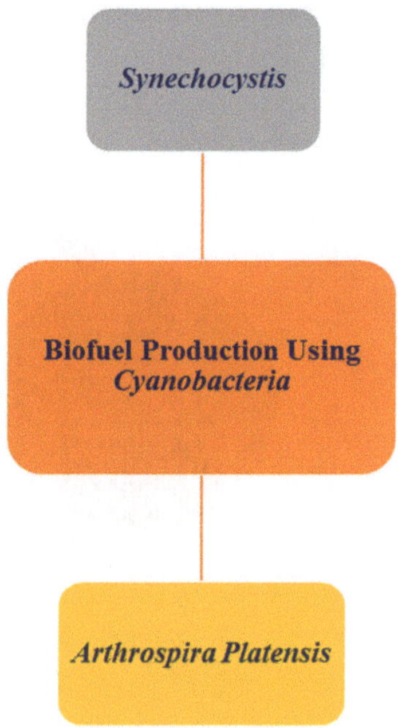

Fig. 17.3 Biofuel production using *Cyanobacteria*

17.2.3.2 *Arthrospira platensis*

Research has proved that free-floating *Arthrospira platensis* is a *filamentous cyanobacterium*, that is extensively used for ethanol production and whose yield could be increased by using $CaCl_2$ and lysozyme (Aikawa et al. 2018). The feedstock of this *cyanobacterium* is subjected to a fermentation process using recombinant yeast *Saccharomyces cerevisiae* strain of BY4741 AASS/GASS that overexpresses α-amylase and glucoamylase gene (Inokuma et al. 2014).

17.2.4 Biofuel Production Using Fungi

17.2.4.1 Saccharomyces

Saccharomyces cerevisiae like any other microorganism can be modified genetically for the efficient production of biofuel. In the Ehrlich pathway isobutanol is produced as a by-product using yeast (Hazelwood et al. 2008).

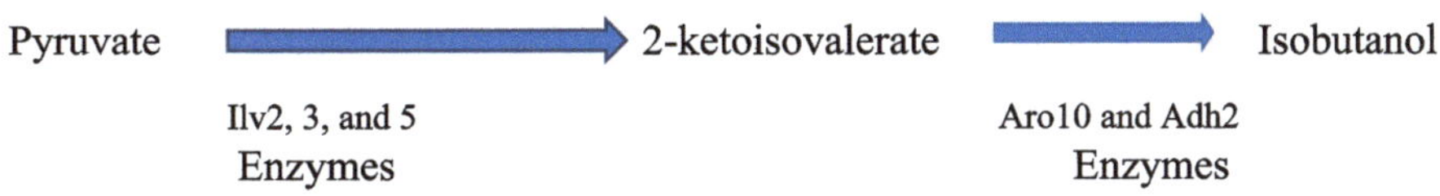

Ehrlich Pathway (Brat et al. 2012)

However, the yield of isobutanol is very low in the conventional pathway but in valine, the biosynthesis gene responsible for coding the enzymes likeILV2, ILV3, ILV5, and BAT2 has been overexpressed leading to increased isobutanol production (Wess et al. 2019).

17.2.4.2 *Chrysoporthe cubensis*

Scientists have reported plant *Chrysoporthe cubensis*is an efficient cellulase and xylanase manufacturer that saccharifies sugarcane and will release 320.8 mg/g monomers of sugar, i.e., glucose and xylulose further utilized for biofuel production (Falkoski et al. 2013).

17.2.4.3 Trichoderma

Sesbania bispinosa which belongs to the plant family of *Fabaceae* this biomass was subjected to hydrolysis by using the enzyme cellulase isolated from *Trichoderma*

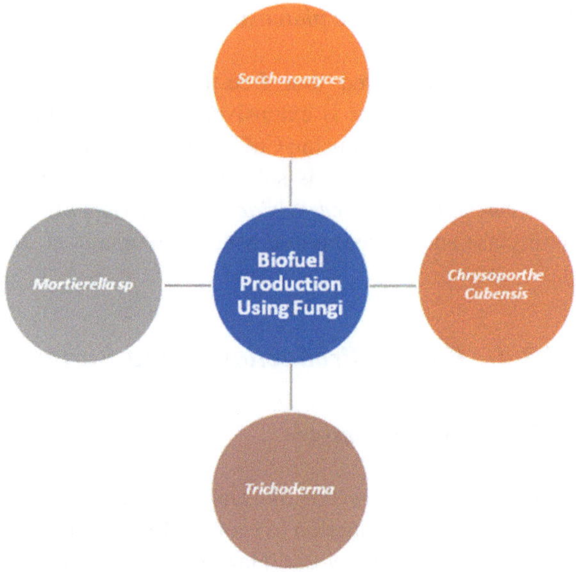

Fig. 17.4 Biofuel production using fungi

longibrachiatum a filamentous fungus was immobilized with a magnetic nanoparticle to accelerate the yield of bioethanol (Baskar et al. 2016) (Fig. 17.4).

17.2.4.4 Mortierella Sp.

Mortierella vinacea, M. isabellina, and *Aspergillus terreus* are filamentous fungi that are actively explored as biodiesel which can be manufactured from the biomass of lignocellulosic as it has the capability to grow and even sustain in acidic dilute sulphuric and hydrolyses the wheat straw and produces lipids (Zheng et al. 2012).

Apart from yeast *Saccharomyces cerevisiae,* there are various molds like Mucorales, *Mucor indicus*, *Mucor hiemalis*, *Mucor circinelloides*, and *Rhizopus oryzae.* They can ferment different types of sugars and produce ethanol (Haghighi Mood et al. 2013; Satari et al. 2015, 2016).

17.3 Genetic Manipulation of Microbes for the Production of Biodiesel

Genetic engineering could be used to manipulate the pathway in such a way that it will transform the microbes into the factories of the cell for the de novo biosynthesis of biodiesel. Many pathways are involved in this process like production of ethanol, metabolism of fatty acid, and transesterification (Fig. 17.5).

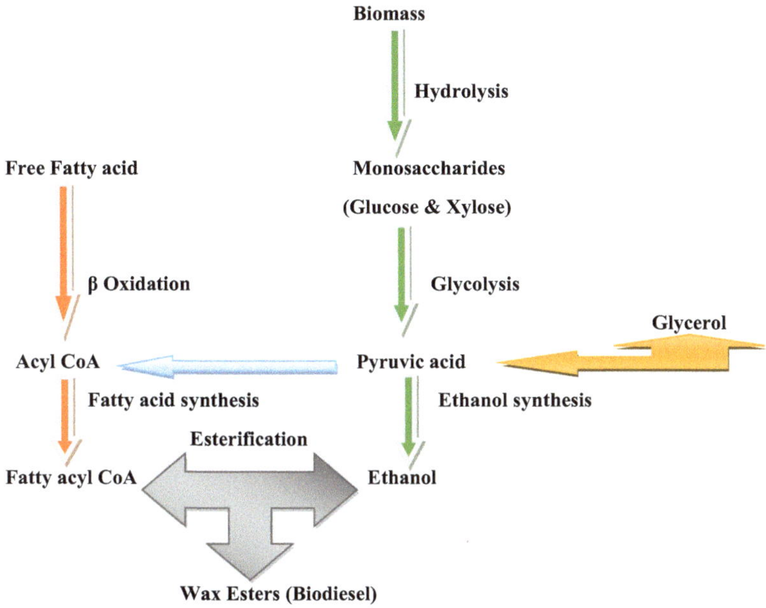

Fig. 17.5 Recombinant microorganisms capable of de novo biosynthesis of biodiesel

Genetic manipulation's best example is *E. coli* which has the ability to synthesize chemicals that will carry a lot of genetic information and participate in the metabolism process (Handke et al. 2011; Huffer et al. 2012). Researchers showed the first proof of a de novo biosynthesis pathway for biodiesel production in *E. coli* was observed by overexpression of the genes of ethanol production from *Z. mobilis* and the WS/DGAT gene extracted by the *Acinetobacter baylyi* strain ADP1 (Kalscheuer et al. 2006). In this process ethanol production was clubbed with esterification followed by cultivation in the presence of monosaccharide and oleic acid under aerobic conditions it was proved that FAEEs biosynthesis depends on the external fatty acid's supplementations (Kalscheuer et al. 2006). Yangkai Duan's co-workers further manipulated the fatty acids pathway of *E. coli* by increasing the content of endogenous fatty acids, which was provided as acyl moieties of coenzyme A which is a thioester of fatty acids for transesterification. Now this new genetically modified strain was able to store FAEEs using hemicellulose as raw materials (Fig. 17.6).

Enzymes were isolated from the microorganism like endoxylanase (Xyn10B) from *Clostridium stercorarium* and a xylanase (Xsa) from *Bacteroides ovatus* they will hydrolyze the hemicellulose to xylose. The pathway of ethanol formation was used for overexpressing of gene pdc and adhB which will code enzyme pyruvate decarboxylase (PDC) and alcohol dehydrogenase (ADH) routed through the fatty acid pathway of E. coli will enhance the production of free fatty acids and similarly fadE gene will code the enzyme acyl-CoA dehydrogenase (FadE) for enhance

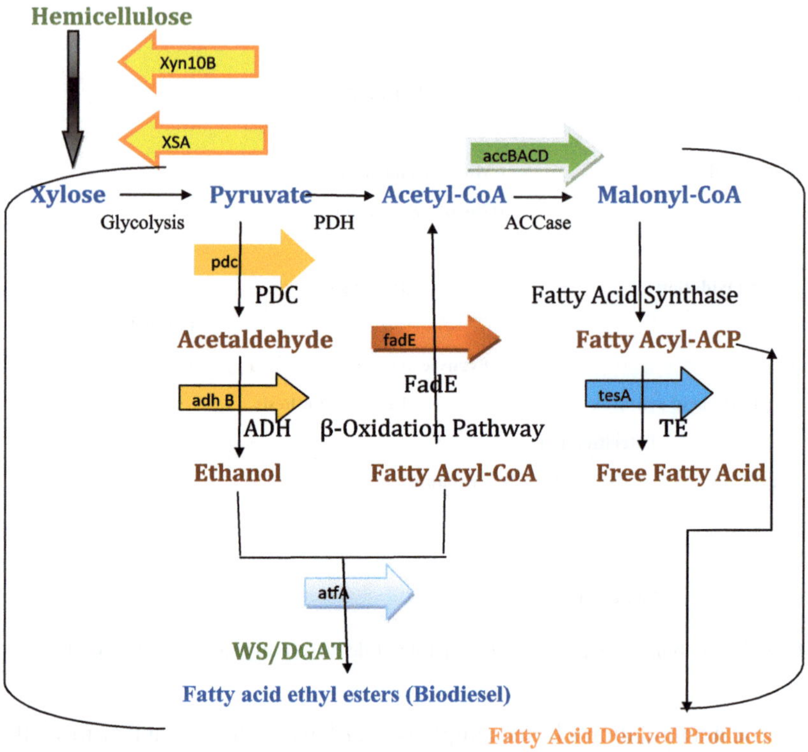

Fig. 17.6 Genetically engineered pathways in E. coli for de novo biosynthesis of biodiesel

synthesis acyl-CoAs which will happen via overexpression of acetyl-CoA carboxylase (ACCase), thioesterases (TE) and fatty acyl-CoA synthase (FadD). The atfA gene from *A. baylyi* coding the wax ester synthase/acyl-coenzyme A diacylglycerol acyltransferase (WS/DGAT) was introduced for fatty acid ethyl esters formation.

17.3.1 Genetic Engineering in Metabolism of Substrates

The current scenario of agro-industrial waste introduction into the cell is the hottest topic as it will decrease the cost of investment for the synthesis of recombinant biodiesel (Li et al. 2011; Weber et al. 2010).

17.3.2 Lignocellulosic Biomass

Genetic engineering can lead to De novo synthesis of Fatty acid ethyl esters from the biomass of lignocellulose which is considered to be the best source of renewable carbon having 40–45% cellulose, 10–40% lignin & 25–50% hemicellulose as this can be degraded by hydrolysis into fermentable sugars in the presence of glycoside hydrolase enzymes which are in huge demand because of this capability (Himmel et al. 2007) (Fig. 17.7).

Consolidated bioprocess (CPB) is nothing but a combination of lignocellulose biomass utilization and the formation of product in a single microorganism which is a potential breakthrough. CPB is efficiently used to convert this mixed sugar and hydrolyze it into FAEEs (Lynd et al. 2005). Cellulose and hemicellulose undergo enzymatic hydrolysis into fermented sugars like hexoses and pentoses which later get converted to biodiesel by the modification into fatty acid pathways.

17.4 Microbes and Biofuel: Toward a Brighter Future

Globalization and industrialization have increased a lot of problems globally like climatic change, overpopulation, and oil are getting extinct. So new innovative technologies need to be researched for microbes to engineer better and more efficient biofuels, but the main problem associated with this is scaling up for these new techniques need to be researched so that we can scale up for commercial purposes.

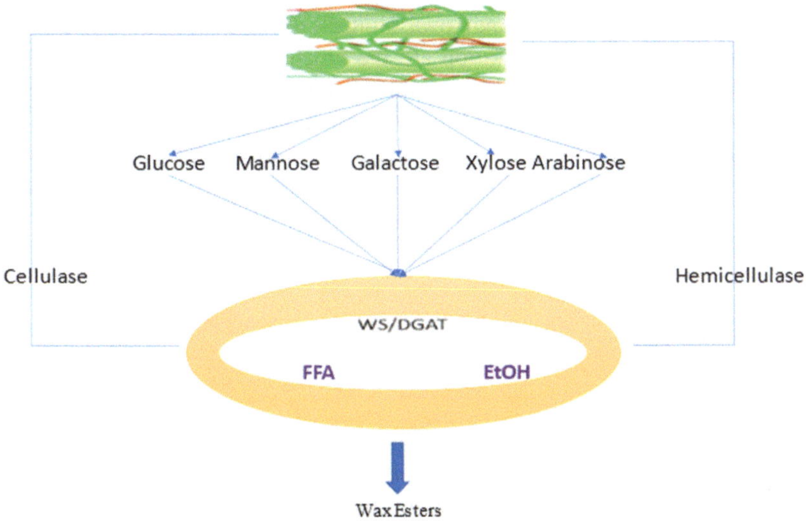

Fig. 17.7 Lignocellulose conversion to biodiesel in a single microbe by consolidated bioprocess (CBP)

17.5 Microbial Technologies for Biofuel Production

The lessening of powers and expansion in the affidavit of nursery gasses has turned into the worldwide worry of today, which is prompting climatic changes and consequently food creation concerns. The development of biofuels (for example alcohols, biodiesels, bioethanol, biobutanol, hydrogen, and biogas) has at the same time occurred to supplant the ordinary powers to diminish the climatic worries.

A few innovations are utilized to deliver different assortments of biofuels from different biomass and rough materials, like microorganisms, creature and horticultural deposits, seeds, vegetable oil, and lignocelluloses. The interest in the creation of different biofuels made by microorganisms has drawn in light of a legitimate concern for the worldwide stage lately. Microorganisms can change the ongoing world view, energy-climate, and open up innumerable open doors for delivering new energy sources, particularly hydrogen, which is an ideal energy hotspot for all frameworks (organic, physical, mechanical) (Talapko et al. 2022).

Bioenergy research is the focal point of logical and mechanical examination in the system of finding savvy biorefineries as an exit from the flow elevated degree of air, water, and soil contamination to find photosynthetic and non-photosynthetic microorganisms that can create clean energy (straightforwardly) or clean hydrogen (Zetterholm et al. 2020).

Growing such energy creation advancements can fundamentally change the all-around accomplished basic degree of ozone-depleting substances that essentially influence the environment (MTA). There are two unique microbial cycles to change over biomass into biofuel: thermochemical and biochemical. Thermochemical change processes, like direct ignition, gasification, and pyrolysis, change over natural parts into biofuels thermochemically. The pyrolysis or gasification process is viewed as the best cycle to change over biomass as fluid fuel (bio-oil). Anaerobic assimilation and alcoholic maturation are the biochemical cycles of energy transformation into biofuels. Right now, biodiesels and bioethanol are the monetarily delivered biofuels that are acquired by the trans-esterification cycle (biochemical interaction) by the handling of plants, creature fats, and vegetable oils. The foundation of further developed innovations will speed up the framework effectiveness and creation of biofuels for a minimal price (Raven et al. 2021).

17.5.1 Thermochemical Change

This interaction is exposed to the warm breaking down of biomass into natural synthetic change, which produces biofuels as the last result (Raheem et al. 2018; Naik et al. 2010). Thermochemical transformation is made out of the accompanying cycles: gasification, pyrolysis, and aqueous liquefaction (HTL). These cycles convert lipids and complete biomass into biofuels.

17.5.1.1 Gasification

The partial oxidation of biomass inside a held measure of air, oxygen, or steam at 7,001,000 °C, which at long last delivers Syngas, this course of development of Syngas is known as gasification. (Chaiwong et al. 2013) The two primary domains of the gasification process are as follows:

1. **Conventional Gasification**:
   In conventional gasification, either a fluidized bed or a fixed bed is used at high temperatures between 800 and 1000 °C.
2. **Supercritical Water Gasification (SCWG)**:
   With the assistance of a catalyst, SCWG is carried out at a temperature of approximately 375–500 °C at 22.1–36 MPa (Chen et al. 2015). It involves the partial combustion of biomass in a controlled environment with a limited supply of oxygen, resulting in the production of syngas (synthesis gas) rich in carbon monoxide (CO), hydrogen (H_2), and other combustible gases.

The mechanism of conventional gasification can be described in several steps:

1. **Drying**:
   The biomass feedstock, such as wood chips, agricultural residues, or energy crops, is first dried to remove moisture content. This is important as excessive moisture can reduce the efficiency of the gasification process. Drying is typically achieved by exposing the biomass to hot air or by utilizing pre-drying techniques.
2. **Pyrolysis**:
   Pyrolysis is achieved by the warm deterioration of biomass by warming it without any oxygen/air. Working circumstances for various kinds of pyrolysis are at climatic strain going with various temperatures (Chen et al. 2015).
3. **Gasification**:
   The pyrolysis products, along with a controlled amount of air or oxygen, are introduced to the gasification reactor. The oxygen-starved environment prevents complete combustion and allows for the partial oxidation of the biomass. At high temperatures, the volatile gases released during pyrolysis react with limited oxygen, resulting in the production of syngas.

The main reactions involved in gasification are:

Reduction reactions: Carbon dioxide (CO_2) reacts with carbon(C) to produce carbon monoxide (CO):

$$CO_2 + C \rightarrow 2CO$$

Water–gas shift reaction: Carbon monoxide reacts with steam (H2O) to produce carbon dioxide and hydrogen gas:

$$CO + H_2O \rightarrow CO_2 + H_2$$

Methanation reaction: Carbon monoxide reacts with hydrogen to produce methane (CH_4):

$$CO + 3H_2 \rightarrow CH_4 + H_2O$$

These reactions result in the conversion of biomass into syngas, which primarily consists of carbon monoxide, hydrogen, methane, and trace amounts of other gases such as nitrogen (N_2) and carbon dioxide.

1. **Gas Cleaning**:
   The syngas produced in the gasification process contains impurities such as particulate matter, tars, sulfur compounds, and trace contaminants. It undergoes a gas cleaning step where these impurities are removed to ensure the quality and stability of the syngas. Cleaning techniques can include processes like scrubbing, filtration, catalytic conversion, and tar cracking.
2. **Utilization**:
   The cleaned syngas can be utilized in various ways for biofuel production. It can be directly burned for heat and power generation, used in combined heat and power (CHP) systems, or further processed to produce liquid biofuels or chemicals. For example, the syngas can undergo additional steps, such as gas-to-liquids (GTL) or Fischer–Tropsch synthesis, to convert them into transportation fuels like gasoline or diesel.

   Conventional gasification offers several advantages, including high energy efficiency, flexibility in feedstock selection, and the ability to produce versatile syngas that can be used in different applications. However, it also has challenges, such as the need for gas cleaning to remove impurities and the potential for tar formation, which can require additional processing steps.

17.5.1.2 Pyrolysis

Pyrolysis is achieved by the warm deterioration of biomass by warming it without any oxygen/air. Working circumstances for various kinds of pyrolysis are at climatic strain going with various temperatures. (Chen et al. 2015) Pyrolysis is a thermal decomposition process that is commonly used in the production of biofuels. It involves the conversion of biomass, such as agricultural residues, wood chips, or energy crops, into biofuels through the application of heat in the absence of oxygen. This method is considered a promising alternative to traditional fossil fuels because it offers several advantages, including the reduction of greenhouse gas emissions and the utilization of renewable resources. The pyrolysis process begins with the collection and preparation of biomass feedstock. The feedstock is typically dried and ground into small particles to increase its surface area and facilitate the pyrolysis reactions. The prepared biomass is then fed into a pyrolysis reactor, which can take various forms such as fluidized beds, rotary kilns, or screw reactors.

Inside the pyrolysis reactor, the biomass undergoes three primary stages: drying, pyrolysis, and char formation. Initially, the biomass is heated to remove any moisture content present. This drying stage typically occurs at temperatures ranging from 100 to 200 °C. Once the biomass is sufficiently dried, the temperature is increased to initiate the pyrolysis process.

During pyrolysis, the biomass is subjected to high temperatures typically between 300 and 800 °C, and the absence of oxygen prevents combustion from occurring. As a result, biomass decomposes into three main products: bio-oil, biochar, and syngas.

1. **Bio-Oil**
   Also known as pyrolysis oil or bio-crude, bio-oil is a dark, viscous liquid that is rich in organic compounds. It contains a wide range of chemical compounds, including aldehydes, ketones, acids, esters, and phenolic compounds. Bio-oil can be further processed and upgraded to remove impurities and increase its stability and energy content. It can be used directly as a fuel or as a feedstock for the production of transportation fuels and other value-added chemicals.
2. **Biochar**
   Biochar is a solid residue that remains after the pyrolysis process. It consists of carbonaceous material and various inorganic compounds. Biochar has a high carbon content and is often used as a soil amendment to improve soil fertility, retain moisture, and sequester carbon in agricultural and environmental applications. It can also be utilized as a renewable energy source through combustion.
3. **Syngas**
   Syngas, or synthesis gas, is a mixture of carbon monoxide, hydrogen, methane, and other light hydrocarbons. It is formed during the pyrolysis process due to the thermal breakdown of biomass. Syngas can be used as a fuel for heating or electricity generation, or they can be further processed through additional conversion technologies, such as gasification or Fischer–Tropsch synthesis, to produce higher-value fuels and chemicals.

   The choice of pyrolysis conditions, such as temperature, residence time, and heating rate, can significantly influence the yield and properties of the bio-oil, biochar, and syngas produced. Optimal process parameters are often determined through experimental investigations and optimization studies to maximize the desired product yields and quality.

   In summary, pyrolysis is a thermochemical conversion method used in biofuel production. It offers a means to convert biomass into valuable bio-oil, biochar, and syngas. These products can be utilized as renewable energy sources, reducing dependence on fossil fuels and contributing to the mitigation of climate change. Additionally, the process produces biochar, which has applications in agriculture and carbon sequestration, further enhancing its environmental benefits.
4. **Aqueous Liquefaction (HTL)**
   HTL of biomass includes physical and synthetic change, which is performed at a high temperature of around 250–500 oC of high auto thermal water strain of around 5–20 bar, either regardless of utilizing an impetus. (Raheem et al. 2015; Matsui et al. 1997).

17.5.1.3 Biochemical Change

Biochemical transformation is completed by the hydrolysis of cell walls utilizing microbes to be changed into aged sugars. Maturation is the course of anaerobic processing of sugars into various biofuel creations, for example, biogas, bioethanol, and biohydrogen (Raheem et al. 2018; Adeniyi et al. 2018).

17.5.1.4 Anaerobic Absorption

Biochemical transformation of natural matter like biomass is changed into biomethane and carbon dioxide with the assistance of catalysts and microorganisms in this course of anaerobic assimilation. This process involves the breakdown of complex organic compounds by a consortium of microorganisms under controlled anaerobic conditions.

The mechanism of anaerobic digestion can be described in several stages:

1. **Hydrolysis**:
   The first step in anaerobic digestion is hydrolysis, where complex organic materials, such as carbohydrates, proteins, and fats, are broken down into simpler compounds by enzymes produced by hydrolytic bacteria. These enzymes degrade large molecules into smaller soluble compounds, such as sugars, amino acids, and fatty acids, which can be easily digested by microorganisms in subsequent stages.
2. **Acidogenesis**:
   In this stage, the hydrolyzed compounds are further broken down by acidogenic bacteria. These bacteria convert the simpler compounds into volatile fatty acids (VFAs), such as acetic acid, propionic acid, and butyric acid, as well as alcohol and carbon dioxide. VFAs are important intermediates in the anaerobic digestion process and serve as a source of energy for the subsequent stages.
3. **Acetogenesis**:
   Acetogenic bacteria convert VFAs produced during acidogenesis into acetic acid, hydrogen, and carbon dioxide through a series of biochemical reactions. Acetic acid is a crucial precursor for the production of methane, which is the main component of biogas.
4. **Methanogenesis**:
   Methanogenic archaea play a vital role in the final stage of anaerobic digestion. These microorganisms convert acetic acid, hydrogen, and carbon dioxide produced in the previous stages into methane gas (CH_4). Methanogens are strict anaerobes and thrive in the absence of oxygen. The produced methane can be used as a biofuel for electricity generation, heating, or transportation purposes.

The overall anaerobic digestion process is dependent on maintaining specific environmental conditions, including temperature, pH, and retention time. The optimal temperature for anaerobic digestion typically ranges between 25 and 55

°C, depending on the type of microorganisms involved. The pH level is usually maintained within a neutral to slightly acidic range to support the growth of various microbial species.

Different types of feedstocks can be used for anaerobic digestion, including animal manure, agricultural residues, food waste, and wastewater sludge. The composition of the feedstock affects the efficiency and performance of the anaerobic digestion process. It is common to use a mixture of feedstocks to ensure a balanced nutrient supply and maximize the production of biogas.

Anaerobic digestion offers several environmental and economic benefits. It helps in the management and treatment of organic waste, reducing its volume and minimizing the release of harmful pollutants into the environment. Additionally, the production of biogas as a renewable energy source helps in reducing greenhouse gas emissions and dependence on fossil fuels.

17.5.1.5 Fermentation

Maturation is one more sort of biochemical transformation that includes the change of carb constituents like cellulose, starch, or sugar into bioethanol. Aging is the enzymatic cycle completed by the hydrolysis of starch into more modest types of sugars, alongside yeast maturation prompting the arrangement of bioethanol (Milano et al. 2016).

17.5.1.6 Photobiological H_2 Creation

In this course of biochemical transformation, hydrogen gas is delivered through utilizing biomass, explicitly with the assistance of hydrogenase proteins, which change over electrons and protons into hydrogen gas (Maness et al. 2009).

$$H^+ + 2e^- \leftrightarrow H2 \uparrow$$

This Mechanism Involves Several Steps:

1. **Selection of Photosynthetic Microorganisms**:
   The first step is to select suitable photosynthetic microorganisms, such as certain strains of cyanobacteria or green algae, which have the ability to utilize sunlight to convert water into hydrogen gas through the process of photosynthesis. These microorganisms possess specialized enzymes called hydrogenases that enable the production of hydrogen gas.
2. **Cultivation of Microorganisms**:
   The selected microorganisms are cultivated in a controlled environment, such as bioreactors or open ponds, providing optimal conditions for growth and hydrogen production. Factors such as temperature, light intensity, nutrient availability, and pH are carefully controlled to ensure the highest productivity.

3. **Light Absorption**:
   The photosynthetic microorganisms utilize light energy from the sun as their primary energy source. Pigments, such as chlorophyll, within the microorganisms, absorb photons of light, which excite electrons and initiate the photosynthetic process.
4. **Photosynthesis**:
   In the presence of light, the microorganisms undergo photosynthesis, a complex biochemical process that consists of two stages: light-dependent reactions and light-independent reactions.
5. **Hydrogen Production**:
   The microorganisms have the ability to produce hydrogen gas under specific conditions. Various mechanisms can contribute to the generation of hydrogen gas:

   1. **Hydrogenase Activity**:
      The microorganisms possess hydrogenase enzymes that can directly convert protons (H^+) from the surrounding environment and electrons (e^-) from the photosynthetic process into molecular hydrogen (H_2). These enzymes play a crucial role in the final step of hydrogen production.
   2. **Depletion of Electron Acceptors**:
      Under certain conditions, the availability of alternative electron acceptors, such as oxygen or other compounds, can be limited. In such cases, the excess electrons produced during photosynthesis are diverted toward the hydrogenase enzymes, favoring the production of hydrogen gas.

6. **Harvesting Hydrogen Gas**:
   The generated hydrogen gas can be harvested from the system through various techniques, such as gas bubbling or membrane separation. The collected hydrogen gas can then be further purified, compressed, and stored for subsequent use as a biofuel.

   Photobiological hydrogen creation offers several advantages, including its potential as a clean and renewable energy source. It utilizes abundant resources such as sunlight and water, and the process itself does not generate carbon dioxide emissions. However, there are challenges to overcome, such as the need for optimizing the efficiency of hydrogen production and scaling up the process for commercial viability.
7. **Transesterification (or Substance Transformation)**:
   This interaction includes the response of fatty oils with liquor, most usually methanol, and ethanol to create biodiesel and glycerol utilizing an acidic or an essential impetus (Machado and Atsumi 2012). As indicated by (Milano et al. 2016), transesterification is depicted by the change of biomass to biodiesel and is additionally capable in the compound transesterification of biomass. Transesterification is of two kinds (Milano et al. 2016) (Fig. 17.8).

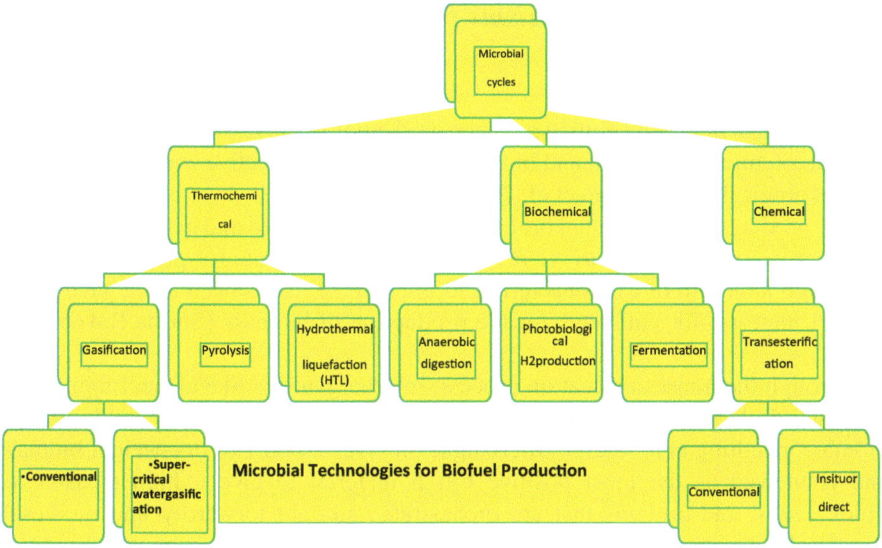

Fig. 17.8 Microbial technologies for biofuel production

17.6 Generations of Biofuel

First-generation biofuels are those produced from agricultural food products like grains, corn, maize, or soybeans. Due to the durability of their polymers and their cost-effectiveness, lignocellulosic feedstocks, which are widely available worldwide, are regarded as second-generation biofuels.

As algae can produce 200 times more biomass per area than first- and second-generation biofuel sources, third-generation biofuels are made from cyanobacteria, algae, and other microorganisms and offer intriguing possibilities for future production (Balan et al. 2013) (Fig. 17.9).

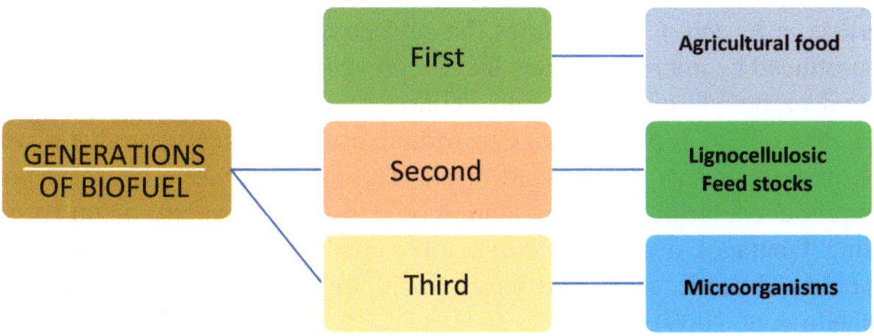

Fig. 17.9 Generation of biofuel

17.6.1 Use of Microorganisms for More Efficient Recovery of Fuel

The creation of microbial biofuel has been marketed, especially biodiesel, bioethanol, biobutanol, microbial lipids, and biogas, and is the main elective energy energizer to limit ecological harm (Balan et al. 2013).

The production of microbial biofuels has been recognized and extensively studied in recent years. According to the research, a promising route for producing biofuel is through a biofilm of cyanobacteria and microalgae. It has been acknowledged as a novel platform for pathways for the production of biomass (Spriha Raven et al. 2021).

Through the transesterification process, microorganisms like algae, fungi, yeast, and bacteria biosynthesize and store a large amount of intercellular lipids in their biomass, resulting in biodiesel. Microalgae oil gives 20-60% of dry-weight biomass and is viewed as the most positive rough material to give a high measure of lipids, carbs, and proteins that are orchestrated straight forwardly into biodiesel. Biodiesel is designed by utilizing *E coli* and *Saccharomyces cerevisiae*, which contain heterologous qualities (Spriha Raven et al. 2021).

The most environmentally friendly product made from renewable biomass of ethanol is bioethanol as a biofuel. Some crude materials, such as corn, wheat, sugar beet, sugarcane, and maize, which provide fermentable sugars and contain a significant amount of carbohydrates, can be used to make bioethanol. Since about half of algae is made up of oil, it is thought to be used to make biofuels. Through the gasification process, the algae species *Spirulina* is utilized in the production of biofuel (Spriha Raven et al. 2021).

Biobutanol is produced through the fermentation process of *Clostridium* species, which converts sugars into solvents like butanol, acetone, and ethanol. Because of its high energy level, high volatility, and hygroscopic nature, biobutanol is regarded as a promising biofuel.

The production of biofuels by microorganisms requires a unique metabolic pathway and catalytic enzymes; for instance, the direct decarboxylation of pyruvate in S. cerevisiae results in the direct formation of ethanol. During the decarboxylation of pyruvate in *E. coli*, CoA activates the acyl group and reduces to ethanol. Pyruvate is produced by microorganisms like *S. cerevisiae*, and *Zymomonas mobilis* uses pyruvate decarboxylase to turn it into alcoho.

At a rate of 3:6:1 proportion Clostridium acetobutylicum produces acetone, butanol, and ethanol as the finished result. The three primary groups of products that are produced during the fermentation process in C. acetobutylicum are solvent (including 1-butanol, acetone, and ethanol), organic acids (including butyric acid, lactic acid, and acetic acid), and gases (including hydrogen and carbon dioxide) (Gheshlaghi et al. 2009).

Aldehyde is made from 2-keto acids, which then turn into alcohol. Iso-butanol is produced on a large scale under anaerobic conditions with *E. coli* and *Lactococcus lactis*. (Atsumi et al. 2008a, b; Bastian et al. 2011, Huo et al. 2011).

The large-scale production of lipids and hydrocarbons has done by many species of microalgae like *Chlorella sp.*, *Crypthecodinium cohnii*, and *Botryococcus braunii*. Chorella has approximately 50% lipids, while *B braunii* has the highest oil content, at about 80%. Commercial travel uses algae-derived green jet fuel (Raven et al. 2021).

17.7 Industrial Trends: Current Status

Ongoing researchers have presented a plethora of opportunities for the production of biofuels using resources that are cost-effective and environment-friendly. However, the feasibility and commercial exploitation of such resources needs to be assessed on the industrial level. Biofuel production has increased only by 4% up to 2017 including sugar and starch-based ethanol, hydrogenated vegetable oil, etc. (2019 report by International Energy Agency). This 4% increase numerically accounts for about 83 million tonnes of oil equivalents (Mtoe). However, a growth of about 2.5% per year is being speculated and a target has been set for 284 Mtoe. Bioethanol industries are vastly located in Brazil that uses sugarcane molasses and starch. These industries generally initiate ethanol production using S. cerevisiae that upon batch fermentation also leads to the generation of products such as methanol. Biofuel production from lignocellulosic biomass is commercially limited due to a lack of enzymes that might efficiently convert the cellulosic components into their reducing sugars. Now days, biomasses are being subjected to specialized biorefineries for the integrated production of food, chemicals as well as biofuels at an industrial scale. Food crops such as Cassava are being used for bioethanol and other biofuel production at the industrial level.

Recent trends in the production of biofuels at the industrial level encompass the development of certain policies in different countries including those of the European Union, the USA, etc. For example, Brazil encourages its citizens to use hydrous bioethanol in place of gasoline while in the USA additional incentives are provided for using cellulosic bioethanol. According to the Energy Policy Act, 2005 the US Congress has also extended biodiesel fuel excise tax credit through 2008.

Following the footsteps of developed countries, some underdeveloped and developing countries such as sub-Saharan African countries are also taking into account the use of l, and industrial advances are being approached in countries including Ghana, Burkina Faso, Mali, etc.. South Africa has implemented a 5-year pilot phase plan and mandates the use of bioethanol and biodiesel together with conventional fuels. On the national level, for biofuel production with low carbon emission, India joined United States Partnership to Advance Clean Energy and majorly uses broken rice, pearl millet, and sorghum at ethanol distilleries of Haryana. These distilleries also make use of *S. cerevisiae* which produces bioethanol by saccharification or fermentation of the feedstocks.

17.8 Future Viewpoint

The greatest test in utilizing microorganisms to deliver biofuels is creating a lot of fuel more economically and productively than conventional petroleum derivatives delivering this sum on a modern scale is testing. Subsequently, future biofuel efficiency ought to be focused on increment microbial biofuel's agreeableness. They are made to most successfully resolve the issues of the energy emergency, contamination, an unnatural weather change, and waste administration. Microorganisms used to be simple biomass decomposers, but since of headways in biotechnology, quality altering, and manufactured science, they currently produce biofuel (Sahni et al. 2021).

17.9 Conclusion

At present, the most challenging issue is the sustainable utilization of energy and to preserve the valuable assets we need to explore some newer avenues for the production of energy in terms of biofuels, bio-products, etc. These days we have our technology on the verge to find some alternative system, and in this context, we came across some advanced technologies that have paved new pathways for utilizing diverse microorganisms to enhance the production of biofuels. Advancement in a couple of years in the field of metabolic engineering has accelerated the production of biofuels like fatty acid, alcohols, and gaseous derivatives, that has the potential to compete with the fossil fuels being used these days. Although there is a huge advancement in the strategies, as well as tools to understand the production mechanism for biofuels or other products but the paradigm has shifted towards the cost-effective approach, and researchers, are still needed in this direction. Synthesis of biofuels requires a huge amount of reducing powers for reduction of carbon residues but still in this process, a large amount of CO_2 is lost and not utilized in ethanol production. We need to minimize the cost that is spent on maintaining the fermentation process because not all the plant biomass is utilized in biofuel production, some biomass goes in vain. Biofuels are being encouraged as a low-carbon alternative to fossil fuels and therefore it would certainly help to cut greenhouse gas (GHG) emissions and the associated climate change impacts and sustainable development.

References

Adeniyi OM, Azimov U, Burluka A (2018) Algae biofuel: current status and future applications. Renew Sustain Energy Rev 90:316–335

Aikawa S, Inokuma K, Wakai S, Sasaki K, Ogino C, Chang JS (2018) Biotechnology for biofuels direct and highly productive conversion of cyanobacteria *Arthrospira platensis* to ethanol with $CaCl_2$ addition. Biotechnol Biofuels 1–9

Alba GL, Torri C, Samorì C, Van Der Spek J, Fabbri D, Kersten SRA, Brilman DWF (2012) Hydrothermal treatment (HTT) of microalgae: evaluation of the process as conversion method in an algae biorefinery concept. Energy Fuels 26(1):642–657

Atsumi S, Cann AF, Connor MR et al (2008a) Metabolic engineering of Escherichia coli for 1-butanol production. Metab Eng 10:305–311

Atsumi S, Hanai T, Liao JC (2008b) Non-fermentative pathways for synthesis of branched-chain higher alcohols as biofuels. Nature 451(7174):86–89

Baez Vasquez MA, Demain AL (2008) Ethanol, biomass and clostridia. In: Harwood CS, Demain AL, Wall JD (eds) Bioenergy. ASM Press, Washington, DC

Balan V, Chiaramonti D, Kumar S (2013) Review of US and EU initiatives toward development, demonstration, and commercialization of lignocellulosic biofuels. Biofuels Bioprod Biorefin 7:732–759

Barreiro LD, Prins W, Ronsse F, Brilman W (2013) Hydrothermal liquefaction (HTL) of microalgae for biofuel production: state of the art review and future prospects. Biomass Bioenergy 53:113–127

Baskar G, Naveen Kumar R, Heronimus Melvin X, Aiswarya R, Soumya S (2016) Sesbaniaaculeate biomass hydrolysis using magnetic nanobiocomposites of cellulase for bioethanol production. Renew Energy 98:23–28

Bastian S, Liu X, Meyerowitz JT et al (2011) Engineered ketol-acid reductoiso-merase and alcohol dehydrogenase enable anaerobic2-methylpropan-1-oil production at the oretical yield in *Escherichia coli*. Metab Eng 13:345–352

Blinová L, Bartošová A, Gerulová K (2015) Cultivation of microalgae (*Chlorella vulgaris*) for biodiesel production. Res Pap Fac Mater Sci Technol Slovak Univ Technol 23(36):87–95

Brat D, Weber C, Lorenzen W, Bode HB, Boles E (2012) Cytosolic re-localization and optimization of valine synthesis and catabolism enables increased isobutanol production with the yeast *Saccharomyces cerevisiae*. Biotechnol Biofuels 5(1):1–16

Chaiwong K, Kiatsiriroat T, Vorayos N, Thararax C (2013) Study of bio-oil and bio-char production from algae by slow pyrolysis. Biomass Bioenergy 56:600–606

Chen WH, Lin BJ, Huang MY, Chang JS (2015) Thermochemical conversion of microalgal biomass into biofuels: a review. Bioresour Technol 184:314–327

Clarke K, Mokomele T, Callanan L, Groenewald J (2017) *Zymomonas mobilis*–towards bacterial biofuel. Energy Environ Climate Change The Nexus 205–219

Doi RH, et al (2008) Cellulases of mesophilic microorganisms: cellulosome and non-cellulosome producers Ann NY Acad Sci 1125:267–279

Falkoski D, Guimarães V, de Almeida M, Alfenas A, Colodette J, de Rezende S (2013) *Chrysoporthe cubensis*: a new source of cellulases and hemicellulases to application in biomass saccharification processes. Bioresour Technol 130:296–305

Filho EXF, Tuohy MG, Puls J, Coughlan MP (1991) The xylan-degrading enzyme systems of *Penicillium capsulatum* and *Talaromyces emersonii*. Biochem Soc Trans 19(1):25S

Gabra FA, Abd-Alla MH, Danial AW, Abdel-Basset R, Abdel-Wahab AM (2019) Production of biofuel from sugarcane molasses by diazotrophic *Bacillus* and recycle of spent bacterial biomass as biofertilizer inoculants for oil crops. Biocatal Agric Biotechnol 19:101112

Gao Z, Zhao H, Li Z, Tan X, Lu X (2012) Photosynthetic production of ethanol from carbon dioxide in genetically engineered cyanobacteria. Energy Environ Sci 5(12):9857–9865

Gheshlaghi R, Scharer JM, Moo-Young M, Chou CP (2009) Metabolic pathways of clostridia for producing butanol. Biotechnol Adv 27:764–781

Gulati M, Kohlmann K, Ladisch MR, Hespell R, Bothast RJ (1996) Assessment of ethanol production options for corn product. Bioresour Technol 58(3):253–264

Haghighi Mood S, Hossein Golfeshan A, Tabatabaei M, Salehi Jouzani G, Najafi G, Gholami M, Ardjmand M (2013) Lignocellulosic biomass to bioethanol, a comprehensive review with a focus on pre-treatment. Renew Sust Energ Rev 27:77–93

Handke P, Lynch SA, Gill RT (2011) Application and engineering of fatty acid biosynthesis in *Escherichia coli* for advanced fuels and chemicals. Metab Eng 13:28–37. https://doi.org/10.1016/j.ymben.2010.10.007

Hazelwood L, Daran J, van Maris A, Pronk J, Dickinson J (2008) The Ehrlich pathway for fusel alcohol production: a century of research on *Saccharomyces cerevisiae* metabolism. Appl Environ Microbiol 74(8):2259–2266

Himmel ME, Ding SY, Johnson DK, Adney WS, Nimlos MR, Brady JW et al (2007) Biomass recalcitrance: engineering plants and enzymes for biofuels production. Science 315:804–807. https://doi.org/10.1126/science.1137016

Huffer S, Roche CM, Blanch HW, Clark DS (2012) *Escherichia coli* for biofuel production: bridging the gap from promise to practice. Trends Biotechnol 30:538–545. https://doi.org/10.1016/j.tibtech.2012.07.002

Huo YX, Cho KM, Rivera JG et al (2011) Conversion of proteins into biofuels by engineering nitrogen flux. Nat Biotechnol 29:346–351

Ingram LO, Conway T, Clark DP, Sewell GW, Preston JF (1987) Genetic engineering of ethanol production in *Escherichia coli.* Appl Environ Microbiol 53(10):2420

Inokuma K, Yoshida T, Ishii J, Hasunuma T, Kondo A (2014) Efficient co-displaying and artificial ratio control of α-amylase and glucoamylase on the yeast cell surface by using combinations of different anchoring domains. Appl Microbiol Biotechnol 99(4):1655–1663

Kalscheuer R, Stölting T, Steinbüchel A (2006) Microdiesel: *Escherichia coli* engineered for fuel production. Microbiology 152:2529–2536. https://doi.org/10.1099/mic.0.29028-0

Klasson KT, Ackerson MD, Clausen EC, Gaddy JL (1992) Bioconversion of synthesis gas into liquid or gaseous fuels. Enzyme Microb Technol 14(8):602–608

Kopke M, Held C, Hujer S, Liesegang H, Wiezer A, Wollherr A, Ehrenreich A, Liebl W, Gottschalk G, Durre P (2010) *Clostridium ljungdahlii* represents a microbial production platform based on syngas. Proc Natl Acad Sci 107(29):13087–13092

Li X, Hu HY, Zhang YP (2011) Growth and lipid accumulation properties of a freshwater microalga *Scenedesmus sp.* under different cultivation temperature. Bioresour Technol 102:3098–30102. https://doi.org/10.1016/j.biortech.2010.10.055

Lynd LR, van Zyl WH, McBride JE, Laser M (2005) Consolidated bioprocessing of cellulosic biomass: an update. Curr Opin Biotechnol 16:577–583. https://doi.org/10.1016/j.copbio.2005.08.009

Machado IM, Atsumi S (2012) Cyanobacterial biofuel production. J Biotechnol 162:50–56

Maness PC, Yu J, Eckert C, Ghirardi ML (2009) Photobiological hydrogen production—prospects and challenges. Efforts to scale up the capacity of green algae and cyanobacteria to use sunlight to convert water into hydrogen gas for energy use. https://doi.org/10.1128/microbe.4.275.1

Matsui TO, Nishihara A, Ueda C et al (1997) Liquefaction of micro-algae with iron catalyst. Fuel 76:1043–1048

Microbial Biofuels: An Economic and Eco-Friendly Approach. https://www.researchgate.net/publication/341431529_Microbial_Biofuels_An_Economic_and_Eco-Friendly_Approach. Accessed 1 May 2023

Microbial Biofuels: An Economic and Eco-Friendly Approach. https://www.researchgate.net/publication/341431529_Microbial_Biofuels_An_Economic_and_Eco-Friendly_Approach. Accessed 30 April 2023

Microbial Biofuels: An Economic and Eco-Friendly Approach. https://www.researchgate.net/publication/341431529_Microbial_Biofuels_An_Economic_and_Eco-Friendly_Approach. Accessed 29 April 2023

Milano J, Ong HC, Masjuki HH et al (2016) Microalgae biofuels as an alternative to fossil fuel for power generation. Renew Sustain Energy Rev 58:180–197

Minowa T, Yokoyama S, Kishimoto M, Okakura T (1995) Oil production from algal cells *Dunaliella tertiolecta* by direct thermochemical liquefaction. Fuel 74(12):1735–1738

Naik SN, Goud VV, Rout PK, Dalai AK (2010) Production of first- and second-generation biofuels: a comprehensive review. Renew Sustain Energy Rev 14:578–597

Nichols NN, Dien BS, Monceaux DA, Bothast RJ (2008) Production of ethanol from corn and sugarcane. In: Harwood CS, Demain AL, Wall JD (eds) Bioenergy. ASM Press, Washington, DC

Outlaw JL, Collins KJ, Duffield JA (2005) Agriculture as a producer and consumer of energy. CABI Publishing, Oxford, UK

Pan P, Hu C, Yang W, Li Y, Dong L, Zhu L, Tong D, Qing R, Fan Y (2010) The direct pyrolysis and catalytic pyrolysis of *Nannochloropsis sp.* residue for renewable bio-oils. Bioresour Technol 101(12):4593–4599

Rabaey K, Boon N, Höfte M, Verstraete W (2005) Microbial phenazine production enhances electron transfer in biofuel cells. Environ Sci Technol 39(9):3401–3408

Raghu S, Anderson RC, Daehler CC, Davis AS, Wiedenmann RN, Sim Berloff D, et al (2006) Adding biofuels to the invasive species fire? Science 313(5794):1742

Raheem A, Azlina WW, Yap YT et al (2015) Thermochemical conversion of microalgal biomass for biofuel production. Renew Sustain Energy Rev 49:990–999

Raheem A, Prinsen P, Vuppaladadiyam AK et al (2018) A review on sustainable microalgae- based biofuel and bioenergy production: recent developments. J Clean Prod 181:42–59

Sahni S, Singh MK, Narang A (2021) Sustainable solution for future energy challenges through mic robes. Energy 231–249

Satari B, Karimi K, Zamani A (2015) Oil, chitosan, and ethanol production by dimorphic fungus *Mucor* indices from different lignocelluloses. J Chem Tech Biotechnol 91(6):1835–1843

Satari B, Karimi K, Taherzadeh M, Zamani A (2016) Co-production of fungal biomass derived constituents and ethanol from citrus wastes free sugars without auxiliary nutrients in airlift bioreactor. Int J Mol Sci 17(3):302

Shivers RP, Sonenshein AL (2004) Activation of the bacillus subtilis global regulator CodY by direct interaction with branched-chain amino acids. Mol Microbiol 53(2):599–611

Raven S, Ekka SB, Chattree SE, Sadanand SS, Rina L, Tiwari A (2021) Microbial technology for biofuel production. Wiley

Talapko J, Talapko D, Matic A, Škrlec I (2022) Microbial technologies applied biofuel production. https://doi.org/10.3390/en15176365

Tilman D, Hill J, Lehman C (2006) Carbon-negative biofuels from low-input high-diversity grassland biomass. Science 314(5805):1598–1600

Wall JD, Harwood CS, Demain AL (2008) Bioenergy. ASM Press, Washington DC

Weber C, Farwick A, Benisch F, Brat D, Dietz H, Subtil T et al (2010) Trends and challenges in the microbial production of lignocellulosic bioalcohol fuels. Appl Microbiol Biotechnol 87:1303–1315. https://doi.org/10.1007/s00253-010-2707-z

Wess J, Brinek M, Boles E (2019) Improving isobutanol production with the yeast Saccharomyces cerevisiae by successively blocking competing metabolic pathways as well as ethanol and glycerol formation. Biotechnol Biofuels 12(1)

Yang S et al (2016) *Zymomonas mobilis* as a model system for the production of biofuels and biochemicals. Microb Biotechnol 9(6):699–717

Zetterholm J, Bryngemark E, Ahlströ m J, Sö derholm P, Harvey S, Wetterlund E (2020) Economic evaluation of large-scale biorefinery deployment: a framework integrating dynamic biomass market and techno-economic models. Sustainability 12:712

Zheng Y, Yu X, Zeng J, Chen S (2012) Feasibility of filamentous fungi for biofuel production using hydrolysate from dilute sulfuric acid pre-treatment of wheat straw. Biotechnol Biofuels 5(3):1–10

Chapter 18
Role of Bioactive Metabolites Produced by Plants and Microbes in Fuel Production

Vishal U. Kalebar and Pritha Chakraborty

Abstract Bioactive metabolites are a diverse group of compounds with specific biological activities and are produced by various living organisms. Plants and microbes produces different bioactive molecules which plays an important role in biofuel production. Bioactive materials enhances the biofuel production by modulating different metabolic pathways. Several enzymes and organic acids produced by microbes helps in degrading the complex sugar to improve the fermentation process which helps in increasing the efficacy and yield of biofuel production. Phytocompounds such as terpenoids, are considered as precursor molecule for bioethanol and biodiesel generation. Additionally, these metabolites also possess antifungal and antibacterial properties, which reduces the contamination occurring during the process of biofuel generation. Several studies also reveal that these phytocompounds also helps in extraction and purification of biofuel. Hence bioactive compounds plays a crucial and multifaceted role in biofuel production. As sustainability and efficacy of biofuel production hold major importance, microbial and plant derived bioactive compounds can redefine the field of renewable energy.

Keywords Bioactive metabolites · Plants · Microbes · Biofuel · Renewable energy · Biomass

V. U. Kalebar · P. Chakraborty (✉)
Department of Molecular Medicine, School of Allied Healthcare and Sciences, JAIN (Deemed to be University), Bangalore, Karnataka 560066, India
e-mail: prithachakraborty7@gmail.com

V. U. Kalebar
Department of Molecular Medicine, school of Allied Healthcare and Sciences, JAIN (Deemed to be University), Bangalore, India

M. Shah and D. Deka (eds.), *Emerging Sustainable Technologies for Biofuel Production*, Environmental Science and Engineering, https://doi.org/10.1007/978-3-031-52167-6_18

18.1 Introduction

The use of fossil fuel for energy has been rising since the industrial revolution and currently the catastrophic effects are experienced worldwide (Solaymani 2019). Global climate change, rising sea levels, destruction of coral reef and weak thermohaline circulations are directly linked to the increasing concentration of greenhouse gases (Verma et al. 2019). According to the recent report by (Li et al. 2021), the atmospheric concentration of CO_2 is 400 ppm and usage of fossil fuel releases approximately 7 Gt CO_2 per year at present. It is also estimated that the rate of CO_2 level will double by 2050 which may increase the global temperature by 2 °C with catastrophic after effects. The long term role of fossil fuel is due to its compatibility with present infrastructure and their higher energy density. The increasing demand for renewable and sustainable sources of fuel is driven by several factors. There is a growing awareness of the need to reduce our reliance on fossil fuels and transition to cleaner and more sustainable energy sources. This is driven in part by concerns about the environmental impact of fossil fuel extraction and transportation, as well as the economic and geopolitical risks associated with reliance on a single source of energy (Ahmed et al. 2022). Advances in technology have made it increasingly feasible to produce energy from renewable sources such as wind, solar, and biofuels. As the cost of these technologies continues to decrease, they are becoming more competitive with fossil fuels, making them an attractive option for businesses and governments looking to reduce their carbon footprint (Chu and Majumdar 2012). Overall, the increasing demand for renewable and sustainable sources of fuel is driven by a combination of environmental, economic, and technological factors, and is expected to continue to grow in the coming years as we work to address the challenges of climate change and create a more sustainable future. Additionally, many countries have set targets and goals to reduce their greenhouse gas emissions, and transitioning to renewable and sustainable sources of fuel is seen as a key strategy to achieve these targets. For example, the Paris Agreement, signed by 196 countries in 2015, aims to limit global warming to well below 2 °C above pre-industrial levels, and calls on countries to achieve net-zero greenhouse gas emissions by the second half of this century (Lewiński and Mohammed 2019).

Furthermore, the use of renewable and sustainable sources of fuel can also have social and economic benefits, particularly in rural areas. For example, the production of biofuels from crops such as sugarcane and corn can provide income for farmers and create job opportunities, while also reducing dependence on imported fossil fuels. The increasing demand for renewable and sustainable sources of fuel is driven by a range of environmental, economic, and social factors, and is expected to continue to grow as we work for transition towards a more sustainable and low-carbon future. The use of plant metabolites for fuel production is one potential strategy to meet this demand, and offers a range of advantages, including the ability to produce energy from abundant and renewable sources, and the potential to reduce greenhouse gas emissions and create economic opportunities for rural communities. Currently, bioethanol is most widely used biofuel, which is produced majorly from corn starch and cane sugar (Chen et al. 2020). Microbes are greatly involved in the production of different biofuels like ethanol, n-butanol, biodiesel and bio-hydrogen. Biofuel generated from bioactive molecules or biomass derived from microbes are referred as advanced biofuel. Biofuel generated using combustion precursors from microalgae or heterotrophic microbes are known as third generation (3G) biofuel (Zabed et al. 2019). Genetically engineered microalgae, bacteria and fungi are used to produce fourth generation (4G) biofuel (Torres-Tiji et al. 2020). Microalgae, bacteria and fungi are metabolically different than each other which enables them to use different substrate for production of biofuel. Lignocellulose substrates can be used by fungi while bacteria and microalgae can fix CO_2 and methane generated from landfill. Lignocellulosic substrates from plant source are first hydrolysed by fungi which is then fermented by *Clostridium thermocellum* to produce biofuel. Cyanobacteria and microalgae can directly use CO_2. Methanogenic bacteria uses methane which is more potent greenhouse gas then CO_2 (Rosentreter et al. 2021). Microbes are found all over the world including most adverse and unfavourable growth conditions. Under stress conditions, the macro and micro molecules with unique properties are produced by these microbes which helps them to survive these adverse situations (de Morais et al. 2015). These unique molecules are identified as bioactive molecules from microbes. Several microbial species produces different types of bioactive compounds only few of which has been identified and studied (Murata et al. 2021). Bioactive compounds like protein, lipid, sterol, polysaccharide, enzymes, vitamins and other value added compounds with promising nutritional and pharmaceutical importance can be used for commercial purposes (Levasseur et al. 2020). Bioactive compounds are derived from either primary or secondary metabolism and exert antimicrobial, antioxidant, anti-inflammatory effects. These bioactive compounds can be stored within the microbial biomass or can be secreted into the medium which are referred to as exo-metabolites. Several microbes are rich source of antioxidants like lutein, ascorbic acid, α and β carotene, α-tocopherol with pharmaceutical and nutraceutical importance. Scientists are keen to explore this enormous biological resources growing in different ecological niches with different physiological potentials (Kumar et al. 2018). The bioactive compounds from both microbes and plant are discussed in detailed in the following sections.

18.2 Biofuels

18.2.1 Bioethanol

Bioethanol is mostly produced in large scale from cane molasses or starch hydrolyzation by *Saccharomyces cerevisiae* (Solaymani 2019). *Zymomonus mobilis* bacteria are capable of using more sugar to produce more ethanol than yeasts. But it will be unfair to use these food crops for solely producing energy while many countries faces hunger problem (Centobelli et al. 2022). Thus, ethanol production from agricultural wastes (husk, straw, wood chips and saw dust) are more environment friendly (Sikarwar et al. 2017). Ethanol production from cellulosic biomass involves several steps like pre-treatment, hydrolysis, saccharification and fermentation. Although enzymatic hydrolysis of sugar is slower than fermentation process by yeast (Osman et al. 2021). A report by (Kumar et al. 2020; Robak et al. 2018) proposed the use of *Clostridium thermocellum* which possess a better rate of hydrolysis and fermentation of cellulosic and hemicellulose biomass. The carbohydrate content of microalgae can reach upto 70% under optimum growth conditions making them a sustainable option for bioethanol production (Markou and Nerantzis 2013).

18.2.2 Biodiesel

Vegetable oil is trans-esterified into mono-alkyl ester of fatty acids, which is known as biodiesel. Cyanobacteria can be a substitute for vegetable oil for biodiesel production as they are rich in lipid content and does not require arable land for their cultivation (Mohsenpour et al. 2021; Ubando et al. 2020). Different microbes like *Mucor* sp. and *Mortierella* sp. are also high in lipid content and can accumulate lipid using wide variety of substrate like whey permeate, industrial glycerol, sugar cane molasses, sewage sludge and rice straw hydrolysate (Chu et al. 2021). Microbes are employed to produce alcohol from glycerol which is a by-product of biodiesel production from vegetable oil. Though conventional biodiesel production does not involve microbes, but studies reported *Escherichia coli*, *Klebsiella* sp., *Clostridium* sp. and *Enterobacter* sp. are used to produce ethanol, butanol and propanol which is used in the production of biodiesel from vegetable oil (Verma et al. 2019; Veza et al. 2021).

18.2.3 Bio-Hydrogen

Cyanobacteria can produce bio-hydrogen through bio-photolysis of water and photo-fermentation of organic substrates derived from other photosynthetic bacteria (Yin et al. 2020). A wild type cyanobacteria *Cyanothece* sp produces hydrogen at a rate of 465 mol/mg of chlorophyll/h from glycerol (Deviram et al. 2020; Sarker et al. 2019). Also anaerobic organisms like acidogenic bacteria can produce biohydrogen from organic wastes through dark fermentation. Immobilized form of thermophilic bacteria can also produce bio-hydrogen through liquid fermentation. Though commercial bio-hydrogen production has not achieved the economically viable level, it has enormous opportunity in the field of renewable energy (Łukajtis et al. 2018; Sridhar et al. 2022).

18.2.4 Bio-Methane

Syngas which is combination of carbon monoxide (CO), carbon dioxide (CO_2) and hydrogen (H_2) is generated by direct or thermochemical gasification of biomass, followed by chemical methanation (Ciliberti et al. 2020). Organic matter like food waste, sewage, manure can be anaerobically fermented or digested to produce biofuel. Chemical upgradation is implied through which contaminants are removed and methane percentage are increased to 90% to produce commercial biogas or bio-methane (Le Borgne and Baquerizo 2019). Microbial methanogenesis is a well-studied process for biofuel production involving both mesophilic and thermophilic organisms (Khan et al. 2021). In the first phase organic polymers present in the feed stock are digested by facultative anaerobic bacteria to produce simple and soluble substrate. Direct metabolism of these substrates produces different gases, alcohols, volatile fatty acids, ketones and organic acids during acidogenic phase. In next step, acetogenic, homoacetogenic and H_2 producing bacteria directs the aceto-genesis process where acetate, formate, CO_2 and H_2 are produced by acetogenic bacteria using the substrates from previous step (acidogenesis). Homoacetogenic bacteria follows acetyl-CoA pathway to produce acetate using CO_2 and H_2. In the last step, methanogenesis is carried out by different methanogenic areheabacteria (methylotrophs, hydrogenotrophs and aceticlasts). Hydrogenotrophs reduces CO_2 to CH_4 while aceticlasts cleaves the acetate to produce CO_2 and CH_4 in presence of hydrogen. Methylotrophs converts methylated substrates like di-methyl sulphide, methylamines and methanol into methane (Qiu et al. 2019).

18.3 Bioactive Molecules from Microbes

18.3.1 Bioactive Molecules from Bacteria

Microorganisms are considered as unique source for biofuel production due to their ubiquitous nature and ability to survive in various habitat like cold desert, saline water, hot spring and drought environment. There is a growing interest about role of microbes in biofuel production as they are cost effective, microbial biomass is renewable and no chemical treatment is required (Yusoff et al. 2019). *Clostridium ljungdahii* and *C. autoethanogenum* are Gram positive, anaerobic, spore forming, rod shaped and motile bacteria which are acetogenic in nature and produces biofuel from carbon monoxide (Ramli et al. 2020). Metabolically engineered *Escherichia coli* are also known for their ability to produce biofuel which is more cost effective than other conventional processes (Chintagunta et al. 2021). A few bacteria like *Acetobacterium woodii* and *C. ljungdahii* are reported to convert synthetic gas or syn gas (CO, CO_2, H_2) in biofuel (ethanol, butanol and acetic acid) (Bertsch and Müller 2015). Butanol is recognized as an alternative supplement to ethanol used in gasoline and diesel. Acetone-butanol-ethanol fermentation is observed and reported in *C. beijerinckii*. C.

acetobutylicum and *Thermoanaerobacterium thermosaccharolyticum* are known for butanol and biohydrogen production (Łukajtis et al. 2018; Veza et al. 2021). Different microbes used in biofuel production is discussed in details in Tables 18.1, 18.2.

Three types of bacterial lipids (triacylglycerols, polyhydroxybutyrate and wax esters) are considered as great resources for biofuel production. Triacylglycerol (TAG) and wax esters are considered as neutral lipids and used for biofuel production. TAG are consists of three fatty acids chain (C16–C20) attached to one glycerol through ester linkage (Klok et al. 2014; Sharma et al. 2018). Fatty acid methyl esters (FAME) and fatty acid ethyl esters (FAEE) are the most used form of TAG for biofuel production through transesterification (Zhang et al. 2021). Wax esters are long chain fatty acids (C12–C24) combined with long chain alcohol with wide industrial applications (Santala et al. 2014). Gaseous hydrocarbons, jet fuels, gasoline are some of the example of stable fuels that can be derived from wax esters via fast deoxygenation in absence of H_2 in addition to transesterification to FAME (Hwangbo and Chu 2020; Shimada et al. 2018). Bacteria are more unlikely to accumulate TAGs and Wax esters than eukaryotes. Gram positive actinomycetes like Nocardia sp., Rhodococcus sp. and Gram negative *Pseudomonas* sp. and *Acinetobacter* sp. are few examples of bacteria which can accumulate TAGs and wax esters (Garay et al. 2014). They are only synthesized when bacteria enters any stressed conditions like nitrogen deficiency using lignocellulosic biomass, sugars, alcohols and organic acids

Table 18.1 Microorganisms used in biofuel production

Organisms	Substrate	Product	References
Escherichia coli	Synthetic medium	Fatty alcohol	Liu et al. (2014)
Clostridium autoethanogenum	Synthetic medium	Ethanol	Liew et al. (2017)
Saccharomyces cerevisiae	Glucose and galactose	2,3-Butanediol	Lian et al. (2014)
Saccharomyces cerevisiae	Synthetic medium	Isobutanol and n-Butanol	Avalos et al. (2013), Schadeweg and Boles (2016)
Corynebacterium glutamicum	Glucose and xylose	3-Hydroxypropic acid	Chen et al. (2017)
Clostridium cellulolyticum	Cellulose	n-Butanol	Gaida et al. (2016)
Clostridium cellulovorans	Cellulose	Ethanol and n-Butanol	Yang et al. (2015)
Clostridium acetobutylicum	Glucose, corn starch, corn stover	n-Butanol	Luo et al. (2017)
Caldicellulosiruptor sp.	Lignocellulose	Hydrogen	Cha et al. (2013)
Klebsiella pneumoniae	Glucose	n-Butanol	Chen et al. (2015)
Escherichia coli	Glucose	1-Butanol, 2-methyl-1-butanol, 3-methyl-1-butanol	Atsumi et al. (2008)

Table 18.2 Types of plant metabolites for fuel production

Plant Metabolite	Potential Fuel	Advantages	Disadvantages	References
Cellulose	Bioethanol, biobutanol, biodiesel	These are cost effective feedstock for biofuel production and do not require any additional economic input	Requires pretreatment to break down the tough cell wall structure	Chandel et al. (2018)
Hemicellulose	Bioethanol,	Abundant and renewable, does not compete with food crops	More difficult to extract and process than cellulose	Huang et al. (2022)
Starch	Bioethanol, biobutanol	Abundant and renewable, compatible with existing infrastructure	Competes with food crops for resources	Marques et al. (2018)
Lipids	Biodiesel, renewable diesel	High energy density, can be produced from various plant sources	Limited availability of suitable plant sources	Tursi (2019)
Terpenes	Biojet fuel, renewable diesel	High energy density, can be produced from various plant sources	Limited availability of suitable plant sources, difficult to extract and process	Mewalal et al. (2017), Wang et al. (2019a)
Lignin	Biofuel Bioplastics, carbon fiber, biocomposites	Abundant and renewable, does not compete with food crops	More difficult to extract and process than other plant metabolites	Ragauskas et al. (2014)
Polyphenols	Biodiesel, renewable diesel	Abundant and renewable, act as catalyst in transesterification process and also as antioxidant in stability and prevent oxidation in biofuel production	More difficult to extract and process than other plant metabolites	Amran et al. (2022), Yusoff et al. (2019)

(Wang et al. 2014). *R. opacus* is an ideal organism that accumulate TAGs as 75% of their dry weight using gluconate under limited nitrogen supply (Gill et al. 2018). While biofuel derived from plants require valuable resources like land and water, biofuel generated from bacterial TAGs and wax esters using lignocellulosic biomass are carbon neutral process. Also these bacteria can be cultivated all year around and much cheaper and easier than microalgae cultivation (De Carvalho and Caramujo 2018). The generation time of *Rhodococcus* sp. is only 3–4 h in compared to *Nanochloropsis* with 2–4 days. The general cell density are higher in bacteria (5.4 g/L) than microalgae (1 g/L) (Thanapimmetha et al. 2017; Wang et al. 2016). With all the above mentioned advantages bacterial lipids are considered as favourable feedstock for biofuel production.

Polyhydroxybutyrate or PHB are intracellular polymeric lipids composed of short chain fatty acids. It reacts with methanol and produces hydroxybutyrate methyl esters as a lipid based biofuel. Mostly all bacteria and some archaea uses PHB as energy and internal energy storage during the stationary phase (Asiri et al. 2020). *Haloarcula* sp. *Haloferax mediterranei* and *H. volcanii* are some of the PHB accumulating archaea while *Haloarcula* sp. showed highest accumulation of PHB (63% of the dry weight) (Koller 2019). *Cupriavidus* sp., *Alcaligenes* sp., *Halomonas* sp. *Bacillus* sp., *Burkholderia* sp., *Vibrio* sp., *Methylobacterium* sp., *Pseudomonas* sp., are some of the examples of bacteria which accumulates PHB. *C. necator* are reported to accumulate PHB using glucose and oil as energy source (Bellini et al. 2022). Glycerol, a by-product of biodiesel production also can be used as carbon source for PHB accumulating bacteria. However, other bacteria like *Bacillus* sp. and *Pseudomonas oleovorans* are reported to produce PHB (81–85%) using synthetic glycerol as carbon source (Saratale et al. 2021). *E. coli* is perfect example of non-oleaginous bacteria accumulating PHB following engineered lipid synthesis pathway. The engineered *E. coli* carries a plasmid which increases the production of TAGs and PHB (Chen et al. 2018; Xu et al. 2018).

18.3.2 Bioactive Molecules from Fungi

Mycodiesel is referred to hydrocarbon derived fuel from endophytic fungi. A number of bioactive compounds like cyclohexane, mono-terpenoids, cyclopentane, alkane, benzenes, alkyl alcohols/ketone and polyaromatic hydrocarbon have been isolated from different endophytic fungi (Khare et al. 2018; Kour et al. 2019). These identified bioactive compounds are structurally identical or similar to the compounds present in diesel. Endophytic fungi uses cellulosic biomass including hemicellulose and other polymers that are found in agricultural wastes to produce hydrocarbon (Mallette et al. 2014; Strobel et al. 2013). Organisms like *Gilocladium* sp., *Annulohypoxylon* sp., *Ascocornye sarcoides*, *Phomopsis* sp and *Daldinia* sp. produces hydrocarbons that are used for biofuel production (Oukala et al. 2021; Strobel 2015). *A. sarcoides* was found to produce acetic acid chain compounds including hectyl, heptyl, octyl, sec-octyl and decyl alcohol. All the diesel fuel available in world has straight chained hydrocarbon as their backbone compound. *Persea indica* is an evergreen tree native to the Canary Island are known to associate with another endophytic fungi *Hypoxylon* sp. high energy density volatile compounds produced by this endophytic fungi includes 1,8-cineole, 1-methyl-1,4-cyclohexadiene with potential to produce mycodiesel fuel (Shaw et al. 2015). *Hypoxylon* sp. is first known fungi to produce 1,8-cineole which is mostly found in plant tissue. Starch or sugar containing media is used for the growth of these fungi and proton transfer mass spectrometry is used to measure the production of these volatile compounds (Schoen et al. 2016). The highest production of volatile compounds (145 ppmv) are recorded around 6th day with a rate of production of 7.65 ppmv/h. 11 new and distinctive terpene synthase genes are identified in *Hypoxylon* sp. in addition to a specific cineole synthase gene which is very important to identify the complete fungal pathway for 1,8-cineole production. Combination of conventional metagenesis techniques and genetic manipulation can be used to solve the yield related problems of these volatile compounds from endophytic fungi to make them a more reliable source of hydrocrabons. 1,8-cineole is definitely possess great potential as a replacement of fossil fuel based hydrocarbons (Tomsheck et al. 2010).

18.3.3 Bioactive Molecules from Algae and Cyanobacteria

Microalgae are dynamic photoautotrophic microorganisms with over 40,000 identified species which can survive and flourish in extreme (volcanic hot springs and Antarctic ice) as well as normal environments like fresh or hypersaline water. Microalgae are suitable for biofuel production as they can be cultivated on non-arable land and in wastewater brackish or saline water (McKinlay et al. 2021). Protein, cellulose and shorter polysaccharide chains are present in microalgal cell wall (Ghassemi et al. 2021). As a result, they contribute to more homogenous biomass simplifying the fermentation process and reducing the post-harvest waste. Another important aspect is that, in compared to terrestrial feedstocks, the lipid content and biomass of microalgae depends on species, location and conditions of culture facility (Nitsos et al. 2020). Studies reported that under lab conditions, 30% of fresh biomass of oleaginous microalga yields biodiesel approximately 52,000 kg/ha/year in comparison to 150 kg/ha/year from maize or corn and 860 kg/ha/year from rape seed and 5000 kg/ha/year from palm oil (Bošnjaković and Sinaga 2020). Tri-acyl-glycerol (TAG) is another interesting precursor of biodiesel. Several oleaginous microalgae like *Chlorella* sp. *Nannochloropsis* sp. and *Scesedemus* sp (Pal et al. 2019). Under nitrogen and phosphate depleted environment, with decreased rate of cell division, microalgae can redirect the extra electrons in photosynthetic chain in TAG and fatty acid synthesis. This process consumes twice NADPH in compared to protein or carbohydrate synthesis (Alishah Aratboni et al. 2019). TAG represents upto 20–50% of the dry weight of the cell and accumulates in liposomes or chloroplast. As TAG level greatly depends on the species, strains, environmental factors and the age of the culture upscaling of biodiesel production can face quality control issues. Thus more focus is required into cultivation, processing technology development and life cycle assessment of microalgae (Mehariya et al. 2021; Villanova and Spetea 2021).

Hydrocarbons (C_{20}–C_{40}) from green algae is another unique source of advanced biofuel production as it can make upto 80% of the dry biomass (Einhaus et al. 2022). Based on metabolic pathways and hydrocarbons, *Botryococcus braunii* is classified into four different phenotypic races of A, B, L and S. Race B of *B. braunii* contains major amount of hydrocarbon in cell wall and extracellular matrix and only 7% of hydrocarbons are stored in liposomes (Li-Beisson et al. 2019). Although it favours easy extraction of hydrocarbons, the major cost of culture, processing and installation makes the process unprofitable for industrial production (Kleinert and Griehl 2021; Yin et al. 2020). A light activated enzymes named fatty acid photodecarboxylase or FAP is identified in *Chlorella variabilis*. FAP directs the decarboxylation of free fatty acids into alkene in presence of blue light. Thioesterase FAP expression in engineered E. coli can produce hydrocarbons (C_{11}–C_{17}) in presence of low-irradiance blue light (Moulin et al. 2019; Sorigué et al. 2017).

Marine cyanobacteria with smaller genome size presents great opportunity to optimize industrial biofuel production by encouraging genetic modifications. Halophilic nature of these cyanobacteria provides the advantage of growing in sea water instead of freshwater which is becoming limited in many parts of the world (Kapoore et al. 2021). USA based companies like Algenol and Joule Unlimited are actively producing biofuel from algal biomass (Debnath et al. 2021). Cyanobacteria is highly efficient in transforming light energy into biomass along with simple cell wall structure and glycogen storage which makes the mobilization easier (Rodionova et al. 2017). Marine cyanobacteria Cyanothece sp can produce hydrogen as a by-product of nitogen fixation in higher rates (Chandrasekhar et al. 2015) (Table 18.1).

18.4 Bioactive Molecules from Plants

18.4.1 Carbohydrates

Carbohydrates from plants can be a valuable source of biofuel. Plants naturally contain carbohydrates in the form of sugars, starches, and cellulose, which can be converted into different forms of biofuel. Some of the plant carbohydrates useful in the production of biofuel are discussed below and Table 18.2.

18.4.1.1 Cellulose

Cellulose is one of the component of the plant cell wall, which plays an important role in structure and stability to cell wall. Increasing production of cellulose in plants by manipulating the biosynthetic pathways for bioethanol production is still a challenge and ongoing research (Chen et al. 2021). It is the most abundant organic material on earth, and it is produced by all photosynthetic organisms. This means that plant cellulose is a highly sustainable resource that can be grown and harvested indefinitely. To produce biofuels from plant cellulose, the cellulose must first be broken down into simple sugars that can be fermented and distilled. This process is called cellulolysis, and it is typically accomplished using enzymes that break down the complex cellulose molecules into smaller sugar molecules (Wang et al. 2019b). One of the main advantages of using plant cellulose as a biofuel feedstock is that it is a highly sustainable resource. Unlike fossil fuels, which are finite and non-renewable, plant cellulose can be continually grown and harvested. Additionally, the use of plant-based biofuels can help reduce greenhouse gas emissions and mitigate climate change (Dheeran and Reddy 2018).

18.4.1.2 Hemicellulose

Hemicellulose is a complex carbohydrate that makes up a significant portion of plant cell walls, along with cellulose and lignin. It is an abundant and renewable resource that can potentially be used as a biofuel feedstock. Hemicellulose can be extracted from plant materials, such as agricultural residues and woody biomass, using various chemical and enzymatic methods (Zabed et al. 2016). One potential application of hemicellulose is as a feedstock for the production of biofuels, such as ethanol. Hemicellulose can be hydrolysed into simple sugars, which can be fermented by microorganisms to produce ethanol. The advantage of using hemicellulose as a biofuel feedstock is that it does not compete with food crops, as is the case with corn-based ethanol production. Another potential application of hemicellulose is as a component of bio-based materials, such as bioplastics. Hemicellulose can be chemically modified to create new materials with different properties, such as improved biodegradability or water resistance. These materials could potentially replace petroleum-based plastics in various applications, reducing the dependence on fossil fuels (Opia et al. 2021). Plant hemicellulose has significant potential as a biofuel feedstock due to its abundance and renewable nature. Unlike fossil fuels, hemicellulose-based biofuels have the advantage of being carbon–neutral, as they only release the same amount of carbon dioxide during combustion as the plants absorbed during their growth. Hemicellulose can also be extracted from a variety of plant materials, including agricultural residues and woody biomass, which can reduce waste and provide a source of revenue for farmers and foresters (Soltanian et al. 2020).

18.4.1.3 Starch

Plant starch is a carbohydrate that is commonly used as a feedstock for the production of biofuels, particularly ethanol. Starch is abundant in many plant sources, including corn, wheat, and potatoes, making it a readily available and renewable resource. Starch is typically extracted from the plant material and hydrolyzed into simple sugars, such as glucose, using enzymes. These sugars can then be fermented by microorganisms, such as yeast, to produce ethanol. The process of starch-based ethanol production is well-established and widely used, particularly in the United States, where corn-based ethanol is a significant contributor to the renewable fuel industry.

18.4.2 Lipids

Among the different types of biomass, plant lipids have gained significant attention as a potential source for biofuel production due to their high energy content and ability to be converted into various types of biofuels, such as biodiesel, bio-jet fuel, and renewable diesel (Dhyani and Bhaskar 2018). Plant lipids are composed of triglycerides, which are esters of glycerol and fatty acids, and they can be obtained from various parts of plants, such as seeds, fruits, leaves, and stems. Significant progress has been made in recent years in utilizing plant lipids for biofuel production. Lipid production from different plant species has been investigated, including oilseed crops such as soybean, canola, sunflower, and oil palm, as well as non-food crops such as jatropha, camelina, and algae (Callegari et al. 2020). Various approaches have been explored to improve lipid yield and quality, including genetic engineering, plant breeding, and cultivation practices. For example, genetic engineering has been used to modify lipid metabolism in plants to increase lipid content, alter fatty acid composition, and reduce undesirable components such as polyunsaturated fatty acids that can lead to oxidative instability of biofuels (Abdullah et al. 2019). Additionally, plant breeding and cultivation practices have been optimized to improve lipid yield and quality, such as the development of high-yielding oilseed crop varieties and the use of agronomic practices to enhance lipid accumulation in plants (Hu et al. 2015).

18.4.3 Polyphenols

Plant polyphenols are a group of naturally occurring compounds found in various plants, including fruits, vegetables, grains, and herbs. They are known for their antioxidant properties and have been studied for their potential health benefits. One of the ways plant polyphenols can be used in biofuel production is through their conversion into biofuels such as bioethanol or biodiesel. Bioethanol is a type of biofuel that can be produced by fermenting plant-based materials, such as cellulose or lignocellulose, which are rich sources of polyphenols. Polyphenols can be extracted from these plant materials and then subjected to fermentation processes using microorganisms that can convert them into ethanol. The resulting bioethanol can be used as a fuel additive or blended with gasoline to reduce the dependence on fossil fuels (Bušić et al. 2018). Polyphenols can also be used in the production of biodiesel, which is a renewable alternative to conventional diesel fuel. Biodiesel is typically produced from plant-based oils, such as soybean oil or canola oil, through a process called transesterification. Plant polyphenols such as oils can be used as catalysts in this process to improve the efficiency and yield of biodiesel production. Additionally, polyphenols can be used as antioxidants in biodiesel to improve its stability and prevent oxidation, which can lead to degradation and reduce the quality of the fuel (Amran et al. 2022; Yusoff et al. 2019). It's worth noting that the use of plant polyphenols in biofuel production is still in the early stages of research and development, and there are challenges to overcome, such as cost-effectiveness, scalability, and sustainability of the production processes. However, the potential of plant polyphenols as a renewable source for biofuel production is promising and warrants further investigation and development as we seek sustainable alternatives to fossil fuels.

18.4.4 Terpenoids

Terpenes are found naturally in plant as a chemically diverse set of secondary metabolites which is often related with physiology development of plant either as mutualistic or antagonistic in plant–herbivore and plant-environment interactions, Terpenes and aromatic compounds make up plant essential oils, with the storage cavities of leaves frequently having the highest terpene content. Terpenes are utilised in the production of various pharmaceuticals compounds, nutraceuticals industry, agrichemicals industry and in production of fragrances (Mewalal et al. 2017).

Terpenes can be found in both cyclic and acyclic forms, but cyclic forms are preferred as feedstocks for biofuels due to their higher energy density and net heat of combustion (NHOC) (Kim et al. 2020). Terpenes can be used as biofuels directly or blended with existing fuels like jet fuel, missile propellant, gasoline, or diesel fuels. For instance, hydrogenated monoterpenes like myrcene and limonene have been shown to be suitable additives to diesel, and β-pinene has been used to synthesize high-density biofuels comparable to JP-10, although with slightly higher viscosity and freezing point (Beller et al. 2015; Wang et al. 2019a). Terpenes can be used as standalone fuels or blended up to 65% (w/w) in diesel or gasoline engines. However, commercial implementation of naturally derived terpenes as biofuels has been limited due to low yields of specific terpenes in plants. Recent advancements in omics technologies and gene editing tools such as CRISPR-Cas9 have provided a potential avenue for large-scale production of terpenes from plants for biofuel production in the future.

Studies suggest that terpenes have the potential to be used as standalone fuels or blended up to 65% (w/w) in diesel or gasoline engines. However, the commercial implementation of naturally derived terpenes as biofuels has been limited due to the low yield of specific terpenes in plants. In recent years, advancements in omics technologies such as DNA sequencing, RNA sequencing, metabolic profiling, and proteomics, as well as genome modification/editing tools like CRISPR-Cas9, have revolutionized plant research. These developments, combined with the inherent ability of many plants to produce and store terpenes, offer a promising platform for large-scale production of terpenes for biofuels (Liu et al. 2016; Ninkuu et al. 2021).

By utilizing the existing variation in plant terpene content and employing high-throughput omics profiling technology, researchers can gain a deeper understanding of the biological processes involved and enable predictive engineering of plants to produce specific terpenes using synthetic biology techniques such as single gene effects or additive effects of gene stacking to optimize the terpene metabolic pathway. This can be achieved through selective breeding approaches as well as a combination of different strategies. When combined with established silviculture and harvesting practices, terpenes synthesized in planta hold significant potential as sustainable specialty biofuels (Mewalal et al. 2017).

18.5 Substrates Used for Biofuel Production

The biofuel production process should be sustainable and efficiently use raw materials, atmospheric CO_2 or carbon that contributes to greenhouse gas emission. Sufficient quantity of these alternative resources is an absolute requirement for replacing the fossil fuel completely.

Plant biomass is readily available in environment and is ideal substrate due to their cellulose (40–50%), hemicellulose (20–40%) and lignin (20–35%) content (Kumar et al. 2020). The complex plant biomass has to be processed into simpler forms like organic acids, syngas, sugar and methane, as cellulose is hydrophobic and remains encapsulated within lignin (crystalline) and hemicellulose (Panahi et al. 2022; Rauch et al. 2014). Thermochemical processes followed by microbial conversion can produce syngas from plant biomass. Organic acids and methane can be produced from mixed fermentation of the substrate (Liao et al. 2016). The pre-treatment process prior to enzymatic hydrolysis of lignocellulose biomass into sugars involves physical, chemical, biological or combination treatment. Cellulolytic microbes and non-complexed cellulase enzymes are then used in hydrolyzation of pre-treated biomass (Wang et al. 2017). Non-complexed cellulase enzymes is a combination of cellobiohydrolases, endoglucanases, and β-glucosidases (Østby et al. 2020). Cellobiose is generated from either reducing or non-reducing end of the polymeric chain of cellulose after cellobiohydrolase breaks the β(1–4)-Dglucosidic linkage. Endogluconases randomly cut the internal bonds to free more cellulose polymers. In next step, cellobiose is converted to glucose by β-gluosidases (Liao et al. 2016). *Penicillium* sp., *Chrysosporium* sp., *Trichoderma reesei* are few fungi which can break down the complex sugar into glucose (Adsul et al. 2020). *Clostridium thermocellum*, *C. cellulovorans* and *C. cellulolyticum* produces CipA scaffolding protein which helps in anchoring on the substrate (Lu et al. 2021a; Lynd et al. 2022). Endoglucanase, exoglucanase, Xylanases, Mannases, lichenase enzymes are produced by these anaerobic bacteria and breaks down the cellulose in cellodextrins. In turn cellodextrins are absorbed by the cell followed by phosphorolytic energy (Liao et al. 2016). Apart from cellulose, lignocellulose comprised of lignin and hemicellulose is structurally diverse. Hemicellulases (arabinofuranosidases, xylosidases) are the key enzyme in hydrolysing the glycosidic bonds or carbohydrate esterases (Wu et al. 2022). Lignin constitute approximately 20–35% of plant biomass. Lignin is mainly used to generate heat or energy by combustion as they structurally very different to convert into fuel (Schutyser et al. 2018). Conversion of lignin into fuel or value added components are also expensive due to the product heterogeneity and product purification. Laccase and peroxidases are ligninolytic oxidative enzymes secreted mainly by fungi in addition to few bacteria (*Pseudomonas autotrophica* and

Comamonas testoteroni) (Janusz et al. 2020; Lee et al. 2019). As lignin hydrolysis is a complex process, the end products are not cost effective.

Direct conversion of CO_2 into liquid fuel using microbes is a great alternative of producing biofuel from biomass. Microalgae and cyanobacteria can fix CO_2 in presence of sunlight which can be exploited in biofuel production with the help of metabolic engineering (Barkia et al. 2019). Microalgae are considered as great resource for biofuel production due to their high lipid content. Next generation sequencing and RNAi can be explored to enhance the capability of microalgae to produce biofuel (Li-Beisson et al. 2019). RNA-sequencing is used in identifying the genes responsible for tri-acyl glycerol accumulation under nitrogen depleted condition thus engineering strains with potential of producing high lipid content. Genetic engineering of more versatile cyanobacteria leads to production of biofuel and value added chemicals (Angermayr et al. 2015). One major shortcoming of this process is requirement of large surface exposed to sunlight. Cultivation in open ponds can be an attractive and inexpensive solution along with problems like water loss, contamination and control of cultures (Faust and Logan 2018). Closed photo-bioreactors are expensive where limited length and intensity radiation is employed. Also the apparatus is sensitive to higher light intensity, high temperature and salt concentrations (Singh et al. 2020).

Methane is a major component of greenhouse gas and it is generated by anaerobic digestion of organic wastes. During last decade, the emission of methane has been drastically increased. Methanotrophs are only known microbiological system that can decrease the methane release from dumping areas (Conrado and Gonzalez 2014). Mainly chemical and biological methods are used for the conversion of methane into different biofuel like ethanol and butanol. Fischer–Tropsch methods is a chemical method which consumes excessive energy due to high temperature and pressure (Strong et al. 2015). On the other hand biological processes are far more cost effective as they require oxidation if methane into methanol using methane monooxygenase enzyme. Membrane-bound particulate monooxygenase is accumulated in the internal cytoplasmic stacks in methanogenic bacteria while soluble monooxygenase carry out the oxidation of methane into methanol (Banerjee et al. 2015; Li et al. 2020). The main drawbacks of this process are lower efficiency of methane activation, higher loss of energy during the oxidation process which is exothermic in nature and maintaining the optimum growth parameters for methanotrophs (Kalyuzhnaya et al. 2015).

18.6 Biofuel Production Pathway

18.6.1 Extraction and Processing of Plant Metabolites for Fuel Production

For the production of biofuel from the plant metabolite, characterization of the various biomass, its intermediate products and final by products is very essential for transforming their biomass to biofuel. Among all the plant metabolites wide variety of research and study has been focused on lignocellulosic biomass because of its potential high yield in biofuel production and greater availability as feedstocks for the process. There are several different techniques involved in the production of biofuel, but majorly it comprises of raw material (biomass) availability, breakdown of the rigid structure of raw materials, finally conversion of the complex sugars components from the raw material into simple glucose, followed by fermentation and processing of the biofuel. Some of these steps are been discussed as follows.

18.6.2 Biomass Feedstock

There are many countries including India have reported for production of biofuel (bioethanol) using plant raw materials such as agriculture residues, forest residues, forest wastes and industrial residues etc., which are known for surplus availability of lignocellulosic biomass. Among these Agricultural residues and forest biomass are widely recognized as promising feedstocks for biofuel production, owing to their ample availability, abundance, and relatively low cost (Wang et al. 2017). Plant cell walls are made up of structural carbohydrates, including cellulose and hemicelluloses, as well as the heterogeneous polymer lignin, which are the primary components. The composition of these components can vary greatly depending on factors such as species, variety, climate, and soil fertility (Dheeran and Reddy 2018).

18.6.3 Pretreatment

This method is essential and uses several combination technique so that cellulosic biomass is transformed into fermentable sugars. Some of these techniques are discussed as follows.

18.6.3.1 Physical Pretreatment

Physical pretreatment methods, such as milling, grinding, or steam explosion, can break down the structural integrity of biomass and increase the surface area, reduce the particle size, and create pores in the biomass material. This leads to an increase in the accessible surface area, which can enhance the contact between the biomass and the subsequent processing steps, such as enzymatic hydrolysis or chemical extraction, making the biomass more amenable to further treatment. The smaller particle size and increased porosity can also improve mass transfer during subsequent reactions, facilitating the extraction or conversion of plant metabolites for fuel production (Abu-Omar et al. 2021). Further the method also employs disrupting the rigid structure cell wall using techniques such as microwave treatment, sonication thereby making the availability of cellulose components for next process (Kan et al. 2016; Ngatcha et al. 2022), it has been reported that comparatively 85% of the cellulose can been converted into monomeric sugars by the process of pyrolysis.

18.6.3.2 Chemical Pretreatment

The pretreatment method described involves using alkali solutions such as NaOH and KOH to remove lignin and some hemicellulose, making the cellulose more accessible to enzymes and increasing saccharification production. This process can be performed at low temperatures but with a longer duration and higher concentration of base. The ester bond between the components is effectively broken without fragmentation (Saravanan et al. 2022). In contrast, pretreatment with different types of acids at normal temperatures enhances anaerobic digestibility, with dilute acid affecting hemicellulose and solubilizing it, making cellulose more accessible to enzymes. Mineral acids like HCl and H_2SO_4 are commonly used for this treatment. Dilute acid pretreatment is a simple process that involves treating biomass with dilute sulfuric acid at suitable concentrations and temperatures for a period of time, with the need for cellulase enzyme for hydrolysis of carbohydrates (Kapoor et al. 2020). Solvent pretreatment is an important process that involves using organic or aqueous organic solvents, with or without catalysts, to deconstruct LCCs. Non-polar solvents have been used to hydrolyze hemicellulose and destruct lignin, rendering cellulose for enzymatic hydrolysis (Schutyser et al. 2018; Yoo et al. 2020).

18.6.3.3 Biological Treatment

Although biological pretreatment offers significant advantages, including reduced energy and chemical usage, there is still no efficient and prompt method available. However, chemical treatment poses several drawbacks, such as the need for corrosion-resistant equipment, proper disposal of chemical waste, and thorough washing procedures. Despite these challenges, biological pretreatment remains a safe and environmentally friendly way to eliminate lignin and break down other components in lignocellulose (Saha et al. 2016).

In order to address the limitations of individual pretreatment methods and enhance the overall efficiency of the treatment process, researchers have been increasingly combining different methods to overcome challenges. Nowadays, there have been numerous studies conducted that utilize a combination of various pretreatment methods. For instance, in the case of biological pretreatment, which tends to have longer operation times as a disadvantage, combining it with other physical and chemical treatment methods has been found to improve the process efficiency (Kumar et al. 2020).

18.6.3.4 Enzymatic Hydrolysis

The enzymatic hydrolysis stage plays a crucial role in producing bioethanol, as it generates the sugars needed for fermentation in the final stage. To make the process more cost-effective, researchers are exploring new and more efficient enzymes for the hydrolysis process. Numerous enzyme strategies are being employed in hydrolysis protocols, using various lignocellulosic biomass materials like wood feedstocks, agricultural wastes, and marine algae to ensure efficient bioethanol production.

The production of ethanol from pretreatment materials involves a second step known as enzymatic hydrolysis. During this process, cellulytic enzymes are used to break down the polymers of cellulose and hemicellulose. Cellulose is made up of glucans while hemicellulose contains polymers of several sugars such as xylan, glucan, galactan, mannan, and arabinan. The main product of cellulose hydrolysis is glucose, while hemicellulose produces several pentoses and hexoses. However, high lignin content in the materials can block enzyme accessibility, cause end-product inhibition, and reduce the rate and yield of hydrolysis. Cellobiose and glucose are also strong inhibitors of cellulases. Enzymatic hydrolysis is carried out under mild conditions, which include low pressure and long retention time especially in relation to the hydrolysis of hemicellulose (Raj et al. 2019). The production of monomeric sugars is influenced by multiple factors including the ratio of liquid to solid components, the type of acid used, the temperature and reaction time, the size of the biomass particles, the length of the macromolecules, the degree of cellulose polymerization, the structure of the cellulose chain, and the way in which cellulose is connected to other polymeric structures present in the feedstock cell wall, such as lignin, pectin, hemicellulose, proteins, and minerals (Lu et al. 2021b).

18.6.4 Fermentation

The success of fermentation depends greatly on the pretreatment and hydrolysis processes, which convert fermentable sugars into alcohol or other end products using microorganisms like industrial yeasts such as *S. cerevisiae*. However, *S. cerevisiae* can only ferment glucose and not pentose sugars like D-xylose, which are commonly found in plant and microbial biomass materials. Other microorganisms like *Z. mobilis* and natural xylose fermenting yeasts can metabolize xylose through different methods (Vasić et al. 2021).

The fermentation process can be performed in different ways such as continuous, batch, or fed-batch processes, depending on the substrate and microorganisms used. Continuous processes involve a steady supply of substrate and culture medium to a reactor with active microorganisms, while batch processes have limited initial nutrients and are inoculated with microorganisms for fermentation. Fed-batch processes combine the advantages of both continuous and batch processes, allowing for increased cell concentration, longer culture lifetimes, and higher product yields (Gutiérrez-Antonio et al. 2017).

The use of different microbial consortia also effects in the yield and the purity of the ethanol produced in the fermentation process such as breakdown of hemicellulosic hydrolysates, Detoxified xylose rich hydrolysate etc., the most commonly reported microbial species used in the process include: *P. stipites, C. tropicalis, C. shehatae, C. parapsilosis. Aspergillus, Fusarium, Monilia, Rhizopus, Neurospora, Trichoderma, Phlebia,* and *Mucor spp.* Apart from these several thermophilic bacteria (*Clostridium thermocellum, T. thermosulfurigenes, Caloramator viberbiensis*) are also been reported for in efficient bioethanol production from feedstock. Recently there many genetically modified microbes have been tested to act efficiently upon the feedstock thereby increasing in the yield of biofuel production (Basak et al. 2018; Bishnu et al. 2011).

There are various techniques and strategies for combining hydrolysis and fermentation in a single reactor, such as Separate Hydrolysis and Fermentation (SHF) and Simultaneous Saccharification and Fermentation (SSF). SHF involves separating enzymatic hydrolysis from the fermentation step to optimize each process, while SSF combines diluted acid with high temperatures in the pretreatment process to enhance the hydrolysis rate and decrease enzyme loading, resulting in higher bioethanol yields with shorter process durations and smaller reactor volumes (Chow et al. 2015; Tavva et al. 2016).

18.6.5 Distillation

The conventional method for the separation of biofuel from the feedstock is by the process of distillation for instance, ethanol separation from the feedstock is carried out by fractional distillation. The by-product from the fermentation process is heated up at 78 °C which is the boiling point of ethanol and collected as a fuel. In order to meet the industrial grade of ethanol several advanced techniques have developed namely vacuum, azeotropic, diffusion, membrane distillation etc.

Currently membrane technology is widely regarded as the most efficient and economically resourceful operation in numerous industrial processes. Moreover, it is recognized for its potential applications in the bioethanol industry (Haider et al. 2022). The separation is based on several factors such as size-exclusion, molecular weight (polymers, proteins, colloidal matters), charge of the particles etc., finally pervaporation of the by product that employs a membrane separation technique driven by vacuum to separate the liquid components of an azeotropic mixture. This technique

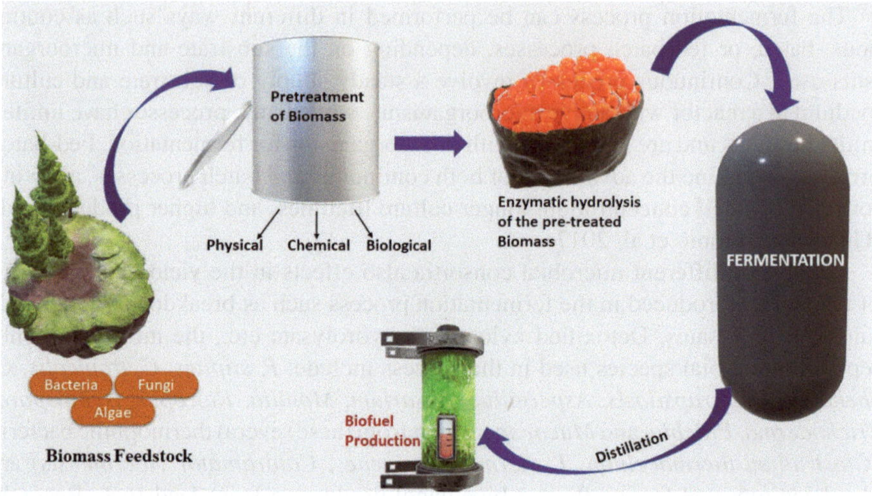

Fig. 18.1 A schematic representation of biofuel production pathway (Gutiérrez-Antonio et al. 2017)

is capable of achieving high levels of separation, enabling the production of pure ethanol while simultaneously separating water from the ethanol–water mixture (Dey et al. 2020, 2021) (Fig. 18.1).

18.7 Challenges and Future Opportunities

Biofuel production is a rapidly evolving field that offers both challenges and future opportunities in the quest for sustainable energy sources. While biofuels have the potential to reduce greenhouse gas emissions and decrease dependence on fossil fuels, several key challenges need to be addressed to unlock their full potential. One significant challenge is feedstock availability and sustainability, future opportunities lie in the development of advanced feedstocks, which can be grown on non-arable land and have higher yields and lower environmental footprints. Another challenge is the efficiency and cost-effectiveness of biofuel production processes. Many biofuel conversion technologies are still in the early stages of development and require further optimization to achieve commercial viability. Future opportunities lie in advancing biochemical and thermochemical conversion methods, as well as developing novel catalytic processes and biotechnological approaches to improve efficiency and reduce costs.

Furthermore, the development of advanced biofuels presents an opportunity for the future. Advanced biofuels, such as cellulosic ethanol and algae-based biofuels, offer the potential to utilize non-food feedstocks and agricultural residues, reducing the competition for land and resources. Research and development in this area aim to improve the efficiency and scalability of these processes. Moreover, technological advancements hold promise for biofuel production. Genetic engineering and synthetic biology can enhance the yield and characteristics of biofuel crops, making them more suitable for commercial production. Additionally, the integration of biofuel production with other industries, such as agriculture and waste management, can create synergies and enhance sustainability. The utilization of waste and biomass residues also presents a significant opportunity. By converting agricultural, forestry, and municipal waste into biofuels, the environmental impact can be minimized, and waste management issues can be addressed simultaneously.

References

Abdullah B, Muhammad SAFS, Shokravi Z, Ismail S, Kassim KA, Mahmood AN, Aziz MMAJR (2019) Fourth generation biofuel: a review on risks and mitigation strategies. Renew Sustain Energy Rev 107:37–50

Abu-Omar MM, Barta K, Beckham GT, Luterbacher JS, Ralph J, Rinaldi R, Román-Leshkov Y, Samec JS, Sels BF, Wang FJE, Science E (2021) Guidelines for performing lignin-first biorefining. Energy Environ Sci 14(1):262–292

Adsul M, Sandhu SK, Singhania RR, Gupta R, Puri SK, Mathur AJE (2020) Designing a cellulolytic enzyme cocktail for the efficient and economical conversion of lignocellulosic biomass to biofuels. Enzyme Microb Technol 133:109442

Ahmed Z, Cary M, Ali S, Murshed M, Ullah H, Mahmood HJE (2022) Moving toward a green revolution in Japan: symmetric and asymmetric relationships among clean energy technology development investments, economic growth, and CO_2 emissions. Energy Environ 33(7):1417–1440

Alishah Aratboni H, Rafiei N, Garcia-Granados R, Alemzadeh A, Morones-Ramírez JR (2019) Biomass and lipid induction strategies in microalgae for biofuel production and other applications. Microb Cell Fact 18:1–17

Amran NA, Bello U, Ruslan MSHJH (2022) The role of antioxidants in improving biodiesel's oxidative stability, poor cold flow properties, and the effects of the duo on engine performance: a review. Heliyon e09846

Angermayr SA, Rovira AG, Hellingwerf KJJ (2015) Metabolic engineering of cyanobacteria for the synthesis of commodity products. Trends Biotechnol 33(6):352–361

Asiri F, Chen C-H, Hwangbo M, Shao Y, Chu K-HJ (2020) From organic wastes to bioplastics: feasibility of nonsterile poly (3-hydroxybutyrate) production by Zobellella denitrificans ZD1. ACS Omega 5(38):24158–24168

Atsumi S, Hanai T, Liao JCJ (2008) Non-fermentative pathways for synthesis of branched-chain higher alcohols as biofuels. Nature 451(7174):86–89

Avalos JL, Fink GR, Stephanopoulos GJ (2013) Compartmentalization of metabolic pathways in yeast mitochondria improves the production of branched-chain alcohols. Nat Biotechnol 31(4):335–341

Banerjee R, Proshlyakov Y, Lipscomb JD, Proshlyakov DAJN (2015) Structure of the key species in the enzymatic oxidation of methane to methanol. Nature 518(7539):431–434

Barkia I, Saari N, Manning SRJ (2019) Microalgae for high-value products towards human health and nutrition. Mar Drugs 17(5):304

Basak B, Fatima A, Jeon B-H, Ganguly A, Chatterjee PK, Dey AJ (2018) Process kinetic studies of biohydrogen production by co-fermentation of fruit-vegetable wastes and cottage cheese whey. Energy Sustain Dev 47:39–52

Beller HR, Lee TS, Katz LJ (2015) Natural products as biofuels and bio-based chemicals: fatty acids and isoprenoids. Nat Prod Rep 32(10):1508–1526
Bellini S, Tommasi T, Fino DJ (2022) Poly (3-hydroxybutyrate) biosynthesis by Cupriavidus necator: a review on waste substrates utilization for a circular economy approach. Bioresour Technol Rep 100985
Bertsch J, Müller VJ (2015) Bioenergetic constraints for conversion of syngas to biofuels in acetogenic bacteria. Biotechnol Biofuels 8(1):1–12
Bishnu J, Bhatt MR, Sharma D, Joshi J, Malla R, Sreerama L (2011) Lignocellulosic ethanol production: current practices and recent developments. Biotechnol Mol Biol Rev 6:(8)172–182
Bošnjaković M, Sinaga NJAS (2020) The perspective of large-scale production of algae biodiesel. Appl Sci 10(22):8181
Bušić A, Marđetko N, Kundas S, Morzak G, Belskaya H, Ivančić Šantek M, Komes D, Novak S, Šantek BJ (2018) Bioethanol production from renewable raw materials and its separation and purification: a review. Food Technol Biotechnol 56(3):289–311
Callegari A, Bolognesi S, Cecconet D, Capodaglio AGJ (2020) Production technologies, current role, and future prospects of biofuels feedstocks: a state-of-the-art review. Crit Rev Environ Sci Technol 50(4):384–436
Centobelli P, Cerchione R, Del Vecchio P, Oropallo E, Secundo GJ (2022) Blockchain technology for bridging trust, traceability and transparency in circular supply chain. Inf Manag 59(7):103508
Cha M, Chung D, Elkins JG, Guss AM, Westpheling JJ (2013) Metabolic engineering of *Caldicellulosiruptor bescii* yields increased hydrogen production from lignocellulosic biomass. Biotechnol Biofuels 6(1):1–8
Chandel AK, Garlapati VK, Singh AK, Antunes FAF, da Silva SSJBT (2018) The path forward for lignocellulose biorefineries: bottlenecks, solutions, and perspective on commercialization. Biores Technol 264:370–381
Chandrasekhar K, Lee Y-J, Lee D-WJ (2015) Biohydrogen production: strategies to improve process efficiency through microbial routes. Int J Mol Sci 16(4):8266–8293
Chen Z, Huang J, Wu Y, Wu W, Zhang Y, Liu DJ (2017) Metabolic engineering of *Corynebacterium glutamicum* for the production of 3-hydroxypropionic acid from glucose and xylose. Metab Eng 39:151–158
Chen H, Simoska O, Lim K, Grattieri M, Yuan M, Dong F, Lee YS, Beaver K, Weliwatte S, Gaffney EMJCR (2020) Fundamentals, applications, and future directions of bioelectrocatalysis. Chem Rev 120(23):12903–12993
Chen Z, Wu Y, Huang J, Liu DJ (2015) Metabolic engineering of *Klebsiella pneumoniae* for the de novo production of 2-butanol as a potential biofuel. Bioresour Technol 197:260–265
Chen J, Li W, Zhang Z-Z, Tan T-W, Li Z-JJ (2018) Metabolic engineering of *Escherichia coli* for the synthesis of polyhydroxyalkanoates using acetate as a main carbon source. Microb Cell Factories 17:1–12
Chen W-H, Lin B-J, Lin Y-Y, Chu Y-S, Ubando AT, Show PL, Ong HC, Chang J-S, Ho S-H, Culaba ABJ (2021) Progress in biomass torrefaction: principles, applications and challenges. Prog Energy Combust Sci 82:100887
Chintagunta AD, Zuccaro G, Kumar M, Kumar SJ, Garlapati VK, Postemsky PD, Kumar NS, Chandel AK, Simal-Gandara JJ (2021) Biodiesel production from lignocellulosic biomass using oleaginous microbes: prospects for integrated biofuel production. Front Microbiol 12:658284
Chow T-J, Su H-Y, Tsai T-Y, Chou H-H, Lee T-M, Chang J-SJ (2015) Using recombinant cyanobacterium (*Synechococcus elongatus*) with increased carbohydrate productivity as feedstock for bioethanol production via separate hydrolysis and fermentation process. Bioresour Technol 184:33–41
Chu R, Li S, Zhu L, Yin Z, Hu D, Liu C, Mo FJR, Reviews SE (2021) A review on co-cultivation of microalgae with filamentous fungi: Efficient harvesting, wastewater treatment and biofuel production. Renew Sustain Energy Rev 139:110689
Chu S, Majumdar AJ (2012) Opportunities and challenges for a sustainable energy future. Nature 488(7411):294–303

Ciliberti C, Biundo A, Albergo R, Agrimi G, Braccio G, de Bari I, Pisano IJP (2020) Syngas derived from lignocellulosic biomass gasification as an alternative resource for innovative bioprocesses. Processes 8(12):1567
Conrado RJ, Gonzalez RJS (2014) Envisioning the bioconversion of methane to liquid fuels. Science 343(6171):621–623
De Carvalho CC, Caramujo MJJ (2018) The various roles of fatty acids. Molecules 23(10):2583
de Morais MG, Vaz BdS, de Morais EG, Costa JAVJ (2015) Biologically active metabolites synthesized by microalgae. BioMed Res Int 2015
Debnath C, Bandyopadhyay TK, Bhunia B, Mishra U, Narayanasamy S, Muthuraj MJ (2021) Microalgae: sustainable resource of carbohydrates in third-generation biofuel production. Renew Sustain Energy Rev 150:111464
Deviram G, Mathimani T, Anto S, Ahamed TS, Ananth DA, Pugazhendhi AJ (2020) Applications of microalgal and cyanobacterial biomass on a way to safe, cleaner and a sustainable environment. J Clean Prod 253:119770
Dey P, Pal P, Kevin JD, Das DBJ (2020) Lignocellulosic bioethanol production: prospects of emerging membrane technologies to improve the process–a critical review. Rev Chem Eng 36(3):333–367
Dey S, Reang N, Das P, Deb MJ (2021) A comprehensive study on prospects of economy, environment, and efficiency of palm oil biodiesel as a renewable fuel. J Clean Prod 286:124981
Dheeran P, Reddy LJ (2018) Biorefining of lignocelluloses: an opportunity for sustainable biofuel production. Biorefining Biomass Biofuels: Oppor Percept 1–23
Dhyani V, Bhaskar TJ (2018) A comprehensive review on the pyrolysis of lignocellulosic biomass. Renew Energy 129:695–716
Einhaus A, Steube J, Freudenberg RA, Barczyk J, Baier T, Kruse OJ (2022) Engineering a powerful green cell factory for robust photoautotrophic diterpenoid production. Metab Eng 73:82–90
Faust JE, Logan JJ (2018) Daily light integral: a research review and high-resolution maps of the United States. HortScience 53(9):1250–1257
Gaida SM, Liedtke A, Jentges AHW, Engels B, Jennewein SJ (2016) Metabolic engineering of *Clostridium cellulolyticum* for the production of n-butanol from crystalline cellulose. Microb Cell Factories 15(1):1–11
Garay LA, Boundy-Mills KL, German JB (2014) Accumulation of high-value lipids in single-cell microorganisms: a mechanistic approach and future perspectives. J Agric Food Chem 62(13):2709–2727
Ghassemi N, Poulhazan A, Deligey F, Mentink-Vigier F, Marcotte I, Wang TJ (2021) Solid-state NMR investigations of extracellular matrixes and cell walls of algae, bacteria, fungi, and plants. Chem Rev 122(10):10036–10086
Gill JJ, Wang B, Sestak E, Young R, Chu K-HJ (2018) Characterization of a novel Tectivirus phage toil and its potential as an agent for biolipid extraction. Sci Rep 8(1):1062
Gutiérrez-Antonio C, Gómez-Castro FI, de Lira-Flores JA, Hernández SJR, Reviews SE (2017) A review on the production processes of renewable jet fuel. Renew Sustain Energy Rev 79:709–729
Haider J, Long NVD, Lee M (2022) Intensified and hybrid distillation technologies for production of high value-added products from lignocellulosic biomass. Biofuels Biorefining, Elsevier, 197–229
Hu G, Ji S, Yu Y, Wang S, Zhou G, Li F (2015) Organisms for biofuel production: natural bioresources and methodologies for improving their biosynthetic potentials. Biotechnol Appl Biodivers 185–224
Huang C, Jiang X, Shen X, Hu J, Tang W, Wu X, Ragauskas A, Jameel H, Meng X, Yong QJR, Reviews SE (2022) Lignin-enzyme interaction: a roadblock for efficient enzymatic hydrolysis of lignocellulosics. Renew Sustain Energy Rev 154:111822
Hwangbo M, Chu K-H (2020) Recent advances in production and extraction of bacterial lipids for biofuel production. Sci Total Environ 734:139420
Janusz G, Pawlik A, Świderska-Burek U, Polak J, Sulej J, Jarosz-Wilkołazka A, Paszczyński A (2020) Laccase properties, physiological functions, and evolution. Int J Mol Sci 21(3):966

Kalyuzhnaya MG, Puri AW, Lidstrom ME (2015) Metabolic engineering in methanotrophic bacteria. Metab Eng 29:142–152
Kan T, Strezov V, Evans TJ (2016) Lignocellulosic biomass pyrolysis: a review of product properties and effects of pyrolysis parameters. Renew Sustain Energy Rev 57:1126–1140
Kapoor R, Ghosh P, Kumar M, Sengupta S, Gupta A, Kumar SS, Vijay V, Kumar V, Vijay VK, Pant D (2020) Valorization of agricultural waste for biogas based circular economy in India: a research outlook. Biores Technol 304:123036
Kapoore RV, Wood EE, Llewellyn CA (2021) Algae biostimulants: a critical look at microalgal biostimulants for sustainable agricultural practices. Biotechnol Adv 49:107754
Khan MU, Lee JTE, Bashir MA, Dissanayake PD, Ok YS, Tong YW, Shariati MA, Wu S, Ahring BK (2021) Current status of biogas upgrading for direct biomethane use: a review. Renew Sustain Energy Rev 149:111343
Khare E, Mishra J, Arora NK (2018) Multifaceted interactions between endophytes and plant: developments and prospects. Front Microbiol 9:2732
Kim T, Song B, Cho KS, Lee I-S (2020) Therapeutic potential of volatile terpenes and terpenoids from forests for inflammatory diseases. Int J Mol Sci 21(6):2187
Kleinert C, Griehl CJ (2021) Identification of suitable *Botryococcus braunii* strains for non-destructive in situ hydrocarbon extraction. J Appl Phycol 33(2):785–798
Klok A, Lamers P, Martens D, Draaisma R, Wijffels RJ (2014) Edible oils from microalgae: insights in TAG accumulation. Trends Biotechnol 32(10):521–528
Koller MJB (2019) Polyhydroxyalkanoate biosynthesis at the edge of water activitiy-haloarchaea as biopolyester factories. Bioengineering 6(2):34
Kour D, Rana KL, Yadav N, Yadav AN, Rastegari AA, Singh C, Negi P, Singh K, Saxena AK (2019) Technologies for biofuel production: current development, challenges, and future prospects. Prospect Renew Bioprocess Futur Energy Syst 1–50
Kumar B, Bhardwaj N, Agrawal K, Chaturvedi V, Verma P (2020) Current perspective on pretreatment technologies using lignocellulosic biomass: an emerging biorefinery concept. Fuel Process Technol 199:106244
Kumar A, Kaushal S, Saraf SA, Singh JS (2018) Microbial bio-fuels: a solution to carbon emissions and energy crisis. Front Biosci-Landmark 23(10):1789–1802
Le Borgne S, Baquerizo G (2019) Microbial ecology of biofiltration units used for the desulfurization of biogas. ChemEngineering 3(3):72
Lee S, Kang M, Bae J-H, Sohn J-H, Sung BH (2019) Bacterial valorization of lignin: strains, enzymes, conversion pathways, biosensors, and perspectives. Front Bioeng Biotechnol 7:209
Levasseur W, Perré P, Pozzobon V (2020) A review of high value-added molecules production by microalgae in light of the classification. Biotechnol Adv 41:107545
Lewiński M, Mohammed D (2019) The 2015 Paris climate conference: arguing for the fragile consensus in global multilateral diplomacy. J Argum Context 8(1):65–90
Li Z, Peng Y, Gao H (2020) Enhanced long-term advanced denitrogenation from nitrate wastewater by anammox consortia: dissimilatory nitrate reduction to ammonium (DNRA) coupling with anammox in an upflow biofilter reactor equipped with EDTA-2Na/Fe (II) ratio and pH control. Biores Technol 305:123083
Li S, Gao Y, Li N, Ge L, Bu X, Feng PJE, Science E (2021) Transition metal-based bimetallic MOFs and MOF-derived catalysts for electrochemical oxygen evolution reaction. Energy Environ Sci 14(4):1897–1927
Lian J, Chao R, Zhao H (2014) Metabolic engineering of a *Saccharomyces cerevisiae* strain capable of simultaneously utilizing glucose and galactose to produce enantiopure (2R, 3R)-butanediol. Metab Eng 23:92–99
Liao JC, Mi L, Pontrelli S, Luo S (2016) Fuelling the future: microbial engineering for the production of sustainable biofuels. Nat Rev Microbiol 14(5):288–304
Li-Beisson Y, Thelen JJ, Fedosejevs E, Harwood JL (2019a) The lipid biochemistry of eukaryotic algae. Prog Lipid Res 74:31–68

Liew F, Henstra AM. Köpke M, Winzer K, Simpson SD, Minton NP (2017) Metabolic engineering of *Clostridium autoethanogenum* for selective alcohol production. Metab Eng 40:104–114

Liu R, Zhu F, Lu L, Fu A, Lu J, Deng Z, Liu T (2014) Metabolic engineering of fatty acyl-ACP reductase-dependent pathway to improve fatty alcohol production in *Escherichia coli*. Metab Eng 22:10–21

Liu D, Hu R, Palla KJ, Tuskan GA, Yang X (2016) Advances and perspectives on the use of CRISPR/Cas9 systems in plant genomics research. Curr Opin Plant Biol 30:70–77

Lu S, Wang Q, Liang Z, Wang W, Liang C, Wang Z, Yuan Z, Lan P, Qi W (2021b) Saccharification of sugarcane bagasse by magnetic carbon-based solid acid pretreatment and enzymatic hydrolysis. Ind Crops Prod 160:113159

Lu J, Li J, Gao H, Zhou D, Xu H, Cong Y, Zhang W, Xin F, Jiang M (2021a) Recent progress on bio-succinic acid production from lignocellulosic biomass. World J Microbiol Biotechnol 37:1–8

Łukajtis R, Hołowacz I, Kucharska K, Glinka M, Rybarczyk P, Przyjazny A, Kamiński MJR (2018) Hydrogen production from biomass using dark fermentation. Renew Sustain Energy Rev 91:665–694

Luo H, Zeng Q, Han S, Wang Z, Dong Q, Bi Y, Zhao Y (2017) High-efficient n-butanol production by co-culturing *Clostridium acetobutylicum* and *Saccharomyces cerevisiae* integrated with butyrate fermentative supernatant addition. World J Microbiol Biotechnol 33:1–10

Lynd LR, Beckham GT, Guss AM, Jayakody LN, Karp EM, Maranas C, McCormick RL, Amador-Noguez D, Bomble YJ, Davison BH (2022) Toward low-cost biological and hybrid biological/catalytic conversion of cellulosic biomass to fuels. Energy Environ Sci 15(3):938–990

Mallette ND, Pankrantz E, Busse S, Strobel GA, Carlson RP, Peyton BM (2014) Evaluation of cellulose as a substrate for hydrocarbon fuel production by *Ascocoryne sarcoides* (NRRL 50072). J Sustain Bioenergy Syst 4(1):33–49

Markou G, Nerantzis E (2013) Microalgae for high-value compounds and biofuels production: a review with focus on cultivation under stress conditions. Biotechnol Adv 31(8):1532–1542

Marques S, Moreno AD, Ballesteros M, Gírio F (2018) Starch biomass for biofuels, biomaterials, and chemicals. Biomass Green Chem: Build Renew Pathw 69–94

McKinlay CJ, Turnock SR, Hudson DA (2021) Route to zero emission shipping: Hydrogen, ammonia or methanol? Int J Hydrogen Energy 46(55):28282–28297

Mehariya S, Goswami RK, Karthikeysan OP, Verma PJC (2021) Microalgae for high-value products: a way towards green nutraceutical and pharmaceutical compounds. Chemosphere 280:130553

Mewalal R, Rai DK, Kainer D, Chen F, Külheim C, Peter GF, Tuskan GA (2017) Plant-derived terpenes: a feedstock for specialty biofuels. Trends Biotechnol 35(3):227–240

Mohsenpour SF, Hennige S, Willoughby N, Adeloye A, Gutierrez T (2021) Integrating micro-algae into wastewater treatment: a review. Sci Total Environ 752:142168

Moulin S, Légeret B, Blangy S, Sorigué D, Burlacot A, Auroy P, Li-Beisson Y, Peltier G, Beisson F (2019) Continuous photoproduction of hydrocarbon drop-in fuel by microbial cell factories. Sci Rep 9(1):13713

Murata MM, Ito Morioka LR, Da Silva Marques JB, Bosso A, Suguimoto HH (2021) What do patents tell us about microalgae in agriculture? AMB Express 11(1):1–12

Ngatcha ADP, Muhammad G, Lv Y, Xiong W, Zhao A, Xu J, Alam MA (2022) Microalgae biomass pre-treatment with deep eutectic solvent to optimize lipid isolation in biodiesel production. Biomass Convers Biorefinery 12(Suppl 1):133–143

Ninkuu V, Zhang L, Yan J, Fu Z, Yang T, Zeng H (2021) Biochemistry of terpenes and recent advances in plant protection. Int J Mol Sci 22(11):5710

Nitsos C, Filali R, Taidi B, Lemaire J (2020) Current and novel approaches to downstream processing of microalgae: a review. Biotechnol Adv 45:107650

Opia AC, Hamid MKBA, Syahrullail S, Abd Rahim AB, Johnson CA (2021) Biomass as a potential source of sustainable fuel, chemical and tribological materials–overview. Mater Today: Proc 39:922–928

Osman AI, Mehta N, Elgarahy AM, Al-Hinai A, Al-Muhtaseb AH, Rooney DW (2021) Conversion of biomass to biofuels and life cycle assessment: a review. Environ Chem Lett 19:4075–4118
Østby H, Hansen LD, Horn SJ, Eijsink VG, Várnai AJ (2020) Enzymatic processing of lignocellulosic biomass: principles, recent advances and perspectives. Off J Soc Ind Microbiol Biotechnol 47(9–10):623–657
Oukala N, Aissat K, Pastor V (2021) Bacterial endophytes: the hidden actor in plant immune responses against biotic stress. Plants 10(5):1012
Pal P, Chew KW, Yen H-W, Lim JW, Lam MK, Show PL (2019) Cultivation of oily microalgae for the production of third-generation biofuels. Sustainability 11(19):5424
Panahi HKS, Dehhaghi M, Guillemin GJ, Gupta VK, Lam SS, Aghbashlo M, Tabatabaei M (2022) Bioethanol production from food wastes rich in carbohydrates. Curr Opin Food Sci 43:71–81
Qiu L, Deng Y, Wang F, Davaritouchaee M, Yao Y (2019) A review on biochar-mediated anaerobic digestion with enhanced methane recovery. Renew Sustain Energy Rev 115:109373
Ragauskas AJ, Beckham GT, Biddy MJ, Chandra R, Chen F, Davis MF, Davison BH, Dixon RA, Gilna P, Keller M (2014) Lignin valorization: improving lignin processing in the biorefinery. Science 344(6185):1246843
Raj K, Krishnan C (2019) Improved high solid loading enzymatic hydrolysis of low-temperature aqueous ammonia soaked sugarcane bagasse using laccase-mediator system and high concentration ethanol production. Ind Crops Prod 131:32–40
Ramli NM, Verreth J, Yusoff FM, Nurulhuda K, Nagao N, Verdegem MC (2020) Integration of algae to improve nitrogenous waste management in recirculating aquaculture systems: a review. Front Bioeng Biotechnol 8:1004
Rauch R, Hrbek J, Hofbauer H (2014) Biomass gasification for synthesis gas production and applications of the syngas. Wiley Interdiscip Rev: Energy Environ 3(4):343–362
Robak K, Balcerek M (2018) Review of second generation bioethanol production from residual biomass. Food Technol Biotechnol 56(2):174
Rodionova MV, Poudyal RS, Tiwari I, Voloshin RA, Zharmukhamedov SK, Nam HG, Zayadan BK, Bruce BD, Hou HJ, Allakhverdiev SI (2017) Biofuel production: challenges and opportunities. Int J Hydrog Energy 42(12):8450–8461
Rosentreter JA, Borges AV, Deemer BR, Holgerson MA, Liu S, Song C, Melack J, Raymond PA, Duarte CM, Allen GH (2021) Half of global methane emissions come from highly variable aquatic ecosystem sources. Nat Geosci 14(4):225–230
Saha BC, Qureshi N, Kennedy GJ, Cotta MA (2016) Biological pretreatment of corn stover with white-rot fungus for improved enzymatic hydrolysis. Int Biodeterior & Biodegrad 109:29–35
Santala S, Efimova E, Koskinen P, Karp MT, Santala V (2014) Rewiring the wax ester production pathway of *Acinetobacter baylyi* ADP1. ACS Synth Biol 3(3):145–151
Saratale RG, Cho S-K, Saratale GD, Kadam AA, Ghodake GS, Kumar M, Bharagava RN, Kumar G, Kim DS, Mulla SI (2021) A comprehensive overview and recent advances on polyhydroxyalkanoates (PHA) production using various organic waste streams. Bioresour Technol 325:124685
Saravanan A, Kumar PS, Jeevanantham S, Karishma S, Vo D-VN (2022) Recent advances and sustainable development of biofuels production from lignocellulosic biomass. Biores Technol 344:126203
Sarker S, Lamb JJ, Hjelme DR, Lien KM (2019) A review of the role of critical parameters in the design and operation of biogas production plants. Appl Sci 9(9):1915
Schadeweg V, Boles E (2016) Increasing n-butanol production with *Saccharomyces cerevisiae* by optimizing acetyl-CoA synthesis, NADH levels and trans-2-enoyl-CoA reductase expression. Biotechnol Biofuels 9:1–11
Schoen HR, Peyton BM, Knighton WB (2016) Rapid total volatile organic carbon quantification from microbial fermentation using a platinum catalyst and proton transfer reaction-mass spectrometry. AMB Express 6:1–10

Schutyser W, Renders T, Van den Bosch S, Koelewijn S-F, Beckham G, Sels BF (2018) Chemicals from lignin: an interplay of lignocellulose fractionation, depolymerisation, and upgrading. Chem Soc Rev 47(3):852–908

Sharma PK, Saharia M, Srivstava R, Kumar S, Sahoo L (2018) Tailoring microalgae for efficient biofuel production. Front Mar Sci 5:382

Shaw JJ, Berbasova T, Sasaki T, Jefferson-George K, Spakowicz DJ, Dunican BF, Portero CE, Narvaez-Trujillo A, Strobel SA (2015) Identification of a fungal 1, 8-cineole synthase from *Hypoxylon sp.* with specificity determinants in common with the plant synthases. J Biol Chem 290(13):8511–8526

Shimada I, Nakamura Y, Kato S, Mori R, Ohta H, Suzuki K, Takatsuka T (2018) Catalytic cracking of wax esters extracted from *Euglena gracilis* for hydrocarbon fuel production. Biomass Bioenerg 112:138–143

Sikarwar VS, Zhao M, Fennell PS, Shah N, Anthony EJ (2017) Progress in biofuel production from gasification. Prog Energy Combust Sci 61:189–248

Singh D, Sharma D, Soni S, Sharma S, Sharma PK, Jhalani A (2020) A review on feedstocks, production processes, and yield for different generations of biodiesel. Fuel 262:116553

Solaymani S (2019) CO_2 emissions patterns in 7 top carbon emitter economies: The case of transport sector. Energy 168:989–1001

Soltanian S, Aghbashlo M, Almasi F, Hosseinzadeh-Bandbafha H, Nizami A-S, Ok YS, Lam SS, Tabatabaei M (2020) A critical review of the effects of pretreatment methods on the exergetic aspects of lignocellulosic biofuels. Energy Convers Manag 212:112792

Sorigué D, Légeret B, Cuiné S, Blangy S, Moulin S, Billon E, Richaud P, Brugière S, Couté Y, Nurizzo D (2017) An algal photoenzyme converts fatty acids to hydrocarbons. Science 357(6354):903–907

Sridhar A, Ponnuchamy M, Kumar PS, Kapoor A, Xiao L (2022) Progress in the production of hydrogen energy from food waste: a bibliometric analysis. Int J Hydrog Energy 47(62):26326–26354

Strobel G, Booth E, Schaible G, Mends MT, Sears J, Geary B (2013) The Paleobiosphere: a novel device for the in vivo testing of hydrocarbon producing-utilizing microorganisms. Biotechnol Lett 35:539–552

Strobel GA (2015) Bioprospecting—fuels from fungi. Biotechnol Lett 37:973–982

Strong PJ, Xie S, Clarke WP (2015) Methane as a resource: can the methanotrophs add value? Environ Sci Technol 49(7):4001–4018

Tavva SMD, Deshpande A, Durbha SR, Palakollu VAR, Goparaju AU, Yechuri VR, Bandaru VR, Muktinutalapati VSR (2016) Bioethanol production through separate hydrolysis and fermentation of *Parthenium hysterophorus* biomass. Renew Energy 86:1317–1323

Thanapimmetha A, Suwaleerat T, Saisriyoot M, Chisti Y, Srinophakun P (2017) Production of carotenoids and lipids by *Rhodococcus opacus* PD630 in batch and fed-batch culture. Bioprocess Biosyst Eng 40:133–143

Tomsheck AR, Strobel GA, Booth E, Geary B, Spakowicz D, Knighton B, Floerchinger C, Sears J, Liarzi O, Ezra D (2010) *Hypoxylon sp.*, an endophyte of *Persea indica*, producing 1, 8-cineole and other bioactive volatiles with fuel potential. Microb Ecol 60:903–914

Torres-Tiji Y, Fields FJ, Mayfield SP (2020) Microalgae as a future food source. Biotechnol Adv 41:107536

Tursi A (2019) A review on biomass: importance, chemistry, classification, and conversion. Biofuel Res J 6(2):962

Ubando AT, Felix CB, Chen W-H (2020) Biorefineries in circular bioeconomy: a comprehensive review. Bioresour Technol 299:122585

Vasić K, Knez Ž, Leitgeb M (2021) Bioethanol production by enzymatic hydrolysis from different lignocellulosic sources. Molecules 26(3):753

Verma S, Lu S, Kenis PJ (2019) Co-electrolysis of CO_2 and glycerol as a pathway to carbon chemicals with improved technoeconomics due to low electricity consumption. Nat Energy 4(6):466–474

Veza I, Said MFM, Latiff ZA (2021) Recent advances in butanol production by acetone-butanol-ethanol (ABE) fermentation. Biomass Bioenergy 144:105919

Villanova V, Spetea C (2021) Mixotrophy in diatoms: molecular mechanism and industrial potential. Physiol Plant 173(2):603–611

Wang B, Rezenom YH, Cho K-C, Tran JL, Russell DH, Gill JJ, Young R, Chu K-H (2014) Cultivation of lipid-producing bacteria with lignocellulosic biomass: effects of inhibitory compounds of lignocellulosic hydrolysates. Biores Technol 161:162–170

Wang Q, Lu Y, Xin Y, Wei L, Huang S, Xu J (2016) Genome editing of model oleaginous microalgae *Nannochloropsis spp.* by CRISPR/Cas9. Plant J 88(6):1071–1081

Wang S, Dai G, Yang H, Luo Z (2017) Lignocellulosic biomass pyrolysis mechanism: a state-of-the-art review. Prog Energy Combust Sci 62:33–86

Wang M, Dewil R, Maniatis K, Wheeldon J, Tan T, Baeyens J, Fang Y (2019a) Biomass-derived aviation fuels: challenges and perspective. Progress Energy Comb Sci 74:31–49

Wang X, Chao N, Zhang M, Jiang X, Gai Y (2019b) Functional characteristics of caffeoyl shikimate esterase in *Larix kaempferi* and monolignol biosynthesis in gymnosperms. Int J Mol Sci 20(23):6071

Wu D, Wei Z, Mohamed TA, Zheng G, Qu F, Wang F, Zhao Y, Song CJC (2022) Lignocellulose biomass bioconversion during composting: mechanism of action of lignocellulase, pretreatment methods and future perspectives 286:131635

Xu L, Wang L, Zhou X-R, Chen W-C, Singh S, Hu Z, Huang F-H, Wan X (2018) Stepwise metabolic engineering of Escherichia coli to produce triacylglycerol rich in medium-chain fatty acids. Biotechnol Biofuels 11(1):1–13

Yang X, Xu M, Yang S-T (2015) Metabolic and process engineering of *Clostridium cellulovorans* for biofuel production from cellulose. Metab Eng 32:39–48

Yin Z, Zhu L, Li S, Hu T, Chu R, Mo F, Hu D, Liu C, Li B (2020) A comprehensive review on cultivation and harvesting of microalgae for biodiesel production: environmental pollution control and future directions. Biores Technol 301:122804

Yoo CG, Meng X, Pu Y, Ragauskas AJ (2020) The critical role of lignin in lignocellulosic biomass conversion and recent pretreatment strategies: a comprehensive review. Biores technol 301:122784

Yusoff FM, Nagao N, Imaizumi Y, Toda TJ (2019) Bioreactor for microalgal cultivation systems: strategy and development. Prospect Renew Bioprocess Futur Energy Syst 117–159

Zabed HM, Akter S, Yun J, Zhang G, Awad FN, Qi X, Sahu JJR, Reviews SE (2019) Recent advances in biological pretreatment of microalgae and lignocellulosic biomass for biofuel production. Renew Sustain Energy Rev 105:105–128

Zabed H, Sahu J, Boyce AN, Faruq GJR (2016) Fuel ethanol production from lignocellulosic biomass: an overview on feedstocks and technological approaches. Renew Sustain Energy Rev 66:751–774

Zhang F, Zhu Y, Lin Q, Zhang L, Zhang X, Wang HJE (2021) Noble-metal single-atoms in thermocatalysis, electrocatalysis, and photocatalysis. Science 14(5):2954–3009

Chapter 19
Sustainability Assessment of Liquid Biofuel Production Technologies-Challenges and Prospects

Jayita Chopra

Abstract Liquid biofuel is one of the most sought-after renewable energy sources that have the potential to substitute fossil fuel primarily because of its greener composition, cleaner combustion, and wide array of available feedstocks from which it can be produced. Sustainability assessment studies, encompassing multiple aspects such as environmental, economic, and societal, are a powerful tool and way forward for the successful commercialisation of biofuels. In this chapter, both conventional and novel biofuel production technologies have been discussed, and their commercialization potential has been assessed with the help of life cycle assessment (LCA) studies previously conducted on them. The adoption of varied technologies, system boundaries, product streams, and economic uncertainties in the production of biofuels remains to be the major challenges in identifying the right technology. Therefore, this chapter aims to critically analyze a set of indicators, including possible valorization routes of wastes, that can offset the high production cost to get a clear picture in identifying the most sustainable and adaptable biofuel production technology.

Keywords Biofuel · Life cycle assessment · Environmental impacts · Sustainability · Renewable energy

19.1 Introduction

The Anthropocene era has witnessed an unprecedented increase in global CO_2 concentration, leading to rising in global temperature. This is primarily attributed to industrialization, rapid growth in population, and lifestyle change, among several other factors. The increasing demand for fossil fuels and their combustion-related emission has led the scientific community to search for renewable and sustainable energy sources. Therefore, to address the issues of energy security and slowly move

J. Chopra (✉)
Department of Chemical Engineering, BITS Pilani K. K. Birla Goa Campus, Zuarinagar, Goa 403726, India
e-mail: jayitac@goa.bits-pilani.ac.in

M. Shah and D. Deka (eds.), *Emerging Sustainable Technologies for Biofuel Production*, Environmental Science and Engineering, https://doi.org/10.1007/978-3-031-52167-6_19

towards a low-carbon economy, zeroing in on a sustainable feedstock is imperative (Isler-Kaya and Karaosmanoglu 2022).

In this regard, liquid biofuels, viz. bioethanol and biodiesel, have gained increased attention because these can be easily blended with fossil-based petrol and diesel, respectively, and can be used in existing engines in planes, marine vessels, and marine vessels and other heavy transport modes that cannot be electrified (Cabrera-Jiménez et al. 2022). Bioethanol production is more widely explored, dominating the global biofuel market. USA and Brazil are the global leaders in bioethanol and biodiesel production, using corn and sugarcane as the primary feedstock. Global bioethanol production was 27 BT in 2021, out of which 15 BT was produced by the USA and 7 BT by Brazil, accounting for 55% and 26%, respectively of global production (Fig. 19.1). In India, sugarcane molasses is one of the widely used feedstock for bioethanol production. However, it cannot meet the increasing demand and the national blending criteria. According to the National Policy on Biofuels, India aims to achieve 20% blending of bioethanol by 2025 and around 5% blending of biodiesel by 2030 (Singh et al. 2017). Therefore, other lignocellulosic biomass, the most abundant resource, such as crop residues, must be explored to achieve the target. India is a tropical and agriculturally active country; the primary crop residue includes rice straw and wheat straw. However, the major deterrent in the commercialization of 2G bioethanol in India is the high production cost primarily contributed by the various energy-intensive pretreatment methods.

Over the years various feedstocks have been tested for production of biofuel. The first-generation biofuels from crops like sugarcane, corn and jatropha has been

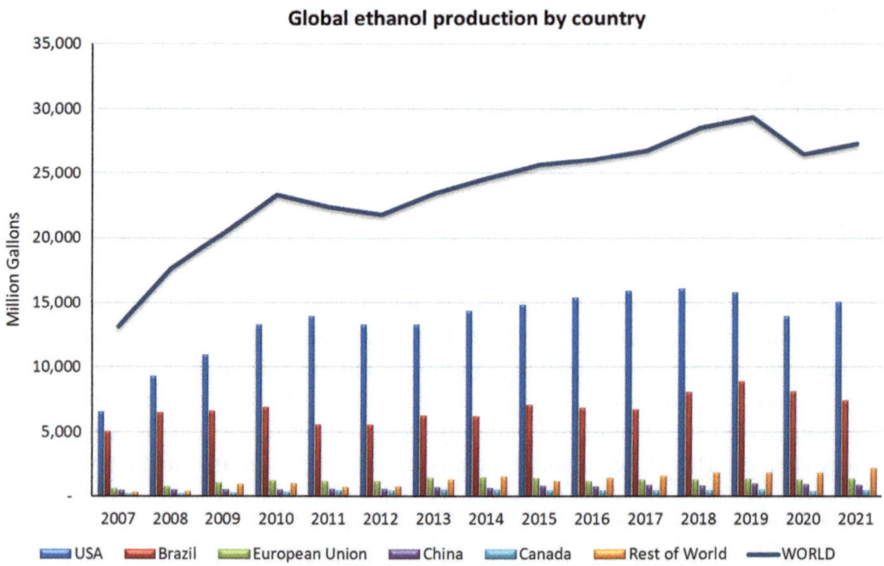

Fig. 19.1 Country-wise global ethanol production (Data source: Renewable Fuels Association (RFA) (https://ethanolrfa.org/markets-and-statistics/annual-ethanol-production))

deemed unsustainable because of the huge requirement in land use change, i.e. growing crops for fuel rather than food (Singh and Olsen 2011). Currently, USA uses around 80% of globally produced corn for bioethanol production, disrupting the international corn market with prices sky rocketing for human consumption (Singh et al. 2017). There are also concerns regarding freshwater usage, deforestation, use of fertilizers and pesticides, leading to severe environmental impacts even more than resulting from the usage of fossil fuels (Holma et al. 2013). The 'food versus fuel' debate can be addressed by using non-edible lignocellulosic biomass, mainly agricultural residues (Zoghlami and Paës 2019) and single-cell microorganisms capable of storing triglycerides, like microalgae, yeasts, etc. (Li et al. 2022; Pragya and Pandey 2016; Chopra et al. 2016). These constitute the second generation biofuels produced from non-edible oil sources and third generation biofuels mostly from algae and yeast, also called advanced biofuels. However, due to the vast diversity in composition of the various lignocellulosic biomass (LCB), it is very challenging to design and develop a single robust biofuel production technology. LCB is very recalcitrant, with lignin tightly bound with the other cell wall constituents such as cellulose and hemicellulose (Nargotra et al. 2018). For efficient conversion of the sugar fraction to bioethanol, it is essential first to hydrolyze these rigid polymers and release the sugar monomers. The enzyme cellulase is responsible for the breakdown of cellulose, and much research is going on worldwide for its efficient production (Kumar et al. 2020). There is no single robust pretreatment method due to the wide variation in LCB composition. Pretreatment methods employ physical, chemical and biological methods in combination (Soam et al. 2016). The overall success of an economically viable biofuel production process depends mainly on the pretreatment method. Therefore, a life cycle assessment study will help us show the feasibility of approaches, as most of the methods require high energy, prolonged reaction time and hazardous chemicals (Rebello et al. 2020). To make the biofuel production economically viable, it is also necessary to utilize the residual fractions of hemicellulose and lignin for either biofuel production or other value-added products in a biorefinery model (Chrysikou et al. 2018; Velvizhi et al. 2022).

3G biofuels, particularly biodiesel, are produced by transesterification of single-cell oil (SCO) accumulated inside yeast, bacteria or microalgae cells. Microalgae species such as *Scenedesmus* sp., *Chlorella* sp., and *Botryococcus* sp. have been well explored in literature for their lipid accumulating capabilities. One major limitation of algal cultivation is the slow growth rate and low biomass density resulting in lower productivity. Oleaginous yeasts perform better in terms of shorter doubling time and higher biomass density than algae, where the growth depends on light. Yeasts such as *Yarrowia lipolytica*, *Rhodotorula glutinis*, *Rhodosporidium toruloides* have been reported to accumulate lipids up to 50–70% of their dry cell weight. However, several energy-intensive techniques are involved in the cultivation of the organism, extraction of intracellular lipids and its conversion to esters, limiting their adoption on a large scale, therefore essential to evaluate the sustainability and potentiality of scale-up of such processes by environmental impact analysis.

There are several reports on biofuel production life cycle assessment from various feedstocks and using different technologies. Most of the studies have focused on the

GHG emissions from the process and its potential impact on other environmental indicators like climate change, marine toxicity, ozone depletion, etc. However, it is not easy to compare different LCA studies because of variations in the functional unit, system boundary and product allocation. In most cases, the results of these studies are conflicting and therefore challenging to come to a general conclusion. The commercialization potential of a biofuel production technology can only be evaluated by a detailed LCA study. Thus, this chapter aims to consolidate the information from LCA review papers and also provide a clear understanding of the environmental, economic and social impact of different biofuel production technologies. It also aims to bridge existing knowledge gaps in biofuel production and suggest a future way forward.

19.2 Biofuel Production Technologies

Biofuel production technology involves a comprehensive combination of processes and can be classified based on the feedstock or conversion route employed. The selection of the economically viable way for biofuel production depends on several factors:

- Nature and availability of feedstock
- Other high-value targeted products
- Biofuel yield
- Energy consumption
- Fuel characteristics or blending properties

The chronological development of biofuel production technologies is depicted in Fig. 19.2.

19.2.1 Conventional or First-Generation Biofuel

19.2.1.1 Sugar-Based Ethanol

First-generation biofuels are commercially produced from sugar and starch-based crops. Sucrose obtained from sugarcane, sugar beet, etc., is fermented to ethanol, while starch, obtained from corn crops, needs to be hydrolyzed first, followed by fermentation. The pretreatment in the latter case is more energy intensive. The overall economy of the process is also dependent on the value of co-products, such as DDGS (dried distiller's grain with solubles).

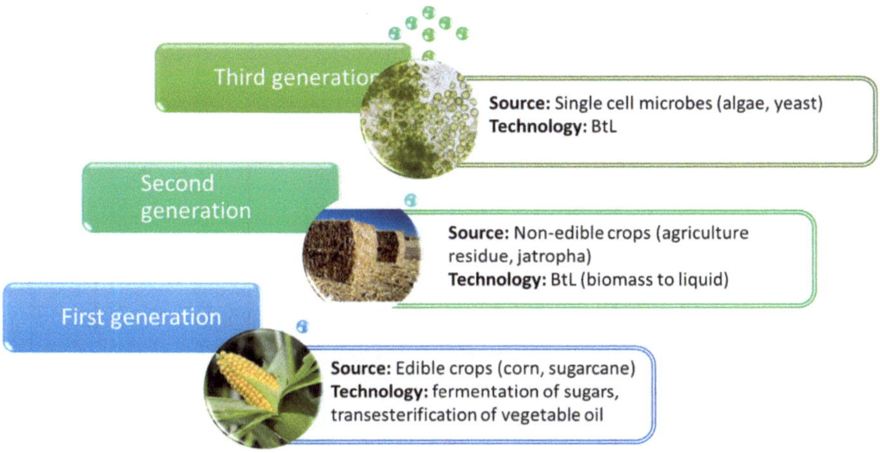

Fig. 19.2 Biofuel production technologies

19.2.1.2 Biodiesel from Oil Crops

Oil crops such as rapeseed, sunflower, oil palm etc., are conventionally used as feedstocks for biodiesel production, wherein the triglycerides are trans-esterified to methyl/ethyl esters in the presence of an acid/base catalyst. Purifying the esters from the reaction mixture containing glycerol, acids, catalyst and residual free fatty acids is a cumbersome process. The high production and purification cost are often balanced by using glycerol as a platform chemical to produce a wide array of glycerol derivatives. Enzyme-catalyzed transesterification is much more sustainable; however, the primary deterrent is the high enzyme cost and slower reaction rate.

The different biomass-to-biofuel conversion routes are shown in Fig. 19.3.

19.2.2 Advanced Biofuel/Jet Fuel/Bio-Crude

19.2.2.1 LCB Bioethanol

Second-generation bioethanol is produced from reducing sugars obtained by breaking down lignocellulosic biomass to individual monomers by hydrolysis, followed by microbial fermentation to ethanol (Fig. 19.3) (Santos et al. 2021). The significant challenges are extracting fermentable sugars, microorganism selection and to utilize all the sugar fractions of LCB (Mishra and Ghosh 2020). Lignin, the most recalcitrant compound, also needs to be valorized. Lignin is an aromatic compound with a complex structure that binds cellulose and hemicellulose together. Valorization of lignin to value-added products will augment the economy of the LCB biorefinery (Alwadani and Fatehi 2018). Therefore, most current research focuses on

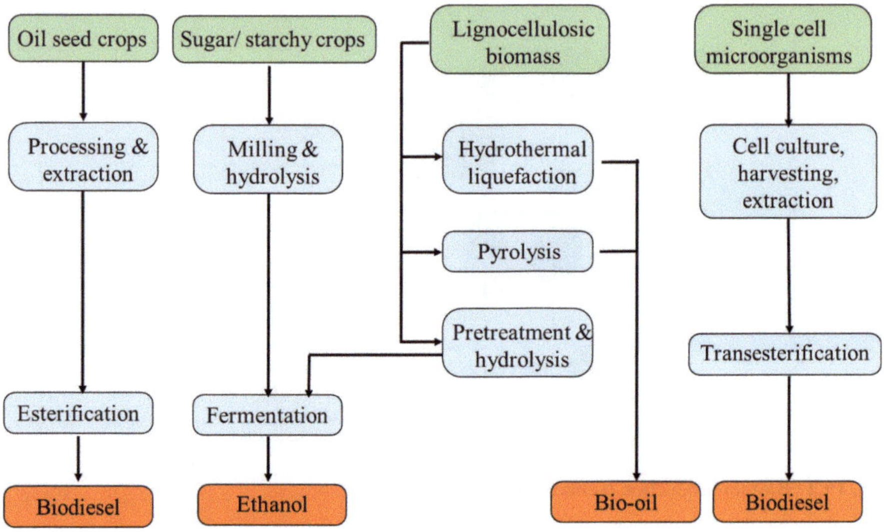

Fig. 19.3 Overview of biomass to various liquid biofuel conversion technologies

identifying the right pretreatment steps and valorising residual fractions to high-value products, typically in a biorefinery model. LCB is primarily composed of cellulose (40–45%), followed by hemicellulose (30–35%) and lignin (20–30%) (Perez et al. 2021). Cellulose, the primary component in LCB, can be easily broken down to glucose by cellulase for subsequent fermentation. However, the challenge lies in using the pentose (C5) sugars obtained from hemicellulose hydrolysis as a feedstock for fermentation. Some C5 sugar fermenting yeasts have been reported, but the ethanol yield is usually significantly less and the performance is also affected by the presence of inhibitors produced during LCB pretreatment.

19.2.2.2 3G Biodiesel

Single-cell microorganisms, particularly microalgae and yeasts, capable of accumulating lipids or triglycerides intracellularly, also known as oleaginous microbes. They can accumulate lipid more than 70% of their dry cell weight under optimal culture conditions. The lipids are thereafter extracted and transesterified to biodiesel (Chopra et al. 2016). Microalgal biofuel productivities per unit area are much higher than other biofuel production crops. Hence, the land footprint will be much lesser than conventional biofuels (Correa et al. 2019).

19.2.2.3 Bio-Crude

Hydrothermal liquefaction is a promising strategy to convert LCB, municipal waste, and sewage sludge to energy-dense liquid fuel called bio-crude (Kapusta 2018; Huang et al. 2018). The reaction is similar to the geological formation of fossil fuel, which takes hundreds of years under the natural condition of high temperature and pressure deep down inside the Earth's surface. The reaction occurs at subcritical condition (<374 °C) with pressure in the range of 120–150 bar to maintain the liquid state of water, where water acts as an organic solvent (Chopra et al. 2019). The heterogeneous biomass gets depolymerized initially, which later repolymerizes to a more homogenous compound. The reaction products are distributed in solid, liquid and gaseous phases, with the distribution yields varying based on the temperature–pressure conditions, feedstock composition and reaction time. The significant advantage is that the feedstock can be used with high moisture content, and the bio-crude has a high calorific value (30–40 MJ/kg) which can be distilled to fractions for blending with diesel or aviation fuel. However, the high energy demand for the technology is considered as the major challenge hindering its large-scale application.

19.2.2.4 Pyrolysis Oil

The reaction takes place at 300–700 °C in the absence of oxygen (Khan et al. 2021). It is a promising technology used to convert crop residues to energy-dense bio-oil and char (Xu et al. 2021). The organic fraction constituting the bio-oil is pulled together from the solid (char), aqueous and gas phases. The bio-oil is separated from the aqueous phase by liquid–liquid extraction using methylene chloride, dimethyl chloride or chloroform. The non-condensable gas called syngas consists of carbon dioxide, carbon monoxide, hydrogen and methane; and can also be used as a fuel. Bio-oil can be used in heavy-duty vehicles and can also be used for biodiesel and hydrogen production. The oxygen content in the bio-oil can be reduced by using catalysts like zeolites or mesoporous molecular sieves, which remove oxygen via deoxygenation reactions, thereby improving the calorific value of the oil. Temperature is the most important criteria in pyrolysis determining the product yields. Temperature ranges are higher in pyrolysis compared to HTL and therefore concerns for input energy source is more. The sustainability of the process depends on the energy balance (output input) of the system. At the same time, biochar finds application in catalyst preparation, as an adsorbent, additive, soil amendment and in electrochemical energy storage (Li et al. 2021).

19.3 Importance of Sustainability Assessment in Biofuel Production/Main Challenges to Biofuel Sustainability

Life cycle assessment is an essential systematic tool used to analyze the environmental impact of a product's life cycle, considering all the inputs and outputs, starting from the extraction of raw materials to its end-of-life (Joseph et al. 2015). LCA of biofuels show the net greenhouse gas savings compared to fossil fuel and quantifies the impact on climate change, land use change, human health, and resource consumption, among other indicators (Gnansounou and Kenthorai Raman 2016). Net GHG emissions depend on the feedstock, technology employed, and assumptions. The GHG emissions depend on a crop's availability or yield, possibilities of by-product production, change in land use, for example, forest or agricultural land for energy crop cultivation and also on input energy source (hydel, solar or thermal). Change in land use patterns is by far one of the significant concerns questioning the sustainability of biofuels from crops grown on converted forest land. Forests act as carbon sinks, and clearing of mass forest lands mobilises the stored carbon and negates biofuel combustion's overall GHG mitigation effects.

The availability of knowledge is limited for some key indicators, such as eutrophication potential, biodiversity and acidification, which are critical in determining the sustainability of biofuels. Application of fertilisers and pesticides for the cultivation of crops for biofuel production has led to increasing in nitrogen, phosphorous and carbon in the water streams in many places, thereby causing severe eutrophication issues. The variations in results from different LCA studies are due to a wide range of available methodologies and assumptions. However, ISO standards have clearly indicated that LCA studies are not designed for comparison studies and should be adapted for every circumstance to generate meaningful data (Collotta et al. 2019).

Another major challenge identified in the existing study includes the complicated biofuel supply chain. Often biofuel production may involve several countries and assembling details like producers of raw materials, chemicals in production processes and process water and emissions from the process becomes a tedious task.

19.4 Approaches to Assess the Sustainability

Most LCA studies reported in the literature are attributional, considering the direct effect of the system of interest defined by the system boundary. It gives an estimate of the environmental impacts of the various sub-processes, compares different methods and identifies the processes having the maximum environmental footprint and therefore helps improvise the process with lesser impact. The sustainability of biofuel production largely depends on the energy utilised in the process (production and transportation) and the energy recovered in the fuel (calorific value and yield). An LCA study is performed based on the ISO 14040-14,044 standard series. A typical

LCA study consists of (1) Goal and scope definition, (2) Inventory analysis, (3) Impact assessment, (4) Interpretation.

19.4.1 Goal and Scope

The first step includes defining the goal and scope of the system, which includes the functional units, system boundaries and allocation procedures (Jeswani et al. 2020). Usually, most of the studies employ an attributional cradle-to-gate approach to estimate the environmental impacts of the biofuel production process. The system boundaries include all the processes starting from the extraction of raw materials needed for the production process to all other inputs (water, energy, etc.) and associated emissions, while the utilisation of biofuel is excluded (products of combustion). The functional unit is chosen as a basis, based on which the inputs and outputs are calculated. Since biofuel is an energy source, the commonly used available unit is MJ, or it can be expressed as a volumetric unit kg. If biofuel is the only product, then no allocation is needed. However, most of the studies have reported the co-production of biofuel and other high-value products and therefore, allocation should be clearly defined in such cases. Distribution can be done based on the ratio of energetic content, economic basis or volumetric yield of the products.

Important impact categories that must be considered include global warming potential (GWP), acidification potential, eutrophication potential, fossil depletion potential and human health. A typical system boundary for biofuel production is shown in Fig. 19.4.

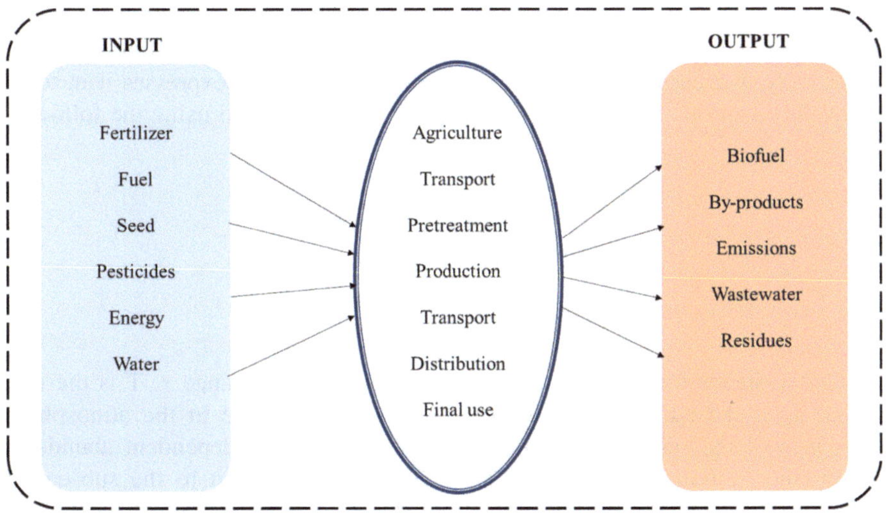

Fig. 19.4 System boundary of the biofuel production process

19.4.2 Inventory Analysis

All lab-scale and pilot-scale data for the cultivation of energy crops, pretreatment, transportation, distribution, production (transesterification or fermentation), and processing is included in the inventory. All secondary data are obtained from associated databases like Ecoinvent, IPCC and USDA (Tibesigwa et al. 2023).

19.4.3 Life Cycle Impact Assessment (LCIA) and Interpretation

The LCIA phase evaluates the potential environmental impacts using the LCI results. It includes the selection of impact categories, assignment of LCI results, characterization (given by all mid-point indicators like eutrophication, acidification, ozone depletion, etc.) and damage assessment (endpoint indicators like human health, ecosystems and resources) (Padilla-Rivera et al. 2019). Most of the studies have compared the environmental impacts of biofuel production to that of fossil fuels. The effects of biofuel production vary widely based on the feedstock and processing technologies employed. Biofuels have been found to perform better in most of the impact categories in comparison with fossil fuels. Table 19.1 shows the environmental impacts of various impact categories for biofuel production from various sources. Various tools exist for LCIA. These include CML2 baseline 20 0 0, IMPACT2002+ , ReCiPe2008, etc. Some of the critical impact categories have been discussed below, along with their significance.

GWP—GHGs are emitted due to change in land use, cultivation of feedstocks, harvesting, transportation and conversion to biofuels (Correa et al. 2019). GWP and climate change are used for methods like Impact 2000+ and Recipe midpoint, respectively. It measures the amount of GHGs released and expresses it in terms of $kgCO_2$ eq/kg oil. The CO_2 equivalency factor is calculated using the following formulae:

$$GWP_{x,T} = \frac{\int_0^T a_x \times [x(t)]dt}{\int_0^T a_r \times [r(t)]dt}$$

$GWP_{x,T}$ stands for the global warming potential of substance x, T is the time period, a_x is the radiative efficiency due to a unit increase in the atmospheric abundance of the substance in question, [x(t)] is the time-dependent abundance of substance x (Goedkoop et al. 2009). GWP also relates it to the subsequent environmental mechanisms affecting the endpoint categories, human health and ecosystem.

Table 19.1 Quantitative estimation of major impact categories for different biofuel production routes

Fuel type	Feedstock	Technology	LCA scope	GWP (kg CO_2 eq/kg oil)	Land use (PDF*m^2y)	AP (kg PO_4 eq/kg oil)	EP (kg PO_4 eq/kg oil)	References
2G Bioethanol	Molasses	Fermentation	Cradle to gate	–	4.89	0.0852	–	Isler-Kaya and Karaosmanoglu (2022)
	LCB	Pretreatment	Cradle to grave	0.14	–	0.0007	0.001	Smullen et al. (2019)
2G Biodiesel	Safflower	Transesterification	Cradle to gate	–	0.294	0.117	–	Isler-Kaya and Karaosmanoglu (2022)
	Jatropha	Transesterification	Cradle to gate	35	–	–	0.005 (kg P eq/GJ)	Tibesigwa et al. (2023)
3G Biodiesel	Microalgae	Transesterification + Anaerobic digestion	–	0.22	0.01	–	–	Gnansounou and Kenthorai Raman, (2016)
Biocrude	Oil palm fruit	HTL	Cradle to grave	3.8	–	0.06	0.008	Chan et al. (2016)
	LCB (lignin-rich)	HTL + APR	Cradle to gate	56.1 gCO_2eq/MJ	–	380 mgSO2 eq/MJ	52.1 mgPO4 eq/MJ	Zoppi et al. (2023)
Bio-oil	Oil palm fruit	Pyrolysis	Cradle to grave	8	–	0.09	0.009	Chan et al. (2016)
Biodiesel + Biocrude	Yeast	Transesterification + HTL	Cradle to gate	47	–	0.37	0.098	Chopra et al. (2020)

*GWP—Global warming potential, AP—Acidification potential, EP—Eutrophication potential, PDF—Potentially disappeared fraction, APR—Aqueous phase reforming, HTL—Hydrothermal liquefaction

Land use—It reflects the damage to ecosystem due to the effects of land use transformation like forest land to agricultural land or for mining activities and also on occupation. It determines how much GHG will be emitted and how much biodiversity loss will occur when native forest land is converted for biofuel production (Correa et al. 2019).

Acidification potential—Atmospheric deposition of inorganic substances like phosphates and sulfates causes acidity in the soil. There is a clearly defined optimum acidity level for most of the plant species. Ocean acidity, on the other hand, is the phenomenon aggravated by the release of large quantities of CO_2 in the atmosphere which eventually dissolves in the ocean to form carbonic acid and poses a threat to aquatic shell-bearing animals (Goedkoop et al. 2009).

Eutrophication potential—The process of nutrient enrichment of aquatic systems leading to severe depletion of dissolved oxygen is known as eutrophication. It results from human activities like the release of agricultural run-offs containing pesticides into water bodies. It typically considers the limiting nutrients (which can limit the growth of aquatic biomass- algae or duckweed), like nitrogen and phosphorus, when there is an excess of other nutrients (Holma et al. 2013).

19.5 Analysis of Sustainability Assessment Studies/ Overview of Major Findings

Available literature shows many LCA reports studying the environmental impact of different biofuel production technologies. Table 19.1 summarizes the impact of a few selected studies on the production of bioethanol, biodiesel, biocrude and bio-oil. The land use change indicator shows higher values for 2G bioethanol and biodiesel because of the extensive land requirement for crop cultivation. In contrast, it is very minimal in the case of 3G biodiesel, biocrude and bio-oil. The GWP values were found to be higher for energy-intensive processes like transesterification and HTL, and the impact can be reduced by using renewable energy for the production process.

19.6 Critical Indicators

The indicators used in a study depend on several factors like data availability, the scope of the LCA study, system boundary and functional unit. This section discusses the most commonly used indicators that comprehensively capture the environmental, social and economic aspects of the selected literature studies. Assessments methods that bring together all these impacts can guide us in choosing sustainable biofuel production technology. It is, therefore crucial to develop standardized assessment methods and selection of critical indicators that can give a holistic picture of the process.

19.6.1 Environmental Indicators

Most of the studies have reported environmental impacts and indicated the cultivation of biomass as one of the significant contributors to the overall impact. Biomass cultivation is dependent on the input energy source, oil yield or monosaccharide yield per hectare of land, climatic conditions, etc. (Collotta et al. 2019). Acidification and eutrophication are the two most widely used impact categories reported in the literature (as discussed in Sect. 4.3) and accounts for carbon sequestration and N_2O emissions and nitrogen losses from the soil in the form of NOx and NO_3-. GWP is another critical indicator as it considers the carbon fluxes through the input and output of the system boundary and quantifies any net CO_2 emission from biofuel production. Table 19.2 shows a comparative air emission upon combustion of different biofuels and fossil fuels for a travel distance of 1 km. Bioethanol and biodiesel show lesser CO_2 and CO emission compared to diesel, compressed gas (CG) and gasoline. In contrast, the NOX emissions were found to be higher in the case of bioethanol blend 85% (E85) because of the presence of protein fractions present in the lignocellulosic biomass feedstock.

Table 19.2 Air emissions over the life cycle of each fuel based on a travel distance of 1 km

	GHG emission					Other emissions (mg/km)			References
	CO_2 (kg/km)	CO (mg/km)	CH_4 (mg/km)	N_2O (mg/km)	NMVOC	NOx	NH_3	SO_2	
CG	0.273	522	135	0.65	107	198	0.43	420	Daylan and Ciliz (2016)
Diesel	0.134	–	0.076	–	–	0.256	–	–	Isler-Kaya and Karaosmanoglu (2022)
Bioethanol (E85)	0.129	341	670	0.84	63	237	15	227	Daylan and Ciliz (2016)
	0.065	248		–	–	124	–	13.9	Kim and Dale (2006)
Biodiesel	0.007	–	0.061	–	–	0.289	–	–	Isler-Kaya and Karaosmanoglu (2022)
Gasoline	0.229	468	–	–	–	62.1	–	42.8	Kim and Dale (2006)

19.6.2 Social Indicators

Social impacts of biofuel production include the benefits at a local/regional level, impact on land, food and feed prices, changes in land usage/ownership and crop cultivation patterns, and socio-political scenario of the country. The most important social factor related to biofuel production is its increased impact on food systems, especially when produced from crops like corn or sugarcane, a major food source (Collotta et al. 2019). A sustainable biofuel production process should not compete with the food supply. The third-generation biofuel produced mainly from algae performs better in terms of food security. Still, it shows impact on employment as it involves a lot of labour-intensive steps like cultivation, lipid extraction and lipid conversion (Yang et al. 2015). However, the US and EU still need to mandate incorporation of social sustainability indicators for biofuel production (Collotta et al. 2019). Few studies have included social impacts such as employment, labour situations, human rights, health and safety, minimum wages, gender equality, injuries, unemployment, forced labour etc. (Ekener-Petersen et al. 2014; Ekener et al. 2018).

19.6.3 Economic Indicators

Economic consideration of biofuel production is crucial in determining its commercial viability and competitiveness with fossil fuels. Currently, the global biofuel market is still at its nascent stage and the production cost is much higher compared to fossil fuel. This high cost is one of the main reasons for the slow growth of the biofuel market since companies and consumers make their choices based on cost. However, emerging biofuel technologies can provide incentives to developing countries and provide job opportunities (Correa et al. 2019). Economic indicators are rarely included in LCA studies and are only sometimes reported alongside environmental indicators. There is a strong interrelation between the two indicators; therefore, combinatorial LCA and techno-economic assessment (TEA) studies will indicate the process's sustainability and economic viability. Some studies have also included the shadow costs involved due to different environmental impacts, for example, the costs incurred for treating diseases caused by environmental emissions in air, water or land. The unit processes/operations contributing to high biofuel production costs include energy intensive pretreatment steps when using lignocellulosic biomass as feedstock. This input energy is usually from fossil sources like natural gas (NG) or coal. Sensitivity analysis by Kim and Dale (2008) showed higher environmental footprints when using coal instead of NG since extraction of 1 MJ of NG requires more energy than mining 1 MJ of coal. Therefore, it also offers more environmental credits in terms of GHG emissions.

19.7 Challenges and Prospects

To meet the future global energy demand, we need to diversify the energy sources and identify conversion technologies with minimal environmental footprint. The viable process can be adopted only after exploring its potential for commercialization. Therefore, a systematic and strategic method incorporating environmental, economic and social indicators will be required to guide the decision-making authorities. One of the significant challenges LCA researchers face is the need for more standardisation in different studies regarding functional units, system boundaries and impact categories. Hence, a standardized global LCA methodology should be followed so that comparisons among different LCA studies can be performed. Although it is understandable that to capture the actual picture (variations in technologies, local aspects), it is not always possible to follow a general method, there should be an equivalent quantitative estimation made available for the same. Another critical challenge in performing an LCA study carried out on small scale or lab scale is data extrapolation. Inventory data for small scale cannot be extrapolated for a given/larger functional unit. Therefore, relevant data should be obtained from pilot scale or commercial scale plants to get a more realistic result. A sensitivity analysis will help us identify the most critical parameter with maximum environmental impact and also help us compare different scenarios.

19.8 Conclusion

The feedstock chosen and the production route opted for biofuel production will determine its overall environmental impact, sustainability, and potential for scalability and commercialization. Hence, a critical analysis of LCB biomass is essential to check its viability (regarding recalcitrance and availability) as a feedstock before indulging in designing a biofuel production system. Using second and third-generation biofuels from lignocellulosic agricultural residue and microalgal feedstock, respectively, can reduce the environmental and socio-economic impacts. Moreover, process intensification and optimization of existing bioprocess systems, preferably in a biorefinery model, are required to achieve more sustainable and economically viable biofuel production technologies. A sustainability assessment of novel biofuel production technologies should be performed before their commercial utilization considering the UN Sustainable development goals (SDGs) such as no poverty, zero hunger, affordable and clean energy, climate action, responsible consumption and production, etc.

References

Alwadani N, Fatehi P (2018) Synthetic and lignin-based surfactants: challenges and opportunities. Carbon Resour Convers 1(2):126–138. https://doi.org/10.1016/j.crcon.2018.07.006

Cabrera-Jiménez R, Mateo-Sanz JM, Gavaldà J, Jiménez L, Pozo C (2022) Comparing biofuels through the lens of sustainability: a data envelopment analysis approach. Appl Energy 307.https://doi.org/10.1016/j.apenergy.2021.118201

Chan YH, Tan RR, Yusup S, Lam HL, Quitain AT (2016) Comparative life cycle assessment (LCA) of bio-oil production from fast pyrolysis and hydrothermal liquefaction of oil palm empty fruit bunch (EFB). Clean Technol Environ Policy 18(6):1759–1768. https://doi.org/10.1007/s10098-016-1172-5

Chopra J, Dineshkumar R, Bhaumik M, Dhanarajan G, Kumar R, Sen R (2016) Integrated: In situ transesterification for improved biodiesel production from oleaginous yeast: a value proposition for possible industrial implication. RSC Adv 6(74):70364–70373. https://doi.org/10.1039/c6ra14003c

Chopra J, Mahesh D, Yerrayya A, Vinu R, Kumar R, Sen R (2019) Performance enhancement of hydrothermal liquefaction for strategic and sustainable valorization of de-oiled yeast biomass into green bio-crude. J Clean Prod 227:292–301. https://doi.org/10.1016/j.jclepro.2019.04.147

Chopra J, Tiwari BR, Dubey BK, Sen R (2020) Environmental impact analysis of oleaginous yeast based biodiesel and bio-crude production by life cycle assessment. J Clean Prod 271:122349. https://doi.org/10.1016/j.jclepro.2020.122349

Chrysikou LP, Bezergianni S, Kiparissides C (2018) Environmental analysis of a lignocellulosic-based biorefinery producing bioethanol and high-added value chemicals. Sustain Energy Technol Assess 28:103–109. https://doi.org/10.1016/j.seta.2018.06.010

Collotta M, Champagne P, Tomasoni G, Alberti M, Busi L, Mabee W (2019) Critical indicators of sustainability for biofuels: an analysis through a life cycle sustainabilty assessment perspective. Renew Sustain Energy Rev 115:109358. https://doi.org/10.1016/j.rser.2019.109358

Correa DF, Beyer HL, Fargione JE, Hill JD, Possingham HP, Thomas-Hall SR, Schenk PM (2019) Towards the implementation of sustainable biofuel production systems. Renew Sustain Energy Rev 107:250–263. https://doi.org/10.1016/j.rser.2019.03.005

Daylan B, Ciliz N (2016) Life cycle assessment and environmental life cycle costing analysis of lignocellulosic bioethanol as an alternative transportation fuel. Renew Energy 89:578–587. https://doi.org/10.1016/j.renene.2015.11.059

Ekener E, Hansson J, Gustavsson M (2018) Addressing positive impacts in social LCA—discussing current and new approaches exemplified by the case of vehicle fuels. Int J Life Cycle Assess 23(3):556–568. https://doi.org/10.1007/s11367-016-1058-0

Ekener-Petersen E, Höglund J, Finnveden G (2014) Screening potential social impacts of fossil fuels and biofuels for vehicles. Energy Policy 73:416–426. https://doi.org/10.1016/j.enpol.2014.05.034

Gnansounou E, Kenthorai Raman J (2016) Life cycle assessment of algae biodiesel and its co-products. Appl Energy 161:300–308. https://doi.org/10.1016/j.apenergy.2015.10.043

Goedkoop M, Heijungs R, Huijbregts M, De Schryver A, Struijs J, Van Zelm R (2009) ReCiPe 2008. Potentials 1–44. http://www.pre-sustainability.com/download/misc/ReCiPe_main_report_final_27-02-2009_web.pdf

Holma A, Koponen K, Antikainen R, Lardon L, Leskinen P, Roux P (2013) Current limits of life cycle assessment framework in evaluating environmental sustainability—case of two evolving biofuel technologies. J Clean Prod 54:215–228. https://doi.org/10.1016/j.jclepro.2013.04.032

Huang Z, Wufuer A, Wang Y, Dai L (2018) Hydrothermal liquefaction of pretreated low-lipid microalgae for the production of bio-oil with low heteroatom content. Process Biochem 69:136–143. https://doi.org/10.1016/j.procbio.2018.03.018

Isler-Kaya A, Karaosmanoglu F (2022) Life cycle assessment of safflower and sugar beet molasses-based biofuels. Renew Energy 201(P1):1127–1138. https://doi.org/10.1016/j.renene.2022.11.041

Jeswani HK, Chilvers A, Azapagic A (2020) Environmental sustainability of biofuels: a review: environmental sustainability of biofuels. Proc R Soc A: Math, Phys Eng Sci 476(2243). https://doi.org/10.1098/rspa.2020.0351

Joseph T, Baah K, Jahanfar A, Dubey B (2015) A comparative life cycle assessment of conventional hand dryer and roll paper towel as hand drying methods. Sci Total Environ 515–516:109–117. https://doi.org/10.1016/j.scitotenv.2015.01.112

Kapusta K (2018) Effect of ultrasound pretreatment of municipal sewage sludge on characteristics of bio-oil from hydrothermal liquefaction process. Waste Manag 78:183–190. https://doi.org/10.1016/j.wasman.2018.05.043

Khan N, Chowdhary P, Gnansounou E, Chaturvedi P (2021) Bioresource technology biochar and environmental sustainability : emerging trends and techno-economic perspectives. 332

Kim S, Dale BE (2006) Ethanol fuels: E10 or E85—life cycle perspectives. Int J Life Cycle Assess 11(2):117–121. https://doi.org/10.1065/lca2005.02.201

Kim S, Dale BE (2008) Life cycle assessment of fuel ethanol derived from corn grain via dry milling. Biores Technol 99(12):5250–5260. https://doi.org/10.1016/j.biortech.2007.09.034

Kumar P, Kumar V, Kumar S, Singh J, Kumar P (2020) Bioethanol production from sesame (Sesamum indicum L.) plant residue by combined physical, microbial and chemical pretreatments. Bioresour Technol 297:122484. https://doi.org/10.1016/j.biortech.2019.122484

Li Y, Yu H, Liu L, Yu H (2021) Application of co-pyrolysis biochar for the adsorption and immobilization of heavy metals in contaminated environmental substrates. J Hazard Mater 420:126655. https://doi.org/10.1016/j.jhazmat.2021.126655

Li P, Luo Y, Yuan X (2022) Life cycle and techno-economic assessment of source-separated wastewater-integrated microalgae biofuel production plant: a nutrient organization approach. Bioresour Technol 344. https://doi.org/10.1016/j.biortech.2021.126230

Mishra A, Ghosh S (2020) Sacchari fi cation of kans grass biomass by a novel fractional hydrolysis method followed by co-culture fermentation for bioethanol production. Renew Energy 146:750–759. https://doi.org/10.1016/j.renene.2019.07.016

Nargotra P, Sharma V, Gupta M, Kour S, Bajaj BK (2018) Application of ionic liquid and alkali pretreatment for enhancing saccharification of sunflower stalk biomass for potential biofuel-ethanol production. Biores Technol 267:560–568. https://doi.org/10.1016/j.biortech.2018.07.070

Padilla-Rivera A, Paredes MG, Güereca LP (2019) A systematic review of the sustainability assessment of bioenergy: the case of gaseous biofuels. Biomass Bioenergy 125:79–94. https://doi.org/10.1016/j.biombioe.2019.03.014

Perez CL, Milessi TS, Sandri JP, Foulqui MR, Giordano RC, Thevelein JM, De Lima R, Giordano C, Zangirolami C (2021) Unraveling continuous 2G ethanol production from xylose using hemicellulose hydrolysate and immobilized superior recombinant yeast in fixed-bed bioreactor. 169. https://doi.org/10.1016/j.bej.2021.107963

Pragya N, Pandey KK (2016) Life cycle assessment of green diesel production from microalgae. Renew Energy 86:623–632. https://doi.org/10.1016/j.renene.2015.08.064

Rebello S, Anoopkumar AN, Aneesh EM, Sindhu R, Binod P, Pandey A (2020) Sustainability and life cycle assessments of lignocellulosic and algal pretreatments. Bioresour Technol 301:122678. https://doi.org/10.1016/j.biortech.2019.122678

Santos JC, Ingle AP, Antunes FAF, Thom LC, Costa CB, Dos V, Bell MJV, Rosa CA, Da Silva SS (2021) Multi-scale study of the integrated use of the carbohydrate fractions of sugarcane bagasse for ethanol and xylitol production. 163. https://doi.org/10.1016/j.renene.2020.08.020

Singh S, Adak A, Saritha M, Sharma S, Tiwari R, Rana S, Arora A, Nain L (2017) Bioethanol production scenario in India: potential and policy perspective. In: Chandel AK, Sukumaran RK (eds) Sustainable biofuels development in India. Springer International Publishing, pp 21–37. https://doi.org/10.1007/978-3-319-50219-9_2

Singh A, Olsen SI (2011) A critical review of biochemical conversion, sustainability and life cycle assessment of algal biofuels. Appl Energy 88(10):3548–3555. https://doi.org/10.1016/j.apenergy.2010.12.012

Smullen E, Finnan J, Dowling D, Mulcahy P (2019) The environmental performance of pretreatment technologies for the bioconversion of lignocellulosic biomass to ethanol. Renew Energy 142:527–534. https://doi.org/10.1016/j.renene.2019.04.082

Soam S, Kapoor M, Kumar R, Borjesson P, Gupta RP, Tuli DK (2016) Global warming potential and energy analysis of second generation ethanol production from rice straw in India. Appl Energy 184:353–364. https://doi.org/10.1016/j.apenergy.2016.10.034

Tibesigwa T, Iezzi B, Lim TH, Kirabira JB, Olupot PW (2023) Life cycle assessment of biodiesel production from selected second-generation feedstocks. Clean Eng Technol 100614. https://doi.org/10.1016/j.clet.2023.100614

Velvizhi G, Balakumar K, Shetti NP, Ahmad E, Kishore Pant K, Aminabhavi TM (2022). Integrated biorefinery processes for conversion of lignocellulosic biomass to value added materials: Paving a path towards circular economy. Bioresour Technol 343. https://doi.org/10.1016/j.biortech.2021.126151

Xu S, Chen J, Peng H, Leng S, Li H, Qu W, Hu Y, Li H, Jiang S, Zhou W, Leng L (2021) Effect of biomass type and pyrolysis temperature on nitrogen in biochar , and the comparison with hydrochar. 291

Yang Y, Zhang B, Cheng J, Pu S (2015) Socio-economic impacts of algae-derived biodiesel industrial development in China: an input-output analysis. Algal Res 9:74–81. https://doi.org/10.1016/j.algal.2015.02.010

Zoghlami A, Paës G (2019) Lignocellulosic biomass: understanding recalcitrance and predicting hydrolysis. Front Chem 7. https://doi.org/10.3389/fchem.2019.00874

Zoppi G, Tito E, Bianco I, Pipitone G, Pirone R, Bensaid S (2023) Life cycle assessment of the biofuel production from lignocellulosic biomass in a hydrothermal liquefaction—aqueous phase reforming integrated biorefinery. Renew Energy 206:375–385. https://doi.org/10.1016/j.renene.2023.02.011

Chapter 20
Microbial Engineering to Design Fuel Production

Himanshi Chaudhary, SanjuktaVidyant, Pooja Sharma, and Shubha Dwivedi

Abstract The rapid increase in population growth has affected the environment in many aspects; one of the aspects is the depletion of renewable sources due to the overconsumption of resources for the population's survival. To overcome this problem research has been focused on finding a solution that is less expensive, more productive, and eco-friendly. One such solution is to use the microorganism available in abundance in the environment for the production of renewable resources. Biofuel production using microorganisms contains the same properties as petroleum-based fuels and is economically beneficial to society. Biofuel can be produced by utilizing biomass of biological waste or microalgae. In this chapter, we discuss how biofuel can be produced by exploring microorganisms through engineering methods by utilization of their genetic materials. Furthermore, we will discuss the latest techniques used for the modification of microbes and will discuss recent research and development in biofuel production.

Keywords Biofuel · Engineered microbes · Biomass · Eco friendly · Techniques

Outline of the Chapter

- Introduction on biofuel
- Role of microbes in biofuel production
- Introduction to microbial engineering
- Role of microbes in 2nd generation biofuel production
- Role of microbes in 3rd generation biofuel production
- Genetically modified microbes for the production of bioethanol
- The role of the CRISPR/Cas9 system in microbial engineering for the production of biofuels

H. Chaudhary
Department of Microbiology, IIMT University, Meerut 250001, Uttar Pradesh, India

SanjuktaVidyant · P. Sharma · S. Dwivedi (✉)
Department of Biotechnology, IIMT University, Meerut 250001, Uttar Pradesh, India
e-mail: shubhadwivedi5@gmail.com

M. Shah and D. Deka (eds.), *Emerging Sustainable Technologies for Biofuel Production*, Environmental Science and Engineering, https://doi.org/10.1007/978-3-031-52167-6_20

- Optimization of microorganisms for high yield of biofuel
- Application of functional genomics to metabolic engineering
- Future Potential
- Conclusion
- References

20.1 Introduction

Rising atmospheric greenhouse gases, like CO_2, N_2O, and CH_4, have been associated with worldwide climate and environmental changes over the past few decades (US Energy Information Administration (EIA) 2019; US Environmental Protection Agency), including coral reef elimination, rising sea levels, and weakened thermohaline circulation. In 2019 around the world, transport was responsible for around 25% of all greenhouse gas emissions (Keasling et al. 2021; US Environmental Protection Agency). Currently, the use of fossil fuels results in a worldwide7 Gigatonne of carbon emission annually in the form of CO_2 (Pacala 2004; Lewis and Nocera 2006). The atmospheric CO_2 concentration is 400 ppm. By 2050, the CO_2 emission rate is anticipated to double if the rising trend continues (Pacala 2004). Without remediation, it is anticipated that doubling CO_2 emission rates will result in a 500 ppmCO_2 concentration, which would result in a 2 °C rise in global warming (O'Neill 2002). The West Antarctic Ice Sheet (WAIS) 2 would be more likely to disintegrate at this rate of temperature rise, among other detrimental consequences. Even if it happened gradually, the WAIS' total collapse would result in an upsetting 4–6-m rise in sea level (O'Neill 2002). Even if these scenarios and the precise relationship among carbon dioxide emission, atmospheric carbon dioxide content, and effect are unclear, there is still a very real risk that they will have detrimental effects on society, finances, and the environment. If we are to keep below the 2 °C threshold, there must be an enormous decrease in carbon dioxide released from the combustion of fossil fuels within a few decades, as well as an increase in renewable energy sources and a decrease in energy consumption (Liao et al. 2016).

Around 84% of all energy and 96% of all consumed transport fuels come from fossil resources, whereas 11% comes from renewable sources and just 4% comes from transport fuels (US Energy Information Administration (EIA) 2019). Global energy utilization is expected to increase by 50% by 2050, with just 25% of the energy coming from renewable sources (Keasling et al. 2021).

Using transportation fuels produced from carbon dioxide in the environment is one way to reduce the amount of fresh carbon discharged into the atmosphere for the purpose of energy, surroundings, and economic security (Keasling et al. 2021). Liquid biofuels have a greater energy density and are more compatible with the present infrastructure when compared to other renewable energy sources, which lays the groundwork for their long-term responsibilities in the future energy mix (Liao et al. 2016).

20.1.1 Biofuel

Fuels produced by biological processes from other renewable resources, such as CO_2 and landfill gas, can be included in this description. Traditionally, biofuels are defined as fuels made from biomass by biological or non-biological processes. Nowadays, ethanol, which is made from cane sugar or corn starch (Bothast and Schlicher 2004), is the most widely used biofuel (Liao et al. 2016).

Biofuels, derived from renewable biological sources, have gained increasing attention as a promising solution to address the challenges of energy security and environmental sustainability. Biofuels can be used as a replacement for fossil fuels in various sectors, including power generation, transportation, and industrial processes, with the ability to reduce greenhouse gas emissions, mitigate climate change, and foster rural development.

The increasing demand for energy, coupled with concerns about climate change and environmental degradation, has led to the search for sustainable and renewable sources of energy. Biofuels, which are derived from organic matter, such as crops, agricultural residues, forest residues, and algae, have emerged as promising solutions to address the challenges of energy security and environmental sustainability. Biofuels offer several advantages, including their capability to reduce greenhouse gas emissions, mitigate climate change, foster rural development, and reduce dependence on fossil fuels. However, biofuel production also presents challenges, such as competition with food crops, land use changes, water usage, and potential negative impacts on biodiversity and ecosystems (Liao et al. 2016).

20.1.2 Production of Biofuel

The chemical energy of the sun is captured during the synthesis of biofuels by the bonds in biologically created molecules. Thus, photosynthesis is the starting point of all pathways to biofuels, and it is also the point at which they split (Fargione et al. 2008). In general, renewable resources can be converted to energy-rich, fuel precursors in three ways:

- The first is the direct production of biofuel through photosynthetic organisms like plants and algae (Fargione et al. 2008). The CO_2 in the atmosphere is fixed by plants to form sugars, which are subsequently fermented by yeast cells to create bioethanol. A costly feedstock deconstruction phase is not required as cane sugar or maize starch is employed as substrates in the manufacture of bioethanol (Fargione et al. 2008). It seems preferable to generate biofuels directly from carbon dioxide through photosynthesis because of its cost-effectiveness and strictly carbon–neutral. But there are many significant obstacles in the way of this strategy, chief among being scalability and land use concerns.
- The second way is to chemically convert biomass to fuels. This conversion involves various processes, such as thermochemical, biochemical, and catalytic

methods, which transform biomass into liquid or gaseous fuels. Biochemical operations involve the enzymatic or microbiological decomposition of biomass into fermentable sugars that can then be turned into biofuels like bioethanol or biodiesel. Thermochemical processes typically entail pyrolysis, gasification, and liquefaction. Catalytic processes, on the other hand, involve the use of catalysts to facilitate the conversion of biomass feedstock into fuels (Liao et al. 2016). Utilizing heat to break down the feedstock of biomass into intermediate goods, that are then further processed to produce liquid or gaseous fuels, is known as the thermochemical conversion of biomass. The three primary thermochemical conversion methods for biomass are gasification, liquefaction, and pyrolysis.

Pyrolysis is a thermochemical method that involves the thermal decomposition of biomass in anaerobic condition, resulting in the production of bio-oil, biochar, and syngas as main products. In this process the biomass is heated at high temperatures, as high as 300–800 °C, in an oxygen-limited or oxygen-free environment. During pyrolysis, biomass undergoes several complex chemical reactions, including depolymerization, vaporization, and condensation, resulting in the formation of different products depending on the process conditions.

- The third is production by heterotrophic microbes like bacteria, yeast, or fungi, which can be fermentative or non-fermentative (Liao et al. 2016).

The spectrum of raw materials utilized for future-oriented biofuels should be expanded, nevertheless, to maximize the impact of biofuels. The carbon that is ultimately employed as the substrate for the manufacture of biofuels should be carbon that is emitted as greenhouse gas emissions or a net gain in global CO_2 fixation. In addition, switching from fossil fuels to biofuels does not automatically lower CO_2 emissions unless all associated emissions are taken into account (Fargione et al. 2008). Therefore, in order to produce next-generation biofuels, it is also necessary to take into consideration the emissions brought on by the use of land for biomass cultivation and harvesting (Searchinger et al. 2009) as well as the impact of that use on biodiversity (Fargione et al. 2008) and additional manufacturing costs (Pacala 2004; Fiorese et al. 2013).

20.1.3 Biofuel Types

First generation biofuels, second generation biofuels, and third generation biofuels are the three basic categories based on the substrates used for production divided as shown in Fig. 20.1.

- First generation biofuels

First generation biofuels are those that are produced from food crops like corn, vegetable oils, wheat, and sugar cane. These biofuels include biodiesel and bioethanol, which are created with the transesterification of vegetable or animal

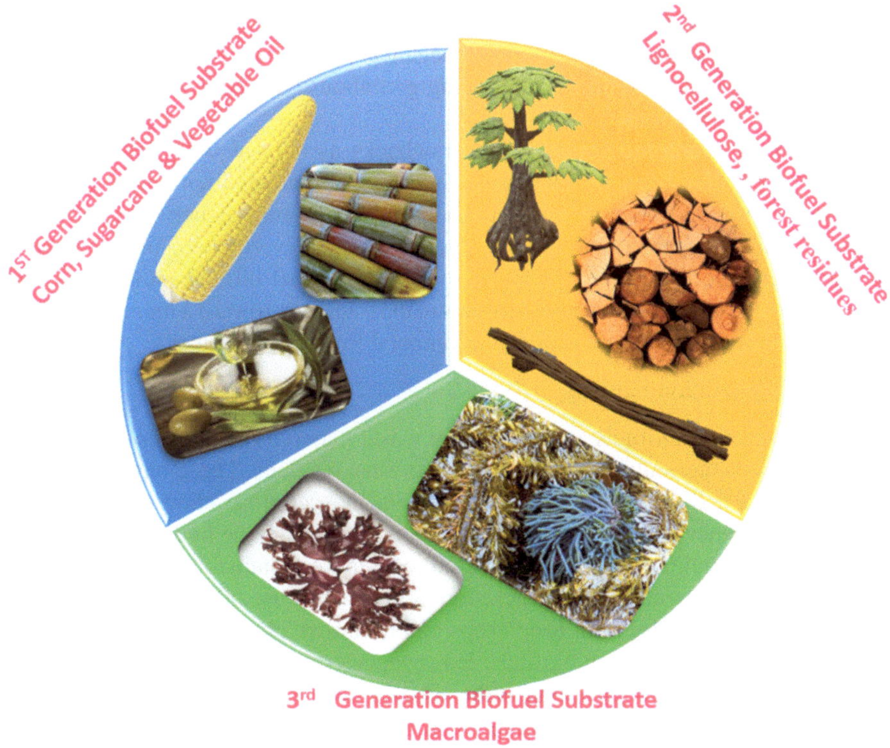

Fig. 20.1 Depicts the various types of substrates used in biofuel production

fats and carbohydrates, respectively. Examples of first-generation biofuels include bioethanol, biogas, and biodiesel.

Bioethanol is the most commonly produced biofuel from the first generation and is typically made from feedstocks such as corn, sugarcane, and wheat. The feedstock is converted into sugar, which is then fermented by microorganisms to produce ethanol. Ethanol can be blended with gasoline and used as a fuel additive or as a standalone fuel for transportation. Contrarily, the process known as transesterification uses vegetable or animal fats to react with an alcohol, usually methanol, to create biodiesel. Diesel fuel can be replaced with biodiesel in diesel engines or combined with diesel fuel in different ratios.

Another first-generation biofuel, biogas is created through the anaerobic breakdown of organic materials like animal dung, food waste, and agricultural wastes. In the absence of oxygen, anaerobic bacteria decompose organic waste, generating a combination of CO_2 and CH_4 that can be utilized as a renewable energy source for heating, power production, or as a fuel for vehicles.

First-generation biofuels have been widely used as alternatives to fossil fuels, as they are renewable, can reduce greenhouse gas emissions, and can be blended with conventional fuels without significant modifications to existing infrastructure.

However, they also face criticisms and limitations, such as competition with food crops for land and resources, potential negative impacts on food prices and food security, and limited potential for reducing greenhouse gas emissions compared to other biofuel generations.

In order to overcome some of the drawbacks associated with first-generation biofuels, there has been an increasing interest in creating improved or second-generation biofuels that are made from non-food biomass, such as agricultural residues, forestry waste, and specific energy crops. Nevertheless, first-generation biofuels continue to play a significant role in the global biofuel industry and are expected to coexist with advanced biofuels in the transition toward a more sustainable and renewable energy future (US Environmental Protection Agency).

- Second-generation biofuels

Second-generation biofuels are made from waste materials or non-food crops such as energy crops, agricultural residues, and forestry residues. These biofuels include cellulosic bioethanol, which is produced from the cellulose component of plant biomass, and renewable diesel, which is produced through the hydro-processing of plant oils or animal fats and are considered to be more sustainable and environmentally friendly compared to first-generation biofuels. The use of non-food substrates, which can lessen rivalry with food crops and minimize changes to land usage, is a benefit of second-generation biofuels. However, second-generation biofuel production also presents challenges, such as the need for advanced conversion technologies and the high cost of production (US Environmental Protection Agency). Second-generation biofuels have the major benefit of not competing with food crops for resources and land, potentially minimizing the effects on prices for food and food security. A variety of non-food biomass feedstocks, such as agricultural and forestry leftovers, which are sometimes regarded as waste or low-value commodities, can be used to make these biofuels. Furthermore, because they are often made from feedstocks that have less carbon intensity and are more plentiful than fossil fuels, second-generation biofuels offer the ability to considerably less greenhouse gas emissions when compared to fossil fuels (Naik et al. 2010).

The production of second-generation biofuels often involves more complex and advanced processes compared to first-generation biofuels. For example, cellulosic or lignocellulosic biomass, which consists of complex carbohydrates such as cellulose and hemicelluloses, requires pre-treatment and enzymatic or thermochemical conversion to convert the biomass into fermentable sugars, which can then be converted into biofuels through processes such as fermentation or gasification as shown in Fig. 20.2. Advanced biofuels can include bioethanol, biodiesel, renewable diesel, renewable jet fuel, and other hydrocarbon-based fuels that are chemically similar to their fossil fuel counterparts (Gomez et al. 2008; Zabaniotou et al. 2008).

Despite their potential benefits, second-generation biofuels face challenges such as high production costs, technological barriers, and limited commercial-scale production. The complex nature of the feedstocks and the conversion processes can pose challenges in terms of scalability, cost-effectiveness, and efficiency. However,

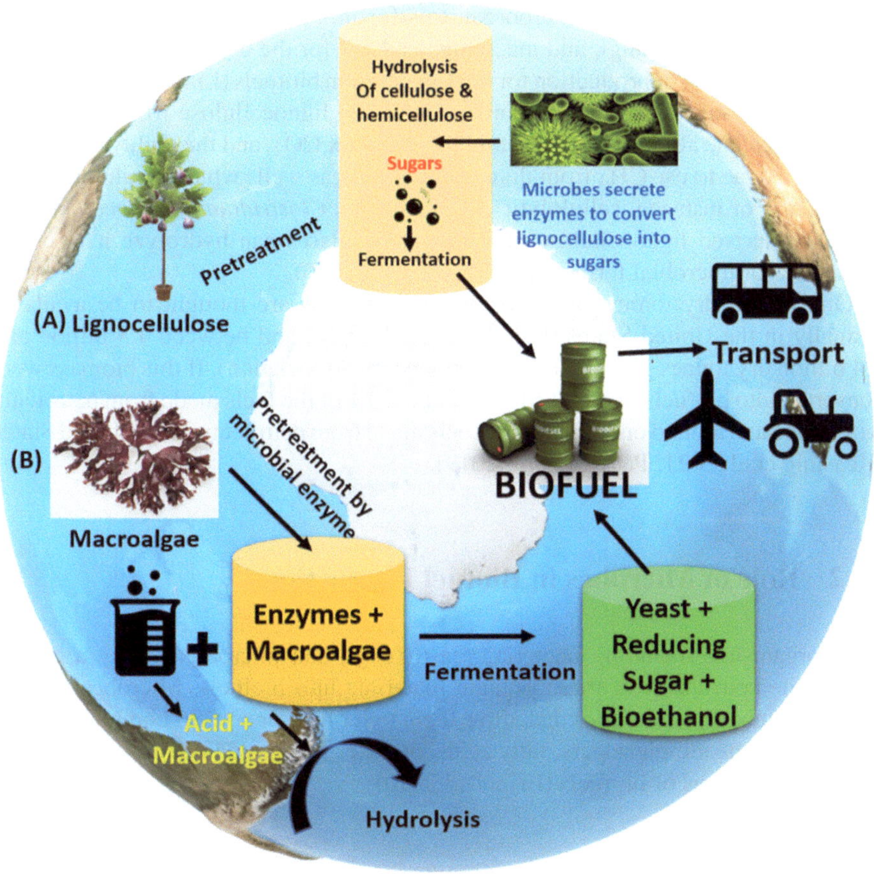

Fig. 20.2 Depicts (A) second-generation biofuel production from lignocellulose, (B) third generation biofuel production from macroalgae

ongoing research and development efforts are focused on overcoming these challenges and advancing the production of second-generation biofuels to achieve commercial viability (Naik et al. 2010).

- Third-generation biofuels

Third-generation biofuels are produced from microorganisms, such as algae or bacteria, which can convert sunlight, carbon dioxide, and nutrients into biofuels. These biofuels include algal biofuels as shown in Fig. 20.2, which are produced from microalgae, and microbial biofuels, which are produced from bacteria or yeast. Third-generation biofuels have the advantage of high productivity, as microorganisms can produce large amounts of biomass and lipid content for biofuel production (Chowdhury and Loganathan 2019).

One advantage of using microorganisms for this purpose is that the metabolic diversity of bacteria, fungi, and macroalgae allows for the use of diverse substrates as the starting point of production for next-generation biofuels (Liao et al. 2016). This includes the capacity of microorganisms to utilize lignocellulose from wastes and energy crops, the ability of bacteria and fungus to fix CO_2, and the ability of bacteria and microalgae to use CH_4 from dumps or natural gas wells which would otherwise be vented. For instance, cellulolytic organisms like *Clostridium thermocellum* may use plant-derived lignocellulose, and fungal cellulases can hydrolyze it to create substrates for microbial fermentation (Liao et al. 2016).

One billion dry tonnes of lignocellulosic biomass are thought to be available annually in the United States (Langholtz et al. 2016) and between 4.3 billion and 9.4 billion dry tonnes globally (Global Bioenergy Association). If this biomass were converted into biofuels, it might substitute 14–70% of the fuels made from petroleum worldwide and around one-third of the fuels made from petroleum in the United States (Keasling et al. 2021; Pattrick et al. 2019).

20.2 Role of Microbes in Biofuel Production

Microorganisms, including bacteria, yeast, fungi, and microalgae, play a crucial role in the conversion of these organic materials into biofuels. Microbes possess unique metabolic capabilities that allow them to ferment or convert complex organic compounds into simple sugars, fatty acids, or gases, which can be further processed into different types of biofuels (Elshahed 2010).

20.2.1 Role of Microbes in Bioethanol Production

Bioethanol, a type of biofuel, is produced by the fermentation of simple sugars derived from various feedstocks, such as corn, sugarcane, and lignocellulosic biomass. Microorganisms especially yeast, and bacteria, are widely used in bioethanol production as shown in Fig. 20.4. Yeast, such as *Saccharomyces cerevisiae*, ferments simple sugars into ethanol through the process of alcoholic fermentation. Bacteria, such as *Zymomonas mobilis* and *Escherichia coli*, can ferment both simple and complex sugars, making them suitable for bioethanol production from a wide range of feedstocks. Microorganisms also play a critical role in lignocellulosic biomass conversion, where they produce enzymes, such as cellulases and hemicellulases, which break down the complex polysaccharides into fermentable sugars as shown in Fig. 20.2 (Fargione et al. 2008). As per the statistics of production shown in Fig. 20.3 USA is the major producer of bioethanol followed by Brazil in the second position.

As per Fig. 20.3, % production of bioethanol worldwide showed that the major producer of bioethanol was the USA followed by Brazil at the second position.

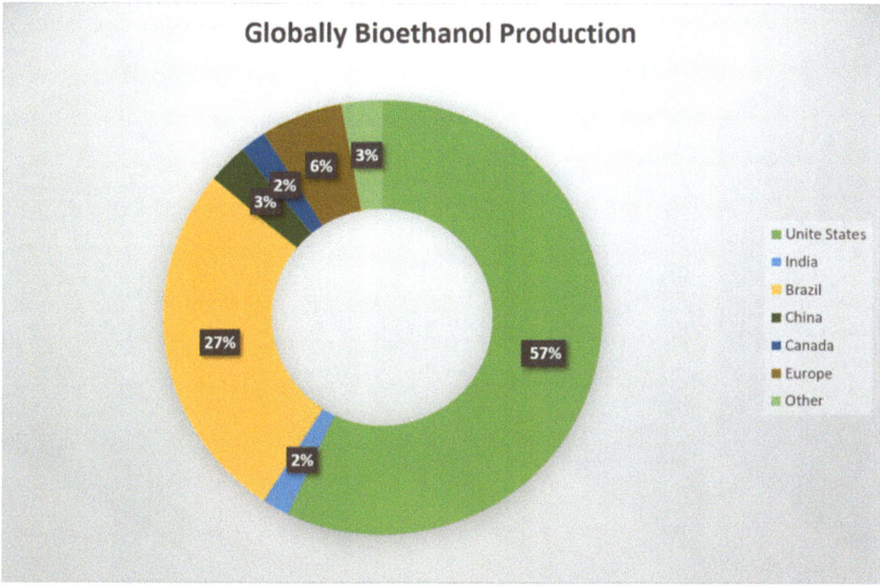

Fig. 20.3 Globally Bioethanol Production

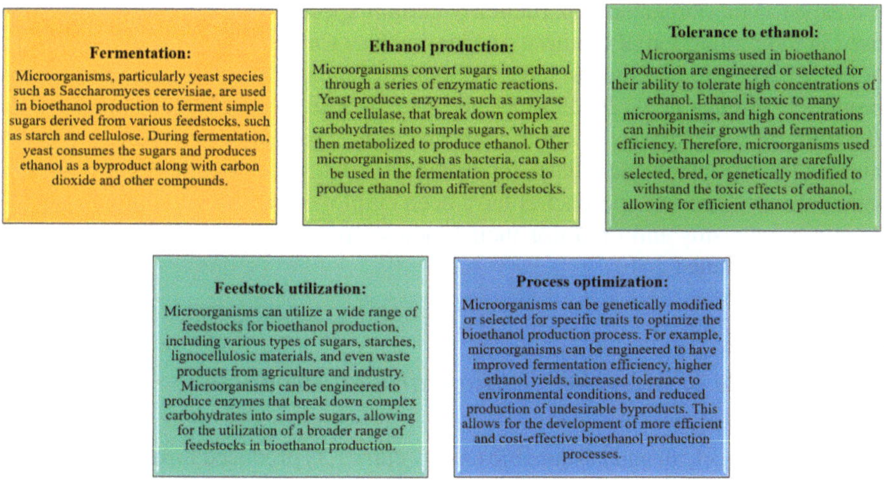

Fig. 20.4 Depicts the key roles of microbes in bioethanol production

20.2.2 Role of Microbes in Biodiesel Production

Another type of biofuel is biodiesel, which is created by transesterifying animal or vegetable lipids with alcohol (usually methanol or ethanol) in the presence of

Fig. 20.5 Depicts some specific roles of microbes in biodiesel production

a catalyst. Microorganisms, particularly certain species of algae and bacteria, are used in the production of biodiesel. Algae, such as *Chlorella, Nannochloropsis*, and *Scenedesmus*, can accumulate large amounts of lipids, which can be converted into biodiesel (Fukuda et al. 2001). Bacteria, such as *Rhodococcus, Pseudomonas*, and *Bacillus*, produce lipases, enzymes that can catalyze the transesterification reaction, making them potential candidates for biodiesel production. Genetic engineering and metabolic engineering approaches have been employed to enhance lipid production in microorganisms and optimize their biodiesel production potential as shown in Fig. 20.5 (Elshahed 2010).

20.2.3 Role of Microbes in Biohydrogen Production

Biohydrogen, a clean and sustainable energy carrier, is produced through the biological process of fermentation or photosynthesis by microorganisms. Various types of bacteria, such as *Clostridium, Escherichia*, and *Rhodobacter*, are known to produce hydrogen through the fermentation of organic compounds. Microalgae, such as *Chlamydomonas* and *Scenedesmus*, can produce hydrogen through the process of photosynthesis, where hydrogen is evolved as a byproduct of their metabolic pathway

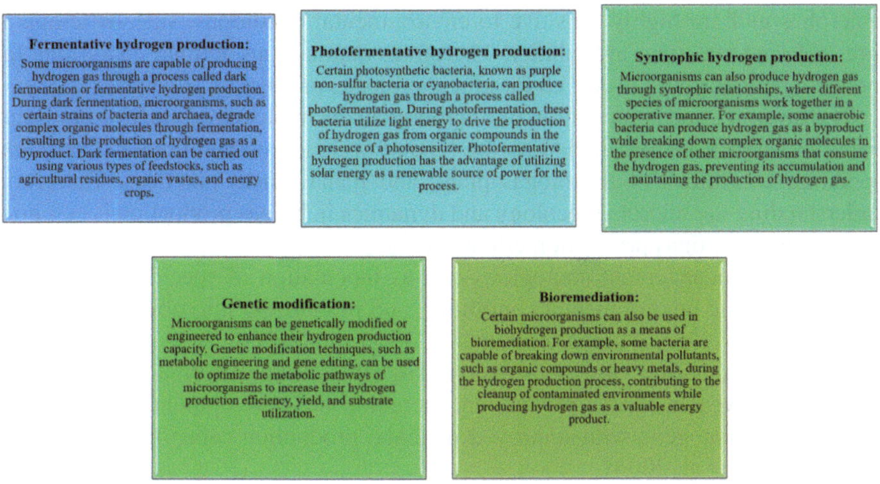

Fig. 20.6 Depicts some specific roles of microbes in biohydrogen production

(Prince and Kheshgi 2005). Genetic engineering and metabolic engineering techniques have been used to enhance the hydrogen production efficiency of microorganisms by manipulating their metabolic pathways and optimizing their growth conditions as shown in Fig. 20.6 (Vignais and Billoud 2007).

20.2.4 Role of Microbes in Biogas Production

The anaerobic breakdown of organic materials, such as crop residue, food scraps, and sewage waste, by microorganisms results in the production of biogas, a renewable energy source. Microbes play a central role in biogas production by breaking down complex organic compounds into simpler compounds, which are then fermented into biogas, primarily consisting of methane (CH_4) and carbon dioxide (CO_2). The key microorganisms involved in biogas production are bacteria, specifically anaerobic bacteria, which thrive in oxygen-free environments (Youssef et al. 2006).

Acidogenic bacteria are the first group of bacteria involved in biogas production. Through a process known as acidogenesis, they convert complex organic substances like proteins, lipids, and carbohydrates into simple organic acids like propionic acid, acetic acid, and butyric acid. These organic acids are then further fermented by acetogenic bacteria into acetic acid, hydrogen, and carbon dioxide in a process called acetogenesis (Singh and Panigrahi 2000).

Finally, methanogenic bacteria convert the acetic acid, hydrogen, and carbon dioxide produced by acidogenic and acetogenic bacteria into methane (CH_4) and carbon dioxide (CO_2) through the process of methanogenesis. Methanogens are strict anaerobes and are highly sensitive to environmental conditions, such as temperature, pH, and substrate availability. Therefore, maintaining optimal conditions for methanogenic bacteria is crucial for efficient biogas production (Elshahed 2010).

The diversity and composition of microorganisms in biogas production are influenced by various factors, including the type of organic material used as feedstock, operating conditions of the biogas plant, and management practices. Therefore, understanding the microbial ecology and dynamics in biogas production is essential for optimizing biogas production efficiency.

Microbial management strategies, such as inoculation of specific microorganisms, optimizing environmental conditions, and monitoring microbial activity, can be employed to enhance biogas production. Genetic engineering and metabolic engineering approaches have also been explored to enhance the performance of microorganisms involved in biogas production, including the development of genetically modified methanogens with improved methane production capabilities (Elshahed 2010; Youssef et al. 2006).

In addition to methane, biogas also contains small amounts of other gases, such as carbon dioxide (CO_2), hydrogen (H_2), and traces of other gases, which can be utilized as energy sources or further processed to obtain desired biogas compositions for specific applications, such as transportation fuel or electricity generation (Singh and Panigrahi 2000).

20.3 Introduction to Microbial Engineering

Microbial engineering is a rapidly advancing field that involves manipulating the genetic makeup of microorganisms to achieve specific outcomes, such as designing microorganisms for biofuel production. Microbial engineering for fuel production is primarily focused on two approaches: bioethanol and biodiesel. In bioethanol production, microorganisms such as yeast or bacteria are engineered to convert sugar into ethanol through a process known as fermentation. In biodiesel production, microorganisms such as algae are engineered to produce lipids that can be converted into biodiesel (Gong et al. 2017).

20.3.1 Benefits of Microbial Engineering for Fuel Production

One of the primary benefits of microbial engineering for fuel production is that it offers a sustainable alternative to fossil fuels as shown in Fig. 20.7. Unlike fossil fuels, which are finite resources and contribute to greenhouse gas emissions, biofuels can be produced from renewable sources such as crops and waste materials. Additionally, microbial engineering offers the potential for increased efficiency and cost-effectiveness compared to traditional fuel production methods.

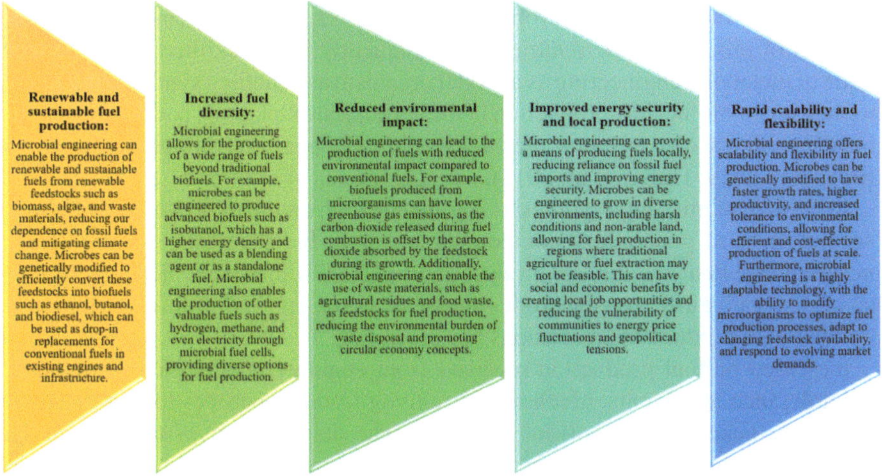

Fig. 20.7 Depicts benefits of microbial engineering for fuel production

20.3.2 Challenges of Microbial Engineering for Fuel Production

One of the primary challenges of microbial engineering for fuel production is the need for high yields and productivity. This requires a deep understanding of the metabolic pathways involved in fuel production and the ability to optimize these pathways for maximum efficiency. Additionally, there are concerns about the potential environmental impact of large-scale biofuel production, including land use changes and the release of harmful byproducts (Srivastava 2019).

20.3.3 Current Research in Microbial Engineering

Current research in microbial engineering for fuel production is focused on improving yields and productivity, as well as developing new microorganisms that can produce a wider range of fuels. For example, researchers are exploring the potential of using bacteria to produce butanol, a fuel with a higher energy density than ethanol. Additionally, there is growing interest in using microorganisms to produce biofuels from waste materials such as agricultural and forestry residues.

The goal of the present study has been to determine the financial advantages of various methods or approaches for utilizing the cheap and plentiful nature of substrate for the synthesis of biofuels (Stephanopoulos 2007). A successful approach for biomass pretreatment operations with adequate selection and the best ability of microbes that can produce fuel has been developed in response to the challenges posed by lignocellulose from maize. Some inhibitors are produced during the pretreatment of biomass, which inhibits microbial growth and results in the generation of small amounts of product. *Escherichia coli* has frequently been modified using the technique of metabolic engineering as a model organism as a biofuel-synthesizing microbial strain, for biofuel bio-production in the appropriate metabolic pathways with flux management that can overcome the numerous challenges during the adoption process (Srivastava 2019; Huffer et al. 2012).

In the area of microbial synthesis showing the high nature of production of biofuel (second or third generation) compounds and bulk biochemicals, the problem of preserving the final metabolism product or product with an exaggerated possibility of having a negative effect on the production of the host microbial strain and a limitation of metabolite product concentrations has emerged.

Increased biofuel production received from engineered microbial strains has a high tolerance to solvents. Mechanisms ranging from chaperones to transcriptional factors are necessary for solvent tolerance. The development of transporters has contributed significantly to the emergence of a potent class of proteins, improving biofuel production and solvent tolerance. Engineering in transporters protein of the strain has increased bacterial solvent tolerance and biofuel production by producing membrane-localized transporters (Srivastava 2019). *Escherichia coli* has improved the production of liberated fatty acids (titer 4.82 g L and yield 29.55%) by up-regulating the production of the enzymes transhydrogenase and NAD+ kinase, which employ glycerol as a substrate or as a precursor. A metabolically designed free fatty acid (FFA) synthesis pathway using acyl-acyl carrier protein thioesterase to employ glycerol as feedstock has been developed using cofactor engineering. This proposed stain has produced the maximum amount (83%) of the projected value (g FFAs/g glycerol) in the route using crude-grade glycerol molecule (as feedstock) acquired from a biodiesel facility. For the production of FFA from crude glycerol, a titer (3.5 g L) and production value (24.13%), which is the pathway's highest potential yield (68%) are reported. And the same microbial strain's contribution to biofuel yield and titer was influenced by the provenance of the substrates (Mukhopadhyay 2015).

Overall, microbial engineering for fuel production offers a promising avenue for sustainable energy production. While there are challenges to overcome, ongoing research in this field is advancing our understanding of microbial metabolism and offering new possibilities for renewable energy production.

20.4 Role of Microbes in 2nd Generation Biofuel Production

The rapid decline of petroleum reserves and the potential for climate change are expected to cause a major increase in the need for alternative biofuels in the coming years. Lignocellulose biomass has come out to be an efficient technique for bioethanol production and other valuable products. Due to their production from affordable and plentiful lignocellulosic feedstocks, second-generation biofuels have received a lot of attention in recent decades. Significant renewable substrates for the production of bioethanol can be found in lignocellulosic biomass, which is derived from agricultural wastes like hay, switch grass, clover, and miscanthus grass, and backwoods items like hardwoods and softwoods (Viikari et al. 2012). According to Prasad et al. (2007), the majority of plant cell walls include roughly 15–40% cellulose, 10–30% hemicellulose and pectin, and 5–20% lignin. According to Van Dyk and Pletschke (2012), the complex chemical makeup of lignocellulosic biomass—which includes lignin, hemicellulose, and acetyl groups as well as its physical characteristics of cellulose crystallinity and degree of polymerization is associated to recalcitrance. Anhydroglucopyranose molecules coupled by β-1,·4-glycosidic connections make up the main structure of cellulose. The complex of powerful intra- and intermolecular hydrogen bonding observed in cellulose leads to the formation of microfibrillar chains, which significantly contribute to the resistance of biomass and make its hydrolysis an expensive operation (Himmel et al. 2007).

Heterogeneous D-xylose, D-mannose, and other substituted sugars make up the majority of the branched and linear polysaccharides that make up hemicellulose (Sánchez 2009). It is possible to hydrolyze both cellulose and hemicellulose into simple sugars, which may then be fermented to produce ethanol (Bhattacharya et al. 2015).

20.4.1 Enzymes for Lignocellulose Decomposition

Microorganisms, such as fungi and bacteria produced variety of extracellular enzymes which may break down cellulose and hemicellulose and use it as a source of carbon. However, to break down every part of lignocellulose, a variety of enzymes are needed. A group of lignocellulolytic enzymes collaborates in nature to digest lignocellulosic biomass.

20.4.2 Cellulases

Exo-1, 4-β glucanases, endo-1,4-β glucanases, cellobiohydrolase, and β glucosidases are at least three enzymes that work together to break down cellulose into glucose.

A cellulose-binding domain (CBD), makes it easier for the enzyme to bind to its substrate, a linker (hinge) area, which connects the two domains, and a lot of serine, proline residues, and threonine make up a cellulase's typical structure.

20.4.3 Fungal and Bacterial Cellulases

The primary microbes responsible for the breakdown of cellulose are fungi, which are further broken down into two subgroups. The first subgroup consists of the white rot and soft rot fungi, such as *Phanerochaete chrysosporium.* A combination of enzymes released by a fungus called *T. reseei* (*Hypocrea jecorina*), which contains 200 GH genes in total, is the primary ingredient in most commercially available cellulase mixtures (Bhattacharya et al. 2015; Kubicek et al. 2009; Martinez et al. 2008). Numerous fungi have been shown to produce two CBHs (Cel6A and Cel7A), for instance, *T. reesei.* The second group consists of fungi that cause brown rot, like *Postia placenta.*

Numerous aerobic bacteria secrete varying levels of free cellulases, which work together to break down cellulose. They contain Carbohydrate Binding Module (CBM), linker peptides, and CDs (Catalytic Domains) that degrade cellulose without the use of processing. Numerous genera of aerobic bacteria are listed in Table 20.1. Different types of cellulases are known to be produced by *Cellulomonas, Erwinia, Pseudomonas, Paenibacillus, Streptomyces, Fibrobacter, Geobacillus,* as well as *Bacillus* (Bhattacharya et al. 2015).

Different types of cellulolytic systems, known as cellulosomes, have emerged in anaerobic bacteria (Doi and Kosugi 2004). These systems feature intricate protein structures that support enzymes that aid in the hydrolysis of cellulose. Many cellulose-digesting enzymes and hemicellulolytic enzymes are arranged on the structural scaffoldin subunits by these bacteria, including *Clostridium, Acetivibrio, Bacteroides,* and *Ruminococcus,* with non-covalent proteins interactions between dockerin and cohesins (Bhattacharya et al. 2015). Scaffoldin, a fibrillar protein, makes up cellulosomes. The CbpA, Cip A, or Cip Cproteins make up the scaffolding proteins, often known as "scaffoldins" (Doi and Kosugi 2004). A small number of anaerobic cellulose-degrading thermophilic bacteria, like *Caldicellulosiruptor* sp., employ a middle method in which they release many free cellulases with various catalytic domains. Two GH 9 and one GH 48 catalytic domain type III cellulose-binding modules make up the *Caldicellulosiruptor bescii* CelA. This CelA speeds up cellulose hydrolysis in addition to the renowned surface removal process by burrowing through the substrate's surface and producing sizable voids (Bhattacharya et al. 2015; Brunecky et al. 2013).

Anaerobic cellulolytic rumen bacteria *Fibrobacter succinogenes* and aerobic soil bacteria *Cytophaga hutchinsonii* have both been found to utilize various forms of cellulolytic breakdown mechanisms. These bacteria's genomic sequences strongly imply that they do not utilize cellulosomes or free cellulases since none of their genes produce scaffoldins, dockerins, or CBMs. Therefore, it is thought that these organisms have acquired a new cellulolytic process to breakdown cellulose fibrils by using a complex on the outer membrane and transporting cellulose molecules to periplasmic regions through super-channels, where they are subsequently degraded by the activity of endoglucanase present in the periplasm (Bhattacharya et al. 2015).

Table 20.1 Depicts engineered microbes for biofuel production

Microbes	Substrate	Substrate concentration (g/L)	Genotype	Ethanol yield (mM/g)	References
Clostridium thermocellum	Cellobiose	5.0	Orotidine-5-phosphate decarboxylase, phosphotransacetylase	0.59	Scully and Orlygsson (2017)
Thermoanaerobacter TD1	Xylose	5.0	Lactate dehydrogenase	0.98	Scully and Orlygsson (2017)
Thermoanaerobacter ethanolicus	Wood HL	8.0	Alcohol dehydrogenases A, B and E	3.30–4.50	Scully and Orlygsson (2014)
Thermoanaerobacter mathranii BG1L1	Glucose, xylose, galactose and mannose	9.0	Lactate dehydrogenase	0.39–0.42	Taylor et al. (2009), Georgieva et al. (2007)
Clostridium thermocellum	Avicel	19.5	Phosphotransacetylase, lactate dehydrogenase	1.26	Scully and Orlygsson (2017)
G. thermoglucosidasius TM242	D-xylose and D-glucose	4.7	Lactate dehydrogenase	0.44	Taylor et al. (2009), Cripps et al. (2009)
Thermoanaerobacter mathranii BG1G1	Wheat straw	30–120	Lactate dehydrogenase	1.53–1.67	Scully and Orlygsson (2017)
Thermoanaerobacter Pentocrobe 411	Wheat straw	65x	Acetate kinase, phosphotransacetylase	1.84 sug eq	Scully and Orlygsson (2017)
Geobacillus thermoglucosidasius	Glucose	34	Lactate dehydrogenase	0.39	Chang and Yao (2011)
Thermoanaerobacter saccharolyticum M1051	Cellobiose	27.5	Lactate dehydrogenase, acetate kinase, phosphotransacetylase	0.39	Chang and Yao (2011)

20.4.4 Hemicellulases

Between 20 and 35% of lignocellulosic biomass is made up of hemicellulose, specifically xylan. To increase the total yield of bioethanol, the hemicellulosic fraction must be converted. Due to its more complex structure than cellulose, hemicellulose necessitates a wide range of enzymes for efficient hydrolysis (Dyk and Pletschke 2012). Firstly, a group of de-polymerizing enzymes referred to as core enzymes that cleave the hemicellulose backbone; and secondly, a class of de-branching enzymes that eliminate alternatives that could cause steric hindrances to de-polymerizing enzymes and then rise the overall production of reducing sugars produced during hydrolysis.

20.4.5 Fungal and Bacterial Hemicellulases

Hemicellulose may be broken down into mono- or disaccharides by a group of hemicellulolytic enzymes that are produced by fungi. *A. niger's* fungal hemicellulases are the most researched which contain one acetyl xylan esterase, two feruloyl esterases, one β-xylosidase, one β-mannosidase, two α-galactosidases, one β-galactosidases, one α-glucuronidases, and two endo-xylanases (Brink and Vries 2011). Similar to this, *T. reesei,* an aerobic fungus, has been found to release a variety of hemicellulases (Bhattacharya et al. 2015). The filamentous fungus *Aspergillus niger* and *Penicillium wortmanii* have been shown to have many β-xylosidases (Juturu and Wu 2012). Similar to this, the filamentous fungus *Cochliobacillus carbonum's* β xylosidase adheres to the inversion of the anomeric center mechanism. Both Gram-positive and Gram-negative types of bacteria have been observed to produce bacterial hemicellulases. Most bacterial xylanases fall within the GH family 10 and GH family 11 categories. Functional characterization of one β xylosidase and two endo-xylanases from *T. maritima* has been done (Zhengqiang et al. 2001). Acetyl xylan esterase and feruloyl esterase modules of multifunctional xylanases, respectively, have been identified in *Clostridium thermocellum* (Prates et al. 2001) and *Clostridium cellulovorans* (Kosugi et al. 2002), respectively. Rakotoarivonina et al. (2012) disclosed *Thermobacillus xylanolytic* hemicellulolytic arsenal, which includes the enzymes xylanase, arabinofuranosidase, feruloyl esterase, acetyl esterase, and xylosidase. The bacteria *Bacillus* sp., *Caldicellulosiruptor* Rt8b, *Streptomyces* sp., *Caldibacillus cellulovorans,* and *Caldocellum saccharolyticum* may all generate mannanases (Bhattacharya et al. 2015; Zhang et al. 2006).

20.4.6 Bacterial and Fungi Lignocellulolytic Enzymes in Synergy

According to Hu et al. (2011), the majority of the commercial lignocellulolytic combinations now in use are based on fungal cellulases and hemicellulases. The topic of whether it is possible to create novel cocktails based on bacterial and fungal glycosyl hydrolases or what more can be done to enhance fungal enzymes is raised.

Since bacterial strains can survive in extreme conditions and in unique habitats, we can add bacterial cellulases and hemicellulases from a tiny gene pool that we have access to, to commercialized cellulase and hemicellulase mixes which are currently dominated by enzymes from fungi, to give them better properties. Using a hybrid blend of bacterial and fungal hemicellulases, Gao et al. (2010) reported that they were able to convert 95% of the xylan and 65% of the glucan in maize stover that had been treated with AFEX (Ammonia Fibre Expansion). Unlike commercial enzymes, a number of auxiliary hemicellulases from various bacterial sources and essential fungal cellulases worked together synergistically to produce significant outputs of glucose (80%) and xylose (70%) with only mild enzyme loading (*20 mg protein/g glucan). Due to the complicated structure of raw substrates, the wide range of pre-treatment techniques, and the financial constraints associated with commercialization, novel methodologies must be employed to supplement and improve the already available commercial cellulase-hemicellulase cocktails. This quest for improved saccharification and fermentation of lignocellulosic biomass must constantly evolve (Bhattacharya et al. 2015).

20.4.7 Amendment of the Cellulosomal Bacteria's Saccharolytic Technique

Cellulosomes are large extracellular complexes with a high molecular weight that are released by anaerobic microbes and include cellulosomal enzymes and scaffoldins that may break down plant cell walls (Doi and Kosugi 2004).

It is possible to redesign the cellulosomes of cellulolytic bacteria to increase their effectiveness for consolidated bioprocessing. Clostridium thermocellum has been suggested as an appropriate candidate for consolidated bioprocessing due to its anaerobic nature and resistance to high temperatures. The main soluble disaccharide end product cellobiose is produced in high quantities as a result of the synergistic destruction of the cellulosome of *C. thermocellum* that is made possible by the presence of the various enzymes. As with most bacterial and fungal cellulase systems, cellobiose is the main byproduct, significantly inhibits *Clostridium thermocellum's* multi-enzyme cellulosome system.

Cellobiose-induced suppression of *Clostridium thermocellum* cellulosomal cellulases and the toxicity of end products like lactic acid and acetic acid, which can impair cell development and metabolism, can be overcome by using β-glucosidases from *Aspergillus niger* or redesigning the cellulosome (Bhattacharya et al. 2015). The term "recombinant cellulolytic technique" refers to the process of modification of non-cellulolytic organisms with high product yields to induce the expression of a heterologous cellulase system that degrades cellulose. For this, a variety of well-reviewed bacterial and yeast strains, including *Escherichia coli, Pichia pastoris, Klebsiella oxytoca, Saccharomyces cerevisiae, Zymomonas abilis*, and *Pichia stipis* have been used (Elkins et al. 2010). Our attention is on synergy models including fungi and bacteria. The expression of the genes cel5 E endoglucanase and bgl3A β-glucosidasefrom *Saccharomycopsis fibuligera* and *Clostridium thermocellum*, respectively, in a strain of *Saccharomyces cerevisiae*, resulted in increased endoglucanase activity and improved PASC (phosphoric acid swollen cellulose) to ethanol conversion (Jeon et al. 2009).

The secret to a greener future is the efficient de-polymerization of lignocellulosic biomass. Innovative methods must be sought in order to get through biomass resistance. Due to the intricacy of the substrates and the popularity of numerous pretreatment approaches, the commercialization of lignocellulolytic enzymes is possible if effective sugar release is achieved with extremely low protein loadings and high fermentative yields. Therefore, equal weight must be paid to improving both saccharification and fermentation strategically. Environment research, genome evaluation, data mining, and proteomic investigation of secretomes (fungal and bacterial) may aid in the identification of new enzymes and the comprehension of enzyme interaction and synergism. It is crucial for us to comprehend that lignocellulose hydrolysis works best in natural circumstances and that our objective should be to mimic such circumstances in order to maximize our chances of producing biofuels at an affordable cost (Bhattacharya et al. 2015).

20.5 Role of Microbes in 3rd Generation Biofuel Production

The first-generation biofuel manufacturing processes have significant financial and environmental limitations due to their requirement for large land areas. They also place higher demands on land for agriculture needed for food production to feed the growing world population, which is expected to reach 7.2 billion in 2050. The main limitations of the second generation of production of biofuel are the necessity for a sizable quantity of agricultural land for cultivation and the woody components of crops that do not compete with food production (Schenk et al. 2008). The drawbacks of first- and second-generation biofuels can be solved by considering microalgae as a potential alternative energy source since they can generate significant amounts of biomass and then biofuels in much smaller areas (Maity et al. 2014; Demirbas 2009).

Seaweed, *cyanobacteria*, and marine resources are intriguing possibilities for third-generation biofuel because they can offer higher results with less resource input. Macroalgae might represent the most potential non-consumable biofuel source since they may develop exponentially in saline water, unfavorable climates, and other circumstances. The benefits of macroalgae over terrestrial biomass include not needing freshwater, arable land, agricultural fertilizer, or pesticides. Due to its low or absent lignin content and high productivity, algal biomass is also very simple to extract carbohydrates from and has little rivalry with agricultural feed and food crops (Kawai and Murata 2016).

The algae biofuel is totally safe, biodegradable, and devoid of sulfur (Fiorese et al. 2013). Each category contains varying amounts of ash (18–55%), carbs (25–60%), proteins (5–47%), and lipids (5%). These amounts vary among species and are strongly impacted by both abiotic and biotic environment growth factors, including temperature and light (Godvin Sharmila et al. 2021).

Brown macroalgae have various characteristics that make them suitable feedstock for the synthesis of renewable commercial chemicals and biofuels. Alginate, mannitol, and glucan are the sugars in brown macroalgae that are most prevalent. When mannitol and glucan are mixed, approximately 0.08–0.12 wt ethanol/wt dried macroalgae is created (Wargacki et al. 2012).

When mannitol is broken down compared to glucose, more reducing equivalents are produced, which leads to an unbalanced redox environment. As a result, only the existence of electron shunts makes the generation of ethanol from mannitol possible. *Zymobacter Palmae* was able to produce 0.38 wt/wt of ethanol from mannitol under semi-fermentative conditions (Horn et al. 2000). Additionally, this strain produced 1.6% v/v of these sugar portions in *Saccharina hyperborea* in microaerobic conditions by co-metabolizing mannitol and glucan (Horn et al. 2000).

Alginate fermentation to ethanol uses two reducing equivalents per mole unit as opposed to one for mannitol. The catabolic pathway of alginate functions as both an additional source of sugars and a counterbalance to the excess-reducing equivalents produced by mannitol catabolism in order to permit simultaneous ethanol fermentation employing all three sugar components in macroalgae. Alginate is a linear block copolymer made up of the uronic acids β-D-mannuronate (M) and α-L-guluronate (G). It has different sequences of uniform sections of either M (poly-M) or G (poly-G) (Wargacki et al. 2012).

Alginate is broken down into oligomers as shown in Fig. 20.8 by a process known as endolytic β-elimination, which is catalyzed by a set of enzymes known as alginate lyases (*Alys*) found in microorganisms (Wargacki et al. 2012; Wong et al. 2000). The exolytic enzyme oligo alginate lyase (Oal) further degrades oligomers into unsaturated monomers. These monomers spontaneously rearrange to form 4-deoxy-L-erythro-5-hexoseulose uronic acid (DEHU) (Wargacki et al. 2012). DEH reductase (DehR) then transforms DEH into 2-keto-3-deoxygluconate (KDG), a common metabolite that enters the Entner-Doudoroff (ED) pathway. Pyruvate and glyceraldehyde-3-phosphate are eventually produced by KDG through the operations of KDG kinase and KDG-6-phosphate aldolase (Wargacki et al. 2012; Takase et al. 2010). In the alginate-metabolizing bacterium *Sphingomonas* sp. A1, a homoethanol route was engineered, according to recent research (Takeda et al. 2011).

Because *E. coli* naturally has the capacity to metabolize mannitol and glucose, Wargacki et al. (2012) decided to use it as a prototype. A wide range of native and

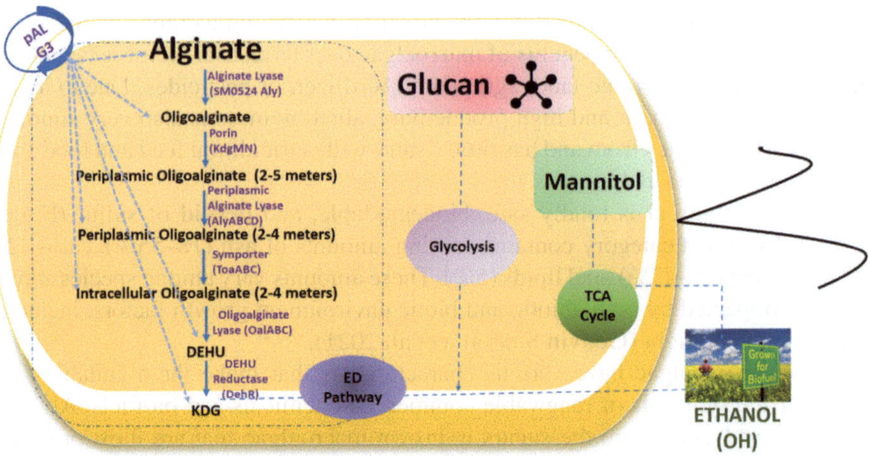

Fig. 20.8 Depicts the creation of the microbial system for biofuel production. An alginate lyase first breaks down the alginate polymer into oligomers. The oligomers are subsequently broken into smaller monomer units (DEH) by oligo alginate lyases (OalABC). DEH reductase (DehR) transforms DEH into KDG, which joins the ED route

heterologous metabolites have been produced in *Escherichia coli* as hosts, including, such as alcohols (Atsumi et al. 2008), plant natural products (Wargacki et al. 2012; Ajikumar et al. 2010), alkanes (Schirmer et al. 2010), fatty acid ethyl esters (Wargacki et al. 2012), polyketides (Atsumi et al. 2008), 1,3-propanediol (Wargacki et al. 2012), and 1,4-butanediol (Yim et al. 2011).

20.5.1 Engineered Secretable Aly *Enzyme for* E. coli *Alginate Degradation*

In order to facilitate the quick and effective breakdown of alginate by the microbial platform, Wargacki et al. (2012) initially designed a secretable *Aly* system. Consolidated bioprocessing (CBP), which enables the immediate production of biofuels and renewable commodity chemical compounds from macroalgae without thermal, chemical, or enzymatic pretreatment or saccharification prior to fermentation, was the aim of combining this designed system with the alginate acceptance and metabolic pathway. Since CBP combines the production of enzymes, feedstock metabolism and breakdown, and product formation in a single process step, it eliminates the requirement for feedstock-degrading enzyme production and purifying costs. This is why CBP is commonly recommended as a cost-effective approach (Stephanopoulos 2007; Lynd et al. 2005). The *Pseudoalteromonas* sp. SM0524 *Aly* (SM0524 *Aly*) is a tiny bifunctional enzyme (32 kD) that primarily generates dimers, trimers, and tetramers from alginate substrates, making it an appealing candidate (Wargacki et al.

2012). Additionally, SM0524 *Aly* functions in the extracellular milieu because its natural producer secretes it. As a result, we selected this enzyme as the model *Aly*. The mechanism that must be developed for the secretion of SM0524 *Aly* was then found by Wargacki et al. (2012). They discovered antigen 43 (Ag43) to be a superb contender that could be able to meet all of these requirements. The majority of *E. coli* strains naturally produce this autotransporter protein, which possesses all the necessary information for localization to the membrane and extracellular secretion (Klemm et al. 2003).

Wargacki et al. (2012) assessed the effectiveness of the *Aly* secretion technique by introducing materials that were appealing to an alginate solution and seeing them destroy alginate. They achieved this by expressing this planned *Aly* secretion creation (N455 + tSM0524 Aly) in an ATCC 8739 *E. coli* strain using a single-copy plasmid vector. The cell culture medium's 48-h time course revealed *Aly* activity to be as high as 1000 U/liter. The initial ten hours of cell development saw 40% of the *Aly* activity, showing that tSM0524 *Aly* may be quickly released for efficient substrate breakdown in the early phases of fermentation (Wargacki et al. 2012).

20.5.2 The Use of Genetically Modified E. coli *for Ethanol Production from Macroalgae*

The effectiveness of this platform for making ethanol out of macroalgae was assessed by Wargacki et al. (2012) after the alginate metabolism cascade genes from pALG3 were inserted into the *E. coli* genome. They added a recombinant homoethanol pathway made up of alcohol dehydrogenase B (AdhB) and pyruvate decarboxylase (PDC) from *Zymomonas mobilis* in order to develop an effective ethanol-production phenotype (Wargacki et al. 2012). By removing pflB-focA, frdABCD, and ldhA from the host strain's genome, they were able to divert carbon flux away from fermentative by-products. Wargacki et al. (2012) were the ones to confirm that the resulting strain (BAL1366) was able to develop on alginate with no enzymatic saccharification step and secrete alginate lyase. Then, they introduced this strain's chromosome with the genetically altered secretable alginate lyase system, N455 + tSM0524 *Aly*. The resultant strain, designated BAL1611, was tested for its ability to ferment a 5% sugar solution made up of alginate, mannitol, and glucose in a ratio of 5:8:1, simulating the sugar ratio seen in brown macroalgae. They experimented with BAL1611 at 25, 30, and 37 °C to examine the modified strain's fermentation ability at various temperatures. Wargacki et al. (2012) discovered that keeping the working temperature between 25 and 30 °C allowed for the achievement of higher ethanol titers (up to about 20 g/liter or 2.4% v/v). Thus, the following laboratory-scale fermentation experiments were conducted within this temperature range. As a representative fermentation substrate, they selected the widely accessible and typical brown macroalga *Saccharina japonica* (kombu). A final titer of 4.7% v/v, which is equivalent to the previously reported standard ethanol titer from lignocellulosic biomass fermentation using *Saccharomyces cerevisiae*, was used to show the synthesis of ethanol from macroalgae using our microbial platform (Lau and Dale 2009). The bioconversion value for this amount of ethanol is 0.28 wt/wt of biomass, and the yield is 0.41 wt/wt of the entire sugar content (alginate, glucan, and mannitol). Over 80% of the theoretical potential yield was obtained by fermentation. 83% of this titer was likewise attained within the first 48 h of fermentation, corresponding to an overall yield of 0.64 g/liter/hour and a sugar utilization ratio of around 2.5 mol mannitol/mol alginate. A final concentration of about 2 g/liter, or a rather tiny amount, of the typical fermentation by-product lactate was identified. The methyl glyoxylate pathway may play a role in this lactate production (Yomano et al. 2009). The amounts of acetate, succinate, and formate were incredibly low and were not quantifiable. These findings demonstrate an effective carbon flow from pyruvate. Our designed microbe breaks down alginate, mannitol, and glucose to produce the ubiquitous metabolic intermediate pyruvate, which may be used to make a wide variety of fuels and common chemical compounds. Thus, our CBP broad initiative to realize the generation of renewable fuels, as well as commodity chemical substances from sustainable biomass resources, would benefit from the development of a microbial platform that is capable of converting marine biomass. This platform would complement microbial platforms currently being researched for lignocellulosic feedstocks and would be mutually advantageous with them (Wargacki et al. 2012).

20.6 Genetically Modified Microbes for the Production of Bioethanol

Many microbes have been modified to specifically produce ethanol over the past 20 years. Complex carbohydrates found in lignocellulosic biomass require the use of microbes to ferment sugars that brewer's yeast cannot. Xylose is the most important carbohydrate among them. Bacteria like *Klebsiella oxytoca, Zymomonas mobilis* and *Escherichia coli*, have seen the highest engineering success. Research has concentrated on designing *Escherichia coli* and *Klebsiella oxytoca* to preferentially make ethanol since these strains can naturally use a range of carbohydrates (Dien et al. 2003).

20.6.1 Modifying Escherichia coli *to Produce Ethanol*

The development of *Escherichia coli* strains that generated ethanol only when necessary was one of the first effective metabolic engineering applications (Dien et al. 2003). *Escherichia coli* offers a number of benefits as a biocatalyst for the synthesis of ethanol, including the capacity to ferment a broad range of sugars, no need for complicated growth agents, and previous commercial application (Dien et al. 2003).

In the process of fermenting carbohydrates, *Escherichia coli* produces ethanol and acids that are organic. Pyruvate formate lyase converts pyruvate to ethanol. For every pyruvate created from carbs, one NADH, H+ is produced; however, two NADH, H+ are required to convert pyruvate to ethanol, making this fermentation process unbalanced. *Escherichia coli* also produces acetic and succinic acids to balance its fermentation. Because yeasts and *Z. mobilis* use pyruvate decarboxylase (PDC), which utilizes just one NADH, H+ for every unit of ethanol produced, to convert pyruvate to ethanol, they are homoethanol fermentative organisms. Therefore, it was anticipated that *Escherichia coli* would only produce ethanol when PDC was expressed in it. The first attempt at creating a yeast-like pathway for pyruvate to ethanol conversion was unsuccessful because *Escherichia coli's* natural alcohol dehydrogenase function was insufficient to create substantial ethanol yields. Anti-diuretic hormone (ADH) II was extracted out of the genomic library of *Zymomonas mobilis* by Ingram et al. (1987) utilizing an entirely novel medium as a biomarker of ADH activity. *Escherichia coli* which has been plasmid-treated containing pyruvate decarboxylase and Anti-diuretic II produced practically all of the ethanol (Ingram et al. 1987). The production of ethanol operon was the construct's designation since the pyruvate decarboxylase and Anti-diuretic IIgenes were expressed together while being regulated by the native *lac* promoter (Ingram et al. 1987). A number of *E. coli* strains were then evaluated as hosts by Ingram and his associates (Alterthum and Ingram 1989). The key screening criteria were enhanced ethanol yield on xylose, plasmid stability on a non-selective medium, and tolerance to challenging growth conditions (such as ethanol endurance). The most promising strain was found to be *Escherichia coli* ATCC11303. To completely characterize the strain, xylose fermentations using ATCC11303 were carried out across a wide range of culture conditions (Beall et al. 1991). The ideal culture environment was found to be above pH 6.0 and below 42 °C. The greatest ethanol threshold was seen in xylose-fermenting cultures under these growth circumstances, which had an average output of 0.72 g for xylose concentrations of 80 g/l (Dien et al. 2003).

20.6.2 Modifying the Metabolism of Arabinose and Xylose in Zymomonas mobilis

Zymomonas mobilis is a peculiar gram-negative bacterium with a number of desirable qualities for use as biological catalysts for the synthesis of ethanol. The organism can withstand around 120 g/l of ethanol and possesses a homoethanol fermentation pathway. It has much higher specific ethanol productivity (2.5×) and yields 5–10% additional ethanol in each fermented gram of glucose than *Saccharomyces sp.* Additionally, *Zymomonas mobilis* has been considered to be safe. Because *Zymomonas mobilis* is so effectively suited to the production of ethanol, some researchers recommended it over *S. cerevisiae* for the transformation of starch to ethanol in the 1970s and 1980s. *Zymomonas mobilis* was successfully used in industrial-level experiments, while because of the resilience of the yeast, the industry continues to select *Saccharomyces cerevisiae* (Dien et al. 2003). Only *Zymomonas* uses the Entner-Doudoroff pathway to anaerobically metabolize glucose, in contrast to the glycolytic pathways. In comparison to the EMP pathway, the Entner-Doudoroff pathway only yields 50% more ATP for every mole of glucose. As a result, *Zymomonas* creates a smaller amount of biomass compared with yeast and directs more carbon to the products of fermentation. In addition, *Zymomonas* retains a high glucose flow through the ED route as a result of poor ATP output. Over fifty percent of the total protein in cells is made up of fermentation enzymes, all of which are produced constitutively (Dien et al. 2003). *Zymomonas* also has basic dietary requirements. For the development of certain strains, just pantothenate and biotin are needed (Rogers et al. 1982).

Zymomonas mobilis exclusively ferments glucose, fructose, and sucrose, which makes it unsuitable for biomass conversion despite its benefits as an ethanologenic. However, over the past 10 years, researchers at the United States National Renewable Resources Laboratory have proved successful in developing strains that can ferment arabinose and xylose. Zhang et al. (1995) claim that the first recombinant strain was made to ferment xylose. Four *Escherichia coli* genes have to be introduced and expressed in order to do this: xylose isomerase, transketolase, xylulose kinase, and transaldolase (Zhang et al. 1995). Xylosekinase and xylulose isomerase transform xylose into xylulose-5-P, a crucial stage in the pentose phosphate cycle. The Entner-Doudoroff pathway intermediates are then produced from xylulose-5-P by transketolase and transaldolase. On a plasmid, the genes are expressed by the glyceraldehyde-3-P dehydrogenase or enolase promoters from *Zymomonas mobilis*, both of which have a significant constitutive promoter effect. The ethanol production was 86% while growing on xylose with the modified strain CP4 (pZB5). Additionally, the strain fermented both xylose and glucose simultaneously (Dien et al. 2003)

20.6.3 **Erwina chrysanthemi** ***and*** **Klebsiella oxytoca** ***Modification for the Production of Ethanol***

Ingram et al. (1987) have used the pet operon to turn gram-negative bacteria *Erwina chrysanthemi* and *Klebsiella oxytoca* along with *Escherichia coli*. Despite the fact that the resultant strains produce less ethanol than *Escherichia coli*, substantial advancements have been achieved in creating enhanced *K. oxytoca* strains that are particularly effective at turning cellulose into ethanol.

An enteric bacteria called *Klebsiella oxytoca* can be discovered growing near different sources of wood as well as in streams used to make paper and pulp. With temperatures as elevated as 35 °C and pH levels as low as 5.0, the bacteria may flourish. *Klebsiella oxytoca* can thrive on a wide range of carbohydrates, including cellobiose and cellotriose as well as hexoses and pentoses. The strain is highly desirable for cellulose fermentations due to the latter quality (Dien et al. 2003).

By fermenting glucose, *Klebsiella oxytoca* produces a range of acids that are organic and neutral byproducts. The Pyruvate formate lyase pathway leads to the formation of ethanol. After plasmids containing the pet operon were used to transform *Klebsiella oxytoca* strain M5A1, ethanol increased to make up over ninety percent of the products of fermentation (on a molar basis) (Dien et al. 2003). The higher ethanol productivity was obtained when the pet operon was expressed on a plasmid with a lower copy number (pLOI555) than on a plasmid with a higher copy number (pLOI297). In contrast to what was seen for *Escherichia coli* strains, this was the case. Based on initial glucose, ethanol generation was 94–98% for M5A1 carrying either plasmid of what was anticipated. It appears that the strain M5A1 (pLOI555) is ideal for fermenting xylose. In comparison to *Escherichia coli* strain KO11 with glucose (2 g/l/h over the first 24 h), it fermented xylose twice as rapidly. Genetic stabilization of the ethanol production operon was achieved in *Klebsiella oxytoca* M5A1 by pairing it using a chloramphenicol acetyltransferase marker on the position of the chromosomal pyruvate formate lyase gene. The first M5A1 transformants discovered had low productivity and output of ethanol, while strains with high Cm (600 mg/l) resistance, as executed before for ethanologenic mutant *Escherichia coli* strains with enhanced ethanol production properties (Dien et al. 2003). The best mutant (P2) quickly produced 44–45 g/l of ethanol from either 100 g/l of glucose or 100 g/l of cellobiose in less than 48 h. P2 findings that were equivalent to regulating fermentations infected with M5A1 demonstrated that the integrated bacterium was producing enough pyruvate decarboxylase and anti-diuretic drugs (Dien et al. 2003).

20.7 CRISPR/Cas9 System Role in Microbial Engineering for the Production of Biofuels

Modern techniques like (CRISPR)/Cas9 defined as frequently clustered interspaced short palindromic repeats, allow for quick and efficient genome editing by swapping out 20-nucleotide regions with chimeric single-guide RNAs that are complementary to the targeted sequence of interest, are speeding up the genetic engineering of microbes (Cho et al. 2018). The Cas9 protein's endonuclease activity cuts the target DNA sequence as soon as the Cas9-sgRNA combination binds to it (Cho et al. 2018). These speeds up both the editing of advantageous genes and the metabolic alteration of proteins that may increase resistance to inhibitors or promote the use of diverse substrates used in the production of biofuels. Numerous studies have been conducted on the Streptococcus pyogenes Type II CRISPR-Cas9 system's mechanism and essential elements. This system consists of auxiliary non-coding trans-activating crRNAs that base-pair to the target-recognizing CRISPR RNA (crRNA) and allow the formation of the Cas9-crRNA complex. The crRNA attaches to the target DNA and directs cleavage. For better performance, it is feasible to introduce new genes, delete undesired genes, or knock-off undesirable genes in biofuel-producing bacterial strains. The CRISPR activation (CRISPR/Cas) mechanism is utilized to activate certain genes. Recently, the CRISPR/Cas 9 technology was used to modify the genetic material of *E. coli* strain MG1655 that generated 5.4 g per litre of n-butanol in a medium using glucose which is the source of carbon. Then, Abdelaal et al. (2019) duplicated this procedure in an ethanologenic strain of *Escherichia coli* strain SSK42 that generates butanol from xylose by means of a redox-balanced route. The CRISPR/Cas9 method was used to introduce an artificial butanol cassette to the *Escherichia coli* SSK42 genome after the gene for endogenous ethanol synthesis was eliminated. During fed-batch fermentation, the recently created strain ASA02 generated 4.32 g per litre of butanol (Abdelaal et al. 2019). In the bacterium *Clostridium saccharoperbutylacetonicum* N1-4, which is known to produce hyper-butanol, two important genes, pta, and buk, which encode for the production of butyrate and acetate, respectively, were selected for editing (Adegboye et al. 2021). Batch fermentation led to increased butanol synthesis, greater yields, and mutant selectivity, but these results were reliant on the fermentation medium employed. With a yield of 19.0 g per litre, the P2 medium produced the most butanol during the batch fermentation process (Adegboye et al. 2021). Efficiency was enhanced by using the PJ23119 promoter to guide RNA production (Adegboye et al. 2021). In order to create mutants to generate an isopropanol-butanol-ethanol combination, Wasels et al. (2017) developed a dual plasmid-inducible CRISPR/Cas9 genome editing approach for the solventogenic bacterium Clostridium acetobutylicum ATCC 824 (Adegboye et al. 2021).

20.8 Optimization of Microorganisms for High Yield of Biofuel

Compared to maize or sugarcane, cellulosic biomass has a substantially larger resource base (Fischer et al. 2008). In the near future, it is anticipated to dominate the production of biofuels. Compared to maize or sugar cane, a cellulosic feedstock is more challenging to convert into fermentable sugars because (i) Typical cellulosic biomasses contain 17–25% of the very resistant matrix polymer of aromatic alcohols known as lignin (Maris et al. 2006); and (ii) Starches and simple oligosaccharides are substantially less tolerant of the breakdown than cellulose. The first challenge can be overcome by choosing and/or developing microorganisms that can convert xylose and other sugars with five carbons to ethanol under anaerobic conditions. Pentose phosphate is the metabolic process used to break down these five-carbon carbohydrates. Although the rate of lignin enzymatic degradation is too slow for industrial use, lignin can be gasified or burned effectively for thermochemical conversion. conversion using thermochemistry. The third barrier is the most crucial. Cellulosic feedstock needs to be chemically "pretreated" due to its resistance in order to create a partly hydrolyzed end product that cellulase enzymes can more easily break down. Different pretreatment approaches have been put forth, but many of them result in fermentation inhibitors. Cellulase delivery and fermentation of sugar to ethanol production have long been researched using two basic metabolic engineering techniques. The first aims to genetically alter *Saccharomyces cerevisiae, Zymomonas mobilis, Escherichia coli*, or other microbes to have pathways for both the synthesis and utilization of ethanol (Fischer et al. 2008).

Recent reviews of promising *Saccharomyces cerevisiae* outcomes have been published (Maris et al. 2006). High ethanol tolerance, low pH fermentation, and the ability to skip medium sterilization are all benefits of *Saccharomyces*. The drawback is that *Saccharomyces* cannot directly utilize cellulose or derived oligosaccharides, unlike *Escherichia coli* and *Zymomonas mobilis*. This necessitates the development of a biorefinery's own specialized on-site cellulase production process, which frequently involves the aerobic growth of microorganisms like *Trichoderma reesei*. Another technique, called consolidated bioprocessing (formerly known as a direct microbial transformation), use thermophilic *Clostridium thermocellum*, a highly cellulolytic bacterium, either alone or in conjunction with other thermophilic, more efficient sugar fermenters (Fischer et al. 2008).

These microbes enable anaerobic cellulose hydrolysis, fermentation, and cellulase synthesis to take place in the same processing vessel. Due to their ability to function at 60 °C, these bacteria' thermophilic nature may make mild sterilization unnecessary, (ii) The majority of clostridia may utilize xylose without recombinant modifications, and (iii) Cellulase synthesis and cellulose hydrolysis can both occur in the same container. The disadvantages of consolidated bioprocessing include (i) decreased solvent resistance of clostridia, (ii) Comparative difficulties with genetic alteration in clostridia, and (iii) In anaerobic settings, cellulase production requires more energy (Fischer et al. 2008).

20.8.1 Desirable Properties

The most apparent requirement of any metabolic engineering endeavor is to find or create an ideal host for the synthesis of biofuel, but this is far from an easy task. The most advantageous host in this condition would degrade lignocellulosic components and rapidly and vigorously ferment the resultant sugars (both pentoses and hexoses), and accept large titers of the finished product at high temperatures and harsh pH in order to avoid the costs of cooling and sterilization. The incredible variety, particularly for microbes, may indicate that an optimal host for the synthesis of biofuels already exists and that the actual difficulty is in locating it. The majority of researchers have chosen to concentrate on the design problem of introducing desirable features into a designed host using recombinant DNA technology since there is no apparent solution to this "needle in a haystack" dilemma. Any such effort typically involves changing phenotypes that are controlled by numerous genes. Since the systematic application of this strategy demands a thorough understanding of the host being modified, it has favored the utilization of "laboratory strains" that are suitable for research but might not be strong enough for industrial applications and survive high titers of the end result at high pH values and temperatures (to prevent paying for cooling and sterilization).

The incredible variety, particularly for microbes, may indicate that an optimal host for the synthesis of biofuels already exists and that the actual difficulty is in locating it. The majority of researchers have chosen to focus on the design challenge of fusing desirable traits into an engineered host utilizing recombinant DNA technology since there is no apparent solution to this "needle in a haystack" dilemma. Any such effort typically involves changing phenotypes that are controlled by numerous genes. The systematic use of this strategy necessitates an in-depth understanding of the host being altered, which has favored utilizing "laboratory strains" that may not be suitable for use in industrial applications but are suitable for study. In order to create production strains with industrial significance, the qualities that have led to the popularity of experimental strains like *Escherichia coli, Bacillus subtilis, Saccharomyces cerevisiae*, and will primarily be necessary. The ability of a strain to accept foreign DNA in a controllable fashion is known as genetic competence and is the first need. For combinatorial libraries to be used effectively for analyzing genes or gene variations that result in a specific trait of interest, significant transformation efficiencies are required. The accessibility of metabolic engineering "modules," with a clear definition that enables diverse genotype modification, is a second quality that attracts researchers to laboratory strains. Examples include repressor-inducer systems, plasmids, termination sequences, chromosomal integration cassettes to produce stable gene replication or knockouts, promoters with various strengths (Alper et al. 2005), and chromosome integration cassettes. Thirdly, the ease of use of "omics" platforms and methods for genome-wide assessment of biological reactions to various medicines and environments. The fourth factor—and most undervalued of all—is the vast body of accumulated information on the physiology of laboratory strains that have been made available by decades of researchers.

20.8.2 Strain Improvement

Given that titer and maximum output are important factors in the successful fermentative production of common chemicals, enhancing environmental tolerances has been a focus of research. Product toxicity issues have been effectively addressed with innovative reactor or procedure design tactics (Fischer et al. 2008), although these methods are often accompanied by strain engineering. Gene alterations for strain enhancement may often be divided into two categories: directed and combinatorial changes. When the sequence of genes that are expected to affect a characteristic of interest is known, the first scenario is simple. The second option, which depends on our capacity to create populations with genetic variation and screen them, is preferable when this information is missing.

These efforts have mostly been directed at "classical strain improvement" (CSI), sometimes referred to as long-lasting adaption. Using physical or chemical mutagens, such as UV or nitrosoguanidine, the mutagenesis stage creates libraries of variations that may subsequently be tested for increased tolerance. The obvious drawback of this technique is the intractability and non-transferability of the resulted in genotypic alterations. Many library-based strain augmentation methods that rely on phenotypic alteration with manageable components have emerged in recent years. These include synthetic global transcription machinery engineering (Alper et al. 2006), siRNA libraries, transcription factors (Fischer et al. 2008), randomized ribozymes, knockout, and over expression libraries (Fischer et al. 2008), and others.

Despite being obvious, when working to improve a strain, it is occasionally overlooked that any genetic modification leads to changes to an already functioning genome. This observation suggests that improvements rely on the host strain's genetic composition depending on the kind of modifications that occur. As a result, the maximum number of modifications that may be tested out sets a limit on how much can be improved.

Finding the correct host as an initial point and selecting which characteristics to change will decide the possibility of creating a strain that is ideal for generating biofuel. This does not, however, mean that the tolerance cannot be increased in comparison to the wild type. Instead, it implies that the phenotype we select to create will determine the practical importance of the threshold for augmentation.

20.8.3 Tolerance for Typical Process Stresses in Engineering

Due to the fact that operating fermentations at temperatures that are higher lowers cooling costs and lowers the danger of contamination, heat tolerance has drawn a lot of attention. Heat increases the membrane's fluidity, which interferes with energy transfer and pH maintenance (Balogh et al. 2005). The majority of cellular processes are also adversely affected by the protein denaturation and clumping it induces. Both random and rational systems have improved thermotolerance. Cell survival following a 4-h shift to 51C was increased when the yeast *Saccharomyces cerevisiae* overexpressed the *Chaetomium globosum*-derived heat shock protein, hsp22.4. This method has been used for a variety of species, including bacteria and extremely complex multicellular systems (Fiocco et al. 2007). It has been demonstrated that preconditioning, excessive synthesis of protective compounds like trehalose, and supplementation with betaine and choline all work to increase thermotolerance (Fischer et al. 2008).

These three chemicals, along with others, modify the interactions between water and proteins, stabilize them, and stop them from aggregating, which has a similar impact on the cytoplasm's osmolarity (Cayley and Record 2003). Combinatorial strategies have also proved productive. For instance, repeated freeze–thaw cycles on a mutagenized *Saccharomyces* population followed by selection have produced mutants with improved tolerance to a variety of stressors, including heat (Fischer et al. 2008). Engineering thermotolerance in *S. cerevisiae* and identifying the gene causing this phenotypic change in *E. coli* have both been accomplished by the screening of synthetic transcription factor libraries. Additionally, osmotolerance has been studied in relation to strain improvement because commercial fermentations sometimes involve high substrate, salt levels, or products that enhance osmotic pressure. Numerous of the methods described for improving thermotolerance can be utilized or have been applied in this situation. For instance, it has been demonstrated that trehalose and betaine encourage osmotolerance in *Escherichia coli* and that trehalose is protective in yeast (Hounsa et al. 1998). Yeast strains improved for development in high salt content and glucose environments have been produced using transcriptional engineering. Similar to how it has enhanced metabolite overproduction, this technique has increased *E. coli's* resistance to a range of stressors (Fischer et al. 2008).

20.8.4 Stresses Particular to Biofuels

For fermentation technology, environmental endurance is a challenge and is especially important for producing biofuels for two different reasons. First, the substrate for fermentation may include hazardous substances if the producing strain is unable to directly break down the lignocellulosic material. Feedstocks derived from lignocellulosic hydrolysates have been extensively researched since pretreatment by acid hydrolysis produces a mixture of sugars, phenol derivatives, organic acids, and furans—all but the first of which are bacterial growth inhibitors—this pretreatment is also known as acid hydrolysis. The second possibility is that the fermentation byproduct itself might be dangerous, in which case the titer may already be limited. Ethanol synthesis in Saccharomyces is a well-known illustration of this (Alper et al. 2006).

ADH (Alcohol Dehydrogenase) 6, a 5-hydroxymethyl furfural (HMF)-reducing enzyme, was overexpressed in *Saccharomyces cerevisiae*, and this resulted in the improved conversion of the inhibitor that might likely be employed in detoxification, but no improvement in ethanol production was noted (Petersson et al. 2006). Higher furfural tolerance was obtained by a similar strategy using ZWF1, which encodes a glucose-6-phosphate dehydrogenase. In the presence of cinnamic acids and ferulic, Saccharomyces strains with increased phenyl acrylic acid decarboxylase production produced more ethanol (Fischer et al. 2008). The fermentation has been manipulated since protonated species are primarily responsible for toxicity; pH has been used to reduce acetic acid tolerance (Fischer et al. 2008). Due to the additional cost and the reality that low pH reduces the danger of contamination; pH control is not preferred, nevertheless. By introducing a gene from the acid-resistant *Oenococcus oeni*, a strain of *E. coli* with improved low pH tolerance was developed. These hosts or ones that are built similarly may be more suitable for fermentations with low pH levels. When the inhibitor is present, acetic acid-resistant mutants of the ethanologenic *Candida krusei* have outperformed the original strain in ethanol fermentations (Wei et al. 2008). It has also been observed that CSI techniques can help yeasts become more tolerant to the components of lignocellulosic hydrolysate. In a manner similar to this, research demonstrates that *Pichia stipitis* exhibited increased ethanol production and titer after becoming long-term acclimated to rising concentrations of hardwood hydrolysate that was either partly neutralized or alkalinized with lime (Fischer et al. 2008).

20.9 Application of Functional Genomics to Metabolic Engineering

More lately, there has been a great deal of interest in both the microbial creation of artificial compounds not present in any creature and the production of natural molecules found in plants and other species. Microorganisms' metabolisms will need to be significantly redesigned in order to produce advanced biofuels (such as alkenes, alkanes and aromatics) (Kour et al. 2019). Understanding metabolism at the systems level and having the capacity to control several factors at once are necessary for manipulating microbial metabolism.

In order to boost the production of a natural product or insert genes that permit the synthesis of synthetic products, it might have detrimental effects on host growth and, as a result, alter product titers. The inability to monitor more than a few, frequently gross, factors made it difficult, if not impossible, to diagnose and fix problems. One such parameter was growth rate. We can now simultaneously monitor hundreds of parameters thanks to the development of functional genomics approaches (Kour et al. 2019). However, detecting and treating such disorders requires system-wide models of genetic control and metabolism as well as computational methods to assess the large volumes of data, this discipline is known as systems biology (Kour et al. 2019).

20.9.1 Strategies for Systems Biology

Since this information gives context for comprehending genomic sequence data, high-density DNA microarray analysis of gene expression levels is becoming a common component of functional genomics pipelines. A staggering amount of array data is now accessible for hundreds of different microbial and non-microbial systems, providing cell-wide details on a dizzying variety of mutations, stressors, toxins, growth conditions and adaptations. More recently, mutation studies and microarray data have been coupled to reveal abnormalities noticed in a better context (Global Bioenergy Association). High-density DNA array technology has also enabled the quick study of microbial diversity (Pattrick et al. 2019) and gene function (Basso et al. 2011) from complex environmental samples. The creation of strict reporting requirements (Naik et al. 2010) was required as a result of attempts to analyze and make use of the enormous quantity of data produced by transcriptional array investigations, which has given rise to completely novel disciplines in the areas of computational biology and bioinformatics (Kour et al. 2019).

20.9.2 Functional Genomics and Metabolic Engineering Applications

Most engineering efforts are primarily motivated by the goal to raise product titers and reduce precursor costs, especially for low-margin goods like commodity chemicals. While commodity chemicals and biofuels can reach titers of hundreds of g/L, pharmaceuticals (Mukhopadhyay et al. 2008) can have economically practical titers of a few mg/L. To achieve these production levels, a significant amount of pathway optimization is required after establishing the route and eliminating or minimizing any detrimental effects the pathway or production may have on the cell. Strain creation is an iterative process, involving pathway modification, system-level analysis, and disclosure of potentially detrimental side effects. After it has been proven that a product of interest can be produced in vivo, it is necessary to reduce the synthesis of less desirable side products and increase carbon flow toward the target product in order to reach economically viable production levels. For instance, several of these procedures have been used to increase the synthesis of bioethanol in *Escherichia coli*, and this development has been widely documented (Dien et al. 2003).

The majority of metabolic engineering projects combine native and heterologous genes, as is the case with the generation of isobutanol in *E. coli*, 1,3-propanediol, and the precursor to the anti-malarial medication artemisinic acid (Mukhopadhyay et al. 2008). It's crucial to comprehend how the addition of a created exogenous pathway affects the host system in order to remove pathway bottlenecks. For instance, the over expression of heterologous enzymes, particularly when the genes originate from an organism with a different codon usage than the generating strain, may result in stress by reducing the supply of amino acids that are available. As an outcome, many heterologous proteins are now codon-optimized to minimize any mRNA structural difficulties and to predominantly draw from the largest tRNA pools available (Mukhopadhyay et al. 2008).

As a result of imbalances in enzyme activity, toxic or inhibitory pathway intermediates may build up and significantly impede both cellular development and output levels. Functional genomics may be crucial for solving these problems. For instance, the formation of is oprenoids in *Escherichia coli* was shown to be constrained by the buildup of 3-hydroxy-3-methylglutaryl-coenzyme A (HMG-CoA), which was remedied by upregulating the synthesis of HMG-CoAor over expressing tHMG1 (Mukhopadhyay et al. 2008).

If the product is not effectively exported out of the cell in metabolic engineering, it may become a bottleneck. Large amounts of the product in the medium can also negatively impact cellular metabolism, which is especially true for biofuels like longer-chain alcohols, ethanol, and long-chain hydrocarbons (Olsson 2007). Developing a strong production host can also benefit from understanding the processes that confer fitness in response to unfavorable production circumstances such as pH fluctuations, the presence of salts, and the buildup of hazardous byproducts. Recent investigations by Alper and Stephanopoulos (2007) highlighted the significance of optimizing strains for industrial use and demonstrated the need to simultaneously change many features in a host organism. Engineering many genes to produce a single characteristic, like ethanol resistance, is common.

A higher number of genes are needed to get strains with different characteristics, making standard sequential screening methods useless. By altering the global transcriptional machinery, Alper and Stephanopoulos (2007) were able to effectively identify strains that were resistant to both large dosages of ethanol and SDS. These findings further highlight the significance of regulatory proteins for the evolution of strains with a variety of desirable features as well as for the maintenance of stable and predictable phenotypes throughout scale-up and long-term growth. A challenging group of genes to analyze using microarrays and proteomics are transcriptional regulators since they frequently exhibit minimal change and are generally present in low abundance. However, a variety of state-of-the-art tools can improve our ability to conduct focused, high-throughput research for certain gene classes. One such device is nCounter (Mukhopadhyay et al. 2008), which employs a freshly developed, direct counting approach to enable extremely sensitive detection of mRNA variations for as many as 524 transcripts. The intrinsic complexity of microbial systems provides a significant obstacle to cellular engineering, despite several attempts to develop the best production hosts. Systems biology offers the chance to examine the cell holistically in order to acquire an overview of the systems that are impacted during production and provide hints as to where route bottlenecks are located. It is crucial to remember that current research using model organisms like *Escherichia coli* and *Saccharomyces cerevisiae* provides a wealth of knowledge that is quite relevant to many of the factors crucial to industrial processes (Mukhopadhyay et al. 2008)

20.10 Future Potential

The usage of various, affordable non-food feedstocks has expanded over the last two years, as has the inventive engineering of enzymes and metabolic pathways for the generation of advanced biofuels. The conversion of syngas, CO_2, algal hydrolysate, and switch grass into higher alcohols, fatty acids, and biofuels produced from isoprenoids has demonstrated the viability of various biofuel production routes. Back in 2012, the only relevant query would have been, "Which advanced biofuel can be produced from glucose in high yield?" Researchers are currently focusing on whether feedstocks may provide high-yield advanced biofuels at low cost. The difficulty is in obtaining a high yield of advanced biofuels, which calls for designing both the routes used to utilize substrates and produce advanced biofuels (Adegboye et al. 2021).

Microbial metabolic engineering is crucial for the growth of the biofuel manufacturing industry, despite the challenge of discovering effective bacterial strains. The relevant enzymes must be identified, and the important metabolic pathways must be fully understood. Metabolic engineers and synthetic biologists are identifying the limiting areas of pathways using various methods. Metabolic engineering is a method for enhancing recombinant protein expression that is constantly evolving and becoming more sophisticated. Industrial microbes have been modified to produce more recombinant metabolites more quickly and cheaply (Adegboye et al. 2021).

Understanding particular gene expression and their function in the metabolic processes involved in lignocellulose-degrading microorganisms' production of biofuel is becoming more common through the use of high throughput technologies, such as transcriptomics, proteomics, and metabolomics (Wang et al. 2016). Computational approaches are routinely used to get an analytical overview of the data collected from this cutting-edge technology. The generation of two terpenes via the heterologous mevalonate pathway in *Escherichia coli* was improved by more than 40% as a result of the application of principle component analysis proteomics-guided engineering. These computational techniques and cutting-edge technologies should be heavily used for strain selection and metabolic engineering to increase the industrial output of biofuel (Adegboye et al. 2021).

20.11 Conclusion

Numerous issues, including resource depletion, climate change, energy security, the cost of oil, and the price of oil are threatening the environment either directly or indirectly in the modern world. These difficulties have spurred important developments in the study and production of fuels and energy produced from biomass. Biofuels are therefore anticipated to be most beneficial in this respect to solving such issues in a highly sustainable manner. Biofuels have been viewed as the most practical solution for lowering carbon dioxide emissions in the transportation industry. Additionally, biofuels can be easily made from locally accessible indigenous resources.

Major contributors to the creation of biofuel are microorganisms. Biodiesel and isobutanol synthesis are two significant fuel production systems that have recently been shown to exist in microorganisms. Although these studies show promise, significant engineering will be required to raise the titers in order to produce commercially viable fuels. If systems-level evaluations identify several solutions that may be coupled to optimize production, it won't be shocking. In reality, a number of recent cell-wide studies have focused on the cellular physiology that creates microorganisms subjected to stress, deprivation, and heterologous metabolic pathways with their potentially damaging intermediates. Many of these findings are currently being applied to enhance hosts. Even while the great bulk of functional genomics research to until has been focused on studying cellular physiology, it is assumed that this information may now be utilized to detect route bottlenecks and enhance cellular circuit design. As the technology advances from the discovery stage into the application stage, the use of data acquired from systems biology research in metabolic engineering holds the possibility of developing strong production host organisms for industrial applications. The goal of recent research has also been to maximize high productivity and energy value at reduced production costs by simulating strain development via metabolic engineering. Database mining may soon lead to the creation of more different metabolic pathways for the production of biofuels. Therefore, by including such routes in commercial fermentation hosts, any barriers to employing lignocellulosic biomass as a sustainable fermentation feedstock may be eliminated. Metabolic engineers must use the most advanced technologies currently available, like the omics technologies and CRISPER/Cas9 system, to design and develop novel strains of microbes with enhanced ability to produce biofuel from a variety of substrates by introducing useful genes into the genome or deleting obstructive ones.

Algal biofuels are now receiving a lot of attention and are seen to be the most viable solution to the world's energy issue. The primary benefits of using algae are the possibility for high productivity and the lack of rivalry for land and freshwater resources with food crops. Around the world, there is a lot of research being done to increase the production of biofuels. Biofuel production technology has advanced significantly, and a thorough understanding of the processes involved in producing biofuel has also been attained. However, fossil fuels cannot yet be entirely replaced by biofuels, and a number of integrated engineering and biological approaches are still needed to optimize the production of biofuels on a commercial scale. Furthermore, in order to create a sustainable biofuel economy, it is essential to comprehend how the production of biofuels will be impacted by future climatic changes. Therefore, biofuels as a replacement for fossil fuels in the future will undoubtedly be a leading supplier of energy in a sustainable way with the ability to increase supply security; this will also undoubtedly result in a decrease in the volume of vehicle emissions.

References

Abdelaal AS, Jawed K, Yazdani SS (2019) CRISPR/Cas9-mediated engineering of Escherichia coli for n-butanol production from xylose in defined medium. J Ind Microbiol Biotechnol 46(7):965–975

Adegboye MF, Ojuederie OB, Talia PM, Babalola OO (2021) Bioprospecting of microbial strains for biofuel production: metabolic engineering, applications, and challenges. Biotechnol Biofuels 14(1)

Ajikumar PK, Xiao W-H, Tyo KEJ, Wang Y, Simeon F, Leonard E, Stephanopoulos G (2010) Isoprenoid pathway optimization for taxol precursor overproduction in Escherichia coli. Science 330(6000):70–74

Alper H, Stephanopoulos G (2007) Global transcription machinery engineering: a new approach for improving cellular phenotype. Metab Eng 9(3):258–267

Alper H, Fischer C, Nevoigt E, Stephanopoulos G (2005) Tuning genetic control through promoter engineering. Proc Natl Acad Sci 102(36):12678–12683

Alper H, Moxley J, Nevoigt E, Fink GR, Stephanopoulos G (2006) Engineering yeast transcription machinery for improved ethanol tolerance and production. Science 314(5805):1565–1568

Alterthum F, Ingram LO (1989) Efficient ethanol-production from glucose, lactose, and xylose by recombinant Escherichia coli. Appl Environ Microbiol 55:1943–1948

Atsumi S, Hanai T, Liao JC (2008) Non-fermentative pathways for synthesis of branched-chain higher alcohols as biofuels. Nature 451(7174):86–89

Balogh G, Horváth I, Nagy E, Hoyk Z, Benkõ S, Bensaude O, Vígh L (2005) The hyperfluidization of mammalian cell membranes acts as a signal to initiate the heat shock protein response. FEBS J 272(23):6077–6086

Basso LC, Basso TO, Rocha SN (2011) Biofuel production. In: Bernardes MAS (ed) Recent developments and prospects, pp 85–100

Beall DS, Ohta K, Ingram LO (1991) Parametric studies of ethanol production from xylose and other sugars by recombinant Escherichia coli. Biotechnol Bioeng 38(3):296–303
Bhattacharya AS, Bhattacharya A, Pletschke BI (2015) Synergism of fungal and bacterial cellulases and hemicellulases: a novel perspective for enhanced bio-ethanol production. Biotech Lett 37(6):1117–1129
Bothast RJ, Schlicher MA (2004) Biotechnological processes for conversion of corn into ethanol. Appl Microbiol Biotechnol 67(1):19–25
Brunecky R, Alahuhta M, Xu Q, Donohoe BS, Crowley MF, Kataeva IA, Bomble YJ (2013) Revealing nature's cellulase diversity: the digestion mechanism of Caldicellulosiruptor bescii CelA. Science 342(6165):1513–1516
Cayley S, Record MT (2003) Roles of cytoplasmic osmolytes, water, and crowding in the response of Escherichia coli to osmotic stress: biophysical basis of osmoprotection by glycine betaine†. Biochemistry 42(43):12596–12609
Chang T, Yao S (2011) Thermophilic, lignocellulolytic bacteria for ethanol production: current state and perspectives. Appl Microbiol Biotechnol 92(1):13–27
Cho S, Shin J, Cho B-K (2018) Applications of CRISPR/Cas system to bacterial metabolic engineering. Int J Mol Sci 19(4):1089
Chowdhury H, Loganathan B (2019) 3rd generation biofuels from microalgae: a review. Curr Opin Green Sustain Chem
Cripps RE, Eley K, Leak DJ, Rudd B, Taylor M, Todd M, Atkinson T (2009) Metabolic engineering of Geobacillus thermoglucosidasius for high yield ethanol production. Metab Eng 11(6):398–408
Demirbas A (2009) Biofuels securing the planet's future energy needs. Energy Convers Manag 50(9):2239–2249
Dien BS, Cotta MA, Jeffries TW (2003) Bacteria engineered for fuel ethanol production: current status. Appl Microbiol Biotechnol 63(3):258–266
Doi RH, Kosugi A (2004) Cellulosomes: plant-cell-wall-degrading enzyme complexes. Nat Rev Microbiol 2(7):541–551
Elkins JG, Raman B, Keller M (2010) Engineered microbial systems for enhanced conversion of lignocellulosic biomass. Curr Opin Biotechnol 21(5):657–662
Elshahed MS (2010) Microbiological aspects of biofuel production: current status and future directions. J Adv Res 1(2):103–111
Fargione J, Hill J, Tilman D, Polasky S, Hawthorne P (2008) Land clearing and the biofuel carbon debt. Science 319(5867):1235–1238
Fiocco D, Capozzi V, Goffin P, Hols P, Spano G (2007) Improved adaptation to heat, cold, and solvent tolerance in Lactobacillus plantarum. Appl Microbiol Biotechnol 77(4):909–915
Fiorese G, Catenacci M, Verdolini E, Bosetti V (2013) Advanced biofuels: future perspectives from an expert elicitation survey. Energy Policy 56:293–311
Fischer CR, Klein-Marcuschamer D, Stephanopoulos G (2008) Selection and optimization of microbial hosts for biofuels production. Metab Eng 10(6):295–304
Fukuda H, Kondo A, Noda H (2001) Biodiesel fuel production by transesterification of oils. J Biosci Bioeng 92(5):405–416
Gao D, Chundawat SPS, Krishnan C, Balan V, Dale BE (2010) Mixture optimization of six core glycosyl hydrolases for maximizing saccharification of ammonia fiber expansion (AFEX) pretreated corn stover. Biores Technol 101(8):2770–2781
Georgieva T, Mikkelsen M, Ahring B (2007) High ethanol tolerance of the thermophilic anaerobic ethanol producer Thermoanaerobacter BG1L1. Open Life Sci 2(3)
Global Bioenergy Association. Global bioenergy statistics 2019
Godvin Sharmila V, Dinesh Kumar M, Pugazhendi A, Bajhaiya AK, Gugulothuf P, Rajesh BJ (2021) Biofuel production from macroalgae: present scenario and future scope. Bioengineerd 12(2):9216–9238
Gomez LD, Steele-King CG, McQueen-Mason SJ (2008) Sustainable liquid biofuels from biomass: the writing's on the walls. New Phytol 178(3):473–485

Gong Z, Nielsen J, Zhou YJ (2017) Engineering robustness of microbial cell factories. Biotechnol J 12(10):1700014
Himmel ME, Ding S-Y, Johnson DK, Adney WS, Nimlos MR, Brady JW, Foust TD (2007) Biomass Recalcitrance: engineering plants and enzymes for biofuels production. Science 315(5813):804–807
Horn SJ, Aasen IM, Østgaard K (2000) Production of ethanol from mannitol by Zymobacter palmae. J Ind Microbiol Biotechnol 24(1):51–57
Hounsa C-G, Brandt EV, Thevelein J, Hohmann S, Prior BA (1998) Role of trehalose in survival of Saccharomyces cerevisiae under osmotic stress. Microbiology 144(3):671–680
Hu J, Arantes V, Saddler JN (2011) The enhancement of enzymatic hydrolysis of lignocellulosic substrates by the addition of accessory enzymes such as xylanase: is it an additive or synergistic effect? Biotechnol Biofuels 4(1):36
Huffer S, Roche CM, Blanch HW, Clark DS (2012) Escherichia coli for biofuel production: bridging the gap from promise to practice. Trends Biotechnol 30(10):538–545
Ingram LO, Conway T, Clark DP, Sewell GW, Preston JF (1987) Genetic engineering of ethanol production in Escherichia coli. Appl Environ Microbiol 53:2420–2425
Jeon E, Hyeon J, Suh DJ, Suh Y-W, Kim SW, Song KH, Han SO (2009) Production of cellulosic ethanol in Saccharomyces cerevisiae heterologous expressing Clostridium thermocellum endoglucanase and Saccharomycopsis fibuligera β-glucosidase genes. Mol Cells 28(4):369–373
Juturu V, Wu JC (2012) Microbial xylanases: engineering, production and industrial applications. Biotechnol Adv 30(6):1219–1227
Kawai S, Murata K (2016) Biofuel production based on carbohydrates from both brown and red macroalgae: recent developments in key biotechnologies. Int J Mol Sci 17(2):145
Keasling J, Garcia Martin H, Lee TS, Mukhopadhyay A, Singer SW, Sundstrom E (2021) Microbial production of advanced biofuels. Nat Rev Microbiol
Klemm P, Hjerrild L, Gjermansen M, Schembri MA (2003) Structure-function analysis of the self-recognizing Antigen 43 autotransporter protein from Escherichia coli. Mol Microbiol 51(1):283–296
Kosugi A, Murashima K, Doi RH (2002) Xylanase and acetyl xylan esterase activities of XynA, a key subunit of the Clostridium cellulovorans cellulosome for Xylan degradation. Appl Environ Microbiol 68(12):6399–6402
Kour D, Rana KL, Yadav N, Yadav AN, Rastegari AA, Singh C, Saxena AK (2019) Technologies for biofuel production: current development, challenges, and future prospects. Plant Long Non-Coding RNAs 1–50
Kubicek CP, Mikus M, Schuster A, Schmoll M, Seiboth B (2009) Metabolic engineering strategies for the improvement of cellulase production by Hypocrea jecorina. Biotechnol Biofuels 2(1):19
Langholtz MH, Stokes BJ, Eaton LM (2016) Billion-ton report: advancing domestic resources for a thriving bioeconomy, Volume 1: Economic availability of feedstocks. Oak Ridge National Laboratory
Lau MW, Dale BE (2009) Cellulosic ethanol production from AFEX-treated corn stover using Saccharomyces cerevisiae 424A(LNH-ST). Proc Natl Acad Sci 106(5):1368–1373
Lewis NS, Nocera DG (2006) Powering the planet: chemical challenges in solar energy utilization. Proc Natl Acad Sci 103(43):15729–15735
Liao JC, Mi L, Pontrelli S, Luo S (2016) Fuelling the future: microbial engineering for the production of sustainable biofuels. Nat Rev Microbiol 14(5):288–304
Lynd L, Zyl W, McBride J, Laser M (2005) Consolidated bioprocessing of cellulosic biomass: an update. Curr Opin Biotechnol 16(5):577–583
Maity JP, Bundschuh J, Chen C-Y, Bhattacharya P (2014) Microalgae for third generation biofuel production, mitigation of greenhouse gas emissions and wastewater treatment: present and future perspectives—a mini review. Energy 78:104–113
Martinez D, Berka RM, Henrissat B, Saloheimo M, Arvas M, Baker SE, Brettin TS (2008) Genome sequencing and analysis of the biomass-degrading fungus Trichoderma reesei (syn. Hypocrea jecorina). Nat Biotechnol 26(5):553–560

Mukhopadhyay A (2015) Tolerance engineering in bacteria for the production of advanced biofuels and chemicals. Trends Microbiol 23(8):498–508
Mukhopadhyay A, Redding AM, Rutherford BJ, Keasling JD (2008) Importance of systems biology in engineering microbes for biofuel production. Curr Opin Biotechnol 19(3):228–234
Naik SN, Goud VV, Rout PK, Dalai AK (2010) Production of first- and second-generation biofuels: a comprehensive review. Renew Sustain Energy Rev 14(2):578–597
O'Neill BC (2002) Climate change: dangerous climate impacts and the Kyoto protocol. Science 296(5575):1971–1972
Olsson L (ed) (2007) Biofuels. Advances in biochemical engineering/biotechnology
Pacala S (2004) Stabilization wedges: solving the climate problem for the next 50 years with current technologies. Science 305(5686):968–972
Pattrick CA, Webb JP, Green J, Chaudhuri RR, Collins MO, Kelly DJ (2019) Proteomic profiling, transcription factor modeling, and genomics of evolved tolerant strains elucidate mechanisms of vanillin toxicity in Escherichia coli. mSystems 4(4)
Petersson A, Almeida JRM, Modig T, Karhumaa K, Hahn-Hägerdal B, Gorwa-Grauslund MF, Lidén G (2006) A 5-hydroxymethyl furfural reducing enzyme encoded by the Saccharomyces cerevisiae ADH6 gene conveys HMF tolerance. Yeast 23(6):455–464
Prasad S, Singh A, Joshi HC (2007) Ethanol as an alternative fuel from agricultural, industrial and urban residues. Resour Conserv Recycl 50(1):1–39
Prates JAM, Tarbouriech N, Charnock SJ, Fontes CMGA, Ferreira LMA, Davies GJ (2001) The structure of the feruloyl esterase module of xylanase 10B from Clostridium thermocellum provides insights into substrate recognition. Structure 9(12):1183–1190
Prince RC, Kheshgi HS (2005) The photobiological production of hydrogen: potential efficiency and effectiveness as a renewable fuel. Crit Rev Microbiol 31(1):19–31
Rakotoarivonina H, Hermant B, Monthe N, Rémond C (2012) The hemicellulolytic enzyme arsenal of Thermobacillus xylanilyticus depends on the composition of biomass used for growth. Microb Cell Fact 11(1):159
Rogers PL, Lee KJ, Skotnicki ML, Tribe DE (1982) Ethanol production by Zymomonas mobilis. Microbial React 37–84
Sánchez C (2009) Lignocellulosic residues: biodegradation and bioconversion by fungi. Biotechnol Adv 27(2):185–194
Schenk PM, Thomas-Hall SR, Stephens E, Marx UC, Mussgnug JH, Posten C, Hankamer B (2008) Second generation biofuels: high-efficiency microalgae for biodiesel production. Bioenergy Res 1(1):20–43
Schirmer A, Rude MA, Li X, Popova E, del Cardayre SB (2010) Microbial biosynthesis of alkanes. Science 329(5991):559–562
Scully S, Orlygsson J (2014) Recent advances in second generation ethanol production by thermophilic bacteria. Energies 8(1):1–30
Scully SM, Orlygsson J (2017) Recent advances in genetic engineering of thermophilic ethanol producing bacteria. In: Engineering of microorganisms for the production of chemicals and biofuels from renewable resources, pp 1–29
Searchinger TD, Hamburg SP, Melillo J, Chameides W, Havlik P, Kammen DM, David Tilman G (2009) Fixing a critical climate accounting error. Science 326(5952):527–528
Singh BP, Panigrahi MR (2000) Review of biomass as a source of energy for India. Energy Sources 22(7):649–658
Srivastava RK (2019) Bio-energy production by contribution of effective and suitable microbial system. Mater Sci Energy Technol
Stephanopoulos G (2007) Challenges in engineering microbes for biofuels production. Science 315(5813):801–804
Takase R, Ochiai A, Mikami B, Hashimoto W, Murata K (2010) Molecular identification of unsaturated uronate reductase prerequisite for alginate metabolism in Sphingomonas sp. A1. Biochimica et Biophysica Acta (BBA) Proteins Proteomics 1804(9):1925–1936

Takeda H, Yoneyama F, Kawai S, Hashimoto W, Murata K (2011) Bioethanol production from marine biomass alginate by metabolically engineered bacteria. Energy Environ Sci 4(7):2575

Taylor MP, Eley KL, Martin S, Tuffin MI, Burton SG, Cowan DA (2009) Thermophilic ethanologenesis: future prospects for second-generation bioethanol production. Trends Biotechnol 27(7):398–405

US Energy Information Administration (EIA) (2019) International energy outlook 2019: with projections to 2050

US Environmental Protection Agency. Sources of greenhouse gas emissions. EPA

Van den Brink J, de Vries RP (2011) Fungal enzyme sets for plant polysaccharide degradation. Appl Microbiol Biotechnol 91(6):1477–1492

Van Dyk JS, Pletschke BI (2012) A review of lignocellulose bioconversion using enzymatic hydrolysis and synergistic cooperation between enzymes—factors affecting enzymes, conversion and synergy. Biotechnol Adv 30(6):1458–1480

Van Maris AJA, Abbott DA, Bellissimi E, van den Brink J, Kuyper M, Luttik MAH, Pronk JT (2006) Alcoholic fermentation of carbon sources in biomass hydrolysates by Saccharomyces cerevisiae: current status. Antonie Van Leeuwenhoek 90(4):391–418

Vignais PM, Billoud B (2007) Occurrence, classification, and biological function of hydrogenases: an overview. Chem Rev 107(10):4206–4272

Viikari L, Vehmaanperä J, Koivula A (2012) Lignocellulosic ethanol: from science to industry. Biomass Bioenergy 46:13–24

Wang Y, Chen L, Zhang W (2016) Proteomic and metabolomic analyses reveal metabolic responses to 3-hydroxypropionic acid synthesized internally in cyanobacterium Synechocystis sp. PCC 6803. Biotechnol Biofuels 9(1)

Wargacki AJ, Leonard E, Win MN, Regitsky DD, Santos CNS, Kim PB, Yoshikuni Y (2012) An engineered microbial platform for direct biofuel production from brown macroalgae. Science 335(6066):308–313

Wasels F, Jean-Marie J, Collas F, López-Contreras AM, Lopes Ferreira N (2017) A two-plasmid inducible CRISPR/Cas9 genome editing tool for Clostridium acetobutylicum. J Microbiol Methods 140:5–11

Wei P, Li Z, He P, Lin Y, Jiang N (2008) Genome shuffling in the ethanologenic yeast Candida krusei to improve acetic acid tolerance. Biotechnol Appl Biochem 49(2):113

Wong TY, Preston LA, Schiller NL (2000) Alginate lyase: review of major sources and enzyme characteristics, structure-function analysis, biological roles, and applications. Annu Rev Microbiol 54(1):289–340

Yim H, Haselbeck R, Niu W, Pujol-Baxley C, Burgard A, Boldt J, Van Dien S (2011) Metabolic engineering of Escherichia coli for direct production of 1,4-butanediol. Nat Chem Biol 7(7):445–452

Yomano LP, York SW, Shanmugam KT, Ingram LO (2009) Deletion of methylglyoxal synthase gene (mgsA) increased sugar co-metabolism in ethanol-producing Escherichia coli. Biotech Lett 31(9):1389–1398

Youssef N, Simpson DR, Duncan KE, McInerney MJ, Folmsbee M, Fincher T, Knapp RM (2006) In situ biosurfactant production by Bacillus strains injected into a limestone petroleum reservoir. Appl Environ Microbiol 73(4):1239–1247

Zabaniotou A, Ioannidou O, Skoulou V (2008) Rapeseed residues utilization for energy and 2nd generation biofuels. Fuel 87(8–9):1492–1502

Zhang M, Eddy C, Deanda K, Finkelstein M, Picataggio S (1995) Metabolic engineering of a pentose metabolism pathway in ethanologenic Zymomonas mobilis. Science 267(5195):240–243
Zhang Q, Yan X, Zhang L, Tang W (2006) Cloning, sequence analysis, and heterologous expression of a β-mannanase gene from Bacillus subtilis Z-2. Mol Biol 40(3):368–374
Zhengqiang J, Kobayashi A, Ahsan MM, Lite L, Kitaoka M, Hayashi K (2001) Characterization of a thermostable family 10 endo-xylanase (XynB) from Thermotoga maritima that cleaves p-nitrophenyl-β-d-xyloside. J Biosci Bioeng 92(5):423–428

Chapter 21
Microbial Fuel Cells for Bioelectricity Generation and Wastewater Treatment—A Review

Pranay Sadawarti and Shishir V. Mendhekar

Abstract Devastating environmental health which is associated with many factors of which one is improper sanitation of water, uncontrolled release of many toxic organic contaminants from the modern industries had exploited the nature a lot. The major causes for environmental pollution and depletion in the quality of surface and ground water bodies are wastewater which comes from many ways. The exhaustible usage of non-renewable sources of energy (coal, oil, natural gas, etc.) are not confined to developing countries alone but are the worldwide along. Recent advancement in the Microbial Fuel Cell (MFC) technology may serve as an eco-friendly, sustainable and the best accepted way to conserve environment and meet the criterion for the water sanitation needs. Ongoing waste water treatment technologies are less effective when it comes to sustainability, as there is rapid growth in population and industrialization and hence sudden increase in the exploitation of resources combinedly, had given researchers insight to brought such a novel technique which can serve the purpose of water sanitation along with the energy yielding conserving of the environment. MFC technology has proven the efficient mode for the generation of energy by transforming the organic waste into the electricity through microbially catalysed anodic, and microbial/enzymatic/abiotic cathodic electrochemical reactions using various redox mediators. To enhance the electron transfer between microorganisms and electrode the redox mediators are majorly used in MFC or microbial electrochemical systems. To produce electricity from the oxidation of organic compounds by using the bio-electrochemical catalytic activity of microbes which in turns uses wastewater as a source for the same, hence reducing the contaminants from wastewater. In some cases, the sources can be industrial wastewater, substrates present in urban sewage, agriculture, food, dairy industries, etc. Thus, scientists are aiming to construct the effective microbial fuel cell or microbial electrochemical systems for generation of electricity and wastewater treatment simultaneously. Besides so many

P. Sadawarti (✉)
Department of Biochemistry, Kamla Nehru Mahavidyalaya, Sakkardara Road, Nagpur 440024, MS, India
e-mail: pranaysadawarti0805@gmail.com

S. V. Mendhekar
Department of Microbiology, Kurukshetra University, Kurukshetra 136119, Haryana, India

M. Shah and D. Deka (eds.), *Emerging Sustainable Technologies for Biofuel Production*, Environmental Science and Engineering, https://doi.org/10.1007/978-3-031-52167-6_21

benefits come along with this technology, it still faces practical barriers such as low production of power and electric current density. To determine the real potential of the MFC technology for sustainable and energy generating wastewater treatment, some remarkable investigation or important discoveries in electrode materials, innovative and integrated methods configurations are urgent requirement. In the present article different parts of MFC such as anode, cathode, membrane, redox mediators and also basic mechanism of the microbial electrochemical systems or MFC for wastewater treatment and production of energy have been reviewed.

Keywords Microbial fuel cell · Wastewater treatment · Power generation · Bioelectricity · Redox mediator · Anode chamber · Cathode chamber · Electrode · Proton exchange membrane

21.1 Introduction

The exponentially increased in the population all over the world had led the tremendous increase in world energy requirement and hence, leading to the severe scarcity of these energy resources. The overconsumption and exploitation of these non-renewable energy resources generated a global crisis resulting to the liberation of various toxic gases and CO_2 to the environment (IPCC 2007; Sharma et al. 2009). It directly results in the environmental pollutions and excessive usage of natural resources, damaging the flora and fauna. Therefore, various methods and strategies are undertaken to resolve this problem and meeting the availability of the energy resources. Industrialization and economic developments had leads to the exhalation of various toxic gases in the environment along with the emission of CO_2 which is one of the major problems that has to be solved soon. Since past few years, scientists are more focusing on the renewable energy resources as an alternative to find solutions on the environmentally related issues. The major alternative sources are solar power, wind energy, tidal energy, hydrogen from biomass, hydropower energy, etc. Among these alternative resources, bioenergy showed a promising and effective alternative way. Microbial fuel cell (MFC) has been considered as one of the efficient and promising alternative renewable bioenergy technologies, the concept came into existence in year 1910 since the demonstration of MFC by Michael C. Potter in 1911 at University of Durham, UK (Potter 1911) and generation of electricity (Kim et al. 1999). Numerous findings and experiments have been conducted in recent decades for increasing power output of MFC and to achieved substantial developments and advancements in MFC. MFC has various applications in addition to bioelectricity production; it could be utilized for industrial wastewater treatment (Du et al. 2007), agricultural wastewater treatment, dairy wastewater treatment (Mohan et al. 2010), leachate treatment (Choi and Ahn 2015), treatment of toxic gases, degradation of petroleum components, production of hydrogen and methane, etc. (Greenman et al. 2009; Yi et al. 2009; Evelyn et al. 2014). The oxidation of organic substrate by metabolic processes of microorganisms to produce electrical energy with high efficiencies for long periods of time is the basic working of the MFCs. MFCs are microorganism-mediated electrochemical devices. The reactions take place in anodic compartment and cathodic compartment, both these compartments are separated by the selective proton exchange membrane. At anode compartment there is a liberation of electrons and hydrogen which moves to the anode and proton/cation exchange membrane (PEM/CEM), respectively. The whole process is summarized further in this chapter.

The basic steps of MFC are as follows:

i. As a first step, there will be the conversion of organic substrate into electrons and protons which takes place in anode compartment. The degradation of organic compound or substrate occurs in the anode compartment by the microbial activity and delivers H^+ ions and electrons.
ii. The transferring of the electrons to electrodes from the bacterial cell either directly or by the help of mediators.
iii. Further the transferred electrons pass through external circuit, from anode to cathode. The protons permeation occurs from anode to cathode simultaneously. The migration of protons in facilitated through PEM/CEM, whereas the flow of electrons from anode chamber to cathode chamber occurs through electrodes. An electrical current is generated as electrons travels through the electrical circuit.
iv. The diffusion of H^+ ions from anode to cathode creates a high electro-chemical gradient close to the anode.
v. The acceptance of electrons and protons takes place in cathode chamber which facilitate more diffusion of H^+ ions from anode to cathode, hence reduction of oxygen into water (Logan et al. 2005; Du et al. 2007; Kumar et al. 2016) (Fig. 21.1).

21.2 Basic Design and Materials of Microbial Fuel Cells

21.2.1 Electrode Materials

The efficiency of MFC for electricity generation and its bioremediation efficacy is solely depends on proper selection of anode and cathode materials. The selection of electrodes is usually depending on electrochemical efficiency, bacterial adhesion, and electron transfer (Mustakeem 2015). Generally, the cost of materials used for electrode must be low and power densities maximized. The criteria for selecting the anodic and cathodic material are as follow:

A. **Electrical Conductivity**: General mechanism for electricity generation through MFCs follows the electron released from microbes and travelling this released electron from the anode to the external circuit. Hence the electrode material having lower resistance will have higher conductivity for the electric current as it does not resist the flow of electron and results in efficient generation of electricity. The low interfacial impedance also plays a key role in facilitating the

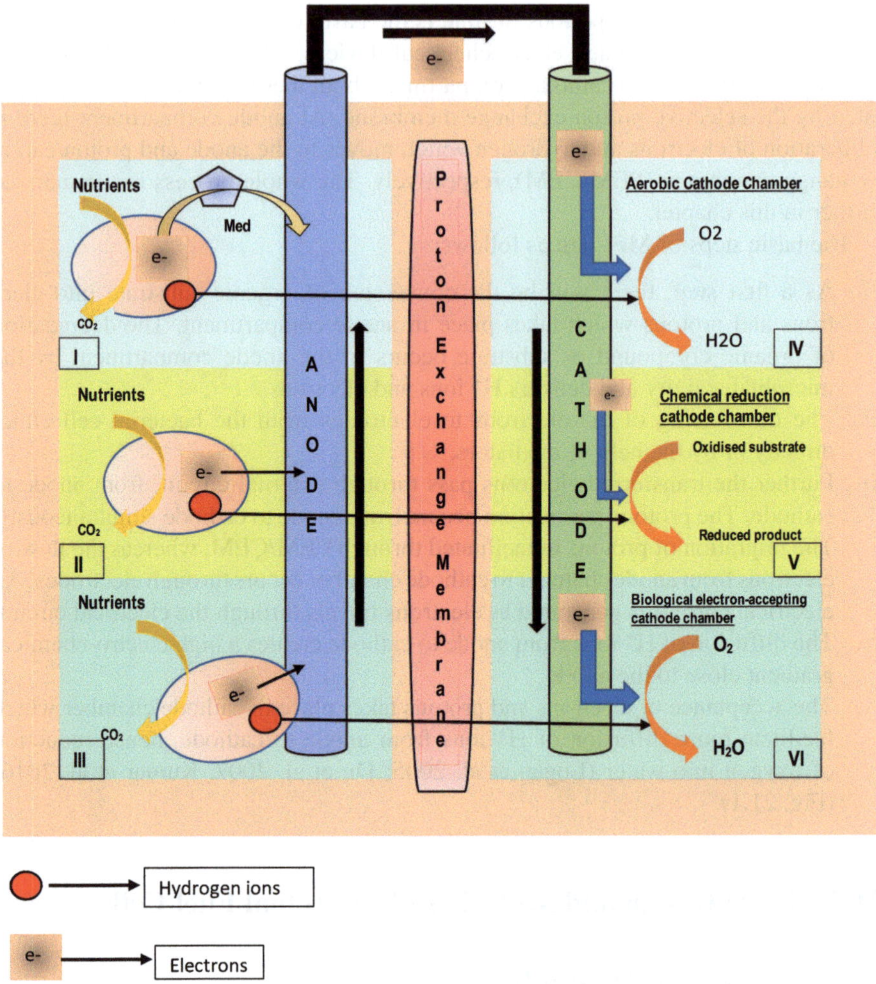

Fig. 21.1 Schematic representation of various types of microbial fuel cell. (I) Mediator-dependent anode chamber, (II) bacterial mediator-dependent anode chamber, (II) mediator-less/direct electron transfer anode chamber, (IV) air cathode chamber, (V) chemical electron-accepting cathode chamber, (VI) biological electron-accepting cathode chamber

electron transfer. The higher ionic conductivity is required for facilitating the triple-phase boundary reaction, in case of cathode (Mustakeem 2015; Natarajan and Van Nguyen 2004).

B. **Surface Area and Porosity**: The surface area of electrode play a very crucial role in the final output power of MFCs. The loss of current is directly proportional to the electrode resistance. The efficiency of the MFC can be enhanced by increasing the effective surface area by keeping the same volume and by decreasing the resistance of the electrode material. The larger surface area will provide large area for reactions that will enhance electrode kinetics (Wang et al. 2011; Rismani-Yazdi et al. 2008). Higher porosity of the electrodes allows bacteria to access and colonize more that help in biofilm formation.

C. **Stability and Durability**: The material used for electrode must be highly durable which can withstand highly reducing and oxidizing environment in MFC that may lead to decomposition of the electrode materials. This can be overcome by increasing the surface roughness of the electrodes however, excessive roughness may decrease overall performance of the MFC in longer term. Therefore, a highly stable and durable electrode are required for overall enhanced performance of MFC (Santoro et al. 2017; Mustakeem 2015)
D. **Cost and Accessibility**: For implementation of the MFC at commercial scale, the electrode material should be of low cost, sustainable, and readily available. The use of such electrode materials which are easily accessible to develop at lab scale and will help in implementation at large scale. The cost of the MFC depends on the cost of the electrode materials to a large extent. For example, platinum is widely used; however, it is expensive. Carbonaceous and non-precious metal materials such as composites might be a substitute to precious metals in electrodes in the future (Santoro et al. 2017; Mustakeem 2015).
E. **Biocompatibility**: A highly biocompatible electrode material will increase the bacterial adhesion and hence ease of current flow and the life of the MFC (Santoro et al. 2017; Mustakeem 2015).

21.2.2 Anode Materials

In general, MFC is a half-biological system in which biological activity takes place in the anode chamber. The anode chamber might be recognised the configuration's heart because it comprises the electricity generated by bacteria, the reduction of organic compounds, the decrease of chemical oxygen demand (COD), different wastewater treatments, biofilm creation, and hydrogen generation. The MFC's potential is found in the anode chamber. Microbial consortia, substrate utilised, mediators, pH, and anode material used in the anode chamber are all essential aspects in MFC efficiency. Extensive research on anode material, arrangement, compatibility, and stability has been conducted in order to increase the MFC's efficiency. The anode material should have numerous properties, including high physical stability, high chemical stability, high conductivity, low resistance, and so on. Carbon paper, graphite sheet, carbon mesh, carbon cloth, carbon felt, carbon foam, carbon brush, graphite rod, graphite plates, and reticulated vitreous carbon (RVC) are some of the most often used nonmetallic anode materials for MFCs (Wei et al. 2011; Zhou et al. 2011). Surface shape and chemistry, affects the interaction between biofilm and anode electrode. Controlling surface chemistry features such as surface charge, hydrophilicity/hydrophobicity, oxygen/nitrogen functional groups, and immobilised mediators can govern bacterial adhesion to the electrode. (Santoro et al. 2015, 2017; Du et al. 2017; Li et al. 2014). The majority of organic contaminants in waste and wastewater are decreased, and the anode serves as a location for electron donor oxidation (Zhang et al. 2016). The chemical oxygen demand (COD) is used to calculate the oxidation power of contaminants. COD is transformed to carbon dioxide and water in the anodic chamber (Zhang et al. 2016; Kim et al. 2015a, b). It has also been used to treat a variety of wastes, including dairy manure, contaminated soil samples, and landfill leachate. The anode's solid waste digesting effectiveness is low, but numerous modifications have been made to increase its efficiency, for example, polluted sediments/soil/groundwater may also be remediated by embedding anodes in the polluted matrix with an external cathode exposed to the air (Table 21.1).

21.2.3 Cathode Materials

In a traditional two-chambered MFC, the cathode chamber serves as an electron and proton acceptor, but in a single-chambered MFC, the air cathode serves as an electron acceptor. Electrons enter the cathode compartment through an external connection, where they are reduced by electron-accepting molecules and the electron cycle is completed. The cathode chamber is a limiting component that can have a significant impact on MFC performance (Rismani-yazdi et al. 2008). The oxygen reduction process (ORR) occurs at the cathode's surface at three phasic interfaces, namely electrode (solid), electrolyte (liquid), and air (gas), to produce water. The ORR is the limiting reaction of MFCs, and a typical MFC cathode is composed of three layers: diffusion, conducting support, and catalyst. To function as a robust cathode, the materials employed must have particular features such as (a) strong mechanical strength, (b) catalytic property, and (c) high electronic and ionic conductivity (Santoro et al. 2017; Mustakeem 2015). To achieve cathodic reduction, various electron acceptors such as chlorate, perchlorate, nitrite, nitrate, dichloroethane, hexavalent chromium, tetrachloroethane, ferricyanide, potassium ferricyanide, ferric ions, fumarate, sodium bromated, and others are used in the cathode chamber (Logan et al. 2006; Jadhav et al. 2014; Yang et al. 2014; Cui et al. 2016; Lian et al. 2017; Kim et al. 2017). Most anode materials may also be used as cathodes; however, carbon-based materials have limited catalytic activity and require an extra catalyst material to improve the reduction process. Table 21.2 lists the various materials utilised as cathodes.

(a) Pt-based catalyst cathode
(b) non-Pt-based catalyst cathode
(c) metal-free catalyst cathode
(d) Biocathode

Table 21.2 contains a list of cathodic materials utilised in MFC, as well as their characteristics.

At the cathode of an MFC, electron acceptors undergo reduction reactions. If the available electric potential exceeds the oxidised substance threshold, the oxidised substrates are reduced at the cathode s (Zhang et al. 2016; Nancharaiah et al. 2015).

Table 21.1 Contains a list of several anodes and their properties

Anode materials	Properties	References
Carbon cloth	High surface area, porosity, electrical conductivity, flexibility and mechanical strength in forming complex 3D structure. But expensive	Guerrini et al. (2014), Santoro et al. (2013)
Carbon brush	Expensive material consists of titanium core with twisted carbon fibres, high surface area that provides an optimal area to volume ratio. The central titanium core guarantee the electrical conductivity	Liao et al. (2015)
Carbon rod	Most affordable anode used in MFC. Surface areas for carbon rod are low. Thus they are more preferred as current collectors rather being used as anode	Jiang and Li (2009), Liu et al. (2004)
Carbon mesh	Commercially available at low cost. Carbon mesh can be folded efficiently to prepare a 3D electrode, low electrical porosity conductivity, mechanical strength and poor durability	Wu et al. (2017), Wang et al. (2009)
Carbon veil	Very cheap carbonaceous material with high electrical conductivity and porosity. It can be easily folded to form a robust and porous 3D electrode. Carbon veil is fragile	Artyushkova et al. (2016), Boghani et al. (2014), Winfield et al. (2014)
Carbon paper	Planar and relatively porous carbonaceous material. Expensive and fragile with its application limited to lab-scale batch system	Santoro et al. (2014), Srikanth et al. (2008)
Carbon felt	High electrical conductivity. Carbon felt has high porosity that allows the microbes to colonize the biofilm by penetrating through the large pores. High mechanical strength owing to thickness of the material and its cost is relatively low	Seviour et al. (2015), Roy et al. (2014), Calignano et al. (2015)
Granular activated carbon (GAC)	The granular activated carbon is used as packaging material rather than using GAC as stand-alone anode. Due to GAC's intrinsic property of very high surface area, it can help in the adsorption of organics pollutants or heavy metals. Low cost and electrical conductivity, high porosity	Yasri and Nakhla (2017), Zhao et al. (2016), Jiang et al. (2011)
Granular graphite	It has properties similar to GAC except the granular graphite has much lower surface area because of lack of activation. It is also used as packing material rather than using a stand-alone anode. High electrical conductivity	Feng et al. (2010)
Carbonized cardboard	3D materials consist of single wall corrugated cardboard from recycled paper. The material is very low cost and has high electrical conductivity and porosity	Kretzschmar et al. (2017), Chen et al. (2012b)

(continued)

Table 21.1 (continued)

Anode materials	Properties	References
Graphite plate	It has relatively lower surface area, surface to volume ratio and electric output. Due to its high mechanical strength, it is often used as support for modified structures. Graphite electrode is simple anode with high electrical conductivity and relative low cost	Dewan et al. (2008), Heijne et al. (2008)
Reticulated vitreous carbon	Reticulated vitreous carbon is highly conductive. The unique property of high porosity allows the biofilm to penetrate through the entire structure and colonize the entire electrode. The major disadvantage is the material is quite fragile and very expensive	Lepage et al. (2012)
Electrospun carbon fibres	Electrospun carbon fibres are prepared by layer-by-layer (LBL) electrospinning of polyacrylonitrile onto thin natural cellulose paper followed by carbonization. High-density layered biofilm propagation and high bioelectrocatalytic anodic current density	Chen et al. (2011), He et al. (2011)
Activated carbon nanofibres	Activated carbon nanofibres have an ultra-thin, porous interconnected structure with high bioaccessible surface area that promotes wellsupported biofilm growth. Application in simultaneous power generation and removal of organics from wastewater	Delord et al. (2017), Karra et al. (2012)
Carbonized plant materials	Several plant materials (e.g. corn stem, different mushrooms) are carbonized and tested as anode. The carbonized plant materials have highly porous architecture support, high electron transfer rate and electroactive biofilm growth. Low-cost alternative to costly anodes. Very conductive, robust and cheap and easily accessible. Major drawbacks are low surface area, biocompatibility issues and corrosion	Karthikeyan et al. (2015), Chen et al. (2012a)
Metallic anode materials	Several metallic materials such as stainless steel (plate, mesh, foam or scrubber) have been used as anodic material. Copper, nickel, silver, gold and titanium have been also successfully tested as anodic material. However, copper and nickel ions can be poisonous for microbes forming biofilm due to its metal toxicity but high and stable performance	Baudler et al. (2017), Guo et al. (2016)

(continued)

Table 21.1 (continued)

Anode materials	Properties	References
Composite material	Carbon nanotubes–polyaniline composite enhanced the electrocatalytic property and adhesion to the bacterial cell. Polypyrole-coated carbon nanotubes have enhanced power density by six times as compared to graphite electrode. PPy-coated CNT showed high electron transfer	Mustakeem (2015), Sharma et al. (2008)
Material treatments	Several surface treatments such as ammonia-treated electrode and graphene over stainless steel mesh. Acid treatment of electrode and electrochemical oxidation treatment helps in protonation of functional group and addition of functional group, respectively	Lim and Wilcox (2012), Cheng and Logan (2007), Lowy et al. (2006)

Because the cathodic chamber environment is highly reducing, it is widely used in waste (landfill leachate) and wastewater treatment (liquid fraction of pig slurry, swine wastewater), organic substances (chlorobenzene and trichloroethylene), heavy metals (Cr6+, V5+), and inorganic substances such as ammonia (Sotres et al. 2015), xenobiotic compounds, etc.

21.2.4 Membrane Material

The separator is located between the anode and cathode chambers and transmits protons from the anode compartment to the cathode compartment. In general, a separator must have a high proton transfer rate, low gas permeability, strong thermal stability, and biofouling resistance (Liu and Logan 2004; Zhao et al. 2009a, b; Li et al. 2011; Hernandez-Flores et al. 2015). Several studies examined towards the effect of different separators on internal resistance, gas diffusion, and cost-effective membrane. Separators are grouped into two types based on their material and exchange mechanism: ion exchange membrane separators and size selective separators. Membranes are categorised into five types based on their materials: proton exchange membrane (PEM), anion exchange membrane (AEM), cation exchange membrane (CEM), bipolar membrane (BPM), and polymer composite porous membrane (PCPM). The most prevalent membrane system is cation-exchange membranes (CEM), such as Naflon (Santoro et al. 2017). Because of its high proton transfer capacity, stability, and low resistance, Nafion is regarded as an excellent separator in terms of power generation. However, it has several drawbacks: the first is the high cost of the membrane; the second is the high permeability of oxygen across the membrane; and the third is the exchange of cation across the membrane (Wang et al. 2010a, b; Leong et al. 2013; Ercelik et al. 2017). As a result, many alternative separators were investigated to overcome these issues. A composite membrane with polyvinyl alcohol borosilicate and polyvinyl Nafion borosilicate as a separator

Table 21.2 List of cathodic materials and their properties used in MFC

Cathode material	Cathode catalyst	Properties	References
Cathode with Pt-based catalyst/ platinum metal based	Carbon-Pt	Platinum is the most successful catalyst used for carbon cathode coating for oxygen reduction reaction because of its high surface area and low over potential for ORR. Limitations such as high cost, pH sensitivity, sulphide poisoning and non-sustainable	Watanabe (2008)
	Double-layer polydimethylsiloxane (PDMS)/carbon loaded with Pt	High-power density and hydrophobic materials like PDMS decrease the water diffusion into a single-chambered MFC	Zhang et al. (2016), Mustakeem (2015)
	Carbon nanotubes (CNTs) loaded with Pt	CNTs loaded with Pt by depositing it over CNTs surface. ORR catalytic activity was improved for CNT–Pt composite. The ORR was not affected considerably by decreasing Pt loading by 20-fold	Ghasemi et al. (2011), Cheng et al. (2006)
Cathode with non-Pt-based catalyst/ platinum group metal-free	Iron phthalocyanine	Rapid electron transfer takes advantage of π–π interaction of metal and carbon	Yuan et al. (2011)
	Iron phthalocyanine-aminofunctionalized multiwalled CNTs	High-power density	Yuan et al. (2011)
	Cobalt tetramethoxyphenylporphyrin (CoTMPP)	Coulombic efficiency is comparable to Pt-based catalyst	Cheng et al. (2006)
	Lead dioxide–carbon	Alternative to Pt as a catalyst, lead oxide has low cost and shows four times higher power densities as compared to conventional Pt electrodes	Morris et al. (2007)

(continued)

Table 21.2 (continued)

Cathode material	Cathode catalyst	Properties	References
	Transition metal-based oxide	The cathode current with MnO_2-catalysed cathode was found to be much larger than graphite. β-MnO_2 deposited over glassy carbon using polyvinylidene fluorine (PVDF) as binder showed high catalytic efficiency and power densities as close to Pt based. CNTs covered with MnO_2 as cathode material showed high-power density increased by two order of magnitude as compared to plain stainless steel cathode	Amade et al. (2015), Cao et al. (2003)
	Palladium	Palladium, a Pt-like transition metal, has excellent catalytic. High stability and low cost as compared to Pt	Huang et al. (2011)
	Lead oxide	Four time higher-power density and eight time less cost compared to commercial Pt/C	Morris et al. (2007)
	Activated carbon	Large surface area, synthesized by electrospinning and pyrolysis of polyacrylonitrile precursor. Activated carbon fibre-treated electrodes was increased by more than 75%	Ghasemi et al. (2011)
Cathode with metal-free catalyst/carbonaceous materials	Chemically modified carbon-based materials	Carbon nanotubes, graphene, and graphite foam can be used as catalyst for ORR	Su et al. (2010), Matter et al. (2006)
		Low cost, relatively high and stable performance	

(continued)

Table 21.2 (continued)

Cathode material	Cathode catalyst	Properties	References
	Nitrogen-doped graphene nanosheets	High catalytic activity, onset potential, higher electrocatalytic activity, lower over potential and long-term stability than commercially available Pt (10 or 20 wt%, E-TEX) for ORR	Lu et al. (2013), Ci et al. (2012), Gong et al. (2009)
Biocathode	Aerobic biocathode	In aerobic biocathode, oxygen acts as terminal electron acceptor, hydrogen peroxide acts as intermediate and transition metals such as iron and manganese act as electron mediators between the electrode and oxygen	Park and Zeikus (2003)
	Anaerobic type biocathode	In anaerobic type biocathode, oxygen is not present, whereas nitrates and sulphates act as terminal electron acceptors. Advantage by preventing the loss of electrons through oxygen also microorganisms at the cathode can act as biosensor for detection of biological oxygen demand (BOD) in water. Further microorganisms at biocathode can be used to produce as methane, ethanol and formic acid	Rabaey and Rozendal (2010), Kim et al. (2003)

was developed, which reduced the cost of the membrane by 11-fold when compared to traditional Nafion Tiwari et al. (2016). Another study discovered that the use of ceramic membrane could increase the efficiency of proton transfer, resulting in high power density output when compared to Nafion Daud et al. (2017). These breakthroughs in membrane production have recently decreased the cost of the technology. However, obstacles remain in producing membranes for large-scale MFC applications.

21.3 Types of Waste Materials Used as Substrates in MFC

MFCs are regarded as an efficient technology that successfully utilises wastewater for energy generation (Winfield et al. 2016). The following waste products are utilised as substrates in MFC.

21.3.1 Biomass from Lignocellulosic Materials

Natural resources such as lignocellulosic materials are plentiful and renewable. However, it cannot be used directly because it must first be converted to monosaccharides or low-molecular-weight compounds before being used by the microorganism to generate electricity (Huang et al. 2011; Ren et al. 2007). Cellulose and chitin are inexpensive, renewable, and widely accessible biopolymeric materials that can be utilised to generate power (Rezaei et al. 2009). To convert cellulose directly, the microorganism(s) must be able to hydrolyze it anaerobically, using an anode as an electron acceptor and oxidising metabolites produced by cellulose hydrolysis.

21.3.2 Synthetic Wastewater

It has also been discovered that synthetic or chemical effluent with a specific composition has been applied (Pant et al. 2010). Synthetic wastewater, on the other hand, may contain redox mediators such as cysteine and sulphur species. These redox mediators can operate as abiotic electron donors and help boost electricity output for a brief time, but they do not represent the real performance of the system, resulting in uncertainty in the results (Aldrovandi et al. 2009). Thus, using minimal salt media with a single electron donor such as glucose or acetate can overcome this.

21.3.3 Wastewater from a Brewery

Brewery wastewater has been employed in MFCs largely due to its low strength, suitability for power production due to the food-derived nature of organic matter, and lack of high quantities of inhibitory chemicals, such as ammonia in animal wastewaters (Feng et al. 2008).

21.3.4 Wastewater from Starch Processing

Starch processing wastewater (SPW) includes a high concentration of carbs, sugars, protein, and starch, which may be transformed into a range of valuable products (Jin et al. 1998). SPW was utilised as a fuel to enhance a microbial consortium that produced energy and current.

21.3.5 Wastewater from Dyes

Azo dyes are the most common type of synthetic dye and are widely prevalent in the effluent of the dye and textile industries. Toxic compounds included in effluents must be removed before disposal because they have a negative impact on the environment. In aquatic systems, the strong colour of dyes has obstructed light and oxygen transport into water bodies, causing harm to aquatic plants and wildlife (Pant et al. 2007). Recently, efforts have been made to use these dyes as a substrate in MFC, resulting in purification and the generation of energy. In the present generation, the concentration of glucose and dyes is crucial; nevertheless, the microorganisms are unable to decolorize high amounts of dye (Sun et al. 2009). Another approach that may be employed is the simultaneous treatment of azo dye-containing wastewater and easily biodegradable organic matter-containing wastewater.

21.3.6 Leachates from Landfills

Landfill leachates are highly contaminated landfill effluents with four primary types of pollutants: dissolved organic matter, inorganic macro-components, heavy metals, and xenobiotic organic chemicals (Kjeldsen et al. 2002). Greenman et al. (2009) has proved that it is feasible to generate energy in addition to treating landfill leachate in MFC columns.

21.3.7 Inorganic and Other Substrates

There are several substrates that have been employed in the MFC as substrates, including carbon monoxide (Kim and Chang 2009), 1,2-dichloroethane (Pham et al. 2009), sulphate, and thiosulphate (Zhao et al. 2009a, b), effluent from paper recycling facilities (Huang and Logan 2008), unaltered wastewater and phenol (Luo et al. 2009). MFC was used to remove fermentation inhibitors such as furfural, 5-hydroxymethylfurfural, vanillic acid, 4-hydroxybenzaldehyde, and 4-hydroxyacetophenone from cellulosic biorefineries while additionally generating energy (Borole et al. 2009).

21.3.8 Heavy Metals

There are several industrial uses for chromium, including wood preservatives, electroplating, leather tanning, and metallurgy (Chaturvedi and Verma 2016). Because of its mutagenic and carcinogenic characteristics, Cr(VI) is more dangerous (Humphries et al. 2004); therefore, hexavalent chromium Cr(VI) detoxification is consequently necessary. The most recommended process is chemical or electrochemical reduction of toxic chromium into harmless chromium. Therefore, chromium reduction can be combined with MFC electricity generation and used for Cr(VI) treatment at the cathode (Tandukar et al. 2009). Pseudomonas aeruginosa and Trichococcus pasteurii are only a couple of the microorganisms involved in the procedure.

Selenium may enter the environment through the mining of metal ores, oil refineries, coal-fired power plants, and sewage sludgem (Tandukar et al. 2009). Selenite accumulates in aquatic plants and is more hazardous than selenate to aquatic invertebrates and fishes. This bioaccumulation in higher species can result in both acute and chronic toxicity in aquatic creatures (Hamilton 2004; Lemly 1997; Rovira et al. 2008). Utilising waste containing selenium to produce electricity and reduce selenium levels while acting as a substrate for microorganisms was investigated (Catal et al. 2009).

21.3.9 Nitrate

The amount of nitrate in water has grown due to the usage of nitrate-based fertilisers and animal manure. After entering the human body, nitrate (which is nontoxic) can change into nitrite (NO_2), which can lead to the "blue baby syndrome" commonly seen in newborns, or it can change into N-nitroso compounds, which are carcinogenic to humans. The use of MFC for the removal of nitrate has gained relevance due to the process's viability because numerous approaches currently in use are costly. In a few investigations, the bio-electrochemical denitrification system's cathode chamber underwent electrochemical denitrification to remove nitrate ions (Kondaveeti and Min 2013).

21.3.10 Marine Sediments Rich in Acetate

The substrate for MFC has been made from acetate-rich marine sediments. Inertness towards alternate microbial conversions, fermentations, and methanogenesis is facilitated by the presence of acetate at ambient temperature (Sun et al. 2009; Aelterman 2009). Acetate is an end product of various metabolic routes, including the Entner-Doudoroff pathway for the metabolism of glucose, and it may be use as carbon source to activate electroactive bacteria (Biffinger et al. 2008; Bond and Lovley 2005).

21.3.11 Animal Wastes

Due to its high daily production, high COD, high nutrient (N and P) concentrations, and high solution conductivity, animal waste can be thought of as a potential substrate for the generation of electricity (Barbosa et al. 2017; Jimenez et al. 2016). Apparently, human faeces, cow dung, chicken droppings, and other animal wastes are used as a substrate for the production of bioelectricity.

21.3.12 Petroleum Hydrocarbons

MFC can be a cost-effective and environmentally friendly approach for bioremediation of petroleum hydrocarbons. As a result, MFC technology can be used effectively for detoxification and enhancing biodegradation of polluted soil/wastewater containing petroleum contaminants in anoxic environments, thereby eliminating the need to adjust terminal electron acceptors like oxygen.

21.4 Types of Microbial Fuel Cells

21.4.1 Classification of Microbial Fuel Cell

The MFC could be classified based on the configuration, mode of nutrition, microbe used, mediators and usage of membrane. The different types of MFC are classified and represented in Fig. 21.2. The types and design of the MFC are temporary and can be change according to the experimental need hence, new compatible designs and types will emerge to overcome the present barriers of the existing type. Therefore, here, a few types based on the above-said factors have been discussed. The various types of MFC configurations, electrodes used, substrate utilized, power output and its applications are summarized in Table 21.3.

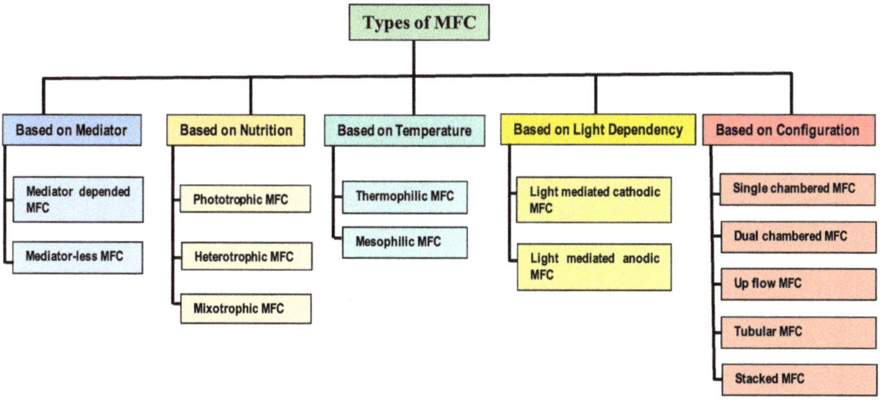

Fig. 21.2 Schematic representations of several MFC types that have been proposed

21.4.1.1 Based on Whether Mediator Use or Not

Electrons can travel to anode in three different ways:

(i) The mediators which are exogenous and soluble accept the electron from microbe and can transfer it to anode (e.g., methylene blue, neutral red, etc.)
(ii) The electron shuttle from microbial membrane to anode through mediators produce by microbes. (e.g., flavin, pyocyanin, etc.)
(iii) Direct transfer of electron to the electrode from microbe via nanowires (e.g., c-Cyts, pili)

Some microbes, such as Desulfobulbus species, Geobacter species, Pseudomonas species, Rhodopseudomonas species, and Shewanella species, have the ability to transfer electrons extracellularly through pilin-related proteins (type IV pil, pil S, and pil R), membrane-bound c-type cytochromes, and cell-secreted substances like flavin and pyocyanin. For instance, Shewanella oneidensis and Geobacter sulfurreducens transmit electrons by forming biofilms on the anode surface. Genes like gene pilA, long PilA, gene pilC, gene pilR, and OmcZ, which in turn regulate the biogenesis of type IV pili, the attachment of biofilm on the electrode surface, and the synthesis of structural proteins and outer membrane c-cytochrome Z, are responsible for the efficiency of electron transfer (Xing et al. 2008; Rollefson et al. 2011; Luo et al. 2013; Smith et al. 2014).

However, some of the microbes synthesize the redox-active compounds such as pyocyanin and flavin which are potent to create an electron gradient in the outer membrane which facilitates the transfer of the electrons (Rollefson et al. 2011). But certain bacteria's and yeast cells require mediator for electron transfer from the outer membrane. It has been reported that secondary substances from wastewater treatment, such as intermediates of azo dyes and malachite green, are used and involved in the transfer of electrons from the membrane to the anode. Also, external mediators such as ferricyanide, humic acid, methylene blue (MB), methyl viologenare,

Table 21.3 Types of MFC, configuration and applications

Type of MFC construction	Microbes utilized	Anode material	Cathode material	Substrate in anode	Catalyst used	Separator material	Power output	Application apart from electricity	References
Single chambered (super capacitive microbial fuel cells)	Electro active bacteria	Carbon brush	Airbreathing cathode	Potassium phosphate buffered saline Potassium chloride Activated sludge	Fe-Amino antipyrine	–	8.1 mW	Hydrogen production	Santoro et al. (2016)
Dual chambered H type	Phototrophic consortium, from a wastewater treatment plant	Graphite felt	Graphite felt	Synthetic media	–	Ultrex membrane	2650 m Wm^{-2}	-	Cao et al. (2008)
Single chamber	Mixed photosynthetic consortia	Plain graphite plates	Plain graphite plates	Designed synthetic wastewater	–	Nafion-117	3.2 mA	Volatile fatty acids removal, treatment of acid-rich effluent, wastewater treatment	Chandra et al. (2012)
Dual chamberd MFC	Intestinal microbes	Carbon fiber cloth	Carbon paper electrode	Stimulated intestinal fluid with fresh faecal samples	Pt catalyst	Carbon paper	73.3 mW/m^2	Power for implantable medical devices	Han et al. (2010)

(continued)

Table 21.3 (continued)

Type of MFC construction	Microbes utilized	Anode material	Cathode material	Substrate in anode	Catalyst used	Separator material	Power output	Application apart from electricity	References
Stacked MFC	Sludge microbial community	Carbon cloth	Carbon cloth	Activated sludge	–	Nafion 117	22.2 mW	Power production	Zhang et al. (2017)
Phototrophic MFC	Rhodo pseudomonas spp.	Carbon felts	Carbon felts	Wetland sediment Anaerobic wastewater	–	Cation exchange membranes	202.9 ± 18.1 mW/m^2	Remove pollutant	Zheng et al. (2017)
Mixotrophic MFC	*Synechocystis* sp. PCC 6803 *Shewanella oneidensis* MR-1	Carbon-based material	Air cathode	LB-broth media	Pt/C	–	8 μA/cm^2	Miniature solardriven MFCs	Liu and Choi (2017)
Thermophilic MFC	Thermophilic bacteria from anaerobic reactor	Carbon fiber felt	Carbon fiber felt	Synthetic medium	–	Cation exchange membrane CMI-7000	437 mWm^{-2}	–	Dai et al. (2017)
Thermophilic MFC	Municipal solid waste as innoculum	Carbon felt	Graphite electrode	Thermophilic anaerobic sludge	–	Cation-exchange membrane	10 mA	Biogas production Anaerobic digestion	Liu et al. (2017)
Tubular MFC	Microbial consortia	Poly vinyl alcohol-coke	Air–cathode	Benzene	–	Poly vinyl alcohol hydrogel	30 mWm^{-2}	Benzene removal	Chang et al. (2017)

neutral red (NR), and thionin has been reported (Chen et al. 2014; Xu et al. 2014). A mediator should be able to easily diffuse into cell membranes, have a high rate of solubility, take electrons from membranes, and not be poisonous to microorganisms (Park et al. 2000; Park and Zeikus 2000; Chen et al. 2014; Zhang et al. 2018). Ferricyanide and methylene blue both have the distinguishing characteristics, however they are expensive and enhance the technique's cost, in addition use of these mediator decreases the viability of the microbe when used. In order to increase efficiency, several research studies are concentrating on secondary metabolite-mediated electron transfer pathways and focusing on mediator-less MFC.

21.4.1.2 Based on Microbial Nutrition Dependency

Phototrophic MFC

The metabolism of the phototrophic MFC is primarily dependent on light. Photosynthetic microorganisms use light energy to generate chemically energy-rich compounds, which are then transformed into electric energy. As a result, light plays a vital role in phototrophic MFCs and may have a considerable influence on microbial processes and metabolism. Recent interest in phototrophic microalgae for electricity generation has grown due to its several uses such as high-power production, waste water treatment, the capacity to manufacture high-potential valuable items, and so on. In the presence of light, microalgae use carbon dioxide to create electrons. The benefits of these phototrophic algal MFC in wastewater treatment, biofuel production, and power generation all work together to make them cost-effective for widespread use. Several microalgae have been studied for the simultaneous generation of energy and wastewater treatment (e.g., *Arthrospira maxima, Chlorella vulgaris, Dunaliella tertiolecta, Scenedesmus obliquus*) (Wang et al. 2010a; Lakaniemi et al. 2012; Inglesby and Fisher 2013; Kakarla and Min 2014).

Heterotrophic MFC

The second category of nutritional-dependent MFC is the use of heterotrophic microorganisms to generate power. The use of heterotrophic microorganisms in MFC might result in quicker removal of organic material in wastewater, energy production, nitrogen removal, and phosphorus removal (Kessel et al. 2015; Wang et al. 2015). In general, most MFC studies are supported by heterotrophic bacteria, which consume various organic carbon sources for growth. For example, in algal MFC under light conditions, photosynthesis supports growth and metabolism, whereas in dark conditions, algal cells uptake organic compounds for metabolism from the medium/substrate (Sanders et al. 2001; Liang et al. 2009; Wang et al. 2015).

Mixotrophic MFC

The third type of nutritional-dependent MFC is mixotrophic MFC, which combines algae and bacteria. The benefits of mixotrophic MFCs over phototrophic and heterotrophic MFCs are numerous. The dead algae replace the need for extra substrate for the bacteria (electrogens) in the mixotrophic mode of operation. The inclusion of substrate improves the MFC's performance. In certain research, algal cells were used as a substrate in the anode chamber. In MFC, many microalgae (e.g., *Chlorella, Scenedesmus, Dunaliella tertiolecta, Microcystis aeruginosa, Arthrospira maxima*, etc.) were used to provide organic substrate to the bacteria (Lakaniemi et al. 2012; Inglesby et al. 2012; Inglesby and Fisher 2013; Nishio et al. 2013; Cui et al. 2014; Wang et al. 2015).

21.4.1.3 Based on Dependency of Light

Although the integration of phototrophic microorganisms in MFC requires light for metabolism, the light source was delivered to either the anode or cathode side of the chamber depending on the type of the microbe utilised in MFC. The phototrophic organism might function as an electron donar or acceptor. Microalgae grown in the anode compartment function as electron donors, while those growing in the cathodic compartment work as electron acceptors. As a result of this circumstance, MFC is further classified into two types:

- Anodic MFC mediated by light
- Cathodic MFC mediated by light

In the anode compartment of light-mediated anodic MFC, either phototrophic or mixotrophic microorganisms are utilised. The light source is located on the anode side of this kind. MFC construction ranges from single-chamber to two-chamber layout. For power generation, the microalga was put in the anode compartment, where it produced 7.3 W/m^3 of power density (Rosenbaum et al. 2005). Similarly, phototrophic microorganisms in the cathode compartment are aided by a light source in light-mediated cathodic MFC (Powell et al. 2009; Wu et al. 2013; El-Mekawy et al. 2014; Mohan et al. 2014).

21.4.1.4 Based on Dependency of Temperature

In recent years several studies integrated thermophiles for the production of electricity and wastewater treatment using MFC. Some unique characteristics of thermophiles make them advantageous over mesophiles. Therefore, the dependency of the temperature for growing microorganism in MFC leads to classify into thermophilic MFC and mesophilic MFC. The mesophilic MFC operated in ambient temperature for bioenergy and wastewater treatment, whereas thermophilic microbe requires high temperature for metabolism; therefore, the MFC operated at elevated temperature.

Carver et al. (2011) conducted a study with thermophilic consortium at 57 °C in anaerobic condition with glucose. The batch mode operation produced 375 mWm^{-2} power density (Carver et al. 2011). In another study Dai et al. (2017) reported a power density of 437 mWm^{-2} by using thermophilic bacterium with ethanol as a substrate.

21.4.1.5 Based on Configuration

Regardless of the mediators, bacteria used, light requirements, or nourishment method, the MFC may be classified into many fundamental categories depending on their setup. Currently, the following five principal varieties are widely utilised for bioenergy generation and wastewater treatment:

- MFC with a single chamber
- MFC with dual chambers
- MFC with upflow
- MFC with tubular tubes
- MFC with stacked tubes

However, MFC configurations vary, and new configurations are continually evolving based on the obstacles that exist in MFC. Several publications documented the setup of the aforementioned basic kinds; thus the general benefits and drawbacks of these types are examined in this section. The single-chambered MFC provides benefits such as low construction costs, the absence of a membrane, low internal resistance, high power output, and ease of fabrication (Liu and Logan 2004; Zhang et al. 2011; Logroño et al. 2017).

However, it has major drawbacks, including strong oxygen diffusion, which reduces coulombic efficiency (Nimje et al. 2012; Yao et al. 2014). The dual-chambered MFC provides benefits over the single-chambered MFC, such as avoiding oxygen diffusion by using an alternating membrane between the anode and the cathode. Upflow and tubular MFC increase the benefits of bioenergy and wastewater treatment. Several studies, however, have found that electricity output is reduced in upflow and tubular MFC. Furthermore, external help is necessary in upflow MFC, which raises the expense of the procedure and necessitates a large volume of area (Zhou et al. 2013; Tee et al. 2016). The main disadvantage of adopting stacked MFC instead of a single big anode and/or cathode chamber in continuous flow mode of wastewater treatment MFC is greater internal resistance. A number of separate models are layered in parallel or vertical position in stacked MFC. This will lower internal resistance caused by the compact set-up and boost total power output (Aelterman et al. 2006; Ledezma et al. 2013; Zhou et al. 2013).

21.5 Application of MFC

The main uses of MFCs are the generation of power, wastewater treatment, chemical compound degradation, biohydrogen synthesis, and the manufacture of biosensors. Since the primary objective of MFC is the production of bioelectricity, numerous studies have concentrated on improving power density. To boost the power production, numerous alterations were made, including changes to the substrate, the pH, the anode and cathode, the membrane components, and various microorganisms. Along with the generation of electricity, concurrent wastewater treatment was also extensively researched using MFC. In MFC, various wastewater types have been treated and used to generate bioelectricity. Li et al. (2016) used coupled system MFC with acetate as the substrate for azo dye decoloration investigations. The system increased 36.52–75.28% more azo dye decolorization than a single MFC, according to an evaluation of the model's performance (Li et al. 2016). Gajda et al. (2014) used MFC for a combined biodegradation and bioenergy investigation as an alternative method. According to the study's findings (Gajda et al. 2014), the synthesis of catholytes improves the efficiency of electricity production and aids in the treatment of wastewater and the biodegradation of organic wastes. A technology's success in the market depends on a number of factors, including affordability, efficiency, environmental friendliness, market size, and, most importantly, profitability. The study's findings made it abundantly evident that using MFC instead of conventional aerated activated sludge is preferable and offers potential cost advantages. Research paths may be focused on the choice of microbial consortia, together with operational consistency and stability of the produced system, while designing for a present state-of-the-art MFC while bearing in mind future issues.

21.6 Challenges and Future Prospects

The development of MFC offers various advantages, however, several limitations are persisting in MFC scale-up. Some of the challenges must be addressed to overcome for successful development in this area. The major limitation of this technique is voltage instability. A stable voltage is required to run a system, but voltage production in MFC is comparatively low, and stability of the production is also quite uncertain. This will lead to power inadequate to run a system like biosensor, etc. (Shantaram et al. 2005; Paitier et al. 2017). In recent years, significant technological advancement and improvement have been observed for waste/wastewater treatments in MFCs. MFCs have multiple applications, including energy generation, waste treatment and production of chemicals, sensors for pollution level in water/soil, etc. It can generate energy, treat waste, produce chemicals, and more. Some of the MFC challenges and their solutions are stated further:

(a) Voltage instability can be overcome by periodical adjustment of operating parameter, microbial load, oxidant concentration.

(b) High internal resistance can be solved through employing stacked type with less distance between electrodes.
(c) Identifying good mediator without toxicity can solve the problem of low electron transfer.
(d) Low power density generation can be resolve by using Identifying potential microbe. Mutant strains and DNA recombinant strains. Using microbial consortia.
(e) Challenge of biofouling of membrane can have solution like using ceramic membrane. Synthesis of antifouling membranes.

21.7 Conclusions

Most intriguingly, the MFCs offer opportunity for the selection of microbes either bacteria or algae, and the recent progress in MFC such as bacterio-algal fuel cell provides additional advantages such as wastewater treatment, toxicity removal, and energy demand reduction. This expanded use of MFC lowers the fabrication cost and makes the technology practical. The main constraints, however, restrict the power output. Therefore, prior to starting up, a specific development and management should be taken care of. If these restrictions and cost-related problems are resolved, MFC will be a crucial technology for the production of electricity. Any sustainable and promising technology requires thorough research. Thus, improving power generation and large-scale operations still require additional research and development.

References

Aelterman P (2009) Microbial fuel cells for the treatment of waste streams with energy recovery. Dissertation, Ghent University, Faculty of Bioscience Engineering

Aelterman P, Rabaey K, Pham HT, Boon N, Verstraete W (2006) Continuous electricity generation at high voltages and currents using stacked microbial fuel cells. Environ Sci Technol 40:3388–3394

Aldrovandi A, Marsili E, Stante L, Paganin P, Tabacchioni S, Giordano A (2009) Sustainable power production in a membrane-less and mediator-less synthetic wastewater microbial fuel cell. Bioresour Technol 100:3252–3260

Amade R, Vila-Costa M, Hussain S, Casamayor EO, Bertran E (2015) Vertically aligned carbon nanotubes coated with manganese dioxide as cathode material for microbial fuel cells. J Mater Sci 50:1214–1220

Artyushkova K, Roizman D, Santoro C, Doyle LE, Fatima Mohidin A, Atanassov P, Marsili E (2016) Anodic biofilms as the interphase for electroactive bacterial growth on carbon veil. Biointerphases 11:31013

Barbosa SG, Peixoto L, Ter Heijne A, Kuntke P, Alves MM, Pereira MA (2017) Investigating bacterial community changes and organic substrate degradation in microbial fuel cells operating on real human urine. Environ Sci Water Res Technol 3:897–904

Baudler A, Langner M, Rohr C, Greiner A, Schröder U (2017) Metal–polymer hybrid architectures as novel anode platform for microbial electrochemical technologies. Chemsuschem 10:253–257

Biffinger JC, Byrd JN, Dudley BL, Ringeisen BR (2008) Oxygen exposure promotes fuel diversity for Shewanella oneidensis microbial fuel cells. Biosens Bioelectron 23:820–826

Boghani HC, Papaharalabos G, Michie I, Fradler KR, Dinsdale RM, Guwy AJ, Ieropoulos I, Greenman J, Premier GC (2014) Controlling for peak power extraction from microbial fuel cells can increase stack voltage and avoid cell reversal. J Power Sources 269:363–369

Bond DR, Lovley DR (2005) Evidence for involvement of an electron shuttle in electricity generation by Geothrix fermentans. Appl Environ Microbiol 71:2186–2189

Borole AP, Mielenz JR, Vishnivetskaya TA, Hamilton CY (2009) Controlling accumulation of fermentation inhibitors in biorefinery recycle water using microbial fuel cells. Biotechnol Biofuels 2:7

Calignano F, Tommasi T, Manfredi D, Chiolerio A (2015) Additive manufacturing of a microbial fuel cell—a detailed study. Sci Rep 5:17373

Cao YL, Yang HX, Ai XP, Xiao LF (2003) The mechanism of oxygen reduction on MnO_2-catalyzed air cathode in alkaline solution. J Electroanal Chem 557:127–134

Cao X, Huang X, Boon N, Liang P, Fan M (2008) Electricity generation by an enriched phototrophic consortium in a microbial fuel cell. Electrochem Commun 10(9):1392–1395

Carver SM, Vuoriranta P, Tuovinen OH (2011) J Power Sour In Press. https://doi.org/10.1016/j.jpowsour.2010.12.088

Catal T, Bermek H, Liu H (2009) Removal of selenite from wastewater using microbial fuel cells. Biotechnol Lett 31(8):1211–1216

Chandra R, Subash GV, Mohan SV (2012) Mixotrophic operation of photo-bioelectrocatalytic fuel cell under anoxygenic microenvironment enhances the light dependent bioelectrogenic activity. Bioresour Technol 109:46–56

Chang SH, Wu CH, Wang RC, Lin CW (2017) Electricity production and benzene removal from groundwater using low-cost mini tubular microbial fuel cells in a monitoring well. J Environ Manag 193:551–557

Chaturvedi V, Verma P (2016) Microbial fuel cell: a green approach for the utilization of waste for the generation of bioelectricity. Bioresour Bioprocess 3:38

Cheng S, Liu H, Logan BE (2006) Power densities using different cathode catalysts (Pt and CoTMPP) and polymer binders (Nafion and PTFE) in single chamber microbial fuel cells. Environ Sci Techn 40(1):364–369

Cheng S, Logan BE (2007) Ammonia treatment of carbon cloth anodes to enhance power generation of microbial fuel cells. Electrochem Commun 9:492–496

Chen S, He G, Carmona-Martinez AA, Agarwal S, Greiner A, Hou H, Schröder U (2011) Electrospun carbon fiber mat with layered architecture for anode in microbial fuel cells. Electrochem Commun 13:1026–1029

Chen S, He G, Hu X, Xie M, Wang S, Zeng D, Hou H, Schröder U (2012a) A three-dimensionally ordered macroporous carbon derived from a natural resource as anode for microbial bioelectrochemical systems. Chemsuschem 5:1059–1063

Chen S, He G, Liu Q, Harnisch F, Zhou Y, Chen Y, Hanif M, Wang S, Peng X, Hou H, Schroder U (2012b) Layered corrugated electrode macrostructures boost microbial bioelectrocatalysis. Energy Environ Sci 5:9769–9772

Chen BY, Xu B, Qin LJ, Lan JCW, Hsueh CC (2014) Exploring redox-mediating characteristics of textile dye-bearing microbial fuel cells: thionin and malachite green. Bioresour Technol 169:277–283

Choi J, Ahn Y (2015) Enhanced bioelectricity harvesting in microbial fuel cells treating food waste leachate produced from biohydrogen fermentation. Bioresour Technol 183:53–60

Ci S, Wu Y, Zou J, Tang L, Luo S, Li J, Wen Z (2012) Nitrogen-doped graphene nanosheets as high efficient catalysts for oxygen reduction reaction. Chin Sci Bull 57:3065–3070

Cui Y, Rashid N, Hu N, Rehman MSU, Han JI (2014) Electricity generation and microalgae cultivation in microbial fuel cell using microalgae-enriched anode and bio-cathode. Energy Convers Manag 79:674–680

Cui H, Yang H, Liu X, Zhao Y, Liang Z (2016) Performance of sodium bromate as cathodic electron acceptor in microbial fuel cell. Bioresour Technol 202:220–225

Dai K, Wena JL, Zhang F, Maa XW, Cui XY, Zhang Q, Zhao TJ, Zeng RJ (2017) Electricity production and microbial characterization of thermophilic microbial fuel cells. Bioresour Technol 243:512–519

Daud SM, Daud WRW, Kim BH, Somalu MR, Bakar MHA, Muchtar A, Jahim JM, Lim SS, Chang IS (2017) Comparison of performance and ionic concentration gradient of two chamber microbial fuel cell using ceramic membrane (CM) and cation exchange membrane (CEM) as separators. Electrochim Acta

Delord B, Neri W, Bertaux K, Derre A, Ly I, Mano N, Poulin P (2017) Carbon nanotube fiber mats for microbial fuel cell electrodes. Bioresour Technol 243:1227–1231

Dewan A, Beyenal H, Lewandowski Z (2008) Scaling up microbial fuel cells. Environ Sci Technol 42:7643–7648

Du H, Li H, Gu T (2007) A state of the art review on microbial fuel cells: a promising technology for wastewater treatment and bioenergy. Biotechnol Adv 25:464–482

Du Q, An J, Li J, Zhou L, Li N, Wang X (2017) Polydopamine as a new modification material to accelerate startup and promote anode performance in microbial fuel cells. J Power Sources 343:477–482

El-Mekawy A, Hegab HM, Vanbroekhoven K, Pant D (2014) Techno-productive potential of photosynthetic microbial fuel cells through different configurations. Renew Sust Energy Rev 39:617–627

Ercelik M, Ozden A, Devrim Y, Colpan CO (2017) Investigation of Nafion based composite membranes on the performance of DMFCs. Int J Hydrog Energy 42:2658–2668

Evelyn LY, Marshall A, Gostomski PA (2014) Gaseous pollutant treatment and electricity generation in microbial fuel cells (MFCs) utilising redox mediators. Rev Environ Sci Biotechnol 13:35–51

Feng Y, Wang X, Logan BE, Lee H (2008) Brewery wastewater treatment using air-cathode microbial fuel cells. Appl Microbiol Biotechnol 78:873–880

Feng Y, Lee H, Wang X, Liu Y, He W (2010) Continuous electricity generation by a graphite granule baffled air-cathode microbial fuel cell. Bioresour Technol 101:632–638

Gajda I, Greenman J, Melhuish C, Santoro C, Li B, Criatiani P, Ieropoulos I (2014) Water formation at the cathode and sodium recovery using microbial fuel cells (MFCs). Sustain Energy Technol Assess 7:187–194

Ghasemi M, Shahgaldi S, Ismail M, Kim BH, Yaakob Z, Wan Daud WR (2011) Activated carbon nanofibers as an alternative cathode catalyst to platinum in a two-chamber microbial fuel cell. Int J Hydrogen Energy 36:13746–13752

Gong K, Du F, Xia Z, Durstock M, Dai L (2009) Nitrogen-doped carbon nanotube arrays with high electrocatalytic activity for oxygen reduction. Science 323:760–764

Greenman J, Galvez A, Giusti L, Ieropoulos I (2009) Electricity from landfill leachate using microbial fuel cells: comparison with a biological aerated filter. Enzym Microb Technol 44:112–119

Guerrini E, Grattieri M, Trasatti SP, Bestetti M, Cristiani P (2014) Performance explorations of single chamber microbial fuel cells by using various microelectrodes applied to biocathodes. Int J Hydrogen Energy 39:21837–21846

Guo K, Hidalgo D, Tommasi T, Rabaey K (2016) Pyrolytic carbon-coated stainless steel felt as a high-performance anode for bioelectrochemical systems. Bioresour Technol 211:664–668

Hamilton SJ (2004) Review of selenium toxicity in the aquatic food chain. Sci Total Environ 326:1–31

Han Y, Yu C, Liu H (2010) A microbial fuel cell as power supply for implantable medical devices. Biosens Bioelectron 25:2156–2160

He G, Gu Y, He S, Schröder U, Chen S, Hou H (2011) Effect of fiber diameter on the behavior of biofilm and anodic performance of fiber electrodes in microbial fuel cells. Bioresour Technol 102:10763–10766

Heijne A, Hamelers HVM, Saakes M, Buisman CJN (2008) Performance of non-porous graphite and titanium-based anodes in microbial fuel cells. Electrochim Acta 53:5697–5703
Hernandez-Flores G, Poggi-Varaldo HM, Solorza-Feria O, Romero-Castanon T, RiosLeal E, Galindez-Mayer J, Esparza-Garcia F (2015) Batch operation of a microbial fuel cell equipped with alternative proton exchange membrane. Int J Hydrog Energy 40(48):17323–17331
Huang L, Logan BE (2008) Electricity generation and treatment of paper recycling wastewater using a microbial fuel cell. Appl Microbiol Biotechnol 80:349–355
Huang YX, Liu XW, Sun XF, Sheng GP, Zhang YY, Yan GM, Wang SG, Xu AW, Yu HQ (2011) A new cathodic electrode deposit with palladium nanoparticles for cost-effective hydrogen production in a microbial electrolysis cell. Int J Hydrogen Energy 36:2773–2776
Humphries AC, Nott KP, Hall LD, Macaskie LE (2004) Continuous removal of Cr (VI) from aqueous solution catalysed by palladised biomass of Desulfovibrio vulgaris. Biotechnol Lett 26:1529–1532
Inglesby AE, Fisher AC (2013) Downstream application of a microbial fuel cell for energy recovery from an Arthrospira maxima fed anaerobic digester effluent. RSC Adv 3:17387–17394
Inglesby AE, Beatty DA, Fisher AC (2012) Rhodopseudomonas palustris purple bacteria fed Arthrospira maxima cyanobacteria: demonstration of application in microbial fuel cells. RSC Adv 2:4829–4838
IPCC (2007) Climate change 2007: synthesis report. In: Pachauri RK, Reisinger A (eds) Contribution of working groups I, II and III to the fourth assessment. Report of the intergovernmental panel on climate change. IPCC, Geneva, p 104
Jadhav DA, Ghadge AN, Mondal D, Ghangrekar MM (2014) Comparison of oxygen and hypochlorite as cathodic electron acceptor in microbial fuel cells. Bioresour Technol 154:330–335
Jiang D, Li B (2009) Granular activated carbon single-chamber microbial fuel cells (GACSCMFCs): a design suitable for large-scale wastewater treatment processes. Biochem Eng J 47:31–37
Jiang D, Curtis M, Troop E, Scheible K, McGrath J, Hu B, Suib S, Raymond D, Li B (2011) A pilot-scale study on utilizing multi-anode/cathode microbial fuel cells (MAC MFCs) to enhance the power production in wastewater treatment. Int J Hydrogen Energy 36:876–884
Jimenez IM, Santoro C, Carbonell SR, Greenman J, Ieropoulos I, Atanassov P (2016) Carbonbased air-breathing cathodes for microbial fuel cells. Catalyst 6:1–13
Jin B, van Leeuwen HJ, Patel B, Yu Q (1998) Utilisation of starch processing wastewater from production of microbial biomass protein and fungal α-amylase by Aspergillus oryzae. Bioresour Technol 66:201–206
Kakarla R, Min B (2014) Photoautotrophic microalgae Scenedesmus obliquus attached on a cathode as oxygen producers for microbial fuel cell (MFC) operation. Int J Hydrog Energy 39:10275–10283
Karra U, Manickam SS, Mccutcheon JR, Patel N, Li B (2012) Power generation and organics removal from wastewater using activated carbon nanofiber (ACNF) microbial fuel cells (MFCs). Int J Hydrogen Energy 38:1588–1597
Karthikeyan R, Wang B, Xuan J, Wong JWC, Lee PKH, Leung MKH (2015) Interfacial electron transfer and bioelectrocatalysis of carbonized plant material as effective anode of microbial fuel cell. Electrochim Acta 157:314–323
Kessel MAHJV, Speth DR, Albertsen M, Nielsen PH, Camp HJMOD, Kartal B, Jetten MSM, Lücker S (2015) Complete nitrification by a single microorganism. Nature 528:555–559
Kim D, Chang IS (2009) Electricity generation from synthesis gas by microbial processes: CO fermentation and microbial fuel cell technology. Bioresour Technol 100:4527–4530
Kim BH, Kim HJ, Hyun MS, Park DH (1999) Direct electrode reaction of Fe(III)-reducing bacterium, Shewanella Putrefaciens. J Microbiol Biotechnol 9(2):127–131
Kim BH, Chang IS, Gil GC, Park HS, Kim HJ (2003) Novel BOD (biological oxygen demand) sensor using mediator-less microbial fuel cell. Biotechnol Lett 25:541–545
Kim B, An J, Fapyane D, Chang IS (2015a) Bioelectronic platforms for optimal bio-anode of bioelectrochemical systems: from nano-to macro scopes. Bioresour Technol 195:2–13

Kim KY, Yang W, Logan BE (2015b) Impact of electrode configurations on retention time and domestic wastewater treatment efficiency using microbial fuel cells. Water Res 80:41–46

Kim C, Lee CR, Song YE, Heo J, Choi SM, Lim DH, Cho J, Park C, Jang M, Kim JR (2017) Hexavalent chromium as a cathodic electron acceptor in a bipolar membrane microbial fuel cell with the simultaneous treatment of electroplating wastewater. Chem Eng J 328:703–707

Kjeldsen P, Barlaz MA, Rooker AP, Baun A, Ledin A, Christensen TH (2002) Present and longterm composition of MSW landfill leachate: a review. Crit Rev Environ Sci Technol 32:297–336

Kondaveeti S, Min B (2013) Nitrate reduction with biotic and abiotic cathodes at various cell voltages in bioelectrochemical denitrification system. Bioprocess Biosyst Eng 36:231–238

Kretzschmar J, Riedl S, Brown RK, Schröder U, Harnisch F (2017) eLatrine: lessons learned from the development of a low-tech MFC based on cardboard electrodes for the treatment of human feces. J Electrochem Soc 164:H3065–H3072

Kumar R, Singh L, Wahid ZA (2016) Exoelectrogens: recent advances in molecular drivers involved in extracellular electron transfer and strategies used to improve it for microbial fuel cell applications. Renew Sust Energy Rev 56:1322–1336

Lakaniemi AM, Tuovinen OH, Puhakka JA (2012) Production of electricity and butanol from microalgal biomass in microbial fuel cells. Bioenergy Res 5:481–491

Ledezma P, Greenman J, Ieropoulos I (2013) MFC-cascade stacks maximise COD reduction and avoid voltage reversal under adverse conditions. Bioresour Technol 134:158–165

Lemly AD (1997) Ecosystem recovery following selenium contamination in a freshwater reservoir. Ecotoxicol Environ Saf 36:275–281

Leong JX, Daud WRW, Ghasemi M, Liew KB, Ismail M (2013) Ion exchange membranes as separators in microbial fuel cells for bioenergy conversion: a comprehensive review. Renew Sust Energy Rev 28:575–587

Lepage G, Albernaz FO, Perrier G, Merlin G (2012) Characterization of a microbial fuel cell with reticulated carbon foam electrodes. Bioresour Technol 124:199–207

Li WW, Sheng GP, Liu XW, Yu HQ (2011) Recent advances in the separators in microbial fuel cells. Bioresour Technol 102(1):244–252

Li Y, Yang HY, Shen JY, Mu Y, Yu HQ (2016) Enhancement of azo dye decolourization in a MFC–MEC coupled system. Bioresour Technol 202:93–100

Li B, Zhou J, Zhou X, Wang X, Li B, Santoro C, Grattieri M, Babanova S, Artyushkova K, Atanassov P, Schuler AJ (2014) Surface modification of microbial fuel cells anodes: approaches to practical design. Electrochim Acta 134:116–126

Lian J, Tian X, Li Z, Guo J, Guo Y, Yue L, Ping J, Duan L (2017) The effects of different electron donors and electron acceptors on perchlorate reduction and bioelectricity generation in a microbial fuel cell. Int J Hydrog Energy 42:544–552

Liang Y, Sarkany N, Cui Y (2009) Biomass and lipid productivities of Chlorella vulgaris under autotrophic, heterotrophic and mixotrophic growth conditions. Biotechnol Lett 31:1043–1049

Liao Q, Zhang J, Li J, Ye D, Zhu X, Zhang B (2015) Increased performance of a tubular microbial fuel cell with a rotating carbon-brush anode. Biosens Bioelectron 63:558–561

Lim DH, Wilcox J (2012) Mechanisms of the oxygen reduction reaction on defective graphene supported Pt nanoparticles from first-principles. J Phys Chem C 116:3653–3660

Liu L, Choi S (2017) Self-sustaining, solar-driven bioelectricity generation in micro-sized microbial fuel cell using co-culture of heterotrophic and photosynthetic bacteria. J Power Sources 348:138–144

Liu H, Logan BE (2004) Electricity generation using an air-cathode single chamber microbial fuel cell in the presence and absence of a proton exchange membrane. Environ Sci Technol 38:4040–4046

Liu H, Ramnarayanan R, Logan BE (2004) Production of electricity during wastewater treatment using a single chamber microbial fuel cell. Environ Sci Technol 38:2281–2285

Liu SY, Charles W, Ho G, Cord-Ruwisch R, Cheng KY (2017) Bioelectrochemical enhancement of anaerobic digestion: comparing single- and two-chamber reactor configurations at thermophilic conditions. Bioresour Technol 245:1168–1175

Logan BE, Murano C, Scott K, Gray ND, Head IM (2005) Electricity generation from cysteine in a microbial fuel cell. Water Res 39:942–952

Logan BE, Hamelers B, Rozendal R, Schroder U, Keller J, Freguia S, Aelterman P, Verstraete W, Rabaey K (2006) Microbial fuel cells: methodology and technology. Environ Sci Technol 40:5181–5192

Logroño W, Pérez M, Urquizo G, Kadier A, Echeverría M, Recalde C, Rákhely G (2017) Single chamber microbial fuel cell (SCMFC) with a cathodic microalgal biofilm: a preliminary assessment of the generation of bioelectricity and biodegradation of real dye textile wastewater. Chemosphere 176:378–388

Lowy DA, Tender LM, Zeikus JG, Park DH, Lovley DR (2006) Harvesting energy from the marine sediment–water interface II: kinetic activity of anode materials. Biosens Bioelectron 21:2058–2063

Lu Z-J, Bao S-J, Gou Y-T, Cai C-J, Ji C-C, Xu M-W, Song J, Wang R (2013) Nitrogen-doped reduced-graphene oxide as an efficient metal-free electrocatalyst for oxygen reduction in fuel cells. RSC Adv 3:3990

Luo H, Liu G, Zhang R, Jin S (2009) Phenol degradation in microbial fuel cells. Chem Eng J 147:259–264

Luo J, Yang J, He H, Jin T, Zhou L, Wang M, Zhou M (2013) A new electrochemically active bacterium phylogenetically related to Tolumonas onensis and power performance in MFCs. Bioresour Technol 139:141–148

Matter PH, Zhang L, Ozkan US (2006) The role of nanostructure in nitrogen-containing carbon catalysts for the oxygen reduction reaction. J Catal 239:83–96

Mohan SV, Mohanakrishnan G, Velvizhi G, Babu VL, Sarma PN (2010) Bio-catalyzed electrochemical treatment of real field dairy wastewater with simultaneous power generation. Biochem Eng J 51(1–2):32–39

Mohan SV, Srikanth S, Chiranjeevi P, Arora S, Chandra R (2014) Algal biocathode for in situ terminal electron acceptor (TEA) production: synergetic association of bacteria–microalgae metabolism for the functioning of biofuel cell. Bioresour Technol 166:566–574

Morris JM, Jin S, Wang J, Zhu C, Urynowicz MA (2007) Lead dioxide as an alternative catalyst to platinum in microbial fuel cells. Electrochem Commun 9:1730–1734

Mustakeem (2015) Electrode materials for microbial fuel cells: nanomaterial approach. Mater Renew Sustain Energy 4:1–11

Nancharaiah YV, Mohan SV, Lens PNL (2015) Metals removal and recovery in bioelectrochemical systems: a review. Bioresour Technol 195:102–114

Natarajan D, Van Nguyen T (2004) Effect of electrode configuration and electronic conductivity on current density distribution measurements in PEM fuel cells. J Power Sources 135:95–109

Nimje VR, Chen CC, Chen HR, Chen CY, Tseng MJ, Cheng KC, Shih RC, Chang YF (2012) A single-chamber microbial fuel cell without an air cathode. Int J Mol Sci 13(3):3933–3948

Nishio K, Hashimoto K, Watanabe K (2013) Digestion of algal biomass for electricity generation in microbial fuel cells. Biosci Biotechnol Biochem 77:670–672

Paitier A, Godaina A, Lyona D, Haddoura N, Vogela TM, Monierb JM (2017) Microbial fuel cell anodic microbial population dynamics during MFC startup. Biosens Bioelectron 92:357–363

Pant D, Singh A, Satyawali Y, Gupta RK (2007) Effect of carbon and nitrogen source amendment on synthetic dyes decolourizing efficiency of white-rot fungus, Phanerochaete Chrysosporium. J Environ Biol 29:79

Pant D, Van Bogaert G, Diels L, Vanbroekhoven K (2010) A review of the substrates used in microbial fuel cells (MFCs) for sustainable energy production. Bioresour Technol 101:1533–1543

Park DH, Zeikus JG (2000) Electricity generation in microbial fuel cells using neutral red as an electronophore. Appl Environ Microbiol 66:1292–1297

Park DH, Zeikus JG (2003) Biotechnol Bioeng 81(3):348–355

Park AH, Kim SK, Shin IH, Jeong YJ (2000) Electricity production in biofuel cell using modified graphite felt electrode with neutral red. Biotechnol Lett 22:1301–1304

Pham H, Boon N, Marzorati M, Verstraete W (2009) Enhanced removal of 1,2-dichloroethane by anodophilic microbial consortia. Water Res 43:2936–2946
Potter MC (1911) Electrical effects accompanying the decomposition of organic compounds. Proc R Soc B 84:260–276
Powell EE, Mapiour ML, Evitts RW, Hill GA (2009) Growth kinetics of Chlorella vulgaris and its use as a cathodic half-cell. Bioresour Technol 100:269–274
Rabaey K, Rozendal RA (2010) Microbial electrosynthesis revisiting the electrical route for microbial production. Nat Rev Microbiol 8:706–716
Ren Z, Ward TE, Regan JM (2007) Electricity production from cellulose in a microbial fuel cell using a defined binary culture. Environ Sci Technol 41:4781–4786
Rezaei F, Richard TL, Logan BE (2009) Analysis of chitin particle size on maximum power generation, power longevity, and Coulombic efficiency in solid-substrate microbial fuel cells. J Power Sources 192:304–309
Rismani-Yazdi H, Carver SM, Christy AD, Tuovinen OH (2008) Cathodic limitations in microbial fuel cells: an overview. J Power Sources 180:683–694
Rollefson JB, Stephen CS, Tien M, Bond DR (2011) Identification of an extracellular polysaccharide network essential for cytochrome anchoring and biofilm formation in Geobacter sulfurreducens. J Bacteriol 193:1023–1033
Rosenbaum M, Schröder U, Scholz F (2005) In situ electrooxidation of photobiological hydrogen in a photobioelectrochemical fuel cell based on Rhodobacter sphaeroides. Environ Sci Technol 39:6328–6333
Rovira M, Gim J, Mart X, de Pablo J, Mart V, Duro L (2008) Sorption of selenium (IV) and selenium (VI) onto natural iron oxides: goethite and hematite. J Hazard Mater 150:279–284
Roy JN, Babanova S, Garcia KE, Cornejo J, Ista LK, Atanassov P (2014) Catalytic biofilm formation by Shewanella oneidensis MR-1 and anode characterization by expanded uncertainty. Electrochim Acta 126:3–10
Sanders R, Caron D, Davidson J, Dennett M, Moran D (2001) Nutrient acquisition and population growth of a mixotrophic alga in axenic and bacterized cultures. Microb Ecol 42:513–523
Santoro C, Li B, Cristiani P, Squadrito G (2013) Power generation of microbial fuel cells (MFCs) with low cathodic platinum loading. Int J Hydrogen Energy 38:692–700
Santoro C, Guilizzoni M, Correa Baena JP, Pasaogullari U, Casalegno A, Li B, Babanova S, Artyushkova K, Atanassov P (2014) The effects of carbon electrode surface properties on bacteria attachment and start up time of microbial fuel cells. Carbon N Y 67:128–139
Santoro C, Artyushkova K, Cornejo JA, Ista L, Bretschger O, Marsili E, Atanassov P, Schuler AJ (2015) Influence of anode surface chemistry on microbial fuel cell operation. Bioelectrochemistry 106:141–149
Santoro C, Soavib F, Arbizzanib C, Serova A, Kabira S, Carpenterc K, Bretschgerc O, Atanassova P (2016) Co-generation of hydrogen and power/current pulses from supercapacitive MFCs using novel HER iron-based catalysts. Electrochim Acta 220:672–682
Santoro C, Arbizzani C, Erable B, Ieropoulos I (2017) Microbial fuel cells: from fundamentals to applications. A review. J Power Sources 356:225–244
Seviour T, Doyle LE, Lauw SJL, Hinks J, Rice SA, Nesatyy VJ, Webster RD, Kjelleberg S, Marsili E (2015) Voltammetric profiling of redox-active metabolites expressed by Pseudomonas aeruginosa for diagnostic purposes. Chem Commun 51:3789–3792
Shantaram A, Beyenal H, Veluchamy RRA, Lewandowski Z (2005) Wireless sensors powered by microbial fuel cells. Environ Sci Technol 39:5037–5042
Sharma A, Bhattacharya A, Singh S (2009) Purification and characterization of an extracellular carbonic anhydrase from Pseudomonas fragi. Process Biochem 44:1293–1297
Sharma T, Mohana AL, Chandra TS, Ramaprabhu S (2008) Development of carbon nanotubes and nanofluids based microbial fuel cell. Int J Hydrogen Energy 33:6749–6754
Smith JA, Tremblay PL, Shrestha PM, Snoeyenbos-West OL, Franks AE, Nevin KP, Lovley DR (2014) Going wireless: Fe(III) oxide reduction without pili by Geobacter sulfurreducens strain JS-1. Appl Environ Microbiol 80:4331–4340

Sotres A, Cerrillo M, Vias M, Bonmat A (2015) Nitrogen recovery from pig slurry in a two chambered bioelectrochemical system. Bioresour Technol 194:373–382
Srikanth S, Marsili E, Flickinger MC, Bond DR (2008) Electrochemical characterization of Geobacter sulfurreducens cells immobilized on graphite paper electrodes. Biotechnol Bioeng 99:1065–1073
Su DS, Zhang J, Frank B, Thomas A, Wang X, Paraknowitsch J, Schlögl R (2010) Metal-free heterogeneous catalysis for sustainable chemistry. Chemsuschem 3:169–180
Sun J, Hu Y-Y, Bi Z, Cao Y-Q (2009) Simultaneous decolorization of azo dye and bioelectricity generation using a microfiltration membrane air-cathode single-chamber microbial fuel cell. Bioresour Technol 100:3185–3192
Tandukar M, Huber SJ, Onodera T, Pavlostathis SG (2009) Biological chromium(VI) reduction in the cathode of a microbial fuel cell. Environ Sci Technol 43:8159–8165
Tee PF, Abdullah MO, Tan IAW, Amin MAM, Nolasco-Hipolito C, Bujang K (2016) Performance evaluation of a hybrid system for efficient palm oil mill effluent treatment via an air-cathode, tubular upflow microbial fuel cell coupled with a granular activated carbon adsorption. Bioresour Technol 216:478–485
Tiwari BR, Noori MT, Ghangrekar MM (2016) A novel low cost polyvinyl alcohol-Nafion borosilicate membrane separator for microbial fuel cell. Mater Chem Phys 182:86–93
Wang X, Cheng S, Feng Y, Merrill MD, Saito T, Logan BE (2009) Use of carbon mesh anodes and the effect of different pretreatment methods on power production in microbial fuel cells. Environ Sci Technol 43:6870–6874
Wang L, Zhou X, Zhong S, Zhou N (2010a) Novel materials and technologies of microbial fuel cell in environmental engineering. ECS Trans 25:311–333
Wang X, Feng Y, Liu J, Lee H, Li C, Li N, Ren N (2010b) Sequestration of CO_2 discharged from anode by algal cathode in microbial carbon capture cells (MCCs). Biosens Bioelectron 25:2639–2643
Wang HY, Bernarda A, Huang CY, Lee DJ, Chang JS (2011) Micro-sized microbial fuel cell: a mini-review. Bioresour Technol 102:235–243
Wang H, Lu L, Liu D, Cui F, Wang P (2015) Characteristic changes in algal organic matter derived from Microcystis aeruginosa in microbial fuel cells. Bioresour Technol 195:25–30
Watanabe K (2008) Recent developments in microbial fuel cell technologies for sustainable bioenergy. J Biosci Bioeng 106:528–536
Wei J, Liang P, Huang X (2011) Recent progress in electrodes for microbial fuel cell. Bioresour Technol 102:9335–9344
Winfield J, Chambers LD, Stinchcombe A, Rossiter J, Ieropoulos I (2014) The power of glove: soft microbial fuel cell for low-power electronics. J Power Sources 249:327–332
Winfield J, Gajda I, Greenman J, Ieropoulos I (2016) A review into the use of ceramics in microbial fuel cells. Bioresour Technol 215:296–303
Wu XY, Song TS, Zhu XJ, Wei P, Zhou CC (2013) Construction and operation of microbial fuel cell with Chlorella vulgaris biocathode for electricity generation. Appl Biochem Biotechnol 171:2082–2092
Wu S, He W, Yang W, Ye Y, Huang X, Logan BE (2017) Combined carbon mesh and small graphite fiber brush anodes to enhance and stabilize power generation in microbial fuel cells treating domestic wastewater. J Power Sources 356:348–355
Xing D, Zuo Y, Cheng S, Regan JM, Logan BE (2008) Electricity generation by Rhodopseudomonas palustris DX-1. Environ Sci Technol 42:4146–4151
Xu B, Chen BY, Hsueh CC, Lj Q, Chang CT (2014) Deciphering characteristics of bicyclic aromatics—mediators for reductive decolorization and bioelectricity generation. Bioresour Technol 163:280–286
Yang Y, Xiang Y, Xia C, Wu WM, Sun G, Xu M (2014) Physiological and electrochemical effects of different electron acceptors on bacterial anode respiration in bioelectrochemical systems. Bioresour Technol 164:270–275

Yao S, He YL, Li YS, Xi H (2014) Effect of the membrane electrode assemble design on the performance of single chamber microbial fuel cells. Energy Procedia 61:1947–1951

Yasri NG, Nakhla G (2017) The performance of 3-D graphite doped anodes in microbial electrolysis cells. J Power Sources 342:579–588

Yi H, Nevin KP, Kim BC, Franks AE, Klimes A, Tender LM, Lovley DR (2009) Selection of a variant of Geobacter sulfurreducens with enhanced capacity for current production in microbial fuel cells. Biosens Bioelectron 24:3498–3503

Yuan Y, Zhao B, Jeon Y, Zhong S, Zhou S, Kim S (2011) Iron phthalocyanine supported on amino-functionalized multi-walled carbon nanotube as an alternative cathodic oxygen catalyst in microbial fuel cells. Bioresour Technol 102:5849–5854

Zhang X, Cheng S, Liang P, Huang X, Logan BE (2011) Scalable air cathode microbial fuel cells using glass fiber separators, plastic mesh supporters, and graphite fiber brush anodes. Bioresour Technol 102:372–375

Zhang Q, Hu J, Lee D (2016) Microbial fuel cells as pollutant treatment units: research updates. Bioresour Technol 217:121–128

Zhang L, Li J, Zhu X, Ye DD, Fu Q, Liao Q (2017) Response of stacked microbial fuel cells with serpentine flow fields to variable operating conditions. Int J Hydrog Energy 42(45):27641–27648

Zhang Y, Jiang J, Zhao Q, Wang K, Yu H (2018) Analysis of functional genomes from metagenomes: revealing the accelerated electron transfer in microbial fuel cell with rhamnolipid addition. Bioelectrochemistry 119:59–67

Zhao F, Rahunen N, Varcoe JR, Roberts AJ, Avignone-Rossa C, Thumser AE, Slade RCT (2009a) Factors affecting the performance of microbial fuel cells for sulfur pollutants removal. Biosens Bioelectron 24:1931–1936

Zhao F, Slade RCT, Varcoe JR (2009b) Techniques for the study and development of microbial fuel cells: an electrochemical perspective. Chem Soc Rev 38:1926–1939

Zhao L, Li J, Battaglia F, He Z (2016) Computational investigation of the flow field contribution to improve electricity generation in granular activated carbon-assisted microbial fuel cells. J Power Sources 333:83–87

Zheng W, Cai T, Huang M, Chen D (2017) Comparison of electrochemical performances and microbial community structures of two photosynthetic microbial fuel. J Biosci Bioeng 124(5):551–558

Zhou M, Chi M, Luo J, He H, Jin T (2011) An overview of electrode materials in microbial fuel cells. J Power Sour 196:4427–4435

Zhou M, Yang J, Wang H, Jin T, Hassett DJ, Gu T (2013) Bioelectrochemistry of microbial fuel cells and their potential applications in bioenergy, Bioenergy Research. Adv Appl 131–153. books.google.com

Chapter 22
Microbial Engineering in Biofuel Production—A Global Outlook, Advances, and Roadmap

Sagnik Mitra, Sonam Paliya, and Ashootosh Mandpe

Abstract The dwindling supply of fossil fuels and environmental issues have raised awareness of alternative energy sources like biofuels. Microbial engineering, on the contrary, presents a viable approach for sustainable biofuel production considering the rising demand for renewable energy sources and the pressing need to mitigate climate change. This technique involves the strategic manipulation of microbes to maximize the production of biofuels, a renewable and sustainable substitute for fossil fuels. However, stepping up biofuel production poses numerous challenges, such as cost-effectiveness, sustainability, and regulatory concerns. Therefore, overcoming challenges of scaling up microbial biofuel production and creating cost-effective downstream processing are need of the hour. Through the exploration of concepts such as techno-feasibility, resource sustainability, resilience, and techno-economic analysis, it is conceivable that the development of technologies aligned with the principles of the circular economy can be achieved. In the present chapter the importance of metabolic engineering, synthetic biology, systems biology, and multi-omics for the design and optimization of microbial strains for biofuel production is reviewed and discussed. Also, a brief overview of the current state of microbial engineering for biofuel production globally is presented. The development of more effective and resistant microbial strains with the integration of an omics knowledge base for comprehension of systems-level knowledge of metabolic networks along with the use of novel bioprocessing methods for large-scale production have also been emphasised in this chapter.

Keywords Biofuel · Climate change · Microbial engineering · Metabolic engineering · Scale-up · Techno-economic feasibility

---

S. Mitra · S. Paliya (✉)
Department of Biosciences and Biomedical Engineering, Indian Institute of Technology Indore, Indore 453552, Madhya Pradesh, India
e-mail: sonampaliya140@gmail.com

A. Mandpe
Department of Civil Engineering, Indian Institute of Technology Indore, Indore 453552, India

M. Shah and D. Deka (eds.), *Emerging Sustainable Technologies for Biofuel Production*, Environmental Science and Engineering, https://doi.org/10.1007/978-3-031-52167-6_22

22.1 Introduction

Despite significant research advancements in the past two decades, the current state of biofuels remains suboptimal in terms of production, utilization, and market penetration when compared to conventional non-renewable energy sources (Sobrino et al. 2010). The first eponymous engine designed by Rudolf Diesel was powered by peanut oil, and Henry Ford designed his equipment to run on ethanol (Garba et al. 2006). The oil crises of 1970 rekindled interest in renewable fuels. Some of the main factors that have sustained this interest to date are rising prices of petroleum products, oil supply uncertainty, and creeping anxiety over the security of crude in the future. Hence, research and development (R&D) efforts in biofuels have focussed primarily on ethanol and biodiesel.

In the study of Lundgren et al. (2008), the potential for energy provision from biomass in general and from solid biofuels has been vividly discussed and reported that there are still untapped potentials for biofuels and these sustainable fuels can contribute significantly to cover the energy demand. The standardization of biofuels is seen as a major key issue to develop the markets, especially the European Union (EU) countries like Belgium and Austria, which have introduced fiscal measures, feed-in tariffs, and investment subsidies for electricity generation from solid biomass.

The various kinds of emissions including CO_2, NOx, SOx, PM, etc. from the combustion of fossil fuels, has paved the way for generation of biofuels around the world. Nevertheless, the impact of biofuel adoption has been limited in developed countries, whereas developing countries have experienced positive effects on their local economies (Chen et al. 2016; Gomiero 2015; Staab and Duffield 2019; The Gazette of India 2018).

22.1.1 Overview of the Status of Biofuel Production

The world's biofuel industry is growing at a faster rate than any other major industry. Intermediate inputs, subsidizing value-adding factors, and granting incentives for achieving target end products are some of the factors where the government has intervened to ramp up the production supply chain (Sorda et al. 2010). To overcome the gap between biomass production and biomass consumption site, a robust, efficient, and smooth supply chain is desirable. Processes like thermo-chemical, physico–chemical, and biochemical conversion are some of the end-use technologies which depend on the characteristics of the biomass getting processed (Kaltschmitt and Weber 2006). The study of Filip et al. (2016) suggested that lignocellulosic biomass which is considered to be second-generation feedstock to some extent can be used as a low-cost feedstock for biofuel production, with costs estimated at US$50–70 per ton for certain types of biomasses as being cost-competitive on a per-gallon basis with current fossil fuels. Cellulose, hemicellulose, and lignin are the most abundant lignocellulosic renewable materials on earth and their subsequent utilization is expected to increase soon (Ryu and Lee 1986). Lignocellulose source with bioethanol potential is a great biomass feedstock for biorefineries due to easy availability, low cost, output/input energy ratio, and high ethanol productivity (Joshi et al. 2022). The rapid creation and deployment of cost-effective and scalable processes for converting this biomass into infrastructure-compatible, low-carbon, and sustainable transportation biofuels are critical to the economic vitality and energy security of the world.

Biodiesel, ethanol, biofuel, food versus fuel, land use change, and indirect land use change have been debatable topics in recent years. Biofuel's impact to create robust and energy-intensive environmental policies is driven by efforts for figuring out ways to simultaneously produce and commercialize biofuels on a large scale to cope with the energy demands. One of the first comprehensive life-cycle inventories for biodiesel production was conducted in the United States (US) using soybean oil in 1998 (Sheehan et al. 1998). It was found that for every unit of fossil fuel used during the life cycle of biodiesel production, 3.2 units of energy are obtained from biodiesel. The use of biodiesel also reduces net emissions of greenhouse gas (GHG) by 78.45%. The agricultural production expansion for meeting the biofuel demand is one of the main factors for indirect land use change. Certain economic and environmental models have been simulated to capture quantitatively the land use change for biofuels as it cannot be captured through direct measurement (Staab and Duffield 2019).

It is important to understand public opinion and the dynamics associated with biofuels. In one of the study, it was found that party identification and risk/benefit perceptions can work in tandem in influencing public opinion and attitude towards the implementation of biofuels and can influence government policy decisions. Recent studies show that mere perceptions about the advantages and disadvantages of biofuels are critical parameters of attitudes people have. People having prior experience with biofuels does influence their attitude (Cacciatore et al. 2012; Wegener et al. 2014). A study conducted by National Biodiesel Board reported that more than half of truck fleet operators used biodiesel in their vehicles which was indicated to be favourable for them, however, 10% displayed a negative impression about biodiesel (Fung et al. 2014). Energy independence, climate change, and rural development combined with political motivation are the main drivers for biofuel usage in the society (Luque et al. 2008).

Biofuel products can also be used for powering internal combustion engines. These are renewable and can recycle CO_2 from their combustion through photosynthetic mechanisms. There can be direct and immediate replacements for liquid fuels that are used in transport which can be easily integrated into logistic systems. The resource utilization that goes into the making and production of biodiesel and ethanol producing chain is an important aspect that requires study as well as the nature of impacts the extensive biofuel production programs have on food supply and the environment (Escobar et al. 2009).

Multi-pronged approaches are being researched for developing technology in conjunction with the use of microorganisms in the production of various biofuels. The direct fermentation and indirect fermentation are the two main approaches used in biofuel research. Direct fermentation involves the degrading of the initial plant material into fermentable sugars and the subsequent conversion of those sugars into alcohol. In indirect fermentation, the starting plant material undergoes pyrolysis followed by conversion of the gas produced into syngas (Osman et al. 2021). Biofuel research involves identifying starting plant material, isolation and development of bacterial and fungal strains, and design of appropriate protocols for conversion of plant material to sugar monomers. Various starting plant materials are available, ranging from molasses to cellulose, hemicellulose, and lignin polymers. A range of different kinds of microorganisms, enzymes, incubation conditions, and engineering schemes are needed for efficient depolymerization. Based on the kind and nature of the initial starting material, biofuel research could be divided into three different generations: first-generation biofuels which comprises of using crops to produce simple sugars and subsequently converting them to ethanol. The second-generation biofuels comprise of using native, perennially growing plants that require no cultivation, or entire prairie system flora for sugar and ethanol production. The third-generation biofuels use microalgae/cyanobacteria/oleaginous microorganism which produces sugars that can be converted into ethanol or biodiesel (Elshahed 2010). Researchers are switching to the fourth generation biofuels, an amalgamation of first, second, and third-generation feedstocks which have been modified genetically to get better yields in terms of quality as well as quantity (Demirbas 2009; Simmons 2011).

Tools such as Life-cycle Assessment (LCA) framework are used for measuring the carbon intensity of biofuels, which focuses on the direct effects such as emissions emanating from growing the feedstock, transport of feedstock to the factory, processing of the feedstock into biofuel, transporting the biofuel from the factory to its point of use, the efficiency of the biofuel compared with standard diesel, and benefits due to the production of useful by-products. Most of the LCAs are dependent on the scope, therefore, defining the scope of the LCA is very important in certain cases which otherwise fail to capture parameters like induced indirect land use change, carbon leakage, changes in crop yield, substitution between fuels, and consumption effects (Ciaian et al. 2015; Passell et al. 2013).

However, the results are also dependent upon the methodological considerations of the assessments, including the reference system, functional unit, impact assessment methods adopted, allocation methods, and important processes such as the agricultural inputs for the system and energy systems employed. Biofuel producers have the potential to improve environmental performance by making use of by-products, waste and diversifying their products, promoting a transition toward biorefinery systems and a bio-based economy for environmental sustainability (Asghar et al. 2020).

22.1.2 Importance of Sustainable and Eco-friendly Sources of Energy

It has been found that the world consumption of marketed energy expanded from 549 quadrillion British thermal units (Btu) in 2012 to 629 quadrillion Btu in 2020 which is estimated to rise up to 815 quadrillion Btu in 2040, a 48% increase from 2012 to 2040. The carbon dioxide in the atmosphere had risen from 278 ppm at the beginning of the industrial revolution to >400 ppm and is projected to triple by 2040 if we are to stay in the present course and emit other significant greenhouse gases which include CH_4, N_2O, CFCs, halons, N_2O, ozone, and peroxy-acetyl-nitrate in the atmosphere (Energy Information Administration 2015). Burning fossil fuels also creates other environmental pollutants such as SOx, NOx, particulate matter (PM), volatile organic compounds, and toxic heavy metals. Factors like increasing heat waves and droughts, torrential downpours, and accompanying floods along with encroaching seas are affecting the greenhouse gas (GHG) emissions enormously.

Solutions are needed to meet such global challenges, which should be clean energy generation, transmission, and distribution, the storage of electrical and chemical energy, energy efficiency, and better energy management systems. Materials science has enabled us to convert CO_2 and H_2O directly into fuels and chemicals. These types of conversions help to capture carbon from power generation, cement, aluminium, plastic, steel production, and the atmosphere. The SunShot initiative by Solar Energy Technologies Office has gained huge success in the US wherein in early 2011, solar power comprised less than 0.1% of US electric supply with an installed capacity of 3 GW but as of 2017, solar now supplies more than 1% of the US electricity demands with an installed capacity of 47 GW. The DOE recently announced their 2030 goals of 0.03–0.05 US\$ kWh^{-1} levelized cost of energy for utility and residential solar power (Chu et al. 2016).

The Intergovernmental Panel on Climate Change (IPCC) of the United Nations (UN) have expressed concern about the future and the main findings that have been enclosed in the IPCC fifth assessment synthesis report. The International Energy Agency recognizes energy security as reliable, affordable access to all fuels and energy sources, and the UN supplements this treatment with the priority connected with the implementation of SDG 7 in the aspect of ensuring access (Jordaan 2007; O'Connor et al. 1999). Fuel and energy together are a part of the core of the environmental economy and have been assigned three strategic functions of providing stability, high effectiveness of the in-use energy, and protecting the environment. The US energy policy has focused on eight traditional objectives: secure, plentiful, diverse energy supply, robust, reliable energy infrastructure, affordable, stable energy price, and environmentally sustainable energy production and use. EU, on the other hand, has adopted ambitious policies for tackling climate change by bringing down greenhouse gas (GHG) emissions in 2020 by 20%. Raising the shares of renewables in final energy consumption to 20%, and realizing energy savings of 20% in 2020 compared to an official baseline energy consumption level that has been achieved.

Some of the energy policies which have been extensively discussed are the Energy-Efficiency Standard (EES), Feed-in-Tariff (FiT), and Building Energy Performance Certification (BEPC) Schemes which provide a unified reference benchmark for future work concerning the development of renewable and sustainable energy policy (Lu et al. 2020).

Renewable energy which can also be termed a sustainable and clean energy source does have the potential to meet domestic energy requirements with almost close to zero emissions of both air pollutants and greenhouse gases leading to minimal depletion of natural resources. These energy sources can be used to improve energy reliability, solve energy and water shortages, increase the standard of living and employment of local populations, and ensure the sustainable development of remote regions. The adoption of decentralized modes for harvesting renewable energy could be one of the sought-after options for meeting small-scale energy needs in a much more reliable, affordable, and environmentally sustainable way (Panwar et al. 2011).

A typical example of sustainable and clean energy could be a fuel cell which is an electrochemical device that converts fuels into energy and it is emerging as one of the clean energy solutions to meet the world's electricity requirements (Krupnov et al. 2022). The world's total installed grid-connected power generation capacity stands at 580.1 GW along with 3.4 GW of off-grid PV, with coal and oil reserves accounting for 80% and renewable energy from large dam-based hydro, biomass, solar, and wind contributing 29%. However, renewables are not much used in power generation as they require high operating costs and efficiency compared to fossil fuels. Concentrating solar power (CSP) or solar thermal electricity (STE) is a technology that can produce electricity based on utility requirements which in turn offers firm capacity and accessible power on demand by integrating thermal energy storage or in hybrid operation. CSP plants are predicted to produce a global electricity contribution of 7% by 2030 and 25% by 2050 (Pira 2020). Certain man-made catalysts are required for solar energy storage and has been one of the crucial topics of research. The development of such man-made catalysts would provide a key enabling technology for a full solar energy conversion and storage system, whether the fuel-forming system is separate or integrated. The key components needed to enable the whole system to remain the same are cost-effectiveness, efficient capture, conversion, and storage of sunlight. Therefore, further advances in basic science and engineering are needed to realize the full potential of solar energy (Chu and Majumdar 2012; Elavarasan 2018; Sastry and Sreenu 2012; Stambouli 2011).

22.1.3 Brief Background on Microbial Engineering and Its Role in Biofuel Production

The importance of microbial engineering lies in the fact that mere repurposing or applying engineering aspects to produce advanced biofuels with similar properties to petroleum-based fuels would help to meet the needs of energy demands possibly a little faster. Engine type, energy content, combustion quality, ignition delay, cloud point, volatility, lubricity, viscosity, stability, odour, toxicity, water miscibility, and cost must be considered when developing an organism or pathway to produce an advanced fuel. Even though the native hosts do convert feedstocks to advanced biofuels, the efficiency is rather poor when compared to engineered ones. The reason of this characteristic is that the native organisms retain the endogenous regulation of biofuel-producing pathways which restrict the organism to provide high yields. Metabolic engineering of the pathways used in the production of advanced biofuels, irrespective of the feedstock and host organism, and synthetic-biology-driven approaches for host and pathway optimization and future directions are strategies that have been adopted or are being explored for the production of biofuels (Oluchukwu et al. 2014; Peralta-Yahya et al. 2012).

Gasoline, diesel, aviation fuel, long-chain alcohols, fatty acid-derived and isoprenoid-derived fuels are some of the most widely established fuels that have been commercialized and have been produced using engineered microbes. Many derivatives of these compounds are produced using different strains of the same microbes (Zhang et al. 2021). Recent advances in systems and synthetic biology provided a framework for tolerance engineering which has been an emerging research area as microbes may possess inherent tolerance which could pose a hindrance to production. Expression of efflux pumps, heat shock protein modifications, membrane protein modifications, and general stress response gene activation are some of the tolerance-based mechanisms that have been adopted. These mechanisms are combined with various kinds of strain developments and improvements along with media supplementation strategies and in situ approaches for recovering biofuel that can be used to reduce the need for highly tolerant strains (Zhang et al. 2011).

Long-chain alcohols are generally more toxic than short-chain alcohols, and toxicity typically increases with solvent hydrophobicity. Certain toxicity levels are associated with several of the next-generation biofuels and not all have antimicrobial activity. Several pumps appear to be specific for solvents, such as the solvent-resistant efflux pumps (srpABC) from *Pseudomonas putida* S12 and the toluene tolerance genes (ttg) from DOT-T1E strain of P.(Dunlop 2011). Several processes had been commercialized which involve the utilization of cellulolytic organisms, photosynthetic organisms, microalgae, cyanobacteria, or lithoautotrophic organisms that can use energy derived from renewable sources. Fuel gases like methane, a more deleterious greenhouse gas than CO_2, can be utilized by methanotrophs or, after activation to methanol, by methylotrophs. All challenges pertain to the use of microbiological tools to improve the carbon assimilation pathways and channelling metabolic flux of the core carbon metabolism and assimilation pathways in the microorganisms themselves to produce the product of interest (Liao et al. 2016).

22.2 Advances in Microbial Engineering for Biofuel Production

Biomass-to-biofuels (B2B) is a cost-efficient biofuel production process that uses reduced biomass in size and is treated to loosen up the lignin-cellulose fibre entanglement. Several methods have been used for this purpose including biomass treatment with saturated steam at 200 °C, exploitation with ammonia, and cooking with warm dilute acid to get usable and available biomass for the production of biofuels. After pre-treatment, the solid suspension is exposed to cellulolytic enzymes that digest the cellulosic and hemicellulosic biomass components to release hydrolysis products, primarily six and five-carbon sugars. The type of pre-treatment defines the optimal enzyme mixture to be used and the composition of the hydrolysis products (Stephanopoulos 2007). Biomass is the only resource that can be converted into liquid transportation fuels that have desirable characteristics for mobility and are likely required for such applications as aviation and long-haul trucking. Advances in genetic engineering and metabolomics have expedited the construction of microbial tools for biofuel production which holds great promise for meeting the growing global demand for renewable energy sources and reducing our dependence on fossil fuels.

22.2.1 Synthetic Biology and Systems Biology Approaches

The profound impact of metabolic engineering and synthetic biology in biocatalyst engineering for bio-renewable fuels and chemical production, such as ethanol, butanol, acetate, lactate, succinate, alanine, and xylitol has been well established and well known. To understand some of the strategies for improving biocatalysts' tolerance to chemical inhibitors the microbial fermentation of biomass-derived sugars to commodity fuels and chemicals were explored. To achieve this, microorganisms with desired traits must be developed, either by modification of existing microbes or by the de-novo design of new microbes. The challenges associated with these strategies include molecule selection, pathway optimization, toxic metabolic intermediates and proteins, dynamic pathway regulation, host engineering, stoichiometric modelling, growth coupling strategies, and successful scale-up of processes to commercial scales. Hence, continual and sustained efforts in improving reproducibility and predictability are needed for further development of bioengineering (Saxena et al. 2009).

Molecular biology tools for the modification of a chassis can be used to formulate an implementation approach, identify pathways and properties, optimize the redesigned system, and assess the system properties. Systems biology, standardization of biological systems, and metabolic evolution have been used to redesign biocatalysts to produce commodity fuels and chemicals, both natural and synthetic such as D- and L- lactate, L-alanine, succinate, ethanol, and butanol (Shanmugam and Ingram 2022).

Erdrich et al. (2015) used a stoichiometric network model of *Synechocystis* sp. PCC 6803 to identify and characterize suitable strain design strategies for biofuel synthesis. The study demonstrated that higher-order knockout strategies exist in the model that led to the coupling of growth with high-yield biofuel synthesis under phototrophic conditions. The knockout strategies seek to reduce the ratio of ATP to NADPH produced by the photosynthetic electron transport chain, such that ATP and NADPH are exclusively synthesized via the linear electron flow. The calculated intervention strategies do not rely on optimal growth and ensure that maintenance of metabolism in the absence of light remains feasible (Buschke et al. 2013).

Bioethanol which also can be used as a source for fuel cells could become a crucial part of long-term solutions to climate change phenomena. Microbial conversion of the sugar residues present in wastepaper and yard trash from US landfills alone could provide more than 400 billion litres of ethanol, 10 times the corn-derived ethanol burned annually as a 10% blend with gasoline. Even though only three steps for efficient ethanol fuel production are important: delignification to liberate cellulose and hemicellulose from their complex with lignin, depolymerization of the carbohydrate polymers to produce free sugars, and fermentation of mixed hexose and pentose sugars to produce ethanol. The development of a feasible biological delignification process should be possible if lignin-degrading microorganisms (e.g., *Phanerochaete chrysosporum* and *Phlebia radiata*) and optimal bioreactor design are effectively coordinated. New fermentation technology for converting 5C sugars like xylose to ethanol also needs to be developed to make the overall conversion process more cost-effective (Aristos and Merja 2000; Bilal et al. 2018).

Although there is a huge demand for biofuels, integration into existing petroleum-based technologies and infrastructure along with rapidly changing economics does not showcase a good ground for building the biofuels industry. The primary cost for producing biofuels is the cost of the feedstock, which is 60% in the case of corn ethanol and 80% for soybean biodiesel. However, yield optimization has been a critical aspect of biochemical engineering processes. Ethanol is a poor fuel due to its hygroscopic nature and low energy content. Biodiesel is a better fuel, but it requires large quantities of petroleum-derived methanol and has only 89% of the energy content of its analog, petrodiesel (Tai and Stephanopoulos 2012).

Photosynthetic microorganisms and microalgae offer an alternative approach, which can be located on cheap, rural land and cultured in seawater rather than freshwater (Harrison et al. 2019). Microalgae have become the prime focus for biofuel production due to their versatility and ability to accumulate high lipid content and biomass in large amounts. Species such as *Chlorella, Dunaliella, Schizochytrium,* and *Tetraselmis* have been reported to accumulate lipid content between 20 and 50% of their cell dry weight. Microalgae can directly convert atmospheric carbon dioxide to fuel molecules or its precursor, but biofuel production technologies from microalgae have yet to be established at scales that are at par with fossil fuels.

Microalgae, although a promising feedstock for the production of biofuels and other bioactive compounds, is still facing challenges towards transition to commercial applications of microalgae-based products. The economic analysis, environmental impact, and industrial potential of biofuel production from microalgae still needs to be successfully figured out for making this technology commercially viable and environmentally friendly. Biofuel production remains debatable fuel sources, but the integration of biorefinery paths with biofuel production for the recovery of value-added products could substantially reduce production costs. Biofuel production using microalgae has been commercialized in the United States of America (USA), Australia, and certain European countries where the year-round weather conditions could permit outdoor cultivation of microalgae. Along with factors like oil content, microalgae yield, conversion cost, and other overhead and miscellaneous costs, economic concerns are amongst the primary concern for this kind of project which happens to be resource-intensive especially when considered for building up a microalgal multi-product biorefinery. The biofuel production process needs to be streamlined by reducing the cost of production through more efficient microalgal cultivation and harvesting to make it cost-competitive with petroleum-based fuels (Chia et al. 2018; G. P. 't Lam et al. 2018).

Some of the technologies used concerning the development of microalgal biorefinery include ionic liquids (ILs), microfiltration (filters/membranes with pore size from 0.1 to 10 mm), photobioreactors (PBRs), Pulsed electric field (PEF), gas medium recycles, cell disruption, extractions, oil phase, water phase, cell debris, and end-products. The proposed technologies include harvesting, centrifugation, filtration, flocculation/sedimentation, flotation, cell disruption, bead mill, high-pressure homogenization, alkali and acid treatment, enzymes, microwave, ultrasound, organic solvents, aqueous two-phase extraction, alkali extraction, and supercritical extraction have been extensively used. Marine macroalgae have rapid growth rates and can produce gases in high amounts, while marine brown and red algae accumulate high levels of carbohydrates and are explored as feedstocks for the production of biofuels. Cyanobacteria is another model organism for the production of biofuel, with *Synechococcus elongatusis* being the most exploited (Ghag et al. 2019).

Escherichia coli and *Saccharomyces cerevisiae* are the most convenient microorganisms that have been widely used for biofuel production over and over through engineering applications on them (Cardoso et al. 2020; Feist et al. 2010; Lacerda et al. 2020; Runguphan and Keasling 2014). *Zymomonas mobilis*, an ethanogenic bacterium was introduced to *E. coli* and increased the growth and production of bioethanol production. By introducing the same genes of *Z. mobilis* in *E. coli* and giving it anaerobic conditions, it produced biodiesel. By deleting the Fad D gene from the chromosome of E. coli colistin, the production of biodiesel was achieved. Genetically modified *E. coli* KO11 is highly tolerant of ethanol and can produce bioethanol from all types of sugars (Ohta et al. 1991). Clostridium is also used for the production of 1-butanol from lignocellulosic biomass. Insertion of PDC I and II genes in *C. cellulolyticum* increases cellulose fermentation for biofuel production

(Konar et al. 2020). These are some of the strategies for the production of biofuel through microorganisms engineering.

Thus, the current chapter provides an understanding as to how synthetic biology and systems biology is making their way into the biofuels research domain and how their impact is affecting the world's economic policies and the context of an imminent shift towards the usage of biofuels has also been realized (Chubukov et al. 2018; Dellomonaco et al. 2010; Martin 2013; Mukhopadhyay et al. 2008; Rupprecht 2009).

22.2.2 Genome Editing Technologies

The CRISPR (clustered regularly interspaced short palindromic repeat)/Cas9 (CRISPR-associated nuclease 9), a prokaryotic molecular immunity system, has emerged as a novel technology for targeted genetic engineering. Microbial strains such as *Saccharomyces cerevisiae, Zymomonas mobilis, Clostridium thermosaccharolyticum, C. thermohydrosulfuricum, Thermoanaerobacter mathranii, T. brockii,* and *T. ethanolicus* are used for fermentation of monomeric sugars for ethanol production on industrial scale (Ozsoz et al. 2019).

Its properties have enabled rapid adoption for genome engineering applications across biology, including therapeutics, xenotransplantation, and uses in livestock, crops, food organisms, and industrial microbes. Genomic cuts are precisely done at the site of Cas9-generated DSBs, and do involve exploiting host cell machinery to repair the genome accordingly at the cut site. CRISPR-Cas9 is being used to treat genetic disorders in animals and humans along with approval for targeted cancer therapies. It is also being used to expedite crop and livestock breeding, engineer new antimicrobials, and control disease-carrying insects. Recent efforts to mine the abundant bacterial and archaeal CRISPR–Cas systems should increase the range of molecular tools available to researchers (Kukreja et al. 2021).

CRISPR-based acetyltransferases and demethylases have also been successful in engineering epigenetic changes in the genome, and orthogonal sgRNA: Cas9 systems are being developed to exploit different CRISPR tools for distinct applications. Some of the recent advances in the development of microbial cell factories for chemical production using various CRISPR-Cas systems had been made public. It provides a toolbox with excellent opportunities for identifying better biocatalysts, multiplexed fine-tuning of metabolic flux, efficient utilization of low-cost substrates, and improvement of metabolic robustness (Barrangou and Doudna 2016). The perspective for further development or applications of the systems for strain improvement have been discussed in various research (Bikard et al. 2013; Shanmugam et al. 2020).

However, some chemicals are non-natural and heterologous pathways must be established in the first step. A focus on the recent advances and immediate challenges for creating efficient cell factories for chemical production is needed. The construction of a superior cell factory is key to the industrial biotechnology process, which requires combinatorial and multiplex modifications in cell metabolisms. CRISPR-Cas system has been demonstrated to implement most of the genetic modifications from the enzyme to the genome level, which was determined by the pool of gRNA used (Shi et al. 2022).

While genome editing techniques have shown great potential in biofuel research, there are also concerns about the safety and ethical implications of these techniques. As the field of biofuels continues to evolve, it will be important to carefully consider the ethical and safety implications of genome editing techniques. The information of the microbes which have been used in some genetics techniques and/or modification has been presented in Table 22.1.

22.2.3 Optimization of Microbial Strains for Biofuel Production

Since biofuels' formal recognition is more than two decades ago, metabolic engineering has evolved into a more systematic and high-throughput discipline, which also termed as systems metabolic engineering. Integration of traditional metabolic engineering approaches with other fields, such as systems biology, synthetic biology, and evolutionary engineering, while continuing to consider upstream (strain development) to midstream (fermentation) to downstream (separation and purification) processes as a whole is defined as systems metabolic engineering. It has resulted in the development of *Escherichia coli* strains overproducing l-valine and l-threonine in 10 person-years. Similar approaches have enabled the successful development of an increasing number of industrial strains capable of efficiently producing various bioproducts, such as 1,4-butanediol, the antimalarial drug artemisinin, succinic acid, lactic acid, and 1,2-propanediol (Javed et al. 2019; Rubin 2008).

The optimization of substrate spectrum and process robustness by rational and evolutive strategies in industrial use microorganisms for biorefinery applications have been reported in various studies (Javed et al. 2019). Bio-based production promises a sustainable route to chemicals, materials, and fuels, but its future success depends on the next level of bioprocesses using raw materials beyond glucose. Around 35 strategies have been applied for this purpose (Lee and Kim 2015). A moderately increased ATP turnover is a promising target to achieve increased biofuel yields. Cyanobacteria are primary producers of hydrocarbons and have the potential to overcome some of the problems associated with plant-derived first-generation biofuels. There has been considerable interest in biotechnological applications of cyanobacteria, such as the production of bioactive compounds, secondary metabolites, bioplastics, and bioremediation purposes, but most applications are still characterized by low product yield. Computational methods can be used to improve product yield (Erdrich et al. 2014).

Metabolic network models have been used to contextualize high throughput experimental data and have shown targets for metabolic engineering approaches. They

Table 22.1 Microbes used for genetic engineering and their functionality post engineering for production of fuels

S. No.	Name of microbe	Genetic engineering strategy	Products	References
1	*Escherichia coli*	• Metabolic pathway engineering • Transcriptional regulation • Reducing competing pathways • Increasing precursor supply • Improving stress tolerance	Ethanol, butanol	Cardoso et al. (2020), LaCroix et al. (2015), Levskaya et al. (2005), Thiele et al. (2009)
2	*Zymomonas mobilis*	• Overexpression of genes in ED pathway • Overexpression and co-expression of PDC and ADH genes • Heterologous expression of genes from other organisms • Modification of regulatory network • Adaptive lab evolution	Ethanol	He et al. (2014)
3	*Pseudomonas putida*	• Metabolic engineering of acetyl-CoA carboxylase gene • Introduction of FAEE synthesis pathway • Elimination of competing pathways • Engineering regulatory pathways	Fatty acid production	Dinamarca et al. (2003), Nogales et al. (2008)
4	*Clostridium acetobutylicum*	• Metabolic engineering of ABE pathway • Modification of solvent tolerance • Usage of alternate Carbon sources • Modulation of regulatory networks • Use of synthetic biology tools like ZFNs, TALENs and CRISPR-Cas9	Butanol, isobutanol	Foster et al. (2021), Girbal et al. (1995), Jin et al. (2009), Senger and Papoutsakis (2008)
5	*Saccharomyces cerevisiae*	• Metabolic pathway engineering including heterologous expression • Reducing byproduct formation • Ethanol tolerance enhancement	Ethanol	Oh and Jin (2020), Ouellet et al. (2011), Rodriguez et al. (2016), Runguphan and Keasling (2014)

(continued)

Table 22.1 (continued)

S. No.	Name of microbe	Genetic engineering strategy	Products	References
6	*Pseudomonas aeruginosa*	• Tweaking rhlAB operon for rhamnolipid production • Redirecting metabolic pathways using CRISPR-Cas9 • Optimization of fermentation conditions	Lipid synthesis	Liu et al. (2015), Rollins et al. (2015), Shi et al. (2013)
7	*Alcanivorax borkumensis*	• Overexpression of beta-oxidation pathway • Genetic modification of regulatory pathways (acyl-CoA synthetase) • Horizontal gene transfer (acyl-CoA thioesterase) • Directed evolution (high salt concentration)	Glucose lipids	Kalscheuer et al. (2007)
8	*Pseudoalteromonas haloplanktis*	• Metabolic engineering • Gene knockout • Heterologous gene expression • Genome shuffling • Directed evolution	Various types of proteins (low temperature)	Ricciardelli et al. (2019)
9	*Azotobacter vinelandii*	• Overexpression of hydrogenase enzyme • Heterologous expression • Regulation of nitrogen fixation	Hydrogen, ethanol, isobutanol	Yoneyama et al. (2015)
10	*Marinobacter aquaeolei*	• Overexpression of acetyl-CoA carboxylase and fatty acid synthase • Upregulation in TCA cycle • Improving tolerance to stress • Upregulation of genes involved in Calvin cycle • CRISPR-Cas9	Wax ester	Lijewski et al. (2021)

(continued)

Table 22.1 (continued)

S. No.	Name of microbe	Genetic engineering strategy	Products	References
11	*Bacillus subtilis*	• Manipulation of central metabolic pathways (TCA cycle and PPP) • Manipulation of stress response pathways • Overexpression of adhA, bdhA, kivD • Manipulation of carbon catabolite repression • Integration of heterologous pathways	Acetone, butanol, ethanol	Fu et al. (2016), Romero et al. (2007)
12	*Synechococcus elongatus*	• Introduction of heterologous pathways • Modification of native pathways • Metabolic engineering • Regulation of gene expression	Isoprenoids, terpenoids	Mashayekhi et al. (2017)
13	*Corynebacterium glutamicum*	• Introduction of heterologous pathways • Manipulation of central metabolism • Engineering of transport systems (PTS, Glycerol-3-phosphate transporter) • Regulation of gene expression • Knockout of competing pathways	Ethanol	Inui et al. (2005)
14	*Cupriavidus necator*	• Manipulation of PHA biosynthesis pathway • Integration of heterologous pathways • Regulation of gene expression • Metabolic engineering • Strain improvement	PHAs	Crépin et al. (2016)
15	*Botryococcus braunii*	• Gene knockout (PDC complex gene) • Gene overexpression • Gene editing • Metabolic engineering	Lipid synthesis	Hirano et al. (2019)

(continued)

Table 22.1 (continued)

S. No.	Name of microbe	Genetic engineering strategy	Products	References
16	*Methanosarcina acetivorans*	• Overexpression of key enzymes (MCR enzyme) • Manipulation of electron transfer pathways • Gene knockouts (acetoclastic and methylotrophic pathways) • Metabolic engineering (methyltransferases and methyl-tetrahydromethanopterin dehydrogenase) • Genome shuffling	Methane	Nazem-Bokaee et al. (2016)
17	*Clostridium thermocellum*	• Cellulase expression • Metabolic engineering • Consolidated bioprocessing • Stress tolerance (high ethanol concentration) • Co-culture with yeast	Ethanol	Tian et al. (2017)
18	*Trichoderma reesei*	• Overexpression of cellulase and hemicellulose genes • Deletion of protease genes • Introduction of transporters for pentose sugars • Optimization of growth conditions	Cellulases, hemicellulases	Chen et al. (2020)
19	*Clostridium cellulolyticum*	• Overexpression of cellulase genes • Engineering of sugar transport systems • Metabolic engineering (acetyl CoA) • Genome shuffling	Ethanol	Gaida et al. (2016)

(continued)

Table 22.1 (continued)

S. No.	Name of microbe	Genetic engineering strategy	Products	References
20	*Clostridium cellulovorans*	• Overexpression of cellulase genes • Metabolic engineering • Optimization of growth conditions • CRISPR Cas9 based genome editing	Ethanol	Yang et al. (2015)
21	*Caldicellulosiruptor bescii*	• Overexpression of key enzymes (adhA) • Genome editing (CRISPR) • Gene deletion (ldh) • Synthetic biology	Ethanol	Straub et al. (2020)
22	*Pseudonocardia autotrophica*	• Metabolic pathway engineering • Gene knockout • Heterologous gene expression (pyruvate decarboxylase) • Promoter engineering • Genome shuffling	Fatty acid production, ethanol, biodiesel	Grumaz et al. (2017)
23	*Comamonas testosteroni*	• Metabolic engineering to increase fatty acid production • Heterologous expression of alkane biosynthesis genes • Modification of cell membrane to increase cell permeability • Optimization of growth conditions	Fatty acid production	Thakor et al. (2002)
24	*Synechocystis* sp. PCC 6803	• Overexpression of key enzymes (RuBISCo) • Expression of heterologous pathways • Modification of photosynthetic pigments • Optimization of carbon flux	Butanol	Kang et al. (2021)

(continued)

Table 22.1 (continued)

S. No.	Name of microbe	Genetic engineering strategy	Products	References
25	*Pyrococcus furiosus*	• Manipulation of the glycolytic pathway • Metabolic engineering of hydrogenases • Genetic manipulation of the cell membranes • Optimization of fermentation conditions	Ethanol, hydrogen	Keller et al. (2017)
26	*Metallosphaera sedula*	• Overexpression of key enzymes (2-keto acid decarboxylase) • Optimization of carbon metabolism • Modification of metal resistance genes (copA) • Construction of synthetic pathways	Isobutanol	Wheaton et al. (2016)
27	*Geobacter* spp.	• Enhancing electron transfer pathways through overexpression • Genetic modification of extracellular electron transfer • Regulation of gene expression	Ethanol, butanol	Choi and Sang (2016)
28	*Sporomusa* spp.	• Introduction of heterologous genes • Knockout of unwanted pathways • Overexpression of key enzymes • Optimization of culture conditions	Ethanol	Madjarov et al. (2022)
29	*Moorella thermoacetica*	• Overexpression of pyruvate decarboxylase • Deletion of competing pathways • Introduction of foreign genes • Optimization of culture conditions	Ethanol, acetic acid, lactic acid	Takemura et al. (2021)

(continued)

Table 22.1 (continued)

S. No.	Name of microbe	Genetic engineering strategy	Products	References
30	*Clostridium beijerinckii*	• Metabolic engineering for butanol production • Tolerance to butanol • Production of non-native biofuels • Genome editing for strain improvement (CRISPR-Cas9) • Genetic regulation and gene expression	Butanol, biohydrogen, isobutanol	Olorunsogbon et al. (2022)
31	*Photorhabdus luminescens*	• Modification of metabolic pathways (PPP and mevalonate pathway) • Engineering of regulatory systems • Heterologous gene expression • Optimization of culture conditions	Isoprenoids, fatty acids, ethanol, butanol, isobutanol	
32	*Nostoc punctiforme*	• Overexpression of nitrogenase • Manipulation of lipid metabolism • Improvement of photosynthesis (psbA) • Enhancing stress tolerance	Lipid production	Erdrich et al. (2014), Moraes et al. (2017)
33	*Yarrowia lipolytica*	• Overexpression of acetyl-CoA carboxylase • Knockdown/Overexpression of diacylglycerol acyltransferase (DGAT) • Overexpression of phospholipid diacylglycerol acyltransferase (PDAT) • Overexpression of sterol regulatory element-binding protein (SREBP) • Overexpression of acyl-CoA synthetase • Knockdown of beta-oxidation pathway	Lipid production, TAGs	Liu et al. (2021)

(continued)

Table 22.1 (continued)

S. No.	Name of microbe	Genetic engineering strategy	Products	References
34	*Synechococcus* sp. PCC 7002	• Engineering the carbon fixation pathway • Engineering the photosynthetic machinery • Engineering the acetyl-CoA pathway • Engineering the hydrogen production pathway	Hydrogen	Kopka et al. (2017)
35	*Streptomyces venezuelae*	• Metabolic engineering • Genome editing using CRISPR-Cas9 • Horizontal gene transfer • Adaptive laboratory evolution	Antibiotics like chloramphenicol	Phelan et al. (2015)
36	*P. kudriavzevii*	• Metabolic engineering • Genome shuffling • CRISPR-Cas9 gene editing • Directed evolution	Ethanol	Chu et al. (2023)
37	*K. marxianus*	• Overexpression of PDC and ADH • Expression of heterologous genes (xylose isomerase) • Genome shuffling • Metabolic engineering • CRISPR-Cas9 genome editing	Ethanol	Bilal et al. (2022)
38	*Pichia stipitis*	• Xylose utilization pathway engineering • Ethanol tolerance engineering • Cellulase expression • Metabolic engineering • Genome editing	Ethanol	Silva et al. (2011)

(continued)

Table 22.1 (continued)

S. No.	Name of microbe	Genetic engineering strategy	Products	References
39	*Clostridium ljungdahlii*	• Enhancing flux in carbon fixation pathway • Increasing efficiency of the Wood-Ljungdahl pathway • Improving tolerance to toxic compounds	Acetic acid and ethanol	Oswald et al. (2018)
40	*Rhodobacter sphaeroides*	• Metabolic engineering • Light harvesting antenna engineering • CO_2 fixation • Overexpression of acetyl CoA carboxylase and fatty acid synthase	Hydrogen, fatty acid, isoprenoids	Orsi et al. (2021)
41	*Chlamydomonas reinhardtii*	• Overexpression of acetyl CoA carboxylase • Overexpression of Delta 12—fatty acid desaturase gene • Optimizing photosynthetic efficiency • Facilitating lipid extraction • Modification of cell wall	EPA	Scranton et al. (2015)
42	*Thermoanaerobacterium saccharolyticum*	• Metabolic engineering of glycolysis pathway • Elimination of competing pathways • Optimization of carbon catabolite repression • Manipulation of regulatory proteins (CcpA, Rex) • Use of plasmids	Hydrogen and ethanol	Currie et al. (2015)
43	*Geobacillus thermoglucosidasius*	• Metabolic engineering • Deletion or overexpression of genes • Promoter engineering • Genome shuffling • Adaptive lab evolution	Riboflavin	Yang et al. (2021)

(continued)

Table 22.1 (continued)

S. No.	Name of microbe	Genetic engineering strategy	Products	References
44	*Rhodosporidium toruloides*	• Metabolic engineering • Gene knockout • CRISPR Cas9 genome editing • Promoter engineering	Lipid production	Lyu et al. (2021)
45	*Methanococcus maripaludis*	• Manipulation of methanogenesis pathway • Utilization of alternative carbon sources • Optimization of gene expression • Modification of cell membrane	Methane	Goyal et al. (2016)

have provided established frameworks for the integration of omics data to allow the formulation and testing of downstream hypotheses. The reconstruction of metabolic network models and major computational tools and pipelines that hold the potential to contribute to the optimization of algal and many such microorganismal strains for biofuel production is reviewed. Several tools and databases like BRENDA and KEGG remain mostly unused by the algal research community, such as 7 algal-based PGDBs (Koussa et al. 2014).

The processes of pre-treatment and fermentation usually generate inhibitors to microbes. These inhibitors are derived from lignocellulose-derived by-products and residues of pre-treatment solvent. Various physical, chemical, and biological methods of detoxification processes along with tolerance and evolutionary engineering have been developed to alleviate the effects of inhibitors. However, detoxification before fermentation is not cost-efficient and product inhibition is one of the main bottlenecks for higher yield. To overcome these inhibitors, improving microbe tolerance has been suggested as a practical approach (Wang et al. 2018). Exposed concentrations and acid chain length are the factors that cause acid toxicity in microbial growth regimes. Longer chain acids are more hydrophobic, while aliphatic acid isomers with branched chains are less toxic due to decreased hydrophobicity. Phenolic compounds, including acids, alcohols, and aldehydes, also have hydrophobic interactions between the metabolite and cell membrane (Callegari et al. 2020; Fischer et al. 2008; Garba et al. 2006; Oh et al. 2018; Srivastava et al. 2020).

The main drivers for a reduction in fossil fuel consumption are global warming, issues related to peak oil, sustainability issues, and mounting concern for national energy security. The development of conversion technologies fed by renewable resources has appeared as a promising option. Biorefinery complexes use biomass as a renewable carbon-based source for the production of bioenergy and biochemicals, and the importance of efficient use of biomass with maximization of useful final products is well-acknowledged (Cherubini and Strómman 2010). Industry-driven initiatives provide a vital role in endorsing biofuel process development using organic wastes. Some of the most well-established and commercially operational wastes into biofuels conversion technologies are: (i) ethanol production from agricultural residues, (ii) biogas production through anaerobic fermentation, and (iii) biodiesel production via the transesterification process (Sheppard 2008; Stephen and Periyasamy 2018; Zaldivar et al. 2001).

Biofuels mostly refer to ethanol and esterified vegetable oil, but new products such as methane, dimethyl ether, Fischer–Tropsch (FT) diesel, and ethanol from lignocellulosic feedstock are coming into existence as a result of active research and development programs. Transport biofuel production pathways are numerous and technology choices are closely linked to the biomass type considered, but most biofuels can have the advantages of being compatible with existing vehicle engines, in contrast to compressed or liquefied natural gas, and being amenable to blending with conventional fuels within existing equipment and infrastructure (Girard and Fallot 2006; Larson et al. 2009; Lynd et al. 2009; Sokhansanj et al. 2009; Wyman and Huber 2009).

22.3 Challenges and Limitations of Microbial Engineering for Biofuel Production

Since the translation and subsequent commercialization is already so much time taking be it in terms of the process, or the microorganism of interest, or the product, or the regulatory concerns, or the socio-economic context, biofuels still have a very long path to traverse before finally blooming into something that can be easily translated and upscaled.

22.3.1 Scalability of Microbial Production

The scalability of combining microalgae-based biofuels with wastewater facilities has been a predominant subject of discussion. It uses municipal wastewater as the source of nutrients and compares the lipid productivities of microalgae. Further considerations are taken into account, such as the uniqueness of the wastewater treatment facility, wastewater characteristics and production, potential inhibitors, and the presence of biological nutrient removal (BNR) operations. Microalgae growth is governed by photosynthesis, oxygen required by aerobic bacteria, and nitrogen uptake energy from wastewater.

There is an increasing interest in growing microalga-based biofuels by taking wastewater as the growth media and the main source of nutrients. Municipal wastes have been identified as the most promising growth media, with the highest net biomass accumulation. However, research conducted at a pilot or in higher-scale plants containing wastewater facilities are scarce. Microalgae genera and strains that are able to grow in wastewater have been extensively studied, with blue-green algae, diatoms, flagellate algae, and green algae identified as the four groups of microalgae found in wastewater stabilization ponds. *Chlorella*, which is a green alga, was found to be more abundant in the South-eastern region of the US, as well as some ponds in Africa. *Scenedesmus* was the most abundant algae at Lancaster and California, while *Chlamydomonas* grew at pH between 7.0–7.8 during mid-day. Interest in the feasibility of growing microalgae-based biofuels on municipal wastewater has led researchers to invest in the production of these biofuels (Quiroz Arita et al. 2015).

The fermentation of glucose-based media using the yeast strain *Rhodosporidium toruloides* has been considered a cost-competitive process compared to the direct conversion of dried yeast cells into fuel. The estimated unitary cost of purified microbial oil is $3.4/kg. Sunflower, rapeseed, soybean, and palm oil seeds are the main oilseeds used to produce biodiesel, but biofuel production should depend on non-food crops, agro-industrial wastes, and renewable resources that do not compete with food and feed production (Koutinas et al. 2014). Recent research initiatives have focused on biodiesel production using microbial oil (MO) that is accumulated by oleaginous microorganisms during fermentation on various renewable resources. MO accumulation with similar fatty acid composition to vegetable oils can be achieved under growth-limiting conditions by various yeast (e.g., *Rhodosporidium, Rhodotorula, Lipomyces, Trichosporon*) and fungal organisms (Koutinas et al. 2014).

Oleaginous yeasts (MOs) can be used as multi-purpose or case-specific feedstocks. They grow faster than microalgae and are not affected by environmental conditions, seasonal production, or geographic location. There is no competition with food or feed production, and scale-up is easier than in the case of microalgae cultivation. Oleaginous yeast can be cultivated in the conventional stirred tank or air-lift bioreactor configurations, but the development of cost-efficient bio-diesel production is dependent upon the optimization and advancement of fermentation processes. Microbial oil accumulation is initiated under nutrient-limiting conditions, and production is carried out in fed-batch mode where yeast cell growth is achieved in the first stage and MO accumulation is induced in the second (Coleman et al. 2014).

The Integrated Assessment Framework (IAF) is an integrated multi-scale modelling, analysis, and data management suite used to address key issues in developing and operating an open-pond microalgae production facility. The costs can be reduced by scaling downstream processing capabilities to fit site-specific growing conditions, available and economically viable resources, and specific microalgal strains. In the study of Chu and Majumdar (2012), one alternative liquid transportation fuel is sourced from photoautotrophic microalgae, which can be used in many different fuel conversion pathways. The results of research enhanced the understanding towards the potential viability and sustainability of a microalgae-based biofuels industry in terms of resource use, feedstock demand, net energy production, system benefits, and economics. A demonstration of a new modelling capability, the IAF, that integrates spatiotemporal-based resource analysis with techno-economic analysis (TEA) capabilities in a high-performance environment. It provides the ability to assess variability and uncertainty in unit area biomass production over time and space followed by evaluation of the effects of site-specific facility scaling on capital and operational expenses (Thomassen et al. 2016).

22.3.2 Feedstock Availability and Processing

In the new biofuels industry, biomass availability and supply chain session had focused on practical supply chain questions for biomass feedstocks, such as how much feedstock is available, properties, and supply chain issues. An update was given on the Billion-Ton Resource Assessment report by Bob Perlack (Turhollow et al. 2014), which estimated the current and potential availability of biomass feedstocks. The report gave a detailed and broad mandate on the types and the kind of biomass feedstocks which included primary sources of forest and agriculture-derived biomass, such as logging residues, fuel treatment thinning, crop residues, and perennially grown grasses and woody crops. Michael Blaylock highlighted corn stover as a potential biomass feedstock for cellulosic ethanol, as its planting, cultivation, and harvesting costs can be shared with a popular cash crop (Parrish 2009).

The characteristics of an ideal biomass feedstock include bioengineering aspects followed by perennial productivity, having minimal antiequality factors, not being weedy/invasive, being readily established, having flexible harvests and no or few pests, being drought and heat-tolerant, having low input requirements, and being economically sound to be harvested and transported to the end user. Switchgrass cultivars have been grown under research conditions for biofuel in the Pacific Northwest (PNW) since 2002. Dacotah and Nebraska 28, the earliest maturity switchgrass cultivars, easily produce two biomass crops per season. Intermediate varieties such as Trailblazer, Forestburg, Cave-in-Rock, Blackwell, and Shawnee will produce two excellent biomass harvests with one harvest slightly less mature. Biomass yields and quality were compared among a full range of switchgrass variety maturities grown in the PNW (Werpy et al. 2004).

The supply chain involves trade-offs of productivity, production, transportation, and capital costs. Forest residues are the only untapped sources of biomass from forestlands and the economics of harvesting and transporting these residues were shown to be very favourable. The impact of expanding dedicated energy crop production to meet expected cellulosic ethanol demand could have negative global impacts if this expanse occurs on lands that are currently used for crop production or if high-carbon lands are converted to produce dedicated bioenergy crops. Several synergistic production options have been suggested that utilize non-arable lands and/or degraded lands (Inman et al. 2010).

In cities, solid municipal waste can be of utmost valuable biomass, while in rural areas, agricultural residues have major potential but face competing uses. Dedicated lignocellulosic energy crops in the future have a greater potential if land availability is made more for food as well as fuel. The present feedstock cost share is 40–70% of total production costs, but this valuation may grow as a decline in capital cost and technology development conversion becomes more efficient. Establishing a well-practical approach and efficient feedstock supply chains at scale is crucial for the success of advanced biofuels. The market size for global liquid biofuels is expected to grow from 2020 to 2027 at a compound annual growth rate (CAGR) of 6.4% (Sandesh and Ujwal 2021).

It is also anticipated to witness marginal growth and 141,592.2 thousand tons of oil equivalent in terms of demand by 2027. Policies in the international markets, trade, and prices for biofuels and various agricultural commodities have been influencing the trends in biofuel production, conversion, and trade. The OECD countries have created unintended consequences and imposed large costs on their taxpayers and consumers. There is a need for international policy to harness the effect on the production sector in the most economically suitable locations. Brazil's Energy policy encourages renewable energy to replace fossil fuel, and the proportion of renewables in the energy matrix is expected to rise to 47% by 2027. The Renewable Energy Production Incentive (REPI) has set goals to increase the blending of ethanol and biodiesel with fossil fuel by 28% and 18% by 2030 respectively (Piemonte et al. 2014).

22.3.3 The Economic Viability of Biofuels

Many nations are setting ambitious targets for increasing the use of biofuels and decreasing fossil fuel consumption. The United States and Brazil are responsible for 77% of global ethanol production, while other countries are also showing increasing interest in ethanol production (Adsul and Gokhale 2012). Several studies exhibited that companies are attracted to the utilization of renewable resources for biofuel production, but due to the overall cost of ethanol produced from biomass, industries slowed down the further steps in production and comprehend the major hurdles in commercialization of biofuels. It is also reported that ethanol cannot compete with gasoline because ethanol generates less energy than gasoline and the return of ethanol as a commodity product will be low if it is not used in large quantities to partially replace gasoline (Kohli 1980).

Consolidated bioprocessing and the utilization of waste by-products for the generation of value-added products may lead to the reduction of final biofuel costs. However, the commercialization of cellulosic ethanol is not cost-effective due to several challenges, including the difficulty of converting lignocellulose into fermentable sugars due to its recalcitrance nature (Vasić et al. 2021; Wadler et al. 2022). Consolidated bioprocessing (CBP) being a new emerging technology involves direct conversion of biomass to fuel using a single organism capable of producing ethanol/butanol using their complete cellulosic complex. The present use of ethanol as a fuel is around 3.8 GL, accounting for 0.7% of the world's oil production and 2% of the gasoline consumption while using less than 1% of the agricultural land in use in the world. Searchinger et al. (2018) calculated that 100,000 km^2 of additional land would have to come into cultivation in Brazil, China, and India, leading to massive deforestation.

To put the problem into perspective, the land in use for ethanol production in 2006 in the United States was 51,000 km^2, and in Brazil 29,000 km^2. This is 0.55% of the agricultural area in use in the world, which has over 14 million km^2. Additionally, the expansion of new crops can generate regional problems, with maize acreage growing by 19% and soybean planting decreasing by 17%. Biofuels have different impacts on food prices, such as in Brazil, where increased production of ethanol from sugarcane did not lead to an increase in sugar prices. However, the production of biofuels such as ethanol from maize and sugarcane could affect the price of cereals, but it is too early to attribute that cause to recent price variations. The reduction in greenhouse gases can be assessed by a life-cycle analysis of the energy balance involved in the preparation of ethanol (Goldemberg and Guardabassi 2009).

Cellulosic biofuels are being promoted for their potential to contribute to energy security, stable energy prices, and climate change mitigation. A key constraint to their expansion is the limited amount of agricultural land available to produce food, feed, and energy crops. To be economically viable, energy crops must compete successfully both as crops and as fuels. Herbaceous energy crops such as miscanthus and switchgrass have been identified as promising crops because they have higher yields than other perennial grasses, provide high nutrient use efficiency, and require growing conditions and equipment like those for corn (Khanna 2008; Ajanovic and Haas 2010).

Biofuels offer a new source of income to farmers and generate employment opportunities in rural areas. Biofuel production may cause net CO_2 emissions, and the term "carbon debt" has been coined to express this concept. Interest in biofuel is still high, with private investments and public subsidies fuelling the sector. To overcome the conflict over land use, marginal land, and non-crop species may be used, such as switchgrass, miscanthus, and jatropha (Godha and Jain 2015; Krauss 2008). Palm oil is also becoming of high interest in the biofuel market. Palm oil plantations may increase the displacement of native forests in tropical countries and replace other food crops without providing any benefits to farmers. International pressure drove the European Parliament's environmental committee to agree that biofuel from food crops should not exceed 6% of final energy use in consumption in transport by 2020. However, there is still 3 million ha of rapeseed cultivated with biodiesel and cereal production used to produce ethanol (Godha and Jain 2015; Jain and Sharma 2010).

The analysis of cost developments is conducted in a dynamic framework till 2030 in the EU for biofuel production. The most important impact factors on biofuel costs are feedstock costs, investment costs, fixed and variable operating and maintenance (O&M) costs, distribution, and retail costs, as well as policies (subsidies, taxation respectively tax exemption of (bio)fuels). Economic analysis is the major focus but issues of potential land availability, ecological performance, and ethical aspects must be considered (Giampietro et al. 1997).

The environmental, economic, and policy literature predicts the possible effects of biofuels and identifies gaps in understanding, and recommends areas for future work. It finds three key conclusions: the current generation is intensive in the land, water, energy, and chemical inputs, the environmental literature is dominated by net carbon offset and net energy gain, and indicators related to the impact on human health, soil quality, biodiversity, water depletion, etc., have received much less attention. The marginal impact of biofuels on welfare or the environment had been realized from the aforementioned issues. The importance of the role which biomass plays in rural areas of developing countries, and the use of biomass for producing fuel for cars can affect access to energy and fodder (Rajagopal and Zilberman 2007).

22.3.4 Regulatory Concerns and Ethical Considerations

An examination of the relationship between environmental sustainability and vulnerability of biofuels, from the perspective of a sociology of networks and flows has been carried out. It argues that current biofuel developments should be understood in terms of the emergence of a globally integrated biofuel network, where environmental sustainability is more easily accommodated than vulnerabilities for marginal, peripheral groups and countries. It also argues that most major players in the global energy markets and most environmental and development non-governmental organizations are supportive of biofuel production and consumption (Morone et al. 2019).

There are four main reasons behind this recent increase in the attention attracted by biofuels and the corresponding increase in biofuel production, R&D programs, policy initiatives, and debates, although not all reasons for this are the same in every country and region (Mol 2007). The regulatory hurdles faced by transgenic biofuel crops, particularly cellulosic ethanol, and the regulatory process in the US that oversees the introduction of new transgenic plants. It argues that the level of regulation for transgenic organisms is not proportional to their potential risk to human health or the environment and that further modifications are necessary to reflect the risk level of transgenic crops and realize their benefits. Transgenic plants with modified cell wall composition and engineered microbes for enhanced fermentation are expected to increase the efficiency of cellulosic biofuel (Jeswani et al. 2020; Saha et al. 2018). Using biotechnology to modify the cell wall structure or lignocellulosic composition of the plant could yield improvements in the ethanol yield per acre and improve the energy balance and economics of cellulosic ethanol production. The key limitation of converting plant biomass to biofuel is the inability of hydrolytic enzymes to access and deconstruct the plant cell wall, which is a complex mixture of polysaccharides cross-linked with lignin, phenolic acids, pectin, and structural proteins (Lee et al. 2009).

In the last decade, numerous national, subnational policies and regulations have incentivized biofuel production in both developed and developing economies. Global production has increased by a factor of four and international trade of feedstock and fuel is rapidly expanding. Large-scale biofuel production can alter landscapes and strain social-ecological relationships, so numerous actors have produced a complex mix of standards, meta-standards, and codes of conduct to minimize the negative impacts. However, sustainability itself is a complex and contested concept, so it should not be surprising that this complexity carries through to biofuels. Biofuel governance has evolved from standards addressing GHG emissions to comprehensive meta-standards that incorporate multiple social issues like labour conditions, land tenure, poverty alleviation, and food security (Bailis and Baka 2011). The aviation industry set targets for reducing CO_2 and flew its first commercial test flight with biofuels in 2008. As of mid-2015, 22 airlines had completed more than 2000 passenger flights with biofuel representing up to half the jet fuel mix. However, near-term plant construction has slowed down, reflecting policy uncertainty in major markets and structural challenges (Araújo et al. 2017; Saravanan et al. 2018; Sentanuhady et al. 2021).

The EU directive on the deliberate release of genetically modified organisms (GMOs) explicitly gives Member States the possibility to take into consideration the ethical aspects of GMOs in decision-making. It is suggested that public participation is an important factor in the successful implementation of the ethics of sustainable development. The new EU directive on the deliberate release of GMOs was adopted in 2001, but despite stricter provisions, a moratorium is still advocated by many non-governmental organizations. The preamble of the directive articulates respect for ethical principles, and gives Members States the possibility to take into consideration ethical aspects when GMOs are deliberately released or placed on the market as or in products (Ormandy et al. 2011).

In Sweden, a similar provision has already existed for several years, which captures in the Swedish regulation that through an empirical analysis of decisions concerning field trials with GMOs, the Gene Technology Act (SFS 1994: 900) was passed by the Swedish parliament and entered into force in 1994. It was based on the EU directives on deliberate release and contained the use of GMOs (Karlsson 2001). The purpose of the legislation was to protect human health, animal health and the environment to ensure that ethical concerns were considered when using GMOs. It also demanded precaution, prevention, and pre-assessments of risks, enforced through a licensing procedure for deliberate release or placing on the market of GMOs. Finally, a special authority, the Gene Technology Advisory Board, was established to oversee and give advice on gene technology (Román-Figueroa and Paneque 2015).

Chile has adopted legislation for the promotion of industry, research, development, and innovation for process optimization and control of the mandated use of biofuels. The regulatory framework for biofuels in Chile includes Circular 30/2007, the Internal Tax Service (SII), Law 18,502/1986, Decree 11/2008, Decree 20,339/2009, Decree-Law 1/1979 of the Ministry of Mining, Exempt 294C. Paneque, Resolution 746/2008 of the SEC, and state participation is high (Román-Figueroa and Paneque 2015). Governments should encourage policies that promote research and ethical production of biofuels according to social development and international treaties.

Target-based policies need to be more responsive to unintended consequences and allow for changing the pace of upscaling. Biofuels should be environmentally sustainable, with a mandatory and enforceable international environmental standard for biofuel production. Single international standards including a clear methodological framework for calculating GHG emissions across the whole biofuel life cycle would be the best way to ensure savings. A global, coordinated response to climate change should address land-use change directly, with strong international and local measures to prevent the destruction of high-carbon stock. Biofuel policies must protect human rights, the environment and should recognize the rights of people. Policymakers should consider the social impacts of mandated biofuel imports and implement strict requirements with strong audit trails for pay and working conditions. Intellectual property (IP) will play a key role in understanding investment in biofuels and securing profits for investors (Buyx and Tait 2011).

22.4 Roadmap for the Future of Microbial Engineering for Biofuel Production

The need to identify potential areas to increase usage of biofuels has already dawned upon us and it is our sole responsibility as to how best we can possibly make use of this industry to maximize the outputs or the benefits that it will reap keeping in mind all the factors that have been taken into consideration. A brief discussion on certain issues pertaining to biofuels' domain has been discussed in the subsequent sections.

22.4.1 Prioritized Research Areas to Address Challenges and Limitations

Analysing past experiments for developing new biofuel policies is used to explain the success and failure of these projects. In 2005, the North-Brabant government took up biofuels as an option to deal with greenhouse gas emissions, along with contributing to a sustainable energy system, creating new market opportunities for the agricultural sector. The level of success of these projects in the Netherlands is analysed by using strategic niche management (SNM) (Msangi et al. 2008; Thompson 2008; van der Laak et al. 2007; Zhang 2011).

A relationship has also been explored between biofuels and sustainable development goals (SDGs). Results show that biofuels contribute to achieving mainly SDG 13 (climate action) and SDG 7 (affordable and clean energy), but also cause negative impacts on other SDGs. Third-generation biofuels are promising alternatives to increasing sustainable development but require more R&D to reduce production costs and facilitate their commercialization. The petroleum chain uses a large quantity of feedstock and inputs, which generates wastes, effluents, and air emissions. This sector contributes to increasing greenhouse gas (GHG) emissions, soil contamination, and modification of surface waters and groundwater quality. Biofuels are a promising alternative to reduce the dependence on fossil fuels and can be classified according to their production technologies (Cheng et al. 2010; Nazari et al. 2021).

EPA's renewable fuel standard assessment found that rape-seed-based biodiesel fuels could meet the European Union Renewable Energy Directive (60% greenhouse gas emission savings) only by taking credit for rape cake (a coproduct) together with the use of glycerol as a heat source. However, farmers and foresters may be reluctant to make long-term land and resource commitments to perennial feedstocks, such as miscanthus, switchgrass, and tree crops, until there are clear market signals and adequate financial security. Without demand growth for biofuels, feedstock suppliers would need a strong incentive to substitute miscanthus or other cellulosic crops for existing corn plots. Repurposing land otherwise productively used could increase the price of the goods in global markets, which could trigger land use conversion elsewhere (McKone et al. 2011).

In conclusion, prioritized research areas in biofuels include feedstock development, metabolic engineering, synthetic biology, systems biology, sustainable production and environmental friendly processing methods. Addressing these research areas can help overcome the challenges and limitations facing biofuels and enable the widespread adoption of this promising alternative to fossil fuels.

22.4.2 Development of New and Efficient Microbial Strains

To develop a sustainable energy supply, cellulosic biomass should be used for bioethanol production instead of grain biomass. Cell surface engineering technology is a promising method, but it has limitations such as low productivity and low solvent tolerance of microorganisms. Biobutanol is produced by the fermentation process of *Clostridium* species, which naturally possess metabolic pathways for the conversion of sugar into solvents such as acetone, butanol, and ethanol. For the industrial-level production of butanol, clostridia fermentation has been widely studied, including pre-treatment of substrates, techniques of butanol recovery, and induction of solventogenesis. The introduction of butanol synthetic pathways into hosts with clostridial metabolic pathways or keto-acid pathways has also been investigated (Gottumukkala et al. 2019; Medfu Tarekegn et al. 2020).

Bioethanol production is a carbon–neutral fuel derived from biomass feedstocks such as grain and wood. The yeast *S. cerevisiae* produces ethanol in batch fermentation, with CO_2 and small amounts of methanol, glycerol, and others as by-products. Modern industrial yeast strains produce up to 20% of ethanol, by using glucose derived from starch, owing to their increased inhibitor resistance, ethanol tolerance, and ethanol-specific productivity. Most of the raw materials currently used for bioethanol production are grain biomass, but a future limitation in the supply of these materials is inevitable due to their use as food crops. To construct an energy-saving and sustainable society, the development of technologies to utilize lignocellulosic biomasses is strongly required so that a shift can be done from food to fuel crops. Cellulose is the most abundant biomacromolecule on earth and is embedded with hemicelluloses and lignin in the plant cell wall matrix and is bound to cellulose by hydrogen bonds and to lignin by ether, ester, and glycoside bonds (Huang and Ramaswamy 2013; Ueda et al. 2011).

22.4.3 Optimization of Feedstock Selection and Processing

Biomass conversion processes including separation and purification technologies in biorefineries are becoming increasingly important due to increased global demand, price, and decrease in the potential availability of crude oil, the need for energy independence and energy security, and the need for reduction in greenhouse gases emission from fossil fuel contributing to global climate change. Biorefineries' facility to convert biomass to bioproducts, such as fuels, heat and power, and a variety of co-products, can be divided into two basic conversion platforms: biochemical and thermo-chemical. Biochemical conversions of biomass using enzymes and microorganisms (yeast and bacteria) are often referred to as sugar-platform conversions. Thermochemical conversion of biomass includes biomass combustion, pyrolysis for bio-oil and biochar, hydrothermal liquefaction to bio-oils as major products, and biomass gasification to syngas. There are also a variety of other chemical conversion processes such as the conversion of oil-containing biomass (Huang and Ramaswamy 2013).

Hydroprocessing of vegetable oils is an effective pathway for processing vegetable oils into biofuels, especially in the aviation industry. Recent literature shows that development, conversion routes, and the role of processing conditions maximize the production of renewable jet fuel (Vásquez et al. 2017). The reaction temperature and acidic strength of the catalyst had a greater influence on the composition of the final products. Wide ranges of feedstocks have been studied for the production of jet biofuel, with jatropha and camelina being promising options due to their sustainability and availability. Current initiatives and companies are trying to boost jet biofuel production.

Alternative jet biofuels are being developed to replace petroleum-based fuels with renewable ones, but there is not a single technology that could reach this goal due to the lack of available raw materials. Hydroprocessing of vegetable oils is the most speedily developing jet biofuel, but its main limitations are high capital and production costs, slow technology development, low productivity, and technical complexities. It can be integrated into existing crude oil refineries, blended with conventional fuels, and overcome some of the drawbacks of other processing methods, resulting in a better-quality final product. HEFA-jet is co-produced with HEFA-Diesel and other lighter compounds at a commercial scale, and its yield depends on the type of feedstock, catalyst possibilities, reaction temperature, partial pressure, and operating conditions (Monteiro et al. 2022).

22.4.4 Integration of Renewable Energy Sources

Carbon trade has been projected as a necessary action by developing countries to reduce carbon emissions. The role of renewable energy for clean energy combustion, model biomass components for the development of bio-oil, policy framework in the development of bioenergy, and carbon trade and carbon emissions have been looked into. The sustainability of the first, second, third, and fourth-generation biofuels coupled with state-of-the-art production technologies has increased the capacity of bioenergy production. However, sustainability in producing bio-based chemicals and biofuels via various bio-refinery technologies is still a major challenge. Biomass feedstocks such as corn and sugarcane contain high levels of starch and glucose, but their use as biofuel sources threaten the food supply. Lignocellulosic feedstocks are still economically uncompetitive due to their conversion to biofuels being laborious and energy intensive. To foster sustainable energy demands, biomass refinery technologies that are environmentally desirable and energy-efficient are needed (Kibet et al. 2021).

The European Council endorsed a mandatory target of a 20% share of energy from renewable sources in overall energy consumption by 2020 and a mandatory 10% minimum target to be achieved by all Member States. To achieve these targets, Member States have applied a variety of measures, such as command and control instruments, standards, quotas, economic and fiscal measures, and communication and collaborative measures. Different countries have adopted different measures to introduce biofuels, and the economics of each are different. Stakeholders have invested in researching and implementing new feedstocks and production technology for biofuels, such as ethanol, biodiesel, green diesel, and dimethyl ether (DME) (Savaliya et al. 2015).

Investment in R&D related to advanced biofuels has risen significantly during the current decade. Biomass technologies, especially those for converting biomass to liquid biofuels for transportation, have cornered a large share of the new investments. Cutting-edge knowledge in genomics and biotechnology, process chemistry, and engineering is being applied to produce new types of energy feedstock and process them into novel biofuels. These investments have the potential to displace a substantial amount of oil over the next few decades, with a limited negative impact on the food supply and the natural habitat. The search for new transportation fuels is also giving rise to the development of new paradigms in innovation, commercialization, and regulation (Savaliya et al. 2015).

Global renewable energy initiatives can also draw inspiration from the Brazilian sugarcane ethanol program, PROALCOOL, which began in the 1970s and continues today. Biofuels, hybrids, plug-in electrics, compressed natural gas, or hydrogen-fuelled vehicles can become significant players under the right economic and policy conditions, but liquid biofuels appear to have a head start in this race due to their simplicity and cost advantages. The International Energy Agency predicts that global oil demand will reach 106 million barrels per day by 2030 (excluding biofuels) (Energy Agency 2021). Large investments are being made by OPEC and non-OPEC nations in costly and increasingly polluting sources of oil such as deep-sea drilling, enhanced oil recovery, oil sands, coal liquefaction, and gas liquefaction. According to New Energy Finance, 98% of energy R&D by private companies is for conventional fuels, but there is significant government support for clean non-fossil technologies to address the energy challenge. Non-fossil fuel resources include nuclear energy, energy efficiency, renewable energy, hydrogen and fuel cell technologies, and energy storage (Rajagopal et al. 2009).

In conclusion, the integration of renewable energy sources with microbial engineering is crucial to increase the yield, quality, and sustainability of biofuels. Solar energy, wind energy, and biogas energy have emerged as promising renewable energy sources for microbial engineering in biofuel production, and several studies have focused on developing efficient systems to integrate these energy sources with microbial fermentation processes. Addressing these research areas can help overcome the challenges facing biofuel production and enable the widespread adoption of this promising alternatives to fossil fuels.

22.4.5 The Potential of Microbial Engineering in the Production of Other Value-Added Products

Biodiesel has gained a significant amount of attention over the past decade as an environmentally friendly fuel, but the production process generates glycerol-containing waste streams which have become a disposal issue for biodiesel plants and generated a surplus of glycerol. Microbial conversions from glycerol to valuable chemicals performed by various bacteria, yeast, fungi, and microalgae have been reported, as well as the possibility of extending these conversions to microbial electrochemical technologies have been explored (Amin et al. 2017; Duraiarasan 2012). Value-added transformation processes can be accomplished by direct application, chemical transformation, or microbial conversion. Direct application refers to processes without any catalytic reaction, while chemical transformation involves three major approaches: oxidation/reduction, synthesis of higher carbon compounds with glycerol and other substrates, and industrial combustion. However, it is difficult to combust glycerol efficiently due to its low energy density.

With the development of the biodiesel industry, breakthroughs have been made each year using different microbial species and bioengineering techniques to convert waste glycerol to value-added products. The glycerol utilization pathways in Enterobacteriaceae have been studied with the goal of industrial production of 1,3-propanediol (1,3-PDO) and may apply to other bacteria that contain similar metabolic grids (Guo et al. 2015; Gupta 2013; Rajagopal et al. 2007; Wang et al. 2015).

Microbial engineering offers tremendous potential for the production of value-added products beyond biofuels. The use of microbial engineering to produce chemicals, pharmaceuticals, and bioplastics offers a sustainable alternative to conventional production methods that rely on non-renewable resources. Addressing these research areas can help unlock the full potential of microbial engineering and enable the widespread adoption of sustainable production methods in a range of industries, including biofuel production. A schematic diagram (Fig. 22.1) has been shown to discuss what are the areas in which microbial engineering of biofuels is having an impact.

22.5 Conclusion and Future Perspectives

Biofuels have emerged as a promising alternative to fossil fuels, as they can reduce greenhouse gas emissions and increase energy security. This chapter on microbial engineering in combination with biofuel production pipeline and technologies associated provides insights into various aspects of biofuel production. Feedstocks, conversion technologies, co-products and by-products, sustainability considerations, policy, and market drivers are the major parameters associated with biofuel production around the world. Microbial engineering on the other hand has played a very important role in the biofuels research domain and has immensely contributed to increased productivity, improved yield, enhanced resistance, cost-effectiveness, reduced environmental impacts, and ability to meet end-user requirements to a certain extent but immense amount of research work and translation is still required to fully realize the true potential of the biofuels' domain.

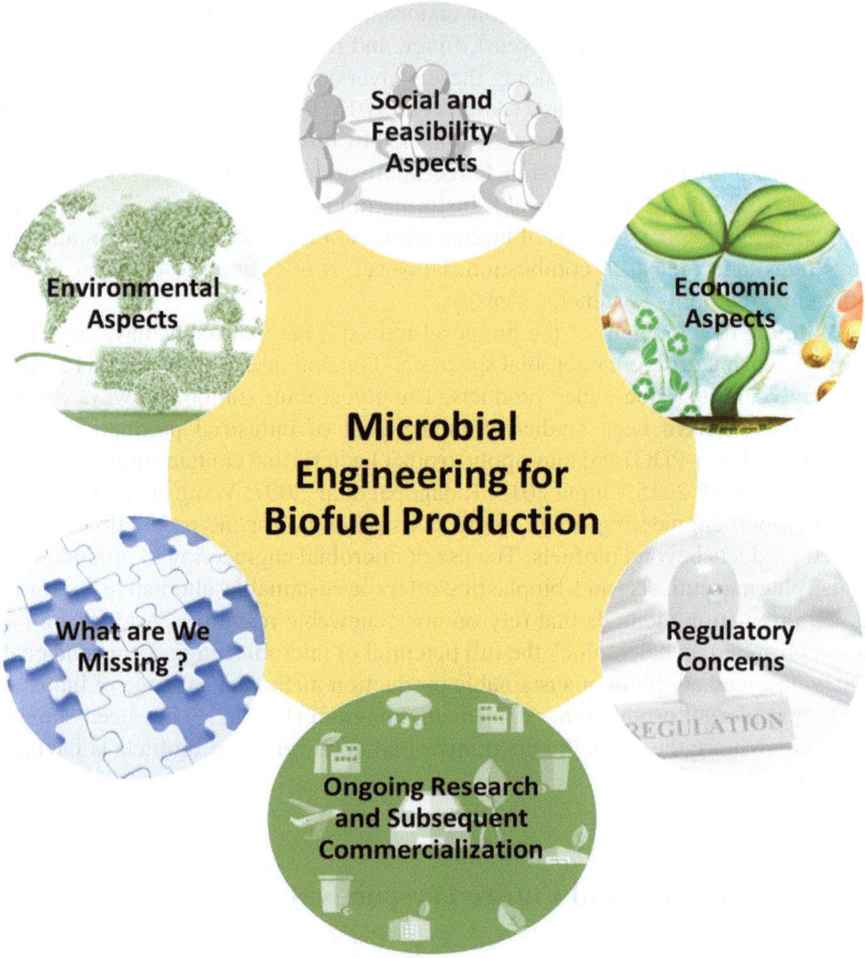

Fig. 22.1 Different aspects of microbial engineering in biofuels wherein researchers have been working to figure out the impact these aspects are having on the biofuel industry

References

Adsul MG, Gokhale DV (2012) The conundrum of making biomass-to-biofuels economic. Biofuels 3(4):383–386. https://doi.org/10.4155/bfs.12.30

Ajanovic A, Haas R (2010) Economic challenges for the future relevance of biofuels in transport in EU countries. Energy 35(8):3340–3348. https://doi.org/10.1016/j.energy.2010.04.020

Amin L, Hashim H, Mahadi Z, Ibrahim M, Ismail K (2017) Determinants of stakeholders' attitudes towards biodiesel. Biotechnol Biofuels 10(1). https://doi.org/10.1186/S13068-017-0908-8

Araújo K, Mahajan D, Kerr R, Da Silva M (2017) Global biofuels at the crossroads: an overview of technical, policy, and investment complexities in the sustainability of biofuel development. Agriculture (Switzerland) 7(4). MDPI AG. https://doi.org/10.3390/agriculture7040032

Aristos A, Merja P (2000) Metabolic engineering applications to renewable resource utilization. Curr Opin Biotechnol 11:187–198
Asghar A, Ajar R, Yadav N, Gupta A (2020) Biofuel and biorefinery technologies 10 prospects of renewable bioprocessing in future energy systems. http://www.springer.com/series/11833
Bailis R, Baka J (2011) Constructing sustainable biofuels: governance of the emerging biofuel economy. Ann Assoc Am Geogr 101(4):827–838. https://doi.org/10.1080/00045608.2011.568867
Barrangou R, Doudna JA (2016) Applications of CRISPR technologies in research and beyond. Nat Biotechnol 34(9):933–941. https://doi.org/10.1038/nbt.3659
Bikard D, Jiang W, Samai P, Hochschild A, Zhang F, Marraffini LA (2013) Programmable repression and activation of bacterial gene expression using an engineered CRISPR-Cas system. Nucleic Acids Res 41(15):7429–7437. https://doi.org/10.1093/nar/gkt520
Bilal M, Iqbal HMN, Hu H, Wang W, Zhang X (2018) Metabolic engineering and enzyme-mediated processing: a biotechnological venture towards biofuel production—a review. Renew Sustain Energy Rev 82:436–447. Elsevier Ltd. https://doi.org/10.1016/j.rser.2017.09.070
Bilal, M., Ji, L., Xu, Y., Xu, S., Lin, Y., Iqbal, H. M. N., & Cheng, H. (2022). Bioprospecting Kluyveromyces marxianus as a Robust Host for Industrial Biotechnology. In Frontiers in Bioengineering and Biotechnology (Vol. 10). Frontiers Media S.A. https://doi.org/10.3389/fbioe.2022.851768
Buschke N, Schäfer R, Becker J, Wittmann C (2013) Metabolic engineering of industrial platform microorganisms for biorefinery applications—optimization of substrate spectrum and process robustness by rational and evolutive strategies. Biores Technol 135:544–554. https://doi.org/10.1016/j.biortech.2012.11.047
Buyx A, Tait J (2011) Ethical framework for biofuels. Science 332(6029):540–541. https://doi.org/10.1126/science.1206064
Cacciatore MA, Scheufele DA, Binder AR, Shaw BR (2012) Public attitudes toward biofuels: effects of knowledge, political partisanship, and media use. Polit Life Sci 31(1–2):36–51. https://doi.org/10.2990/31_1-2_36
Callegari A, Bolognesi S, Cecconet D, Capodaglio AG (2020) Production technologies, current role, and future prospects of biofuels feedstocks: a state-of-the-art review. Crit Rev Environ Sci Technol 50(4):384–436. https://doi.org/10.1080/10643389.2019.1629801
Cardoso VM, Campani G, Santos MP, Silva GG, Pires MC, Gonçalves VM, de Giordano R, Sargo CR, Horta ACL, Zangirolami TC (2020) Cost analysis based on bioreactor cultivation conditions: production of a soluble recombinant protein using Escherichia coli BL21(DE3). Biotechnol Rep 26.https://doi.org/10.1016/j.btre.2020.e00441
Chen H, Xu M, Guo Q, Yang L, Ma Y (2016) A review on present situation and development of biofuels in China. J Energy Inst 89(2):248–255.https://doi.org/10.1016/j.joei.2015.01.022
Chen Y, Wu C, Fan X, Zhao X, Zhao X, Shen T, Wei D, Wang W (2020) Engineering of Trichoderma reesei for enhanced degradation of lignocellulosic biomass by truncation of the cellulase activator ACE3. Biotechnol Biofuels 13(1). https://doi.org/10.1186/s13068-020-01701-3
Cheng J et al (2010) Advanced biofuel technologies status and barriers. http://econ.worldbank.org
Cherubini F, Strómman AH (2010) Production of biofuels and biochemicals from lignocellulosic biomass: estimation of maximum theoretical yields and efficiencies using matrix algebra. Energy Fuels 24(4):2657–2666. https://doi.org/10.1021/ef901379s
Chia SR, Chew KW, Show PL, Yap YJ, Ong HC, Ling TC, Chang JS (2018) Analysis of economic and environmental aspects of microalgae biorefinery for biofuels production: a review. Biotechnol J 13(6). Wiley-VCH Verlag. https://doi.org/10.1002/biot.201700618
Choi O, Sang BI (2016) Extracellular electron transfer from cathode to microbes: application for biofuel production. Biotechnol Biofuels 9(1). https://doi.org/10.1186/s13068-016-0426-0
Chu Y, Li M, Jin J, Dong X, Xu K, Jin L, Qiao Y, Ji H (2023) Advances in the application of the non-conventional yeast fichia kudriavzevii in food and biotechnology industries. J Fungi 9(2). MDPI. https://doi.org/10.3390/jof9020170

Chu S, Cui Y, Liu N (2016) The path towards sustainable energy. Nat Mater 16(1):16–22. Nature Publishing Group. https://doi.org/10.1038/nmat4834

Chu S, Majumdar A (2012) Opportunities and challenges for a sustainable energy future. Nature 488(7411):294–303. https://doi.org/10.1038/nature11475

Chubukov V, Mukhopadhyay A, Petzold CJ, Keasling JD, Martín HG (2018) Synthetic and systems biology for microbial production of commodity chemicals. npj Syst Biol Appl 2. Nature Publishing Group. https://doi.org/10.1038/npjsba.2016.9

Ciaian P, Kancs A, Pirolix G, Rajcaniova M, Piroli G (2015) LICOS discussion paper series from a rise in B to a fall in C? SVAR analysis of environmental impact of biofuels *. http://www.econ.kuleuven.be/licos

Coleman AM, Abodeely JM, Skaggs RL, Moeglein WA, Newby DT, Venteris ER, Wigmosta MS (2014) An integrated assessment of location-dependent scaling for microalgae biofuel production facilities

Crépin L, Lombard E, Guillouet SE (2016) Metabolic engineering of Cupriavidus necator for heterotrophic and autotrophic alka(e)ne production. Metab Eng 37:92–101. https://doi.org/10.1016/j.ymben.2016.05.002

Currie DH, Raman B, Gowen CM, Tschaplinski TJ, Land ML, Brown SD, Covalla SF, Klingeman DM, Yang ZK, Engle NL, Johnson CM, Rodriguez M, Joe Shaw A, Kenealy WR, Lynd LR, Fong SS, Mielenz JR, Davison BH, Hogsett DA, Herring CD (2015) Genome-scale resources for Thermoanaerobacterium saccharolyticum. BMC Syst Biol 9(1). https://doi.org/10.1186/s12918-015-0159-x

Dellomonaco C, Fava F, Gonzalez R (2010) The path to next generation biofuels: successes and challenges in the era of synthetic biology. http://www.microbialcellfactories.com/content/9/1/3

Demirbas A (2009) Political, economic and environmental impacts of biofuels: a review. Appl Energy 86(Suppl 1). https://doi.org/10.1016/j.apenergy.2009.04.036

Dinamarca MA, Aranda-Olmedo I, Puyet A, Rojo F (2003) Expression of the Pseudomonas putida OCT plasmid alkane degradation pathway is modulated by two different global control signals: evidence from continuous cultures. J Bacteriol 185(16):4772–4778. https://doi.org/10.1128/JB.185.16.4772-4778.2003

Dunlop MJ (2011) Engineering microbes for tolerance to next-generation biofuels. Biotechnol Biofuels 4. https://doi.org/10.1186/1754-6834-4-32

Duraiarasan S (2012) Biofuel and bioactive compounds production from algae view project microalgal biodiesel—a comprehensive review on the potential and alternative biofuel. Res J Chem Sci 2(11). www.isca.in

Elavarasan RM (2018) The motivation for renewable energy and its comparison with other energy sources: a review. Eur J Sustain Dev Res 3(1). https://doi.org/10.20897/ejosdr/4005

Elshahed MS (2010) Microbiological aspects of biofuel production: current status and future directions. J Adv Res 1(2):103–111. https://doi.org/10.1016/j.jare.2010.03.001

Energy Information Administration (2015) World energy demand and economic outlook Table 1–1. World energy consumption by country grouping, 2012–40 (quadrillion Btu). https://www.eia.gov/analysis/

Energy Agency I (2050) (2021). Net Zero by 2050—a roadmap for the global energy sector. www.iea.org/t&c/

Erdrich P, Knoop H, Steuer R, Klamt S (2014) Cyanobacterial biofuels: new insights and strain design strategies revealed by computational modeling. Microbial Cell Factor 13. http://www.microbialcellfactories.com/content/13/128

Erdrich P, Steuer R, Klamt S (2015) An algorithm for the reduction of genome-scale metabolic network models to meaningful core models. BMC Syst Biol 9(1). https://doi.org/10.1186/s12918-015-0191-x

Escobar JC, Lora ES, Venturini OJ, Yáñez EE, Castillo EF, Almazan O (2009) Biofuels: environment, technology and food security. Renew Sustain Energy Rev 13(6–7):1275–1287. https://doi.org/10.1016/j.rser.2008.08.014

Feist AM, Zielinski DC, Orth JD, Schellenberger J, Herrgard MJ, Palsson BO (2010) Model-driven evaluation of the production potential for growth-coupled products of Escherichia coli. Metab Eng 12(3):173–186. https://doi.org/10.1016/j.ymben.2009.10.003

Filip O, Janda K, Kristoufek L, Zilberman D (2016) Dynamics and evolution of the role of biofuels in global commodity and financial markets. Nat Energy 1(12). https://doi.org/10.1038/nenergy.2016.169

Fischer CR, Klein-Marcuschamer D, Stephanopoulos G (2008) Selection and optimization of microbial hosts for biofuels production. Metab Eng 10(6):295–304. https://doi.org/10.1016/j.ymben.2008.06.009

Foster C, Charubin K, Papoutsakis ET, Maranas CD (2021) Modeling growth kinetics, interspecies cell fusion, and metabolism of a Clostridium acetobutylicum/Clostridium ljungdahlii syntrophic coculture. https://doi.org/10.1128/mSystems

Fu J, Huo G, Feng L, Mao Y, Wang Z, Ma H, Chen T, Zhao X (2016) Metabolic engineering of Bacillus subtilis for chiral pure meso-2,3-butanediol production. Biotechnol Biofuels 9(1). https://doi.org/10.1186/s13068-016-0502-5

Fung TKF, Choi DH, Scheufele DA, Shaw BR (2014) Public opinion about biofuels: the interplay between party identification and risk/benefit perception. Energy Policy 73:344–355. https://doi.org/10.1016/j.enpol.2014.05.016

Gaida SM, Liedtke A, Jentges AHW, Engels B, Jennewein S (2016) Metabolic engineering of Clostridium cellulolyticum for the production of n-butanol from crystalline cellulose. Microb Cell Fact 15(1). https://doi.org/10.1186/s12934-015-0406-2

Garba MU, Alhassan M, Kovo AS (2006) A review of advances and quality assessment of biofuels. http://ljs.academicdirect.org

Ghag SB, Vavilala SL, D'Souza JS (2019) Metabolic engineering and genetic manipulation of novel biomass species for biofuel production. In: Advanced bioprocessing for alternative fuels, biobased chemicals, and bioproducts: technologies and approaches for scale-up and commercialization. Elsevier, pp 13–34. https://doi.org/10.1016/B978-0-12-817941-3.00002-4

Giampietro M, Ulgiati S, Pimente D (1997) Feasibility of large-scale biofuel production does an enlargement of scale change the picture? https://academic.oup.com/bioscience/article/47/9/587/222660

Girard P, Fallot A (2006) Review of existing and emerging technologies for the production of biofuels in developing countries. Energy Sustain Dev 10(2):92–108. https://doi.org/10.1016/S0973-0826(08)60535-9

Girbal L, Croux C, Vasconcelos I, Soucaille P (1995) Regulation of metabolic shifts in Clostridium acetobutylicum ATCC 824. FEMS Microbiol Rev 17(3):287–297. https://doi.org/10.1111/j.1574-6976.1995.tb00212.x

Godha A, Jain P (2015) Sustainability reporting trend in Indian companies as per GRI framework: a comparative study. South Asian J Bus Manag Cases 4(1):62–73. https://doi.org/10.1177/2277977915574040

Goldemberg J, Guardabassi P (2009) Are biofuels a feasible option? Energy Policy 37(1):10–14. https://doi.org/10.1016/j.enpol.2008.08.031

Gomiero T (2015) Are biofuels an effective and viable energy strategy for industrialized societies? A reasoned overview of potentials and limits. Sustainability (Switzerland) 7(7):8491–8521. https://doi.org/10.3390/su7078491

Gottumukkala LD, Mathew A, Abraham A, Sukumaran RK (2019) Biobutanol production: microbes, feedstock, and strategies. In: Biomass, biofuels, biochemicals: biofuels: alternative feedstocks and conversion processes for the production of liquid and gaseous biofuels. Elsevier, pp 355–377. https://doi.org/10.1016/B978-0-12-816856-1.00015-4

Goyal N, Zhou Z, Karimi IA (2016) Metabolic processes of Methanococcus maripaludis and potential applications. Microb Cell Fact 15(1). https://doi.org/10.1186/s12934-016-0500-0

Grumaz C, Rais D, Kirstahler P, Vainshtein Y, Rupp S, Zibek S, Sohn K (2017) Draft genome sequence of Pseudonocardia autotrophica strain DSM 43083, an efficient producer of peroxidases for lignin modification. Genome Announcements 5(5). https://doi.org/10.1128/genomeA.01562-16

Guo M, Song W, Buhain J (2015) Bioenergy and biofuels: history, status, and perspective. Renew Sustain Energy Rev 42:712–725. Elsevier Ltd. https://doi.org/10.1016/j.rser.2014.10.013

Gupta VK (2013) Applications of microbial engineering. CRC Press, Taylor & Francis Group

Harrison MJ, Thompson B, Du Z-Y, Zienkiewicz K, Vande PN, Ostrom NE, Benning C, Bonito GM (2019) Algal-fungal symbiosis leads to photosynthetic mycelium. https://doi.org/10.7554/eLife.47815.001

He MX, Wu B, Qin H, Ruan ZY, Tan FR, Wang JL, Shui ZX, Dai LC, Zhu QL, Pan K, Tang XY, Wang WG, Hu QC (2014) Zymomonas mobilis: a novel platform for future biorefineries. In: Biotechnology for biofuels, vol 7, issue 1. BioMed Central Ltd. https://doi.org/10.1186/1754-6834-7-101

Hirano K, Hara T, Ardianor, Nugroho RA, Segah H, Takayama N, Sulmin G, Komai Y, Okada S, Kawamura K (2019) Detection of the oil-producing microalga Botryococcus braunii in natural freshwater environments by targeting the hydrocarbon biosynthesis gene SSL-3. Sci Rep 9(1). https://doi.org/10.1038/s41598-019-53619-y

Huang H-J, Ramaswamy S (2013) Overview of biomass conversion processes and separation and purification technologies in biorefineries. http://www.nrel.gov/biomass/biorefinery.html

Inman D, Nagle N, Jacobson J, Searcy E, Ray AE (2010) Feedstock handling and processing effects on biochemical conversion to biofuels. Biofuels Bioprod Biorefin 4(5):562–573. https://doi.org/10.1002/bbb.241

Inui M, Kawaguchi H, Murakami S, Vertès AA, Yukawa H (2005) Metabolic engineering of Corynebacterium glutamicum for fuel ethanol production under oxygen-deprivation conditions. J Mol Microbiol Biotechnol 8(4):243–254. https://doi.org/10.1159/000086705

Jain S, Sharma MP (2010) Prospects of biodiesel from Jatropha in India: a review. Renew Sustain Energy Rev 14(2):763–771. https://doi.org/10.1016/j.rser.2009.10.005

Javed MR, Noman M, Shahid M, Ahmed T, Khurshid M, Rashid MH, Ismail M, Sadaf M, Khan F (2019) Current situation of biofuel production and its enhancement by CRISPR/Cas9-mediated genome engineering of microbial cells. Microbiol Res 219:1–11. Elsevier GmbH. https://doi.org/10.1016/j.micres.2018.10.010

Jeswani HK, Chilvers A, Azapagic A (2020) Environmental sustainability of biofuels: a review: environmental sustainability of biofuels. Proc R Soc A: Math Phys Eng Sci 476(2243). https://doi.org/10.1098/RSPA.2020.0351

Jin YL, Jang YS, Lee J, Papoutsakis ET, Lee SY (2009) Metabolic engineering of Clostridium acetobutylicum M5 for highly selective butanol production. Biotechnol J 4(10):1432–1440. https://doi.org/10.1002/biot.200900142

Jordaan SM (2007) Ethical risks of attenuating climate change through new energy systems: the case of a biofuel system. Ethics Sci Environ Polit 7(1):23–29. https://doi.org/10.3354/esep00081

Joshi A, Verma KK, Rajput V, Minkina T, Arora J (2022) Recent advances in metabolic engineering of microorganisms for advancing lignocellulose-derived biofuels. Bioengineered 13(4):8135–8163. Taylor and Francis Ltd. https://doi.org/10.1080/21655979.2022.2051856

Kalscheuer R, Stöveken T, Malkus U, Reichelt R, Golyshin PN, Sabirova JS, Ferrer M, Timmis KN, Steinbüchel A (2007) Analysis of storage lipid accumulation in Alcanivorax borkumensis: evidence for alternative triacylglycerol biosynthesis routes in bacteria. J Bacteriol 189(3):918–928. https://doi.org/10.1128/JB.01292-06

Kaltschmitt M, Weber M (2006) Markets for solid biofuels within the EU-15. Biomass Bioenergy 30(11):897–907. https://doi.org/10.1016/j.biombioe.2006.06.009

Kang MJ, Hong SJ, Yoo D, Cho BK, Lee H, Choi HK, Kim DM, Lee CG (2021) Photosynthetic production of biodiesel in Synechocystis sp. PCC6803 transformed with insect or plant fatty acid methyltransferase. Bioprocess Biosyst Eng 44(7):1433–1439. https://doi.org/10.1007/s00449-021-02520-y

Karlsson M (2001) Ethics of sustainable development—a study of Swedish regulations for genetically modified organisms

Keller MW, Lipscomb GL, Nguyen DM, Crowley AT, Schut GJ, Scott I, Kelly RM, Adams MWW (2017) Ethanol production by the hyperthermophilic archaeon Pyrococcus furiosus by expression of bacterial bifunctional alcohol dehydrogenases. Microb Biotechnol 10(6):1535–1545. https://doi.org/10.1111/1751-7915.12486

Khanna M (2008) The magazine of food, farm, and resource issues cellulosic biofuels: are they economically viable and environmentally sustainable? Choices 23(3). http://www.choicesmagazine.org

Kibet JK, Korir BK, Mosonik BC (2021) A review of the current trends in the production and consumption of bioenergy. Prog Chem Biochem Res J 2021(2):109–133. https://doi.org/10.22034/pcbr.2021.126117

Kohli H (1980) Renewable energy: alcohol from biomass

Konar D, Saha R, Bhattacharya D, Mukhopadhyay M (2020) Present status and future prospect of genetic and metabolic engineering for biofuels from lignocellulosic biomass. In: Genetic and metabolic engineering for improved biofuel production from lignocellulosic biomass. Elsevier, pp 37–46. https://doi.org/10.1016/B978-0-12-817953-6.00003-8

Kopka J, Schmidt S, Dethloff F, Pade N, Berendt S, Schottkowski M, Martin N, Dühring U, Kuchmina E, Enke H, Kramer D, Wilde A, Hagemann M, Friedrich A (2017) Systems analysis of ethanol production in the genetically engineered cyanobacterium Synechococcus sp. PCC 7002. Biotechnol Biofuels 10(1). https://doi.org/10.1186/s13068-017-0741-0

Koussa J, Chaiboonchoe A, Salehi-Ashtiani K (2014) Computational approaches for microalgal biofuel optimization: a review. BioMed Res Int 2014. Hindawi Limited. https://doi.org/10.1155/2014/649453

Koutinas AA, Chatzifragkou A, Kopsahelis N, Papanikolaou S, Kookos IK (2014) Design and techno-economic evaluation of microbial oil production as a renewable resource for biodiesel and oleochemical production. Fuel 116:566–577. https://doi.org/10.1016/j.fuel.2013.08.045

Krauss C (2008) Taking flight on jatropha fuel

Krupnov YA, Krasilnikova VG, Kiselev V, Yashchenko AV (2022) The contribution of sustainable and clean energy to the strengthening of energy security. Front Environ Sci 10. https://doi.org/10.3389/fenvs.2022.1090110

Kukreja S, Gunarathne SMS, Giri T, Goutam U, Gautam S (2021) CRISPR-CAS9: a genome editing tool for improvement of biofuel production in diatoms: a review. Plant Archiv 21(Suppl-1):202–209. https://doi.org/10.51470/plantarchives.2021.v21.s1.035

LaCroix RA, Sandberg TE, O'Brien EJ, Utrilla J, Ebrahim A, Guzman GI, Szubin R, Palsson BO, Feist AM (2015) Use of adaptive laboratory evolution to discover key mutations enabling rapid growth of Escherichia coli K-12 MG1655 on glucose minimal medium. Appl Environ Microbiol 81(1):17–30. https://doi.org/10.1128/AEM.02246-14

Lacerda MP, Oh EJ, Eckert C (2020) The model system Saccharomyces cerevisiae versus emerging non-model yeasts for the production of biofuels. Life 10(11):1–20. MDPI AG. https://doi.org/10.3390/life10110299

Larson ED, Jin H, Celik FE (2009) Large-scale gasification-based coproduction of fuels and electricity from switchgrass. Biofuels Bioprod Biorefin 3(2):174–194. https://doi.org/10.1002/bbb.137

Lee D, Nair R, Chen A (2009) Regulatory hurdles for transgenic biofuel crops. Biofuels Bioprod Biorefin 3(4):468–480. https://doi.org/10.1002/bbb.158

Lee SY, Kim HU (2015) Systems strategies for developing industrial microbial strains. Nat Biotechnol 33(10):1061–1072. Nature Publishing Group. https://doi.org/10.1038/nbt.3365

Levskaya A, Chevalier AA, Tabor JJ, Simpson ZB, Lavery LA, Levy M, Davidson EA, Scouras A, Ellington AD, Marcotte EM, Voigt CA (2005) Engineering Escherichia coli to see light. Nature 438(7067):441–442. https://doi.org/10.1038/nature04405

Liao JC, Mi L, Pontrelli S, Luo S (2016) Fuelling the future: microbial engineering for the production of sustainable biofuels. Nat Rev Microbiol 14(5):288–304. Nature Publishing Group. https://doi.org/10.1038/nrmicro.2016.32

Lijewski AM, Knutson CM, Lenneman EM, Barney BM (2021) Evaluation of two thioesterases from Marinobacter aquaeolei VT8: relationship to wax ester production. FEMS Microbiol Lett 368(1). https://doi.org/10.1093/femsle/fnaa206

Liu Y, Gong AJ, Qiu LN, Li JR, Li FK (2015) Biodegradation of decabromodiphenyl ether (BDE-209) by crude enzyme extract from Pseudomonas aeruginosa. Int J Environ Res Public Health 12(9):11829–11847. https://doi.org/10.3390/ijerph120911829

Lu Y, Khan ZA, Alvarez-Alvarado MS, Zhang Y, Huang Z, Imran M (2020) A critical review of sustainable energy policies for the promotion of renewable energy sources. Sustainability (Switzerland) 12(12). MDPI. https://doi.org/10.3390/su12125078

Lundgren T, Marklund PO, Brännlund R, Kriström B (2008) The economics of biofuels. Int Rev Environ Resour Econ 2(3):237–280. https://doi.org/10.1561/101.00000017

Luque R, Herrero-Davila L, Campelo JM, Clark JH, Hidalgo JM, Luna D, Marinas JM, Romero AA (2008) Biofuels: a technological perspective. Energy Environ Sci 1(5):542–564. https://doi.org/10.1039/b807094f

Lynd LR, Larson E, Greene N, Laser M, Sheehan J, Dale BE, McLaughlin S, Wang M (2009) The role of biomass in America's energy future: framing the analysis. Biofuels Bioprod Biorefin 3(2):113–123. https://doi.org/10.1002/bbb.134

Lyu L, Chu Y, Zhang S, Zhang Y, Huang Q, Wang S, Zhao ZK (2021) Engineering the oleaginous yeast Rhodosporidium toruloides for improved resistance against inhibitors in biomass hydrolysates. Front Bioeng Biotechnol 9. https://doi.org/10.3389/fbioe.2021.768934

Madjarov J, Soares R, Paquete CM, Louro RO (2022) Sporomusa ovata as catalyst for bioelectrochemical carbon dioxide reduction: a review across disciplines from microbiology to process engineering. Front Microbiol 13. Frontiers Media S.A. https://doi.org/10.3389/fmicb.2022.913311

Martin M (2013) Industrial symbiosis in the biofuel industry: quantification of the environmental performance and identification of synergies. Linköping University, Department of Management and Engineering

Mashayekhi M, Sarrafzadeh MH, Tavakoli O, Soltani N, Faramarzi MA (2017) Potential for biodiesel production and carbon capturing from Synechococcus Elongatus: an isolation and evaluation study. Biocatal Agric Biotechnol 9:230–235. https://doi.org/10.1016/j.bcab.2017.01.005

McKone TE, Nazaroff WW, Berck P, Auffhammer M, Lipman T, Torn MS, Masanet E, Lobscheid A, Santero N, Mishra U, Barrett A, Bomberg M, Fingerman K, Scown C, Strogen B, Horvath A (2011) Grand challenges for life-cycle assessment of biofuels. Environ Sci Technol 45(5):1751–1756. https://doi.org/10.1021/es103579c

Medfu Tarekegn M, Zewdu Salilih F, Ishetu AI (2020) Microbes used as a tool for bioremediation of heavy metal from the environment. Cogent Food Agric 6(1). Informa Healthcare. https://doi.org/10.1080/23311932.2020.1783174

Mol APJ (2007) Boundless biofuels? Between environmental sustainability and vulnerability. Sociol Rural 47(4):297–315. https://doi.org/10.1111/j.1467-9523.2007.00446.x

Monteiro RRC, Dos Santos IA, Arcanjo MRA, Cavalcante CL, de Luna FMT, Fernandez-Lafuente R, Vieira RS (2022) Production of jet biofuels by catalytic hydroprocessing of esters and fatty acids: a review. Catalysts 12(2). MDPI. https://doi.org/10.3390/catal12020237

Moraes LE, Blow MJ, Hawley ER, Piao H, Kuo R, Chiniquy J, Shapiro N, Woyke T, Fadel JG, Hess M (2017) Resequencing and annotation of the Nostoc punctiforme ATTC 29133 genome: facilitating biofuel and high-value chemical production. AMB Express 7(1). https://doi.org/10.1186/s13568-017-0338-9

Morone P, Strzałkowski A, Tani A (2019) Biofuel transitions: an overview of regulations and standards for a more sustainable framework. In: Biofuels for a more sustainable future: life

cycle sustainability assessment and multi-criteria decision making. Elsevier, pp 21–46. https://doi.org/10.1016/B978-0-12-815581-3.00002-6
Msangi S, Sulser T, Rosegrant M, Valmonte-santos R, Ringler C (2008) Global scenarios for biofuels: impacts and implications
Mukhopadhyay A, Redding AM, Rutherford BJ, Keasling JD (2008) Importance of systems biology in engineering microbes for biofuel production
Nazari MT, Mazutti J, Basso LG, Colla LM, Brandli L (2021) Biofuels and their connections with the sustainable development goals: a bibliometric and systematic review. Environ Dev Sustain 23(8):11139–11156. Springer Science and Business Media B.V. https://doi.org/10.1007/s10668-020-01110-4
Nazem-Bokaee H, Gopalakrishnan S, Ferry JG, Wood TK, Maranas CD (2016) Assessing methanotrophy and carbon fixation for biofuel production by Methanosarcina acetivorans. Microb Cell Fact 15(1). https://doi.org/10.1186/s12934-015-0404-4
Nogales J, Palsson B, Thiele I (2008) A genome-scale metabolic reconstruction of Pseudomonas putida KT2440: iJN746 as a cell factory. BMC Syst Biol 2. https://doi.org/10.1186/1752-0509-2-79
Oh EJ, Jin YS (2020) Engineering of Saccharomyces cerevisiae for efficient fermentation of cellulose. FEMS Yeast Res 20(1). Oxford University Press. https://doi.org/10.1093/femsyr/foz089
Oh YK, Hwang KR, Kim C, Kim JR, Lee JS (2018) Recent developments and key barriers to advanced biofuels: a short review. Bioresour Technol 257:320–333. Elsevier Ltd. https://doi.org/10.1016/j.biortech.2018.02.089
Ohta K, Beall DS, Mejia JP, Shanmugam KT, Ingram LO (1991) Genetic improvement of Escherichia coli for ethanol production: chromosomal integration of Zymomonas mobilis genes encoding pyruvate decarboxylase and alcohol dehydrogenase II. Appl Environ Microbiol 57(4):893. https://doi.org/10.1128/AEM.57.4.893-900.1991
Olorunsogbon T, Adesanya Y, Atiyeh HK, Okonkwo CC, Ujor VC, Ezeji TC (2022) Effects of Clostridium beijerinckii and medium modifications on acetone–butanol–ethanol production from switchgrass. Front Bioeng Biotechnol 10. https://doi.org/10.3389/fbioe.2022.942701
Oluchukwu A, Innocent A, Ogochukwu E (2014) Sustainable sources of energy and the expected benefits to Nigerian economy. Int J Sustain Energy Environ Res 3(2). http://pakinsight.com/?ic=journal&journal=13
Ormandy EH, Dale J, Griffin G (2011) Genetic engineering of animals: ethical issues, including welfare concerns. Can Vet J 52(5):544. /pmc/articles/PMC3078015/
Orsi E, Beekwilder J, Eggink G, Kengen SWM, Weusthuis RA (2021) The transition of Rhodobacter sphaeroides into a microbial cell factory. Biotechnol Bioeng 118(2), 531–541. John Wiley and Sons Inc. https://doi.org/10.1002/bit.27593
Osman AI, Mehta N, Elgarahy AM, Al-Hinai A, Ala' Al-Muhtaseb H, David, Rooney W (2021) Conversion of biomass to biofuels and life cycle assessment: a review. Environ Chem Lett 19:4075–4118. https://doi.org/10.1007/s10311-021-01273-0
Oswald F, Zwick M, Omar O, Hotz EN, Neumann A (2018) Growth and product formation of Clostridium ljungdahlii in presence of cyanide. Front Microb 9(JUN). https://doi.org/10.3389/fmicb.2018.01213
Ouellet M, Datta S, Dibble DC, Tamrakar PR, Benke PI, Li C, Singh S, Sale KL, Adams PD, Keasling JD, Simmons BA, Holmes BM, Mukhopadhyay A (2011) Impact of ionic liquid pretreated plant biomass on Saccharomyces cerevisiae growth and biofuel production. Green Chem 13(10):2743–2749. https://doi.org/10.1039/c1gc15327g
Ozsoz M, Au I, Pp C (2019) Application of Crispr technology for the generation of biofuels: a review
O'Connor RE, Bord RJ, Fisher A (1999) Risk perceptions, general environmental beliefs, and willingness to address climate change. Risk Anal 19(3):461–471. https://doi.org/10.1111/j.1539-6924.1999.tb00421.x

Panwar NL, Kaushik SC, Kothari S (2011) Role of renewable energy sources in environmental protection: a review. Renew Sustain Energy Rev 15(3):1513–1524. https://doi.org/10.1016/j.rser.2010.11.037

Parrish DJ (2009) New biofuels industry: biomass availability and supply chain. Appl Biochem Biotechnol 154(1–3):268–270. https://doi.org/10.1007/s12010-009-8612-1

Passell H, Dhaliwal H, Reno M, Wu B, Ben Amotz A, Ivry E, Gay M, Czartoski T, Laurin L, Ayer N (2013) Algae biodiesel life cycle assessment using current commercial data. J Environ Manag 129:103–111. https://doi.org/10.1016/j.jenvman.2013.06.055

Peralta-Yahya PP, Zhang F, Del Cardayre SB, Keasling JD (2012) Microbial engineering for the production of advanced biofuels. Nature 488(7411):320–328. https://doi.org/10.1038/nature11478

Phelan RM, Sekurova ON, Keasling JD, Zotchev SB (2015) Engineering terpene biosynthesis in streptomyces for production of the advanced biofuel precursor bisabolene. ACS Synth Biol 4(4):393–399. https://doi.org/10.1021/sb5002517

Piemonte V, Paola LD, Russo V (2014) An LCA study on feedstocks and processes for biofuels production. Chem Eng Trans 37:517–522. https://doi.org/10.3303/CET1437087

Pira S (2020) The importance of renewable energies with emphasize on wind power. www.ijert.org

Quiroz Arita CE, Peebles C, Bradley TH (2015) Scalability of combining microalgae-based biofuels with wastewater facilities: a review. Algal Res 9:160–169. Elsevier B.V. https://doi.org/10.1016/j.algal.2015.03.001

Rajagopal D, Sexton S, Hochman G, Zilberman D (2009) Recent developments in renewable technologies: R&D investment in advanced biofuels. Annu Rev Resource Econ 1(1):621–644. https://doi.org/10.1146/annurev.resource.050708.144259

Rajagopal D, Sexton SE, Roland-Holst D, Zilberman D (2007) Challenge of biofuel: filling the tank without emptying the stomach? Environ Res Lett 2(4). https://doi.org/10.1088/1748-9326/2/4/044004

Rajagopal D, Zilberman D (2007) Review of environmental, economic and policy aspects of biofuels. http://econ.worldbank.org

Ricciardelli A, Casillo A, Vergara A, Balasco N, Corsaro MM, Tutino ML, Parrilli E (2019) Environmental conditions shape the biofilm of the Antarctic bacterium Pseudoalteromonas haloplanktis TAC125. Microbiol Res 218:66–75. https://doi.org/10.1016/j.micres.2018.09.010

Rodriguez S, Denby CM, Vu T, Baidoo EEK, Wang G, Keasling JD (2016) ATP citrate lyase mediated cytosolic acetyl-CoA biosynthesis increases mevalonate production in Saccharomyces cerevisiae. Microb Cell Fact 15(1). https://doi.org/10.1186/s12934-016-0447-1

Rollins MF, Schuman JT, Paulus K, Bukhari HST, Wiedenheft B (2015) Mechanism of foreign DNA recognition by a CRISPR RNA-guided surveillance complex from Pseudomonas aeruginosa. Nucleic Acids Res 43(4):2216–2222. https://doi.org/10.1093/nar/gkv094

Romero S, Merino E, Bolívar F, Gosset G, Martinez A (2007) Metabolic engineering of Bacillus subtilis for ethanol production: Lactate dehydrogenase plays a key role in fermentative metabolism. Appl Environ Microbiol 73(16):5190–5198. https://doi.org/10.1128/AEM.00625-07

Román-Figueroa C, Paneque M (2015) Ethics and biofuel production in Chile. J Agric Environ Ethics 28(2):293–312. https://doi.org/10.1007/s10806-015-9535-1

Rubin EM (2008) Genomics of cellulosic biofuels. Nature 454(7206):841–845. Nature Publishing Group. https://doi.org/10.1038/nature07190

Runguphan W, Keasling JD (2014) Metabolic engineering of Saccharomyces cerevisiae for production of fatty acid-derived biofuels and chemicals. Metab Eng 21:103–113. https://doi.org/10.1016/j.ymben.2013.07.003

Rupprecht J (2009) From systems biology to fuel-Chlamydomonas reinhardtii as a model for a systems biology approach to improve biohydrogen production. J Biotechnol 142(1):10–20. https://doi.org/10.1016/j.jbiotec.2009.02.008

Ryu DDY, Lee SB (1986) Enzymatic hydrolysis of cellulose: determination of kinetic parameters. Chem Eng Commun 45(1–6):119–134. https://doi.org/10.1080/00986448608911377

Saha S, Sharma A, Purkayastha S, Pandey K, Dhingra S (2018) Bio-plastics and biofuel: is it the way in future development for end users? Plastics to energy: fuel, chemicals, and sustainability implications, pp 365–376. https://doi.org/10.1016/B978-0-12-813140-4.00014-5
Sandesh K, Ujwal P (2021) Trends and perspectives of liquid biofuel—process and industrial viability. Energy Convers Manag X 10. https://doi.org/10.1016/j.ecmx.2020.100075
Saravanan AP, Mathimani T, Deviram G, Rajendran K, Pugazhendhi A (2018) Biofuel policy in India: a review of policy barriers in sustainable marketing of biofuel. J Clean Prod 193:734–747. Elsevier Ltd. https://doi.org/10.1016/j.jclepro.2018.05.033
Sastry SVAR, Sreenu P (2012) New energy sources and its sustainability. In: AICERA 2012—Annual international conference on emerging research areas: innovative practices and future trends. https://doi.org/10.1109/AICERA.2012.6306705
Savaliya ML, Dhorajiya BD, Dholakiya BZ (2015) Recent advancement in production of liquid biofuels from renewable resources: a review. Res Chem Intermed 41(2):475–509. https://doi.org/10.1007/s11164-013-1231-z
Saxena RC, Adhikari DK, Goyal HB (2009) Biomass-based energy fuel through biochemical routes: a review. Renew Sustain Energy Rev 13(1):167–178. https://doi.org/10.1016/j.rser.2007.07.011
Scranton MA, Ostrand JT, Fields FJ, Mayfield SP (2015) Chlamydomonas as a model for biofuels and bio-products production. Plant J 82(3):523–531. https://doi.org/10.1111/tpj.12780
Searchinger TD, Wirsenius S, Beringer T, Dumas P (2018) Assessing the efficiency of changes in land use for mitigating climate change. Nature 564(7735):249–253. https://doi.org/10.1038/s41586-018-0757-z
Senger RS, Papoutsakis ET (2008) Genome-scale model for Clostridium acetobutylicum: Part I. Metabolic network resolution and analysis. Biotechnol Bioeng 101(5):1036–1052. https://doi.org/10.1002/bit.22010
Sentanuhady J, Atmaja GPSG, Muflikhun MA (2021) Challenges of biofuel applications in industrial and automotive: a review. J Eng Sci Technol Rev 14(4):119–134. https://doi.org/10.25103/jestr.144.16
Shanmugam S, Ngo HH, Wu YR (2020) Advanced CRISPR/Cas-based genome editing tools for microbial biofuels production: a review. Renew Energy 149:1107–1119. Elsevier Ltd. https://doi.org/10.1016/j.renene.2019.10.107
Shanmugam KT, Ingram LO (2022) Principles and practice of designing microbial biocatalysts for fuel and chemical production. J Ind Microbiol Biotechnol 49(2). Oxford University Press. https://doi.org/10.1093/jimb/kuab016
Sheehan J, Camobreco V, Duffield J, Graboski M, Shapouri H (1998) Life cycle inventory of biodiesel and petroleum diesel for use in an urban bus
Sheppard N (2008) Glenn Beck exposes ethanol's connection to rising food prices
Shi G, Yin H, Ye J, Peng H, Li J, Luo C (2013) Aerobic biotransformation of decabromodiphenyl ether (PBDE-209) by Pseudomonas aeruginosa. Chemosphere 93(8):1487–1493. https://doi.org/10.1016/j.chemosphere.2013.07.044
Shi S, Qi N, Nielsen J (2022) Microbial production of chemicals driven by CRISPR-Cas systems. Curr Opin Biotechnol 73:34–42. Elsevier Ltd. https://doi.org/10.1016/j.copbio.2021.07.002
Silva JPA, Mussatto SI, Roberto IC, Teixeira JA (2011) ETHANOL PRODUCTION FROM XYLOSE BY Pichia stipitis NRRL Y-7124 IN A STIRRED TANK BIOREACTOR. Braz J Chem Eng 28:151–156. https://www.abeq.org.br/bjche
Simmons BA (2011) Opportunities and challenges in advanced biofuel production: the importance of synthetic biology and combustion science. Biofuels 2(1):5–7. https://doi.org/10.4155/bfs.10.75
Sobrino FH, Monroy CR, Luís Hernández Pérez J (2010) Biofuels in Spain: market penetration analysis and competitiveness in the automotive fuel market. https://doi.org/10.1016/j.rser.2010.06.017
Sokhansanj S, Mani S, Turhollow S, Kumar A, Bransby D, Lynd L, Laser M (2009) Large-scale production, harvest and logistics of switchgrass (Panicum virgatum L.)—current technology

and envisioning a mature technology. Biofuels Bioprod Bioref 3(2):124–141. https://doi.org/10.1002/bbb.129

Sorda G, Banse M, Kemfert C (2010) An overview of biofuel policies across the world. Energy Policy 38(11):6977–6988. https://doi.org/10.1016/j.enpol.2010.06.066

Srivastava RK, Shetti NP, Reddy KR, Aminabhavi TM (2020) Biofuels, biodiesel and biohydrogen production using bioprocesses. A review. Environ Chem Lett 18(4):1049–1072. Springer. https://doi.org/10.1007/s10311-020-00999-7

Staab B, Duffield JA (2019) Biofuel impact on food prices index and land use change. https://www.elsevier.com/open-access/userlicense/1.0/

Stambouli AB (2011) Fuel cells: the expectations for an environmental-friendly and sustainable source of energy. Renew Sustain Energy Rev 15(9):4507–4520. https://doi.org/10.1016/j.rser.2011.07.100

Stephanopoulos G (2007) Challenges in engineering microbes for biofuels production. Science 315(5813):801–804. https://doi.org/10.1126/science.1139612

Stephen JL, Periyasamy B (2018) Innovative developments in biofuels production from organic waste materials: a review. Fuel 214:623–633. Elsevier Ltd. https://doi.org/10.1016/j.fuel.2017.11.042

Straub CT, Bing RG, Wang JP, Chiang VL, Adams MWW, Kelly RM (2020) Use of the lignocellulose-degrading bacterium Caldicellulosiruptor bescii to assess recalcitrance and conversion of wild-type and transgenic poplar. Biotechnol Biofuels 13(1). https://doi.org/10.1186/s13068-020-01675-2

't Lam GP, Vermuë MH, Eppink MHM, Wijffels RH, van den Berg C (2018) Multi-product microalgae biorefineries: from concept towards reality. Trends Biotechnol 36(2):216–227. Elsevier Ltd. https://doi.org/10.1016/j.tibtech.2017.10.011

Tai M, Stephanopoulos GN (2012) Metabolic engineering: enabling technology for biofuels production. Wiley Interdiscip Rev: Energy Environ 1(2):165–172

Takemura K, Kato J, Kato S, Fujii T, Wada K, Iwasaki Y, Aoi Y, Matsushika A, Murakami K, Nakashimada Y (2021) Autotrophic growth and ethanol production enabled by diverting acetate flux in the metabolically engineered Moorella thermoacetica. J Biosci Bioeng 132(6):569–574. https://doi.org/10.1016/j.jbiosc.2021.08.005

Thakor NS, Patel MA, Trivedi UB, Patel KC (2002) Production of poly(b-hydroxybutyrate) by Comamonas testosteroni during growth on naphthalene

The Gazette of India (2018) National_biofuel_policy_of_India_2018. Bharat Ka Rajpatra

Thiele I, Jamshidi N, Fleming RMT, Palsson BO (2009) Genome-scale reconstruction of Escherichia coli's transcriptional and translational machinery: a knowledge base, its mathematical formulation, and its functional characterization. PLoS Comput Biol 5(3):e1000312. https://doi.org/10.1371/journal.pcbi.1000312

Thomassen G, Egiguren Vila U, Van Dael M, Lemmens B, Van Passel S (2016) A techno-economic assessment of an algal-based biorefinery. Clean Technol Environ Policy 18(6):1849–1862. https://doi.org/10.1007/S10098-016-1159-2

Thompson PB (2008) The agricultural ethics of biofuels: a first look. J Agric Environ Ethics 21(2):183–198. https://doi.org/10.1007/s10806-007-9073-6

Tian L, Perot SJ, Hon S, Zhou J, Liang X, Bouvier JT, Guss AM, Olson DG, Lynd LR (2017) Enhanced ethanol formation by Clostridium thermocellum via pyruvate decarboxylase. Microb Cell Fact 16(1). https://doi.org/10.1186/s12934-017-0783-9

Turhollow A, Perlack R, Eaton L, Langholtz M, Brandt C, Downing M, Wright L, Skog K, Hellwinckel C, Stokes B, Lebow P (2014) The updated billion-ton resource assessment. http://www.elsevier.com/open-access/userlicense/1.0/

Ueda M, Sakuragi H, Kuroda K (2011) Molecular breeding of advanced microorganisms for biofuel production. J Biomed Biotechnol 2011. https://doi.org/10.1155/2011/416931

Vasić K, Knez Ž, Leitgeb M (2021) Bioethanol production by enzymatic hydrolysis from different lignocellulosic sources. Molecules 26(3). MDPI AG. https://doi.org/10.3390/molecules26030753

van der Laak WWM, Raven RPJM, Verbong GPJ (2007) Strategic niche management for biofuels: analysing past experiments for developing new biofuel policies. Energy Policy 35(6):3213–3225. https://doi.org/10.1016/j.enpol.2006.11.009

Vásquez MC, Silva EE, Castillo EF (2017) Hydrotreatment of vegetable oils: a review of the technologies and its developments for jet biofuel production. Biomass Bioenergy 105:197–206. Elsevier Ltd. https://doi.org/10.1016/j.biombioe.2017.07.008

Wadler CS, Wolters JF, Fortney NW, Throckmorton KO, Zhang Y, Miller CR, Schneider RM, Wendt-Pienkowski E, Currie CR, Donohue TJ, Noguera DR, Hittinger CT, Thomas MG (2022) Utilization of lignocellulosic biofuel conversion residue by diverse microorganisms. Biotechnol Biofuels Bioprod 15(1). https://doi.org/10.1186/s13068-022-02168-0

Wang J, Guleria S, Koffas MA, Yan Y (2015) Microbial production of value-added nutraceuticals, pp 1–2. http://www.elsevier.com/open-access/userlicense/1.0/

Wang S, Sun X, Yuan Q (2018) Strategies for enhancing microbial tolerance to inhibitors for biofuel production: a review. Bioresour Technol 258:302–309. Elsevier Ltd. https://doi.org/10.1016/j.biortech.2018.03.064

Wegener DT, Kelly JR, Wallace LE, Sawicki V (2014) Public opinions of biofuels: attitude strength and willingness to use biofuels. Biofuels 5(3):249–259. https://doi.org/10.1080/17597269.2014.921011

Werpy T, Petersen G, Aden A, Bozell J, Holladay J, White J, Manheim A (2004) Top value added chemicals from biomass Volume I-Results of screening for potential candidates from sugars and synthesis gas produced by the staff at Pacific Northwest National Laboratory (PNNL) National Renewable Energy Laboratory (NREL) Office of Biomass Program (EERE) for the Office of the Energy Efficiency and Renewable Energy. http://www.osti.gov/bridge

Wheaton GH, Mukherjee A, Kelly RM (2016) Transcriptomes of the extremely thermoacidophilic archaeon Metallosphaera Sedula exposed to metal “shock” reveal generic and specific metal responses. Appl Environ Microbiol 82(15):4613–4627. https://doi.org/10.1128/AEM.01176-16

Wyman CE, Huber G (2009) What could be possible with mature biofuels technologies? Biofuels Bioprod Biorefin 3(2):105–107. Wiley. https://doi.org/10.1002/bbb.139

Yang Z, Sun Q, Tan G, Zhang Q, Wang Z, Li C, Qi F, Wang W, Zhang L, Li Z (2021) Engineering thermophilic Geobacillus thermoglucosidasius for riboflavin production. Microb Biotechnol 14(2):363–373. https://doi.org/10.1111/1751-7915.13543

Yoneyama F, Yamamoto M, Hashimoto W, Murata K (2015) Production of polyhydroxybutyrate and alginate from glycerol by azotobacter vinelandii under nitrogen-free conditions. Bioeng 6(4):209–217. https://doi.org/10.1080/21655979.2015.1040209

Zaldivar J, Nielsen J, Olsson L (2001) Fuel ethanol production from lignocellulose: a challenge for metabolic engineering and process integration. Appl Microbiol Biotechnol 56(1–2):17–34. https://doi.org/10.1007/s002530100624

Zhang T, Datta S, Eichler J, Ivanova N, Axen SD, Kerfeld CA, Chen F, Kyrpides N, Hugenholtz P, Cheng JF, Sale KL, Simmons B, Rubin E (2011) Identification of a haloalkaliphilic and thermostable cellulase with improved ionic liquid tolerance. Green Chem 13(8):2083–2090. https://doi.org/10.1039/c1gc15193b

Zhang J, Chen Y, Fu L, Guo E, Wang B, Dai L, Si T (2021) Accelerating strain engineering in biofuel research via build and test automation of synthetic biology. Curr Opin Biotechnol 67:88–98. Elsevier Ltd. https://doi.org/10.1016/j.copbio.2021.01.010

Zhang YHP (2011) What is vital (and not vital) to advance economically-competitive biofuels production. Process Biochem 46(11):2091–2110. Elsevier Ltd. https://doi.org/10.1016/j.procbio.2011.08.005

Chapter 23
Evolution, Challenges and Benefits of Biofuel Production and Its Potential Role in Meeting Global Energy Demands

Deepshikha Datta, Soheli Biswas, Dhriti Barman, Esha Mandal, and Bimal Das

Abstract Biofuel has gained significant attention in recent years due to their potential to mitigate climate change, reduce dependence on fossil fuels, and provide economic opportunities for rural communities. The current work explores the evolution of biofuels and their potential role in meeting global energy demands. The work further describes the various feedstocks used to produce biofuels, such as agricultural crops, forest residues, algae, and municipal waste, and the various processes involved in converting them into usable fuel and emphasizes on the evolution of first, second and third generation biofuels. The work also highlights on various biofuel production techniques with a focus on the current state of the industry, its importance in the global energy mix, and the challenges and opportunities associated with its growth. It also emphasizes on the benefits of biofuels, including their potential to reduce greenhouse gas emissions, increase energy security, and promote rural development and explores on the challenges associated with biofuel production, such as land-use competition, water scarcity, and the potential for negative environmental impacts. However, the production of biofuels has the potential to meet global energy demands while reducing greenhouse gas emissions and promoting rural development. Additionally, the chapter covers the environmental and economic benefits and its potential role in meeting energy demands and reducing greenhouse gas emissions.

D. Datta · B. Das (✉)
Department of Chemistry, Brainware University, Barasat, Kolkata 700125, West Bengal, India
e-mail: bimal.das@che.nitdgp.ac.in

S. Biswas
Department of Biotechnology, Brainware University, Barasat, Kolkata 700125, West Bengal, India

D. Barman
Department of Biotechnology, National Institute of Technology Durgapur, Durgapur 713213, India

E. Mandal
Department of Chemical Engineering, National Institute of Technology Durgapur, Durgapur 713213, India

M. Shah and D. Deka (eds.), *Emerging Sustainable Technologies for Biofuel Production*, Environmental Science and Engineering, https://doi.org/10.1007/978-3-031-52167-6_23

Keywords Evolution · Feedstocks · Production techniques · Benefits · Challenges · Global energy demand · Environmental impact · Economic benefit · Emissions

23.1 Introduction

The evolution of biofuels is considered as a remarkable upgradation in the contemporary world. In the present scenario, various modern technologies and techniques are being adopted for the production of ecologically sound energy. In the past days, fossil fuels were used from which different form of energies were extracted and as these are non-renewable sources, these fossil fuels came to near its depletion without fulfilling the demand of the world. The biofuels are the best alternative to fossil fuels not only for its renewable capability but also for its eco-friendly nature.

Since the beginning human-created fire up to the use of clumped wood as a fuel for thermal plants, biomass has always been a dependable source of energy. Even though the idea of using materials which are made up of lignin and cellulose, as a solid bio-fuel is well-informed, turning bio-mass materials into liquid state fuel is a difficult task. When the complexity of the biomass (in terms of chemical composition) increases, the more the difficulties and generally expensiveness of the alternation process is increases. In the last 10–20 years, interest in biofuels has surged due to the depletion of oil reserves and rising global energy consumption, despite the fact that during the majority of the twentieth century, scientific-research over biofuel deeply tracked the price-rate of petroleum. Despite occasional criticism, using biofuels is frequently a more environmentally friendly option given the rising concerns regarding greenhouse gas emissions because their carbon balance is nearly neutral compared to that of petroleum-derived fuels like petrol, diesel, and kerosene (Lee and Lavoie 2013).

Biofuels, as the term has lately been condensed to mean renewable carbon sources, are defined as fuels produced from biological material. Before Samuel Morey conducted the first internal combustion engine test on ethanol in the early nineteenth century, it had been known to be used for lamp oil and cooking for many years (at the time, it was known as spirit oil). Whale oil was later replaced by ethanol, which was then followed by petroleum distillate (beginning with kerosene for lighting) (Lee and Lavoie 2013).

Biofuels have gone through the evolution cycle up to fourth generations. One of the most popular bio-fuels in the USA is ethanol, while other bio-fuels, including biodiesel, are also widely used in the other regions of all around the world, including Europe, Asia, and Brazil too. While cellulosic ethyl alcohol and dimethyl ether (bio-DME), which are gaining popularity and might be categorised as 2nd and 3rd-generation biofuels, are emerging, ethanol and biodiesel are both regarded as first-generation biofuels. The fourth generation is the most advanced one in which the branch of science known as Biotechnology comes into action. By 2022, the Energy Independence and Security Act of 2007 in the United States required the usage of 136 billion litres (36 billion gallons) of biofuels annually, an increase of more than six times over production levels in 2006. The law maintains some government subsidies and tax breaks for the development of biofuels and mandates, subject to certain restrictions, that 79 billion litres (21 billion gallons) of the total amount be biofuels other than ethanol derived from corn (Lee and Lavoie 2013; Mathews 2007).

One unique promise of biofuels is that these may be able to permanently remove carbon dioxide from the environment when used in conjunction with a cutting-edge technology known as carbon capture and storage. According to this concept, carbon dioxide would be removed from the atmosphere as biofuel crops grew and energy facilities would catch the carbon dioxide released when burning biofuels to produce electricity (Mathews 2007).

23.2 Evolution of Biofuels

Biofuels are categorised as first, second, third and fourth-generation in the 2012 EASAC report, depending on the source and the technology used to produce them (Sumathi et al. 2008). The functional properties of biofuels do not change from generation to generation; rather, the source from which the fuel is extracted changes. The first generation of biofuels are made by extracting starch, sugars, fats, and oils from various food crops. The second-generation fuel is derived from non-edible biomass, this generation of biofuels primarily addresses the Food versus Fuel controversy created by the first-generation biofuels. The first two generations of biofuels involve the use of important resources like water, land, and intense labour. To address this issue, the third generation of biofuels derived from algae requires fewer resources and produces more quickly. The fourth generation of biofuels—which primarily consists of solar fuels—has emerged as a result of technological development over time and could prove to be a game-changer for the biofuel industry (Barrios et al. 2020) (Fig. 23.1). The various processes of the generation of biofuels are discussed below.

a. First-Generation Biofuels

First-generation biofuels are produced by extracting sugars from sugarcane and sugar beetroot, respectively, and edible biomass such as starch from potato, wheat, barley and maize, vegetable oils like, mustard oil, coconut oil, soybean oil, palm oil, corn oil, rice oil, olive oil, rapeseed oil, etc. The fermentation of these foods results in the production of biofuels. It initially displayed a promising ability to reduce the combustion of fossil fuels and lower the atmospheric levels of CO_2, which is utilized by crops during growth. The main benefits of the first-generation raw material are its yield accessibility and relatively simple transformation method. However, the risk of shortage of food supply is the basic issue for raising the cost of food items produced from these raw materials. The use of these edible crops as feedstocks has also raised concerns over the effects on cropland, biodiversity, and the food supply (Fig. 23.4).

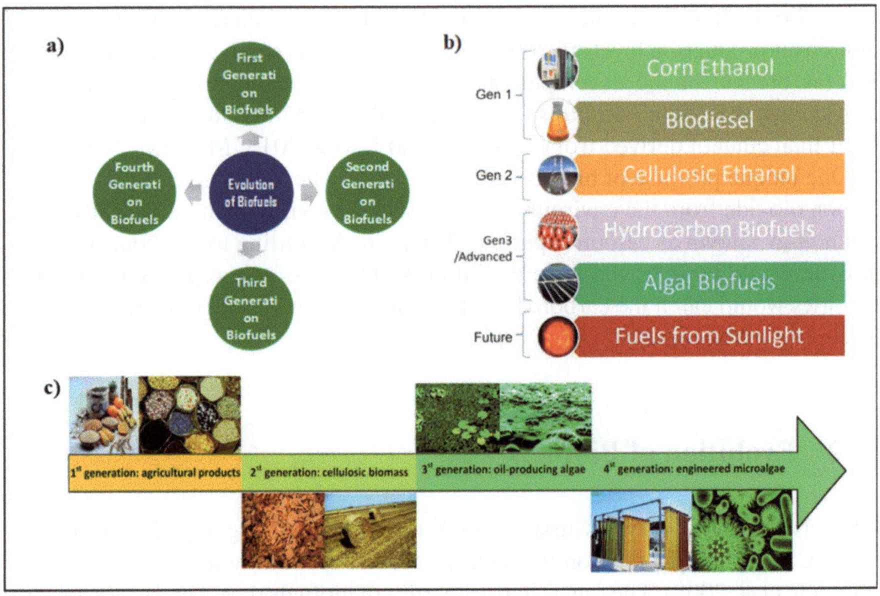

Fig. 23.1 Evolution of biofuels on the basis of their source of extraction (Shen 2014; Pillar-Little 2018)

The most common examples of this generation biofuels are Bioethanol and Biodiesel (Barrios et al. 2020; Karmakar et al. 2010; Paul Abishek et al. 2014).

First-generation bioethanol is the most notable biofuel, produced by fermentation of sugars as well as starches obtained from agricultural crops such as maize (*Zea mays*), sugar beets and sugar cane (*Saccharum* sp.) using yeast (*Saccharomyces cerevisiae*) (Fig. 23.2). The extraction of bioethanol from sugarcane is relatively simple: sugarcane is first passed through an extractor where it is crushed in water and then purified with the help of fermentation yeast to extract bioethanol. Corn is high in carbohydrates, so it must go through preliminary hydrolysis of starch, which necessitates the use of expensive enzymes such as a-amylase to extract sugars, which result in the production of biofuels after fermentation. Its vital diligence can be found in engine vehicles. It is commonly used as the transportation energy as its pure form and also through blending along gases within the conventional combustion engines. Also, generally combined along with gasoline, or E10, at a 10% bioethanol content. It tends to react effectively with the gases, producing slightly aromatic gas known as green gas as a by-product. 85% ethanol and 54% ethanol which are usually known as E85 and E54, respectively, are both components of the other bio gasoline combination that is energising. US along with Brazil together contributed to 89% of the world's capacity for producing bioethanol. By replacing petroleum derivatives in the transportation sector, bioethanol can help to reduce global warming. Since its inception, the use of ethanol fuel has grown. The primary benefit of ethanol from biomass is that, it can replace not renewable power sources like petroleum or ignitable

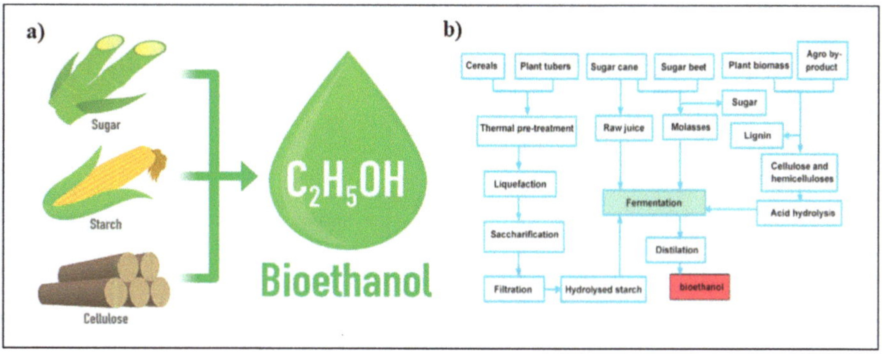

Fig. 23.2 **a** Different sources of bioethanol **b** different stages for first-generation bioethanol production technology (Miskat et al. 2020; Hussain et al. 2020)

gases in combustion engines. Bioethanol has outperformed non-renewable energy sources in terms of productivity (Alalwan et al. 2019; Azhar et al. 2017; Malode et al. 2022; Sims et al. 2010).

Apart from bioethanol, the only other first-generation biofuel produced on an industrial scale is biodiesel. Oil and fats derived from oily plants (Soyabean and Palm oil) and seeds; alcohol, which is usually methanol in used oil or ethanol in crude oil; and a catalyst, which is usually potassium hydroxide (KOH) or sodium hydroxide (NaOH) (Fig. 23.3). The majority of oils and fats are made up of triglycerides, which are molecules made up of glycerol esterified with three fatty acids. Triglycerides are reacted with an alcohol, typically methanol, via a chemical process known as transesterification or esterification reaction, which uses sodium hydroxide (NaOH) as a catalyst. This reaction produces glycerin and three fatty acid esters. Because esters have a lower density than glycerol, they float and separate from it. Biodiesel is made up of molecules of fatty acid esters. The first generation of biofuels appeared to be a viable solution to the global energy crisis because they are carbon neutral, functionally equivalent to fossil fuels, and compatible with existing energy infrastructure—until researchers discovered that producing these fuels on an industrial scale is time, energy, and resource intensive (Barrios et al. 2020) (Fig. 23.4).

b. Second-Generation Biofuels

Learning from the shortcomings of the first generation, second-generation biofuels used non-edible biomass, specifically lignocellulose-rich food residues such as coconut husks, cotton wastes, switchgrass, sugar bagasse, forestry wood residues, and waste papers (Fig. 23.5). The extraction process for second generation biofuels is divided into two distinct categories: biochemical and thermochemical (Fig. 23.6). The biochemical conversion process uses enzymes and microorganisms to convert biomass into sugars and those sugars into biofuels. Firstly, biomass undergoes a pretreatment process so that it can be decomposed easily in the subsequent steps. To break down and separate biomass into cellulose, hemicellulose, and lignin compounds—Steam explosion, Classical pulping, and organosolv pulping methods

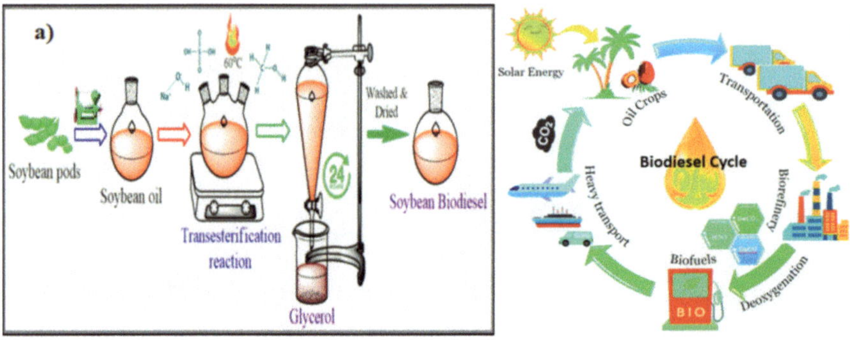

Fig. 23.3 **a** Source and different stages of first-generation biodiesel production technology **b** production of biodiesel and use of it through biodiesel cycle (Jin et al. 2019; Pan 2014)

Fig. 23.4 Food versus fuel controversy of first-generation biofuels (Verma et al. 2019)

are typically used. After pretreatment, biomass goes through an expensive process called steam.

Saccharification, in which enzymes like pullulanase break the cellulose chains into glucose and the hemicellulose chains into xylose, primarily because these simple sugars are easily fermented into ethanol or other biofuels. Following that, biofuel is produced during the fermentation stage, where bacteria or microbes, such as yeast,

Fig. 23.5 Various sources of second-generation biofuels which includes—**a** wood pulps, **b** coconut husks, **c** forestry wood residues, **d** switchgrass, **e** sugar bagasse, **f** straws (Ghosh et al. 2009; Hays et al. 2005; Miesel et al. 2012; Bielen et al. 2013; Baldino et al. 2019; Zhao et al. 2021)

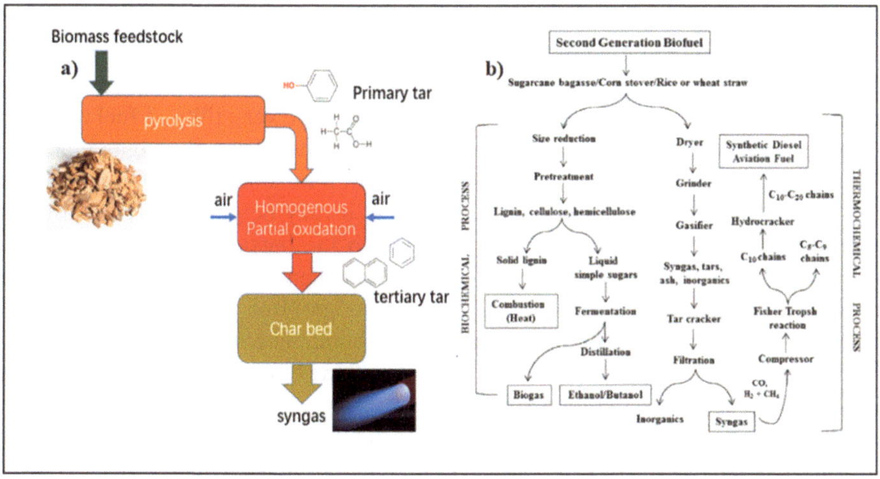

Fig. 23.6 **a** Production method of syngas **b** biochemical and thermochemical processes of 2nd generation biofuel production (Ben-Iwo et al. 2016; Tushar et al. 2014)

can break down sugars into secreted compounds that are used. Heat is used in the thermochemical conversion process to break down biomass into intermediates such as gas or bio-oil, which can then be upgraded to biofuels. Purification or gasification is the first step in the thermochemical process. This method uses heat and pressure to convert biomass, typically wood material such as forest residue, into a hydrogen and carbon monoxide rich gas known as synthesis gas (syngas). Gasification occurs in a precisely controlled environment that typically includes oxygen or The syngas is then conditioned to produce the desired ratio of carbon monoxide to hydrogen in order to maximize yield from subsequent chemical reactions. Pressurized hot syngas is then passed through a catalyst contained within a reactor to convert the gas to liquid. At this point, the carbon monoxide and hydrogen molecules are combined to form larger hydrocarbon molecules. The molecules are then cooled, condensed, and refined into biofuels. While second-generation biofuels are clearly an improvement, owing to the use of inexpensive food residues rather than food crops, the extraction process is still time-consuming and costly. However, second-generation biofuels increase the sustainability of fuel production by using lants that are not edible and thus are not in direct competition with food production (Barrios et al. 2020; Oh et al. 2012).

c. Third-Generation Biofuels

Third-generation biofuels are more advanced biofuels that outperform first- and second-generation biofuels in a variety of ways. The fact that third-generation biofuels are made from microalgae is their primary advantage (Fig. 23.7). Third-generation biofuels are also referred to as "algae fuel" or "oilage." Algae are aquatic plants that can grow in both salt and fresh water. They grow quickly and don't require any land or freshwater to thrive. In terms of quantity and diversity, no feedstock can compete with algae in terms of fuel production potential. The Chlorophyceae family of microalgae is the most promising species because of its relatively fast growth rate, high lipid content, and ease of cultivation. Other microalgae classifications that are crucial for the production of biofuels include Bacillariophyceae, Eustigmatophyte, Chrysophyceae, Haptophyceae, and Cyanophyceas. The versatility of fuel produced by algae is due to the microorganism has two features. First, the algae make oil, which can be simply processed to diesel fuel as well as some amount of petroleum. However, the most significant advantage is that it may be genetically modified in order to generate all through ethanol along with the butanol to petrol and diesel fuels. Butanol is of particular interest because it is chemically identical to gasoline. In fact, it has nearly the same energy density as gasoline and a better emissions profile. Scientists had a difficult time producing butanol before the advent of genetically modified algae. In terms of fuel potential, algae have more than just diversity. It can also generate exceptional amounts. Moreover, up to 9,000 gallons of biofuel are made from algae each acre, which is ten times more than the best traditional feedstock (Alam et al. 2015; Shokravi et al. 2019; Bošnjaković and Sinaga 2020).

Another advantage of using algae is that, there are variety of systems through which it can be grown. The systems are listed below (Fig. 23.8):

- ***Open ponds*** are the most basic systems in which algae grows in a pond which is exposed to the open air. Although they are easy to construct and affordable, they are not as effective as other systems. There are also a source of concern because the pond may become contaminated by other organisms as well, which could harm or kill the algae (Shokravi et al. 2019).
- ***Closed-loop systems*** are like open ponds system, but employ a sterile source of carbon dioxide and are not exposed to the atmosphere. Such systems have potential since they might be able to be directly connected to carbon dioxide sources (like smokestacks) and use the gas before exposed to the air (Shokravi et al. 2019).

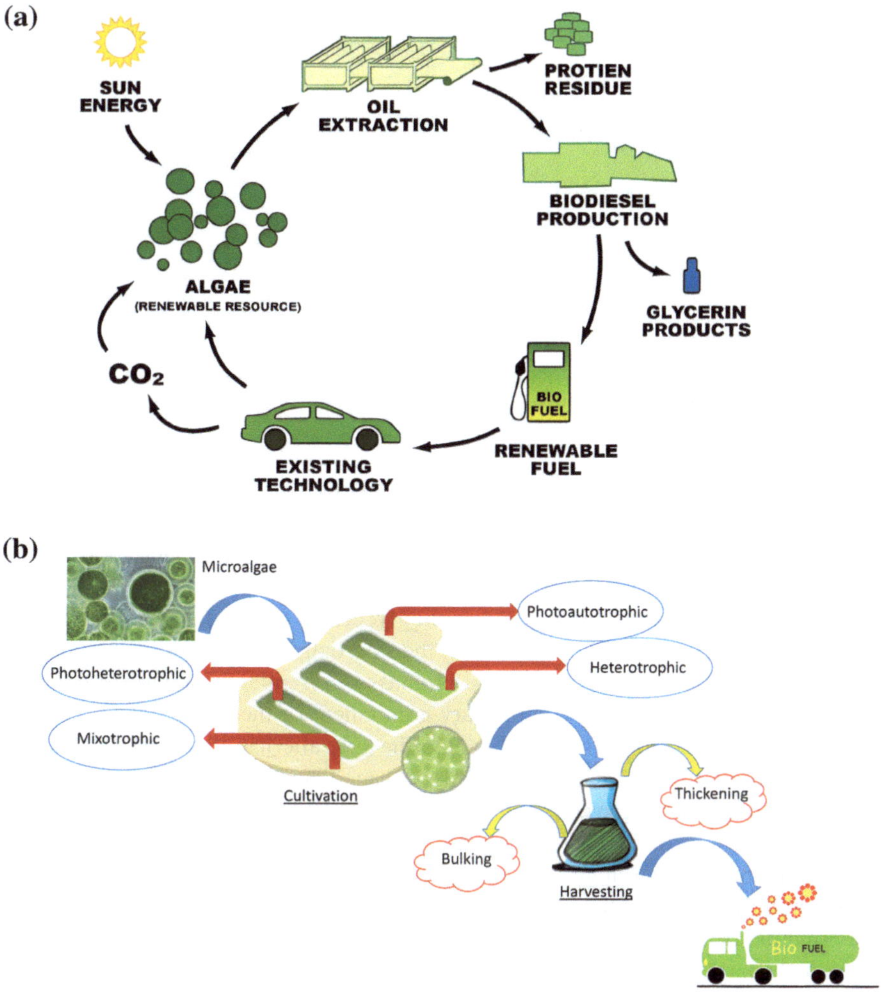

Fig. 23.7 **a** Algae used as the source for 3rd generation biofuel production **b** open ponds system are used to grow microalgae for 3rd generation biofuels (Javed et al. 2019; Slade and Bauen 2013)

- ***Photobioreactors*** are particularly more sophisticated and challenging system to set up, leading to significant capital expenditures. But, when it comes to yield and control, they have unparalleled benefits. Photobioreactors are closed systems (Shokravi et al. 2019).

d. Fourth-Generation Biofuels

The fourth-generation biofuel is accomplished by introducing genetic modification to raise the quality and production of microalgae followed by processes like pyrolysis, gasification, digestion, etc. (Fig. 23.9). Lipids, proteins, carbohydrates, and

Fig. 23.8 Various ways for algae production—**a** open ponds system **b** closed loop system **c** photobioreactors (Shokravi et al. 2019; Jerney and Spilling 2020; Zhou et al. 2015)

nucleic acids are the substances that make up microalgae. Microalgae's biochemical composition contains lipids in any form, whether polar or neutral. By adhering to organelle membranes or the bilayer structure of cell membranes, polar lipids form the structural lipid. Storage lipids known as neutral lipids comprise free fatty acids, triglycerides, diglycerides, and monoglycerides (FFAs). As lipids have the highest energy content (37.6 kJ g1) of all the microalgal biochemical components, increasing the lipid content of the algal biomass is essential to maximize the yield of fourth generation biofuel. The primary component of the microalgae utilized to produce biodiesel is lipid. Biodiesel is now the sole alternative to fuel that may be burnt directly in engines that are currently in use. Trans-esterification of the accumulated lipid in microalgae cells can result in the production of biodiesel. A number of genomes editing techniques, including gene knockdown, knockout, and genetic alterations, are included in genetic engineering. The selection of microalgae strains plays a significant role in the overall performance of the production of algal biofuel. The principle behind strain selection is that some microalgae are more suitable than others for a particular task.

The productivity and lipid content of microalgae are frequently increased by the application of genome editing techniques. The genome editing of microalgae strains is currently carried out using three different types of tools: clustered regularly interspaced palindromic sequences like CRISPR/Cas9, Transcription Activator-Like Effector Nucleases (TALEN), and Zinc-Finger Nuclease (ZFN). The first known microalgae genome editing experiment with ZFN was performed on Chlamydomonas reinhardtii. Phaeodactylum tricornutum (using TALEN and CRISPR/Cas9), Nannochloropsis oceanica (using CRISPR/Cas9), and Chlamydomonas reinhardtii (using TALEN and CRISPR/Cas9) have successfully achieved the genome editing experiment (Fig. 23.9). The potential to alter the metabolism of microalgae has greatly increased through the application of CRISPR/Cas9. In this way, modern genetic techniques are used for the production of fourth generation biofuels (Sharma et al. 2022).

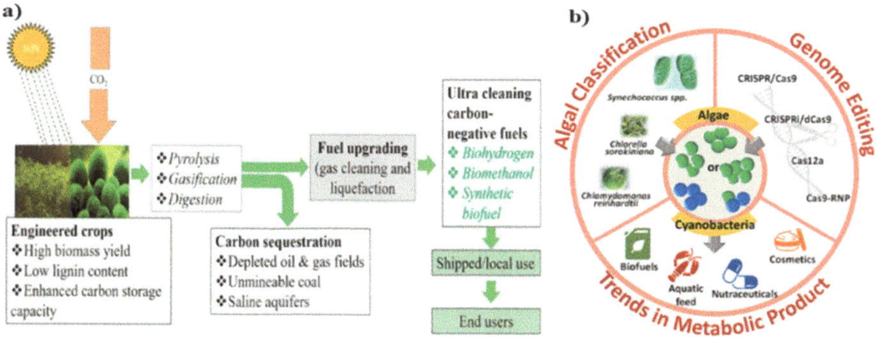

Fig. 23.9 **a** Production methods of 4th generation biofuels, **b** algal classification, genome editing and the final products obtained after genomic modification (Ng et al. 2020; Aryal et al. 2021)

23.3 Potential Role of Biofuel in Meeting Global Energy Demands

Fossil fuels are the most prevalent energy source, and there is a rising need for energy worldwide to meet the needs of the expanding human population. The production of greenhouse gases (GHGs), lead to the phenomenon of global warming, and is one of the most critical environmental problems associated with the utilisation of energy from fossil fuels and also concerns about climate change. By increasing the availability of renewable energy sources, fossil fuels would be replaced and their dominance as carbon-intensive fuels in the future energy system would be greatly reduced. To reduce the environmental issues caused by GHG emissions, it is crucial to develop and use renewable energy sources like solar, bioenergy, wind, hydro, and geothermal energy. The most common type of renewable energy is bioenergy, which is produced when biomass produced through photosynthesis is turned into biofuel (biogas), heat, and electricity (Tomes et al. 2010). The below mentioned graph depicts the production of biofuels by different countries according to the passing years. The graph shows the production of Brazil ethanol, USA ethanol, other ethanol, Europe biodiesel and other biodiesel production rates in billion liters from the year of 2000–2010 (Fig. 23.10).

Biofuels play a particularly crucial role in decarbonising transport by providing a low-carbon option for existing technologies, such as light-duty vehicles in the near term and heavy-duty trucks, ships and aircraft with few viable solutions in the long run. Expanding the use of advanced feedstock is also necessary; by 2030, biofuels made from waste and residue resources will account for 45% of all biofuel demand, up from about 8% in 2021. Biofuels accounted for 3.6% of the world's transportation energy needs in 2021, primarily for road transportation. According to the Net Zero Scenario, by 2030, biofuels will have contributed 15% of the transportation sector's fuel needs, or about one-fifth of it. Despite the fact that the pandemic caused a fall in global demand for biofuels, it has now rebounded to 2019 levels, the recovery has

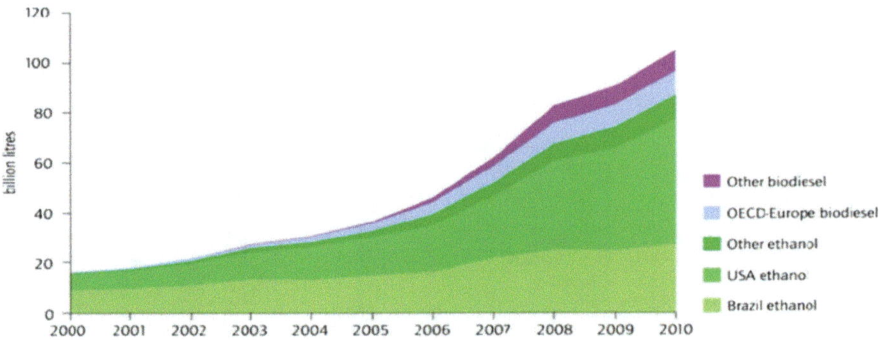

Fig. 23.10 Global biofuel production (Gordon and Szalay 2022)

been unequal for different biofuels. Demand for ethanol increased by 6% between 2020 and 2021 but remained 7% lower than in 2019. FAME (biodiesel) surpassed 2020 demand by 0.3% to reach 1.4 EJ. Nonetheless, renewable diesel (referring to HVO) consumption increased exponentially to over 0.3 EJ in 2021, which is 65% more than in 2019. In order to comply with the Net Zero Scenario, aviation biofuels, commonly known as biojet kerosene, would need to increase from 0.1% of aviation fuel demand in 2021 to more than 5% in 2030 (Fig. 23.11). A number of critical elements must come together for biojet kerosene to take off successfully, including narrowing of the cost gap between it and fossil jet fuel, the implementation of clear regulatory frameworks and policies by governments, and the diversification of sustainable feedstock sources beyond waste and edible oils (Ahorsu et al. 2018).

The Net Zero Scenario assumes that by 2030, almost 50% of all biofuels consumed will be made from wastes, leftovers, and special crops that don't compete with food crops (such as crops cultivated on marginal land), up from an estimated 8% in 2021. The bulk of non-food crop feedstocks used in the production of biofuels today come from used cooking oil and animal fat waste. Due to the scarcity of these feedstocks, new technologies must be commercialised in order to increase the production of non-food crop biofuels. For instance, non-food feedstocks can be used in the production of low-carbon biofuels for use in the transportation sector through cellulosic ethanol and biomass-based Fischer–Tropsch (bio-FT) technologies. Although the average production cost of these biofuels is still two to three times higher than that of their fossil fuel equivalents, over the next ten years it may decrease by as much as 27%, with any remaining cost gaps being filled by policy initiatives that encourage production and demand.

There are numerous biofuel production methods (Fig. 23.12) that are now commercially viable, such as ethanol from corn and sugarcane, **FAME** (Fatty Acid Methyl Esters) biodiesel, **HVO** (Hydrotreated Vegetable oil) sustainable diesel, and **HEFA** (Hydro processed Esters and Free Fatty Acid) bio jet kerosene made from waste and vegetable oils. Yet, there is still a lack of innovation when it comes to turning woody and grassy biomass into liquid biofuels, for instance through thermochemical processes such as biomass gasification followed by bio-FT synthesis,

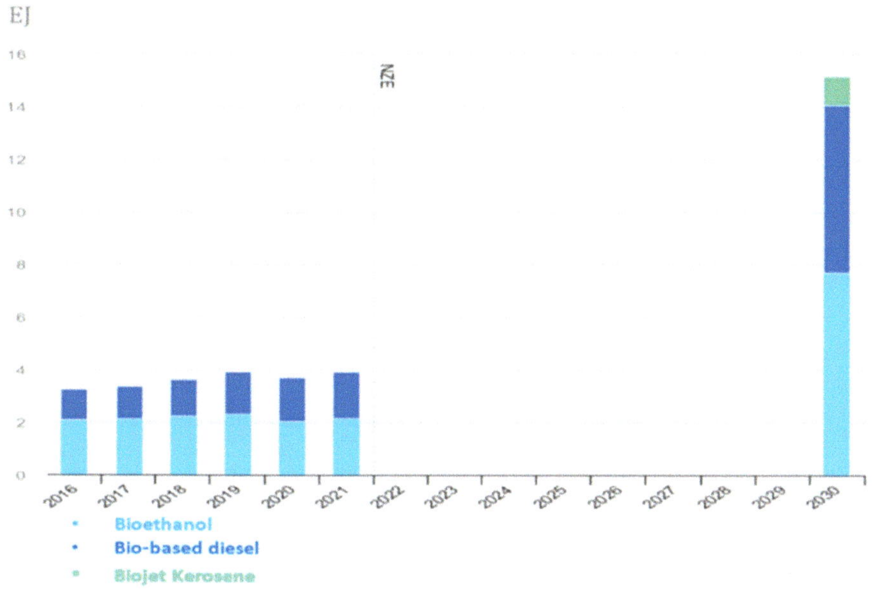

Fig. 23.11 Global biofuel demand in transport in the Net Zero Scenario, 2016–2030 (Ahorsu et al. 2018)

hydrothermal liquefaction, and rapid pyrolysis with upgrading. Both hydrothermal liquefaction and rapid pyrolysis with upgrading have lower levels of technological readiness than bio-FT due to difficulties in pre-treating the bio-oils for subsequent hydroprocessing into renewable fuel. However, after pre-treatment, the bio-oil can be processed alongside petroleum products (up to around 10%) at already-existing oil refineries, avoiding the need for expensive scaling-up-related capital expenditures (Ahorsu et al. 2018).

Several biofuel manufacturing processes naturally release a stream of CO_2 that is almost completely clean. These pathways include cellulosic and crop-based ethanol fermentation, as well as bio-FT. Due to the high CO_2 concentration, CO_2 capture is inexpensive because only dehydration is necessary for further purification. Once the CO_2 has been gathered, it can be compressed and sent, carried by vehicle, or stored using a pipeline. Since 2010, ethanol facilities have begun capturing CO_2, with the first projects selling the CO_2 for use in food and beverage production or in increased oil recovery. The world's first bioenergy carbon capture and storage (BECCS) plant, which can capture 1 Mt of CO_2 annually, was installed in the United States in 2017. A number of ethanol plants were capturing carbon as of 2021, with a total annual capture capacity of 2.21 Mt CO_2 divided almost equally between CO_2 storage and usage, and 1.65 Mt CO_2 coming from initiatives intended for increased oil recovery or storage (Fig. 23.13). A total of 40 ethanol production facilities with a capacity to capture over 15 Mt of biogenic CO_2 are expected to begin doing so by 2030, including about 30 as part of the Midwest Carbon Express project in the United States. The

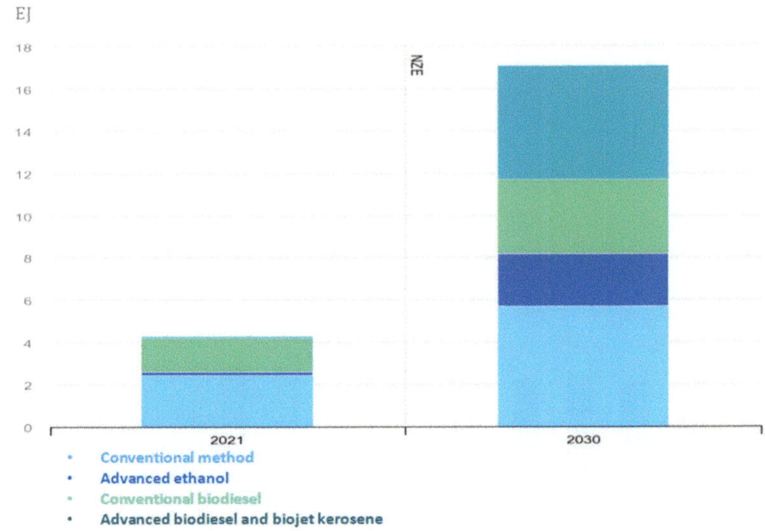

Fig. 23.12 Production of liquid biofuel using current technology and feedstock in the Net Zero Scenario between 2021 and 2030 (Ahorsu et al. 2018)

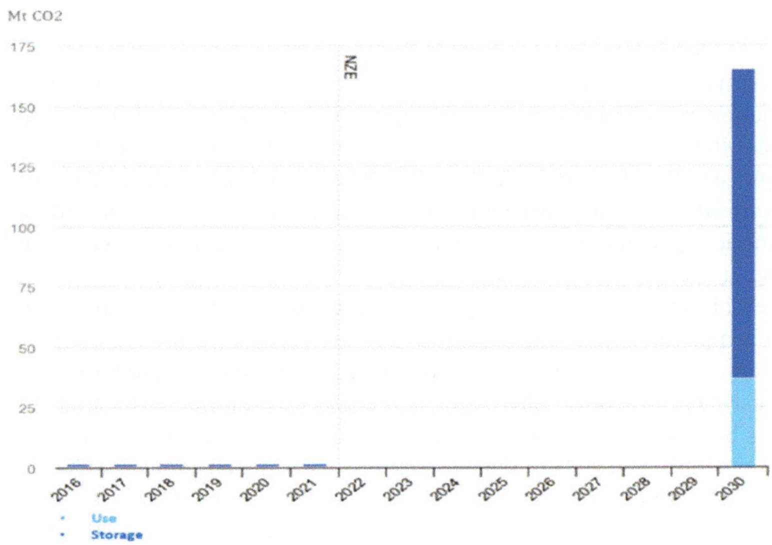

Fig. 23.13 CO_2 captured from biofuels production in the Net Zero Scenario, 2016–2030 (Ahorsu et al. 2018)

Net Zero Scenario, however, calls for well over 50 times as much CO_2 to be caught by 2030 as it does today, leaving a significant gap that needs to be filled (Ahorsu et al. 2018).

On an energy basis, biofuels collectively prevent 4.4% of the world's oil use in road transport. OECD nations account for over 60% of biofuel demand, while non-OECD nations account for 40%. By the end of 2022, existing and new policies are predicted to increase the demand for biofuel by 5% from 2021 levels. For instance, the Inflation Reduction Act, which includes incentives for biodiesel, renewable, sustainable aviation fuel, and advanced fuels as well as support for biofuels infrastructure and production, was enacted by the US in August 2022. Further supporting increased biofuel consumption in the upcoming years are India (greater ethanol blending), Canada (clean fuel standard), Brazil (RenovaBio), Europe (new Fit for 55 objectives), and the United States (California and Oregon low carbon fuel standards [LCFS] and other states exploring LCFSs). However, slower GDP growth and increased oil prices will hinder growth in 2022 (Ahorsu et al. 2018).

23.4 Feedstocks for Biofuel

The term "feedstock" describes the raw material needed for a particular industrial process. Any biomass intended for conversion into energy or biofuel is referred to as a feedstock. For instance, soybean oil and corn are both feedstocks for the manufacturing of ethanol and biodiesel, respectively.

In the below mentioned table, the feedstocks are being shown according to the generations of biofuel (Table 23.1; Fig. 23.14).

23.5 Different Types of Biofuels Production Techniques

23.5.1 Bio-ethanol

Bio-ethanol (also known as ethanol or ethyl alcohol (C_2H_5OH)) is an alcohol derived from starch and sugar cane. It is commonly mixed with gasoline. Bioethanol is primarily produced by fermentation and the reaction of ethylene and steam. A large part of bio-ethanol in the world is yield from corn (resources of starch) by two techniques—wet and dry milling (Onuki 2006). In industry, mainly two types of amylases are used in hydrolysing starch: α-amylase and gluco-amylase. Basic difference in between them is, Alpha amylase can break the 1–4 bonds along a chain, but in order to break the 1–6 side-chain bonds glucoamylase enzyme is needed, which is not normally present in malted grain but can be found on sale in some home brew shops (Maarel et al. 2002; Mollahoseini et al. 2015). Bioethanol is biodegradable, has low toxicity, and low impact on the environment. Ethanol is a high-octane fuel and has replaced lead as an octane booster for gasoline. By mixing ethanol with gasoline, the fuel mixture can also be oxygenated for more complete combustion and reduced pollutant emissions (Da Silva Trindade and dos Santos 2017).

Table 23.1 Flowchart of the feedstocks required for biofuel production in various generations

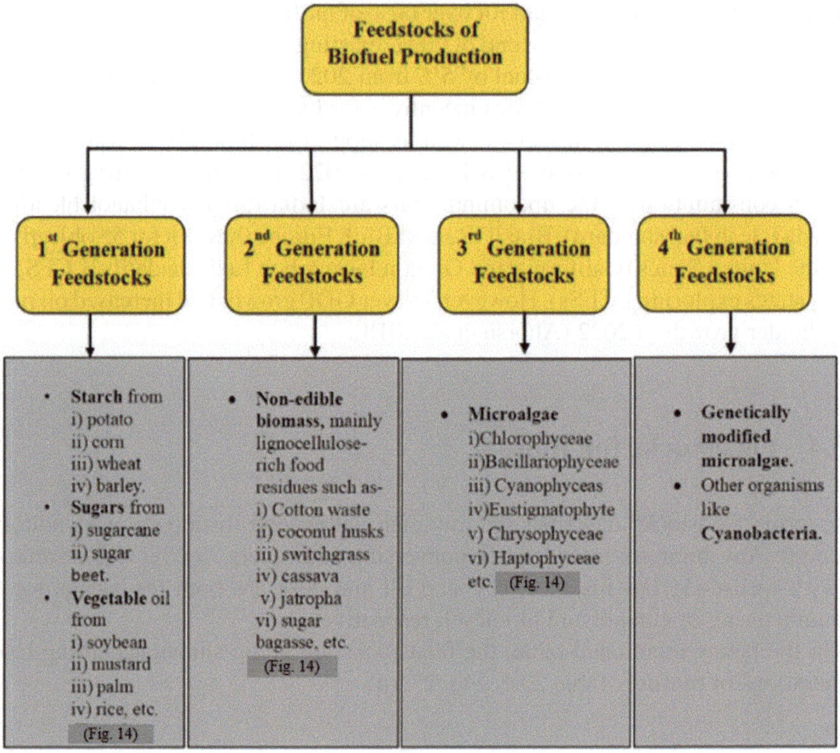

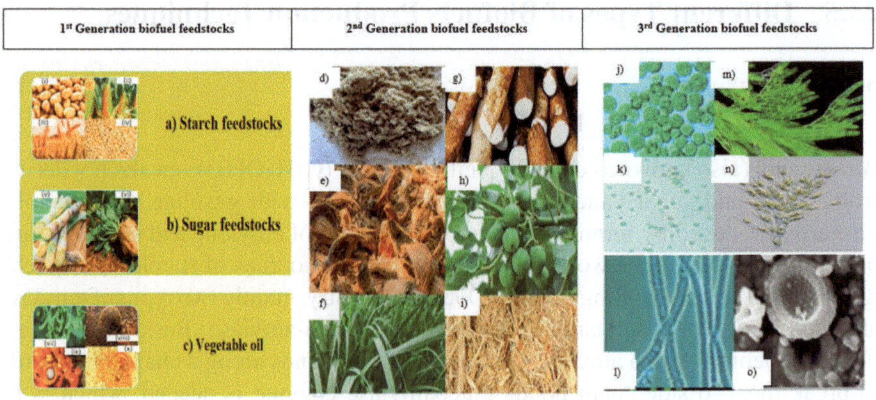

Fig. 23.14 First, second and third generation biofuel feedstock (Moodley 2021)

23.5.2 Biobutanol

Bio-butanol: Biobutanol is 4-carbon alcohol produced by the fermentation of biomass. Biobutanol can be produced in ethanol production plants, but it leads to competition over food resources. So, to overcome this problem, it is needed to find fresh strains and genetically-engineer nowadays used microbes to enhanced the proficiency of the techniques (Lütke-Eversloh and Bahl 2011; Nielsen et al. 2009; Shen and Liao 2013; Steen et al. 2008; Kang et al. 2020). Implementation of high skilled and lower energy needed purification techniques for e.g., membrane separation technologies or nontoxic liquid–liquid extraction can enhance the financial durability and durability of bio-butanol generation (Gnansounou and Dauriat 2005). Biobutanol is mainly used as a fuel for internal combustion engines. Its properties are similar to gasoline. Some gasoline vehicles can even use biobutanol without modification and can be added to gasoline at concentrations up to 11.5% by volume. However, the main drawback of biobutanol is its lower energy content than gasoline, averaging 10–20%. Biobutanol can to decrease CO_2 emissions by 85% compared to gasoline (Habeeb 2012).

23.5.3 Biodiesel

Bio-diesel: It is made by a bio-chemical process called "transesterification". Transesterification is a catalytic pathway to get bio-diesel and the catalyst might be acidic, alkyl, or can be both in nature in either identical or different forms, and enzymes (Melero et al. 2009; Sakai et al. 2009; Mohadesi et al. 2014). Improvements in the area of bio-fuel technology have driven to many commercial bio-diesel generation processes. Most of the established outturn plants, handled either in batch or continuous mode, by using trans-esterification reactions and homogeneous catalysts. Basic alike catalysts such as CH_3NO (sodium methoxide), CH_3KO (potassium methoxide), NaOH (sodium hydroxide), and KOH (potassium hydroxide) are favourable because of their large-scale out comes and low prefecture time (Akia et al. 2014; Antczak et al. 2009; Du et al. 2004). Whereas Identical or same composition-based acid catalysts, defend the high free fatty acid content of the feedstock and could catalyse both transesterification and esterification reactions synchronously. Most common acids that are used in wide range, are H_2SO_4 and HCl. The superior outturn of alkaline catalysts and digestion troubles caused by acid catalysts, as well as the high oil to alcohol ratio requirement, have increased the alkaline-catalysed technique demand to industrial plants (Lam et al. 2010; Felizardo et al. 2006; Jacobson et al. 2008). Biodiesel is made from renewable resources like plant/vegetable/animal oils for use in diesel engines. Vegetable oils are chemically called triglycerides (fat). Biodiesel consists of esters of long-chain fatty acids executed from these oils. To produce biodiesel, the fats (triglycerides) in vegetable oil are reacted with an alcohol (usually methanol). In this reaction, glycerol (in triglycerides) is replaced with methanol to produce a methyl ester (biodiesel). Biofuel development in India is focused on the cultivation of seeds (40%) of the oil-rich Jatropha plant (Grieder et al. 2012).

23.5.4 Biohydrogen

Bio-hydrogen: Biohydrogen can be produced through various processes such as pyrolysis and gasification. Biohydrogen can be viewed as a valuable energy source to replace fossil fuels. However, economically viable large-scale biohydrogen production is essential to replace fossil fuels with biohydrogen (Habeeb 2012). Biohydrogen is generated by two diverse pathways, photo-synthesis technique and anaerobic degradation technique by acidogenic bacteria (acidogenesis fermentation). These two individual techniques entangle some microbes and feedstock (Chong et al. 2009; Teng et al. 2010; Hallenbeck et al. 2009). Dark fermentative bio-hydrogen generation (which is the most commonly used technique for bio-hydrogen generation than with light fermentation and photosynthesis) mainly contains two major steps: hydrolysis and degradation technique by acidogenic bacteria producing CO_2 and organic acids as by products. Complex cultures of microbes are serviceable than with mono-culture ones according to the prospect of hydrogen production. Hydrogen in the aqueous state is generated and eaten up by methanogenic microbes during the process of production by anaerobic mode. To generate hydrogen effectively in the gaseous state, either these bio-hydrogen eaters must be removed or reduction in reaction time should be done (Aftab et al. 2019; Chuang et al. 2012).

23.5.5 Biogas and Biomethane

Bio-gas: Biogas is a by-product of the decomposition of plant and animal waste in low-oxygen environments such as landfills, waste treatment plants, and dairy products. It is formed by the anaerobic decomposition of organic matter. Since biogas consists primarily of methane and carbon dioxide (greenhouse gases), the natural incentives to keep biogas out of the atmosphere are strong. There are four prime stages present in bio-methane generation: hydrolysis, acidogenesis, dehydrogenation, and methanation (Vrije and Claassen 2003; Weiland 2010). Hydrolysing and fermenting microbe onslaught the monomers and polymers. Microbes taking part in this stage are anaerobes such as Bactericides, Clostridia, and Bifidobacterium. In acidogenesis stage other anaerobic bacteria alter the resultant complexes to short-chain organic acids, alcohols, H, and CO_2. At the dehydrogenation step, products of the acidogenic phase are seized as substrate by acetogenic bacteria, which generate H as a by-product while ethanoic acid (CH_3COOH) is also resulted. The last stage entangles methanogens activity which is a group of strictly anaerobic archaea, that take part in methanation. Methanogenic fermentation can be fulfilled using two modes of techniques: wet (substrate with less than 10% solid content) and dry (solid content of 15–35%) digestion (Grieder et al. 2012; Tabatabaei et al. 2011; Deublein and Steinhauser 2011).

23.5.6 Current State of Industry

Biofuels are used globally, most commonly to power vehicles and cooking stoves. However, environmentally sound processes which minimize the waste disposal problem as they produce marketable value-added products are developed. In April 2008 with the help of imaging technology, they noticed that vigorous mixing helps microorganisms turn farm waste into alternative energy, providing farmers with a simple way to treat their waste and convert it into energy (Business Line 2005). Today, with the enhancing usage of biorefineries, especially in the United States, the European Union, and the BICS [Brazil; India, China and South Africa] group, agriculture is now the new energy sector as well as the fulcrum of a "nouvelle économie" (Sebastian et al. 2006).

United States of America

In 2006, the United States president George W. Bush said in a State of the Union speech that the US is "addicted to oil" and should replace 75% of imported oil by 2025 by alternative sources of energy including biofuels (Fig. 23.15). Essentially all the ethanol fuel in the US is generated from corn. Corn is a great energy intensive crop-plant, which requires one unit of fossil-fuel energy to create just 0.9–1.3 energy units of ethanol. The US Energy Independence and Security Act of 2007-12-19 of 2007 requires US "fuel producers to use at least 36 billion gallons of biofuels in 2022. This is a nearly fivefold increase over current levels." This is causing a significant agricultural resource shift away from food production to biofuels. American food exports have decreased (increasing grain prices worldwide), and US food imports have increased significantly. General Motors (GM) is starting a project to produce E85 fuel from cellulose ethanol for a projected cost of $1 a gallon. This is optimistic however, because $1/gal equates to $10/MBTU which is comparable to woodchips at $7/MBTU or cord wood at $6–$12/MBTU, and this does not account for conversion losses and plant operating and capital costs which are significant. The raw materials can be as simple as corn stalks and scrap petroleum-based vehicle tires, but used tires are an expensive feedstock with other more-valuable uses. GM has over 4 million E85 cars on the road now, and by 2012 half of the production cars for the U.S. will be capable of running on E85 fuel, however by 2012 the supply of ethanol will not even be close to supplying this much E85. Coskata Inc. is building two new plants for the ethanol fuel. Theoretically, the process is claimed to be five times more energy efficient than corn-based ethanol, however, is still under development and has not proven cost-effective on the open market. Greenhouse gas emissions are reduced by 86% for pulp compared to a 29% reduction for corn (Okonko et al. 2009).

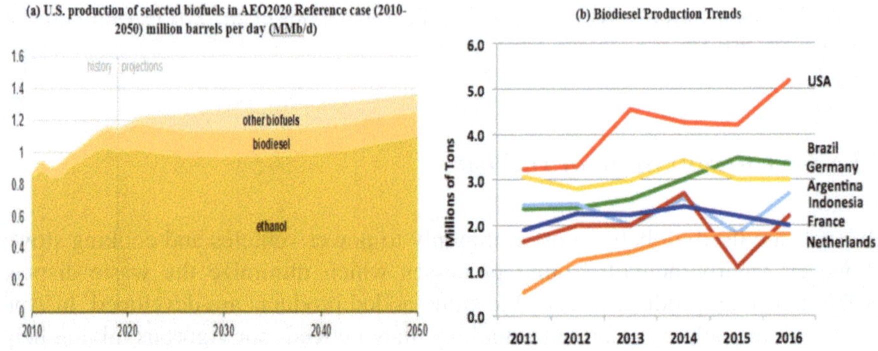

Fig. 23.15 **a** U.S. biofuel production to slowly increase through 2050, **b** comparison between other countries in biodiesel production trends (Bastani et al. 2012; Toumi et al. 2017)

Brazil

Typical Brazilian "flex" models from several car makers, that run on any blend of ethanol and gasoline. In Brazil, the government hopes to build on the success of the Proálcool ethanol program by expanding biodiesel production, which is set to contain 2% biodiesel by 2008 and increase to 5% by 2013 (Shahare et al. 2017). Brazil has ethanol fuel available throughout the country. Brazil is the world's second-largest generator of ethanol and the world's greatest exporter, and it is taken into account the world's first sustainable biofuel economy and a chief in biofuels (Arshad 2017).

Together, Brazil and the United States command the industrialized world in global ethanol generation, together accounting for 70% of global generation (Gebremariam and Marchetti 2021) and nearly 90% of ethanol used as fuel (Kumar 2008). In 2006, Brazil generated 16.3 billion liters (4.3 billion US fluid gallons), representing around 34% of the world's entire ethanol generation and 42% of the world's ethanol used as fuel (Kumar 2008). Brazil's ethanol program provided nearly 700,000 jobs in 2003 and reduced oil imports between 1975 and 2002 by a total of US$50 billion (Okonko et al. 2009). Ethanol production is concentrated in the interior and southeastern areas of the country, along with the primary producer—the state of São Paulo. These two areas were accountable for almost 90% of ethanol production in Brazil in 2004 (Regnier 2007). According to government statistics, Brazil produced around 17.470 billion liters of ethanol in 2006, 23 billion liters in 2007 and 2008. Companhia Nacional de Abastecimento expects production to grow by about around 15 and 20%, bringing the total ethanol generation will range from 26.50 to 28 billion liters (Okonko et al. 2009) (Fig. 23.16).

China

The Chinese government mandates E10 (Ethanol biofuel mixtures have this 'E' numbers which describes the percentage of ethanol) blends in 5 regions that account for 16% of the country's passenger cars. In Southeast Asia, Thailand has an ambitious introduction of 10% ethanol in gasoline since 2007. For similar reasons, the palm

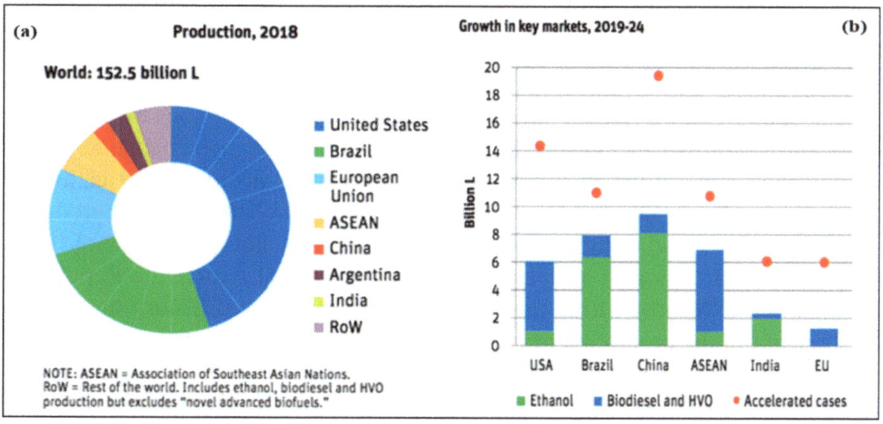

Fig. 23.16 **a** Pie chart and **b** bar chart of biofuel production of different countries (Johansen 2023)

oil industry plans to supply an enhancing share of national diesel fuel needs in Malaysia and Indonesia. In Canada, the govt. aims for 46% of the country's gasoline expenditure to hold 11% ethyl alcohol by 2010 (Okonko et al. 2009).

India

In India, a bioethanol program calls for E5 blends throughout most of the country aiming to raise this necessity to E10 and then E20. Jatropha biodiesel incentives in India are a part of India's goal to achieve energy independence by the year 2012. Jatropha oil is generated from the seeds of the *Jatropha curcas*, a plant that can grow in waste-lands across India and the oil is considered to be a great source of biodiesel. Large-scale plots of wasteland have been chosen for *Jatropha* harvesting and will give much required employment to the rural impoverished of India (Debnati and Bisen 2009). Businesses are also seeing the planting of Jatropha as a good business opportunity. Ninety-eight million acres of land has been identified for the production of Jatropha to replace the diesel consumption in India by 20% by 2011 (Shuit et al. 2009). Figure 23.16 shows the production of biofuel by different countries.

Europe

The European Union in its biofuel directive (updated in 2006) has fixed the goal that for 2010 that each member state should acquire at least 6% biofuel usage of all applied traffic fuel. By 2020 the figure should be near about 10%. As of 2007, the European Union region, led by the Netherlands, is the main importer of Brazilian ethanol, with 265.3 million gallons (1,004.2 million liters) (Shamsul et al. 2014). As of January 2008, these aims are being reconsidered in light of certain environmental and social concerns associated with biofuels such as rising food prices and deforestation (Fig. 23.16).

Biofuel industries are becoming established in many developing countries. Already, sub-Saharan Africa is considered as a major geo-source for bio-energy trade and regional development and production of bio-fuels (Johnson and Matsika 2006). African expectations are that Africa will be the new Biofuel continent (Monkey 2007), because Trees and crops are greening the Sahara (Haglund et al. 2011). Several African countries are pursuing independent biotechnological options to reduce their dependence on fossil fuel energy imports to conserve foreign-exchange resources, minimize pollution of the environment, and improve the quality of human, animal, and plant life. Ghana, Kenya, Malawi, South Africa and Zimbabwe—sugar producing countries—are exploring the option of largescale production of bioethanol (Pillay and Silva 2009). Amongst all the countries in continental Africa, the Republic of South Africa (RSA) by virtue of its past and present shared history has been at the forefront in the promotion of biotechnology to achieve national goals [Department of Science and Technology (DST) 2001, 2002]. However, the southernmost country of Africa has been described by Pillay and Da Silva (2009) as having veins of diamond and arteries of gold, and as a land of bio- and unique cultural diversity in its fauna and flora (features characteristic of 4 other countries: Brazil, China, India and Indonesia), its geography and topography, people, and political history. The potential of *Jatropha curcas*, an alternative bio-fuel source, has enthused bio-economists and policy-decision-makers in Burkina Faso, Cameroon, Ghana, Lesotho, Madagascar, Malawi and South Africa (Johansen 2023). Countries such as India and China are developing both bioethanol and biodiesel programs.

23.5.7 Importance of Biofuel in Global Meet

- Factors like the finite petroleum reserves and constantly rising demands for energy by the industrialised as well as the highly populated countries (on their Way to industrialisation) like India and China have made it absolutely necessary to look into alternate and efficient methods to replace these fuels in future. Also, concerns like steep rise in fossil fuel prices in the recent years, increasing concerns about climate change like global warming, insecurity and unrest among governments due to their depleting natural reserves are just a few factors that define an urgent need for a sustainable way to develop renewable fuel. Lignocellulosic biomass has evolved as one of the most clean and sustainable alternative to fossil fuels essentially because of their cost competitiveness as opposed to the current expensive methods of ethanol production from sugarcane and corn (Bułkowska and Pawłowski 2016).
- Alliance will promote the use of biofuels which will ensure energy security for the country and also reduce import dependence on crude oil.
- The government envisages that biofuel can play a strategic role in this regard by diversifying India's energy basket (Karmakar et al. 2012).
- Biofuels will also help in saving forex reserves due to the diversification of the energy basket.

- India is one of the world's largest producers and exporters of sugarcane. The sugarcane farmers face distress due to a demand–supply mismatch in the market and wastage (Gopinathan and Sudhakaran 2009).
  - In this context, the government promoted the diversion of sugarcane towards ethanol production under the ethanol blending programme.
  - Biofuels also aid in doubling farmers' income and boost price discovery.
- Government think tank NITI Aayog also believes that an Alliance will further the role of biofuel in the global fuel basket.

23.6 Benefits

Biofuels have cited advantages, like increased creation of jobs, enhanced domestic energy security, economic growth, decreased greenhouse gas (GHG) emissions, compared to their counterpart fossil fuels, etc. Accrued to the economy from the perspective of consumption of local resources to decreased crude oil imports, to the environment from the concern of reduced GHG emissions (Huang et al. 2009; Laurentiis et al. 2016), biofuels have drastically shown sustainable benefits and pose to be an alternative to currently used fuels.

23.6.1 *Potential to Reduce Greenhouse Gas Emission*

Being one of the pivotal objectives, biofuels have significantly improved in reducing GHGs. Compared to crude diesel, biodiesel burns up to 75% cleaner fuel, with reduced unburned hydrocarbons, particulate matter, sulfur dioxide emissions etc. (Kumar et al. 2018). Since 1960, an average increase of 3.2% per annum has been observed in CO_2 emissions per capita, which is further projected to increase by almost 80% between 2010–2030 (International Energy Agency 2009; Koizumi 2014). As mentioned in a study by Hoefnagels et al., the global warming potential for CO_2, CH_4 and N_2O is 1, 25 and 298, respectively, based on the fourth assessment report of the Intergovernmental Panel on Climate Change (IPCC) (Hoefnagels et al. 2010). In fact, in their 2011 special report, IPCC have stated that, 'most bioenergy systems can contribute to climate change mitigation if they replace fossil-based energy', citing the importance of biofuels against fossil fuels (Edenhofer et al. 2011).

23.6.2 *Increase Energy Securing*

Renewable energy sources including biofuels have been supported by varied governmental agencies, quoting its advantages in increased domestic energy security, employment generation, reduced GHG emissions etc. Quoting several advantages, one of the major problems faced by developing countries to adopt biofuels as a primary fuel is the energy access and affiliated social and economic developments (Waisman et al. 2012). Though, enhancing national energy security and reducing the dependence of the traditional fossil fuels, have been gradually adopted by several countries, dependence on fossil fuels has been growing tremendously and will increase substantially in the next 20 years. In fact, it is anticipated, by 2030, net oil imports to China and India will make up to 74% and 92%, respectively, of total national demand (International Energy Agency 2009). In order to diversify their domestic energy supplies and partially reduce their reliance on energy imports, a number of nations in the region have adopted bioenergy targets and other supporting policies. Nevertheless, though, biofuel industries depend on imported fuel and feedstock, will reduce the dependency of crude oil.

23.6.3 Promote Rural Development

Studies have implicated that biofuel industry has a great impact on job creation and employability (Owusu and Asumadu-Sarkodie 2016). In fact, biofuel industries require a work force of 100 times more than that of fossil fuel industries (Gwilliam et al. 2004). Exporting biofuels has already helped some countries create a significant amount of jobs. By increasing palm biodiesel production in Indonesia, which employs 1.7–3 million people in the oil palm sector, could create 2.5 million more jobs (Flach et al. 2012; Cassman and Liska 2007; Zen et al. 2006). However, concerns have been raised about the widespread use of land tenure and ownership, which might get mitigated with further research.

23.7 Challenges

Growing attention of biofuels as substitutes to fossil fuels, has been reflected in terms of energy security, economic stability, climate change policies, business opportunities, political concerns, environmental concerns, etc. Though economic stability, business opportunities and political concerns are quite relative with respect to personal perspectives, energy security, climate change policies and environmental concerns play a major role.

23.7.1 Land Use Competition

Since the production of biofuels is based on agricultural products, where land is the primary input, it has a different dynamic than the production of fossil fuels. Since

2005, there has been a significant increase in agricultural production specifically in the United States and Brazil, first through increased production of corn and sugarcane for ethanol production, respectively, and then soybeans for biodiesel production. The necessity to include new arable land in order to quickly increase the production of liquid biofuels has turned out to be a global trend (Mathews 2007; Goldemberg et al. 2008; Wright 2006).

This shows a shift in the application of land from agricultural outturn and sustains a dilemma for the world, placing the need to feed people against farmers' increased financial gains from using land for agro-energy (Simpson et al. 2012). Evident in the 1970s and 1980s, adoption of this practice in some countries (Brazil as the prime example), research has been made to study the effects of warfare between cultivation of food and bio-fuels outturn (Brown 1980; Destek et al. 2021; Meekhof et al. 1980; Silva et al. 1978). In fact, a lot of existing crop-land that is underutilized in many nations around the world and might be used for meal, feed, or agroenergy without creating a land-use warfare concerning biofuel production. For instance, it is estimated that there are over 300 million hectares of potentially arable land available in some African countries alone for the production of both ethanol and biodiesel. Consequently, land use warfare is a matter primarily observed in the major countries producing crops (i.e., the US and Brazil) (Gokcol et al. 2009). Due to this dynamic, the relationship between the quality and use of land has always been hampered. In this regard, national governments could play a significant role in reducing the effects of land use competition. Further, in order to demonstrate that the production of biofuels can be sustained over time, implementing certification standards for biofuels that are based on both origin and the farming methods employed is undoubtedly the best course of action.

23.7.2 Water Scarcity

Besides the adoption of biofuels via microorganisms, the water demand for such products has always stayed high. Microalgae cultivation (in a suitable growth system, such as an open pond, photobioreactor, etc.) is followed by drying, harvesting, lipid extraction, and esterification in a typical process for producing microalgal biomass for various industrial purposes (Batan et al. 2016; Lardon et al. 2009). Given the above cited processes, water loss is mainly observed in processes like drying, harvesting and cultivation (Delrue et al. 2012; Feng et al. 2016). Terms like "water footprint", as coined by UNESCO in 2002, have also been introduced in this regard to the amount of freshwater being consumed for processes like, to produce a biofuel. Particularly, in regard of microalgal biofuel production, water footprint is comprised of three major components, namely, grey water, blue water and white water (Pugazhendhi et al. 2020).

However, several strategies are practiced concerning to reduce water footprint, like recycling, high density cultivation, heterotrophic cultivation etc. About 600%

reduction in water consumption can be achieved if water is recycled during cultivation of microalgae (Huang et al. 2016). Aside from the benefit of less water being used, recycling water also reduces the need for nutrients as recycled medium already contains unused nutrients. The production of *Chlorella vulgaris* biodiesel using recycled medium is predicted to reduce the need for 55% of the total amount of nutrients (Yang et al. 2011). Though water recycling has gots its own benefits and demerits, as in change in salinity, pH, flocculant concentration, and secretions of microalgae, if controlled well, water recycling in addition to reduction in production costs decreases the water footprint (Pugazhendhi et al. 2020). Non-suspended cultivation is yet another method which can be devoured, pertaining to its simple harvesting and easy-dewatering advantages.

23.7.3 Negative Environmental Impacts

Impacts on the environment evaluate how various paths react to stressors such as resource depletion, acidification, and climate change. Keeping in concern the feedstock, microalgae reduce carbon footprint via their respiration process. However, because the reduced emissions must be transferred throughout the entire value chain, the process of turning microalgae into goods is equally crucial. In the "business as usual" scenario, one MJ of bio-oil generates 80 g of CO_2, whereas in the "optimal scenario," with increased productivity and crude yield, it generates 44 g. The Net Energy Ratio (NER), which expresses the ratio of energy saved to energy required in a process, is an important consideration in a life cycle assessment (LCA). Depending on the reactor used for growing the algae, the NER can increase or decrease. In contrast to flat reactors, which have a NER of 4.5, and tubular reactors, which have a NER of 0.2, Raceway Pond, for example, has a NER of 8.3. This net energy ratio affects the entire energy balance, which in turn affects other environmental impact elements like global warming, acidification risk, and resource potential (Barsanti and Gualtieri 2018). The environmental criterion known as global warning potential gives details about the emissions of a system. The likelihood of global warming is typically expressed in equivalent units of CO_2. The potential of the different greenhouse gases is changed to CO_2 equivalency, to use a common expression. A carbon isolation method that lowers overall emissions is algae biorefinery. However, the energy-intensive nature of biomass drying reduces its negative effects.

23.8 Opportunities

Due to the erratic supply of fossil fuels, rising pollution risks, expanding rural economies, and the requirement to reduce carbon emissions, there is an increasing demand for biofuels. To reduce carbon emissions, it is crucial to look for alternative fuel sources. In the modern era, it is easy to see how fossil fuel refineries are giving way to sustainable biorefineries. The search for alternate and sustainable nonfood feedstocks, such as lignocellulosic biomass, has addressed the dependence on food-based crops, which results in the food versus fuel issue. The process for utilising lignocellulosic biomass to its full potential is still in its infancy. In addition, although macro- and microalgae have potential applications in biotechnology, industry, and the environment, producing algal fuel is still more expensive than producing fossil fuels. Wastewater and flue gases have decreased the cost of nutrients, but other mechanical tools and processes are pricey. Or it can be said that the basic procedures applied for producing and selling algae bio-fuels sustain to be a major hurdle. The agricultural system must be enhanced to accessible the outturn of a sizable quantity of algae biomass which are rich in lipid substances, and more work and ingredients are required for the harvesting techniques for algal biodiesel to be more economically viable.

23.9 Environmental and Economic Benefit

23.9.1 Environmental Impacts

The environmental impact of growing crops for biofuel use is complex and widespread (a brief description is depicted in Table 23.2). Pollution level with environmental degradation due to direct production of biofuel, is became more meaningful, especially the fact that while promoting how "green" bio-fuels actually are, few take these problems into account (Gasparatos et al. 2018).

23.9.2 Economic Benefits

Fossil fuels are considered to be the biggest environmental hazards of our time. The biggest objective of several global organizations is to reduce the use of these fossil fuels and replace them with something more organic. This is where biofuels can help. Governments all over the world are giving subsidies to biofuels (Table 23.3). Many government agencies procure biofuels from vendors. They are used to meet the needs of sectors like railways which have high energy requirements. Other economic benefits like-

- Lower Import Bills—Rising oil prices have had a severe economic impact on the current account position of many nations. If they import too much oil, they often have to cut the import of other essential commodities. Lower import bills would provide a wide range of benefits to these governments. Governments experience a lot of flexibility and freedom in decision making after breaking the shackles of fossil fuel consumption (Ewing and Msangi 2009).

Table 23.2 Environmental influences

Environmental influences	Brief description	References
Habitat devastation	• In terms of ethanol production, on the contrary, sugar cane is more efficient; when carbon-rich rainforests are cleared for sugarcane production, this leads to a significant increase in GHG emissions; therefore, the overall competency of sugarcane cultivations is quickly declining • When all the ecological benefits of rainforests are embodied, for example for biodiversity preservation, hydrological operations and soil conservation, then such comparisons are even more one-sided	Sahoo et al. (2022)
Vestige Green House Gas (GHG) emanations from enrichers	• The most efficient biofuel crop-plants, like corn and canola, need a lot of fertilizer. As they penetrate the soil and enter groundwater, at the end they touch streams and water bodies, the latter being irradiated in flight; they will be altered into a large source of N_2O, an important smog which clears stratospheric ozone. This may be recognised as the emission of trace gas • Again, if the real residual greenhouse gases were combined, the measured greenhouse gas emissions used in development, and such crop-plants may be worse than fossil fuels for global warming	Lemoine et al. (2010), Fargione et al. (2008), Melillo et al. (2009)
Aquatic influences	• As water demands grow with population growth and urbanization, and with increasing household and commercial habits, the situation is worsening • Meanwhile, the use of water, particularly for agricultural production, causes the greatest demand for water in the world • This shortfall could be significantly exacerbated if a dramatic increase in biofuel production is supported	Nelson and Robertson (2008)

(continued)

Table 23.2 (continued)

Environmental influences	Brief description	References
Outsized balance farming	• Harvesting of biofuel crop-plants must usually take place on relatively large, modern plantation granary, to have a money-worthy fuel supply • The heavy machinery needed to harvest such large annual crop-plants must run on other alternative sources except fossil fuels • Husbandry cultivation methods leave the soil incredibly depleted, leading to continual deforestation as more and more land is needed to generate uniform amounts of crops	Rosegrant et al. (2008), Woods (2008)
Lower emissions	• Firstly, biofuels emit considerably lower amounts of greenhouse gases in the environment • Comparisons done over a thirty-year period show that biofuels emit less than 50% of greenhouse gases as compared with fossil fuels • Also, there are many second-generation biofuels which drastically cut emissions from day one. Except being good from an environmental point of view, this can also from an economic standpoint • Higher emission levels mean the government has to spend more money trying to control pollution. The usage of biofuels, therefore, provides financial advantages	Bessou et al. (2011)

Table 23.3 Economic benefits based on different countries

<table>
<tr><th>Country</th><th>GDP (Gross Domestic Product)</th><th>SDG (Sustainable Development Goal)</th><th>References</th></tr>
<tr><td>India</td><td>The service sector is the biggest sector of India with the Gross Value Added at current prices as 96.54 lakh crore in 2020–21. Today the service sector accounts for almost 54% of Indian GVA (Gross Value Added) of 179.15 lakh crores. India's total GDP is 3.469 trillion USD (2022)</td><td rowspan="5">Meet the goals-
Goal 6—ensure availability and sustainable management of water and sanitation for all
Goal 7—ensure access to affordable, reliable, sustainable and modern energy for all
Goal 8—promote sustained, inclusive and sustainable economic growth, full and productive employment and decent work for all
Goal 11—make cities and human settlements inclusive, safe, resilient and sustainable
Goal 14—conserve and sustainably use the oceans, seas and marine resources for sustainable development
Goal 15—protect, restore and promote sustainable use of terrestrial ecosystems, sustainably manage forests, combat desertification, and halt and reverse land degradation and halt biodiversity loss</td><td>Agarwal and Saini (2023)</td></tr>
<tr><td>USA</td><td>In 2021, U.S. ethanol production totalled about 15 billion gallons and combined biodiesel/renewable diesel production totalled about 2.5 billion gallons. USA's total GDP is 21.43 trillion USD (2019)</td><td>Hoekman (2009)</td></tr>
<tr><td>Brazil</td><td>In 2019, the production of biofuels in Brazil reached the highest figure of the decade, at 411 thousand barrels of oil equivalent per day. However, figures decreased to 376 thousand barrels of oil equivalent per day in 2021. Brazil is the second largest biofuel producer world-wide, ranking only behind the U.S. Brazil's total GDP is 1.833 trillion USD (2022)</td><td>Hernandes et al. (2021)</td></tr>
<tr><td>China</td><td>The domestic production of crude oil generation is likely to stabilize at around 200 million tonnes by 2020 (Tong et al. 2009), the gap between domestic demand and supply creates opportunities for the development of alternative fuels, including biofuels. China's total GDP is 18.32 trillion USD (2022)</td><td>Shiyan et al. (2012)</td></tr>
<tr><td>Europe</td><td>In 2021, the production of biofuels in Europe was highest in Germany, at 54 thousand barrels of oil equivalent per day. This was six thousand barrels of oil equivalent daily more than France, which produced the second highest volume that year. Germany has around 3.4 percent share of global biofuel production. The overall production of biofuels in EU has increased dramatically since the turn of the century, reaching 236 thousand barrels of oil equivalent per day in 2021. Europe's total GDP is 386.830 billion of USD (2021)</td><td>Næss et al. (2023), Morone et al. (2023)</td></tr>
</table>

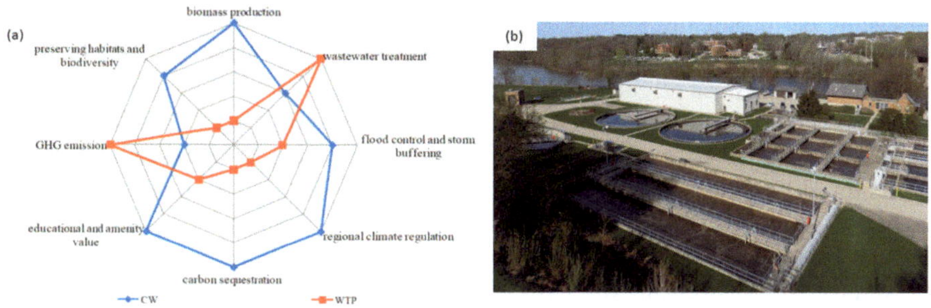

Fig. 23.17 **a** Conceptual framework comparing CW (blue line, left image) and WTP (treatment plant) ecosystem services (red line, right image). **b** Wastewater treatment plant (Liu et al. 2019)

- Terrorism Financing—The proceeds from the sale of fossil fuels are the number one source of financing for terrorist organizations. ISIS derives a large portion of its revenue from selling its oil in the black market. Radical elements in the governments of Saudi Arabia and Iran also send oil money to terrorist sympathizers. If the world were to embrace the use of biofuels, the dependence on fossil fuels would be reduced considerably. This would severely curtail the flow of money to terrorist organizations (Valentine 2011).
- Sustainable Model—Last but not the least biofuels are a sustainable way to meet the energy needs of the planet. Production of biofuels does not take thousands of years. Hence, the production of biofuels can be made faster than its consumption. This would create a sustainable business model. The world will no longer have to live with the fear that the energy sources are going to be depleted. This will lead to stability in the price of these fuels. Biofuels are unlikely to be as volatile as fossil fuels. This will help the economy since it will make production planning a lot easier (Guo et al. 2015).

In addition, CWs (constructed wetlands) can provide additional ecosystem service benefits such as biomass generation, carbon sequestration, multiple water supply, regional climate adjustment, habitat protection, educational and recreational use (Fig. 23.17). The conditions for each service in the Fig. 23.17 are indicated along each "petal" to illustrate the provision of multiple ecosystem services under different land-use regimes. The emission point value of each ecosystem service increases from the inside out.

23.10 Conclusion

Biofuels are becoming an increasingly important alternative to petroleum diesel. Biofuels are valued by some policymakers in developing countries as a durable sake of strengthening regional economics, enhancing energy retreat or acquiring unfamiliar trade, and reducing greenhouse gas emissions, as well as businesses as a way to encourage promising prospects. This concept provides various concrete examples of how biofuel production can lead to positive, neutral, or negative environmental consequences relying on the individual environmental outcomes of interest, the crop used, the land on which the crop is grown and its past use, the practices used management. As well as various factors such as environmental impacts due to agricultural land market intervention and vegetation change. As mentioned above, there are feedstocks (four generations of feedstocks are present) available from where biofuel production becomes easier. It helped to overcome many challenges like land or water scarcity (overlong discussion has been done in this book chapter). Biofuel helps to fulfil environmental and economic needs, with new sustainable development goal. Social, economic and environmental aspects need to be considered judiciously in policies for the managed development of bio-fuels.

References

Aftab MN, Iqbal I, Riaz F, Karadag A, Tabatabaei M (2019) Different pretreatment methods of lignocellulosic biomass for use in biofuel production. In: Biomass for bioenergy-recent trends and future challenges, pp 1–24

Agarwal RN, Saini I (2023) Advances in sustainability management. Paarth Publications

Ahorsu R, Medina F, Constantí M (2018) Significance and challenges of biomass as a suitable feedstock for bioenergy and biochemical production: a review. Energies 11(12):336. https://doi.org/10.3390/en11123366

Akia M, Yazdani F, Motaee E, Han D, Arandiyan H (2014) A review on conversion of biomass to biofuel by nanocatalysts. Biofuel Res J 1(1):16–25

Alalwan HA, Alminshid AH, Aljaafari HA (2019) Promising evolution of biofuel generations. Subject review. Renew Energy Focus 28:127–139

Alam F, Mobin S, Chowdhury H (2015) Third generation biofuel from algae. Procedia Eng 105:763–768

Antczak MS, Kubiak A, Antczak T, Bielecki S (2009) Renew Energy 34:1185–1194

Arshad M (ed) (2017) Perspectives on water usage for biofuels production: aquatic contamination and climate change. Springer

Aryal N, Ottosen LDM, Kofoed MVW, Pant D (eds) (2021) Emerging technologies and biological systems for biogas upgrading

Azhar SHM, Abdulla R, Jambo SA, Marbawi H, Gansau JA, Faik AAM, Rodrigues KF (2017) Yeasts in sustainable bioethanol production: a review. Biochem Biophys Rep 10:52–61

Baldino C, Berg R, Pavlenko N, Searle S (2019) Advanced alternative fuel pathways: technology overview and status. Int Council Clean Transp 13:1–31

Barrios E, Gemmill-Herren B, Bicksler A, Siliprandi E, Brathwaite R, Moller S, Batello C, Tittonell P (2020) The 10 elements of agroecology: enabling transitions towards sustainable agriculture and food systems through visual narratives. Ecosyst People 16(1):230–247

Barsanti L, Gualtieri P (2018) Is exploitation of microalgae economically and energetically sustainable? Algal Res 31:107–115

Bastani P, Heywood JB, Hope C (2012) The effect of uncertainty on US transport-related GHG emissions and fuel consumption out to 2050. Transp Res Part A: Policy Pract 46(3):517–548

Batan LY, Graff GD, Bradley TH (2016) Techno-economic and Monte Carlo probabilistic analysis of microalgae biofuel production system. Biores Technol 219:45–52

Ben-Iwo J, Manovic V, Longhurst P (2016) Biomass resources and biofuels potential for the production of transportation fuels in Nigeria. Renew Sustain Energy Rev 63:172–192

Bessou C, Ferchaud F, Gabrielle B, Mary B (2011) Biofuels, greenhouse gases and climate change. Sustain Agric 2:365–468
Bielen AA, Verhaart MR, Van der Oost J, Kengen SW (2013) Biohydrogen production by the thermophilic bacterium Caldicellulosiruptor saccharolyticus: current status and perspectives. Life 3(1):52–85
Bošnjaković M, Sinaga N (2020) The perspective of large-scale production of algae biodiesel. Appl Sci 10(22):8181
Brown LR (1980) Food or fuel: new competition for the world's Cropland. Worldwatch Paper 35
Bułkowska K, Pawłowski A (2016) Biomass feedstock for biofuels production. In: Biomass for biofuels. CRC Press, pp 37–61
Cassman G, Liska AJ (2007) Food and fuel for all: realistic or foolish? Biofuels Bioproducts Biorefining Innov Sustain Econ 1(1):18–23
Chong ML, Sabaratnam V, Shirai Y, Hassan MA (2009) Biohydrogen production from biomass and industrial wastes by dark fermentation. Int J Hydrogen Energy 34(8):3277–3287
Chuang YS, Huang CY, Lay CH, Chen CC, Sen B, Lin CY (2012) Fermentative bioenergy production from distillers grains using mixed microflora. Int J Hydrogen Energy 37(20):15547–15555
Da Silva Trindade WR, dos Santos RG (2017) Review on the characteristics of butanol, its production and use as fuel in internal combustion engines. Renew Sustain Energy Rev 69:642–651
Da Silva JG, Serra GE, Moreira JR, Conçalves JC, Goldemberg J (1978) Energy balance for ethyl alcohol production from crops. Science 201(4359):903–906
De Laurentiis V, Hunt DV, Rogers CD (2016) Overcoming food security challenges within an energy/water/food nexus (EWFN) approach. Sustainability 8(1):95
De Vrije T, Claassen PAM (2003) Dark hydrogen fermentations. In: Bio-methane & bio-hydrogen, pp 103–123
Debnati MO, Bisen PS (2009) Jatropha Revolution in India
Delrue F, Setier PA, Sahut C, Cournac L, Roubaud A, Peltier G, Froment AK (2012) An economic, sustainability, and energetic model of biodiesel production from microalgae. Biores Technol 111:191–200
Department of Science and Technology (DST), RSA (2001) A national biotechnology strategy for South Africa, June, p 82
Destek MA, Sarkodie SA, Asamoah EF (2021) Does biomass energy drive environmental sustainability? An SDG perspective for top five biomass consuming countries. Biomass Bioenergy 149:106076
Deublein D, Steinhauser A (2011) Biogas from waste and renewable resources: an introduction. Wiley
Du W, Xu Y, Liu D, Zeng J (2004) Comparative study on lipase-catalyzed transformation of soybean oil for biodiesel production with different acyl acceptors. J Mol Catal B Enzym 30(3–4):125–129
Edenhofer O, Pichs-Madruga R, Sokona Y, Seyboth K, Kadner S, Zwickel T, Eickemeier P, Hansen G, Schlömer S, von Stechow C, Matschoss P (eds) (2011) Renewable energy sources and climate change mitigation: special report of the intergovernmental panel on climate change. Cambridge University Press
Ewing M, Msangi S (2009) Biofuels production in developing countries: assessing tradeoffs in welfare and food security. Environ Sci Policy 12(4):520–528
Fargione J, Hill J, Tilman D, Polasky S, Hawthorne P (2008) Land clearing and the biofuel carbon debt. Science 319(5867):1235–1238
Felizardo P, Correia MJN, Raposo I, Mendes JF, Berkemeier R, Bordado JM (2006) Production of biodiesel from waste frying oils. Waste Manag 26(5):487–494
Feng PZ, Zhu LD, Qin XX, Li ZH (2016) Water footprint of biodiesel production from microalgae cultivated in photobioreactors. J Environ Eng 142(12):04016067
Flach B, Bendz K, Lieberz S (2012) USDA foreign agricultural service global agriculture information network report. EU Biofuels Annu
Gasparatos A, Romeu-Dalmau C, von Maltitz GP, Johnson FX, Shackleton C, Jarzebski MP, Jumbe C, Ochieng C, Mudombi S, Nyambane A, Willis KJ (2018) Mechanisms and indicators for

assessing the impact of biofuel feedstock production on ecosystem services. Biomass Bioenergy 114:157–173

Gebremariam SN, Marchetti JM (2021) Process simulation and techno-economic performance evaluation of alternative technologies for biodiesel production from low value non-edible oil. Biomass Bioenergy 149:106102

Ghosh SK, Samanta S, Hirani H, da Silva CRV (2009) Effective waste management and circular economy

Gnansounou E, Dauriat A (2005) Ethanol fuel from biomass: a review

Gokcol C, Dursun B, Alboyaci B, Sunan E (2009) Importance of biomass energy as alternative to other sources in Turkey. Energy Policy 37(2):424–431

Goldemberg J, Coelho ST, Guardabassi P (2008) The sustainability of ethanol production from sugarcane. Energy Policy 36(6):2086–2097

Gopinathan MC, Sudhakaran R (2009) Biofuels: opportunities and challenges in India. In Vitro Cell Dev Biol Plant 45:350–371

Gordon J, Szalay J (2022) Potato nutrition facts & health benefits

Grieder C, Mittweg G, Dhillon BS, Montes JM, Orsini E, Melchinger AE (2012) Kinetics of methane fermentation yield in biogas reactors: genetic variation and association with chemical composition in maize. Biomass Bioenergy 37:132–141

Guo M, Song W, Buhain J (2015) Bioenergy and biofuels: history, status, and perspective. Renew Sustain Energy Rev 42:712–725

Gwilliam KM, Kojima M, Johnson T (2004) Reducing air pollution from urban transport. World Bank, Washington, DC

Habeeb SA (2012) The influence of temperature and types of filter media on the palm oil mill effluent (POME) treatment using the hybrid up-flow anaerobic sludge blanket (HUASB) reactor. Doctoral dissertation, Universiti Tun Hussein Onn Malaysia

Haglund E, Ndjeunga J, Snook L, Pasternak D (2011) Dry land tree management for improved household livelihoods: farmer managed natural regeneration in Niger. J Environ Manag 92(7):1696–1705

Hallenbeck PC, Ghosh D, Skonieczny MT, Yargeau V (2009) Microbiological and engineering aspects of biohydrogen production. Indian J Microbiol 49:48–59

Hays MD, Fine PM, Geron CD, Kleeman MJ, Gullett BK (2005) Open burning of agricultural biomass: physical and chemical properties of particle-phase emissions. Atmos Environ 39(36):6747–6764

Hernandes TAD, Duft DG, dos Santos Luciano AC, Leal MRLV, Cavalett O (2021) Identifying suitable areas for expanding sugarcane ethanol production in Brazil under conservation of environmentally relevant habitats. J Clean Prod 292:125318

Hoefnagels R, Smeets E, Faaij A (2010) Greenhouse gas footprints of different biofuel production systems. Renew Sustain Energy Rev 14(7):1661–1694

Hoekman SK (2009) Biofuels in the US—challenges and opportunities. Renew Energy 34(1):14–22

Huang J, Yang J, Qiu H, Rozelle S, Sombilla MA (2009) Global and regional development and impact of biofuels: a focus on the greater Mekong subregion. Asian Development Bank

Huang Y, Sun Y, Liao Q, Fu Q, Xia A, Zhu X (2016) Improvement on light penetrability and microalgae biomass production by periodically pre-harvesting Chlorella vulgaris cells with culture medium recycling. Biores Technol 216:669–676

Hussain F, Soudagar MEM, Afzal A, Mujtaba MA, Fattah IR, Naik B, Mulla MH, Badruddin IA, Khan TY, Raju VD, Gavhane RS (2020) Enhancement in combustion, performance, and emission characteristics of a diesel engine fueled with Ce-ZnO nanoparticle additive added to soybean biodiesel blends. Energies 13(17):4578

International Energy Agency (2009) World energy outlook. OECD/IEA, Paris, p 17

Jacobson K, Gopinath R, Meher LC, Dalai AK (2008) Solid acid catalyzed biodiesel production from waste cooking oil. Appl Catal B 85(1–2):86–91

Javed MR, Bilal MJ, Ashraf MUF, Waqar A, Mehmood MA, Saeed M, Nashat N (2019) Microalgae as a feedstock for biofuel production: current status and future prospects. Top 5:1–39

Jerney J, Spilling K (2020) Large scale cultivation of microalgae: open and closed systems. Biofuels from algae: methods and protocols, pp 1–8
Jin W, Pastor-Pérez L, Villora-Pico JJ, Pastor-Blas MM, Sepúlveda-Escribano A, Gu S, Charisiou ND, Papageridis K, Goula MA, Reina TR (2019) Catalytic conversion of palm oil to bio-hydrogenated diesel over novel N-doped activated carbon supported Pt nanoparticles. Energies 13(1):132
Johansen BE (2023) Climate crisis: code red for humanity and our home planet: a scorecard for an apocalypse. Global warming and the climate crisis: science, spirit, and solutions. Springer International Publishing, Cham, pp 97–197
Johnson FX, Matsika E (2006) Bio-energy trade and regional development: the case of bio-ethanol in southern Africa. Energy Sustain Dev 10(1):42–53
Kang Y, Yang Q, Bartocci P, Wei H, Liu SS, Wu Z, Zhou H, Yang H, Fantozzi F, Chen H (2020) Bioenergy in China: evaluation of domestic biomass resources and the associated greenhouse gas mitigation potentials. Renew Sustain Energy Rev 127:109842
Karmakar A, Karmakar S, Mukherjee S (2010) Properties of various plants and animals feedstocks for biodiesel production. Biores Technol 101(19):7201–7210
Karmakar A, Karmakar S, Mukherjee S (2012) Biodiesel production from neem towards feedstock diversification: Indian perspective. Renew Sustain Energy Rev 16(1):1050–1060
Koizumi T (2014) Biofuels and food security: biofuel impact on food security in Brazil, Asia and major producing countries. Springer
Kumar A (2008) Bioengineering of crops for biofuels and bioenergy. In: Recent advances in plant biotechnology. IK International, New Delhi
Kumar A, Ogita S, Yau YY (eds) (2018) Biofuels: greenhouse gas mitigation and global warming: next generation biofuels and role of biotechnology. Springer
Lam MK, Lee KT, Mohamed AR (2010) Homogeneous, heterogeneous and enzymatic catalysis for transesterification of high free fatty acid oil (waste cooking oil) to biodiesel: a review. Biotechnol Adv 28(4):500–518
Lardon L, Hélias A, Sialve B, Steyer JP, Bernard O (2009) Life-cycle assessment of biodiesel production from microalgae
Lee RA, Lavoie JM (2013) From first-to third-generation biofuels: challenges of producing a commodity from a biomass of increasing complexity. Anim Front 3(2):6–11
Lemoine DM, Plevin RJ, Cohn AS, Jones AD, Brandt AR, Vergara SE, Kammen DM (2010) The climate impacts of bioenergy systems depend on market and regulatory policy contexts
Line B (2009) India Inc eyes jatropha farming in a big way. Business Line 6–15. Accessed 2 Sept 2005
Liu D, Zou C, Xu M (2019) Environmental, ecological, and economic benefits of biofuel production using a constructed wetland: a case study in China. Int J Environ Res Public Health 16(5):827
Lütke-Eversloh T, Bahl H (2011) Metabolic engineering of Clostridium acetobutylicum: recent advances to improve butanol production. Curr Opin Biotechnol 22(5):634–647
Malode SJ, Gaddi SAM, Kamble PJ, Nalwad AA, Muddapur UM, Shetti NP (2022) Recent evolutionary trends in the production of biofuels. Mater Sci Energy Technol 5:262–277
Mathews JA (2007) Biofuels: what a biopact between north and south could achieve. Energy Policy 35(7):3550–3570
Meekhof RL, Tyner WE, Holland FD (1980) US agricultural policy and gasohol: a policy simulation. Am J Agr Econ 62(3):408–415
Melero JA, Iglesias J, Morales G (2009) Heterogeneous acid catalysts for biodiesel production: current status and future challenges. Green Chem 11(9):1285–1308
Melillo JM, Reilly JM, Kicklighter DW, Gurgel AC, Cronin TW, Paltsev S, Felzer BS, Wang X, Sokolov AP, Schlosser CA (2009) Indirect emissions from biofuels: how important? Science 326(5958):1397–1399
Miesel JR, Renz MJ, Doll JE, Jackson RD (2012) Effectiveness of weed management methods in establishment of switchgrass and a native species mixture for biofuels in Wisconsin. Biomass Bioenergy 36:121–131

Miskat MI, Ahmed A, Chowdhury H, Chowdhury T, Chowdhury P, Sait SM, Park YK (2020) Assessing the theoretical prospects of bioethanol production as a biofuel from agricultural residues in Bangladesh: a review. Sustainability 12(20):8583

Mohadesi M, Hojabri Z, Moradi G (2014) Biodiesel production using alkali earth metal oxides catalysts synthesized by sol-gel method. Biofuel Res J 1(1):30–33

Mollahoseini A, Tabatabaei M, Najafpour GD (2015) Biofuel production. In: Biochemical engineering and biotechnology, 2nd edn. Elsevier, Amsterdam, pp 597–630

Monkey G (2007) In Fuel-Tech, Bio-Fuel Vehicles, March 4

Moodley P (2021) Sustainable biofuels: opportunities and challenges. Sustain Biofuels 1–20

Morone P, Cottoni L, Giudice F (2023) Biofuels: technology, economics, and policy issues. In: Handbook of biofuels production. Woodhead Publishing, pp 55–92

Næss JS, Hu X, Gvein MH, Iordan CM, Cavalett O, Dorber M, Giroux B, Cherubini F (2023) Climate change mitigation potentials of biofuels produced from perennial crops and natural regrowth on abandoned and degraded cropland in Nordic countries. J Environ Manag 325:116474

Nelson GC, Robertson RD (2008) Green gold or green wash: environmental consequences of biofuels in the developing world. Rev Agric Econ 30(3):517–529

Ng IS, Keskin BB, Tan SI (2020) A critical review of genome editing and synthetic biology applications in metabolic engineering of microalgae and cyanobacteria. Biotechnol J 15(8):1900228

Nielsen DR, Leonard E, Yoon SH, Tseng HC, Yuan C, Prather KLJ (2009) Engineering alternative butanol production platforms in heterologous bacteria. Metab Eng 11(4–5):262–273

Oh PP, Lau HLN, Chen J, Chong MF, Choo YM (2012) A review on conventional technologies and emerging process intensification (PI) methods for biodiesel production. Renew Sustain Energy Rev 16(7):5131–5145

Okonko IO, Egwame RA, Fajobi EA, Nkang AO, Iheakanwa CI, Ogunnusi TA, Onajobi BI (2009) Current trends in biofuel production and its use as an alternative energy security. Electron J Environ Agric Food Chem 8(12)

Onuki S (2006) Bioethanol: industrial production process and recent studies. J Chem Technol Biotechnol 5:1–12

Owusu PA, Asumadu-Sarkodie S (2016) A review of renewable energy sources, sustainability issues and climate change mitigation. Cogent Eng 3(1):1167990

Pan N (2014) Alternative and renewable energy. Doctoral dissertation, Worcester Polytechnic Institute

Paul Abishek M, Patel J, Prem Rajan A (2014) Algae oil: a sustainable renewable fuel of future. Biotechnol Res Int 2014

Pillar-Little Jr TJ (2018) Carbon quantum dots: bridging the gap between chemical structure and material properties

Pillay D, Da Silva EJ (2009) Sustainable development and bioeconomic prosperity in Africa: biofuels and the South African gateway. African J Biotechnol 8(11)

Pugazhendhi A, Nagappan S, Bhosale RR, Tsai PC, Natarajan S, Devendran S, Al-Haj L, Ponnusamy VK, Kumar G (2020) Various potential techniques to reduce the water footprint of microalgal biomass production for biofuel—a review. Sci Total Environ 749:142218

Regnier E (2007) Oil and energy price volatility. Energy Econ 29(3):405–427

Rosegrant MW, Zhu T, Msangi S, Sulser T (2008) Global scenarios for biofuels: impacts and implications. Rev Agric Econ 30(3):495–505

Sahoo G, Wani AM, Swamy SL, Sharma A (2022) Environmental impact and economic benefits of biofuel production. In: Bio-clean energy technologies, vol 1. Springer Nature Singapore, Singapore, pp 349–378

Sakai T, Kawashima A, Koshikawa T (2009) Economic assessment of batch biodiesel production processes using homogeneous and heterogeneous alkali catalysts. Biores Technol 100(13):3268–3276

Sebastian A, Madsen AM, Martensson L, Pomorska D, Larsson L (2006) Assessment of microbial exposure risks from handling of biofuel wood chips and straw-effect of outdoor storage. Ann Agric Environ Med 13(1)

Shahare VV, Kumar B, Singh P (2017) Biofuels for sustainable development: a global perspective. In: Green technologies and environmental sustainability, pp 67–89

Shamsul NS, Kamarudin SK, Rahman NA, Kofli NT (2014) An overview on the production of bio-methanol as potential renewable energy. Renew Sustain Energy Rev 33:578–588

Sharma S, Singh S, Sarma SJ, Brar SK (2022) Feedstocks for green diesel. In: Green diesel: an alternative to biodiesel and petrodiesel. Springer Nature Singapore, Singapore, pp 41–53

Shen Y (2014) Carbon dioxide bio-fixation and wastewater treatment via algae photochemical synthesis for biofuels production. RSC Adv 4(91):49672–49722

Shen CR, Liao JC (2013) Synergy as design principle for metabolic engineering of 1-propanol production in Escherichia coli. Metab Eng 17:12–22, 48

Shiyan C, Lili Z, Timilsina GR, Xiliang Z (2012) Development of biofuels in China: technologies, economics and policies. World Bank Policy Research Working Paper 6243

Shokravi Z, Shokravi H, Aziz MA, Shokravi H (2019) The fourth-generation biofuel: a systematic review on nearly two decades of research from 2008 to 2019. In: Fossil free fuels, pp 213–251

Shuit SH, Tan KT, Lee KT, Kamaruddin AH (2009) Oil palm biomass as a sustainable energy source: a Malaysian case study. Energy 34(9):1225–1235

Simpson RD, Toman MA, Ayres RU (2012) Scarcity and growth revisited: natural resources and the environment in the new millennium. Routledge

Sims RE, Mabee W, Saddler JN, Taylor M (2010) An overview of second generation biofuel technologies. Biores Technol 101(6):1570–1580

Slade R, Bauen A (2013) Micro-algae cultivation for biofuels: cost, energy balance, environmental impacts and future prospects. Biomass Bioenergy 53:29–38

Steen EJ, Chan R, Prasad N, Myers S, Petzold CJ, Redding A, Ouellet M, Keasling JD (2008) Metabolic engineering of Saccharomyces cerevisiae for the production of n-butanol. Microb Cell Fact 7:1–8

Sumathi S, Chai SP, Mohamed AR (2008) Utilization of oil palm as a source of renewable energy in Malaysia. Renew Sustain Energy Rev 12(9):2404–2421

Tabatabaei M, Sulaiman A, Nikbakht AM, Yusof N, Najafpour G (2011) Influential parameters on biomethane generation in anaerobic wastewater treatment plants. In: Alternative fuel. INTECH Open Access Publisher, Rijeka

Teng SX, Tong ZH, Li WW, Wang SG, Sheng GP, Shi XY, Liu XW, Yu HQ (2010) Electricity generation from mixed volatile fatty acids using microbial fuel cells. Appl Microbiol Biotechnol 87:2365–2372

Tomes D, Lakshmanan P, Songstad D (eds) (2010) Biofuels: global impact on renewable energy, production agriculture, and technological advancements. Springer Science & Business Media

Tong H, Jin Y, Liu W, Li F, Fang J, Yin Y, Qian Q, Zhu L, Chu C (2009) DWARF AND LOW-TILLERING, a new member of the GRAS family, plays positive roles in brassinosteroid signaling in rice. Plant J 58(5):803–816

Toumi O, Le Gallo J, Rejeb JB (2017) Assessment of Latin American sustainability. Renew Sustain Energy Rev 78:878–885

Tushar MS, Dutta A, Xu C (2014) Effects of reactor wall properties, operating conditions and challenges for SCWG of real wet biomass. In: Near-critical and supercritical water and their applications for biorefineries, pp 207–228

Valentine SV (2011) Emerging symbiosis: renewable energy and energy security. Renew Sustain Energy Rev 15(9):4572–4578

Van Der Maarel MJ, Van der Veen B, Uitdehaag JC, Leemhuis H, Dijkhuizen L (2002) Properties and applications of starch-converting enzymes of the α-amylase family. J Biotechnol 94(2):137–155

Verma D, Fortunati E, Jain S, Zhang X (eds) (2019) Biomass, biopolymer-based materials, and bioenergy: construction, biomedical, and other industrial applications. Woodhead Publishing

Waisman H, Guivarch C, Grazi F, Hourcade JC (2012) The I maclim-R model: infrastructures, technical inertia and the costs of low carbon futures under imperfect foresight. Clim Change 114:101–120

Weiland P (2010) Biogas production: current state and perspectives. Appl Microbiol Biotechnol 85:849–860

Woods J (2008) Sustainable biofuels: prospects and challenges. The Royal Society, London

Wright L (2006) Worldwide commercial development of bioenergy with a focus on energy crop-based projects. Biomass Bioenergy 30(8–9):706–714

Yang J, Xu M, Zhang X, Hu Q, Sommerfeld M, Chen Y (2011) Life-cycle analysis on biodiesel production from microalgae: water footprint and nutrients balance. Biores Technol 102(1):159–165

Zen Z, Barlow C, Gondowarsito R (2006) Oil palm in Indonesian socio-economic improvement: a review of options

Zhao S, Bi X, Pan X, Su Y, Wu W (2021) The optimization of in-situ tar reduction and syngas production on a 60-kW three-staged biomass gasification system: theoretical and practical approach. Biomass Convers Biorefin 11:1835–1846

Zhou X, Yuan S, Chen R, Ochieng RM (2015) Sustainable production of energy from microalgae: review of culturing systems, economics, and modelling. J Renew Sustain Energy 7(1):012701

Chapter 24
Global Status of Microbial Engineering for Biofuel Production - Advances, Challenges, and Roadmap

Shrey Semalty, Ashutosh Bhardwaj, and Santosh Kumar Karn

Abstract Microbial engineering is a crucial technology in the production of biofuels. It involves the manipulation of microorganisms, such as yeast or bacteria, to produce fuels such as ethanol or bio-hydrogen. The global status of microbial engineering has advanced in recent years, with improvements in metabolic engineering and synthetic biology. However, the production of biofuels still faces several challenges, including low productivity, high production costs, and limited scalability. To overcome these challenges, researchers are exploring new metabolic pathways, developing more efficient microorganisms, and improving the production process. A roadmap for the future of microbial engineering for biofuels production would include continued research into more efficient production methods, the development of cost-effective production processes, and the integration of renewable energy sources. Overall, the future of microbial engineering for biofuels production is promising, with the potential for significant advancements in the coming years.

Keywords Biofuel · Microbes · Bacteria · Challenges · Energy

24.1 Biofuels: In General

Biofuels are fuels that are produced from biomass which is organic matter derived from plants or animals. Biomass can be used directly as a source of heat or electricity, or it can be converted into liquid or gaseous fuels that can be used for transportation or other purposes (Hill 2013). Biofuels are renewable energy sources because they can be replenished in a relatively short time span, unlike fossil fuels that take millions of years to form (Zhang 2013). There are different types of biofuels, depending on the feedstock and the conversion process used. Some of the most common biofuels are described in the Table 24.1.

S. Semalty · A. Bhardwaj · S. K. Karn (✉)
Department of Biochemistry & Biotechnology, Sardar Bhagwan Singh University, Balawala, Dehradun, Uttarakhand 248001, India
e-mail: Santoshkarn@gmail.com; santoshkarn.sbsu@gmail.com

M. Shah and D. Deka (eds.), *Emerging Sustainable Technologies for Biofuel Production*, Environmental Science and Engineering, https://doi.org/10.1007/978-3-031-52167-6_24

Biofuels has been categorized based on the source from which that had been extracted like first generation biofuels, second generation biofuels, third generation biofuels, and fourth generation biofuels (Sadeghinezhad et al. 2014) as shown in Fig. 24.1.

Biofuels have several advantages over fossil fuels, they may improve energy security and lessen reliance on imported oil (Hughes, Moser, and Gibbons 2014), cut greenhouse gas emissions and slow climate change (Creutzig et al. 2015), and more.

Table 24.1 Major types of biofuels utilized

S.No	Type of fuel	Description
1	Biogas	A mixture of methane and carbon dioxide that is produced by anaerobic digestion of organic waste, such as manure, sewage, or landfill. Biogas can be used as a fuel for heating, cooking, or electricity generation (Naik et al. 2010)
2	Bioethanol	An alcohol that is produced by fermenting sugars or starches from crops suchas corn, sugarcane, or sweet sorghum. During the fermentation process hydrolysed carbohydrates (like glucose) are transformed into ethanol (bioethanol) with the assistance of microorganisms (like yeast). A high-octane, water-free alcohol is then created through distillation and dehydration. This substance is also referred to as hydrated ethanol or anhydrous ethanol (Escobar et al. 2009)
3	Biodiesel	A fuel that is produced by transesterifying oils or fats in the presence of alcohol or alkaline from plants or animals, such as soybean, rapeseed, palm, orwaste cooking oil (Du et al. 2016)
4	Biohydrogen	A gas that is produced by splitting water molecules using biological catalysts, such as algae or bacteria. Biohydrogen can be used as a clean and efficient fuel for fuel cells or combustion engines (Raturi et al. 2022)

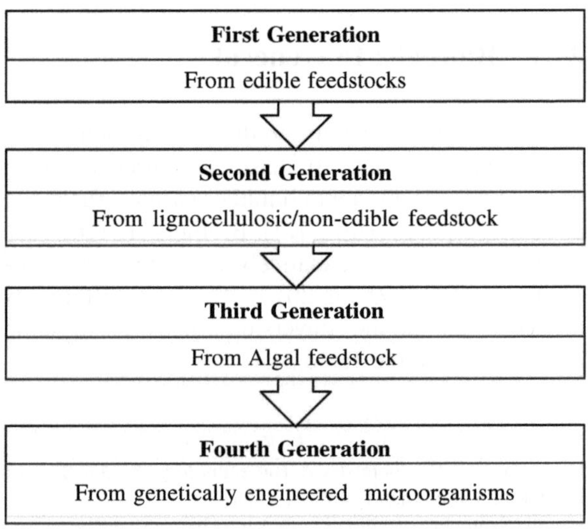

Fig. 24.1 Different generation of biofuels

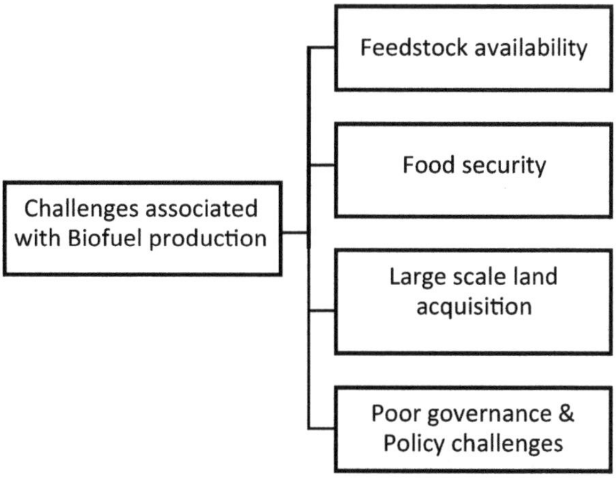

Fig. 24.2 Challenges associated with biofuel production

They may diversify the energy mix and improve energy efficiency. They can also provide new employment and revenue possibilities for farmers and rural communities (Balat 2011). However, biofuels also face some challenges and limitations, they may rise food prices and food security by competing with food production (Pimentel et al. 2008), and they may have greater production costs than fossils fuels (Kern et al. 2017). Therefore, biofuels need to be developed and used in sustainable way, considering the environmental, social, and economic impacts of their production and consumption. Biofuels also need to be integrated with other renewable energy sources and technologies to achieve a low-carbon and resilient energy system.

24.2 Global Scenario in Biofuel Production

The Covid-19 crisis has significantly altered the global context of biofuels. Restrictions imposed around the world and stalling of economic activity have reduced the demand for transportation fuel. Since a large proportion of biofuels are used to combine with fossil fuels, this decrease in demand often affects the demand for biofuels. Ethanol production is expected to decline by 18% (720 million tons) and biodiesel and hydro treated vegetable oil (HVO) production by 3% (21 million tons) as demand for fossil fuels declines due to the global pandemic (IEA 2020). Biofuel demand in 2021 reached 4 EJ (159 200 million litres), returning to near 2019 levels after a decline due to the Covid-19 pandemic. Global demand for biofuels is set to grow by 41 billion litres, or 28%, over 2021-2026 in the main case. The recovery to pre-Covid-19 demand levels accounts for one- fifth of this demand growth (IEA 2021). There are many reasons for this post-pandemic recovery like overall transport fuel demand and cost, different government policies to encourage biofuel production and consumption. Globally, biofuels are produced predominantly from traditional feedstocks i.e., 1st generation biofuels due to multiple challenges being faced to extract biofuel from algae, cyanobacteria i.e., 3rd generation biofuels. Asian countries are driving biofuel supply and demand by favoring domestic production though subsidies, tax credits and ow interest loans for investments. Feedstocks for biofuel products vary from country to country (OECD and Nations 2022). Many jurisdictions are making moves that suggest they see a significant long-term role for bioenergy in the energy transition like Brazil, India, and the United States, as leading biofuel producers and consumers, will work together in the coming years towards the development of a Global Biofuels Alliance along with other interested countries. This Alliance will be aimed at facilitating cooperation and intensifying the use of sustainable biofuels, including in the transportation sector. Additionally, the United States passed the inflation Reduction Act in August 2022, which includes extended and new policy support for biofuels, particularly advanced biofuels, and sustainable aviation fuels. Only bioenergy that reduces lifecycle GHG emissions while avoiding unacceptable social, environmental, and economic impacts should only receive policy support (IEA 2022).

24.3 Challenges Associated with Biofuel Production

Even though they are thought of as alternatives to conventional fuel sources, biofuel manufacturing faces a number of difficulties. Significant sustainability issues, such as feedstock supply, economic feasibility, environmental and social impact, etc., plague the first generation of biofuels. One significant factor contributing to the constant increase in the price of staple foods since 2000 is the growing use of food crops to make biofuels in the United States, the European Union, Japan, and Brazil (Renzaho, Kamara, and Toole 2017). The controversy over using food as a fuel source has given rise to a number of objections to the use of biofuels, many of which are related to the problem of food security in poor nations. One of the biggest obstacles to producing biofuels is poor governance. Large-scale land acquisition for biomass sourcing and a steady supply of feedstock, both of which call for effective government regulations, are necessary for the development of viable biofuels. Environmental restrictions also apply to the production of biofuels.

Although biofuels are thought to be environmentally friendly, their widespread production has brought to light a variety of environmental issues related to their use. Additional acreage is required to generate the necessary oil crops for biofuel production and to meet the demand for biofuel. As a result, there has been significant deforestation to provide the extra acreage needed. Numerous negative consequences of large-scale deforestation include soil erosion, loss of the ecological system, loss of biodiversity, etc. The technology surrounding biofuel is also changing. To be used as an alternative to conventional fuels, they need to have their engines and fuels modified. The generation of biofuel has these key difficulties, which are covered in more depth below.

24.4 Feedstock Availability and Food Security

The conflict between using land for food crops and biofuel feedstock has brought up the "food versus fuel" discussion. In 1991, (D. O. Hall 1991) acknowledged the complexity of the food vs. fuel debate and said it required serious consideration since agricultural and export policies, as well as the politicization of food supply, were more important determinants in the creation of biofuels. A number of moral issues have been brought up by the first generation of biofuels, including the threat to biodiversity posed by their inability to reduce carbon dioxide and greenhouse gas emissions and the threat to food supplies posed by the potential for rising biofuel demand to divert crops from global food markets, raising food prices (Ewing and Msangi 2009). Given that many low- and middle-income nations struggle with problems like hunger and malnutrition, the issue of food security is a significant barrier to the production of biofuels. If the problem of food security can be resolved responsibly, the production of biofuel could present these nations with enormous economic prospects.

24.5 Large-Scale Land Acquisition

Land acquisition on a wide scale to produce biofuels is related to the problem of maintaining domestic food and energy security, particularly when done across and within national borders, and the uneven sustainability effects this produces. The use of land in the global south to grow biomass for fuel in the global north or for domestic urban populations has come under fire. Energy crop farming has been argued to have both positive and negative effects in developing nations, as it may help the rural poor by providing them with a source of income, reduce the country's reliance on foreign imports of fossil fuels, and provide prospects for export earnings (Mohr and Raman 2013). The key factor influencing the purchase of extensive farms is still biofuel feedstock. Large-scale land acquisition by foreign firms in the developing world, particularly in African nations, has expanded dramatically. For instance, in the Democratic Republic of the Congo, over 10 million hectares have been given to various international organizations and governments for the production of biofuel and food for export (R. Hall 2011). Food insecurity will increase, especially in the most vulnerable areas, as more land, water, fertilizers, and other resources are diverted from food production to biofuel production in order to meet the rising energy demands and lessen the dependency on non-renewable fossil fuels (Renzaho, Kamara, and Toole 2017). The fragile food crisis is likely to worsen if smallholder farmers are forced from their traditional lands in favor of biofuels feedstock.

In the Tonkali and Bombali areas of Sierra Leone, an Action Aid (2013) survey indicated that 99% of the respondents linked their experience of hunger to the loss of their land at the hands of foreign corporations. Families or individuals losing their lands did not consent to having their land leased to foreign businesses; they were not adequately informed about the advantages and difficulties of biofuel farming;

the payment for their land lease was insufficient to secure alternative land; the jobs offered were insufficient to cover all households that lost their land; and the wages from the biofuel plantation were too low to keep up with the rising cost of food (Isabirye et al. 2013). Therefore, poorly implemented large-scale biofuel initiatives hurt smallholder farmers and drive them into poverty by denying them access to their ancestral lands and limiting their capacity to grow their own food.

24.6 Poor Governance and Policy Challenges

The creation of a framework for regulating the practice of land acquisition in the global South is a major task for policy given these issues. However, reliable land acquisition monitoring is expected to be challenging in nations with inadequate social and environmental governance given the significant power gaps between big businesses and locals (Van Der Horst and Vermeylen 2011). Murphy along with other researcher (Murphy et al. 2011) makes the case that in order to balance and encourage synergies between various uses of land for food and fuel, we need better land management regulations. The use of land for fuel needs to be evaluated within a larger examination of land use for diverse purposes, and how land is valued. Farmlands are used for a variety of purposes, including agriculture, grazing areas for animals, etc. (Gamborg et al. 2012). Policy makers must evaluate broader (industrial, residential, recreational, etc.) land use and management practices in addition to analyses of current agriculture management practices. Even in the bioenergy sector, however, there are not many discussions on the "sustainability of agriculture" and "sustainability of land use." We would suggest that, given these complexities, there is a need for such a debate. However, we would caution that the potential of such a broader debate is hampered by opposing social, economic, environmental, and policy arguments (and the values embedded therein, both explicit and implicit) (Mohr and Raman 2013).

24.7 Environmental Impacts

A growing need for biofuels for transportation is expected to make biodiversity and water preservation "grand challenges," with far-reaching social repercussions that frequently affect local and regional communities. The results of a study on the land and water implications of global 1G biofuel production in 2030 showed that policymakers would be reluctant to pursue biofuel options based on traditional food and oil crops in areas where traditional agricultural production already faces severe water limitations (such as in India and China, two of the world's largest agricultural producers and consumers) (De Fraiture, Giordano, and Liao 2008). Arable land that was previously set aside by the EU under a programme to protect agricultural ecosystems may experience a decline in biodiversity if it is planted with monoculture energy crops (Romijn and Caniëls 2011). Deforestation and the development of biofuels have also been related. To supply the biofuel business, more than 13 million hectares of South East Asia's rich biodiversity were removed and replaced with palm oil crops (Scharlemann and Laurance 2008). According to some research, the greenhouse gas emissions caused by the destruction of forests for biofuel feedstock can be up to 420 times greater than the savings from using fewer fossil fuels (Fargione et al. 2008). For many nearby people, forests provide a source of livelihood and serve as carbon sinks, habitats for flora and animals, and watersheds (Streck and Scholz 2006). Therefore, the ecosystem would suffer greatly because of deforestation brought on by the purchase of land for the production of biofuels.

24.8 Other Local Impacts

Many of the direct effects of biofuel production are also felt locally, especially when biomass is grown in the less developed Southern regions for use in biofuelelse where (Van Der Horst and Vermeylen 2011). It is widely believed that intensification of 1G or 2G energy crops will have detrimental ecological effects, such as deforestation, habitat loss, and decreased soil fertility, which will negatively impact rural lives (Mohr and Raman 2013). Increased energy crop production has been linked to long-term losses in livelihoods, local food production, and energy output due to the eviction of local subsistence farmers and adverse ecological effects (Van Eijck and Romijn 2008).

24.9 Microbial Engineering for Biofuel Production

High productivity and yields are essential for the cost-effective generation of biofuels. Almost all metabolic engineering techniques begin with the overexpression of the target pathway genes, then include the elimination of rival pathways that deplete precursors, products, or cofactors. Functional-genomics methods are particularly suited for determining the rate- limiting stages in biosynthetic pathways. For example, focused proteomics was used to pinpoint two weakly expressed enzymes in the heterologous mevalonate pathway (Redding- Johanson et al. 2011). Large-scale biomass production places a constraint on the field of biofuel research. Future hurdles will be in the identification, characterization, and overproduction of lipids once this is accomplished. Finally, the extraction of lipids from microalgal cells requires cost-effective procedures. Wastewater treatment and microalgal culture can be combined to produce biomass that is both affordable and sustainable (Hasunuma, Matsuda, and Kondo 2016). Below is a list of the many biotechnological strategies that may be used to optimize the production of biofuels.

24.10 Role of Genomics

The study of the structure, function, evolution, mapping, and editing of the genome is the focus of the interdisciplinary discipline of biology known as genomics. Numerous research using the genomics platform have been carried out to describe model organisms for effective biofuel generation. "Genomic techniques include nucleotide sequences, next-generation sequencing approaches, expressed sequence tags (ESTs), sequencing by ligation pyrosequencing, recombinant cDNAs, real-time sequencing, and full-length cDNAs. Functional genomics includes genome, genes, and protein expression as well as transcriptomic analysis. Comparative genomics focuses on phylogenetic analysis for genes and genomes" (Rai et al. 2016). The ability to design microalgal strains either by disrupting genes or expressing transgenes has been standardized thanks to advances in the identification of the proper genes involved for bioenergy and processes in microalgae (Beer et al. 2009). Genomic information pertinent to the generation of biofuels has been made available for *Geitlerinema* sp. FC II (Batchu et al. 2019), *Coccomyxa* sp. strain KJ (Yoshimitsu, Abe, and Harayama 2018), and many other microalgae strains by complete or partial genome sequencing.

24.11 Role of Proteomics

The field of biotechnology known as proteomics uses genetics, biochemistry, and molecular biology methods to study the structure, interactions, and functions of the proteins made by a specific cell, tissue, or organism. It also organizes the information in databases and uses the data. Numerous proteins that are involved in the increase of lipids have been discovered through proteomics investigations. 60% FAME was found in *Nannochloropsis oculata* development under nitrogen limitation, and nitrogen from the degradation of proteins and pigments was largely funneled towards fatty acid synthesis, according to a proteomic analysis (Tran et al. 2016). Purified oil bodies comprising diverse proteins involved in vesicle trafficking and transport were shown to include these proteins in another publication on the proteome profiling investigation of *C. reinhardtii* under nitrogen restriction (Nguyen et al. 2011). Proteomics techniques for biofuel generation need thorough knowledge of the biochemical, biophysical, and structural characterization of the enzymes involved in the biosynthetic pathway of lipids, as well as the technology needed for separating lipids from biomass.

24.12 Strain Improvement for Biofuel Production

The fourth or most recent generation of biofuels is made from modified microorganisms and engineered algae that are designed to endure stressors and generate more ethanol than naturally occurring algae (Klinthong et al. 2015). One of the fundamental techniques in microbial engineering for biofuel generation is strain improvement. There are a variety of methods for strain improvement, such as the classical strain improvement techniques that focus on obtaining a large number of mutants, followed by screening for desired phenotypes. These techniques are very time- and labor-intensive, but one benefit is that strains developed in this way are currently seen as non-genetically modified organisms, which increases their acceptance by both regulatory bodies and consumers (Bharadwaj et al. 2020). Genome shuffling is an alternative way of strain enhancement that is superior to traditional techniques since it makes use of whole- genome recombination's ability to select for desirable characteristics However, because genome shuffling does not need any genomic data and can be simply applied to any strain with a measurable goal phenotype (Biot-Pelletier and Martin 2014) it is comparable to traditional strain enhancement strategies.

Numerous microbial strains have been subjected to a variety of biotechnological techniques to maximize biofuel production. Such as in the case of *Phaeodactylum tricornutum* overexpression of DGAT2 gene leads to 35% increase in lipid content (Niu et al. 2013). Similarly in the case of *Chlorella ellipsoidea* expression of GmDof4 TFs leads to 52% increase in total lipids (Zhang et al. 2014). Overexpression of acetyl-CoA synthetase in *Schizochytrium sp.* results in 11.3% higher fatty acid production (Yan et al. 2013). Meanwhile in *C. reinhardtii* expression of citrate synthase leads to 169% increase in TAG level (Deng, Cai, and Fei 2013).

24.13 Conclusion

The Covid-19 problem has had an influence on the worldwide scenario for biofuel production, which has decreased demand for transportation fuel and therefore decreased demand for biofuels. However, because of factors including the general need for transportation fuel, governmental regulations, and the formation of coalitions to support sustainable biofuel usage, the demand for biofuels is predicted to increase in the upcoming years. Environmental and social implications, feedstock availability, economic viability, and sustainability concerns are only a few of the difficulties facing the development of biofuels. Concerns regarding food security and higher food costs have been brought up by the use of food crops for the production of biofuel. Large-scale land acquisition for the purpose of obtaining biomass has unfavorable effects, including deforestation and the eviction of smallholder farmers, which results in food insecurity. It takes effective laws and improved land management techniques to generate sustainable biofuels because of issues with governance and policy. Deforestation, habitat loss, and greenhouse gas emissions are some environmental effects. The creation of biofuel has negative repercussions on the local community as well, such as reduced food production and losses to livelihoods.

Optimizing the production of biofuels involves a large amount of microbial engineering. The creation of modified microbes with improved biofuel production capacities is made possible by strain improvement techniques as well as genomics and proteomics, which help uncover important genes and proteins involved in bioenergy pathways.

References

Balat M (2011) Potential alternatives to edible oils for biodiesel production – a review of current work. Energy Convers Manage 52(2):1479–1492. https://doi.org/10.1016/j.enconman.2010.10.011

Batchu NK, Khater S, Patil S, Nagle V, Das G, Bhadra B, Sapre A, Dasgupta S (2019) Whole genome sequence analysis of geitlerinema Sp. FC II Unveils competitive edge of the strain in marine cultivation system for biofuel production. Genomics 111(3):465–472. https://doi.org/10.1016/j.ygeno.2018.03.004

Beer Laura L, Eric S Boyd, John W Peters, Matthew C Posewitz (2009) Engineering Algae for biohydrogen and biofuel production. Current Opinion in Biotechnology, Energy Biotechnology / Environmental Biotechnology 20(3):264–271. https://doi.org/10.1016/j.copbio.2009.06.002

Bharadwaj SV Vamsi, Shristi Ram, Imran Pancha, Sandhya Mishra (2020) Chapter 14 - recent trends in strain improvement for production of biofuels from microalgae. In *Microalgae Cultivation for Biofuels Production*, edited by Abu Yousuf, pp 211–225. Academic Press. https://doi.org/10.1016/B978-0-12-817536-1.00014-X

Biot-Pelletier D, Martin VJJ (2014) Evolutionary engineering by genome shuffling. Appl Microbiol Biotechnol 98(9):3877–3887. https://doi.org/10.1007/s00253-014-5616-8

Creutzig F, Ravindranath NH, Berndes G, Bolwig S, Bright R, Cherubini F, Chum H et al (2015) Bioenergy and climate change mitigation: an assessment. GCB Bioenergy 7(5):916–944. https://doi.org/10.1111/gcbb.12205

De, Fraiture, Charlotte MG, Liao Y (2008) Biofuels and implications for agricultural water use: blue impacts of green energy. Water Policy 10(S1):67–81. https://doi.org/10.2166/wp.2008.054

Deng X, Cai J, Fei X (2013) Effect of the expression and knockdown of citrate synthase gene on carbon flux during triacylglycerol biosynthesis by green algae chlamydomonas reinhardtii. BMC Biochem 14(1):38. https://doi.org/10.1186/1471-2091-14-38

Du C, Zhao X, Liu D, Lin CSK, Wilson K, Luque R, Clark J (2016) Introduction. In Handbook of Biofuels Production, pp 3–12. Elsevier. https://doi.org/10.1016/B978-0-08-100455-5.00001-1

Escobar José C, Electo S Lora, Osvaldo J Venturini, Edgar E Yáñez, Edgar F Castillo, Oscar Almazan (2009) Biofuels: environment, technology and food security. Renew Sustain Energy Rev 13(6–7):1275–1287. https://doi.org/10.1016/j.rser.2008.08.014

Ewing M, Msangi S (2009) Biofuels production in developing countries: assessing tradeoffs in welfare and food security. Environ Sci Policy 12(4):520–528. https://doi.org/10.1016/j.envsci.2008.10.002

Fargione J, Hill J, Tilman D, Polasky S, Hawthorne P (2008) Land clearing and the biofuel carbon debt. Science 319(5867):1235–1238. https://doi.org/10.1126/science.1152747

Gamborg C, Millar K, Shortall O, Sandøe P (2012) Bioenergy and land use: framing the ethical debate. J Agric Environ Ethics 25(6):909–925. https://doi.org/10.1007/s10806-011-9351-1

Hall DO (1991) Biomass energy. Energy Policy 19(8):711–737. https://doi.org/10.1016/0301-4215(91)90042-M

Hall R (2011) Land grabbing in southern Africa: the many faces of the investor rush. Rev Afr Polit Econ 38(128):193–214. https://doi.org/10.1080/03056244.2011.582753

Hasunuma Tomohisa, Mami Matsuda, Akihiko Kondo (2016) Improved sugar-free succinate production by synechocystis Sp. PCC 6803 following identification of the limiting steps in glycogen catabolism. Metab Eng Commun 3(December):130–141. https://doi.org/10.1016/j.meteno.2016.04.003

Hill Jason (2013) Life cycle analysis of biofuels. In *Encyclopedia of Biodiversity*, pp 627–630. Elsevier. https://doi.org/10.1016/B978-0-12-384719-5.00365-8

Hughes Stephen R, Bryan R Moser, William R Gibbons (2014) Moving toward energy security and sustainability in 2050 by reconfiguring biofuel production. In Convergence of food security, energy security and sustainable agriculture, David D Songstad, Jerry L Hatfield, Dwight T Tomes (eds), 67:15–29. Biotechnology in Agriculture and Forestry. Berlin, Heidelberg: Springer Berlin Heidelberg. https://doi.org/10.1007/978-3-642-55262-5_2

Isabirye Moses, Raju DVN, Kitutu M, Yemeline V, Deckers J, Poese J (2013) Sugarcane biomass production and renewable energy. In *Biomass Now - Cultivation and Utilization*, edited by Miodrag Darko Matovic. InTech. https://doi.org/10.5772/56075

Kern JD, Hise AM, Characklis GW, Gerlach R, Viamajala S, Gardner RD (2017) Using life cycle assessment and techno-economic analysis in a real options framework to inform the design of algal biofuel production facilities. Biores Technol 225(February):418–428. https://doi.org/10.1016/j.biortech.2016.11.116

Klinthong W, Yang Y-H, Huang C-H, Tan C-S (2015) A review: microalgae and their applications in CO_2 capture and renewable energy. Aerosol Air Qual Res 15(2):712–742. https://doi.org/10.4209/aaqr.2014.11.0299

Mohr A, Raman S (2013) Lessons from first generation biofuels and implications for the sustainability appraisal of second generation biofuels. Energy Policy 63(December):114–122. https://doi.org/10.1016/j.enpol.2013.08.033

Murphy R, Woods J, Black M, McManus M (2011) Global developments in the competition for land from biofuels. Food Policy 36(January):S52-61. https://doi.org/10.1016/j.foodpol.2010.11.014

Naik SN, Goud VV, Rout PK, Dalai AK (2010) Production of first and second generation biofuels: a comprehensive review. Renew Sustain Energy Rev 14(2):578–597. https://doi.org/10.1016/j.rser.2009.10.003

Nguyen HM, Baudet M, Cuiné S, Adriano J-M, Barthe D, Billon E, Bruley C et al (2011) Proteomic profiling of oil bodies isolated from the unicellular green microalga chlamydomonas reinhardtii: with focus on proteins involved in lipid metabolism. Proteomics 11(21):4266–4273. https://doi.org/10.1002/pmic.201100114

Niu Ying-Fang, Meng-Han Zhang, Da-Wei Li, Wei-Dong Yang, Jie-Sheng Liu, Wei-Bin Bai, Hong-Ye Li (2013) Improvement of neutral lipid and polyunsaturated fatty acid biosynthesis by over-expressing a type 2 diacylglycerol acyltransferase in marine diatom phaeodactylum tricornutum. Marine Drugs 11(11):4558–5469. https://doi.org/10.3390/md11114558

OECD, and Food and Agriculture Organization of the United Nations (2022). "Biofuels." https://www.oecd-ilibrary.org/content/component/c90c6fa0-en

Pimentel D, Marklein A, Toth M, Karpoff M, Paul G, McCormack R, Kyriazis J, Krueger T (2008) Biofuel impacts on world food supply: use of fossil fuel, land and water resources. Energies 1(2):41–78. https://doi.org/10.3390/en1010041

Rai Vineeta, Anbarasu Karthikaichamy, Debasish Das, Santosh Noronha, Pramod P Wangikar, Sanjeeva Srivastava (2016) Multi-Omics frontiers in algal research: techniques and progress to explore biofuels in the postgenomics world. OMICS: A J Integ Biol 20(7):387– 399. https://doi.org/10.1089/omi.2016.0065

Raturi Shubham, Pallavi Singh, Saurabh Kumar Jha, Santosh Kumar Karn (2022) Biohydrogen: opportunities and challenges as an alternative energy resource. J Appl Nat Sci 14(2):692–701. https://doi.org/10.31018/jans.v14i2.3480

Redding-Johanson AM, Batth TS, Chan R, Krupa R, Szmidt HL, Adams PD, Keasling JD, Lee TS, Mukhopadhyay A, Petzold CJ (2011) Targeted proteomics for metabolic pathway optimization:

application to terpene production. Metab Eng 13(2):194–203. https://doi.org/10.1016/j.ymben.2010.12.005

Renzaho AMN, Kamara JK, Toole M (2017) Biofuel production and its impact on food security in low and middle income countries: implications for the post-2015 sustainable development goals. Renew Sustain Energy Rev 78(October):503–516. https://doi.org/10.1016/j.rser.2017.04.072

Romijn HA, Caniëls MCJ (2011) The jatropha biofuels sector in Tanzania 2005–2009: evolution towards sustainability? Res Policy 40(4):618–636. https://doi.org/10.1016/j.respol.2011.01.005

Sadeghinezhad E, Kazi SN, Foad Sadeghinejad A, Badarudin MM, Sadri R, Safaei MR (2014) A comprehensive literature review of bio-fuel performance in internal combustion engine and relevant costs involvement. Renew Sustain Energy Rev 30(February):29–44. https://doi.org/10.1016/j.rser.2013.09.022

Scharlemann, Jörn PW, William F Laurance (2008) How green are biofuels? Science 319(5859):43–44. https://doi.org/10.1126/science.1153103

Streck C, Scholz SM (2006) The role of forests in global climate change: whence we come and where we go. Int Aff 82(5):861–879. https://doi.org/10.1111/j.1468-2346.2006.00575.x

Tran N-A, Padula MP, Evenhuis CR, Commault AS, Ralph PJ, Tamburic B (2016) Proteomic and biophysical analyses reveal a metabolic shift in nitrogen deprived nannochloropsis oculata. Algal Res 19(November):1–11. https://doi.org/10.1016/j.algal.2016.07.009

Van Der Horst, Dan, Saskia Vermeylen (2011) Spatial scale and social impacts of biofuel production. Biomass Bioenergy 35(6):2435–2443. https://doi.org/10.1016/j.biombioe.2010.11.029

Van Eijck, Janske, Henny Romijn (2008) Prospects for Jatropha biofuels in Tanzania: an analysis with strategic niche management. Energy Policy 36(1):311–325. https://doi.org/10.1016/j.enpol.2007.09.016

Yan J, Cheng R, Lin X, You S, Ke, Li, Rong H, Ma Y (2013) Overexpression of Acetyl-CoA synthetase increased the biomass and fatty acid proportion in microalga Schizochytrium. Appl Microbiol Biotechnol 97(5):1933–1939. https://doi.org/10.1007/s00253-012-4481-6

Yoshimitsu Y, Abe J, Harayama S (2018) Cas9-Guide RNA Ribonucleoprotein-Induced Genome Editing in the Industrial Green Alga Coccomyxa Sp. Strain KJ. Biotechnol Biofuels 11(1):326. https://doi.org/10.1186/s13068-018-1327-1

Zhang (2013) Determination of the exergy of four wheat straws. Am J Biochem Biotechnol 9(3):338– 47. https://doi.org/10.3844/ajbbsp.2013.338.347

Zhang J, Hao Q, Bai L, Xu, Jin, Yin W, Song L, Xu, Ling et al (2014) Overexpression of the Soybean transcription factor GmDof4 significantly enhances the lipid content of chlorella ellipsoidea. Biotechnol Biofuels 7(1):128. https://doi.org/10.1186/s13068-014-0128-4

Chapter 25
Advances and Future Prospective of Plant-Based Biofuels

R. V. Hemavathy, D. Kaviya, and K. Parinitha

Abstract The rising energy demand, combined with growing concern about environmental health, has resulted in a significant increase in global biofuel production. Plant-based biofuels are the most abundant renewable fuel source, capable of producing ethanol, butanol, and long-chain hydrocarbons from starch, cellulose, hemicellulose, and oils. First-generation biofuels are made from agricultural food crops such as corn, wheat, sugar beets, and oil seeds. Biodiesel and bioethanol are the primary first-generation biofuel products. Triglyceride-enriched biomass is transetherified to produce biodiesel. Biomass is fermented to produce alcohol. To reduce agricultural land utilization, lignocellulosic biomass has been identified as the feedstock for second-generation biofuels. However, the cost of plant-based biofuel is expensive, compared to oil prices. In addition to other approaches, like boosting plant polysaccharide content and overall biomass, genetically engineering plants to produce cellulases and hemicellulases and lessen the pretreatment procedures through lignin modification are promising ways to overcome the cost issue.

Keywords Biofuel · Plant · Lignocellulose · Sustainability · Bioethanol

25.1 Introduction

The world's population is expected to exceed 9 billion by 2050. Natural energy reserves have been under immense strain due to rising global prices and increased energy consumption, which has resulted in their depletion. The combustion of fossil fuels has a number of environmental consequences, including a rise in greenhouse gas (GHG) emissions, specifically carbon dioxide (CO_2). Global primary energy consumption has sharply expanded in recent decades as a result of rapid industrialization and higher standards of living. To maintain a reasonable quality of living, developing countries such as Brazil, the South Asian region, and South Africa require 12–24 gigajoules (GJ)/cap of energy per year. Around 98% of the energy used in the

R. V. Hemavathy (✉) · D. Kaviya · K. Parinitha
Department of Biotechnology, Rajalakshmi Engineering College, Chennai 602105, India
e-mail: hemavathy.rv@rajalakshmi.edu.in

M. Shah and D. Deka (eds.), *Emerging Sustainable Technologies for Biofuel Production*, Environmental Science and Engineering, https://doi.org/10.1007/978-3-031-52167-6_25

world today is produced by carbon emissions from fossil fuels, such as natural gas, oil, and coal, which account for over 80% of the world's energy supply. Throughout the twenty-first century, it is anticipated that droughts would last longer and be more intense, resulting in a five-fold reduction in water reserves. The surface temperature rise of 1.5 °C over pre-industrial levels will be stabilized with an increase in the proportion of renewable energy in the world's energy supply. Urbanization and industrialization have led to excessive fossil fuel use and an increase in the concentration of polluting gases. Because of the unequally large concentration of fossil fuels in some regions, there is a disparity in the availability of fuel among the countries, which places a financial burden on some nations by forcing them to import fuel to meet their energy needs. Alternative fuels are the most plausible option to fulfill the expanding need due to a lack of resources readily accessible, particularly water, and a growing demand for energy globally. To meet this demand, researchers have already identified alternatives. Furthermore, potential solutions to minimize the effects of climate change and lessen reliance on fossil fuels are urgently needed and are already in the works. Due to their potential to reduce GHG emissions and the accompanying climate change impact from transportation, biofuels are gaining interest on a national and international level as a low-carbon substitute for fossil fuels. As a result of their lower cost of synthesis, using bioenergy and biofuels is one of the potential renewable energy solutions (Neupane 2023).

The main factor causing climate change is the emission of greenhouse gases from the burning of fossil fuels. Its effects on the environment include heatwaves, floods, ozone depletion, air pollution, and global warming. Indeed, since humans are the main offenders, combating climate change has gained international attention. Diversifying their energy sources is one way that nations try to lessen the effects of climate change. Growing energy demands in the United States of America have revealed that energy resources are finite, and the quest for alternative energy sources has increased (Azni et al. 2023).

In contrast to the limited supply, geopolitical unpredictability, and negative global repercussions of fossil fuel energy, it is becoming increasingly obvious that biofuels may be an effective form of renewable energy The rapid depletion of fossil fuels increases the demand for sustainable and renewable fuels, including biofuel to be generated globally. All energy-enhanced substances produced purely biologically or obtained through chemical synthesis from the biomass of once-living organisms are collectively referred to as biofuels. Biofuels are mostly made from photosynthetic organisms such as vascular land plants, micro- and macroalgae, and photosynthetic microorganisms. The major products of biofuel can be in the form of gas, liquid, or solid. These products can be further transformed through physical, thermochemical, and biological processes (Rodionova et al. 2017).

Utilizing biofuels made from sustainable sources is a crucial technique for reducing the environmental effect of fossil fuels. The production of ethanol and butanol (as petrol additives), as well as long-chain hydrocarbons (for diesel additives or jet fuels) from starch, cellulose, hemicellulose, and oils, makes up the majority of renewable fuels. The sun, which will continue to be a reliable supply of energy for the next several billion years, is the origin of the energy that plants absorb. Unlike fossil petroleum and natural gas, which release their carbon from ancient fixed carbon sources, the carbon released during the burning of biofuels is continuously cycled (Hood 2016).

Bioethanol and biodiesel are the two most significant biofuels produced from biomass. Bioethanol is primarily produced by fermenting plant carbohydrates with yeast; it is made from sugar or the starch fraction of plants (such as sugar beetroot, sugar cane, corn, cassava, and wheat); it can also be made from lignocellulosic materials. Whereas, Microalgae, animal fats, and vegetable oils including palm, sunflower, soybean, and rapeseed are used to make biodiesel. In general, there are various feedstocks and conversion technologies that can be used in the production of biofuels. Despite the fact that fossil fuels are considered a non-sustainable resource with an inevitable depletion feature, environmental pollution is increasing day by day due to the widespread use of low-grade petroleum oil, which emits some amounts of SOx, COx, and NOx into the atmosphere. As a result, biofuel has seen significant increases in production in recent years, and this trend is expected to continue. This has led to alternative fuel and bio taking on a more significant role as the global energy supply for future life. Biofuel is the fuel of the future because of its favorable environmental characteristics (Bamerni 2018). However, intensive research is required to make this cost-effective. Innovations are being made in feedstock structure, processing simplicity, conversion effectiveness, co-product production, and sustainability. These findings can be integrated into new industrial applications as they come together (Hood 2016).

25.2 Evolution of Plant-Based Technologies

Biofuels, in addition to their positive carbon balance in comparison to fossil fuels, represent a significant potential for industrialized countries' sustainability and economic growth because they can be generated from locally available renewable material.

Biofuels are usually classified as follows:

1. First-generation biofuels, which are directly related to edible biomass.
2. Second-generation biofuels, which fuels made from a diverse range of feedstocks, ranging from lignocellulosic feedstocks to municipal solid waste.
3. Third-generation biofuels, that are currently associated with algal biomass but may be linked to the use of CO_2 as a feedstock to some extent (Lee and Lavoie 2013).
4. Fourth-generation biofuels, which involve genetically modified organisms (Siddiki et al. 2022).

25.2.1 *First Generation Biofuels*

First-generation biofuels refer to fuels produced directly from food crops, For example, biofuels made from biomass starch, sugar, vegetable oil, and fats. The structure of the biofuel remains identical across the different generations of biofuels, but the classification is based on the source from which the biofuel is derived. The most common feedstocks for the production of first-generation biofuels are wheat, maize, and sugarcane. Sugar crops such as sugar cane and sugar beet contain significant amounts of sucrose, which can be extracted and fermented into bioethanol. From an economic and environmental perspective, sugar cane is a preferred feedstock for ethanol production, but is limited to certain regions due to weather and soil conditions (Naqvi and Yan 2015).

25.2.1.1 Types of First-Generation Biofuels

Bio Alcohol

The fermentation of sugars and starches through the action of enzymes and microorganisms produces Bio alcohols. Bio alcohols mainly refer to bioethanol and to a lesser extent bio propanol, bio methanol, and biobutanol (also known as biofuel). Bio alcohols are an interesting choice for fuels because the chemical properties of the four alcohols can be used in ICEs based on their carbon content. Among the bio alcohols, bioethanol is the most widely used first-generation biofuel, especially in Brazil. Bioethanol is produced through the fermentation of wheat, sugar beet, molasses, sugar cane, corn, and other sugar- and starch-containing raw materials (Naqvi and Yan 2015).

Biodiesel

First-generation biodiesel may be produced from vegetable oils, soybeans, palm oil, sunflower oil, etc. by transesterification. Biodiesel is a liquid with a complementary composition to fossil diesel, but outside sulfur, and consists of fatty acid methyl esters (FAMEs). Biodiesel consists mainly of methyl or ethyl esters of fatty acids that come from herb and animal fats. First-generation biodiesel has an energy density of about 9% below petroleum-based diesel and demands modifications to diesel engines to prevent damage to rings, fuel lines, and rubber components. Biodiesel may be used neat or integrated with fossil diesel in any diesel engine without modification (Naqvi and Yan 2015).

Biogas

Anaerobic digestion of organic materials, such as manure, sewage sludge plant material, wastes, and crops, results in the production of first-generation biogas. The feedstock gets broken down during anaerobic digestion without the use of oxygen. Biogas is a general term because the gas is made up of different chemicals (Naqvi and Yan 2015).

Bio Ether

Bio ether is another first-generation biofuel that can be produced from wheat, waste glycerol, and sugar beet. Bio ether has a potential to replace petroleum-based ether, which is used as an additive to conventional fossil fuels. As bio ether has low energy density, it is unlikely to utilize bio ether as 100% fuel (Naqvi and Yan 2015).

25.2.1.2 First-Generation Feedstock to Biofuels

Among various first-generation biofuels, large quantities of bioethanol, biodiesel, and biogas have been produced commercially worldwide and the production processes are considered as "established technologies." Bioethanol can be used as a substitute for gasoline and is derived from sugar and starch through fermentation. Biodiesel is an alternative to fossil diesel and can be produced through the transesterification of vegetable oils, residual oils, and fats. Whereas, biogas can be used in gasoline engines after slight modification and is produced through anaerobic digestion of organic substrates such as liquid manure and food wastes (Naqvi and Yan 2015).

25.2.1.3 Processes for Converting First-Gen Biofuels

The detailed flow chart on conversion processes for first-generation biofuels, including physical, chemical, biological, and thermal processes or combinations of these processes with their respective products, is shown in Fig. 25.1 (Naik et al. 2010).

25.2.1.4 Impacts of First-Generation Biofuels

There are several significant impacts associated with the production and use of first-generation biofuels. These include the emission of greenhouse gases, atmospheric pollution, water overconsumption and pollution, deforestation, loss of biodiversity, impacts on rural development and food security, effects on energy security, impacts on public health, and the potential for social conflicts. The impacts of biofuels depend

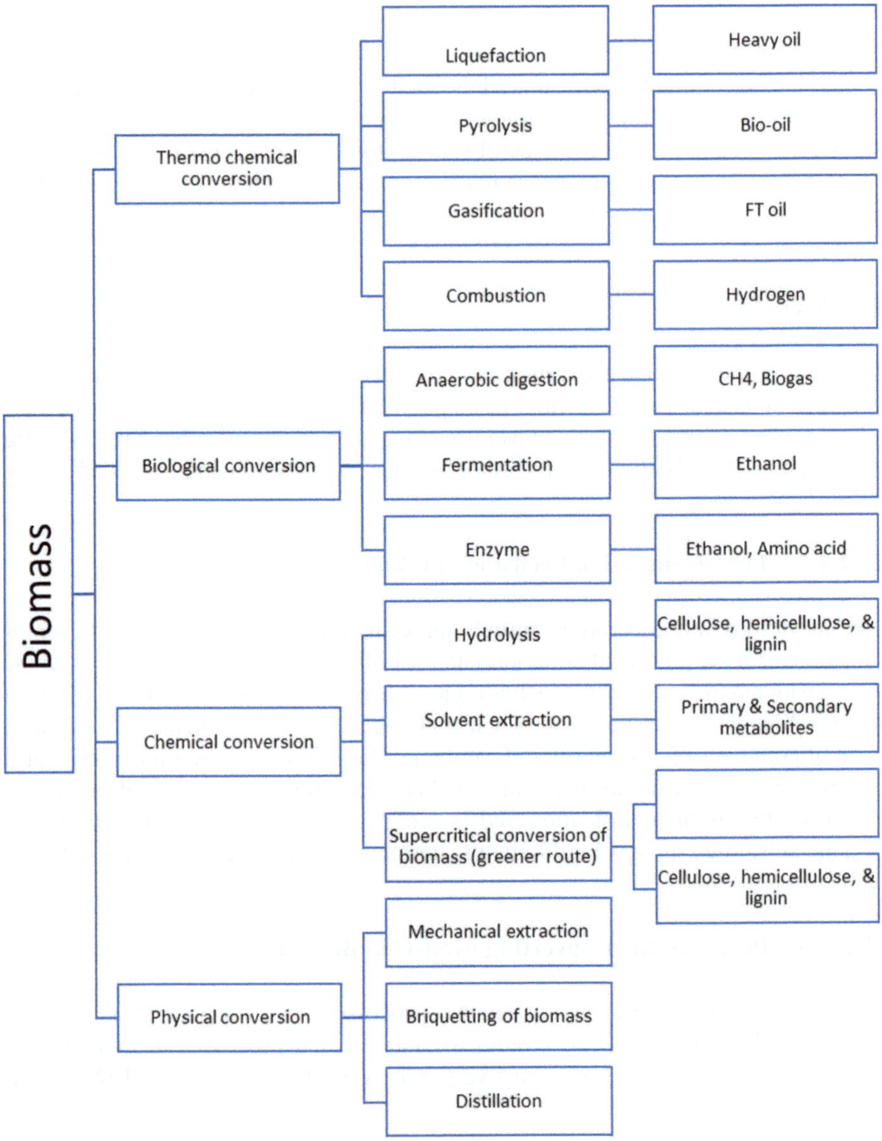

Fig. 25.1 Different methods for Conversion of biomass into biofuel (Naik et al. 2010)

on various factors such as feedstock, agricultural practices, production methods, environmental and socioeconomic contexts, and policies. While trade-offs are unavoidable, careful planning can mitigate some of the negative impacts associated with biofuels. Large-scale sugarcane bioethanol production in Brazil has different drivers and environmental and socioeconomic impacts compared to small-scale jatropha biodiesel production for rural electrification in Sub-Saharan Africa. Moreover, the impacts and potential of the same biofuel practice can significantly vary across different regions of the world (Gasparatos et al. 2013).

It is critical to avoid generalizing the impacts and sustainability of biofuels without taking into account the specific environmental and socioeconomic contexts in which they are produced. Understanding these contexts is crucial for assessing the trade-offs associated with biofuels and their acceptability by various stakeholders. However, the multidisciplinary and constantly expanding biofuel literature poses a significant challenge in obtaining a comprehensive understanding of these trade-offs. Despite the knowledge that trade-offs, careful planning can help mitigate negative impacts. Focusing only on a subset of biofuel impacts can result in inappropriate resource use decisions and support narrow economic interests. Therefore, a comprehensive assessment of biofuel impacts is necessary to ensure responsible resource use (Gasparatos et al. 2013).

25.2.2 Second Generation Biofuels

First-generation biofuels have been criticized because they avoid relatively few greenhouse gas emissions, because their production is unsustainable in terms of deforestation, water use, and land management, because they compete with food, driving up food prices, and because they need generous government support programs to remain competitive even when the technologies are mature. As a result, much hope has been placed in 2nd generation biofuels. Generation. To the extent that these are based on crop and forest residues or high-yielding non-food energy crops grown specifically for feedstock production, they are considered to be more sustainably produced than some 1st generation fuels (Sims et al. 2010).

Current research efforts are focused on lignocellulosic biomass from sources that do not compete with food crops: For example, agricultural waste products such as sugarcane bagasse, wheat straw, rice stalks, cotton linters, and forest thinning, or novel crops that can be grown in environments too marginal for food production, such as switchgrass and eucalyptus trees (Doherty et al. 2011).

Biofuels produced from lignocellulosic biomass are known as Second-generation biofuels. Lignocellulose is highly represented in plant biomass compared to oils and starch; it is the main component of cell walls. Lignocellulose consists of three components: Cellulose (40–50%), Lignin (15–20%), and Hemicellulose (25–35%). All of these components have some value to humans, but extraction of any particular component, especially cellulose, is quite difficult. Lignin is an amorphous polymer consisting of phenylpropane structures. It gives cellulose rigidity and stability to many hydrolytic enzymes. Lignin can be removed by dissolving it in alkaline-alcoholic solutions. Lignin by itself can be burned to obtain heat or electricity; it is also possible to obtain useful chemical compounds from it by chemical treatment.

Hemicellulose is an amorphous polymer containing several different types of sugars: D-xylose, L-arabinose, D-galactose, D-glucose, and D-mannose. It is possible to hydrolyze hemicellulose to obtain monomeric sugars after they have been fermented into alcohol. Cellulose is a straight glucose polymer with a certain rigidity that makes its hydrolysis difficult (Voloshin et al. 2016).

25.2.2.1 Biofuel Production Routes

Biofuels can be produced from lignocellulosic feedstocks via two very different processing routes.

Biochemical routes**:** Biochemical routes are where enzymes and other microorganisms are used to convert the cellulosic and hemicellulosic components of raw materials into sugars before they are fermented to produce ethanol (Sims et al. 2010). Pretreatment is an important process that makes the cellulose accessible to enzymes that convert complex hydrocarbon polymers into fermentable sugars (see Figure). The aim is to break down the lignin polymer and destroy the crystal structure of the cellulose. In many current biomass treatments (pretreatment with ethanol-organic solution, pretreatment with dilute acid, pretreatment with ammonia recovery, pretreatment with filtration), the different components of the biomass (lignin, hemicellulose, and cellulose) are fractionated into separate process streams. The removal of lignin and/or hemicellulose can significantly reduce the resistance of the biomass to enzymatic hydrolysis (Balan et al. 2009).

Thermochemical routes: Thermochemical routes are (also known as biomass to liquids BTL) where pyrolysis/gasification technology produces syngas (CO and H_2) that can be reformed into synthetic diesel jet fuel or long carbon chain biofuels such as ethanol based on the Fischer–Tropsch transformation (Sims et al. 2010).

The advantage of biomass enriched with lignocellulose is the efficiency in Land usage. The impact on the food sector is lower, at the expense of using non-food crops (Voloshin et al. 2016a). However, the full commercialization of biochemical or thermochemical conversion processes to produce 2nd generation biofuels still appears to be some years away. This is despite decades of research and development and recent investments in several pilot and demonstration plants in the US, Europe, and elsewhere. Even with generous government subsidies, commercial risks remain high, especially as recent wildly fluctuating oil prices and global financial turmoil add to investment uncertainty. Unless there is a technical breakthrough by biochemical or thermochemical means that significantly lowers production costs and accelerates investment and deployment, the successful commercialization of 2nd generation bioethanol and other biofuels is likely to take another ten years or so. During this period, demonstration and industrial-scale 2nd generation plants will be continuously improved to make the products competitive with both petroleum fuels and first-generation biofuels.

The main focus for the use of biofuels must be on the aviation, shipping, and heavy transport sectors, for which there will be limited alternatives. From around 2020 onwards, 2nd generation bioethanol could play a much more important role in a global biofuel market characterized by a balance between 1st and 2nd generation technologies. Policies aimed at rewarding the environmental performance and sustainability of biofuels and encouraging the provision of a more abundant and geographically extended feedstock supply could lead to 2nd generation products replacing 2nd generation alternatives in the medium to long term. Overall, it is assumed that 2nd generation technologies for the production of liquid biofuels for transport will not be able to compete with petroleum products in the near future as long as the oil price does not remain permanently above 100 USD/barrel. Therefore, the potential of biofuels should be considered in the long term, but without delaying the necessary investments needed to bring these technologies closer to the market. International cooperation is crucial and cooperation through international organizations should be strengthened, with both the public and private sectors playing an active role to support the 2nd generation biofuel industry (Sims et al. 2010).

The Fig. 25.2 process flow diagram shows how lignocellulosic feedstocks are converted to ethanol and byproducts. Processing steps are presented in grayscale while inputs and outputs are presented in white.

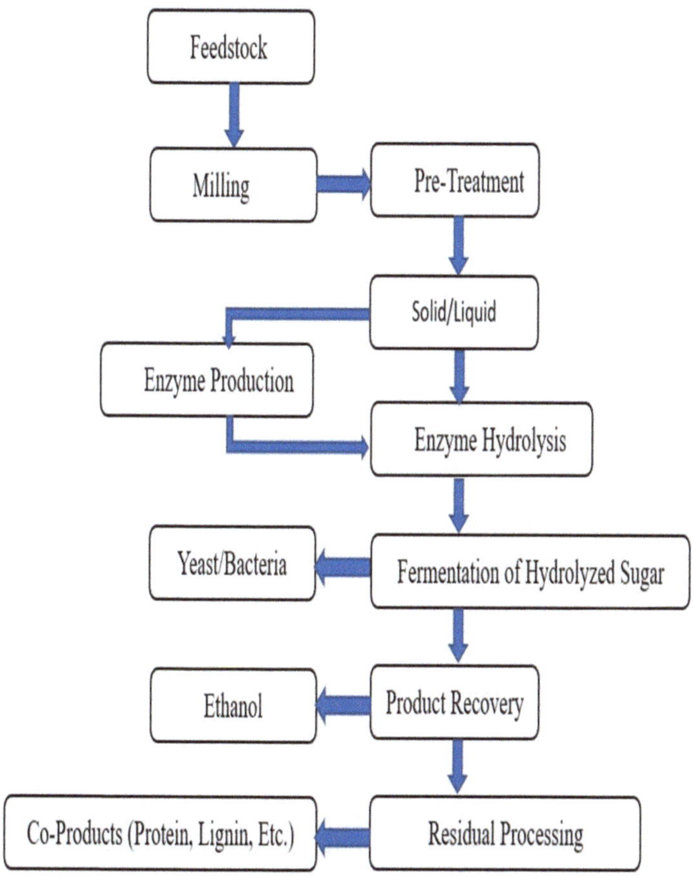

Fig. 25.2 Conversion of lignocellulosic feedstocks to bioethanol and by-products (Balan et al. 2009)

25.2.3 Third-Generation Biofuel

One of the oldest known living species on Earth is the microalgae. They can double their biomass in less than a day and grow at a rate 100 times quicker than terrestrial plants. In addition, some strains of microalgae have the capacity to store huge amounts of lipids inside their cells, where the lipid can be turned to biodiesel. A recent study claims that a reasonable estimate for the annual production of microalgae biomass is between 15 and 25 tonnes/ha. Assuming a 30% lipid content in microalgae cells, this translates to a lipid production rate of 4.5–7.5 tonnes/ha/year (without optimizing the growth conditions). Compared to the production of oil from soybean (0.4 tonne/ha/year), rapeseed (0.68 tonne/ha/year), oil palm (3.62 tonne/ha/year), and jatropha (4.14 tonne/ha/year), this amount is higher. In other words, growing microalgae for biodiesel contains a significant feature for efficient land use and requires the least amount of land (Lam and Lee 2012).

In addition, microalgae are an excellent feedstock for the synthesis of bioethanol. Some microalgae contain carbohydrates (typically not cellulose) that can be used as a carbon source or substrate for fermentation in addition to their high lipid content. Even though there is still little research on the fermentation of microalgae biomass to bioethanol in the literature, this process has many benefits. Furthermore, microalgae have the ability to fix CO_2 from the atmosphere, flue gases, or soluble carbonate into their cells during growth while also absorbing solar energy with efficiency 10–50 times greater than terrestrial plants; this presents a golden opportunity for carbon credit programs. Microalgae have successfully positioned themselves as one of the most promising biofuel feedstocks based on these evidences (Lam and Lee 2012).

Nitrogen (often in the form of nitrate) and phosphorus (generally in the form of ortho-phosphate) are the two main nutrients needed when growing microalgae on an industrial scale for the manufacture of biofuels. These nutrients typically come from chemical or organic fertilizers that are utilized to boost microalgae's growth rate and produce a bulk amount of biomass (Lam and Lee 2012).

Microalgae-based biofuels are among the third generation of biofuels, which completely expand the field of renewable energy. However, it should be noted that there are other feedstocks that can be used to produce third generation biofuels besides microalgae. Other third generation biofuels include bioethanol produced through direct cellulose fermentation and biodiesel made from yeast and fungi (consolidated bioprocessing) (Lam and Lee 2012).

Simple photosynthetic organisms known as microalgae thrive in aquatic environments. Microalgae don't have proper roots, stalks, leaves, or embryos like aquatic plants do. Through photosynthesis, they can generate a large amount of algal biomass using only CO2, water, and sunlight. In flow-through ponds, closed ponds, photobioreactors, fermenters, and sea systems, microalgae can be manually produced. Although there are more than 50,000 different species of microalgae, only about 30,000 have been thoroughly examined. There are several clear benefits to using microalgae as a feedstock for biofuels. First, the photosynthesis and growth rates of microalgae are very high. Microalgae can produce more biomass and fat per unit of cultivated area than terrestrial plants can. The fastest-growing terrestrial plant, switchgrass, is predicted to be outdone by microalgae in terms of biomass production by 50 times. During exponential growth, biomass can double in just 3.5 h, and the average productivity in ponds can reach 20–22 g dry weight–m–2–day-1. Second, because water can be recycled into the cultivation system for microalgae to grow, it uses less water than land crops do. Third, the algae biofuel sector requires little space since it can be grown on unproductive soils such as dry or semi-arid soils, saline soils, infertile croplands, contaminated soils, and other soils (such as deserts) with poor economic value. Given that it does not compete with food crops for arable land, this makes a lot of sense. Fourth, any wastewater can be used as a source of nutrients. This is a novel wastewater treatment technique since waste water's inorganic and organic content can both be broken down physically and by microbial activity. Fifth, since some species of microalgae can withstand CO_2, NOx, SOx, dust, and other elements in the gas, flue gas (which is high in CO_2) can be used as a carbon source for microalgae. Sixth, the process of producing biofuel is sped up by the ease with which microalgae may be collected and pretreated. Seventh, the refining of biofuels allows for the co-production of numerous economical compounds. For instance, more than 400,000 tonnes of glycerol and 1 billion gallons of algae biodiesel may both be produced simultaneously. Eighth, the amount of biomass and algal oil content can be significantly increased by using technological techniques to modify and mutate the algal cell genes. Finally, a single biomass source can be used to produce a variety of biofuels, including biodiesel, bioethanol, and biogas (Zhu et al. 2012).

25.2.3.1 Algal Biomass Production

Ongoing research has shown how impressively algal biomass can be produced on a wide scale using wastewater for biofuel and other purposes. However, wastewater-based green growth development is vulnerable to a number of threats and challenges, such as a wide range of wastewater synthesis due to source, foundation, weather patterns, pre-treatment strategies, inappropriate supplement proportions (such as C/N and N/P), high turbidity due to the presence of shades and suspended strong particles that affect light transmission, the presence of competing microflora and harmful mixtures. There are numerous methods for growing microalgae. However, there are two commonly used development frameworks: (1) suspended societies, which include open lakes and closed reactors, and (2) immobilized societies, which include biofilms and matrix immobilized frameworks. High-rate algal lakes or raceway lakes are, practically speaking, the most well-known large-scale creation frameworks. Raceway lakes are shallow and open, and they include paddle wheels to facilitate the flow of nutrients and green growth. Raceways are often easy to build and operate, although they usually perform poorly for a variety of reasons. The primary type of closed frameworks used for the industrial-scale production of green growth I cylindrical photobioreactors. The photoreactor framework can also go by the names vertical, level, or even, and helical photoreactors. The least labor-intensive method of increasing production is believed to be the helical photoreactor. Round photobioreactors can provide superior pH and temperature control, greater protection against culture contamination, better mixing, less evaporative misfortune, and higher cell densities when compared to open lakes. Every framework, however, has proportional advantages and disadvantages. Utilizing raceways and cylindrical photobioreactors has some significant drawbacks, including biomass recuperation. Algal cycles that have been attached or immobilized societies have somewhat mitigated this problem. Algal biofilms may play a crucial role in overcoming the significant challenges associated with the cultivation and harvesting of microalgae. Green-growth biofilm development can be higher than suspended development if a suitable surface area is provided (Alam et al. 2015).

Algae are currently harvested using a range of methods, including chemical, mechanical, biological, and, to a lesser extent, electrical processes. However, various combinations or orders of these techniques are also frequently used. Algae contain incredibly tiny cells. In order to lower the particle size of the algae, chemical-based flocculation is usually used before applying another technique, such as flotation, to harvest the algae. Centrifugation is the most dependable and efficient method for mechanically recovering suspended algae. In an electric-based method, the negative charge properties of algal cells are used to separate the cells. These cells can be concentrated by the movement of an electric field. There are several processes that can be used to turn microalgae biomass into energy sources, including (a) biochemical conversion; (b) chemical reaction; (c) direct burning; and (d) thermochemical conversion. Other compounds with valuable applications in various industrial sectors may also be extracted from microalgae, in addition to producing biofuel, depending on the species. Biofuel generation is a difficult process. These procedures, however, are time-consuming, expensive, and technologically difficult (Alam et al. 2015).

The Table 25.2 lists the advantages and disadvantages of different types of photobioreactors used for cultivation of microalgae for biofuel production.

25.2.4 Fourth-Generation Biofuels

Fourth-generation biofuels mix genetically designed feedstock with genomically synthesized microorganisms such as cyanobacteria to efficiently generate bioenergy. Similar to third-generation biofuels, they are produced on non-arable land. In order to create biofuel from photosynthetic organisms, the procedure makes use of bioengineering techniques to alter the metabolism and characteristics of algae. They are designed to capture and store carbon dioxide (CO_2) while simultaneously producing bioenergy. The fourth-generation biofuel production is carbon negative due to the geo-sequestered CO_2 that is stored in depleted oil and gas fields, mineral storage (as carbonates), and saline aquifers. This reduces the number of stages between the sun's energy production and its conversion into biofuel, reducing the expense of fermentation and other processing steps. As this is a new field, extensive research

Table 25.1 Overview of 4 generations of biofuels (Modified from (Siddiki et al. 2022))

First generation biofuel	Second generation biofuel	Third generation biofuel	Fourth generation biofuel
Sources are food crops based	Sources are Non-food crops based	Microalgae based source	Synthetic biology-based source (Genetically modified organisms are used)
Poses adverse effect on food security	No impact on food security	No impact on food security	No impact on food security
Monoculture of feedstocks results in loss of biodiversity	No impact on biodiversity	No impact on biodiversity	No impact on biodiversity
Requires cultivable land for feedstock generation	Cultivable lands or forests needed	Non-arable land such as waste lagoons, sea etc., can be utilized	Non-arable land can be utilized
Brackish and saline water is not suitable for feedstock growth	Brackish and saline water is not suitable for feedstock growth	Brackish and saline water can be used for feedstock growth and thus have little impact on the water footprint	Brackish and saline water can be used for feedstock growth
Negative impact on the environment due to pesticide and fertilizer use	Less dependent on fertilizers but deforestation is the concern	Environmental friendly but marine eutrophication is the concern	Environmental friendly but the Release of genetically modified strains is the concern
Feedstock growth is largely dependent on environmental factors such as Temperature, humidity, precipitation etc.,	Feedstock growth is dependent on environmental factors	Feedstock can be grown in rough conditions	Feedstock can be grown in rough conditions
CapEx and OpEx are relatively small	CapEx and OpEx are fairly small	CapEx and OpEx are high	CapEx and OpEx are high
High energy efficiency due to wide and existing commercial production	Commercial production exists	Commercial production is not yet in place; extensive research is ongoing	This technology is still in its developmental stage

is being done to find novel strategies for the environmentally friendly conversion of energy to fuel (Ale et al. 2019).

The goal of genetic engineering is to increase algal biofuel output. However, disposing of GMOs is one of the main issues with their use. To reduce the possibility of lateral gene transmission, the applicable disposal methods should destroy the microorganism and the genetic component. In microorganisms, cell lysis typically results in DNA release. Even after an organism dies, its mechanisms remain active. Chromosome or plasmid DNA may be released as a result of this occurrence. There

Table 25.2 Comparison of advantages and disadvantages of different culture systems (Voloshin et al. 2016)

Culture systems	Advantages	Disadvantages
Open-air systems	Relatively economical, easy to clean up, easy maintenance, utilization of non-agricultural land, low energy inputs	Little control of culture conditions, poor mixing, light and CO2 utilization, difficult to grow algal cultures for long periods, poor productivity, limited to few strains, cultures are easily contaminated
Tubular photobioreactors	Relatively cheap, large illumination surface area, suitable for outdoor cultures, good biomass productivities	Gradients of pH, dissolved oxygen and CO2 along the tubes, some degree of wall growth, requires large land space, photoinhibition
Flat photobioreactors	Relatively cheap, easy to clean up, large illumination surface area, suitable for outdoor cultures, low power consumption, good biomass productivities, good light path, readily tempered, low oxygen build-up, shortest oxygen path	Difficult scale-up, difficult temperature control, some degree of wall growth, hydrodynamic stress to some algal strains, low photosynthetic efficiency
Column photobioreactors	Low energy consumption, readily tempered, high mass transfer, good mixing, best exposure to light/dark cycles, low shear stress, easy to sterilize, reduced photoinhibition, reduced photo-oxidation, high photosynthetic efficiency	Small illumination surface area, sophisticated construction materials, shear stress to algal cultures, a decrease of illumination surface area upon scale-up

are tight rules for the disposal of these products because the intentional or unintentional release of chromosomal or plasmid DNA at specific amounts could lead to transformation and horizontal gene transfer. By replacing hazardous byproducts with more environmentally acceptable alternatives, waste reduction practices strive to reduce the negative environmental effects of GM residue. Several methods (such as waste separation and concentration, waste reduction, energy/material recovery, waste exchange, incineration/treatment, and secure land disposal) could be used to reduce waste. One of the safest methods for getting rid of GM organisms is composting. To avoid cross-pollination with wild species in neighboring ecosystems, the by-products of genetically altered biomass must be decontaminated. Decontaminating GM algae requires a simultaneous breakdown of the organism and the genetic material. In genetically engineered organisms, temperature and pH play key roles in DNA deterioration and slowing the pace of horizontal transfer. Because of the large concentrations of bacteria in active compose and subsequent increases in nuclease concentration, DNA degradation in compost is greater than in soil. For the decomposition of GMO strains, open-turned windrow systems and enclosed bioreactor systems are typically used. It is advised to utilize an enclosed bioreactor since it has a substantially shorter composting period and guarantees that the materials are composted in the best possible way. Concerns about horizontal gene transfer can be allayed by digesting the lipid, carbohydrate, and protein components of algal biomass and reusing the leftovers. Burning dried waste products from the biofuel extraction process has been suggested as a way to recover energy and CO_2. A gas diffuser was used to incorporate the CO_2 produced during the combustion process into the culture media. The operating energy required for the process can be covered by the energy produced by the combustion of the dried residue. Therefore, the suggested method has three benefits: energy recovery, CO_2 recovery, and reuse, and waste reduction (Abdullah et al. 2019).

25.2.4.1 Genetic Engineering for Enhanced Plant-Based Biofuel Production

The presence of different kinds of sugars frequently makes the microbial conversion of carbohydrates to fuel molecules challenging. In order to create the optimal biofuel feedstocks, it is currently being researched to increase the cellulose content due to the difficulty of digesting mixed sugars. Several different cellulose synthases from a large family are involved in the biosynthesis of cellulose (Endler and Persson 2011). Manipulating these enzymes can result in a higher cellulose concentration. The expression of the mutant *Arabidopsis* cellulose synthase gene *CES3A* in tobacco led to more effective enzymatic saccharification of cellulose. More studies on the alteration of cellulose biosynthesis will be necessary to provide possibilities for improved cellulose biosynthesis without reducing biomass production (Furtado et al. 2014).

Increases in cellulose as a result of carbon being diverted from other important cellwall components like lignin and xylan have frequently been documented. The targeted ecrease of xylans has also been investigated as a method to make a feedstock more favorable for the generation of biofuel. Following a change in the spatial and temporal distribution of xylan from the secondary cell wall to the xylem vessels, the xylose content of *irx7, irx8,* and *irx9* mutant *Arabidopsis* plants was subtly decreased by up to 23%. These mutants are known to have lower xylose concentrations than wild-type levels, but their growth is limited, which lowers their cellulose content. The transgenic *Arabidopsis*'s saccharification sugar yields were dramatically increased by up to 42% after a hot water pretreatment (Furtado et al. 2014).

RNA interference (RNAi) of *PoGT47C*, a glycosyltransferase, caused hemicellulose disruption in poplar, which led to a considerable decrease in secondary cell wall thickness and an increase in enzymatic saccharification and glucose yield after pretreatment. Following a 1 N KOH extraction, the amount of xylose in the supernatant was reduced to 43%-77% of the wild type, whereas the amount of xylose still present in the plant cell wall was reduced to 54%-79% of the wild type. In the transgenic trees, there was a slight increase in the amounts of extractable S and G lignin. The amount of glucose remaining in the plant cell wall was 56%-87% lower than in the wild type, which is consistent with prior research that found impairments in cellulose production with glucuronoxylan down-regulation. Transgenic trees had up to a 48% increase in glucose concentration as compared to wild-type poplar. These tests highlight the difficulty of changing cell wall structure without causing downstream pleiotropic growth consequences (Furtado et al. 2014).

The exogenous addition of *Arabidopsis RabG3bCA* to poplar resulted in a 9.8% rise in glucan content without an increase in other biomass constituents. The increased glucan concentration resulted in higher enzymatic hydrolysis yields. Poplar polysaccharide content rose after an endoglucanase cell from *Arabidopsis thaliana* was over-expressed. There was a 10.8% rise in cellulose and a 5.7% increase in non-cellulosic polysaccharide content, but no change in lignin (Furtado et al. 2014).

Lignin biosynthesis involves several enzymes including phenylalanine ammonia-lyase (PAL), cinnamate 4-hydroxylase (C4H), coumarate 3-hydroxylase (C3H), caffeic acid/5-hydroxyferulic acid O-methyltransferase (COMT), ferulate 5-hydroxylase (F5H), 4-hydroxycinnamate CoA ligase (4HCL), cinnamoyl-CoA reductase (CCR), caffeoyl-CoA O-methyltransferase (CCoAOMT) and cinnamyl alcohol dehydrogenase (CAD). Depending on the commitment of the enzymes to the lignin biosynthesis pathway, manipulation of these enzymes may result in changes in lignin production and other pathways that may alter plant growth and development. COMT and CAD are terminal enzymes in the pathway, and changing them has little or no influence on plant growth and development, whereas changing CCR, CCoAOMT, and others may have pleiotropic effects. The quality and composition of the lignin in plant cell walls have been reported to be affected by CAD. Down-regulation of lignin synthesis enzymes such as CAD (to modify lignin content) and COMT (to modify lignin composition and content), reduction of p-coumarate esters and ferulate ethers, reduction of S and G lignin deposition, and increase of phenylpropane units or aldehydes in lignin (Furtado et al. 2014).

Some progress has been achieved in the alteration of the lignin components of grass biomass in biofuel species. In sugarcane, for example, the lignin pathway contains ten different enzymes, including those involved in hydroxylation, methylation, and side-chain reduction of monolignol precursors. COMT alone involves at least 31 distinct expressed sequence tags (ESTs) in the complex sugarcane genome, some of which may be homologous genes/alleles. 14 CAD-like genes have been found in sorghum at seven different genomic regions. Reduced expression of lignin production enzymes was first achieved using antisense oligonucleotides, although RNAi technology has lately become more widely used. The lignin content of sugarcane was decreased by 3.9 to 13.7% by using RNAi suppression to down-regulate COMT by 67-97%. In maize, down-regulating COMT improved the biomass's digestibility, and in switchgrass, down-regulating CAD caused either an overall reduction in lignin or an altered structure. The composition of lignin monomers (S, G, and H lignin) and their ratios have an impact on conversion efficiency. The lignin S/G ratio was decreased from 1.47 in the wild type to values ranging between 1.27 and 0.79 using RNAi suppression of COMT in sugarcane, which increased the sugar yield by up to 29% without pretreatment (Furtado et al. 2014). The S/G ratio was decreased and the ethanol yield rose by 38% in switchgrass as a result of down-regulating COMT (Fu et al. 2012).

The expression of *PvMYB4*, a crucial transcript factor from *Panicum virgatum*, can be altered to reduce the expression of the genes involved in lignin production. *PvMYB4* overexpression in switchgrass increased the wall's phenolic content, decreased recalcitrance in transgenic biomass, and decreased lignin concentration by 40–70%. Several grass COMT and CAD mutants have been reported, including sugarcane mutants with altered lignin content and maize bm1 mutants with altered CAD enzyme, brown midrib sorghum, and a COMT gene in maize with altered lignin subunit composition (Furtado et al. 2014).

Coniferyl ferulate was incorporated into the maize lignin biosynthetic pathway, which resulted in increased yields of enzymatic saccharification sugar and greater degradability when alkaline pretreatments were used. Coniferyl ferulate was added to replace some of the coniferyl alcohol precursor. The proportion of alkali-soluble lignin increased from 12 to 18% as the coniferyl ferulate concentration rose from 20 to 60%. It was reasoned that the more biodegradable lignin should improve the removal of pectin and hemicellulose. After initial drops in lignin concentration, carbohydrate yields after enzymatic saccharification dramatically increased, but finally plateaued regardless of further lignin reduction, showing that lignin was not the only factor influencing cell wall degradation (Furtado et al. 2014).

The level of lignin polymerization was decreased as a result of the expression of a bacterial 4-hydroxycinnamoyl-CoA hydratase/lyase (4HCHL) in Arabidopsis. Hydroxybenzaldehyde and hydroxybenzoate moieties accumulated as a result of the cleavage of the propanoid side chain of the precursors of hydroxycinnamic lignin. This led to a reduction in the molecular weight of lignin and an improvement in saccharification yields (Furtado et al. 2014).

Reduced lignin content was caused by the down-regulation of 4-coumarate: CoA ligase (Pt4CL1) in aspen (*Populus tremuloides*). In order to decrease the amount of lignin and increase the release of sugar in grass or woody species, it may be beneficial to target this enzyme in addition to HCT and CH3. Monolignol ferulate transferase can be utilized to add chemically labile links to the lignin backbone, according to a study (Furtado et al. 2014).

Recent research has demonstrated that altering Mediator, a transcriptional regulatory complex, can change the lignin in Arabidopsis. Plants with typical growth patterns were produced, but they lacked the guaiacyl or syringyl lignin subunits and had a higher potential for saccharification. This implies that lignin may be subject to radical alterations, and these modifications may not necessarily have an impact on plant performance (Furtado et al. 2014).

The cost of converting biomass to biofuel can be decreased by in planta production for producing the enzymes needed to convert polysaccharides in biomass to sugars. Enzymes can be produced at a low cost in specialized enzyme crops. The expression of in planta enzymes in grasses is still in its infancy. However, sugarcane, switchgrass, miscanthus, and eriathus are examples of high biomass-yielding biofuel plants that could serve as potential targets for enzyme expression. To separate the enzymes produced from their substrates, recombinant protein enzymes may be directed to organelles such as chloroplasts, vacuoles, and the endoplasmic reticulum.

The expression of enzymes in leaves could be exploited in sugarcane due to its well-established transformation procedures via Agrobacterium-mediated transformation to break down the bagasse for ethanol. In genetically modified sugarcane leaves, exoglucanases (CBH I and CBH II) and EGs have been overexpressed by employing a maize *PepC* promoter to express at 0.05% of total soluble proteins (EG, in the chloroplast), and lesser levels of CBH I and CBH II without changing the phenotype of the plant (Furtado et al. 2014).

It has been demonstrated that the expression of esterases, such as ferulic acid esterase, enhances cell wall digestibility. Ferulic acid esterase was expressed in Festuca arundinacea, resulting in lignin-polysaccharide ester bond breaking and increased digestibility and enzymatically released monomeric sugars. The production of several enzymes has been engineered into tobacco plants. Endoglucanase, exoglucanase, lipase, pectate lyases, cutinase, endoglucanases, swollenin, xylanase, acetyl xylan esterase, or beta-glucosidase from fungi or bacteria were expressed in tobacco chloroplasts at high levels, producing enzyme preparations that were extremely effective in the biomass conversion process. The expression of chimeric enzymes, which combine domains for various enzyme actions in a single protein, has proved successful in tobacco (Furtado et al. 2014).

The majority of plants with perfected transformation systems, including Arabidopsis, tobacco, potato, maize, and rice, are used for PCDE in-planta expression. the levels of accumulation of enzymes in the plants were as high as 26% of total soluble protein when Enzymes such as EG (E1, E2, Cel6A, Egl), exoglucanase (Cel6B, CelA/Cel6G, CBH I, E3), xylanase (XynA, XyIII, XynII, xynB, xynZ) were expressed in Arabidopsis, tobacco, potato, rice, corn, alfalfa, and barley As the plant serves as both the host and the substrate for sugar conversion, in-planta expression permits the autodegradation of the plant biomass, which lowers the cost of enzymatic pretreatment (Furtado et al. 2014).

Direct alteration of composition with the use of transgenic techniques will be possible by changing particular genes. However, the genomic analysis will aid in the identification of loci for conventional selection or targeted mutation in plant improvement. Advances in the analysis of the genomes of major bioenergy species will hasten these advancements. The creation of selection strategies to target desired changes in cell wall composition will be facilitated by a thorough understanding of the genes involved in cell wall biosynthesis in each species. Technology advancements in targeted plant mutagenesis provide opportunities for significant changes in biomass composition (Furtado et al. 2014).

Synthetic biology developments in plants should enhance the accuracy and effectiveness of genetic engineering for sustainability. Advances in synthetic biology including genome editing, gene circuit design, synthetic promoter development, gene stacking technologies, and the development of environmental sensors can be applied for the plant biofuel production process. Additionally, recent developments in developing spatially resolved and single-cell omics support the identification and characterization of cell-type-specific mechanisms as well as spatiotemporal gene controls in various plant tissues for the expression of cell- and tissue-specific genes, leading to enhanced biofuel production (Yang et al. 2022).

25.3 Limitations, Advances and Future Prospects of Plant-Based Biofuel

Despite the potential of lignocellulosic biomass, the poor photosynthesis efficiency of terrestrial land plants still places a restriction on the solar-to-molecule energy efficiency for biomass-based products. Despite a higher anticipated potential, most land plants have photosynthetic rates of less than 1%, which significantly reduces the efficiency of converting solar energy to end products. Considering the relatively greater photosynthetic productivity and efficiency, algae biofuels and bioproducts have been suggested to address these issues. Tan focuses on the potential of algal biomass and bioproducts as a method of carbon sequestration, while Ma focuses on the synthetic biology design of cyanobacteria to convert CO_2 to a wide range of products. Algal bioproduction has the potential to support sustainability, but due to several intrinsic difficulties such as low light penetration, diluted cell concentrations, and expensive harvesting, algal biofuels, and commodities items are not currently commercially feasible. Some of these issues have recently been solved by advances in artificial intelligence-guided process optimization, although commercial platforms have not been created. Recent developments seek to overcome this obstacle by separating bioproduction from light-driven carbon fixation. The reduction of CO_2 to intermediates like methanol, ethanol, or acetate, which can then be transformed into end products, has been proposed for both electrocatalysis and thermo-catalysis. Recent research has shown that mixing the electrochemical CO_2 reduction process with downstream bioconversion to create more biocompatible ethanol and acetate can result in a solar-to-biomass conversion rate of 4.6%, which is significantly greater than any photosynthetic systems (Yuan et al. 2022).

For the past 20 years, cane sugar and corn starch have been the main raw materials used to produce ethanol. In 2014, the USA produced 14 billion gallons of ethanol from corn starch. The Fig. 25.3 shows that corn-based ethanol production has reached a plateau. Since the remaining corn is used for human food and animal feed, the current corn crop in the USA, which makes up around 40% of the total, is not anticipated to grow any further. In 2014, Brazil generated 7.3 billion gallons of ethanol from cane sugar. More than 90% of the ethanol consumed worldwide is produced by Brazil and the USA together.

In order to produce biodiesel commercially, feedstock oils or fats are typically treated with alcohols and chemical catalysts like salt or potassium hydroxide. Commercial success for this approach is possible, but only under specific circumstances. First off, recycled cooking oil or other used oils are typically more cost-effective feedstocks because virgin plant oils are typically too expensive for the procedure. Second, the feedstock oil needs to have fewer than 2% free fatty acids (FFAs) or those FFAs need to be eliminated before chemical reactions. If left in place, they saponify (i.e., produce soap) and emulsify with the catalyst, making it more challenging to separate the biodiesel from the pollutants. Third, as they generate foaming or emulsions that are challenging to remove from the finished biodiesel,

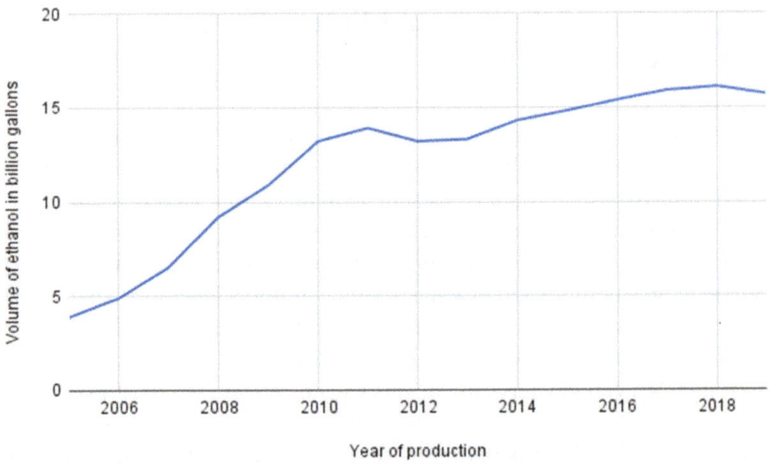

Fig. 25.3 Volumes of ethanol produced from corn starch in the United States over the nine years. Production has risen from over 4 billion gallons (15.1 billion liters) in 2005 to more than 14 billion gallons (53 billion liters) in 2018

phospholipids must also be eliminated. The process is energy-intensive, and the alkaline waste water needs to be treated before it can be used again. Fatty acid methyl esters are the main end products of fats and oils, and they must be distinguished from the contaminating glycerol, water, and acid or base catalyst. Several of these problems can be solved by using enzymes as opposed to non-enzymatic catalysts. Biological processes in any type of industrial reaction can promote more particular products from the feedstock while avoiding co-product production. Furthermore, the end products are regiospecific, meaning that the enzyme selectively produces one constitutional isomer rather than the other. Additionally, they provide an environmentally friendly means of producing biodiesel that generates methyl esters from FFAs and from the DAG released from phospholipids by phospholipases, improving the overall production of biodiesel from the feedstock oil (Hood 2016).

According to the study conducted by Fanar Bamerni, *Camelina sativa* biodiesel production has been shown to be superior to other biodiesel crops such as Canola and Jatropha in terms of anticipated greenhouse emissions and net energy ratio. Additionally, it increases the amount of carbon in the soil while causing no harm. Furthermore, Camelina cultivation does not harm the soil because of its ability to offer more carbon to the soil, which is subsequently transformed to carbon dioxide, in contrast to other crops like Jatropha, which causes soil poisoning. Camelina can be utilized as a raw material in the production of biodiesel because it is not a food crop and does not compete with other crops for agricultural land. Camelina offers several advantages over other conventional oil crops, including a high oil content, a short growing season, the ability to endure drought, and the ability to tolerate low-quality, nutrient-poor soils. It can also be grown on marginal lands. One of the future perspectives of this DIC-transesterification-based study will be a complete investigation of glycerol, a byproduct of the transesterification process, as well as research into its reuse in some new applications. Manufacturing biodiesel from Camelina seeds via the hydrocracking process is also one of the future possibilities (Bamerni 2018).

The large-scale sustainable production of biofuels and biomaterials remains a technical challenge. The usage of vast amounts of land and water may conflict with food production and environmental protection, particularly biodiversity conservation. Biofuel production may fail to achieve the intended greenhouse gas reductions if significant emissions are related to biomass transportation and production, as well as conversion to or transit of biofuels. Minimizing the environmental impact of biofuel production will necessitate the careful selection of biomass production systems and biofuel conversion technologies that consume less land and water and emit less greenhouse gas. The overall biomass production and the efficiency of conversion to fuel are the two aspects that define the long-term yield of biofuel. Plant biotechnology has the ability to contribute to both the amount and type of biomass produced (determining potential conversion efficiency). Improvements in biomass composition have the potential to significantly increase the sustainability of biofuel production by making biomass conversion and fuel yield much easier. The land area necessary to replace fossil fuels at any level of biomass yield per hectare and conversion efficiency to biofuel can be simply determined. This analysis will help in defining the biofuel yields required from current or enhanced biomass to meet fuel requirements with production from available land resources. Because of the cell wall structure, lignocellulosic biomass is resistant to the process of deconstruction to liberate sugars. Large levels of enzymes are frequently needed because lignin limits the ability of hydrolytic enzymes to access polysaccharides and can additionally inhibit these enzymes. Lignocellulosic biomass generated from different tissues or plant species may be structurally diverse, necessitating distinct deconstruction methods to yield fermentable sugars for biofuel production. The cost of lignocellulosic biomass and its deconstruction, which often entails some sort of pretreatment followed by polysaccharide hydrolysis (enzymatic and/or chemical), are the two key factors that determine the final cost of fuel production for a specific conversion method. Plant variety development with a more acceptable biomass composition should help to a more reliable supply of usable biomass and support biofuel production investments. Biomass with a more desirable composition will be more valuable as a substrate for these processes. The composition may be a crucial quality for both biomass producers justifying output due to increased value and processors ensuring a more efficient process and perhaps giving higher biofuel yields. To aid in the selection/generation of the appropriate biomass combination for any conversion process, a full understanding of the chemical and structural components of lignocellulosic biomass will be required (Furtado et al. 2014).

The enormous diversity of structures encountered complicates the analysis of plant cell wall polysaccharides and their structures. Lignin is difficult to quantify due to the variety of chemical structures contained within this fraction, as well as the heterogeneity in the contributions of various moieties in lignin from different sources. Rapid screening of numerous potential biofuel feedstocks is now possible due to the development of high-throughput analytical techniques like vibrational spectroscopy.

More extensive examination of the chemical structures found in specific biomass types, on the other hand, may offer to fresh insights into biomass upgrading choices as a substrate for biofuel production. Further advancements in chemical analysis technologies will be critical for the development of biofuel crops (Furtado et al. 2014).

The ideal natural feedstock would have chemical properties such as high cellulose and low lignin concentration, an appropriate lignin S/G ratio, and low ash content. While high-throughput technologies enable rapid screening of biomass to discover which plants have these traits, many plants do not meet these standards. Manipulation of the structure of plant cell walls in species produced for biofuel production is a key technique for improving biofuel production efficiency. The amount and composition of cell wall polysaccharides (including cellulose and non-cellulosic polysaccharides), lignin (amount and composition), total amount of cell wall, and cross-linking of cell wall components should all be regarded as goals for future genetic improvements in biomass (Furtado et al. 2014).

The yields of first-generation biofuel crops can be increased technologically. Crop variety, soil quality, rainfall and irrigation, nitrogen supplementation, and cultural practices all influence yield. Fermentation and distillation for ethanol generation, as well as extraction and esterification, are commercially established and feasible biofuel conversion technologies. Because the technology is well established and big production programs exist, first-generation crops are anticipated to dominate biofuel production for many years to come. The feedstock for next-generation biofuel crops is lignocellulosic biomass, which is collected for its total biomass. Woody biomass, tall grasses, agricultural residues (bagasse, shells, straw, stocks, leaves, and reproductive parts, etc.), and plantation residues (leaves, sawdust, nut shells, reproductive parts, etc.) are abundant and can be harvested at a much lower cost. Perennial grasses (such as switch grass and miscanthus), short-rotation willows, hybrid poplar, and Eucalyptus are some of the promising crops for next-generation biofuel production. Many lignocellulosic feedstocks, such as crop and forest residue, wood processing waste, and the organic part of municipal solid waste, can be produced with little additional land required. Furthermore, some biofuel crops, such as grasses and tree crops, can be grown in poor soils. The fundamental advantage of next-generation biofuel crops over first-generation crops is the ability to utilize a wide range of feedstocks, leftovers and wastes, and even complete plants, potentially resulting in higher biofuel output per hectare. If the potential of the residues and waste products is low, then dedicated cultivation of vegetative grasses or short-rotation woody crops in degraded or marginal areas should be attempted (Ravindranath et al. 2011).

25.3.1 Impact on Land Usage

The conversion of forest land to farmland is prohibited in India under the National Forest Policy. Furthermore, cropland is unlikely to be transformed for biofuel production because it has been stable for more than 20 years, and government support in

India is mostly for cultivating biodiesel crops on wastelands. As a result, degraded grassland or wasteland are expected to provide the land demand for biofuels. Due to overgrazing, fires, and soil erosion, India's grasslands and wastelands are in a degraded state. This will reduce carbon emissions from biofuel production to a relatively low level when compared to global scenarios involving forest and productive grassland conversion. According to a study conducted in India, if the yield of a biofuel crop falls by 50%, emissions will inevitably increase due to increased transport distances to collect the required raw material to create the same quantity of biofuel. In other words, the yield of a biofuel crop, the distance of a biofuel manufacturing unit from the source of biomass, and the manner of raw material transit are all critical factors in determining the amounts of emissions associated with biofuel (Ravindranath et al. 2011).

25.3.2 *Impact on Food Security*

Plant-based Biofuel production is projected to have both direct and indirect implications on food security, as more land in developing nations is shifted to biofuel production due to decreased production costs, notably labor. Several studies have found that rising demand for biofuel feedstock has already contributed to rising food costs. Increasing worldwide biofuel demand contributed to 30% of the increase in weighted average grain prices between 2000 and 2007; the greatest influence was on maize prices, where increasing biofuel demand accounted for 39% of the price increased (Ravindranath et al. 2011).

Water stress and diminishing groundwater levels are currently threatening food production in several tropical locations, particularly India. In most parts of the developing world, food shortage is intrinsically linked to water scarcity. According to FAO, water shortage, rather than land scarcity, maybe a major limiting issue for biofuel production in many countries. Many of the existing first-generation biofuel crops, such as sugarcane, oil palm, and maize, have rather high requirements for water. As a result, such crops may only be cultivated in irrigated conditions or in areas with significant rainfall, especially if big yields are desired. Extensive commercial development of biofuel crops may result in water conflict between biofuel production and subsistence food production (Ravindranath et al. 2011).

25.3.3 *Impact on Biodiversity*

One of the major concerns in plant-based biofuel production is its impact on biodiversity. Large-scale biofuel production may have both positive and negative impacts on biodiversity, depending on the land category converted for biofuel production, the biodiversity state of the land prior to conversion, the biofuel crop, and farming practices. Increased biofuel production is likely to have major consequences for biodiversity, which is defined as species richness and measured as the number of plant, animal, and microbe species per unit area. Increased biofuel production will have a negative impact on biodiversity because of (i) habitat conversion and loss; (ii) agricultural intensification; (iii) invasive species; and (iv) pollution (Ravindranath et al. 2011).

25.3.4 Impact on Socio Economics

Plant-based Biofuel production may be useful for creating rural jobs and promoting rural development. Biofuel production will result in the development of a huge number of employment across the supply chain, including production (cultivation practices), harvesting, transportation, processing, and marketing (Ravindranath et al. 2011).

The primary obstacle to the adoption of more environmentally friendly biofuel manufacturing methods is economic profitability. Currently, first-generation biofuels have the lowest production costs, particularly sugarcane bioethanol in Brazil and maize bioethanol in the United States, owing in part to government subsidies. The high costs of converting lignocellulosic feedstocks into biofuels, as well as the high capital and operational expenses of establishing microalgal production systems, diminish the economic competitiveness of lignocellulosic and microalgal biofuel production systems. While sugarcane production costs in Brazil and maize production costs in the United States have been calculated to be between US$ 5–9 and US$ 9–20 GJ^{-1}, respectively, estimated production costs for lignocellulosic feedstocks and microalgal systems can range between US$ 19–62 and US$ 13–8949 GJ^{-1}, respectively. In comparison, a US$ 100 $barrel^{-1}$ oil price, a value that has only been exceeded in only 20% of months during the last decade is comparable to US$ 17.12 GJ^{-1} (Low Heat Value LHV). To be competitive, more sustainable biofuel production is required (Correa et al. 2019).

25.3.5 Advancements in Routes of Conversion

Conversion of bio-oil, the product of biomass pyrolysis, which might be routed through a normal petrochemical refinery to generate various chemicals, is an emerging idea in the biorefinery sector. The advantage of this technique is that all essential infrastructure for product separation and purification is already in place. This approach makes perfect sense because most petroleum refineries are well-equipped to handle varied feedstock, assuming that no two batches of crude oil are identical. The chemical characteristics of bio-oil vary depending on the feedstock, but woody biomass typically yields 30% water, 30% phenolics, 20% aldehydes and ketones, 15% alcohols, and 10% miscellaneous chemicals. To replace oxygen with hydrogenation in the unprocessed bio-oils, a procedure called hydrodeoxygenation (HDO) could be used. The bio-oil might be converted into a liquid hydrocarbon with characteristics similar to petroleum crude oil after many HDO treatment processes. Deoxygenated bio-oils may be refined in existing petroleum refineries with only slight modifications to the current petroleum sector refinery infrastructure that is built up for the hydrodesulfurization (HDS) process (Naik et al. 2010).

The new technology, hydrodeoxygenation (HDO) of vegetable oils, normally occurs at reaction conditions of 300° C-450° C, hydrogen pressure above 3 MPa, lowest space velocity, and higher H2: oil ratio. These vegetable oils can be transformed into saturated long-chain hydrocarbons with boiling points comparable to diesel fuel (180-360 °C). Neste Oil Corporation, UOP Honeywell, ConocoPhillips, Valero Energy Corporation, and HP Innovations are among the industrial applications that have successfully implemented HDO technology. They have effectively commercialized the tremendous potential of catalytic hydroprocessing in the generation of green diesel from vegetable oil. Due to economic and social issues, the use of edible vegetable oils for biodiesel synthesis should be replaced with non-edible plant oils. As a result, oil from tropical plants such as *Hevea brasiliensis*, Jatropha, *Ceiba pentandra*, and others is an appealing prospective feedstock. Researchers always place a high priority on finding an easy, environmentally friendly, and affordable approach to making biodiesel. The transesterification reaction should be improved in order to boost ester yield, minimize reaction time, and simplify the procedure. To improve biodiesel production, several novel methods have been used, including in situ transesterification processes, the use of supercritical methanol, ultrasonic irradiation, hydrodynamic cavitation, microwave heating, microwave-assisted transesterification, and cosolvent (Yusup et al. 2019).

The current promising technologies for producing gasoline are catalytic cracking and HDO. This is because these technologies generate high-octane fuel in the presence of a suitable catalyst, which favors the production of the necessary gasoline range hydrocarbons. The biofuel produced has qualities similar to conventional fuel. As a result, it could be immediately employed in current automobile engines. The process of catalytic cracking, however, is superior to HDO since it may operate in atmospheric conditions and does not require a supply of hydrogen. This drastically lowers operating expenses. Furthermore, this evolved technology is already present in existing oil refineries. As a result, no major infrastructure changes are required (Yusup et al. 2019).

The disadvantages of existing bio-gasoline manufacturing processes via thermochemical, chemical, and biochemical routes are summarized. The disadvantages of using biochemical technology (enzyme hydrolysis) in ethanol to gasoline catalysis are as follows: varying enzymes depending on the type of feedstock utilized; short life period and costliness of enzymes; Longer reaction time, inability to store enzymes for a longer period of time; reduced enzyme activity due to product inhibition. The disadvantages of the chemical technology (acid hydrolysis) in ethanol to gasoline catalysis are as follows: the requirement of High temperature; expensive treatment of the hazardous wastewater produced, and the presence of traces of acid in ethanol, resulting in a corrosive fuel. The drawbacks of the Fischer–Tropsch thermochemical technique include polymerization-like nature, which limits selectivity to high-quality gasoline, catalyst deactivation, and heat and mass transport constraints. The following are the drawbacks of Pyrolysis: the bio-oil is highly oxygenated, thermally unstable, corrosive, and viscous; the high content of oxygenated chemicals and water reduces the calorific value of the oil and significantly affects the combustion properties; the type of feedstock used determines product quality. The downsides of the hydrothermal method of thermochemical technology are the production of high oxygen content bio-oil, high operating pressure, rapid corrosion in subcritical water, acidic and oxidizing conditions, and deposits of salt in the reactor, which causes plugging problems. The thermal cracking method has two major drawbacks: it is not very selective in terms of producing gasoline-range hydrocarbons, and it produces mostly paraffinic compounds, which limit fuel octane number (Yusup et al. 2019).

Despite the importance of kinetic mechanisms in commercial reactor scale-up and modeling, there has been far too little emphasis dedicated to developing a viable kinetic model in catalytic cracking and hydroprocessing of crude vegetable oils. Many previously published kinetic studies in catalytic cracking and hydroprocessing were primarily concerned with studying and mapping out hypothetical reaction paths of single-model chemicals such as oleic acid, palmitic acid, and triolein using simple chemical conservation equations. However, the underlying interactions between triglycerides, chemical intermediates, and hydrocarbon products are yet unknown and must be determined in the future. The scarcity of such empirical studies in the literature can be explained by the inherent high complexity of parallel reaction processes and a large number of reaction steps and chemical species involved in developing such a model. Because natural triglycerides are made up of hundreds of chemical components, it is impossible to recreate such a burst of chemical species sizes in a chemical kinetic model. A surrogate triglyceride model generated from selected representative components found in practical vegetable oil can be used to develop more holistic predictive chemical kinetic models. The chemical kinetic mechanism's complexity and scale can be considerably reduced and simplified for inclusion in multidimensional computational fluid dynamic models, or employed in full detail for purely kinetic modeling. A significant number of published papers have included new bimodal pore systems in the construction of hybrid composite catalysts to improve catalyst stability and deactivation resistance. The system took full advantage of the advantages that microporous and mesoporous phases provided. In comparison to traditional sulfide metallic-supported catalysts, bimetallic-supported catalysts such as palladium, platinum, and other transition metals have shown outstanding HDO activity. However, one of the primary challenges in the field of catalytic cracking and hydroprocessing is the catalyst economics of such advanced synthetic catalysts, as high manufacturing costs limit their potential for commercial applications. Exploring the idea of a renewable materials-based catalyst developed from waste materials can greatly enhance catalyst economics. The mesoporous carbon support materials of the catalysts, for example, can be synthesized through the thermal decomposition of RSO shells, rice husks, and coconut husks. Much less is known about the stability of these unique synthetic catalysts in the continuous deoxygenation activity of various triglyceride feedstocks to date, and this needs to be researched. Metal and support system optimization with one-step deoxygenation and isomerization capabilities remains an important challenge in the catalysis industry. The ultimate goal of high activity, and selectivity for gasoline and diesel hydrocarbons with small branched hydrocarbons, coupled with high catalyst stability and deactivation resistance, can be pursued further by tailoring the morphology, structure, and acidity level of mesoporous support materials, as well as introducing rare earth elements as promoters in stimulating higher affinity towards oxygenate. The use of steam as a gasifying agent in new biomass gasification technology has not only boosted the hydrogen content in product gas but also reduced the problem of excess steam removal. The use of PKS, which is waste from the palm oil industry, for syngas production is a less expensive source that also provides a renewable source. The use of a catalyst in biomass gasification has a beneficial impact on system efficiency since it increases the required content in product gas such as hydrogen through tar reduction and reaction acceleration. The most often utilized catalysts for syngas production and CO_2 removal from product gas are dolomite, alkaline metal, and Ni. Nonetheless, these catalysts have some drawbacks, such as short active life, higher cost, and regeneration issue. Coal bottom ash, which is a waste product of the coal power industry, can be used as a catalyst for the production of syngas, which is not only cheaper but also provides a sustainable solution for coal bottom ash disposal. The key obstacles should be addressed in the commercialization of the gasification process in the future. The process should be integrated to use waste heat for power generation, making the process economically viable (Yusup et al. 2019).

The deep eutectic solvent (DES) is a promising new green solvent that can be used as an alternative to ionic liquids. It is a mixture of hydrogen bond acceptors (HBAs) like choline chloride and hydrogen bond donors (HBDs) like amines, alcohols, lactic acid, and carboxylic acids. The typical DES is mixed with choline chloride (ChCl) and lactic acid (LA). In comparison to ionic liquids, DESs exhibit similar physical and chemical properties, however, they exhibit a lower melting point, greater stability, and lower cost. DESs also have some benefits for the subsequent treatment, such as reduced toxicity and biocompatibility. Additionally, the barely volatile components of DESs make them simple to recycle and reuse without affecting the effectiveness of the treatment. High-purity lignin extraction produced by DES pretreatment favors the fractionation and commercialization of lignocellulose because it exhibits specific selectivity to lignin. As a result, recent research has turned its attention to DES and demonstrated various unique methods based on DES as well as the characteristics of DES pretreatment (Ning et al. 2021).

According to the study conducted by Stephanie Searle and Chris Malins, even if practically all 'unused' grassland and savannah are converted, it is shown that the maximum probable limit to sustainable energy crop production in 2050 is 40–110 EJ yr^{-1}. When combined with forests, crop leftovers, and wastes, the maximum long-term total biomass availability in primary energy is 60–120 EJ yr^{-1}. Maximum potentials of 10–20 EJ yr^{-1} of biofuel, 20–40 EJ yr^{-1} of electricity, and 10–30 EJ yr^{-1} of heating in 2050 are estimated after accounting for present trends in bioenergy allocation and conversion losses. In 2050, the IEA predicts a global need for 32 EJ yr^{-1} of biofuel and 60 EJ yr^{-1} of electricity/heating. Given that our projected potential is a maximum value, it is evident that reaching bioenergy utilization at the level envisioned by the IEA would necessitate considerable government commitment, and even then, demand would likely be insufficient to meet sustainably. Furthermore, as policymakers continue to consider expanded use of end-use bioenergy policies, and as researchers consider the role of biomass as a feedstock for the road vehicle, aviation, chemicals, and/or power sectors, adequate recognition of likely resource constraints and the inevitability of intersectoral competition for the resources is critical. Policymakers all throughout the world would benefit enormously from a better knowledge of the supply limits for sustainably produced biomass. Policies that encourage the use of bioenergy may result in unfulfilled targets, unsustainable practices, compromised climate mitigation aims, unintentional deforestation, encroachment on the world's food supply, or some combination of these results. Future bioenergy targets and scenarios should be commensurate with a global capacity, not just in terms of biomass production but also in terms of sustainable governance (Searle and Malins 2015).

Decentralized national or regional regulatory rules that demand or assume bioenergy use (for example, for vehicles, aviation, and power) should expressly address these limits, as they are drawing from the same finite pool of biomass supplies (Searle and Malins 2015).

25.3.6 Potential Feedstocks

Agricultural waste, such as straw or stover, is a potential lignocellulose-based feedstock with a favorable greenhouse gas balance for the manufacture of ethanol and, consequently, bioenergy. They are abundant agricultural byproducts with little commercial value. Since straw is lignocellulosic by nature, it must first undergo some sort of physical, chemical, or biological pretreatment in order to improve accessibility to its three main biopolymer compounds: cellulose, hemicellulose, and lignin. Wheat straw and maize stover both have the potential to be successfully investigated for the commercial production of ethanol, biogas, and biohydrogen. For the manufacture of biobutanol, biodiesel, and biochemicals utilizing this feedstock, decent research efforts have been made, but more advanced scientific insights and technologies are required for their commercialization. The higher operational cost needed for biomass feedstock pretreatment is the main barrier to producing industrial biofuels from lignocellulosic raw materials. A few viable ways to establish cereal straw as a feedstock for effective and affordable biofuel and chemical production include the development of genetically modified microbial strains, enhanced production and extraction processes, coupled with the generation of valuable byproducts (Bakala et al. 2023).

25.3.7 Use of Nanoparticles in Biofuel Production

Due to their unique physicochemical characteristics, such as a high surface area to volume ratio, good selectivity, strong dispersibility, and high reactivity, nanoparticles are proven to be capable of improving biofuel production. Increased substrate digestibility after pretreatment with nanoparticles leads to better biofuel production yields. The discovery of alternative methods for boosting biofuel production can be facilitated by combining nanotechnology and biotechnology. This technology may not be widely available because it is unique and takes a high level of expertise, but if invested in, it can help mitigate the global energy crisis. The price of naked modified magnetic nanoparticles can be significantly lower after nanoparticle reactivation, ranging from £3–500 to £0.5–108 per kg. Several technical issues must be resolved before the improved biofuel production process based on nanotechnology can be commercialized successfully. To solve these problems, scientists must create nanoparticles that are not harmful to microorganisms, use less expensive nanoparticles that are not dangerous to the environment, and adapt chemical synthesis techniques to biological nanoparticle synthesis, which requires extremely strict operating conditions (Ahmed et al. 2023).

To summarize the different generations of biofuels, their advantages, and sources, Table 25.3 is illustrated.

Table 25.3 Summary of the first, second, and third generations of biofuel

Biofuel generation	Feedstock	Production method	Advantages	Challenges
First generation	Food crops (e.g. corn, sugarcane)	Fermentation or transesterification	Established technology, high yield potential	Competes with food production, high land use, high water use, can lead to deforestation
Second generation	Non-food crops (e.g. switchgrass, wood chips)	Cellulose conversion (e.g. gasification, pyrolysis)	Utilizes non-food crops, potential for high yield, reduces competition with food production	High capital costs, limited commercial-scale production, technical challenges
Third generation	Algae, cyanobacteria	Photosynthesis	High yield potential, does not require arable land, can utilize waste water or CO_2 emissions	Technical challenges, limited commercial-scale production, high energy input

25.4 Conclusion

Many scientists believe that the most pressing single issue confronting humanity is climate change caused by anthropogenic emissions of greenhouse gases such as CO_2, N_2O, and methane into the atmosphere. The combustion of fossil fuels, which accounts for approximately 85% of total energy consumption, contributes significantly to CO_2 emissions. When considering ways to decarbonize energy production, one is unavoidably drawn to the fact that the earth's surface receives approximately 9000 times as much energy from the sun as all human energy uses. The sun's energy is used to generate power or fuels in four ways: through wind turbines to generate electricity, through photovoltaic conversion to electricity, through mirrors to heat sterling engines that eventually produce electricity, and through harvesting plant biomass that can be burned as solid or liquid fuels. A fifth possibility, photoelectrochemical to electrochemical hydrogen production, is promising but requires significant further development. At the moment, none of these approaches provide a clear answer to our energy needs. As a result, rather than relying on a single technology, it is critical to approach energy production through a basket of complementary technologies. Because biomass is currently the most cost-effective route from photons to fuels (as opposed to power), much of the focus in the biofuels field has been on its applications (Somerville et al. 2010). The use of biomass for energy production as a replacement for fossil energy is frequently viewed as a promising strategy for reducing fossil-fuel dependency and helping to reduce greenhouse gas (GHG) emissions. It has been argued that large-scale biomass combustion, with subsequent carbon capture and storage (CCS), could be an important option for achieving the negative GHG emissions required to limit global warming to 2 degrees Celsius until 2100, a goal thought to be required to reduce the risk of catastrophic runaway events as the earth's system approaches certain "tipping points" (Haberl et al. 2011).

"Biomass" is defined as "material of biological origin, excluding material embedded in geological formations and transformed to fossils, such as energy crops, agricultural and forestry wastes and byproducts, manure, or microbial biomass." Biofuels are usually categorized as first, second, or third-generation biofuels based on the feedstocks and conversion technology used in their production: First-generation biofuels are created through the cultivation of bioenergy crops (food-crop feedstock). They are primarily composed of sugar, starch, and vegetable oil as feedstocks. Second-generation biofuels are bio-based products derived from non-food feedstock, specifically lignocellulosic biomass such as agricultural and forestry feedstock. They include agricultural and food processing system residues (discarded biomass). Third-generation biofuels are made from aquacultured feedstock (algae) (Saladini et al. 2016). Rapid increases in energy consumption and human reliance on fossil fuels have resulted in greenhouse gas accumulation and, as a result, climate change. As a result, significant efforts have been made to develop, test, and implement clean renewable fuel alternatives. Crop-based bioethanol and biodiesel production is well established. Other feedstock resources and processes, such as landfill and plastic waste conversion, algal photosynthesis, and electrochemical carbon fixation, have also demonstrated high potential to provide efficient and cost-effective alternatives. Furthermore, downstream microbial fermentation can be further engineered to not only increase product yield but also to expand the chemical space of biofuels via rational design and fine-tuning of biosynthetic pathways towards the realization of ''designer fuels'' and diverse future applications (Liu et al. 2021).

25.4.1 Acronyms

SO_X: Sulfur oxides.
CO_X: Carbon oxides.
NO_X: Nitrogen oxides.
GHG: Greenhouse gas.
GMO: Genetically modified organism.
DNA: Deoxyribonucleic acid.
CO_2: Carbon dioxide.

References

Abdullah B, Syed Muhammad SAF ad, Shokravi Z, Ismail S, Kassim KA, Mahmood AN, Aziz MMA (2019) Fourth generation biofuel: a review on risks and mitigation strategies. Renew Sustain Energy Rev 107, 37–50. https://doi.org/10.1016/j.rser.2019.02.018

Ahmed SF, Debnath JC, Mehejabin F, Islam N, Tripura R, Mofijur M, Hoang AT, Rasul MG, Vo DVN (2023) Utilization of nanomaterials in accelerating the production process of sustainable biofuels. Sust Energy Technol Assess 55:102894. https://doi.org/10.1016/J.SETA.2022.102894

Alam F, Mobin S, Chowdhury H (2015) Third generation biofuel from algae. Proc Eng 105:763–768. https://doi.org/10.1016/J.PROENG.2015.05.068

Ale S, Femeena PV, Mehan S, Cibin R (2019) Environmental impacts of bioenergy crop production and benefits of multifunctional bioenergy systems. Bioenergy Carbon Capture Stor Using Nat Res Sust Devel 195–217. https://doi.org/10.1016/B978-0-12-816229-3.00010-7

Azni A., Md Khalid R, Hasran UA, Kamarudin SK (2023) Review of the effects of fossil fuels and the need for a hydrogen fuel cell policy in Malaysia. Sustainability 15(5):4033. https://doi.org/10.3390/SU15054033

Bakala HS, Devi J, Ankita Sarao LK, Kaur S (2023) Utilization of Wheat and Maize Waste as Biofuel Source, pp 27–66. https://doi.org/10.1007/978-981-19-6230-1_2

Balan V, Bals B, Chundawat SPS, Marshall D, Dale BE (2009) Lignocellulosic biomass pretreatment using AFEX. In *Methods in molecular biology (Clifton, N.J.)* (Vol. 581, pp. 61–77). https://doi.org/10.1007/978-1-60761-214-8_5

Bamerni F (2018) Plant-based (Camelina Sativa) biodiesel manufacturing using the technology of Instant Controlled pressure Drop (DIC) : process performance and biofuel quality. https://theses.hal.science/tel-02009827

Correa DF, Beyer HL, Fargione JE, Hill JD, Possingham HP, Thomas-Hall SR, Schenk PM (2019) Towards the implementation of sustainable biofuel production systems. Renew Sustain Energy Rev 107:250–263. https://doi.org/10.1016/j.rser.2019.03.005

Doherty WOS, Mousavioun P, Fellows CM (2011) Value-adding to cellulosic ethanol: Lignin polymers. Ind Crops Prod 33(2):259–276. https://doi.org/10.1016/J.INDCROP.2010.10.022

Endler A, Persson S (2011) Cellulose synthases and synthesis in Arabidopsis. Mol Plant 4(2):199–211. https://doi.org/10.1093/MP/SSQ079

Fu C, Sunkar R, Zhou C, Shen H, Zhang JY, Matts J, Wolf J, Mann DGJ, Stewart CN, Tang Y, Wang ZY (2012) Overexpression of miR156 in switchgrass (Panicum virgatum L.) results in various morphological alterations and leads to improved biomass production. Plant Biotechn J 10(4), 443. https://doi.org/10.1111/J.1467-7652.2011.00677.X

Furtado A, Lupoi JS, Hoang NV, Healey A, Singh S, Simmons BA, Henry RJ (2014) Modifying plants for biofuel and biomaterial production. Plant Biotechnol J 12(9):1246–1258. https://doi.org/10.1111/PBI.12300

Gasparatos A, Stromberg P, Takeuchi K (2013) Sustainability impacts of first-generation biofuels. Anim Front 3(2):12–26. https://doi.org/10.2527/AF.2013-0011

Haberl H, Erb KH, Krausmann F, Bondeau A, Lauk C, Müller C, Plutzar C, Steinberger JK (2011) Global bioenergy potentials from agricultural land in 2050: sensitivity to climate change, diets and yields. Biomass Bioenerg 35(12):4753–4769. https://doi.org/10.1016/J.BIOMBIOE.2011.04.035

Hood EE (2016) Plant-based biofuels. F1000Research, 5. https://doi.org/10.12688/F1000RESEARCH.7418.1

Lam MK, Lee KT (2012) Microalgae biofuels: a critical review of issues, problems and the way forward. Biotechnol Adv 30(3):673–690. https://doi.org/10.1016/J.BIOTECHADV.2011.11.008

Lee RA, Lavoie JM (2013) From first- to third-generation biofuels: challenges of producing a commodity from a biomass of increasing complexity. Anim Front 3(2):6–11. https://doi.org/10.2527/AF.2013-0010

Liu Y, Cruz-Morales P, Zargar A, Belcher MS, Pang B, Englund E, Dan Q, Yin K, Keasling JD (2021) Biofuels for a sustainable future. Cell 184(6):1636–1647. https://doi.org/10.1016/J.CELL.2021.01.052

Naik SN, Goud VV, Rout PK, Dalai AK (2010) Production of first and second generation biofuels: a comprehensive review. Renew Sustain Energy Rev 14(2):578–597. https://doi.org/10.1016/j.rser.2009.10.003

Naqvi M, Yan J (2015) First-generation biofuels. Handbook of Clean Energy Syst 1–18. https://doi.org/10.1002/9781118991978.HCES207

Neupane D (2023) Biofuels from renewable sources, a potential option for biodiesel production. Bioengineering, 10(1). https://doi.org/10.3390/BIOENGINEERING10010029

Ning P, Yang G, Hu L, Sun J, Shi L, Zhou Y, Wang Z, Yang J (2021) Recent advances in the valorization of plant biomass. Biotechnology for Biofuels 14(1):1–22. https://doi.org/10.1186/S13068-021-01949-3

Ravindranath NH, Sita Lakshmi C, Manuvie R, Balachandra P (2011) Biofuel production and implications for land use, food production and environment in India. Energy Policy 39(10):5737–5745. https://doi.org/10.1016/J.ENPOL.2010.07.044

Rodionova MV, Poudyal RS, Tiwari I, Voloshin RA, Zharmukhamedov SK, Nam HG, Zayadan BK, Bruce BD, Hou HJM, Allakhverdiev SI (2017) Biofuel production: challenges and opportunities. Int J Hydrogen Energy 42(12):8450–8461. https://doi.org/10.1016/J.IJHYDENE.2016.11.125

Saladini F, Patrizi N, Pulselli FM, Marchettini N, Bastianoni S (2016) Guidelines for emergy evaluation of first, second and third generation biofuels. Renew Sustain Energy Rev 66:221–227. https://doi.org/10.1016/J.RSER.2016.07.073

Searle S, Malins C (2015) A reassessment of global bioenergy potential in 2050. GCB Bioenergy 7(2):328–336. https://doi.org/10.1111/GCBB.12141

Siddiki SYA, Mofijur M, Kumar PS, Ahmed SF, Inayat A, Kusumo F, Badruddin IA, Khan TMY, Nghiem LD, Ong HC, Mahlia TMI (2022) Microalgae biomass as a sustainable source for biofuel, biochemical and biobased value-added products: an integrated biorefinery concept. Fuel 307:121782. https://doi.org/10.1016/J.FUEL.2021.121782

Sims REH, Mabee W, Saddler JN, Taylor M (2010) An overview of second generation biofuel technologies. Biores Technol 101(6):1570–1580. https://doi.org/10.1016/j.biortech.2009.11.046

Somerville C, Youngs H, Taylor C, Davis SC, Long SP (2010) Feedstocks for lignocellulosic biofuels. Science 329(5993):790–792. https://doi.org/10.1126/SCIENCE.1189268

Voloshin RA, Rodionova MV, Zharmukhamedov SK, Nejat Veziroglu T, Allakhverdiev SI (2016) Review: Biofuel production from plant and algal biomass. Int J Hydrogen Energy 41(39):17257–17273. https://doi.org/10.1016/j.ijhydene.2016.07.084

Yang Y, Chaffin TA, Ahkami AH, Blumwald E, Stewart CN (2022) Plant synthetic biology innovations for biofuels and bioproducts. Trends Biotechnol 40(12):1454–1468. https://doi.org/10.1016/J.TIBTECH.2022.09.007

Yuan JS, Pavlovich MJ, Ragauskas AJ, Han B (2022) Biotechnology for a sustainable future: biomass and beyond. Trends Biotechnol 40(12):1395–1398. https://doi.org/10.1016/j.tibtech.2022.09.020

Yusup S, Bokhari A, Trinh H, Shahbaz M, Patrick DO, Cheah KW, Azizan MT, Ramli A, Ameen M, Osman N, Shuhaili AFA, Singh HKG (2019) Emerging technologies for biofuels production. Biomass, Biofuels, Biochemicals: Biofuels: Alternative Feedstocks and Conversion Processes for the Production of Liquid and Gaseous Biofuels 45–76. https://doi.org/10.1016/B978-0-12-816856-1.00002-6

Zhu L, Hiltunen E, Takala J (2012) Microalgal biofuels beat the first and second generation biofuels. Appl Mech Mater 197:760–763. https://doi.org/10.4028/WWW.SCIENTIFIC.NET/AMM.197.760

Chapter 26
Advances, Applications, Challenges and Future Prospects of Recent Technologies in Biofuel Production

B. Vijaya Geetha, K. S. Shreenidhi, R. Nivaas, T. Raghasree, and B. Ram Kumar

Abstract Fossil fuel is one of the most pernicious and expensive fuels. The release of the greenhouse gas carbon dioxide and methane from vehicles into the environment can cause major effects like global warming, weather change and depletion of the ozone layer etc. As we are in need to find an alternative source of fuel for future use, the major significance of biofuel started there. Biofuel is obtained from bio-based products such as bio-diesel and bio -compressed natural gas. This product is obtained from sustainable lignocellulose biomass, agriculture, woodland residues, cyanobacteria, algae and ammonia. These biofuels are toxic-free, sulfur-free, less expensive, more stable and renewable sources that cause no pollution to the environment. Economically biofuels are more efficient than fossil fuels. The major reason for the introduction of biofuel is to preserve the environment. Nowadays there are more advances and progression updated in biofuel production. This review prioritizes the importance of biofuel and gives alternative solutions for fossil fuels.

26.1 Introduction to Biofuel

Biofuels are forms of gaseous, liquid, and solid energy that are consistent with or generated from biomass. Because they are made from plant-derived polysaccharides, biofuels generally do not result in an increase in CO_2 emissions when they are utilised (burned). For non-food biomass to be converted into biofuels, there are two technical obstacles. (2) alleviation of fermentation inhibition by organic mixtures (aromatics, organic acids, and furans) produced from non-food biomass. (1) technique for efficient conversion of pentose to biofuel. Thermochemical processing of nonfood biomass results in the production of inhibitory mixtures, although this step is necessary for efficient enzymatic saccharification of biomass. Low yield and

B. Vijaya Geetha (✉) · K. S. Shreenidhi · R. Nivaas · T. Raghasree · B. R. Kumar
Department of Biotechnology, Rajalakshmi Engineering College (Autonomous), Affiliated to Anna University, Thandalam, Chennai, Tamilnadu 602105, India
e-mail: vijayageetha.b@rajalakshmi.edu.in

M. Shah and D. Deka (eds.), *Emerging Sustainable Technologies for Biofuel Production*, Environmental Science and Engineering, https://doi.org/10.1007/978-3-031-52167-6_26

productivity of the bioprocess are produced by these inhibitory mixtures. There has been an increase in interest in finding practical biofuel product ways due to the increasing demand for necessary renewable energy sources from emigrations. In some biotechnological operations, membranes are used to retain cells and/or benefits sought by operation of membrane bioreactors for biofuel product mainly concentrated and natural conversion rate due to high cell attention, low energy demand, and ease of conventional MBRs for natural treatment of different By using standard MBR technologies, sources with a high concentration of cell inhibitory mixtures or multiple different prioritised substrate sources are constrained. Feeds containing a lot of suspended solids (SS) can cause membrane fouling in MBR-supported bioconversion systems. impede the operation of the membrane bioreactor (rMBR) in biotechnology. The rMBR is a hybrid design that combines cell technology and traditional MBR. This is where rMBR offers the chance to have a membrane bioreactor system operating on the basis of cell encapsulation principles, serving from the bioconversion of complex feed courses microorganisms in between membrane layers to give high original cell density, rather than on the basis of diffusive mass transfer as for cell encapsulation (Ruan et al. 2019).

Biodiesel may more effectively lower CO_2 emigration, cover the natural environment, and preserve the ecological balance as compared to the usage of reactive energy. Because acid rain poses a serious threat to the environment and terrestrial life, including the acidification of the soil, face, and groundwater, the destruction of wood and vegetation, the erosion of man-made structures, and the literal erosion of monuments made of calcium-containing materials, using biodiesel instead of regular diesel oil will significantly reduce acid rain. Additionally, since biodiesel's ester composites support clean burning and include oxygen, less CO, HC, and particulate matter will be generated. When biodiesel is used in a traditional diesel engine, carbon monoxide, sweet hydrocarbons, alkenes, aldehydes, ketones, and particulate matter are all significantly reduced. On a life cycle basis, it has been calculated that soy biodiesel decreases carbon dioxide by 78. Additionally, biodiesel is the first essential energy to successfully pass all of the US Clean Air Act's health goods testing requirements.

By reducing global warming and saving reactive energy, bioethanol energy plays a significant role in environmental conservation. One method of reducing both the usage of crude oil painting and environmental pollution is the production of bioethanol from biomass or trash. The physical and chemical properties of bioethanol are comparable to those of ethanol; however, their product coffers differ (Eloka-Eboka et al. 2021).

In ponds or tanks on land or in the water, algae can be grown. Algal energies have a high ignition point, can be created using saline water and wastewater, can be cultivated with little negative influence on fresh water supplies, have a high yield, and are biodegradable and relatively non-aggressive to the environment if exposed. Large amounts of energy and toxins are needed to make the product, and the energy it produces degrades more quickly than other biofuels and flows poorly in freezing temperatures. Most efforts to manufacture electricity from algae had been abandoned or switched to other endeavours by 2017 due to financial considerations (Pereira da Silva and Ribeiro 2019).

In order to transform light into chemical energy, either protons are reduced to hydrogen or carbon dioxide to organic compounds. The term "fourth-generation biofuels" also refers to fuels produced by bioengineered organisms like cyanobacteria and algae. Biofuels will be produced by algae and cyanobacteria using water, carbon dioxide, and solar energy. This biofuel production method is still in the exploratory stage. In comparison to older generations of biofuels, it is projected that the bioengineered organisms' buried biofuels will have improved photon-to-energy conversion efficiency. This class of biofuels has the benefit of not requiring pastoralist land for the civilisation of the organisms that make them. The cost of raising the organisms that produce biofuel is one of the drawbacks (Ababneh and Hameed 2022).

26.2 Advances in Biofuel

The use of organic waste from soil and subsurface terrain inventories for renewable energy products for life-sustaining circumstances, maintaining a clean and healthy environment. A renewable source of bioenergy is provided by methane, hydrogen, ethanol, bioelectricity, algal diesel, butanol, or other kinds of energy that can be produced by natural systems. The approaches for converting waste into energy (thermal and biochemical) using agricultural wastes are covered (Jin et al. 2015). The new developments in "coetaneous waste mitigation with energy development" methods are the key topic of this study. Biofuels have been taken into consideration since they are nontoxic, sulfur-free, and irreplaceable for petroleum.

The utilisation of accessible manufacturing materials has been cited as a countersign to the advancement of biofuels. Biofuels are a type of energy that is derived from organic sources that are produced by businesses and living things and are generally represented as biomass. They are divided into four categories: first-, second-, third-, and fourth-generation biofuels.

The first generation of biofuels is made from oil painting-base shops, sugar, and bounce yields, but the second generation is made from non-food yields, primarily agricultural and forestland remnants (Piotrowska et al. 2013). Algae-based third generation biofuels are easily upgraded, can be produced on a massive scale, and absorb CO_2. The fourth generation of biofuels uses cyanobacterial development that has been tweaked. The biomass of lignocellulosic biomass is made up of crops, agricultural waste, and scrap from manufacturing facilities. To improve the volume of the intended product through enzymatic hydrolysis and turbulence, the pre-treatment stage is crucial. Biomass from microalgae can be transformed into biofuels like biodiesel, biooil, biomethane, and crude oil, among other things. In order to localise microalgae growth development within the manufacturing unit installations using wastewater-rich exhaust gas and sewage water are the nutrient budgets to induce enough biomass to meet the rising needs of power consumption, a broad biomass metamorphosis system is being abandoned.

The crucial ways for genetically modifying microalgae include adding the objectification of light into microalgae by limiting the size of specific chlorophyll antenna and colour administration. Microalgal metabolic technology can ameliorate the content of lipids and carbohydrates, adding the performance of microalgae biomass yields. The primary supplements in microalgae are nitrogen and phosphorus, which constitute 10–20 of the biomass. Biofuels of the fourth generation integrate feedstock of inheritable engineering, genomically synthesised micro-organisms, similar as cyanobacteria, and microalgae to produce bioenergy effectively. To develop a massive-scale biofuel system for managing organic trash, a well-determined biomass trash collection and depository system is important (Vallinayagam et al. 2014).

Effective non-supervision measures must be put into place to improve the garbage multifariousness and depository system, and government subsidies to build biofuel businesses and lower operating costs will aid in managing the biofuels produced from trash. Modern technologies are essential for producing biofuel efficiently, and to address the ongoing problems with waste management, a creative strategy is required. Solid, liquid, and gaseous biofuels can replace conventional energy sources and reduce carbon emissions. Bioethanol is produced from bounce and sugar yields, while biodiesel is derived from oil painting yields for traditional fluid biofuels. Modern biofuels are substantially more environmentally friendly and can lower pollution by reducing carbon emissions (Röder et al. 2019). In addition to algal feedstock, lignocellulosic biomass from agrarian and woodland environments can also be used to create advanced renewable energy sources. Rapidly produced liquid biofuels are paraffin energy or hydrocarbons that are manufactured through the chemical or microbiological catalysis of industrial carbohydrates.

Due to their carbon–neutral liquid transportation energy, biofuels products like bioethanol, biobutanol, and biodiesel have attracted a lot of attention. While lignocellulose is an economically viable option because of its abundance and low cost, ethanol is frequently produced using solventogenic provocations and certain microorganisms. The direct conversion of hemicellulose into ethanol was achieved using an artificial Escherichia coli double culture, with a predicted yield of 2.84 g/L. In a microbiological institution, cellulase systems can also be built, comparable to a binary Bacillus microbe/incentive system. Recombinant B (Deng et al. 2015).

Eight cellulosomal genes derived from C. thermocellum, one scaffolding protein gene, and two hemicellulase genes are present in S. subtilis. Lignocellulose can be fermented into ethanol by co-culturing cellulolytic bacteria (cipA, sdbA, two exoglucosidase genes (celK and celS), two endo-glucanase genes (celA and celR), and two xylanase genes (xynC and xynZ). To manufacture ethanol from lignocellulose using CBP, co-culturing cellulolytic microbes with ethanol-producing bones is a practical and adaptable method.

26.3 The Fuel for Future—Ammonia

Fritz Haber and Carl Bosch created the Haber–Bosch method to combine ambient nitrogen and hydrogen under pressure to create ammonia. The production of urea, ammonium nitrate, and almost all of the ammonia in the world still comes from this technique. Climate change is the current concern, and ammonia could help by absorbing, storing, and transporting hydrogen for use in emission-free fuel cells and turbines (Wimmerstedt 1999). Chemical companies are investing in carbon capture and storage to reduce the carbon effect of producing traditional ammonia as well as creating green ammonia, a method of producing ammonia in which hydrogen obtained from water electrolysis powered by alternative energy substitutes hydrogen derived from hydrocarbons.

Compared to hydrogen and natural gas, ammonia has a higher energy density, enabling it to provide greater mileage or smaller tank sizes for the same mileage. Additionally, because it has a higher octane number than fossil fuels, it may be used in engines with higher compression ratios to achieve higher cycle thermal efficiency. According to Liu et al., utilising ammonia as fuel often increases an engine's cycle thermal efficiency to more than 50%, as opposed to only 30 to 37% for a regular engine. Ordinary hydraulic cylinders can store ammonia without the need for more expensive storage equipment and transportation systems (Herbinet et al. 2022).

The autoignition can also benefit from preheating for ammonia. Ammonia, on the other hand, offers a major improvement to cut emissions because the auxiliary fuel generates CO_2 and NOx emissions. The image's caption provides more information.

Ammonia's slower flame speed compared to other conventional fuels is the second obstacle to its application in internal combustion engines. Both compression engines and spark ignition systems exhibit this issue (Fan et al. 2022).

Higher compression ratios, warmed ammonia, and ammonia-fuel blends are a few solutions that can boost flame velocity. By infusing ammonia and petrol separately into the intake manifold in liquid phase, ammonia can be employed in spark ignition engines. Ammonia and air are combined before being delivered into compression ignition engines through the intake manifold. A little amount of diesel fuel or another promoter is then injected within the cylinder to ignite the ammonia-air mixture. Ammonia, fuel, and air mixture ratios need to be adjusted and automatically controlled because this is a serious problem.

The pace at which a chemical reaction takes place is known as its rate. Internal combustion engines employ ammonia as a fuel, but due to its slow reaction time, ammonia is released from the exhaust without being burned. Ammonia-air mixture, conventional fuels, and hydrogen are employed as promoters to speed up the chemical reaction rate of ammonia combustion. As catalysts, NaCl, $BaCL_2$, and NaF improve the chemical kinetics of ammonia combustion. By adding more ammonia to the mixture, CO_2 and NOx emissions can be lowered for various ammonia-fuel mixes with respect to the mixture ratio. Ammonia oxidation catalyst and catalytic reduction can be utilised to reduce NOx emissions. Internal combustion engines that burn hydrogen and use ammonia as the carrier are a quick and practical solution to switch to a carbon-free atmosphere. Ammonia can be used as a temporary substitute for the hydrogen economy that is both economical and environmentally friendly, but it comes with substantial usage issues such difficult igniting, slow flame speed, and increased compression requirements. The challenging ignition issue can be improved with the use of ammonia-fuel mixes, which can also improve engine performance. To reduce NOx emissions, an efficient and automatically controlled ammonia-fuel blender should be created. Ammonia oxidation catalysts and catalytic reduction can also be used (Kalyuzhnaya 2016).

The cost of a metric tonne of "green" ammonia may be as low as US$386 if small-scale NH_3 generation based on renewables produced under good geographic conditions (such as solar electricity in the Middle East region). The price of anhydrous ammonia, however, fell to its lowest level in ten years in the spring of this year as a result of the oil price shocks of early 2020, rendering green NH_3 generation at USD386 uncompetitive on a global fertiliser market where "conventional" ammonia producers would have an advantage due to the low feedstock prices at the time. The cost of a metric tonne of "green" ammonia may be as low as US$386 if small-scale NH_3 generation based on renewables produced under good geographic conditions (such as solar electricity in the Middle East region). The price of anhydrous ammonia, however, fell to its lowest level in ten years in the spring of this year as a result of the oil price shocks of early 2020, rendering green NH_3 generation at USD386 uncompetitive on a global fertiliser market where "conventional" ammonia producers would have an advantage due to the low feedstock prices at the time (Sagel et al. 2022).

Large quantities of hydrogen and ammonia are typically stored under high pressure at low temperatures in specialised tanks or (salt) caves for lengthy periods of time. However, due to its severe toxicity, the subsurface storage of ammonia in salt caverns is frequently overlooked. It is physically possible to store hydrogen underground in salt caverns, despite the fact that the majority of the world's H_2 is now kept in tanks as a compressed gas or cryogenic liquid. Due to the fact that the majority of feasible underground storage facilities are used for the preservation of natural gas, there are currently only a small number of projects in the UK and the USA that involve the underground storage of hydrogen in salt caverns. Economical considerations are important because an H_2 cavern holds around four times less energy than an equal natural gas cavern (de Jong et al. 2017).

The practise of storing H_2 in salt caves, however, might become more widespread in the future. The most important details in this text are that high-pressure tanks of sufficient size that store enough hydrogen (as a gas) to be useful to energy systems on a nation scale are impractical, and that compressed ammonia in a transportable vessel (tank) is cheaper than preserving either compressed or liquefied hydrogen in a tank (120–180 USD/mt for compressed NH_3 vs. 145–234 USD/mt for compressed H_2 or 820–1,990 USD/mt for liquefied H_2). Though the total costs (production + transportable storage) would be higher for hydrogen (506–566 USD/mt-H_2 for NH_3 vs. 366–455 USD/mt for H_2), they would still be lower overall. However, if investors additionally factor in capital expenses, the odds will once more be in favour of "green" ammonia since the capital cost of an ammonia storage facility is roughly 25 times lower than that of a hydrogen storage facility for every unit of stored energy.

Decarbonized energy and power systems can benefit greatly from the long-term storage of huge amounts of energy in a transportable form. Only a few electrochemical, mechanical, and chemical energy storage methods now on the market allow for the retention of energy for lengths longer than a few weeks or months. Ammonia, hydrogen, and natural gas are significant options for chemical energy storage. Compared to hydrogen and ammonia, natural gas has a lot of advantages, but being a fossil fuel is a significant disadvantage. In comparison to ammonia, hydrogen has a higher vapour pressure and lower volumetric energy, making its handling more difficult, costly, and dangerous.

Green ammonia production may increase as a result of government incentives for decarbonization, such as emission limits, carbon levies, subsidies, or investor tax benefits. Only when done on a limited scale in specific geographical areas where the weather and environment are conducive to the creation of ultra-cheap power does green ammonia manufacturing appear to be economically viable. In order to promote the widespread adoption of this storage system, policy and regulatory assistance are required.

26.4 Production of Biohydrogen and Other Biofuels

Due to the scientific, economical, and environmental challenges it faces, the generation of microalgal biohydrogen is becoming more and more important (Mohan and Pandey 2013). In order to achieve long-term sustainability, it is important to optimise experimental parameters, metabolic engineering, and genetic modification. The cost of producing biohydrogen ranges from \$10 GJ-1 to \$20 GJ1.

A significant competitor in the energy market and a potential substitute for many conventional energy sources is the production of biohydrogen. One of the various types of bioenergy products that can be made from the biomass of microalgae is biohydrogen, which is among the most promising. Direct biohydrogen synthesis by microalgae and the use of microalgal biomass as a fermentation substrate by other bacteria are the two ways that microalgae can make biohydrogen (Mathews and Wang 2009). The first approach may produce biohydrogen in the range of 0.015 to 1.084 mmol/L.h, whereas the second process can do so in the range of 0.35 to 10.26 mmol/L.h, depending on the feedstocks, growth conditions, and microorganisms used. The routes used by microalgae to produce biohydrogen are processed. The following subsections discuss various approaches to using microalgae to generate biohydrogen.

The process of manufacturing biohydrogen through fermentation is safe for the environment, uses less fossil fuel, and emits less pollutants. It can be divided into two categories: dark fermentation and photo fermentation, both of which hold tremendous promise as alternatives to fossil fuels.

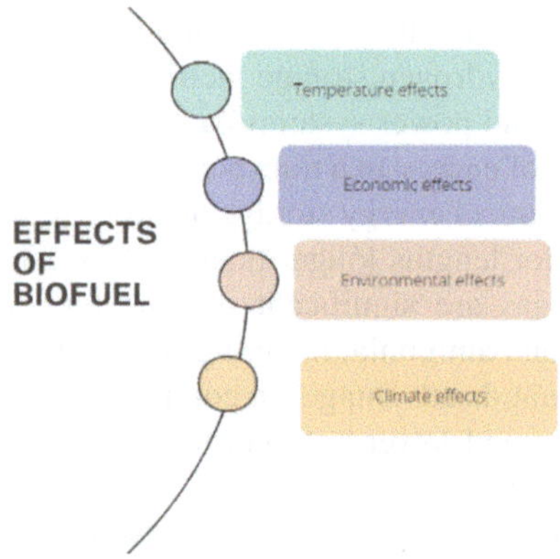

Effects of biofuel (Lundgren 2008)

In Photosystem I (PSI) and Photosystem II (PSII), these electrons are transferred to the plastoquinone (PQ) pool. Nicotinamide adenine dinucleotide phosphate (NADPH)-plastoquinone oxidoreductase (PQOR) is involved in the transport of electrons to the PQ pool in the chloroplast of different kinds of vascular plants. The production of biohydrogen benefits from the incubation of microalgae cultures when DCMU (3-(3,4-dichlorophenyl)-1,1-dimethylurea) is present.

In the dark fermentation pathway, a heterotrophic fermentation technique, green algae are employed to produce biohydrogen. Materials including glucose, sucrose, starch, and cellulose are used to hasten the creation of hydrogen (Soares et al. 2020).

Compared to photo and dark fermentation, in terms of the production of biohydrogen. The most successful photo fermentation was discovered to have a high hydrogen content.

Utilising synergistic microbial conversion assisted by light, modern technology known as the microbial photoelectrochemical cell (PEC) creates sustainable biohydrogen from organic waste. Water is split into hydrogen and oxygen using sunlight, and the device is comprised of semiconductors that are separated by a membrane in an electrolyte (Margaritondo 2005). Using photo-bioanodes with two sides, biocarbon material, and photocatalyst (a-Fe_2O_3), it has been possible to accelerate the production of biofilms, improve extracellular electron transport, and enrich electrogens.

Two of the biggest producers in the world, Malaysia and Indonesia, produce 80% of the world's palm oil. In India, non-edible sources including jatropha, mahua, Karanja, neem, etc. are generally used in the production of ethanol, but palm oil is primarily used in Malaysia to make biodiesel. In the People's Republic of China (PRC), maize is the primary agricultural product utilised to make ethanol. Nigeria has a lot of potential for biofuel production in terms of land availability and biomass supplies. Brazil is another country on the list of those engaged in the production of biofuels.

Brazil could use a variety of crops to make biofuels, including physic nuts, sunflower, soybean, castor bean, African palm, cotton, peanut, linseed, sesame, and canola. Indian rosewood, Perkia biglobosa (Jacq.) R. Br. ex G. Don (locust bean), Delonix regia (Boj. ex Hook.) Raf. and Delonix regia (Boj. ex Hook.), locust beans, and Perkia biglobosa (Jacq.) R. Br. ex G. Don. RAF (locust bean) is a source of carbohydrates that can be utilised as a substrate in the fermentation process. Monocots, which comprise plants like corn, maize, wheat, sugarcane, sorghum, miscanthus, etc. as a source of bioenergy, contain a single cotyledon. Although maize is a bioenergy source, there are significant costs associated with using it as a fuel (Carpita and McCann 2020). A substance known as maize stover is left on the soil's surface after the harvest of maize grains. It can be utilised directly as an energy source in biomass cofiring applications oas a raw material in thermochemical or biochemical conversion processes to create liquid fuels. It can be used to make bioenergy without immediately affecting food output.

The production of biofuels utilising food oils as a feedstock is restricted due to its cost and source. Mahua, Pongamia, Jatropha, and Pongamia are examples of non-edible oils that can assist in lowering the cost of biodiesel production. Several methods, such as micro-emulsification, pyrolysis, and transesterification, can be used to lower the viscosity of vegetable oils so they can be used as fuels. Because of its short reaction time and large production, transesterification is the process that is most usually utilised to create biodiesel for commercial use.

A microwave is an electromagnetic radiation with a frequency range of 0.3 to 300 GHz. They use a rapid heating process and heat generation within the materials, which results in greater yields with a quick response. Biodiesel has been produced using microwaves from feedstocks high in free fatty acids, including jatropha oil. Despite being a well-known technology, the transesterification process used to produce biodiesel has high production costs due to the high cost of heating methods. Among these procedures, microwave transesterification is a quick and effective approach to turn diverse feedstocks into biodiesel (Bart et al. 2010).

Despite being a well-known technology, the transesterification process used to produce biodiesel has high production costs due to the high cost of heating methods (Fayyazi et al. 2015). Among these procedures, microwave transesterification is a quick and effective approach to turn diverse feedstocks into biodiesel. When ultrasound is used as a pre-treatment for lignocellulosic biomass, sugar dissolution and maize starch hydrolysis can both be expedited for fermentation.

Genetic engineering facilitates the development of bioenergy crops and the subsequent production of biofuels. It can be used to alter characteristics such the production of cellulases and other hydrolytic enzymes, the presence of cellulose, and the presence of lignin. Genetic engineering also entails cell wall manipulation, which modifies and lowers the amount of lignin in plant cell walls to make them suitable for use as a source of bioenergy (Rocha-Meneses et al. 2020). Although lignin occasionally does so, there is no increase in pathogen susceptibility following lignin modification or when the amount of lignin in plant cell walls is decreased. Plant cell walls have been altered, which has led to an increase in pathogen resistance in plants.

The primary components of the cell wall are lignin, pectin, cellulose, and hemicellulose. The enzymes cellulases and hemicellulases can convert these polysaccharides into fermentable sugars. Lignin, a monoolignol composed of alcohols like pcoumaryl, coniferyl, and sinapyl, is one of the most common aromatic polymers found in plant cell walls. Although eliminating lignin from plant cell walls has drawbacks, some lignin levels have been lowered through genetic engineering, which promotes better biomass digestion. Cellulose is the most common polymer found in plant cell walls and is composed of cellulose, hemicellulose, pectin, and lignin. The most important topic in this book is the advantages and disadvantages of biofuels, which are made from plants and crops. Cellulose is the world's largest organic carbon reservoir, making up between 15 and 30% of its dry mass in both main and secondary cell walls. Pectin, which accounts for 30 to 35% of the dry weight in nongraminaceous monocots, is a component of plant cell walls. To effectively produce biofuels from raw feedstocks, the proper pectin extraction and degradation optimisation procedures must be used. Two advantages and disadvantages of using biofuels include reduced carbon emissions and help in reducing greenhouse gas emissions.

26.5 Implementation of Biofuels

The restricted water consumption of many first generation biofuels becomes a more tangible issue for nations when their water force rapidly declines. However, it is projected that the production of biofuel will keep rising because nations are looking for sustainable energy sources to substitute nonrenewable energy sources that are running out. The decision over which biofuels to employ becomes in fact increasingly pertinent as developing nations try to scale up biofuel production in order to have a practical source of renewable energy (Ziolkowska and Simon 2014) Many of these nations struggle with poverty, which would be made worse if implicit food sources were used to make first generation biofuels. Countries should think about how to gauge up product while concurrently conserving as much of their water force as possible, using the available land efficiently, and using feedstocks that are water-effective and which won't deplete the available food force. Biofuel production is expanding encyclopedically on a rapid-fire scale.

The key recommendation made by Mission 2017 is a 3-step overall plan to make the production of biofuels more water-efficient. As soon as practicable, switch from heavily water-consuming feedstocks to those that are more water-efficient, such as sweet sludge for ethanol production and won oil painting for biodiesel production. Make the switch to lignocellulosic biofuels and other alternate generation biofuels during the next 20–30 years. It would be impolite to start committing these crimes without knowing enough about the economics of these biofuels. To make the production of lignocellulosic biofuel more affordable and hence more competitive economically, further research must be done. Over time, make an effort to switch to biofuels that are mostly water-effective, like algae. Algae-based biofuel production is still in its infancy, and price barriers must be removed to make the process more commercially viable. The following fashions are also advised for investigation, testing, and use inorder to decrease water use (Gerbens-Leenes and Hoekstra 2011). Using agrarian and horticultural remainders and external solid wastes for biofuel product. Shifting the amalgamation used in colorful different styles of transportation depending on the water consumption of biodiesel. Emphasize the significance of water consumption in all biofuel exploration. Advantages of using sweet sludge for ethanol product lately, numerous developing countries have been giving high precedence to the use of sweet sludge in ethanol product. Advantages of sweet sludge over sludge number advanced photosynthetic effectiveness, advanced stress forbearance, lower toxin demand, rapid-fire growth rate, ease of planting, wide rigidity and further ethanol per unit area than sludge in addition to food and fodder for cattle.

In order to further expand oil painting win colonies, native rainforests have been destroyed in the major win oil painting-producing countries of Malaysia and Indonesia due to a lack of available land. Mission 2017 urges these nations to halt deforestation since it has a negative impact on biodiversity and the carbon cycle that is incalculable (M. O. S. Dias et al.). It is important to remember the worth of the earth's biodiversity for future generations. Using a combination of private and governmental funds, we strongly advised that countries protect their timbers from the growth of the win oil painting assiduity by setting aside timber tracks as reserves. This would effectively protect the timbers from the assiduity's growth.

In many underdeveloped nations, cooking using energy-efficient oil paintings is common. However, if won oil painting were to be used to produce biodiesel rather than food, the price would rise significantly. However, if a country has a water shortage, it should be prepared to pay this price (T. Hasunuma et al.)

Significant amounts of carbon dioxide and trapped carbon are released into the atmosphere as a result of the won oil painting industry's growth into peat bog territory in nations like Indonesia and Malaysia. These nations should stop developing in peatland because one of the key goals of generating biofuels has been to lower hothouse gas emigrations. It is explosively advised that nations use the identical recommendations put forth for the issue of deforestation to cover peat bogs to prevent the expansion of the win oil painting assiduity. Win oil painting can still be used as a feedstock for biodiesel if countries can produce it without significantly affecting the terrain by destroying peat bogs and rainforests (Y. Ge et al.) Oil painting win colonies are capable of expanding in the tropics of Africa, the Americas, and Asia. 90% of the merchandise marketed globally comes from Malaysia and Indonesia, the two countries that manufacture the most won oil paintings. Environmental worries about deforestation are currently present, nevertheless, due to the restricted amount of land that may be used for expansion. This issue serves as a reminder that, despite the incalculable ecosystem services given by peat bogs and rain forests, there is much less pressure to make biodiesel than there is to preserve the ecological legacy of the globe.

Agricultural and horticultural leftovers, wood leftovers, external solid waste, animal manure, indestructible meadows, bioenergy crops, submerged shops, and paper and cotton wastes are all sources for lignocellulosic-based biofuels. In its 2008 study, the International Energy Agency predicted that marketable biofuels would be produced by 2015, widespread biofuels by 2020, and competitive biofuels before 2030 (S. A. Scott et al.). Therefore, it is reasonable to expect that by 2020, countries will increase their investments in lignocellulosic biofuels. Mission 2017 advises that until these sources of energy are economically viable, governments should plan to use sweet sludge for ethanol production and won oil painting for biodiesel production.

This only applies to situations where trees are cultivated to provide lignocellulose as feedstock for biofuels. Because doing so would upset the carbon balance in the landscape and have other detrimental effects on the environment, there shouldn't be a significant change in the amount of lignocellulose collected (S. N. Naik et al.)

To aid with waste/burning implied biofuel sources, timbers need to be controlled (J. Z. Lee et al.) If a country hasn't already done so, it is crucial that they create associations to address the state of its forests.

26.6 Engineering of Plant Cell Walls for Enhanced Biofuel Product

To put this into perspective, consider that in just two days, land businesses produce the entirety of the chemical assiduity produced globally (1.2 billion tons/year; BASF report 2011). During this process, shops absorb large amounts of atmospheric carbon dioxide through photosynthesis, deposit it in the wall, and then store it there, making the wall the planet's primary carbon-insulation system. However, the wall should not be viewed as a simple carbon storing structure but rather as a complex material compound consisting of several polymer networks that serve numerous vital roles during a factory's lifespan (Pauly and Keegstra 2008) Glucans, which form cellulose microfibrils, substituted polysaccharides, which represent colourful hemicelluloses, and a polyanionic polysaccharide network, the pectins, are the main components of a primary wall that surrounds factory cells while they are still protracting. Some cell types deposit a secondary wall after cell expansion stops. Secondary walls also consist of polysaccharides, such as coloured hemicelluloses and cellulose microfibrils, but the water content is significantly reduced (5) because of the presence of lignin, a hydrophobic polyphenol.

According to the factory species, towel type, and experimental condition of the towel (3), the wall's precise composition differs. While the top characteristics of poplar wood and lawn straw, for example, are comparable, as illustrated in Fig. 26.1, there are differences in the comparable abundance of cellulose, non-cellulosic polysaccharides, and lignin as well as fresh distinct differences. The abundance of mannose-rich polymers and glucuronosylated xylans in hardwood as opposed to arabinosylated xylans, and the existence of p-hydroxy-cinnamic acids akin to p-coumaric acid and ferulic acid in the lignin of grass straw are a few examples of these distinctive structural components (Loqué et al. 2015) Due to their "cornucopia factory," or "lignocellulosics," cell walls are permitted.

To serve as a renewable resource for the production of biofuels and/or other commodity chemicals, particularly in light of the fact that lignocellulosic feedstocks don't serve as a human food source (4). Lignocellulosic material is broken down

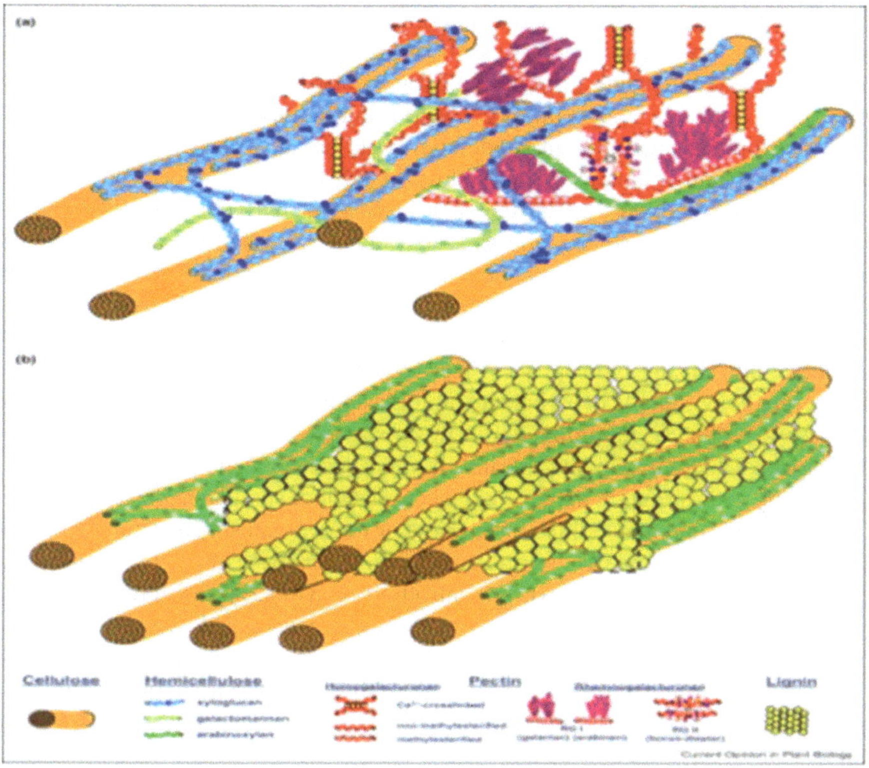

Fig. 26.1 Schematic models of plant cell walls. The emphasis is placed on wall composition rather than architecture. (**a**) Pectinaceous, aqueous primary cell walls found in dicots. (**b**) Lignified secondary walls with only minor water content (Loqué et al. 2015)

using a chemical pretreatment at a high temperature in current marketable biorefinery processes, and the polysaccharides are transformed to monosaccharides by a combination of enzymes. The sugars may also undergo chemical catalysis or microbial instigation to undergo reformation in order to produce the required chemical or energy. Several obstacles must be cleared in order for the biorefining procedures to be economically viable. One chain involves selecting appropriate feedstocks that are adapted to particular environmental rainfall/growth circumstances in order to increase the output of lignocellulosic biomass per ha, as an example, in order to optimise quantitative agronomic traits (4). Another chain comprises the lignocellulosic biomass's qualitative characteristics, such as lowering resistance to declination by making the carbohydrates more palatable to enzymes or optimising the composition for the upheaval step (5, 6).

The defiance has recently been successfully addressed by a number of factory biotechnology methods without sacrificing factory development or biomass accumulation (7, 8). The development of these strategies is then discussed, starting with the first step of reducing resistance by altering the lignin polymer network, followed by the second step of removing processing obstacles that arise from the biomass, and the third step of enhancing turmoil yields by incorporating the plethora of readily fermentable sugars in lignocellulosics. Modifying the network of lignin polymers The Tracheophyte area's lignin production route has historically been extremely well described and maintained (Ciesielski et al. 2014).

Lignin is substantially composed of p-hydroxyphenyl (H), guaiacyl (G), and syringyl (S) units deduced from the arbitrary polymerization of p-coumaryl alcohol, coniferyl alcohol and sinapyl alcohol, independently. Significant progress on the revision of lignin structure, content or distribution has been achieved within the last times to reduce overall biomass defiance and ameliorate saccharification yield without compromising biomass yield (see recent reviews (9, 10, 11)). The lignin-engineering field is fleetly evolving and multiple approaches are presently developed to reduce its defiance, minimize lignin waste sluice, and/ or valorize it (7, 12, 13, 14). Rapid advances in genomics, transcriptomics and synthetic biology contributes significantly to the progress in lignin engineering. Omics studies enable the discovery of new enzymes and metabolites from shops or microorganisms that can latterly be used to manipulate lignin content and composition. Likewise, they grease the reconstruction of nonsupervisory networks controlling biosynthesis of different secondary cell wall factors and the identification of promoters with asked expression biographies that can latterly be used to wangle shops and optimize towel specific expression of transgenes. An illustration of largely successful engineering involved switching the protagonist driving the expression of the Arabidopsis cinnamate 4–hydroxylase (C4H) with a vessel-specific protagonist to circumscribe C4H expression to vessel cells and dissociate it from the fiber nonsupervisory network.

This method stressed the significance of vessel cell wall integrity for optimal factory development and the role of synthetic biology in lignin engineering. the method depended on vacuity of a sheepish lignin biosynthesis mutant and could thus be delicate to restate to a range of crops. the method allowed reducing lignin content by more than 20 without compromising factory growth. The restoration of lignin technology from model shops to crops would be facilitated by the development of dominant inherited characteristics. In order to lessen the risk of introducing unwanted phenotypes, techniques for engineering should precisely target particular apkins and/or be switched on/off only at specified experimental stages of crops. The detrimental effects of lignin modifications on the phenylpropanoid-derived secondary metabolites, which are used for functions other than lignification, such as biotic stress tolerance, might be limited by fine-tuning lignin engineering pathways.

By using a secondary cell wall cellulose synthase protagonist (pAtCesA7) to control the expression of a microRNA raised against CCR1 (cinnamoyl-CoA reductase 1), a towel-specific dominant strategy was developed that successfully reduced lignin accumulation in interfascular filaments while maintaining vessel integrity. Recently, the first successful and enzymatically supported method for cutting the lignin content in half without reducing biomass yield was revealed. In order to decrease the vacuity of shikimate, which is required for the activity of hydroxycinnamoyl-CoA shikimate hydroxycinnamoyltransferase (HCT), a critical enzyme of the lignin pathway, the method focused on the development of a competitive metabolic route in cells forming secondary walls (Eudes et al. 2015).

Lowering obstacles to endogenous processes. The polysaccharides in lignocellulosic biomass comprise acetyl esters, methyl esters, and ethers in addition to carbohydrates. Because they prevent enzymes from accessing the polysaccharides, these changes to the carbohydrate boost defiance. Additionally, the esters are eventually released and help reduce back pain. By combining the acetylated polysaccharides, genetic snooping can be used to lower the lignocellulosic acetyl content. Reduced Wall Acetylation (RWA) proteins, Altered XYloglucan 9 proteins, and Trichome Birefringence/Trichome Birefringence—Like (TBR/ TBL) proteins are three families of enzymes that have been implicated as being necessary for wall polysaccharide O–acetylation. Based on research with Arabidopsis, decreasing the exertion of RWA proteins leads to stores with decreased wall acetylation. In knock-eschewal alleles of AXY9, the O–acetylation location is 70% lower in the stem and 30 percent lower in the splint towel. It is unclear exactly how the Arabidopsis factory-specific gene AXY9 contributes to the polysaccharide O-acetylation pathway. As they influence the acetylation of several polysaccharides, the RWA proteins and AXY9 appear to function upstream of the TBR–TBL proteins. TBL proteins have been demonstrated to be distinct for a certain kind of polysaccharide, in contrast. Nevertheless, inactivation of TBL29 causes a significant slowdown in factory growth and reduced biomass accumulation. However, it is plausible that this may be changed, for instance, by restricting the inactivation to fibre cells, as was done in recent research with xylan and lignin. Putting wall sugars in.

The main components of the wall now employed for biorefining are carbohydrates. The majority of these carbohydrates are used by agitating bacteria in the form of monosaccharides, particularly hexoses comparable to glucose. However, the wall also contains a sizeable amount of pentoses that are analogous to xylose and were derived from hemicellulose xylan. Pentoses cannot be initiated as effectively as hexoses, hence current efforts focus on decreasing xylan and increasing the hexose/ pentose rate in wall accoutrements.

Engineering glycan synthases and glycosyltransferases, which are directly involved in wall polymer aggregation, is a possible method for improving particular wall sugars. A multiple transmembrane gauging protein called CesA forms complexes in the tube membrane and is responsible for synthesising cellulose. A Rhodobacter cellulose synthase's protein structure has only recently been revealed, providing invaluable insight into how this enzyme functions (59). De novo and homology modelling of a cotton cellulose synthase's catalytic sphere reveals good agreement with the bacterial one, demonstrating that cellulose conflation is processive and does not, contrary to what was previously believed, bear a manual. In contrast to expectations, overexpression of CesAs in poplar resulted in a decrease in wall cellulose because indigenous CesAs was silenced. The shops displayed severe, abnormal, pleiotropic traits, indicating that changing CesAs' expression conditions would not be a wise course of action. Additionally, two mutant Arabidopsis plants with CesAs that had point mutations in various locations showed reduced phenotypes or a decrease in cellulose content, separately, which made biomass accumulation difficult. A reduction in cellulose crystallinity results from both mutations.

26.7 Challenges and Opportunities

A prospect of micro-algal civilization technologies for direct energy conversion to make biofuels has been suggested. Recent approaches for microbial biofuel product are well investigated and honoured. As an illustration, the novel platform of biomass product routes that will eventually be investigated for the manufacturing of biofuels may be the biofilm civilization of microalgae or cyanobacteria. The biofilm civilization concept holds promise for the production of biofuel from microalgae or cyanobacteria. Factory biomass has been the most well-known source of biofuels for a number of decades. Algal biomass is currently being attested to as a good source for biofuel products. The capacity of shops and algae to photosynthesize is a key factor in separating them from other sources. The conversion of sugars from atmospheric carbon dioxide using sun energy is known as photosynthesis. It is the unique way that green algae and factories focus on carbon in nature that shows that printing-conflation is the only process of growing factory and algal biomass, which is the starting point for the production of biofuels.

The initial molecular substrate for the creation of bioethanol and biomethanol is sugar. Because sunlight provides energy and atmospheric CO_2 provides carbon, using photosynthetic organisms as a source of biofuel is both affordable and feasible. The two stages of photosynthesis are light-dependent and light-independent.

Light energy is received, transformed into a charge separation, and ultimately transformed into the conflation of ATP and NADPH during the light-dependent stage. Chlorophyll in plants and bacteriochlorophyll in bacteria are the main photosynthetic pigments involved in light absorption. The energy and electrons from ATP and NADPH are used separately to create sugars during the light-independent stage. The electron transport chain, which consists of a series of electron carriers bedded in thylakoid membrane, is a crucial component of the photosynthetic apparatus. Light initiates electron transfer, which is carried out by electron carriers in vast membrane complexes connected by mobile electron carriers such plastoquinone, plastocyanin, cytochrome c, or ferredoxin. [A. Demirbas et al.).

The NADPH molecules are reduced to NADPH by the electrons from the water and the ferredoxin NADP oxidoreductase. In case of oxygenic photosynthesis, water acts as an electron source. Molecular oxygen and protons are evolved as a result of the water corruption. Under certain conditions, protons come upon hydrogenase, not upon ferredoxinNADP oxidoreductase, and molecular hydrogen is generated. This miracle indicates a medium for biohydrogen product in the photosynthetic outfit (M. Seibert et al.). The product of molecular hydrogen by the culture ofOne of the most promising methods for the creation of renewable energy is the production of molecular hydrogen through the cultivation of photosynthetic microorganisms (28e30). The process runs at room temperature and needs sunlight, water, and the bare minimum of macro- and micronutrients. Since the production of hydrogen through photosyn—thesis has no impact on the emigration of greenhouse gases or other environmental pollutants, it is predicted that future research will lead to the development of an environmentally beneficial platform for the production of artificial renewable energy. The only byproduct in both uses of biohydrogen is water, whether it be for powering energy cells or internal combustion engines directly. The poor yields of the available methods for sustainable biohydrogen production remain a problem and difficulty.

Many photosynthetic organisms are capable of producing hydrogen, and the list of hydrogen producers includes species from various prokaryote and eukaryotic taxa, particularly ii, the photosynthetic bacterial, cyanobacterial, and green algal species. Water cannot be used as an electron patron by photosynthetic microorganisms. Cyanobacteria and green algae can convert scalding water into molecule oxygen and protons in the light, in contrast. In both instances, the solar energy is transformed into chemical energy in compounds like ATP and NADPH, which are then used to fuel cellular metabolism, build biomass, and store sugars and other energy-added molecules. The production of biohydrogen can be done in two ways. Through turbulence and/or photofermentation, the collected biomass, particularly carbohydrates, is further transformed to biohydrogen. This strategy incorporates two independent stages that are either interspersed between the photosynthetic and turbulent ages or separated by response spatially in two different bioreactors. The alternative strategy, the main topic of this section, uses direct or circular water biophotolysis processes to use sunlight to convert water to hydrogen and oxygen. For many years, applied research and development has focused on the H_2 production by the green alga Scenedesmus obliques during light biophotolysis. Additionally, it has been observed that filamentous cyanobacteria Anabaena cylindrica exposed to argon (Ar) Atmosphere showed simultaneous H_2 product and O_2 elaboration.

In fact, traditional methods have been used for many years to utilise bioalcohol as an energy source. Currently, non-fossil essential transportation is viewed as being memoir alcohol. Currently, manufacturing material rich in sugars and carbohydrates, such as grain crops and sugarcanes, is the primary source of bioalcohols. However, in recent years, interest in indestructible meadows like switchgrass and Miscanthus has increased. These crops don't compete with food production, but the turbulence and distillation processes demand that the cellulosic biomass be first converted into sugars. The most prevalent bioalcohol is ethanol, while biopropanol and biobutanol are less prevalent. These alcohols are created when microbes ferment carbohydrate-modified feedstock. R. S. Poudyal and others. The amount of industrial biomass depends on how well plants use photosynthetic energy and the environment. Ethanol was traditionally utilised to produce alcohol-containing beverages. However, with the advancement of sanctification practises, ethanol use has mostly been expanded to other uses. To reduce carbon dioxide emissions, for instance, bioethanol can be used in place of petrol. Because oxygen exists in its molecular form, burning at relatively low temperatures is made possible while limiting the escape of various toxic byproducts such carbon monoxide (CO), nitric oxide (NO), and unexpected organic composites (Nigam et al. 2011).

Although they can be used to produce biohydrogen, algae are a different source of bioethanol. According to recent reports, algae serve as an implicit feedstock for the production of biofuels because they include 50 of the lipids needed to produce biodiesel and the remaining proteins and elemental sugars needed to produce bioethanol. Currently, fossil fuels are mostly used to make methanol. The chemical process that turns natural gas (CH4, or methane) into methanol in three steps—brume reforming, conflation, and distillation—is the basis for the production of methanol. Methanol is much simpler to recover than ethanol because ethanol and water combine to produce an azeotrope, which is difficult to purify. Because methanol doesn't form an azeotrope, recycling is simpler in this scenario. In spite of this, biomethanol is chosen for use in biodiesel (P. S. Nigam et al.).

Additionally, a biomethanol product created by gasifying carbohydrates and partially oxidising them with oxygen and water has been pictured. By swirling or distilling the crops that contain sugars and bounce, it is possible to create oxidation response in biomass yields in addition to partial oxidation of carbs. Bounce is transformed by this process into simple sugars, which are then transformed into bioethanol and biomethanol. From biomass and biodegradable waste, biomethanol can be made. It is possible to see biomass as an implicit source of energy for gasification, syngas production, and methanol conflation (79). All the biomass is transformed to biomethanol when sufficient hydrogen is added to the synthesised gas (80). The only type of photosynthetic organism that grows the fastest and is appropriate to produce a variety of bio-oils is microalgae. From biomass and biodegradable waste, biomethanol can be created. Biomass can be viewed as an implicit energy source for gasification, syngas production, and the subsequent methanol conflation process. All the biomass is transformed to biomethanol when sufficient hydrogen is added to the synthesised gas(80). The only type of photosynthetic organism that grows the fastest and is appropriate to produce a variety of bio-oils is microalgae. Oil painting makes up up to 50% of algae's weight, so using algae to manufacture biomethanol is a possibility. Similar to Spirulina sp., microalgae may also gasify microorganisms to make biomethanol. Due to its acute and toxic components, bioethanol has received more attention than biomethanol (Khan et al. 2009).

26.8 Economics of Biofuels

Due to the scientific, economical, and environmental challenges it faces, the generation of microalgal biohydrogen is becoming more and more important (Mohan and Pandey 2013). In order to achieve long-term sustainability, it is important to optimise experimental parameters, metabolic engineering, and genetic modification. The cost of producing biohydrogen ranges from \$10 GJ-1 to \$20 GJ1. A significant competitor in the energy market and a potential substitute for many conventional energy sources is the production of biohydrogen (Condon et al. 2015). One of the various types of bioenergy products that can be made from the biomass of microalgae is biohydrogen, which is among the most promising. Direct biohydrogen synthesis by microalgae and the use of microalgal biomass as a fermentation substrate by other bacteria are the two ways that microalgae can make biohydrogen (Mathews and Wang 2009). The first approach may produce biohydrogen in the range of 0.015 to 1.084 mmol/L.h, whereas the second process can do so in the range of 0.35 to 10.26 mmol/L.h, depending on the feedstocks, growth conditions, and microorganisms used. The routes used by microalgae to produce biohydrogen are processed. The following subsections discuss various approaches to using microalgae to generate biohydrogen. The process of manufacturing biohydrogen through fermentation is safe for the environment, uses less fossil fuel, and emits less pollutants. It can be divided into two categories: dark fermentation and photo fermentation, both of which hold tremendous promise as alternative fossil fuels.

In Photosystem I (PSI) and Photosystem II (PSII), these electrons are transferred to the plastoquinone (PQ) pool. Nicotinamide adenine dinucleotide phosphate (NADPH)-plastoquinone oxidoreductase (PQOR) is involved in the transport of electrons to the PQ pool in the chloroplast of different kinds of vascular plants. The production of biohydrogen benefits from the incubation of microalgae cultures when DCMU (3-(3,4-dichlorophenyl)-1,1-dimethylurea) is present.

In the dark fermentation pathway, a heterotrophic fermentation technique, green algae are employed to produce biohydrogen. Materials including glucose, sucrose, starch, and cellulose are used to hasten the creation of hydrogen (Soares et al. 2020).

Compared to photo and dark fermentation, in terms of the production of biohydrogen. The most successful photo fermentation was discovered to have a high hydrogen content.

Utilising synergistic microbial conversion assisted by light, modern technology known as the microbial photoelectrochemical cell (PEC) creates sustainable biohydrogen from organic waste. Water is split into hydrogen and oxygen using sunlight, and the device is comprised of semiconductors that are separated by a membrane in an electrolyte (Margaritondo et al. 2005). Using photo-bioanodes with two sides, biocarbon material, and photocatalyst (a-Fe_2O_3), it has been possible to accelerate the production of biofilms, improve extracellular electron transport, and enrich exoelectrogens.

Two of the biggest producers in the world, Malaysia and Indonesia, produce 80% of the world's palm oil. In India, non-edible sources including jatropha, mahua, Karanja, neem, etc. are generally used in the production of ethanol, but palm oil is primarily used in Malaysia to make biodiesel. In the People's Republic of China (PRC), maize is the primary agricultural product utilised to make ethanol. Nigeria has a lot of potential for biofuel production in terms of land availability and biomass supplies. Brazil is another country on the list of those engaged in the production of biofuels.

Brazil could use a variety of crops to make biofuels, including physic nuts, sunflower, soybean, castor bean, African palm, cotton, peanut, linseed, sesame, and canola.

Indian rosewood, Perkia biglobosa (Jacq.) R. Br. ex G. Don (locust bean), Delonix regia (Boj. ex Hook.) Raf. and Delonix regia (Boj. ex Hook.), locust beans, and Perkia biglobosa (Jacq.) R. Br. ex G. Don. RAF (locust bean) is a source of carbohydrates that can be utilised as a substrate in the fermentation process.

Monocots, which comprise plants like corn, maize, wheat, sugarcane, sorghum, miscanthus, etc. as a source of bioenergy, contain a single cotyledon. Although maize is a bioenergy source, there are significant costs associated with using it as a fuel (Carpita and McCann et al. 2020). A substance known as maize stover is left on the soil's surface after the harvest of maize grains. It can be utilised directly as an energy source in biomass cofiring applications or as a raw material in thermochemical or biochemical conversion processes to create liquid fuels. It can be used to make bioenergy without immediately affecting food output.

The production of biofuels utilising food oils as a feedstock is restricted due to its cost and source. Mahua, Pongamia, Jatropha, and Pongamia are examples of non-edible oils that can assist in lowering the cost of biodiesel production. Several methods, such as micro-emulsification, pyrolysis, and transesterification, can be used to lower the viscosity of vegetable oils so they can be used as fuels. Because of its short reaction time and large production, transesterification is the process that is most usually utilised to create biodiesel for commercial use.

A microwave is an electromagnetic radiation with a frequency range of 0.3 to 300 GHz. They use a rapid heating process and heat generation within the materials, which results in greater yields with a quick response. Biodiesel has been produced using microwaves from feedstocks high in free fatty acids, including jatropha oil. Despite being a well-known technology, the transesterification process used to produce biodiesel has high production costs due to the high cost of heating methods. Among these procedures, microwave transesterification is a quick and effective approach to turn diverse feedstocks into biodiesel (Bart et al. 2010).

Despite being a well-known technology, the transesterification process used to produce biodiesel has high production costs due to the high cost of heating methods (Fayyazi et al. 2015). Among these procedures, microwave transesterification is a quick and effective approach to turn diverse feedstocks into biodiesel. When ultrasound is used as a pre-treatment for lignocellulosic biomass, sugar dissolution and maize starch hydrolysis can both be expedited for fermentation.

Genetic engineering facilitates the development of bioenergy crops and the subsequent production of biofuels. It can be used to alter characteristics such the production of cellulases and other hydrolytic enzymes, the presence of cellulose, and the presence of lignin. Genetic engineering also entails cell wall manipulation, which modifies and lowers the amount of lignin in plant cell walls to make them suitable for use as a source of bioenergy (Rocha-Meneses et al. 2020). Although lignin occasionally does so, there is no increase in pathogen susceptibility following lignin modification or when the amount of lignin in plant cell walls is decreased. Plant cell walls have been altered, which has led to an increase in pathogen resistance in plants.

The primary components of the cell wall are lignin, pectin, cellulose, and hemicellulose. The enzymes cellulases and hemicellulases can convert these polysaccharides into fermentable sugars. Lignin, a monoolignol composed of alcohols like pcoumaryl, coniferyl, and sinapyl, is one of the most common aromatic polymers found in plant cell walls. Although eliminating lignin from plant cell walls has drawbacks, some lignin levels have been lowered through genetic engineering, which promotes better biomass digestion. Cellulose is the most common polymer found in plant cell walls and is composed of cellulose, hemicellulose, pectin, and lignin.

The most important topic in this book is the advantages and disadvantages of biofuels, which are made from plants and crops. Cellulose is the world's largest organic carbon reservoir, making up between 15 and 30% of its dry mass in both main and secondary cell walls. Pectin, which accounts for 30 to 35% of the dry weight in nongraminaceous monocots, is a component of plant cell walls. To effectively produce biofuels from raw feedstocks, the proper pectin extraction and degradation optimisation procedures must be used. Two advantages and disadvantages of using biofuels include reduced carbon emissions and help in reducing greenhouse gas emissions.

Fossil fuels will be mostly obsolete. The use of biofuels in nature is more effective because the majority of supplies, including manure, corn, soybeans, switchgrass, waste from crops and plants, are renewable resources. These plants can be planted, grown, and utilised again.

Sugar, starch, oilseeds, and animal fats are the main ingredients of first-generation biofuels. Ethanol is the most prevalent bioalcohol.

While the mitigation of global warming is projected to provide income for the community, global warming has a significant role in affecting the economy of the society. The availability of environmental benefits that reduce carbon dioxide emissions is a crucial factor in the rising use of biofuels. the broad perspective of biofuels as the source of net energy gain, acquiring more energy than that used in the production of fossil fuels. According to recent study, biofuels won't have any favourable climate effects (Lundgren et al. 2008).

The usage of biofuels will aid in the US and the world's transition away from oil, along with long-term alternatives including increasing fuel economy standards, tax incentives, and renewable fuels.

Vehicles fueled by biodiesel have a 30% greater fuel economy than those powered by petrol, which reduces emissions and maintenance costs while also saving drivers money.

Oil refineries transform crude oil into useful goods, but they also discharge millions of pounds of pollutants, including several carcinogens, into the environment.

Biofuels will take the place of fossil fuels and offer a number of advantages. Biofuels are created from organic renewable sources, as opposed to fossil fuels, which are finite resources. As a result, their manufacture and use will continue forever. Because they can be produced domestically, biofuels may have lower transportation costs.

To lessen our reliance on fossil fuels as crude oil prices climb, we need a number of other alternative energy solutions. Some pollution and emissions may be reduced by biofuels. On full combustion, ethanol can lower carbon monoxide emissions. Due to its high saturated fat content, biodiesel can operate in existing diesel engines without modification, however it may perform poorly in cold areas. Canola biodiesel, on the other hand, is lower in saturated fat and increases engine longevity (Prieler et al. 2012).

Although government support for the biofuel sector can aid in Tanzania's fight against poverty, switching from food to fuel production has drawbacks. Governments must guarantee a sufficient supply of crops for both food and biofuels. As they release less pollutants and are cleaner to operate, biofuel refineries are more environmentally benign than ethanol plants that burn coal for energy. Nearly 10,000 individuals perish each year as a result of petrol engine pollution. Petrol releases acetaldehyde and nitrogen oxide in addition to other pollutants and vulnerables. When aldehyde molecules interact with sunlight, smog will result. More people get malignancies and respiratory illnesses each year as a result of these pollutants. Compared to fossil fuels, biofuels emit fewer pollutants into the atmosphere. Biodiesel reduces the production of smog-creating particulate matter when compared to regular diesel, which lowers the incidence of asthma and other respiratory ailments. Additionally, biodiesel doesn't release any sulfites or sulphur oxides, which may be responsible for acid rain (Huang et al. 2013).

Increased land use, polluting inputs, higher food prices, and resource competition may result from the development of biofuels. By releasing terrestrial carbon stocks into the environment, GHG emissions may be raised as a result of changes in land use patterns. The biofuel feedstocks, such as soybeans, oil palm from the Amazon Forest and southeast Asia, that are cultivated in land that has been cut from forests especially produce GHG emissions (Melillo et al. 2009).

GHG emissions can be produced during the preparation of biofuels. Nitrous oxide, a tiny amount of greenhouse gases, are released when fertiliser is applied to the soil. The majority of biorefineries will run on fossil fuels.

According to (Mosnier et al. 2013), the production of biofuel may result in worsened air quality, depletion of aquifers, and increased water pollution. The claim that biofuels, especially ethanol, need more energy to create than they produce is frequently made in opposition to them. Additionally, fossil fuels are needed for biofuels. This net energy balance is primarily improved for ethanol produced in Brazil that is grounded with sugar and soy. Despite the fact that it is extremely difficult to grow this crop in terms of both finances and land, American biofuel is predominantly sludge-based. Fossil fuels still cost less to generate than biofuels, despite their rising use, raising concerns about the use of biofuels from a business standpoint. Over the past ten years, the price of feedstock has climbed by 37 for bagasse in Brazil and by 40–50 for sludge in the United States. Similar to how the ratio of crude oil painting to biofuel feedstock has remained fairly stable, any decrease in the price of biofuel typically corresponds with a rise in the price of crude oil painting. Biofuels are more attractive from a financial standpoint as oil painting becomes even more limited and only available in unstable locations. Due to these price changes, oil painting can no longer be considered a sustainable source of energy. An enormous amount of pressure has been placed on the biodiesel assiduity as a result of the circumstance where energy demand is significantly lower as a result of force chain and distribution disruptions as well as the dropped product in feedstocks (Pereira da Silva and Ribeiro 2019).

26.9 Future Perspectives

India's agriculture is one of the fastest-growing in the world. Despite being a coal-rich and renewable energy-rich country India, like many other emerging nations, is a net energy importer. India, like many other emerging nations, is a net energy importer. significances in the form of 25 of the fundamental energy requirements are being satisfied. India is the fifth-ranked country in the world for primary per capita consumption, which is still incredibly low when compared to other emerging nations. India needs to heavily expand its energy industry in order to support its planned development's lucrative growth rate of over 6% annually. The majority of India's energy needs are met by conventional and commercial energy sources, such as energy In rural households, the possibility of similar energy sources is projected to be 155 MTOE annually. Homes, the demand for energy has also increased, particularly for electricity and for electricity product and consumption, which has improved the investment prospects in the electricity sector and has momentarily widened openings to power Indian electricity generation policy (Rosegrant et al. 2008).

India is the fourth-largest producer of edible oil, accounting for 9.3% of the world's total oil seed production. As a result, the possibility of creating biodiesel from edible oil resources in India is almost nonexistent. To meet the needs of the domestic market, edible oil is imported. India's use of and consumption of edible oil. According to a Government of India report, 60 Mha of the total land area are classified as being used for crops such Jatropha curcas and Pongamia, the oil from which can be used to make biodiesel (Kerckow et al. 2010). India also has the third-largest road network in Asia.

Jatropha curcas as a substitute for petro-diesel in india. Jatropha curcas is an imperishable plant that thrives in marshy environments and is failure-resistant. It is simple to establish, develops quickly, and has a 50-time lifespan. The miracle plant, Jatropha curcas, yields seeds with a 37 percent oil painting content. Without meliorating, oil painting can be burned as energy. India is receiving the future source of biodiesel from jatropha curcas. Jatropha curcas is being grown as part of a biodiesel production programme on waste land. Jatropha curcas, one of the vibrant oil painting seeds, has been developed to be more appropriate for biodiesel (Searchinger et al. 2008).

Jatropha curcas is an imperishable plant that thrives in marshy environments and is failure-resistant. It is simple to establish, develops quickly, and has a 50-time lifespan. The miracle plant, Jatropha curcas, yields seeds with a 37% oil painting content. Without meliorating, oil painting can be burned as energy. India is receiving the future source of biodiesel from jatropha curcas. Jatropha curcas is being grown as part of a biodiesel production programme on waste land. Jatropha curcas, one of the vibrant oil painting seeds, has been developed to be more appropriate for biodiesel (Searchinger et al. 2008). Currently unutilized Mha of land will be planted with jatropha in order to produce 13 MT of biodiesel by 2013, which will be blended with petro-diesel. In favour of widespread use of jatropha curcas. There are other motivations for using Jatropha curcas on a wide basis as well. A first phase of the

National Programme on Jatropha, approved with a volume of Rs. 1.500 Crore and 400,000 ha, was released in March 2004 with Rs. 800 Crore to assist the civilization of Jatropha. By 2006, 2.6 MT of Jatropha biodiesel was expected to be generated on 2.2 Mha, based on yields. An arrangement of peripheral plutocrat, subvention, and loan was envisaged for 2007, when the process was intended to move in a tone-sustaining manner. The programme addressed worldwide environmental enterprises and their advancements, thus the expansion of processing capacity is intended to run on a 30 new support for significantly Funding Agencies (Nichols 1986).

Programmes promoting biofuels have an impact on the demand for goods related to biofuel production because they encourage the use of sugarcane, sugar beet, sludge, and oil painting seeds. The main factors driving countries to choose green energy products are the desire (or necessity) to lessen reliance on fossil fuels (energy security), to reduce GHG emigrations (climate change mitigation), and to do so. Energy security, for instance, could be achieved not only by promoting the use of biofuels but also through other domestically produced renewable energy sources, similar to how the use of biofuels has helped to reduce GHG emigrations. Since biofuels are primarily produced from agricultural sources, they are marketed as a fashionable way to meet the demand for increased energy. The idea for biofuels first emerged in the 1970s. According to Rosegrant et al. (2008), biofuels can aid on a global scale because they are renewable and their treasuries are dispersed more unevenly than those from nuclear other reactive sources.

26.10 Conclusion

For poor net buyers of food masses in both civic and pastoral areas, advanced food prices performing in part from increased biofuel demand will pose an immediate trouble to their food security. Indeed if biofuels are only one of several sources of the recent sharp increases in food prices, expanded biofuel products can still continue to ply upward pressure on food prices for considerable time to come. The immediate impact of high food prices on the poor can be eased through meetly designed and targeted safety nets that support access to food. At the same time, it's important to allow rising prices to feed through to growers so as to spark a possible force response. Assessing price controls and import bans, as numerous countries have done in 2008 in sweats to cover consumers from high prices, prevents requests from conforming and, while furnishing an apparent short-term relief, may actually protract and consolidate the food-security crisis. However, advanced prices will give an incitement for increased product and increased employment, which may palliate food-security enterprises over the longer term, If requests are allowed to serve and price signals are effectively transmitted to directors.

It is only reasonable to assume that liquid biofuels derived from agricultural crops will contribute minimally to the world's supply of transport energies and even less to the world's total energy reserves. The expansion of biofuel products sharply drives up the price of agricultural feedstocks and renders them non-competitive versus petroleum-based energies because agrarian demands are minimal in comparison to energy demands. However, nations with abundant natural resources that can generate feedstocks at a price that is competitive and utilise them effectively may be able to grow a viable biofuel industry. The financial viability of biofuels may also vary due to unanticipated changes in energy demands. Technological advancements, such as the creation of alternative-generation biofuels based on cellulosic feedstocks, may increase the number of countries, both explicitly and implicitly, where biofuels may be useful.

Acknowledgements We are thankful for the DEPARTMENT OF BIOTECHNOLOGY, RAJALAKSHMI ENGINEERING COLLEGE and our Mentors Dr B. VIJAYA GEETHA (Associate Professor), K. S. SHREENIDHI (Assistant Professor) for their patience and guidance throughout this book chapter.

References

Ababneh H, Hameed B (2022). Electrofuels as emerging new green alternative fuel: a review of recent literature. Energy Convers Manag 254:115213. https://doi.org/10.1016/j.enconman.2022.115213

Aden A, Foust T (2009) Technoeconomic analysis of the dilute sulfuric acid and enzymatic hydrolysis process for the conversion of corn stover to ethanol. Cellulose 16(4):535–545. https://doi.org/10.1007/s10570-009-9327-8

Bart JC, Palmeri N, Cavallaro S (2010) Feedstocks for biodiesel production. Biodiesel Sci Technol 130–225. https://doi.org/10.1533/9781845697761.130

Carpita NC, McCann MC (2020) Redesigning plant cell walls for the biomass-based bioeconomy. J Biol Chem 295(44):15144–15157. https://doi.org/10.1074/jbc.rev120.014561

Ch Venkata MS, Kakarla R, Shetti RR, NPMSR, Anjanapura VR (2019) Novel Co and Ni metal nanostructures as efficient photocatalysts for photodegradation of organic dyes. Mater Res Exp 6(12):125502. https://doi.org/10.1088/2053-1591/ab5328

Ciesielski PN, Resch MG, Hewetson B, Killgore JP, Curtin A, Anderson N, Chiaramonti AN, Hurley DC, Sanders A, Himmel ME, Chapple C, Mosier N, Donohoe BS (2014) Engineering plant cell walls: tuning lignin monomer composition for deconstructable biofuel feedstocks or resilient biomaterials. Green Chem 16(5):2627. https://doi.org/10.1039/c3gc42422g

Condon N, Klemick H, Wolverton A (2015) Impacts of ethanol policy on corn prices: a review and meta-analysis of recent evidence. Food Policy 51:63–73. https://doi.org/10.1016/j.foodpol.2014.12.007

de Jong S, Hoefnagels R, Wetterlund E, Pettersson K, Faaij A, Junginger M (2017) Cost optimization of biofuel production—the impact of scale, integration, transport and supply chain configurations. Appl Energy 195:1055–1070. https://doi.org/10.1016/j.apenergy.2017.03.109

Demirbas A (2009) Political, economic and environmental impacts of biofuels: a review. Appl Energy. https://doi.org/10.1016/j.jclepro.2014.01.006

Deng A, Ren J, Li H, Peng F, Sun R (2015) Corncob lignocellulose for the production of furfural by hydrothermal pretreatment and heterogeneous catalytic process. RSC Adv 5(74):60264–60272. https://doi.org/10.1039/c5ra10472f

de Vries SC, van de Ven GW, van Ittersum MK, Giller KE (2010) Resource use efficiency and environmental performance of nine major biofuel crops, processed by first-generation conversion techniques. Biomass Bioenergy 34(5):588–601. https://doi.org/10.1016/j.biombioe.2010.01.001

Dias MOS et al. (2009) Production of bioethanol and other bio-based materials from sugarcane bagasse: integration to conventional bioethanol production process. Chem Eng Res Des. https://doi.org/10.1002/bbb.1662

Eloka-Eboka AC, Maroa S, Behera S (2021) Algal biofuels—technologies, scope, opportunities, challenges, and applications. Sustainable Biofuels 449–470. https://doi.org/10.1016/b978-0-12-820297-5.00016-5

Ensinas AV et al (2007) Analysis of process steam demand reduction and electricity generation in sugar and ethanol production from sugarcane. Energy Convers Manag. https://doi.org/10.1016/j.energy.2010.08.011

Eudes A, Sathitsuksanoh N, Baidoo EEK, George A, Liang Y, Yang F, Singh S, Keasling JD, Simmons BA, Loqué D (2015) Expression of a bacterial 3-dehydroshikimate dehydratase reduces lignin content and improves biomass saccharification efficiency. Plant Biotechnol J 13(9):1241–1250. https://doi.org/10.1111/pbi.12310

Fan Q, Liu X, Xu L, Subash AA, Brackmann C, Aldén M, Bai XS, Li Z (2022) Flame structure and burning velocity of ammonia/air turbulent premixed flames at high Karlovitz number conditions. Combustion and Flame 238:111943. https://doi.org/10.1016/j.combustflame.2021.111943

Fargione J, Hill J, Tilman D, Polasky S, Hawthorne P (2008) Land clearing and the biofuel carbon debt. Science 319(5867):1235–1238. https://doi.org/10.1126/science.1152747

Fayyazi E, Ghobadian B, Najafi G, Hosseinzadeh B, Mamat R, Hosseinzadeh J (2015) An ultrasound-assisted system for the optimization of biodiesel production from chicken fat oil using a genetic algorithm and response surface methodology. Ultrason Sonochem 26:312–320. https://doi.org/10.1016/j.ultsonch.2015.03.007

Ge Y et al. (2011) Growth characteristics of Botryococcus braunii 765 under high CO2 concentration in photobioreactorBioresour Technol. https://doi.org/10.1016/j.biortech.2013.04.036

Gerbens-Leenes W, Hoekstra AY (2011) The water footprint of biofuel-based transport. Energy Environ Sci 4(8):2658. https://doi.org/10.1039/c1ee01187a

Hasunuma T et al (2013) A review of enzymes and microbes for lignocellulosic biorefinery and the possibility of their application to consolidated bioprocessing technology. Bioresour Technol. https://doi.org/10.1016/j.biortech.2012.10.047

Herbinet O, Bartocci P, Grinberg Dana A (2022) On the use of ammonia as a fuel – a perspective. Fuel Commun 11:100064. https://doi.org/10.1016/j.jfueco.2022.100064

Hertel TW, Golub AA, Jones AD, O'Hare M, Plevin RJ, Kammen DM (2010) Effects of US maize ethanol on global land use and greenhouse gas emissions: estimating market-mediated responses. BioScience 60(3):223–231. https://doi.org/10.1525/bio.2010.60.3.8

Huang H, Khanna M, Önal H, Chen X (2013) Stacking low carbon policies on the renewable fuels standard: Economic and greenhouse gas implications. Energy Policy 56:5–15. https://doi.org/10.1016/j.enpol.2012.06.002

Jin M, Slininger PJ, Dien BS, Waghmode S, Moser BR, Orjuela A, Sousa LDC, Balan V (2015) Microbial lipid-based lignocellulosic biorefinery: feasibility and challenges. Trends Biotechnol 33(1):43–54. https://doi.org/10.1016/j.tibtech.2014.11.005

Kalyuzhnaya M (2016) Methane biocatalysis. Biotechnol Biofuel Prod Optim 353–383. https://doi.org/10.1016/b978-0-444-63475-7.00013-3

Kerckow B, Hervouet V, Coombs R (2010) Conference report: european biofuels technology platform third stakeholder plenary meeting. Biofuels 1(4):527–530. https://doi.org/10.4155/bfs.10.36

Khan SA et al (2009) Prospects of biodiesel production from microalgae in India. Renew Sust Energy Rev. https://doi.org/10.1016/j.rser.2010.09.018

Kudakasseril Kurian J, Raveendran Nair G, Hussain A, Vijaya Raghavan G (2013) Feedstocks, logistics and pre-treatment processes for sustainable lignocellulosic biorefineries: A comprehensive review. Renew Sustain Energy Rev 25:205–219. https://doi.org/10.1016/j.rser.2013.04.019

Lee JZ et al (2007) The effect of butyrate concentration on hydrogen production via photofermentation for use in a Martian habitat resource recovery process. Int J Hydrogen Energy. https://doi.org/10.1016/j.procbio.2010.08.017

Loqué D, Scheller HV, Pauly M (2015) Engineering of plant cell walls for enhanced biofuel production. Curr Opin Plant Biol 25:151–161. https://doi.org/10.1016/j.pbi.2015.05.018
Margaritondo G (2005) Valence photoemission. Encyclop Cond Matter Phys 279–286. https://doi.org/10.1016/b0-12-369401-9/00682-3
Mathews J, Wang G (2009) Metabolic pathway engineering for enhanced biohydrogen production. Intern J Hydr Energy 34(17):7404–7416. https://doi.org/10.1016/j.ijhydene.2009.05.078
Melillo JM, Reilly JM, Kicklighter DW, Gurgel AC, Cronin TW, Paltsev S, Felzer BS, Wang X, Sokolov AP, Schlosser CA (2009) Indirect Emissions from Biofuels: How Important? Science 326(5958):1397–1399. https://doi.org/10.1126/science.1180251
Mohan SV, Pandey A (2013) Biohydrogen production. Biohydrogen 1–24. https://doi.org/10.1016/b978-0-444-59555-3.00001-5
Mosnier A, Havlík P, Valin H, Baker J, Murray B, Feng S, Obersteiner M, McCarl B, Rose S, Schneider U (2013) Alternative U.S. biofuel mandates and global GHG emissions: the role of land use change, crop management and yield growth. Energy Policy 57:602–614. https://doi.org/10.1016/j.enpol.2013.02.035
Naik SN et al (2010) Production of first and second generation biofuels: a comprehensive review. Renew Sust Energy Rev. https://doi.org/10.1016/j.rser.2009.10.003
National Academies of Sciences, Engineering, and Medicine (2011) Renewable fuel standard: potential economic and environmental effects of U.S. biofuel policy. Washington, DC: The National Academies Press. https://doi.org/10.17226/13105
Nichols JP (1986) Armbruster Walter J, Lester H Myers (eds) Research on effectiveness of agricultural commodity promotion, Oak Brook IL: Farm Foundation, 1985, 213 pp., free or charge. Am J Agricult Econ 68(3):747–748. https://doi.org/10.2307/1241563
Nigam PS et al (2011) Production of liquid biofuels from renewable resources. Prog Energy Combust Sci. https://doi.org/10.1016/j.pecs.2010.01.003
Pauly M, Keegstra K (2008) Cell-wall carbohydrates and their modification as a resource for biofuels. Plant J 54(4):559–568. https://doi.org/10.1111/j.1365-313x.2008.03463.x
Pereira da Silva P, Ribeiro LA (2019) Assessing microalgae sustainability as a feedstock for biofuels. Adv Bioproc Altern Fuels, Biobased Chem Bioprod 373–392. https://doi.org/10.1016/b978-0-12-817941-3.00019-x
Piotrowska P, Zevenhoven M, Hupa M, Giuntoli J, de Jong W (2013) Residues from the production of biofuels for transportation: characterization and ash sintering tendency. Fuel Proc Technol 105:37–45. https://doi.org/10.1016/j.fuproc.2011.09.020
Poudyal RS et al. Current insights to enhance hydrogen production by photosynthetic organisms. https://doi.org/10.1016/j.ijhydene.2019.01.092
Prieler S, Fischer G, van Velthuizen H (2012) Land and the food-fuel competition: insights from modeling. Wiley Interdiscipl Rev Energy Environ 2(2):199–217. https://doi.org/10.1002/wene.55
Röder M, Welfle A (2019) Bioenergy. Manag Glob Warm 379–398. https://doi.org/10.1016/b978-0-12-814104-5.00012-0
Rocha-Meneses L, Ferreira JA, Mushtaq M, Karimi S, Orupõld K, Kikas T (2020). Genetic modification of cereal plants: A strategy to enhance bioethanol yields from agricultural waste. Indust Crops Prod 150:112408. https://doi.org/10.1016/j.indcrop.2020.112408
Rosegrant MW, Zhu T, Msangi S, Sulser T (2008). Global scenarios for biofuels: impacts and implications*. Rev Agricult Econ 30(3):495–505. https://doi.org/10.1111/j.1467-9353.2008.00424.x
Ruan R, Zhang Y, Chen P, Liu S, Fan L, Zhou N, Ding K, Peng P, Addy M, Cheng Y, Anderson E, Wang Y, Liu Y, Lei H, Li B (2019) Biofuels: Introduction. Biofuels: Alternative Feedstocks and Conversion Processes for the Production of Liquid and Gaseous Biofuels, pp 3–43. https://doi.org/10.1016/b978-0-12-816856-1.00001-4
Sagel VN, Rouwenhorst KH, Faria JA (2022) Green ammonia enables sustainable energy production in small island developing states: a case study on the island of Curaçao. Renew Sustain Energy Revi 161:112381. https://doi.org/10.1016/j.rser.2022.112381

Scott SA et al (2010) Biodiesel from algae: challenges and prospects. Bcurr Opin Biotechnol. https://doi.org/10.1016/j.copbio.2010.03.005

Searchinger T, Heimlich R, Houghton RA, Dong F, Elobeid A, Fabiosa J, Tokgoz S, Hayes D, Yu TH (2008) Use of U.S. croplands for biofuels increases greenhouse gases through emissions from land-use change. Science 319(5867):1238–1240. https://doi.org/10.1126/science.1151861

Soares JF, Confortin TC, Todero I, Mayer FD, Mazutti MA (2020) Dark fermentative biohydrogen production from lignocellulosic biomass: technological challenges and future prospects. Renew Sust Energy Rev 117:109484. https://doi.org/10.1016/j.rser.2019.109484

Seibert M, Applied photosynthesis for biofuels production. https://doi.org/10.1039/c1ee02514g

Wimmerstedt R (1999) Recent advances in biofuel drying. Chem Eng Proc Process Intensif 38(4–6):441–447. https://doi.org/10.1016/s0255-2701(99)00041-0

Yüzbaşıoğlu AE, Avşar C, Gezerman AO (2022) The current situation in the use of ammonia as a sustainable energy source and its industrial potential. Curr Res Green Sustain Chem 5:100307. https://doi.org/10.1016/j.crgsc.2022.100307

Ziolkowska JR, Simon L (2014) January). Recent developments and prospects for algae-based fuels in the US. Renew Sustain Energy Rev 29:847–853. https://doi.org/10.1016/j.rser.2013.09.021

MIX
Papier aus verantwortungsvollen Quellen
Paper from responsible sources
FSC® C105338

If you have any concerns about our products,
you can contact us on
ProductSafety@springernature.com

In case Publisher is established outside the EU,
the EU authorized representative is:
Springer Nature Customer Service Center GmbH
Europaplatz 3, 69115 Heidelberg, Germany

Printed by Libri Plureos GmbH
in Hamburg, Germany